THE CULTURAL LANDSCAPE

An Introduction to

Human

Geography

Sixth Edition

James Rubenstein

Miami University, Oxford, Ohio

PRENTICE HALL, Upper Saddle River, NJ 07458

Library of Congress Cataloging-in-Publication Data

Rubenstein, James M.
      The cultural landscape : an introduction to human geography /
   James M. Rubenstein.—6th ed.
         p.   cm.
      Includes bibliographical references and indexes.
      ISBN 0-13-079778-2
      1. Human geography.   I. Title
   GF41.R82   1999   98-7272
   304.2—dc21   CIP

Acquisition Editor: Daniel Kaveney
Editor in Chief: Paul Corey
Editorial Director: Tim Bozik
Director of Production and Manufacturing: David W. Riccardi
Executive Managing Editor: Kathleen Schiaparelli
Assistant Managing Editor: Lisa Kinne
Production Editor: Edward Thomas
Development Editor: Barbara Muller
Editor in Chief of Development: Ray Mullaney
Marketing Manager: Leslie Cavaliere
Marketing Assistant: Cheryl Adam
Creative Director: Paula Maylahn
Art Director: Joseph Sengotta
Photo Editors: Lorinda Morris-Nantz and Melinda Reo
Photo Researcher: Tobi Zausner
Copy Editor: James Tully
Assistant Editor: Wendy Rivers
Editorial Assistant: Margaret Ziegler
Cover/Interior Designer: Maureen Eide
Manufacturing Manager: Trudy Pisciotti
Buyer: Benjamin Smith
Text Composition: Better Graphics, Inc.
Illustrator: GeoSystems Global Corporation
Cover Photo: Woman weaving rugs near Lake Atklan,
Guatemala. Photo by Michele and Tom Grimm/Tony
Stone Images.
Acknowledgments for figures appear on pp. 527–528.

Printed in the United States of America
10 9 8 7 6 5 4 3

ISBN 0-13-079778-2

Prentice-Hall International (UK) Limited, *London*
Prentice-Hall of Australia Pty. Limited, *Sydney*
Prentice-Hall Canada Inc., *Toronto*
Prentice-Hall Hispanoamericana, S.A., *Mexico*
Prentice-Hall of India Private Limited, *New Delhi*
Prentice-Hall of Japan, Inc., *Tokyo*
Pearson Education Asia Pte. Ltd., *Singapore*
Editora Prentice-Hall do Brasil, Ltda., *Rio de Janeiro*

Contents

GLOBALIZATION AND LOCAL DIVERSITY
Intervening Obstacles   99

Chapter 4
Folk and Popular Culture   119

GLOBALIZATION AND LOCAL DIVERSITY
Revival of Lacrosse, a Folk Custom   127

Chapter 5
Language   153

GLOBALIZATION AND LOCAL DIVERSITY

Afghanistan: A Cold War Conflict Becomes a Local Civil
War   292–93

Chapter 9
Development   303

GLOBALIZATION AND LOCAL DIVERSITY

Eastern Europe: An "Un-Developing" Region   309

Chapter 10
Agriculture   335

Chapter 11
Industry 369

Chapter 12
Services 407

Preface

What is geography? Geography is the study of where things are located on Earth's surface and the reasons for the location. The word geography, invented by the ancient Greek scholar Eratosthenes, is based on two Greek words. Geo means "Earth," and graphy means "to write." Geographers ask two simple questions: where and why. Where are people and activities located across Earth's surface? Why are they located in particular places?

Geography as a Social Science

Recent world events lend a sense of urgency to geographic inquiry. Geography's spatial perspectives help to relate economic change in Europe, the Middle East, and other regions to the spatial distributions of cultural features such as languages and religions, demographic patterns such as population growth and migration, and natural resources such as energy and food supply.

Does the world face an overpopulation crisis? Geographers study population problems by comparing the arrangements of human organizations and natural resources across Earth. Given these spatial distributions, geographers conclude that some locations may have more people than can be provided for, whereas other places may be underpopulated.

Similarly, geographers examine the prospects for an energy crisis by relating the spatial distributions of energy sources and consumption. Geographers find that the users of energy are located in places with different social, economic, and political institutions than the producers of energy. Geographers seek first to describe the distribution of features such as the production and consumption of energy, and then to explain the relationships between these distributions and other human and physical phenomena.

The main purpose of this book is to introduce students to the study of geography as a social science by emphasizing the relevance of geographic concepts to human problems. It is intended for use in college-level introductory human or cultural geography courses. The book is written for students who have not previously taken a college-level geography course and have had little, if any, geography in high school.

Divisions within Geography

Because geography is a broad subject, some specialization is inevitable. At the same time, one of geography's strengths is its diversity of approaches. Rather than being forced to adhere rigorously to established disciplinary laws, geographers can combine a variety of methods and approaches. This tradition stimulates innovative thinking, although students who are looking for a series of ironclad laws to memorize may be disappointed.

Human versus Physical Geography. Geography is both a physical and a social science. When geography concentrates on the distribution of physical features, such as climate, soil, and vegetation, it is a natural science. When it studies cultural features, such as language, industries, and cities, geography is a social science. This division is reflected in some colleges, where physical geography courses may carry natural science credit and human and cultural geography courses social science credit.

While this book is concerned with geography from a social science perspective, one of the distinctive features of geography is its use of natural science concepts to help understand human behavior. The distinction between physical and human geography reflects differences in emphasis, not an absolute separation.

Topical versus Regional Approach. Geographers face a choice between a topical and a regional approach. The topical approach, which is used in this book, starts by identifying a set of important cultural issues to be studied, such as population growth, political disputes, and economic restructuring. Geographers using the topical approach examine the location of different aspects of the topic, the reasons for the observed pattern, and the significance of the distribution.

The alternative approach is regional. Regional geographers start by selecting a portion of Earth and studying the environment, people, and activities within the area. The regional geography approach is used in courses on Europe, Africa, Asia, and other areas of the world. Although this book is organized by topics, geography students should be aware of the location of places in the world. A separate index section lists the book's maps by location. One indispensable aid in the study of regions is an atlas, which can also be used to find unfamiliar places that may pop up in the news. Partly for this reason, the publisher has chosen to offer an atlas to accompany this textbook at an additional nominal cost.

Descriptive versus Systematic Method. Whether using a topical or a regional approach, geographers can select either a descriptive or a systematic method. Again, the distinction is one of emphasis, not an absolute separation. The descriptive method emphasizes the collection of a variety of details about a particular location. This method has been used primarily by regional geographers to illustrate the uniqueness of a particular location on Earth's surface. The systematic method emphasizes the identification of several basic theories or techniques developed by geographers to explain the distribution of activities.

This book uses both the descriptive and systematic methods because total dependence on either approach is unsatisfactory. An entirely descriptive book would contain a large collection of individual examples not organized into a unified structure. A completely systematic approach suffers because some of the theories and techniques are so abstract that they lack meaning for the student. Geographers who depend only on the systematic approach may have difficulty explaining important contemporary issues.

Features

This book is sensitive to the study needs of students. Each chapter is clearly structured to help students understand the material and effectively review from the book.

Outline.　　The book discusses the following main topics:

- What basic concepts do geographers use? Chapter 1 provides an introduction to basic geographic concepts, as well as a brief summary of the development of the science of geography. Geographers employ several concepts to describe the distribution of people and activities across Earth, to explain reasons underlying the observed distribution, and to understand the significance of the arrangements.

- Where are people located in the world? Chapters 2 and 3 examine the distribution and growth of the world's population, as well as the movement of people from one place to another. Why do some places on Earth contain large numbers of people or attract newcomers while other places are sparsely inhabited?

- How are different cultural groups distributed? Chapters 4 through 8 analyze the distribution of different cultural traits and beliefs and the problems that result from those spatial patterns. Important cultural traits discussed in Chapter 4 include food, clothing, shelter, and leisure activities. Chapters 5 through 7 examine three main elements of cultural identity: language, religion, and ethnicity. Chapter 8 looks at political problems that arise from cultural diversity. Geographers look for similarities and differences in the cultural features at different places, the reasons for their distribution, and the importance of these differences for world peace.

- How do people earn a living in different parts of the world? Human survival depends on acquiring an adequate food supply. One of the most significant distinctions in the world is whether people produce their food directly from the land or buy it with money earned by performing other types of work. Chapters 9 through 12 look at the three main ways of earning a living: agriculture, manufacturing, and services. Chapter 13 discusses cities, the centers for economic as well as cultural activities.

- What issues result from using Earth's resources? The final chapter is devoted to a study of three issues related to the use of Earth's natural resources: energy, pollution, and food supply. Geographers recognize that cultural problems result from the depletion, destruction, and inefficient use of the world's natural resources.

Chapter Organization.　　To help the student use the material in this book, each chapter is organized with these study aids:

- *Case Study.* Each chapter opens with a case study that illustrates some of the key concepts presented in the text. The case studies are generally drawn from news events or from daily experiences familiar to residents of North America.

- *Key Issues.* Each chapter contains a set of three or four key issues around which the chapter material is organized. These questions reappear as major headings within the chapter. All questions include one of the two key geographic concerns: where or why.

- *Key Terms.* The key terms in each chapter are indicated in bold type when they are introduced. These terms are also defined at the end of each chapter.

- *Globalization and Local Diversity Box.* Each chapter has a one- or two-page box that explores in depth a particular topic related to the subject of the chapter. The Globalization and Local Diversity boxes relate principles and concepts to applied, practical issues. All boxes include the key terms where, why, globalization, and local diversity.

- *Summary.* The key issues are repeated at the end of the chapter with a brief review of the important concepts covered in detail in the text.

- *Case Study Revisited.* Additional information related to the chapter's case study may be used to reinforce some of the main points.

- *Thinking Geographically.* This section offers five questions based on concepts and themes developed in the chapter. The questions help students apply geographic concepts to explore issues more intensively.

- *Further Readings.* A list of books and articles is provided for students who wish to study the subject further.

Appendix.　　A special appendix on scale and major projections enhances the discussion of the subject in Chapter 1 of the text. We are grateful to Phillip C. Muehrcke, Professor of Geography at the University of Wisconsin-Madison, and former president of the American Cartographic Association, for his clear explanation of the subject.

Instructional Package

In addition to the text itself, the author and publisher have been pleased to work with a number of talented people to produce an excellent instructional package. This package includes the traditional supplements that students and professors have come to expect from authors and publishers, as well as new kinds of components that utilize electronic media.

For the Student

- Companion Website: *The Cultural Landscape: An Introduction to Human Geography* web site by Robert E. Nunley, George W. Ulbrick, Daniel L. Roy, and Severin M. Roberts, all of the University of Kansas, gives students the opportunity to further explore topics presented in the book using the Internet. The site contains numerous review exercises (from which students get immediate feedback), exercises to expand students' understanding of human geography, and resources for further exploration. This web site provides an excellent platform from which to start using the Internet for the study of human geography. Please visit the site at http://www.prenhall.com/rubenstein

- *Geosciences on the Internet: A Student's Guide*, by Andrew T. Stull and Duane Griffin, is a guide to the Internet specifically for geography students. *Geosciences on the Internet* is available at no cost to qualified adopters of *The Cultural Landscape*.

- *Study Guide:* Written by experienced educators Robert E. Nunley, George W. Ulbrick, Severin M. Roberts, and Daniel L. Roy, the study guide helps students identify the important points from the text, and then provides them with review exercises, study questions, self-check exercises, and vocabulary review.

For the Professor

- *Slides and Transparencies:* More than 100 full color illustrations from the text are available free of charge to qualified adopters. In order to accommodate instructor preference, these images are available both on transparency acetates and 35 millimeter slides.

- *Presentation Manager:* This user-friendly navigation software enables professors to custom build multimedia presentations. *Prentice Hall Presentation Manager 3.0* contains several hundred images from the text. The CD-ROM allows professors to organize material in whatever order they choose; preview resources by chapter; search the digital library by keyword; integrate material from their hard drive, a network, or the Internet; edit lecture notes; and annotate images with an overlay tool. This powerful presentation tool is available at no cost to qualified adopters of the text.

- *The New York Times* **Themes of the Times—Geography:** This unique newspaper-format supplement features recent articles about geography from the pages of the *New York Times.* This supplement, available at no extra charge from your local Prentice Hall representative, encourages students to make connections between the classroom and the world around them.

- *Instructor's Manual:* Written by Tarek Joseph of Michigan State University, the instructor's manual is intended as a resource for both new and experienced instructors. It includes a variety of lecture outlines, additional source materials, teaching tips, advice about how to integrate visual supplements (including the web-based resources), and various other ideas for the classroom.

- *Test Item File:* The test item file, by Robert E. Nunley, George W. Ulbrick, Severin M. Roberts, and Daniel L. Roy, provides instructors with a wide variety of test questions.

- *PH Custom Test:* Based on the powerful testing technology developed by Engineering Software Associates, Inc. (ESA), *Prentice Hall Custom Test* allows instructors to create and tailor exams to their own needs. With the online testing program, exams can also be administered online and data can then be automatically transferred for evaluation. A comprehensive desk reference guide is included along with online assistance.

Suggestions for Use

This book can be used in an introductory human or cultural geography course that extends over one semester, one quarter, or two quarters. An instructor in a one-semester course could devote one week to each of the chapters, leaving time for examinations. In a one-quarter course, the instructor might need to omit some of the book's material.

A course with more of a cultural orientation could use Chapters 1 through 8, plus Chapter 14. If the course has more of an economic orientation, then the appropriate chapters would be 1 through 3 and 8 through 14.

A two-quarter course could be organized around the culturally oriented Chapters 1 through 8 during the first quarter and the more economically oriented Chapters 9 through 14 during the second quarter. Topics of particular interest to the instructor or students could be discussed for more than one week.

Changes

This edition of the book is organized around a tension between two important themes—globalization and cultural diversity. In many respects we are living in a more unified world economically, culturally, and environmentally. The actions of a particular corporation or country affect people around the world. This book argues that after a period when globalization of the economy and culture has been a paramount concern in geographic analysis, local diversity now demands equal time. People are taking deliberate steps to retain distinctive cultural identities. They are preserving little used languages, fighting fiercely to protect their religions, and carving out distinctive economic roles.

This edition contains one entirely new chapter and several substantially rewritten ones.

The new chapter is Ethnicity (Chapter 7). Ethnicity, like language and religion, is increasingly a source of pride to people and a link to the cultural traditions of ancestors.

Ethnicity may help to explain demographic, health, and economic conditions and patterns of inequality and discrimination. Some of the material in Chapter 7 was found elsewhere in previous editions, including U.S. urban patterns,

South Africa's history of apartheid, and ethnic cleansing in the Balkans.

Chapter 4, Folk and Popular Culture, deals with topics previously contained in the fifth edition's Chapter 6, Social Customs, although the material has been restructured to emphasize cultural patterns. For the first time in a major introductory human geography text, this edition devotes an entire chapter to services (Chapter 12). Two-thirds of North Americans work in services, but this sector of the economy receives minimal treatment in introductory (or even intermediate-level) geography books. Urban Patterns (Chapter 13) combines material that previously had been spread between two chapters.

The organization of material within most of the chapters has been substantially changed to fit more explicitly into the book's where and why framework. For example, Chapter 5 (Language) starts with describing where languages are distributed, then discusses why languages are distributed as they are, then explains why the distribution of languages can cause problems. Similarly, Chapter 10 (Agriculture) has been reorganized around where the major types of agriculture are practiced, why the differences exist, and why problems may arise. Similar frameworks are used for other cultural and economic features.

Maps have also been added to reinforce the tension between globalization and cultural diversity. For example, new maps show ethnic distributions by U.S. state and within the cities of Chicago, Detroit, and Los Angeles. World population has been shown in a cartogram, and new world maps show countries by percent service workers, percent urban dwellers, and total fertility rate.

Finally, given the enormous amount of material now available electronically, through CD-ROM, Internet, and so on, why should an instructor continue to make students buy an expensive textbook? In the computer age, is a textbook an anachronism? A book is a slow way to communicate: by the time this book is in your hands, something in it will be outdated, perhaps a new war, peace treaty, or United Nations member. The information superhighway is filled with information that can be quickly retrieved, but the information is poorly organized and written. In contrast, a high-quality book is crafted carefully by the author, editors, and publisher. For example, the author rewrote this sentence five times to convey a precise meaning. Editors then change many of the words and punctuation to assure that the author's intended meaning is successfully communicated. A book allows an author to lay out a more careful and clear route to explanation and understanding than is possible electronically. For now, computers are tools for retrieval of facts and for advanced analysis, but they can not yet compete with books in explaining a discipline's basic concepts and themes.

Acknowledgments

The successful completion of a book like this requires the contribution of many people. I would like to gratefully acknowledge the help I received. At Prentice Hall, Geography Editor Daniel Kaveney has proved to be a level-headed common-sense leader in the field of geography publishing. Ed Thomas, Production Editor for the second time, sweats the details to make this a well-crafted book. Barbara Muller, Development Editor, played a sensitive role in developing new material for this edition. Joseph Sengotta, department manager Paula Maylahn, and the rest of the design team played an especially important role in creating an outstanding design for the book's new size this edition. I will always be grateful to Paul Corey, Editor-in-Chief for Science, for his long-time support and friendship.

Outside Prentice Hall, the production staff at GeoSystems (formerly Maryland CartoGraphics), led by Tracy Morrill, continue to produce outstanding maps and line drawings for this book, as they have for more than a decade. Tobi Zausner, Photo Researcher, assembled an especially strong collection of engaging photographs. Better Graphics, led by Linda Wachowski, pulled together the page layout and composition far more smoothly and attractively than in the old days of galleys and paste-ups. I am also grateful to the outstanding work done on a variety of ancillaries by the University of Kansas crowd, led by Robert E. Nunley, Severin M. Roberts, Daniel L. Roy, and George W. Ulbrick, as well as by Tarek Joseph, Andrew T. Stull, and Duane Griffin. Finally, I would like to thank my Introduction to Human Geography students at Miami, who make this work worthwhile.

SIXTH EDITION REVIEWERS

Roger Hunt, Grand Valley State University; Robert E. Nunley, University of Kansas; Paul B. Frederic, University of Maine, Farmington; Vernon Domingo, Bridgewater State College; Steve Levy, Portland Community College; Roger Selya, University of Cincinnati; Henry W. Bullamore, Frostburg State University; Hal Jackson, University of New Mexico; Craig Campbell, Youngstown State University; Tarek Joseph, Michigan State University; Richard Pillsbury, Georgia State University; Brock Brown, Southwest Texas State University

About the Author

Dr. James M. Rubenstein received his Ph.D. from Johns Hopkins University in 1975. His dissertation on French urban planning was later developed into a book entitled **The French New Towns** (Johns Hopkins University Press). In 1976 he joined the faculty at Miami University, where he is currently Professor of Geography. Besides teaching courses on Urban and Human Geography and writing textbooks, Dr. Rubenstein also conducts research in the automotive industry and has published a book on the subject entitled **The Changing U.S. Auto Industry: A Geographical Analysis** (Routledge). Originally from Baltimore, he is an avid Orioles fan and follows college lacrosse. Stormy, a lab-pointer mix, takes Dr. Rubenstein for a long walk in the woods every day.

This book is dedicated to Bernadette Unger, Dr. Rubenstein's wife, who has stuck with him through thick and thin. Dr. Rubenstein also gratefully thanks the rest of his family for their love and support.

The world map at right reveals one of the most significant elements of the cultural landscape—the political boundaries that separate its five billion inhabitants. The numerous states range in size from Russia, which occupies one-sixth of the world's land area, to microstates such as Singapore, Malta, or Grenada. The names of these states evoke images of different environments, peoples, cultures, and levels of well-being. However, the political boundaries are only one of the many patterns that geographers observe across Earth's surface. Geographers study the distribution of a wide variety of cultural and environmental features—social customs, agricultural patterns, the use of resources—many of which transcend political boundaries. As scientists, geographers also try to explain why we can observe these patterns on the landscape. The facing map and chapters that follow are intended to begin the student on a journey toward understanding our exciting and complex world.

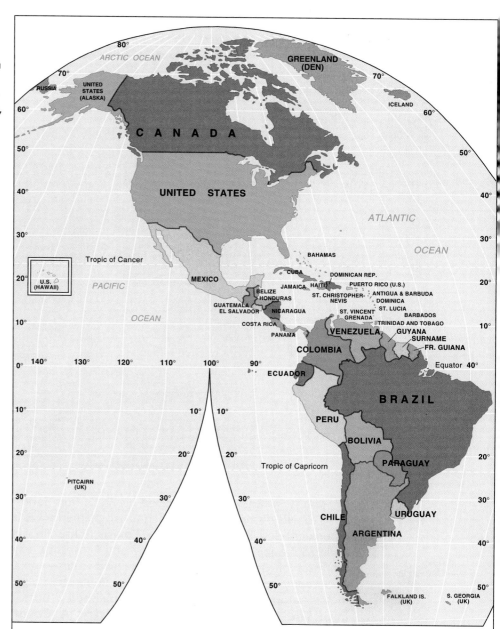

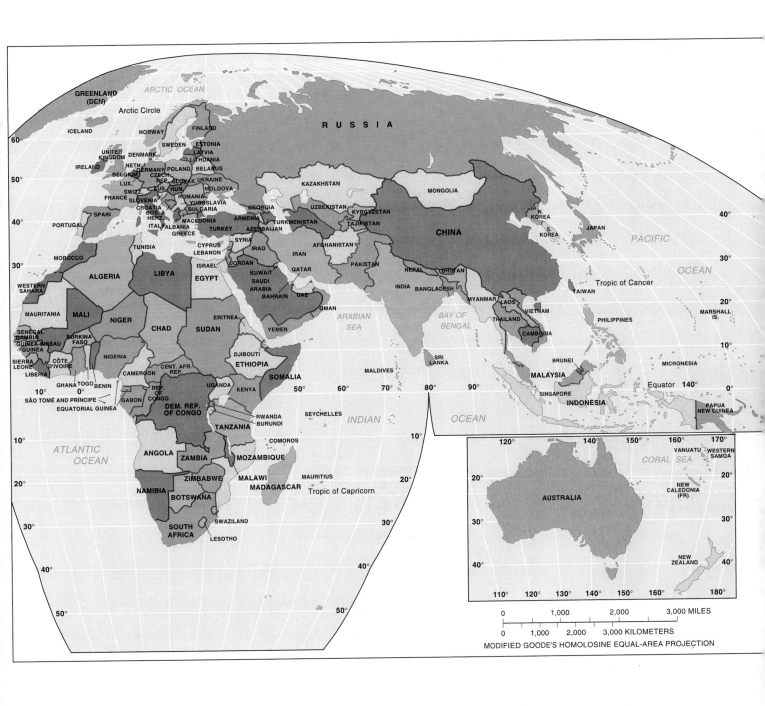

GREENLAND
(DEN)

Arctic Circle

ICELAND

ARCTIC OCEAN

NORWAY FINLAND

RUSSIA

60°

UNITED
KINGDOM
IRELAND

SWEDEN ESTONIA
DENMARK LATVIA
LITHUANIA
NETH. GERMANY POLAND BELARUS
BELGIUM CZECH.
LUX. REP. SLOVAK. UKRAINE
SWITZ. AUS. HUN. MOLDOVA
FRANCE SLOVENIA ROMANIA
CROATIA YUGOSLAVIA
BOS. & BULGARIA
HERZ. MACEDONIA
ITALY ALBANIA
GREECE

KAZAKHSTAN

MONGOLIA

50°

SPAIN

GEORGIA
TURKEY ARMENIA
AZERBAIJAN
SYRIA

UZBEKISTAN
KYRGYZSTAN
TURKMENISTAN TAJIKISTAN

N.
KOREA
S.
KOREA

JAPAN

PACIFIC

40°

PORTUGAL

CYPRUS
LEBANON
ISRAEL
JORDAN

IRAQ
IRAN

AFGHANISTAN

CHINA

OCEAN

30°

MOROCCO

TUNISIA

ALGERIA LIBYA EGYPT

KUWAIT
SAUDI
ARABIA
BAHRAIN UAE

QATAR

PAKISTAN

NEPAL BHUTAN

Tropic of Cancer

TAIWAN

20°

WESTERN
SAHARA

MAURITANIA MALI NIGER CHAD SUDAN

ERITREA

OMAN

ARABIAN
SEA

INDIA BANGLADESH

BAY OF
BENGAL

MYANMAR LAOS
THAILAND VIETNAM

PHILIPPINES

MARSHALL
IS.

SENEGAL
GAMBIA BURKINA
GUINEA-BISSAU FASO
GUINEA
SIERRA CÔTE
LEONE D'IVOIRE
LIBERIA GHANA TOGO BENIN
0°
SÃO TOMÉ AND PRÍNCIPE
EQUATORIAL GUINEA

NIGERIA

CENT. AFR.
REP.
CAMEROON

DJIBOUTI

ETHIOPIA

YEMEN

MALDIVES

SRI
LANKA

CAMBODIA

BRUNEI

MICRONESIA

10°

REP.
OF
CONGO
GABON

UGANDA

SOMALIA

KENYA

50° 60° 70° 80° 90°

MALAYSIA

SINGAPORE

Equator 140°

0°

10°

DEM. REP.
OF CONGO

TANZANIA

RWANDA
BURUNDI

SEYCHELLES

INDONESIA

PAPUA
NEW GUINEA

ATLANTIC
OCEAN

ANGOLA ZAMBIA

COMOROS

INDIAN

OCEAN

10°

10°

ZIMBABWE
NAMIBIA

MOZAMBIQUE

MALAWI

MAURITIUS

20°

BOTSWANA

MADAGASCAR

Tropic of Capricorn

20°

SOUTH
AFRICA

SWAZILAND

LESOTHO

30°

40°

30°

40°

50°

50°

120° 140° 150° 160° 170°

VANUATU WESTERN
SAMOA

CORAL SEA

20°

AUSTRALIA

NEW
CALEDONIA
(FR)

20°

30°

30°

40°

NEW
ZEALAND

40°

110° 120° 130° 140° 150° 160° 180°

0 1,000 2,000 3,000 MILES

0 1,000 2,000 3,000 KILOMETERS

MODIFIED GOODE'S HOMOLOSINE EQUAL-AREA PROJECTION

CHAPTER 1

Basic Concepts

What do you expect from this geography course? You may think that geography involves memorizing lists of countries and capitals, climates and crop types, or exports and imports. Perhaps you associate geography with photographic essays of exotic places in popular magazines.

But contemporary geography is much more. It is a fascinating science, and one in which everyone can participate. Geography is the scientific study of the location of people and activities across Earth's surface, and the reasons for their distribution. Geographers ask "where" things are and "why" they are there.

Like all sciences, the study of geography requires you to understand some basic concepts. For example, the definition of geography in the previous paragraph included the words "location" and "distribution." We use these words commonly in daily speech, but geographers give them precise meanings. This first chapter is a short tour through the interesting basics of geography.

Jim Richardson, Woodfin Camp & Associates.

Case Study

Where Is Miami?

Consider the following conversation between two students during winter vacation:

First student: Where do you go to school?

Second student: Miami University.

First student: I'll bet you enjoy the warm weather and nearby ocean.

Second student: No way. I don't go to school in Florida. We have snow and hills.

First student: Then where is your Miami?

Second student: In Oxford, Ohio.

First student: Where is Oxford, Ohio?

Second student: About 35 miles northwest of Cincinnati and 2 miles from the Indiana border.

First student [overwhelmed]: Oh.

Second student [conversation gets less realistic]: Not only that, Miami, Ohio, is located at 39°30′40″ north latitude and 84°44′40″ west longitude, in township T5N R1E.

The conversation between the two students presents a key question that geographers ask: Where is something located? Geographers study the arrangement of people and activities across Earth's surface. But geography is much more than a description of place-names. It is a scientific study of the reasons why people and activities are arranged in a particular way. Further, geographers seek to know the significance of where something is located. Like other scientists, geographers try to solve problems, in this case those that arise from the location of people and activities. In other words, geographers ask two key questions: *Where?* and *Why?*

The dialogue in the Case Study illustrates the first question geographers ask: *Where?* The other question they ask is *Why?* Where are people and activities located across Earth's surface? Why are they located in particular places?

Geography divides broadly into two categories—*human* geography and *physical* geography—and they ask slightly different "where" and "why" questions. Human geography is the study of where and why human activities are located where they are—for example, religions, businesses, and cities. Physical geography studies where and why natural forces occur as they do—for example, climates, landforms, and types of vegetation.

This text is an introduction to human geography; it concentrates on two main features of human behavior: culture and economy. The first half of the book explains why the most important cultural features, such as major languages, religions, and ethnicities, are arranged as they are across Earth. The second half of the book looks at important economic activities, including agriculture, manufacturing, and services.

This first chapter introduces basic concepts that geographers employ to address their "where" and "why" questions. The first key issue in this chapter looks at two concepts that help geographers to describe where people and activities are found around the world—location and distribution. Everything occupies a unique **location**, or position, on Earth's surface, and geographers have many ways to identify location. The various locations of a collection of people or objects form a **distribution**, or regular arrangement across Earth's surface. In a given area, particular objects may be distributed close together or far apart; they may be numerous or scarce.

The second and third sections of this chapter look at basic concepts geographers use to ask two principal "why" questions. First, geographers want to know why each place on Earth is in some ways unique. For example, why do people living close to each other speak different languages and employ different methods of agriculture?

Geographers use two basic concepts to explain why every place is unique—culture and region. Geographers identify the distinctive **culture** of a group of people, their body of beliefs and traditions, as well as their political and economic practices. Distinctive cultural groups occupy particular **regions**, which are areas of the world distinguished by a collection of distinctive cultural as well as physical features.

The third key issue in this chapter looks at geography's other main "why" question. Geographers want to know why different places on Earth have similar features. For example, why do people living far apart from each other practice the same religion and earn a living in similar ways?

Two basic concepts—spatial interaction and diffusion—help geographers explain why these similarities do not result from coincidence. Geographers study the relationships among people and objects that form across the barrier of space, known as **spatial interaction**, as well as

the various means by which the interaction occurs, known as **diffusion**. Diffusion can involve the movement either of people or—increasingly in an age of electronic communications—of ideas.

Geography matters in the contemporary world because it can explain human actions at all scales, from local to global. At the national and international scales, geography is concerned with such questions as where the population is growing rapidly, where the followers of different religions live, and where corporations place factories. And geography studies why these arrangements can cause problems: Why rapid population growth can exceed available food supply, why different religious groups are unable to live in peace with each other, why some places are unable to attract or retain industries.

Pursuing the "why" question further, geographers observe that people are being pulled in opposite directions by two factors—*globalization* and *local diversity*. On the one hand, modern communications and technology have fostered globalization, pulling people into greater cultural and economic interaction with others. At the same time, people are searching for more ways to express their unique cultural traditions and economic practices. Tensions between the simultaneous geographic trends of globalization and local diversity underlie many of the world's problems that geographers study, such as political conflicts, economic uncertainty, and pollution of the environment.

Geography can be about personal space, not just world issues. On a tiny scale, geography can describe where students sit in a classroom, and why they select their seats. Some students sit in the front of the room for maximum interaction with the instructor. Students near the front can more easily read the chalkboard, hear what the instructor and other students say during lectures and discussions, and make eye contact with the instructor. Other students choose a location in the rear of the room to avoid interaction with the instructor. Perhaps they have not done the assignment, or they wish to spend class time doing other things. Whether in the front or rear, students quickly acquire a sense of place in the classroom: Having selected seats at the beginning of the term, they tend to sit in or near the same location every day, even if the instructor does not require it.

Whether the area of analysis is a room or a planet, location matters. Historians organize material by time, because they understand that action at one point in time can result from past actions and can affect future ones. Geographers organize material by place, because they understand that something happening at one place can result from something that happened elsewhere and can affect conditions at other places. Historians study the logical sequence of human activities in time, whereas geographers study the logical arrangement of human activities in space.

This book focuses on human geography, but it never forgets Earth's atmosphere, land, water, vegetation, and other living creatures. Because geographers are trained in both social and physical sciences, they are particularly well equipped to understand interactions between people and their environment. For example,

- To explain the problem of hunger in Somalia and neighboring countries in Africa, geographers examine relations among population growth, farming practices, political unrest, drought, and environmental degradation.
- To explain unrest in the Middle East, geographers study the distribution of differences in religious beliefs, alternative strategies for modernizing economies, and energy resources.

Human–natural relationships such as these will be examined throughout this book. For example, in Chapter 2 we will see how humans are more likely to live in the areas of Earth where land is flat, water abundant, and climate mild, and they avoid living in mountains, deserts, and areas with harsh climates. The final chapter of the book will explicitly tie human activities to the physical environment.

Key Issue 1

How Do Geographers Describe *Where* Things Are?

- Maps: Scale models that show where something is
- How geography grew as a science by asking "Where?"
- Location: Where something is
- Distribution: Spatial regularities

In ancient times, geographers learned through exploration where people, human activities, and natural features were found across Earth's surface. With the rise of geography as a science in the past couple of hundred years, contemporary geographers employ scientific tools and concepts to describe where objects are found more accurately than was possible in the ancient world. The development of geography from ancient exploration to modern science is briefly reviewed later in this section of Chapter 1, followed by discussions of geography's two basic concepts for addressing "where" questions: location and distribution.

Maps: Scale Models That Show Where Something Is

Geography's most important tool is the map. As you turn the pages of this book, the first thing you probably notice is the large number of maps, more than 200. These maps range in size from two-page spreads of the entire world to tiny boxes covering part of a city. Some are highly detailed, with complex colors, lines, points, and shadings, whereas others seem highly generalized and "unrealis-

tic." Geography is immediately distinguished from other disciplines by its reliance on maps to communicate information.

A **map** is a two-dimensional or flat scale model of Earth's surface, or some portion of it. Maps are scale models of the real world, just like a model automobile or ship, made small enough to work with on a desk or display on a wall. Maps range from hasty sketches (here's-how-to-get-to-the-party) to precise, sophisticated, computer-generated works of art.

Earth is very nearly a sphere, and therefore is quite accurately represented in the form of a globe. However, a globe is an extremely limited tool to communicate information about Earth's surface. A small globe does not have enough space to display detailed information, while a large globe is too bulky and cumbersome to use. And a globe is difficult to write on, photocopy, mail, or carry in the glove box of a car. Consequently, most maps—including all in this book—are flat. Maps are normally made flat because three-dimensional models are expensive and difficult to reproduce.

A map serves two main purposes, as a tool for storing reference material and as a tool for communicating geographic information. As a reference source, a map helps us to find the shortest route between two places and to avoid getting lost along the way. We consult maps to learn where in the world something is found, especially in relation to a place we know, such as a town, body of water, or highway. The maps in an atlas or a road map are especially useful for this purpose.

As a communications tool, a map is often the best means for extracting information about human activities or physical features. A series of maps of the same area over several years can reveal dynamic processes at work, such as human migration or the spread of a disease. Patterns on maps may suggest interactions among different features of Earth. Placing information on a map is a principal way that geographers share data or results of scientific analysis. To communicate specific pieces of information more effectively, the maps in this book omit details such as cities and rivers that are extraneous to their principal purposes.

A map is different from a photograph because it is a less-literal representation of Earth, an artistic creation constrained by scientific principles. For centuries, geographers have worked to perfect the science of mapmaking, called **cartography**. To communicate geographic concepts effectively through maps, cartographers must design them properly and assure that users know how to read them. Cartographers must make two especially important decisions in creating a map—scale and projection.

Scale

The first decision a cartographer faces is how much of Earth's surface to depict on the map. Is it necessary to show the entire globe, or just one continent, or a country, or a city? To make a scale model of the entire world, many details must be omitted because there simply is not enough space. Conversely, if a map shows only a small portion of Earth's surface, like a street map of a city, it can provide a wealth of detail about a particular place.

The level of detail and the amount of area covered on a map depends on its scale. The **scale** of a map is the relation of a feature's size on a map and its actual size on Earth's surface. For example, if one inch of roadway on a map is actually 24,000 inches on the ground, the map scale is 1:24,000.

Cartographers usually present scale in one of three ways: a fraction (1/24,000) or ratio (1:24,000), a written statement ("1 inch equals 1 mile"), or a graphic bar scale (Figure 1–1).

A fractional scale shows the numerical ratio between distances on the map and Earth's surface. A scale of 1:24,000 or 1/24,000 means that one unit (inch, centimeter, foot, finger length) on the map represents 24,000 of the same unit (inch, centimeter, foot, finger length) on the ground. The unit chosen for distance can be anything, as long as the units of measure on both the map

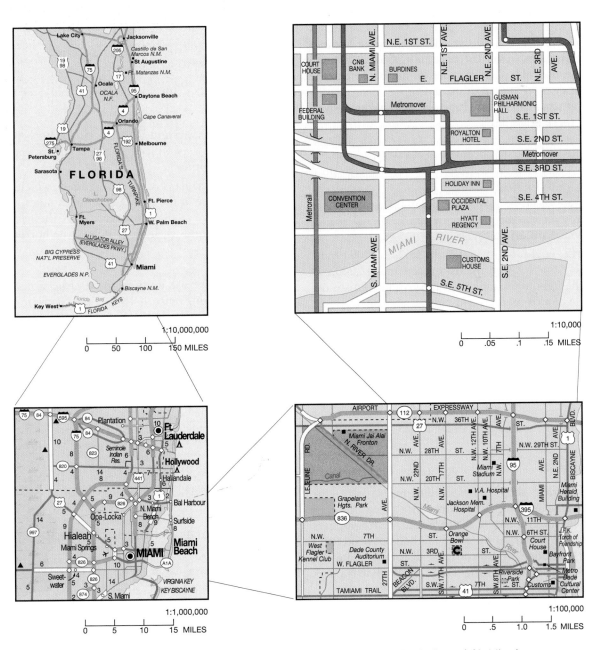

Figure 1–1 Scale. The four maps show Florida (upper left), south Florida (lower left), Miami (lower right), and downtown Miami (upper right). The map of Florida (upper left) has a fractional scale of 1:10,000,000. Expressed as a written statement, 1 inch on the map represents 10 million inches (about 158 miles) on the ground. The bar line below the map displays the scale in a graphic form. Look what happens to the scale on ge other three maps. As the area covered gets smaller, the maps get more detailed, and 1 inch on the map represents smaller distances.

and the ground are the same. The 1 on the left side of the ratio always refers to a unit of distance *on the map*, and the number on the right always refers to the *same unit* of distance *on Earth's surface*.

The written scale describes this relation between map and Earth distances in words. For example, the statement "1 inch equals 1 mile" on a map means that one inch on the map represents one mile on Earth's surface. Again, the first number always refers to map distance, and the second to distance on Earth's surface. (Here the units are different—inch and mile—for ease of use.)

A graphic scale usually consists of a bar line marked to show distance on Earth's surface. To use a bar line, first determine with a ruler the distance on the map in inches or centimeters. Then hold the ruler against the bar line and read the number on the bar line opposite the map distance on the ruler. The number on the bar line is the equivalent distance on Earth's surface.

A map's scale can be any ratio the mapmaker desires. Here are three examples:

1. A flower bed could be mapped at 1:1 scale by drawing the flower bed at exactly the same size on a huge sheet of paper.
2. A state could be mapped at 1:250,000 scale, where one inch on the map represents 250,000 inches (about 4 miles) on the ground.
3. Earth could be mapped at 1:42,000,000 scale, where one inch on the map represents 42,000,000 inches (about 665 miles) on the ground. This is the scale of a common world map.

When comparing map scales, remember that the smaller the fractional scale, the larger the overall area represented (example 3 above), and the larger the fractional scale, the smaller the area covered (example 1 above). A world map uses a smaller scale than a city map, because it covers a larger area. A large-scale map is suitable for detailed information about a small area.

Projections

Earth's spherical shape poses a dilemma to cartographers, because drawing Earth on a flat piece of paper unavoidably produces some distortion. Cartographers have invented hundreds of clever methods of producing flat maps, but none are free of some distortion. The scientific method of transferring locations on Earth's surface to a flat map is called **projection**.

The problem of distortion is especially severe for world-scale and other small-scale maps. Four types of distortion can result:

1. The *shape* of an area can be distorted, so that it appears more elongated or squat than in reality.
2. The *distance* between two points may become increased or decreased.

3. The *relative size* of different areas may be altered. One area may appear larger than another on a map but in reality be smaller.
4. The *direction* from one place to another can be distorted.

Most of the world maps in this book, such as Figure 1–12, are *equal area projections*. The primary benefit of this type of projection is that the relative size of the landmasses on the map are the same as in reality. The projection also minimizes distortion in the shape of most landmasses, although areas toward the North and South poles—such as Greenland and Australia—become more distorted. These areas are sparsely inhabited, so distorting their shape usually is not important.

To largely preserve the size and shape of landmasses, however, the projection in Figure 1–12 forces other distortions:

- The Eastern and Western hemispheres are separated into two pieces, a characteristic known as interruption.
- The meridians (the vertical lines), which in reality converge at the North and South poles, do not converge at all on the map. Also, they do not form right angles with the parallels (the horizontal lines).

In contrast, we use uninterrupted projections to display information in Figure 1–18 and the figure in the Globalization and Local Diversity box. The figure in the box uses the Mercator projection, one of the most common. It has several advantages: Shape is distorted very little, direction is consistent, and the map is rectangular. Its greatest disadvantage is that area is grossly distorted toward the poles, making high-latitude places look much larger than they actually are. For example, compare the sizes of Greenland and South America in the maps on pages 17 and 25. The map on page 25 shows their size accurately.

The Appendix presents some of the decisions that must be made when developing a map.

Contemporary High-Tech Mapping

Having largely completed the great task of accurately mapping Earth's surface, which required several centuries, geographers have turned to new technologies to learn more about the characteristics of places. Two important technologies developed during the past quarter-century are remote sensing from satellites (to collect data) and geographic information systems (computer programs for manipulating geographic data). These technologies help geographers create more accurate and complex maps and measure changes over time in the characteristics of places.

Remote Sensing. The acquisition of data about Earth's surface from a satellite orbiting Earth or other long-distance methods is known as **remote sensing**. Geographic applications of remote sensing include mapping of vegetation and other surface cover, gathering data for large unpopulated areas such as the extent of winter ice cover on the oceans, and monitoring changes such as weather patterns and deforestation.

Remote-sensing satellites scan Earth's surface, much like a television camera scans an image in the thin lines you can see on a TV screen. At any moment, a satellite sensor records the image of a tiny area, an area called a picture element or pixel. A map created by remote sensing is essentially a grid containing many rows of pixels.

The smallest feature on Earth's surface that can be detected by a sensor is the resolution of the scanner. Some can sense objects as small as 10 meters across. Future satellites will improve resolution to 1 meter or less.

Weather satellites take a broader view, looking at several kilometers at a time. This enables them to map a large area such as a continent rapidly. Weather forecasters need data about large areas very quickly, because weather systems change so rapidly.

GIS. A **geographic information system (GIS)** is a high-performance computer system that processes geographic data. Many pieces of information about a location are stored in computer files. Each type of information (topography, political boundaries, population density, manufacturing, soil type, earthquake faults, and so on) is stored as an "information layer." A single layer can be displayed by itself, but the GIS is most powerful when it is used to combine several layers to show relations among different kinds of information (Figure 1–2). Powerful desktop microcomputers have speeded the diffusion of GIS technology worldwide.

Geographers use GIS to analyze both environmental and social phenomena. A human-geography example is combining a street map with a population map to determine the number of people living within walking distance of a proposed bus route.

GPS. The Global Positioning System (GPS) is an example of applying new technology to an old human habit: consulting a map to get to a desired destination. The GPS navigation system takes signals from a series of satellites that can pinpoint the current location of a car within 15 meters (50 feet). As a cross-check, sensors in the wheels pinpoint movement of the vehicle.

The driver programs the desired destination, and the monitor calls up the appropriate map, stored on CD-ROM. An antenna receives signals from GPS satellites and feeds them into a computer stored in the trunk of the car. The navigation system provides oral instructions for where to turn, because reading the map while driving can be dangerous.

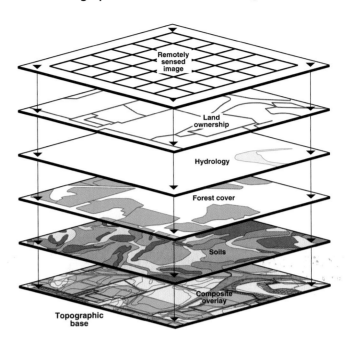

Figure 1–2 A geographic information system. GIS involves storing information about a location in layers. Each layer represents a different piece of human or environmental information. The layers can be viewed individually or in combination.

How Geography Grew as a Science by Asking "Where?"

Long before geography was recognized as a distinct science, humans asked geographic questions, wondering where things were found in the world. The first "geographer" probably was the unknown ancient who crossed a river or climbed a hill, observed what was on the other side, and returned home to tell about it—and to scratch the first rough map in the dirt. The second "geographer" probably was the person who had to find a way from one place to another—in other words, to navigate.

Historical Development of Geography

In early history, geography was synonymous with navigation. As early as 800 B.C., Mediterranean sailors and traders made charts of useful information for finding their way, noting distinctive landmarks such as rock formations, islands, and the direction of ocean currents. For millennia, Polynesian peoples have navigated thousands of kilometers over the South Pacific islands, using three-dimensional maps, called "stick charts," made of strips from palm trees and sea shells. The shells represented islands, and the palm strips represented patterns of waves between the islands (Figure 1–3).

Geography in the Ancient World. The word *geography* was invented by the ancient Greek scholar Eratosthenes (276?–194? B.C.); it is based on two Greek words, *geo* meaning "Earth," and *graphy*, meaning "to write." But the ancient Greeks were concerned with geographic con-

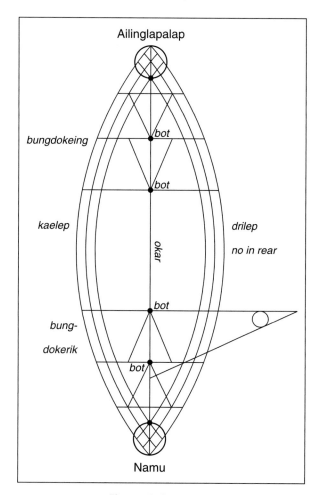

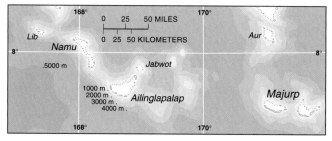

Figure 1–3 Polynesian "stick chart," a type of ancient map, made of shells and strips of palm trees. Islands were shown with shells, and patterns of swelling of waves with palm strips. Curved palms represented different wave swells than straight strips. This ancient example depicted the sea route between Ailinglapalap and Namu, two islands in the present-day Marshall Islands, in the South Pacific Ocean.

cepts for hundreds of years before the name *geography* was invented. In the sixth century B.C., the philosopher Thales of Miletus (624?–546? B.C.) applied principles of geometry to measuring land area. (Miletus was an ancient port city in Turkey.) Thales' student, Anaximander (610–546? B.C.), argued that the world was shaped like a cylinder and made a world map based on information from sailors.

Aristotle (384–322 B.C.) was the first to demonstrate that Earth was spherical, noting that matter falls together toward a common center, that during an eclipse Earth's shadow on the Moon is circular, and that the groups of stars visible at night change as one travels north or south. The Greek astronomer and geographer Pytheas sailed to Iceland in 325 B.C. and worked out a method for determining latitude by observing the position of stars. Hipparchus (190?–125? B.C.) drew lines on maps of Earth's surface to create reference points for the location of places. To this day, we depend on his concept of north-south meridians and east-west parallels (longitude and latitude).

Eratosthenes, the first person of record to use the word *geography*, accepted that Earth was round, as few

did in his day; he also calculated its circumference within an amazing 0.5 percent accuracy. In one of the first geography books, he described the known world and correctly divided Earth into five climatic regions—a *torrid* zone across the middle, two *frigid* zones at the extreme north and south, and two *temperate* bands in between. Eratosthenes also prepared one of the earliest maps of the known world.

Roman geographers made their contribution, too. Strabo (63? B.C.–A.D. 24?) exhaustively described the known world in his 17-volume work *Geography*. Strabo regarded Earth as a sphere at the center of a spherical universe. In the second century A.D., the Roman Empire controlled an extensive area of the known world, including much of Europe, northern Africa, and western Asia. Taking advantage of information collected by Roman merchants and soldiers, another Greek, Ptolemy (A.D. 100?–170?), wrote an eight-volume *Guide to Geography*. He prepared numerous maps, which were not improved upon for more than a thousand years.

Geography also developed in China, independent of European studies. The oldest Chinese geographical writing, from the fifth century B.C., describes the economic resources of the country's different provinces. Phei Hsiu (or Fei Xiu), the "father of Chinese cartography," produced an elaborate map of the country in A.D. 267.

Geography in the Middle Ages.   After Ptolemy, little progress in geographic thought was made in the ancient world. Following the collapse of the Roman Empire in the fifth century A.D., the word *geography* disappeared from European vocabulary. During the Middle Ages (roughly A.D. 1100–1500), geographic inquiry continued outside of Europe. Beginning in the seventh century, Muslim armies controlled much of northern Africa and southern Europe and eventually reached as far east as present-day Indonesia in Southeast Asia. The Muslim geographer al-Edrisi (1100–1165?) prepared a world map and geography text in 1154. Ibn-Battutah (1304–1368?) wrote *Rihlah* ("Travels") based on his journeys of more than 120,000 kilometers through the Muslim world over three decades.

Europeans did continue to explore portions of Earth's surface previously unfamiliar to them. Vikings sailed west from Scandinavia to Iceland in 860. Erik Thorvaldson (Erik the Red), having been banished from Iceland, sailed to Greenland in 982, returned to Iceland to collect colonists, and established a permanent settlement on Greenland in 986. Bjarni Herjulfsson left Iceland in 985 to join Erik's colony, but he sailed too far south and reached Newfoundland instead, probably the first European to sight mainland North America. Bjarni did not land in Newfoundland, preferring to find his way eventually to Greenland. In 1001, Leif Eriksson, son of Erik the Red, sailed off course from Norway to Greenland and landed in Newfoundland, where he set up a camp.

Christopher Columbus sailed west across the Atlantic Ocean in 1492 to try to discover a sea passage between Europe and Asia that would eliminate the long and difficult trip around the southern tip of Africa. Columbus made four voyages across the Atlantic, and died in 1506 believing that he had reached Asia. But other explorers soon realized Columbus's error, and within a year of his death, the first European map was published showing the existence of a landmass in the Western Hemisphere.

The first European explorer to see the Pacific Ocean was Vasco Núñez de Balboa, who viewed it from a mountain in Panama in 1513. The first ship to sail around the world was the *Victoria*, captained for most of its voyage by Ferdinand Magellan. The ship left Spain in 1519, passed from the Atlantic to the Pacific through what is now known as the Estrecho de Magallanes (Straits of Magellan) in southern Chile, and reached the Philippines in 1521. Magellan was killed in a fight with the inhabitants of the Philippines, but another member of the crew, Juan Sebastián del Cano, sailed across the Indian Ocean and around the southern tip of Africa, to complete the around-the-world voyage back to Spain in 1522.

Geographic thought enjoyed a resurgence in Europe in the seventeenth century, inspired by exploits of European explorers to establish trading routes and gain control of resources elsewhere in the world. *Geographia Generalis*, written by the German Bernhardus Varenius (1622–1650), stood for more than a century as the standard treatise on systematic geography. Varenius also wrote a description of Japan, but he died before he could complete a more comprehensive work on regional geography.

How Geography Grew as a Science

The German philosopher Immanuel Kant (1724–1804) placed geography within an overall framework of scientific knowledge. He argued that all knowledge can be classified logically or physically. For example, a logical classification organizes plants and animals into a systematic framework of species, based on their characteristics, regardless of when or where they exist. A physical classification identifies plants and animals that occur together in particular times and places. Descriptions according to time comprise history, and descriptions according to place comprise geography. History studies phenomena that follow one another chronologically, whereas geography studies phenomena that are located beside one another.

As modern geography developed, two opposing views emerged. One group of geographers believed that our physical environment *causes* human behavior. Another group believed that everything in the landscape is interrelated, but physical factors do not necessarily cause human actions. Let us look at each school of thought.

Does the Physical Environment Cause Human Actions?
Modern geography began with two nineteenth-century German geographers, Alexander von Humboldt (1769–1859) and Carl Ritter (1779–1859). Prior to their work, geographers described the physical and social characteristics of places in great detail, but they did not explain their observations systematically. Humboldt and Ritter argued that geography should move beyond describing Earth's surface to explaining *why* certain phenomena were present or absent. This is the origin of our "where" and "why" approach.

Humboldt and Ritter urged human geographers to adopt the methods of scientific inquiry used by natural scientists. They argued that the scientific study of social and natural processes is fundamentally the same. Natural scientists have made more progress in formulating general laws than have social scientists, so an important goal of human geographers is to discover general laws.

According to Humboldt and Ritter, human geographers should apply laws from the natural sciences to understanding relationships between the physical environment and human actions. Humboldt and Ritter concentrated on how the physical environment *caused* social development, an approach called **environmental determinism**.

Other influential geographers adopted environmental determinism in the late nineteenth and early twentieth

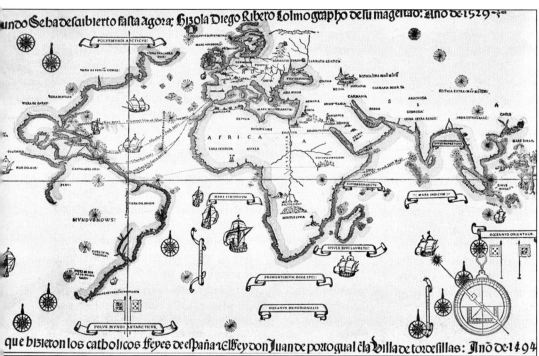

(Top.) Nautical map of Diego Ribero, 1529. This was the first map to show the vast extent of the Pacific Ocean. Survivors of Ferdinand Magellan's around-the-world journey provided Ribero with much of the information when they returned to Spain in 1522. (The Bettmann Archive)

(Bottom.) Compare the accuracy of the coastlines on Ribero's map with the recent image of the world based on satellite photographs. The composite image was assembled by the Geosphere Project of Santa Monica, California. Thousands of images were recorded over a ten-month period by satellites of the National Oceanographic and Atmospheric Administration. The images were then electronically assembled, much like a jigsaw puzzle. (Tom Van Sant/The Stock Market)

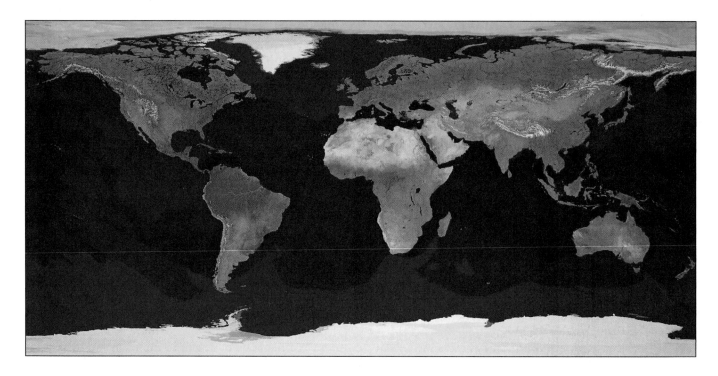

centuries. Friedrich Ratzel (1844–1904) and his American student, Ellen Churchill Semple (1863–1932), claimed that geography was the study of the influences of the natural environment on people. Another early American geographer, Ellsworth Huntington (1876–1947), argued that climate was a major determinant of civilization. For instance, according to Huntington, the temperate climate of maritime northwestern Europe produced greater human efficiency as measured by better health conditions, lower death rates, and higher standards of living.

The geographic approach that emphasizes human-environment relationships is now known as **cultural ecology**. To explain the relationship between human activities and the physical environment, modern geographers reject environmental determinism in favor of possibilism. According to **possibilism**, the physical environment may limit some human actions, but people have the ability to adjust to their environment. People can choose a course of action from many alternatives in the physical environment.

For example, the climate of any location influences human activities, especially food production. From one generation to the next, people learn that different crops thrive in different climates—rice requires plentiful water, whereas wheat survives on limited moisture, and it actually grows poorly in very wet environments. On the other hand, wheat is more likely than rice to be grown successfully in colder climates. Thus, under possibilism, it is possible for people to choose the crops they grow, to be compatible with their environment.

In a Region, Everything Is Related. A second school of geographic thought, regional studies, developed in France during the nineteenth century. The **regional studies** approach—sometimes called the cultural landscape approach—was initiated by Paul Vidal de la Blache (1845–1918) and Jean Brunhes (1869–1930). It was later adopted by several American geographers, including Carl Sauer (1889–1975) and Robert Platt (1880–1950).

These geographers rejected the idea that physical factors simply determine human actions. Instead, they argued that each place has its own distinctive landscape that results from a unique combination of social relationships and physical processes. Therefore, geographers should start by closely observing the physical and social characteristics of a place. They called this the *regional studies* approach, stating that the work of human geography is to discern the relationships among social and physical phenomena in a particular study area. Everything in the landscape is interrelated, so physical factors do not simply cause human actions, as environmental determinists had argued.

Today, contemporary geographers reject the extreme position of the environmental determinists that the physical environment causes human actions. They also have considerably modified the regional studies approach. However, these two traditions of geographic thought—human-environment relationships and regional studies—remain fundamental to the scientific study of geography.

Location: Where Something Is

The most fundamental concept in geography is location, which is the position that something occupies on Earth's surface. Geographers identify the location of something in four ways—by place-name, site, situation, and mathematical location—to answer the "where" question.

The dialogue about "Miami" that opened this chapter illustrates all four methods. The student's first response to the "where" question was the *place-name* "Miami." When this response failed accurately to indicate the location of Miami, the student then referred to its *site* characteristics, such as vegetation, topography, and climate. The next response drew on Miami's *situation*, in the city of Oxford and the state of Ohio and near the city of Cincinnati and the state of Indiana. Finally, the student gave two examples of Miami's *mathematical location*.

Place-names

Because all inhabited places on Earth's surface have been named, the simplest way to describe a particular location is by referring to its name. Geographers call the name given to a portion of Earth's surface its **toponym** (literally, place-name).

The name of a place may give us a clue about its founders, physical setting, social customs, or political changes. Some communities take the name of an otherwise obscure founder or early leader, such as the West Virginia communities of Jenkinjones (named for a mine operator) and Gassaway (named for Senator Henry Gassaway Davis). Others adopt the name of a famous person who had no connection with the community. George Washington's name has been selected for one state, counties in 30 other states, and dozens of cities, including the national capital. Most states also contain a place named after James Madison or Thomas Jefferson.

Places may be named after important historical events. One of the most straightforward is found in England. The key victory in the Norman (French) conquest of England in 1066 was the Battle of Hastings. The actual battle site, 10 kilometers (6 miles) from the town of Hastings, is now simply known as "Battle." After the assassinations of President John F. Kennedy in 1963 and the Rev. Martin Luther King, Jr., in 1968, many communities renamed streets, parks, and buildings after them.

Some place-names derive from features of the physical environment. Trees, valleys, bodies of water, and other natural features appear in the place-names of most languages. The capital of the Netherlands, called *'s Gravenhage* in Dutch (in English, The Hague), means "the prince's forest." *Aberystwyth*, in Wales, means "mouth of the River Ystwyth," while 22 kilometers (13 miles) upstream lies the tiny village of *Cwmystwyth*, which means "valley of the Ystwyth." The name of the river, *Ystwyth*, in turn is the Welsh word for "meandering," descriptive of a stream that bends like a snake.

The name of a place can tell us a lot about the social customs of its early inhabitants. Some settlers select place-names associated with religion, such as Saint Louis, whereas other names derive from ancient history, such as Athens, Attica, and Rome. A place-name may also indicate the origin of its settlers. Place-names commonly have British origins in North America and Australia, Portuguese origins in Brazil, Spanish origins elsewhere in Latin America, and Dutch origins in South Africa.

Repeated use of the same name can cause confusion, as in the introductory case study. Hundreds of streets in London, England, are called High Street, a relic of medieval times, when each neighborhood was an independent town. The most important shopping street in each town was known as the High Street.

Confusion and Change. Confusion may also arise if local residents commonly employ names other than the official ones. New York City has an abundance of unofficial names. The Avenue of the Americas is almost universally known by its former name, Sixth Avenue, and the Queensboro Bridge is generally called the 59th Street

Bridge. A place having two or more local names presents a quandary to cartographers who need to give the place a label.

The Board of Geographical Names, operated by the U.S. Geological Survey, was established in the late nineteenth century to be the final arbiter of names on U.S. maps. In recent years, the board has been especially concerned with removing offensive place-names, such as racial or ethnic slurs.

Places can change names. The city of Cincinnati was originally named Losantiville. The name was derived as follows: *L* is for Licking River; *os* is Latin for mouth; *anti* is Latin for opposite; *ville* is Latin for town—hence, "town opposite the mouth of the Licking River." The name was changed to Cincinnati in honor of a society of Revolutionary War heroes named after Cincinnatus, an ancient Roman general.

Names can also change as a result of political upheavals. For example, following World War II, Poland gained control over territory that was formerly part of Germany and changed many of the place-names from German to Polish. Among the larger cities, Danzig became Gdańsk, Breslau became Wrocław, and Stettin became Szczecin.

After the fall of communism in the early 1990s, names throughout Eastern Europe were changed in the 1990s, in many cases reverting to those used before the Communists had gained power a few decades earlier. For example, after the demise of communism, Olomouc, a Czech city of 100,000, changed Lenin Street to Liberty Street, Red Army Square to Lower Square, and Liberation Street to Masaryk Street (for the first president of democratic Czechoslovakia between 1919 and 1935). Gottwaldov, a Czech city named for a Communist president of Czechoslovakia, reverted to its former name Zlín, and Leningrad, the second largest city in the Soviet Union, reverted to St. Petersburg, Russia, which was the city's name prior to Communist rule.

Someone unfamiliar with foreign languages might have difficulty in identifying the English name for these European countries: Civitas Helvetia, Österreich, Magyarorszag, and Suomi. These are the official names for Switzerland, Austria, Hungary, and Finland, respectively.

Money and Politics. Pioneers lured to the American West by the prospect of finding gold or silver placed many picturesque names on the landscape. Place-names in Nevada selected by successful miners include Eureka, Lucky Boy Pass, Gold Point, and Silver Peak. Unsuccessful Nevada pioneers sadly or bitterly named other places Battle, Disaster Peak, and Massacre Lake. In 1959 the Elko, Nevada, county commissioners gave the name Jackpot to a town near the Idaho state border, in recognition of the importance of legalized gambling to the local economy.

What may be the longest community name in the world has an economic origin—the Welsh town of *Llanfairpwllgwyngyllgogerychwyrndrobwllllantysiliogogogoch*.

The 58-letter name means "the Church of St. Mary's in the grove of the white hazelnut tree near the rapid whirlpool and the Church of St. Tisilio near the red cave." The town's name originally encompassed only the first 20 letters (*Llanfairpwllgwyngyll*), but when the railway was built in the nineteenth century, the townspeople lengthened it. They decided that signs with the longer name in the railway station would attract attention and bring more business and visitors to the town.

Sometimes a place-name is so symbolic that its use can cause great political difficulty—and lost revenue. When Yugoslavia's southernmost republic declared independence in 1991, its leaders wished to call the new country *Macedonia*, the same name it had as a republic of Yugoslavia. But Greece felt threatened by this use of the name, because Macedonia is also the name of Greece's northernmost region. As the home of Aristotle and Alexander the Great, ancient Greek Macedonia was an important cultural hearth for Greece and Western civilization, and the new country adopted Alexander's star-burst symbol for its new flag.

Both countries suffered financially from the dispute. Impoverished Macedonia was prevented by Greece from receiving aid from international organizations, and the Greek port of Thessaloniki lost revenue because Macedonia was not allowed to ship goods through it. In 1995, Greece lifted the economic blockade when Macedonia agreed to adopt a new flag and change its constitution to eliminate any hint of claims on Greek territory.

However, the 1995 agreement did not include a new name for the country. Greece suggested that it be called the Slavic Republic of Macedonia, but this was rejected because only 64 percent of the inhabitants of the new country were Slavs. In an already volatile area of the world, naming the country "Slavic" would inflame Albanians and Turks, who comprised 21 percent and 5 percent of Macedonia's population, respectively. Through the 1990s, the international community sidestepped the dispute by calling the new country by the clumsy name FYROM, an acronym for "Former Yugoslav Republic of Macedonia."

Site

The second way to indicate location is by **site**, which is the physical character of a place. Important site characteristics include climate, water sources, topography, soil, vegetation, latitude, and elevation. The combination of physical characteristics gives each location a unique character.

Site factors always have been essential in selecting locations for settlements, although people have disagreed on the attributes of a good site, depending on cultural values. Some have preferred a hilltop site for easy defense from attack. Other people located settlements near convenient river-crossing points to facilitate communication with people in other places.

An island combines the attributes of both hilltop and riverside locations, because the site provides good defense and transportation links. The site of the country of Singapore, for example, is a small, swampy island approximately 1 kilometer off the southern tip of the Malay Peninsula at the eastern end of the Strait of Malacca. The city of Singapore covers nearly 20 percent of the island.

In general, the characteristics of a site do not change over time, but human preferences do. The relatively warm, humid climate in the southeastern United States traditionally retarded population growth, but in recent years it has become an attraction. People increasingly prefer the climate in the Southeast because they can participate in outdoor recreational activities throughout the year and do not have to shovel snow in winter. At the same time, technological change, especially the invention of air conditioning, has increased the Southeast's attractiveness by enabling people to escape the region's high heat and humidity.

However, humans do have the ability to modify the characteristics of a site. The southern portion of New York City's Manhattan Island is twice as large today as it was in 1626, when Peter Minuit bought the island from its native inhabitants for the equivalent of $23.75 worth of Dutch gold and silver coins. The additional land area was created by filling in portions of the East River and Hudson River. In the eighteenth century, landfills were created by sinking old ships and dumping refuse on top of them.

Because of poor health conditions, the city decided in 1797 to cover all the landfills with soil and gravel and to lay out a new street, called South Street, to halt further dumping in the river. Today, South Street is two blocks from the river. More recently, New York City permitted construction of Battery Park City, a 57-hectare (142-acre) site designed to house more than 20,000 residents and 30,000 office workers (Figure 1–4). The central areas of Boston and Tokyo have also been expanded through centuries of landfilling in nearby bays, substantially changing these sites.

Situation

Situation is the location of a place relative to other places. Situation is a valuable way to indicate location, for

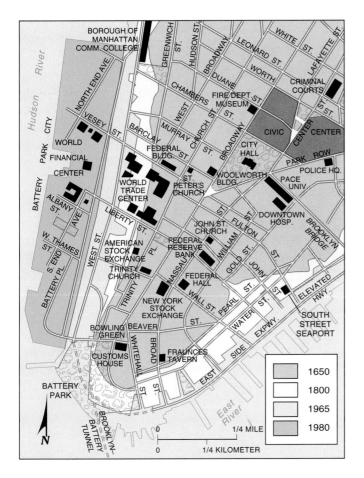

Figure 1–4 Site of New York City. Much of the southern part of New York City's Manhattan Island was built on landfill. Several times in the past 200 years, the waterfront has been extended into the Hudson and East rivers to provide more land for offices, homes, parks, warehouses, and docks. Battery Park City and the 110-story World Trade Center towers (at left in the photograph) were built on landfill in the Hudson River. The office buildings at right, behind Battery Park, were built on landfill in the East River. (Manfred Gottschalk/West Light)

two reasons—finding an unfamiliar place, and understanding its importance.

First, situation helps us find an unfamiliar place by comparing its location with a familiar one. We give directions to people by referring to the situation of a place: "It's down past the court house, on Locust Street, after the third traffic light, beside the yellow brick bank." We identify important buildings, streets, and other landmarks to direct people to the desired location.

For example, even long-time residents of Paris might have difficulty finding the Marmottan Museum by its address, 2 rue Louis-Boilly, because the street is only one block long. The museum, which contains one of the world's largest collections of paintings by Claude Monet, can be found by referring to its situation: one block east of the city's largest park, the Bois de Boulogne, near the Muette stop on the Métro (subway).

Second, situation helps us understand the importance of a location. Many locations are important because they are accessible to other places. For example, because of its location, Singapore has become a center for the trading and distribution of goods for much of Southeast Asia. Singapore is situated near the Strait of Malacca, which is the major passageway for ships traveling between the South China Sea and the Indian Ocean (Figure 1–5).

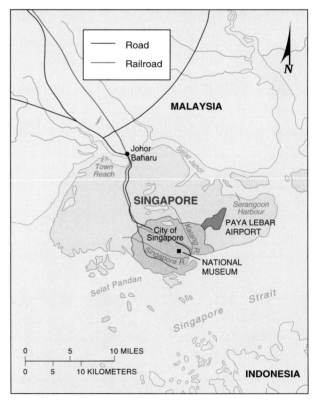

Figure 1–5 Situation of Singapore. The small country of Singapore, less than one-fifth the size of Rhode Island, has an important situation for international trade. The country is situated at the confluence of several straits that serve as major passageways for shipping between the South China Sea and the Indian Ocean. Downtown Singapore is situated near where the Singapore River flows into the Singapore Strait. In the foreground, the dome at right is the National Museum; the spire at center is St. Andrew's Cathedral. (Jose Fuste Raga/The Stock Market)

Mathematical Location

Sometimes it is necessary to pin down the location of something with precise accuracy. In using an atlas, you probably have used the familiar "A–1" system of finding a place on a map. Typically, a row of numbers runs across the top and bottom of the map, and a column of letters runs down each side. When you look up Cactus, Texas, in the atlas, its map coordinates are given as C–7. You simply find C in the column and look across under the 7, and there is Cactus. But this general location is accurate only within several kilometers.

Latitude and Longitude. The location of Cactus, or any other place on Earth's surface, can be described very precisely by drawing an imaginary grid on the globe, and then describing the place's location on the grid, using a set of numbers called latitude and longitude. The universally accepted numbering system of latitude and longitude consists of imaginary arcs drawn on the globe.

A **meridian** is an arc drawn between the North and South poles. Every meridian has the same length and the same beginning and end points. The location of each meridian is identified on Earth's surface according to a numbering system known as **longitude** (Figure 1–6). One meridian, which passes through the Royal Observatory at Greenwich, England, has been designated by international agreement as the "starting point" for numbering the meridians. It is labeled as *0 degrees longitude* and is also called the **prime meridian**.

The meridian on the opposite side of the globe from the prime meridian is 180° longitude. All other meridians have numbers between 0° and 180° and are designated "east" or "west" to show that they are either east or west of the prime meridian. For example, New York City is located at 74° west longitude, whereas Lahore, Pakistan, is 74° east longitude. San Diego is located at 117° west longitude; Tianjin, China, is at 117° east longitude.

The second set of imaginary arcs drawn on Earth's surface are **parallels**. These are circles drawn around the globe parallel to the equator at right angles to the meridians. The numbering system used to indicate the location of parallels is called **latitude**. The equator is 0° latitude, the North Pole 90° north, and the South Pole 90° south. New York City is located at 41° north latitude, whereas Wellington, New Zealand, is at 41° south. San Diego is located at 33° north latitude; Santiago, Chile, is at 33° south. Latitude and longitude are used together to identify locations. For example, Cactus, Texas, is at the intersection of 36° north latitude and 102° west longitude.

The determination of latitude and longitude is a good example of how geography is partly a natural science and partly a study of human behavior. Latitudes are scientifically derived by Earth's shape and its rotation around the Sun. The equator (0° latitude) is the parallel with the largest circumference and is the place where every day has 12 hours of daylight. On the other hand, 0° longitude is a human creation. It runs through Greenwich because England was the world's most powerful country in the eighteenth century, when meridians were first measured.

We can find the mathematical location of a place even more precisely, if necessary. Each degree is divided into 60 minutes (′), and each minute in turn is divided into 60 seconds (″). For example, the official mathematical location of Denver, Colorado, is 39°44′ north latitude and 104°59′ west longitude. The State Capitol building in Denver is located at 39°42′52″ north latitude and 104°59′04″ west longitude. The latitude/longitude system is especially useful for navigation on the sea. (See the box Globalization and Local Diversity: Time Zones and the International Dateline.)

U.S. Land Ordinance of 1785. In addition to the global system of latitude and longitude, other mathematical indicators of locations are used in different parts of the world. In the United States, the **Land Ordinance of 1785** divided much of the country into a system of townships and ranges to facilitate the sale of land to settlers in the West. The initial surveying was performed by Thomas Hutchins, who was appointed geographer to the United States in 1781. After Hutchins died in 1789, responsibility for surveying was transferred to the Surveyor General.

In this system, a **township** is a square 6 miles on each side. Some of the north-south lines separating townships are called **principal meridians**, and some east-west lines

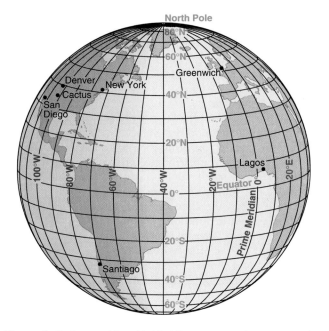

Figure 1–6 Geographic grid. Meridians are arcs that connect the North and South poles. The meridian through Greenwich, England, is the prime meridian or 0° longitude. Parallels are circles drawn around the globe parallel to the equator. The equator is 0° latitude; the North Pole is 90° north latitude.

Globalization and Local Diversity

Time Zones and the International Dateline

Where we are located in the world affects the time of day. One element of **local diversity** that we take for granted is that elsewhere in the world the time of day may be earlier or later than at home. In a world of instantaneous communications, we take account of local time differences in the **globalization** of the economy or culture, such as placing long-distance telephone calls or watching the Olympics on television.

Although time of day varies locally, the world as a whole is organized into 24 standard time zones. At first glance, the reason **why** the world has 24 standard time zones seems straightforward. Earth as a sphere is divided into 360° of longitude (the degrees from 0° to 180° west longitude, plus the degrees from 0° to 180° east longitude). As Earth rotates daily, these 360 imaginary lines of longitude pass beneath the cascading sunshine. If we let every fifteenth degree of longitude represent one time zone, and divide the 360° by 15°, we get 24 time zones, or one for each hour of the day.

Standard time zones were established at the urging of the railroads in the United States in 1883 and in the rest of the world following the international meridian conference in Washington, DC, in 1884. Before standard time zones were created, each locality set its own time, usually that kept by a local jeweler. Railroads were the main cross-country transportation of the time, and each rail company kept its own time, normally that of the largest city it served. Train timetables listed two sets of arrival and departure times, one for local time and one for railroad company time. Railroad stations had one clock for local time and a separate clock for each of the railroad companies using the station.

At noon on November 18, 1883, time stood still in the United States so that each locality could adjust to the new standard time zones. In New York City, for example, time stopped for 3 minutes and 58 seconds to adjust to the new Eastern Standard Time. However, for many years, Chicago resisted the change and continued to be 17 minutes ahead of Central Standard Time.

The international agreement designated the time at the prime meridian or 0° longitude as **Greenwich Mean Time** (GMT), or Universal Time (UT). It is the master reference time for all points on Earth. Earth rotates eastward, so any place to the east of you always passes "under" the Sun earlier. Thus, as you travel eastward from the prime meridian, you are "catching up" with the Sun, so you must turn your clock ahead from GMT, by 1 hour for each 15°. If you travel westward from the prime meridian, you are "falling behind" the Sun, so you turn your clock back from GMT, by 1 hour for each 15°.

The eastern United States, which is near 75° west longitude, is therefore 5 hours earlier than Greenwich Mean Time (the 75° difference between the prime meridian and 75° west longitude, divided by 15° per hour, equals 5 hours). Thus, when the time is 11 A.M. GMT, the time in the eastern United States is 5 hours earlier, or 6 A.M. (Figure 1).

Each 15° band of longitude is assigned to a standard time zone. The 48 contiguous U.S. states and Canada share four standard time zones, known as Eastern, Central, Mountain, and Pacific:

- The Eastern Standard Time Zone is near 75° west longitude, which passes close to Philadelphia, and is 5 hours earlier than GMT.
- The Central Standard Time Zone is near 90° west longitude, which passes through Memphis, Tennessee, and is 6 hours earlier than GMT.
- The Mountain Standard Time Zone is near 105° west longitude, which passes through Denver, Colorado, and is 7 hours earlier than GMT.
- The Pacific Standard Time Zone is near 120° west longitude, which passes through Lake Tahoe in California and Nevada, and is 8 hours earlier than GMT.

Most of Alaska is in the Alaska Time Zone, which is 9 hours earlier than GMT, Hawaii and some of the Aleutian Islands are in the Hawaii-Aleutian Time Zone, which is 10 hours earlier than GMT; and eastern Canada is in the Atlantic Time Zone, which is 4 hours earlier than GMT.

Suppose today is Sunday. Is it Sunday all over the world? No. When you cross the **International Date Line** (shown on the figure) heading eastward toward America, the clock moves back 24 hours, or one entire day. When you go westward toward Asia, the calendar moves ahead one day.

To see the need for the International Date Line, try counting the hours around the world from the time zone in which you live. As you go from west to east, you add 1 hour for each time zone. When you return to your starting point, you will reach the absurd conclusion that it is 24 hours later in your locality than it really is.

Therefore, when the time in New York City is 2 P.M. Sunday (as shown in the figure), it is 7 P.M. Sunday in London, 8 P.M. Sunday in Rome, 9 P.M. Sunday in Jerusalem, and 10 P.M. Sunday in Moscow, 3 A.M. Monday in Singapore, and 5 A.M. Monday in Sydney, Australia. Continuing farther east, it is 7 A.M. *Monday* in Wellington, New Zealand—but when you get to Honolulu, it is 9 A.M.

Sunday, because the International Date Line lies between New Zealand and Hawaii.

Some places have asserted local diversity by tinkering with standard time zones. For example, Newfoundland is $3\frac{1}{2}$ hours earlier than GMT, India $5\frac{1}{2}$ hours later, and central Australia $9\frac{1}{2}$ hours later. The residents of Newfoundland assert that their island, which lies between 53° and 59° west longitude, would face dark winter afternoons if it were 4 hours earlier than GMT like the rest of eastern Canada and dark winter mornings if it were 3 hours earlier than GMT.

In 1997, the country of Kiribati took an even bolder step at altering the world map of time zones—it moved part of the International Date Line 3,000 kilometers west.

Despite the protests of other Pacific island countries, the Royal Observatory in England ruled that Kiribati was within its rights to move the International Date Line.

Why did Kiribati move the International Date Line, and why did neighboring countries care? A collection of small islands extending from about 169° east longitude to 150° west longitude, Kiribati moved the International Date Line to its eastern boundary so that it would be the westernmost country lying to the east of the line, and therefore the first country to see each day's sunrise. Kiribati hoped this feature would attract tourists to celebrate the start of the new millennium on January 1, 2000—or January 1, 2001, when sticklers point out the millennium really begins.

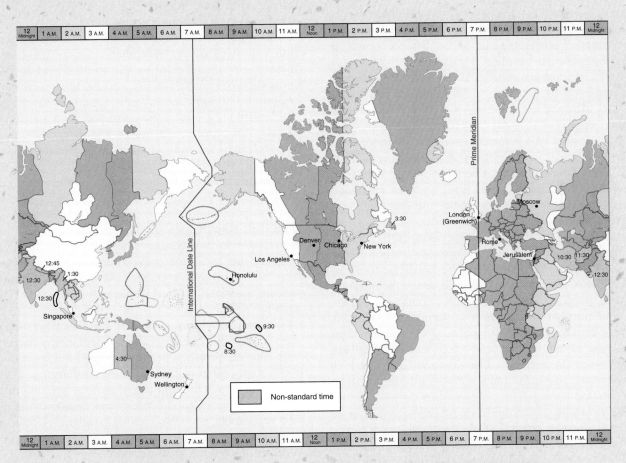

Time zones. The world is divided into 24 standard time zones, each of which represents 15° of longitude. Greenwich Mean Time (GMT) is the time near the prime meridian, or 0° longitude. The Pacific Time Zone, which encompasses the western part of the United States and Canada, is 8 hours behind GMT because it is situated near 120° west longitude.

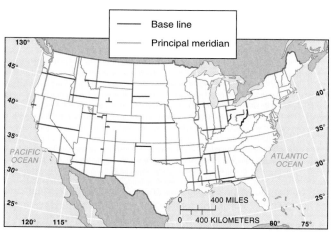

Figure 1–7 Principal meridians and baselines. To facilitate the numbering of townships, the U.S. Land Ordinance of 1785 designated several north-south lines as principal meridians and several east-west lines as baselines. As territory farther west was settled, additional lines were delineated.

are designated **base lines** (Figure 1–7). Each township has a number corresponding to its distance north or south of a particular base line. Townships in the first row north of a baseline are called T1N (Township 1 North), the second row to the north is T2N, the first row to the south is T1S, and so on. Each township has a second number, known as the *range*, corresponding to its location east or west of a principal meridian. Townships in the first column east of a principal meridian are designated R1E (Range 1 East). The Tallahatchie River, for example, is in township T23N R1E, north of a baseline that runs east-west across Mississippi and east of a principal meridian along 90° west longitude.

A township is divided into 36 **sections**, each of which is 1 mile by 1 mile (Figure 1–8). Sections are numbered in a consistent order, from 1 in the northeast to 36 in the southeast. Each section is divided into four quarter-sections, designated as the northeast, northwest, southeast, and southwest quarters of a particular section. A quarter-section, which is 0.5 mile by 0.5 mile, or 160 acres, was the amount of land many western pioneers bought as a homestead. The Tallahatchie River is located in the southeast and southwest quarter-sections of Section 32.

The township and range system is still important in understanding the location of objects across much of the United States. It explains the location of highways across the Midwest, farm fields in Iowa, and major streets in Chicago.

Distribution: Spatial Regularities

Each object has a unique location on Earth, but geographers seek patterns in the arrangement across Earth of collections of objects. On Earth as a whole, or within an area of Earth, objects may be numerous or scarce, close

together or far apart. The arrangement of something across Earth's surface is known as its *distribution*. Geographers identify three main properties of distribution: density, concentration, and pattern.

Density

The frequency with which something occurs in an area is its **density**. The feature being measured could be people, buildings, dwelling units, cars, volcanoes, or anything. The area could be measured in square kilometers, square miles, hectares, acres, or any other unit of area (Figure 1–9).

The *arithmetic density* is the total number of objects in an area, such as people. The arithmetic density of the United Kingdom, for example, is 240 persons per square kilometer (622 persons per square mile). This is simply the total population (58.6 million people) divided by the U.K.'s area (244,110 square kilometers or 94,251 square miles). Arithmetic density is a useful measure of living arrangements in different places.

Remember that a large *population* does not necessarily lead to a high *density*. Arithmetic density involves two measures: the number of people and the land area. The most populous country in the world, China, with approximately 1.2 billion inhabitants, by no means has the highest density. The arithmetic density of China is approximately 126 persons per square kilometer (327 persons per square mile), only one-half as high as the United Kingdom. Although China has about 20 times more inhabitants than the United Kingdom, it also has nearly 40 times more land.

High population density is also unrelated to poverty. The Netherlands, one of the world's wealthiest countries, has an arithmetic density of approximately 377 persons per square kilometer (976 persons per square mile). One of the poorest countries, Mali, has an arithmetic density of only 8 persons per square kilometer (21 persons per square mile).

Geographers measure density in other ways, depending on the subject being studied. Geographers concerned with the relationship between population growth and food supply often calculate two other densities. Physiological density is the number of persons per unit of area suitable for agriculture. Agricultural density is the number of farmers per unit area of farmland. Urban geographers frequently use housing density, which is the number of dwelling units per unit of area.

Concentration

The extent of a feature's spread over an area is its **concentration**. If the objects in an area are close together, they are considered to be *clustered*. If they are relatively far apart, they are considered to be *dispersed*. To compare the level of concentration clearly, two areas need to have the same number of objects and the same size area, or else be adjusted to correspond.

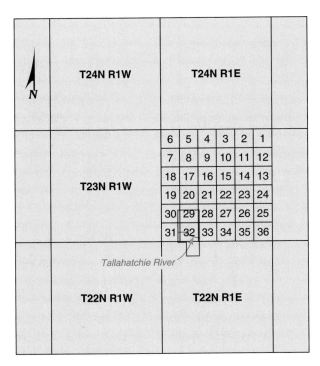

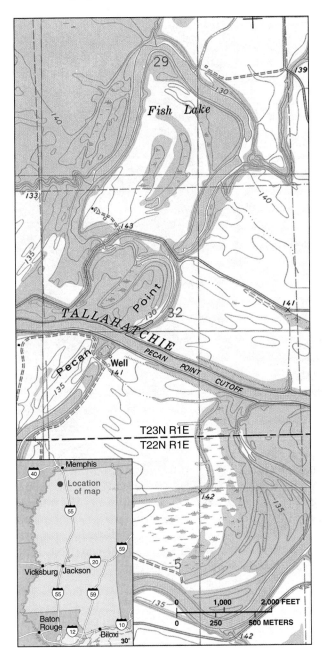

Figure 1–8 (Above) Townships. Townships are typically 6 miles by 6 miles, although physical features, such as rivers and mountains, result in some irregularly shaped ones. The Tallahatchie River, for example, is located in the twenty–third township north of a baseline that runs east-west across Mississippi, and in the first range east of the principal meridian at 90° west longitude. Townships are divided into 36 sections, each 1 square mile. Sections are divided into four quarter-sections. The Tallahatchie River is located in the southeast and southwest quarter-sections of Section 32, T23N R1E. (Right) Topographic map. This map, published by the United States Geological Survey, has a fractional scale of 1:24,000. Expressed as a written statement, one inch on the map represents 24,000 inches (2,000 feet) on the ground. The bar line below the map displays the scale in a third way. The map displays portions of two townships, shown on the above map. The brown lines on the map are contour lines, which show the elevation of any location.

Geographers use the concept of concentration in a number of ways. For example, one of the major changes in the distribution of the U.S. population is increasing dispersion. The total number of people living in the United States is growing slowly—less than 1 percent per year—and the land area is essentially unchanged. But the population distribution is changing from *relatively clustered* in the Northeast to more *evenly dispersed* across the country.

Concentration is not the same as density. One area with relatively high density could have a dispersed population, whereas another area with the same density could have a clustered population. We can illustrate the difference between density and concentration by the change in the distribution of major league baseball teams in North

America (Figure 1–10). In 1900, the major leagues had 16 teams, a distribution that remained unchanged for more than half a century. Beginning in 1953, the following 6 of the 16 teams moved to other cities:

- Braves—Boston to Milwaukee in 1953, then to Atlanta in 1966
- Browns—St. Louis to Baltimore (Orioles) in 1954
- Athletics—Philadelphia to Kansas City in 1955, then to Oakland in 1968
- Dodgers—Brooklyn to Los Angeles in 1958
- Giants—New York to San Francisco in 1958
- Senators—Washington to Minneapolis (Minnesota Twins) in 1961

Figure 1–9 Spatial distribution. Distribution is represented in three ways: density, concentration, and pattern. Assume that the area in all six figures represents 1 acre. The density (left) is 6 houses per acre in A and 12 houses per acre in B. The concentration (center) is dispersed in A and clustered in B. Note that while the concentration changes in the center sketches, the density is the same in both. The pattern (right) is linear in A and alternating setbacks in B.

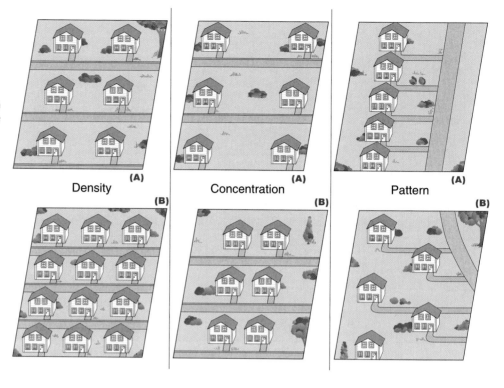

Density (A) (B)　Concentration (A) (B)　Pattern (A) (B)

These moves result in a more dispersed distribution. Before the moves, seven teams were clustered in the three northeastern cities of Philadelphia, New York, and Boston, compared to only three teams after the moves. In 1953, no team was located south or west of St. Louis, but after the moves, teams were located on the West Coast and in the Southeast for the first time.

In addition to the shifts by established teams, the major leagues expanded between 1960 and 1998 from 16 to 30 teams. The new teams selected the following locations:

- Angels—Los Angeles in 1961, then to Anaheim (California) in 1965

- Senators—Washington in 1961, then to Dallas (Texas Rangers) in 1971
- Mets—New York in 1962
- Astros—Houston (originally Colt .45s) in 1962
- Royals—Kansas City in 1969
- Padres—San Diego in 1969
- Expos—Montreal in 1969
- Pilots—Seattle in 1969, then to Milwaukee (Brewers) in 1970
- Blue Jays—Toronto in 1977
- Mariners—Seattle in 1977
- Marlins—Miami (Florida) in 1993
- Rockies—Denver (Colorado) in 1993

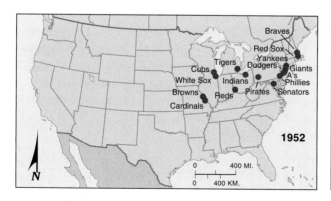

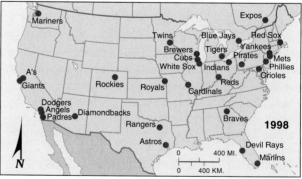

Figure 1–10 Density and concentration of baseball teams. The changing distribution of North American baseball teams illustrates the difference between density and concentration. The density of ball teams in North America increased from 16 in 1952 to 30 in 1998. At the same time, the distribution changed from a clustered arrangement in the northeastern part of the United States to a more dispersed arrangement across the United States and southern Canada.

- Devil Rays—Tampa Bay in 1998
- Diamondbacks—Phoenix (Arizona) in 1998

Thus, the density of major league teams in North America increased from 16 to 30, and at the same time the distribution became more dispersed.

Pattern

The third property of distribution is the **pattern**, which is the geometric arrangement of the objects. Some phenomena are organized in a regular (geometric) pattern, whereas others are distributed irregularly. Geographers observe that many objects form a linear distribution, such as the arrangement of houses along a street or stations along a subway line.

Objects are frequently arranged in a square or rectangular pattern. Many American cities contain a regular pattern of streets, known as a grid pattern, which intersect at right angles at uniform intervals to form square or rectangular blocks. The system of townships, ranges, and sections established by the Land Ordinance of 1785 is another example of a square or grid pattern. The distribution of baseball teams also follows a regular pattern—the teams are located in North America's largest metropolitan areas.

Not all objects are distributed in a regular pattern. The streets in the older parts of European cities are arranged in a random pattern, following centuries of haphazard development.

Key Issue 2

Why Is Each Place Unique?

- Culture: People's beliefs and traits
- Region: Areas of unique characteristics

Geographers recognize that each place in the world is in some respects unique and in other respects similar to other places. The interplay between the uniqueness of each place and the similarities among places is the basis for geographic inquiry into why things are located and distributed where they are.

Two basic concepts help geographers to explain why every place is in some ways unique: culture and region. Culture is the body of customary beliefs, material traits, and social forms that together constitute the distinct tradition of a group of people. Geographers distinguish groups of people according to important cultural characteristics, describe where particular cultural groups are distributed, and offer reasons to explain the observed distribution. Geographers divide the world or a portion of the world into regions, areas distinguished by unique combinations of distinctive characteristics, which in human geography are primarily cultural, although physical features can also be important.

Culture: People's Beliefs and Traits

In everyday language, we think of *culture* as the collection of novels, paintings, symphonies, and other works produced by talented individuals. A person with a taste for these intellectual outputs is said to be "cultured." We commonly distinguish intellectually challenging culture from *popular* culture, such as television programs.

Culture also refers to small living organisms, such as those found under a microscope or in yogurt. Agri-culture is a term for the growing of living material at a much larger scale than in a test tube.

The origin of the word *culture* is the Latin *cultus*, which means "to care for." Culture is a complex concept in modern English, because "to care for" something has two very different meanings:

- To care *about*—to adore or worship something, as in the modern word *cult*.
- To take care *of*—to nurse or look after something, as in the modern word *cultivate*.

Geography looks at both of these facets of the concept of culture to see why each location in the world is unique. As discussed in the third key issue of this chapter, geographers also explain why culture has contributed to creating similarities among places in the modern world. But traditionally, geography has emphasized the role of culture in creating unique places.

Geographic Approaches to Studying Culture

When geographers think about culture, they may be referring to either one of the two main meanings of the concept. Some geographers study what people care about (their ideas, beliefs, and values), while other geographers emphasize what people take care of (their tangible objects and material artifacts).

What People Care About. Geographers study why the customary ideas, beliefs, and values of a people produce a distinctive culture in a particular area of the world. Especially important cultural values derive from a group's language, religion, and ethnicity. These three cultural traits are both an excellent way of identifying the location of a culture and the principal means by which cultural values become distributed around the world.

Language is a system of signs, sounds, gestures, and marks that have meanings understood within a cultural group. People communicate the cultural values they care about through language, and the words themselves tell something about where different cultural groups are located. The distribution of speakers of different languages and reasons for the distinctive distribution are discussed in Chapter 5.

Religion is an important cultural value, because it is the principal system of attitudes, beliefs, and practices through which people worship in a formal, organized way what is important to the group. As discussed in Chapter 6, geographers look at the distribution of religious groups around the world and the different ways that the various groups interact with their environment.

Ethnicity encompasses a group's language, religion, and other cultural values, as well as its physical traits. A group possesses these cultural and physical characteristics as a product of its common traditions and heredity. As discussed in Chapter 7, geographers find that problems of conflict and inequality tend to occur in places where more than one ethnic group inhabits and seeks to organize the same territory.

What People Take Care Of. The second element of a culture is its material artifacts, the visible objects that cultural groups possess and leave behind for the future. Geographers are especially interested in two distinctive kinds of material artifacts—daily necessities of survival (including food, clothing, and shelter) and leisure activities (such as artistic expressions and recreation).

All people consume food, wear clothing, build shelter, and create art, but different cultural groups provide these material artifacts in different ways. As discussed in Chapter 4, geographers classify the various material expressions, document where they are found, and explain why some have widespread distribution across Earth's surface while others are restricted to isolated areas.

Cultural Institutions. In addition to cultural values and material artifacts, geographers are interested in a third aspect of culture, the political institutions that maintain the values and protect the artifacts. The world is organized into a collection of countries, or states, controlled by governments put in place through various representative and unrepresentative means. A major element of a group's cultural identity is its citizenship, the country or countries where it inhabits, pays taxes, votes, and otherwise participates in the administration of space.

As discussed in Chapter 8, in the modern world cultural groups are increasingly asserting their right to organize their own affairs, rather than submit to the control of other cultural groups. Political problems are found in places where the area occupied by a cultural group does not coincide with the boundaries of a country.

Cultural Identity

A pet dog doesn't care if you are male or female, black or white, gay or straight. As long as you feed it and take care of it, your dog will respond with total, unquestioned devotion.

Although dogs don't care about these human traits, people do. They are key characteristics to which people refer in order to identify who they are. Cultural identity is a source of pride and an inspiration for personal values.

Even more important than self-identification, these traits matter to other people. They are the means by which other people classify us. Whatever biological basis may or may not exist for distinguishing among humans, differences in gender, race, and sexual orientation are first and foremost constructed by the attitudes and actions of others. Geographers consider the study of cultural identity to be important because humans repeatedly demonstrate that these factors are important in explaining why they sort themselves out in space and move across the landscape in distinctive ways.

Consider first the "all-American" family of mother, father, son, and daughter. Leave aside for the moment that this family constitutes a very small percentage of American households. Consider the daily movement of this family across a city. In the morning, Dad gets in his car and drives from home to work, where he parks the car and spends the day; then in the late afternoon, he collects the car and drives home. The location of the home was selected in part to ease Dad's daily commute to work.

The mother's travel patterns are likely to be far more complex than the father's. Mom takes the children to school and returns home. She also drives to the supermarket, visits grandmother, and walks the dog. In between, she organizes the several thousand square feet of space that the family calls home. In the afternoon, she picks up the youngsters at school and takes them to Little League or ballet lessons. Later, she brings them home, just in time for her to resume her responsibility for organizing the home.

Most American women are now employed at work outside the home, adding a substantial complication to an already complex pattern of moving across urban space. Where is her job located? The family house was already selected largely for access to Dad's place of employment, so Mom may need to travel across town. Who leaves work early to drive a child to a doctor's office? Who takes a day off work when a child is at home sick?

The role of gender in organizing space is learned as a child. Which child—the boy or girl—went to Little League, and which went to ballet lessons? To which activity is substantially more land allocated in a city— ballfields or dance studios?

If the family described above consisted of persons of color, its occupancy of space would change. The most prominent difference is the location of the family's home. In most neighborhoods, the residents are virtually all whites or virtually all persons of color. Segregation by color was legal in the United States until the 1950s and in South Africa until the 1990s. Although illegal to discriminate against people of color, segregation persists in part because people want to reinforce their cultural identity by living near persons of similar background, and in part because persons of color have lower average incomes than Caucasians. But many still practice discrimination because of a deep-seated fear of occupying a piece of Earth's surface near persons of color.

The effects of race on the organization of space can be seen across America. Stand near the exit from a supermarket or other store in the City of Oak Park or Cicero, Illinois, near the border with the City of Chicago. A mix of African-Americans and whites leave the stores, but nearly all of the whites turn west into the heart of Oak Park or Cicero while nearly all the African-Americans turn east into Chicago. Not surprisingly, most neighborhoods in Oak Park and Cicero are virtually all white, while the nearby neighborhoods inside the city of Chicago are virtually all African-American (Figure 1–11).

In downtown Dayton, watch the people at the bus stops along the main east-west street, Third Street. In the afternoon, when office workers are heading home, persons of color are waiting on the north side of Third Street for west-bound buses, while whites are waiting on the south side for east-bound buses. As discussed in Chapter 13, virtually all African-Americans in Dayton live on the west side, while the east side is virtually all white.

Openly homosexual men and lesbians may be attracted to some locations to reinforce their self-identity, and they may also perceive the aversion of some heterosexuals to living near them. San Francisco reinforces its reputation as a sympathetic home for homosexuals and lesbians through such practical means as preventing the city from doing business with companies that do not provide their employees with domestic-partner benefits. Specific neighborhoods in other cities may be known to have large gay populations.

Geographers face distinctive challenges in studying the distribution of homosexuals and lesbians. The U.S. census identifies the number of persons of different races living in specific neighborhoods or cities, as well as the number of men and women, but it does not identify the number of homosexuals and lesbians. Gays may hide rather than proclaim their cultural identity, and the cultural identity of each generation of gays is renewed afresh rather than transmitted from parents to children.

All academic disciplines, all workplaces have proclaimed sensitivity to issues of cultural diversity. For geographers, concern for cultural diversity is not merely a politically correct expediency; it lies at the heart of geography's subject matter. Nor for geographers is deep respect for the dignity of all cultural groups merely a politically correct expediency; it lies at the heart of geography's explanation of important differences that exist across Earth's surface.

Cultural Ecology: Relations Between Cultures and Environment

Many social sciences, not just geography, study culture. Distinctive to geography is the importance given by the discipline to relationships between culture and the natural environment. Different cultural groups modify the natural environment in distinctive ways to produce

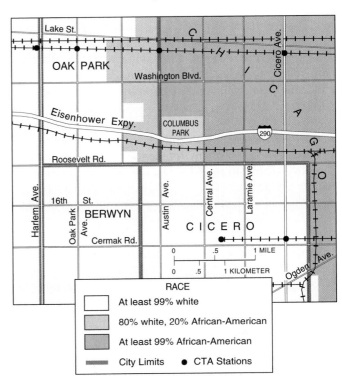

Figure 1–11 Cultural identity. A rather sharp "color" line runs through the Chicago area, as in most U.S. cities. The neighborhoods to the east are predominantly African-American, to the west predominantly white. The two groups interact at stores located close to the boundary between the two areas.

unique cultural landscapes. According to geographer Peirce Lewis, "the cultural landscape is our unwitting autobiography," because it reflects in tangible form our tastes, values, aspirations, and fears.

Humans endow the physical environment with cultural values by regarding it as a large collection of **resources**, or substances with usefulness. As discussed in Chapter 14, humans are depleting some of these valuable resources, especially those used for energy, and through pollution humans are altering or damaging other valuable resources, especially air and water.

Some human impacts on the environment are casual, and some are based on deep-seated cultural values. Why do we plant our front yards with grass, water it to make it grow, mow it to keep it from growing tall, and impose fines on those who fail to mow often enough? Why not let dandelions grow, or pour concrete instead? Why does one group of people consume the fruit from deciduous trees and chop down the conifers, while another group chops down the deciduous trees for furniture while preserving the conifers as religious symbols?

A people's level of wealth can influence its attitude toward modifying the environment. A farmer who possesses a tractor may regard a hilly piece of land as an obstacle to avoid, whereas a poor farmer with a hoe may regard hilly land as the only opportunity to produce food for survival through hand cultivation.

Human geographers use this cultural ecology, or human-environment, approach to explain many global issues. For example, world population growth is a problem if the number of people exceeds the capacity of the physical environment to produce food. However, people can adjust to the capacity of the physical environment by controlling their numbers, adopting new technology, consuming different foods, migrating to new locations, and other actions.

Physical Processes. Human geographers need some familiarity with global environmental processes to understand the distribution of human activities, such as where people live and how they earn a living. Important physical processes include climate, vegetation, soil, and landforms.

Climate is the long-term average weather condition at a particular location. Geographers frequently classify climates according to a system developed by German climatologist Vladimir Köppen. The modified Köppen system divides the world into five main climate regions, which are identified by the letters A through E, as well as by names (Figure 1–12).

- A Tropical Climates
- B Dry Climates
- C Warm Mid-Latitude Climates
- D Cold Mid-Latitude Climates
- E Polar Climates

The modified Köppen system divides the five main climate regions into several subtypes. For all but the B climate, the basis for the subdivision is the amount of precipitation and the season in which it falls. For the B climate, subdivision is on the basis of temperature and precipitation.

Humans have a limited tolerance for extreme temperature and precipitation levels and thus avoid living in places that are too hot, too cold, too wet, or too dry. Compare the map of global climate to the distribution of population (see Figure 2–1). Relatively few people live in the Dry (B) and Polar (E) climate regions.

The climate of a particular location influences human activities, especially production of the food needed to survive. People in parts of the A climate region, especially southwestern India, Bangladesh, and the Myanmar (Burma) coast, anxiously await the annual monsoon rain, which is essential for successful agriculture and provides nearly 90 percent of India's water supply. For most of the year, the region receives dry, somewhat cool air from the northeast. In June, the wind direction suddenly shifts, bringing moist, warm southwesterly air, known as the *monsoon*, from the Indian Ocean. The monsoon rain lasts until September.

In years when the monsoon rain is delayed or fails to arrive—in recent decades, at least one-fourth of the time—agricultural output falls and famine threatens in the countries of South Asia, where nearly 20 percent of the world's people live. The monsoon rain is so important in India that the words for "year," "rain," and "rainy season" are identical in many local languages.

Vegetation, or plant life, covers nearly the entire land surface of Earth. Earth's land vegetation includes four major forms of plant communities, called biomes. Their location and extent are influenced by both climate and human activities. Vegetation and soil, in turn, influence the types of agriculture that people practice in a particular region. The four main biomes are forest, savanna, grassland, and desert.

In the *forest biome*, trees form a continuous canopy over the ground. Although trees are the dominant vegetation, grasses and shrubs may grow beneath the cover. The forest biome covers a large percentage of Earth's surface, including much of North America, Europe, and Asia, as well as tropical areas of South America, Africa, and Southeast Asia.

The *savanna biome* is a mixture of trees and grasses. The trees do not form a continuous canopy, and the resultant lack of shade allows grass to grow. Savanna covers large areas of Africa, South Asia, South America, and Australia.

As the name implies, the *grassland biome* is covered by grass rather than trees. Few trees grow in the region because of low precipitation. Early explorers from northern Europe and eastern North America regarded the American prairies—the world's most extensive grassland area—to be uninhabitable, because of the lack of trees with which to build houses, barns, and fences. However, modern cultivation of wheat and other crops has turned the grasslands into a very productive region.

The *desert biome* is not completely bereft of vegetation. Although many desert areas have essentially no vegetation, the region contains dispersed patches of plants adapted to dry conditions. Vegetation is often sufficient for the survival of small numbers of animals.

Soil is the material that forms on Earth's surface, in the thin interface between the air and the rocks. Not merely dirt, soil contains the nutrients necessary for successful growth of plants, including those useful to humans.

The U.S. Comprehensive Soil Classification System divides global soil types into ten *orders*, according to the characteristics of the immediate surface soil layers and the subsoil. The orders are subdivided into suborders, great groups, subgroups, families, and series. More than 12,000 soil types have been identified in the United States alone.

Human geographers are concerned with the destruction of the soil that results from a combination of natural processes and human actions. Two basic problems contribute to the destruction of soil: erosion and depletion of nutrients. Erosion occurs when the soil washes away in the rain or blows away in the wind. To reduce the erosion problem, farmers reduce the amount of plowing, plant crops whose roots help bind the soil, and avoid planting on steep slopes.

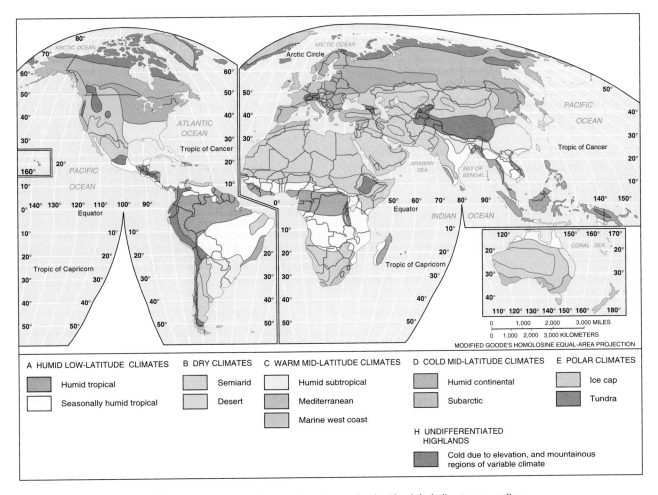

Figure 1–12 Climate regions. Geographers frequently classify global climates according to a system developed by Vladimir Köppen. The modified Köppen system divides the world into five main climate regions, indicated on the map by the letters A, B, C, D, and E.

Nutrients are depleted when plants withdraw more nutrients than natural processes can replace. Each type of plant withdraws certain nutrients from the soil and restores others. Repeated harvesting of the same type of crop year-after-year can remove certain nutrients and reduce the soil's productivity.

To minimize the depletion problem, farmers in more developed countries sometimes plant crops that offer no economic return but restore nutrients to the soil and keep the land productive over a longer term. Farmers also restore nutrients to the soil by adding fertilizers, either natural or synthetic. Farmers in less developed countries may face greater problems with depletion of nutrients because they lack knowledge of proper soil management practices and funds to buy fertilizer.

Landforms, which are Earth's surface features, can vary from relatively flat to mountainous. Geographers find that the study of Earth's landforms—a science known as geomorphology—helps to explain the distribution of people and the choice of economic activities at different locations. People prefer living on flatter land, which generally is better suited for agriculture. Great concentrations of people and activities in hilly areas may require extensive effort to modify the landscape.

Topographic maps, published (for the United States) by the U.S. Geological Survey (USGS), show a remarkable detail of physical features, such as bodies of water, forests, mountains, valleys, and wetlands. They also show cultural features, such as buildings, roads, parks, farms, and dams. "Topos," as they are called, are used by engineers, hikers, hunters, people seeking a homesite, and anyone who really needs to see the lay of the land.

Geographers use topographic maps to study the relief and slope of localities. Relief is the difference in elevation between any two points, and it measures the extent to which an area is flat or hilly. The steepness of hills is measured by slope, which is the relief divided by the distance between two points. Figure 1–8 shows a portion of a USGS map for northern Mississippi, at the scale of 1:24,000. The brown lines on the map are contour lines, which connect points of equal elevation above or below sea level. Contour lines are closer together to show steeper slopes and farther apart in flatter areas.

The Netherlands: Sensitive Environmental Modification. Modern technology has altered the historic relationship between people and the environment. Humans now can modify the physical environment to a greater extent than in the past. For example, air conditioning has increased the attractiveness of living in warmer climates, and better insulation now permits living in colder climates.

Geographers are concerned that people sometimes use modern technology to modify the environment insensitively. Human actions can deplete scarce environmental resources, destroy irreplaceable resources, and use resources inefficiently. The refrigerants in the air conditioners that have increased the comfort of residents of warmer climates have also increased the amount of chlorofluorocarbons in the atmosphere, damaging the ozone layer that protects living things from X rays and contributing to global warming. We explore the consequences of such use, abuse, and misuse of the environment in more detail in Chapter 14.

Few lands have been as thoroughly modified by humans as the Netherlands. Because more than half of the Netherlands lies below sea level, most of the country today would be under water if it were not for massive projects to modify the environment by holding back the sea. The Dutch have a saying that "God made Earth, but the Dutch made the Netherlands." The Dutch have modified their environment with two distinctive types of construction projects: polders and dikes.

A **polder** is a piece of land that is created by draining water from an area. Polders, first created in the thirteenth century, were constructed primarily by private developers in the sixteenth and seventeenth centuries and by the government during the past 200 years. Altogether, the Netherlands has 6,500 square kilometers (2,600 square miles) of polders, comprising 16 percent of the country's land area.

The first step in making a polder is to build a wall encircling the site, which is still under water. Then the water inside the walled area is pumped from the site into either nearby canals or the remaining portion of the original body of water. Before the invention of modern engines, windmills performed the pumping operation. Many of these windmills remain as a picturesque element of the Dutch landscape, although they were originally built for a practical purpose (Figure 1–13).

Once dry, the site—now known as a polder—can be prepared for human activities. The Dutch government has reserved most of the polders for agriculture, to reduce the country's dependence on imported food. Some of the polders are used for housing, and one contains Schiphol, one of the largest airports in Europe.

The second distinctive modification of the landscape in the Netherlands is the construction of massive dikes to prevent the North Sea, an arm of the Atlantic Ocean, from flooding much of the country. The Dutch have built dikes in two major locations, the Zuider Zee project in the north and the Delta Plan project in the southwest.

The Zuider Zee, an arm of the North Sea, once protruded into the heart of the Netherlands. For centuries, the Dutch unsuccessfully attempted to prevent the Zuider Zee from flooding much of the country. Then, in the late nineteenth century, a Dutch engineer named Cornelis Lely proposed an ambitious project to seal off the Zuider Zee permanently from the North Sea, the ultimate source of the floodwaters.

In accordance with Lely's plan, a dike was built, 32 kilometers (20 miles) long, across the mouth of the Zuider Zee to block the flow of North Sea water. When completed in 1932, the dike caused the Zuider Zee to be converted from a saltwater sea to a freshwater lake. The newly created body of water was named the IJsselmeer, or Lake IJssel, because the IJssel River now flows into it. Some of the lake has been drained to create several polders, encompassing an area of 1,600 square kilometers (620 square miles).

A second ambitious project in the Netherlands is the Delta Plan in the southwestern part of the country. Several important rivers flow through the Netherlands, including the Rhine (Europe's busiest river), the Maas (known as the Meuse in France), and the Scheldt (known as the Schelde in Belgium). As these rivers flow into the North Sea, they split into many branches and form a low-lying delta that is vulnerable to flooding.

After a devastating flood in January 1953 killed nearly 2,000 people, the Delta Plan called for the construction of several dams to close off most of the waterways from the North Sea. Together these dams shorten the coastline of the Netherlands by approximately 700 kilometers (435 miles). Because Rotterdam, Europe's largest port, is located nearby, some of the waterways were kept open. The project took 30 years to build and was completed in the mid-1980s.

With these two massive projects finished, attitudes toward modifying the environment have changed in the Netherlands. The Dutch have scrapped plans to build additional polders in the IJsselmeer, in order to preserve the lake's value for recreation. A plan adopted in 1990 calls for returning 263,000 hectares (650,000 acres) of farms to wetlands or forests. Widespread use of insecticides and fertilizers on Dutch farms contributes to contaminated drinking water, acid rain, and other environmental problems. The Dutch are deliberately breaking some of the dikes to flood fields.

But modifying the environment will still be essential to the survival of the Dutch. Global warming could raise the level of the sea around the Netherlands by between 20 and 58 centimeters (8 and 23 inches) within the next 100 years.

Florida: Not-so-sensitive Environmental Modification. Humans do not always modify the environment as sensitively as the Dutch. In Florida, the rechanneling of the Kissimmee River and construction on coastal barrier islands demonstrate the negative consequences that

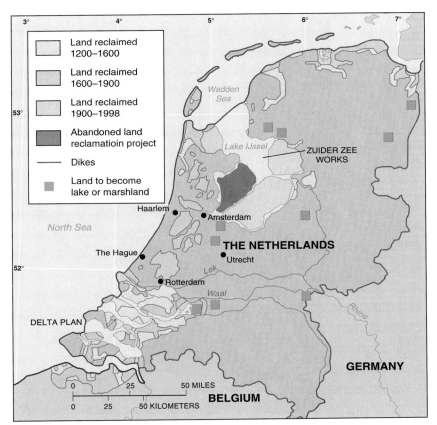

Figure 1–13 Environmentally sensitive cultural ecology in the Netherlands. The Dutch people have considerably altered the site of the Netherlands through creation of polders and dikes. Since the thirteenth century the Dutch have reclaimed more than 2,600 square miles (6,500 square kilometers) of polders, more than three-fourths of which have been reclaimed in the past 200 years. The Zuider Zee and Delta Plan have altered the coastline of the Netherlands and enabled the Dutch to reduce the amount of destruction caused by flooding.

The Dutch still use windmills to pump water on polders, although the windmills may have a more modern appearance, such as this one on a polder near Friesland, near Makkum. (Wolfgang Kaehler Photography)

result from ill-conceived human actions to modify the environment.

In 1961, the state of Florida asked the U.S. Army Corps of Engineers to straighten the course of the Kissimmee River, which meandered for 160 kilometers (98 miles) from near Orlando to Lake Okeechobee (Figure 1–14). In years with heavy rains, the river flooded 117,000 hectares (45,000 acres) of nearby land, creating an obstacle to cattle grazing and construction of buildings in areas with potential for rapid population growth. The Corps channeled the river into a canal 90 meters wide (300 feet) and 9 meters deep (30 feet), which ran in a straight line for 84 kilometers (52 miles).

The opening of the canal in 1971 changed the region's environment. Tens of thousands of gallons of polluted water, mainly from cattle grazing along the banks, ran into the canal and flowed into Lake Okeechobee, which is the source of fresh water for half of Florida's population. Fish in the lake began to die from the high levels of mercury, phosphorous, and other contaminants. The polluted water then continued to flow south, from Lake Okeechobee to the Everglades, where wildlife habitats were also disturbed.

The Corps is now spending hundreds of millions of dollars to restore the Kissimmee River to its meandering course and to buy the nearby grazing land, which will again be subject to flooding. In an ironic reminder of the Dutch saying quoted earlier, Floridians say "God made the world in six days, and the Army Corps of Engineers has been tinkering with it ever since."

The rechanneling of the Kissimmee River is not the only example of insensitive environmental management in Florida. Barrier islands extend for several hundred kilometers along Florida's eastern and western coasts, as well as the rest of the Atlantic and Gulf coasts between Maine and Texas. These barrier islands are essentially large sandbars that shield the mainland from flooding and storm damage. They are constantly being eroded and shifted from the force of storms and pounding surf, and after a major storm large sections can be washed away and disappear.

Despite the fragile condition of the barrier islands, hundreds of thousands of people live on them. They are increasingly attractive locations for constructing homes and recreational facilities to take advantage of proximity to the seashore. Two-thirds of the barrier islands are linked with the mainland by bridge, causeway, ferry service, or airplane flights.

To fight erosion along the barrier islands, people build seawalls and jetties, which are structures extending into the sea, but these projects result in more damage than protection. A seawall or jetty can prevent sand from

Miami Beach, Florida, is built on a series of barrier islands along the Atlantic Ocean. The beach is constantly being eroded along some of these barriers islands, such as the one containing Miami Beach's largest hotels. (James Blank/The Stock Market)

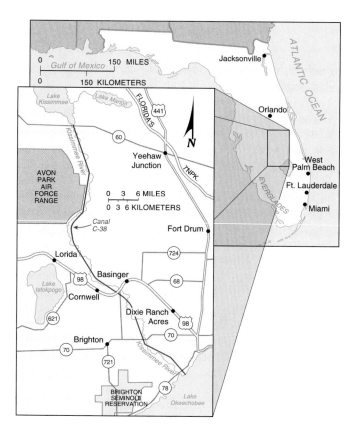

Figure 1–14 Environmentally insensitive cultural ecology in Florida. The U.S. Army Corps of Engineers straightened the course of the Kissimmee River to control flooding in central Florida. After the canal, known as C-38, opened in 1971, millions of gallons of polluted water—mainly runoff from cattle grazing—began pouring into Lake Okeechobee, which is the major source of fresh water for about half of Florida's population. Now the state wants the corps to return the river to its original course.

drifting away, but by trapping sand along the up-current side, it causes erosion on the barrier islands on the down-current side.

Region: Areas of Unique Characteristics

A region is an area of Earth defined by one or more distinctive characteristics, including cultural features such as language and religion, economic features such as agriculture and industry, and physical features such as climate and vegetation. Geographers identify regions to study differences among Earth's peoples and environments, as well as regularities. A combination of cultural, economic, and environmental features gives a region its unified character and distinguishes it from other areas of Earth's surface.

In the past, geographers who used the regional studies approach identified an area of Earth's surface and described in careful detail as many of its characteristics as they could uncover. When Julius Caesar wrote that "All Gaul is divided into three parts," he gave an example of the traditional regional studies approach to geographic explanation.

Some introductory geography courses are organized around regions of the world, such as Latin America, East

Asia, and sub-Saharan Africa. This approach selects a portion of Earth and studies the environment, people, and activities within the region. Although this book is organized by key issues rather than regions, you should be aware of the importance of distinctive characteristics of regions in the world.

Today, regional analysis may start by identifying an important characteristic, such as population growth, level of wealth, or energy consumption. Then, geographers search for reasons to explain why that characteristic is greater or more intense in one area and less so elsewhere. To build a model of explanation, geographers regionalize the world or a portion of it according to these characteristics and explain why the various characteristics are integrated within regions and vary among regions.

Region is a vague concept, because it can apply to any area larger than a point and smaller than the entire planet. Regions can vary in scale from a neighborhood to a large percentage of Earth's surface. Two regions could overlap, or some territory could be excluded from either.

Types of Regions

Geographers employ the concept of region primarily to explain why an area of Earth's surface is distinctive. Within a region, the people, activities, and environment will display similarities and regularities. A region's cultural, economic, and physical characteristics will differ in some way from those of other regions. Geographers identify three types of regions: formal, functional, and vernacular.

Formal Region.   A **formal region**, also called a uniform region or a homogeneous region, is an area within which everyone shares in common one or more distinctive characteristics. The shared characteristic could be a cultural value such as a common language, an economic activity such as production of a particular crop, or an environmental property such as climate. In a formal region, the selected characteristic is present throughout.

Some formal regions are easy to identify, such as countries or local government units. Montana is an example of a formal region, characterized by a government that passes laws, collects taxes, and issues license plates with equal intensity throughout the state. We can easily identify the formal region of Montana, because it has clearly drawn and legally recognized boundaries, and everyone living within them shares the status of being subject to a common set of laws.

In other kinds of formal regions, not everyone possesses the shared characteristic. Not every farmer living in the U.S. or Canadian wheat belt grows wheat, nor does every farmer living in the U.S. ranching area raise cattle. Nonetheless, we can distinguish the wheat belt as a formal region in which the shared characteristic is the predominance of one type of agriculture (wheat growing), whereas in the ranching area the shared characteristic is the predominance of another type of agriculture (cattle raising).

Similarly, we can distinguish formal regions within the United States characterized by a predominant voting for Republican candidates. Republicans may not get 100 percent of the votes in these regions, nor in fact do they always win. However, in a presidential election, the candidate with the largest number of votes receives all of the electoral votes of a state, regardless of the margin of victory. Consequently, a state that usually has Republican electors can be considered a Republican state.

Geographers typically identify formal regions to help explain broad global or national patterns, such as variations in religions and levels of economic development. The characteristic selected to distinguish a formal region often illustrates a general concept, rather than representing a precise mathematical distribution (although it may be quantifiable as well).

A cautionary step in identifying formal regions is the need to recognize the *diversity* of cultural, economic, and environmental factors, even while making a generalization. Problems may arise because a minority of people in a region speak a language, practice a religion, or possess resources different from those of the majority. People in a region may play distinctive roles in the economy and hold different positions in society based on their gender or ethnicity.

Functional Region.   A **functional region**, also called a nodal region, is an area organized around a node or focal point. The characteristic chosen to define a functional region dominates at a central focus or node and diminishes in importance outward. The region is tied to the central point by transportation or communications systems, or by economic or functional associations.

Geographers often use functional regions to display information about economic characteristics. The region's node may be a shop or service, and the boundaries of the region mark the limits of the trading area of the activity. People and activities may be attracted to the node, and information may flow from the node to the surrounding area.

An example of a functional region is the circulation area of a newspaper. A newspaper dominates circulation figures in the city in which it is published. Farther away from the city, fewer people read that newspaper, whereas more people read a newspaper published in a neighboring city. At some point between the two cities, the circulation of the newspaper from the second city equals the circulation of the original newspaper. That point is the boundary between the nodal regions of the two newspapers.

A more contemporary example is television broadcasting. The United States is divided into several hundred functional regions based on television networks. Every television market has an *area of dominant influence (ADI)*, the region in which the preponderance of viewers are tuned to that market's stations. The United States is divided into several hundred functional regions, according to the distribution of the ADIs. The culture disseminated by TV stations diffuses to the surrounding region.

An ADI is a good example of a functional region, because the characteristic—people who are viewing a particular station—is dominant at the center and declines toward the periphery. For example, everyone in Des Moines, Iowa, who wishes to watch a program on NBC tunes to Channel 13. In Omaha, Nebraska, 225 kilometers (140 miles) to the west, everyone watching NBC is tuned to Channel 6. With increasing distance eastward from Omaha, Channel 6's signal gets weaker and Channel 13's gets stronger. The percentage of people watching NBC declines for Channel 6 and increases for Channel 13.

The boundary between the Omaha and Des Moines ADIs is the point where an equal number of people watch Channel 13 and Channel 6, near the Cass/Adair county line. Other functional regions in Iowa centered around NBC affiliates include Sioux City's Channel 4 to the northwest, Davenport's Channel 6 to the east, Waterloo's Channel 7 to the northeast, and Rochester, Minnesota's, Channel 10 to the north (Figure 1–15).

Vernacular Region. A **vernacular region**, or perceptual region, is one that people believe to exist as part of their cultural identity. Such regions emerge from concepts that people use informally in daily life, rather than from scientific models developed through geographic thought. (In language, the term "vernacular" means everyday language that is used by ordinary people.)

As an example of a vernacular region, Americans frequently refer to the "South" as an area with environmental, cultural, and economic features perceived to be quite distinct from the rest of the United States. Many of these perceptions are accurate and can be measured to delineate the boundary between the South and the North.

In terms of climate, the South can be defined as a region where the last winter frost occurs in March and rainfall is more plentiful in winter than in summer. Economically, the South can be defined as a region of high cotton production and low high-school graduation rates. Culturally, the South can be defined as the states that joined the Confederacy during the Civil War, and the region where Baptist is the most prevalent religion (Figure 1–16).

Analysts have difficulty fixing the precise boundary of the South. Kentucky did not join the Confederacy (although both Jefferson Davis, the only president of the Confederate States, and Abraham Lincoln were born in Kentucky). Northern Virginia has April frosts. Southern Florida, Louisiana, and Texas have a majority Roman Catholic. When several environmental, economic, and cultural factors are examined, much of the territory thought to be part of the perceptual region of the South is actually excluded.

Similar studies identifying the Midwest show relatively strong consensus on the location of three boundaries—near the Ohio–Pennsylvania state line on the east, the U.S.–Canada border on the north, and the Ohio River on the south—but wide disagreement on the western boundary, which varies from the Mississippi River to the Rocky Mountains. To target potential subscribers and advertisers more effectively, *Midwest Living* magazine asked several thousand people where they thought the Midwest was located. Only one state made every

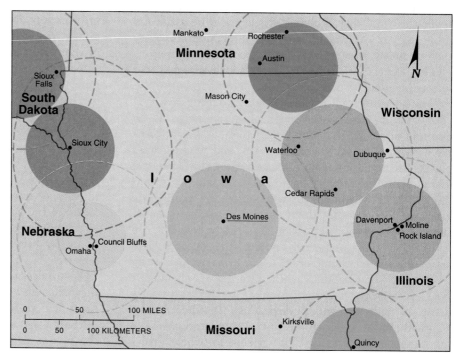

Figure 1–15 Formal and functional regions in Iowa. The state of Iowa is an example of a formal region. The areas of dominance influence (ADIs) for different television stations within Iowa are examples of functional regions. In several of the functional regions, the node—the TV station—is in an adjacent state. Functional regions frequently overlap the formal regions delineated by state or national boundaries.

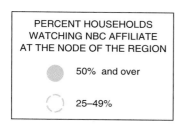

PERCENT HOUSEHOLDS WATCHING NBC AFFILIATE AT THE NODE OF THE REGION

50% and over

25–49%

The frostbelt and sunbelt are vernacular regions distinguished by a number of features, such as differences in winter climate. In a frostbelt city like Chicago, average winter temperatures are generally 40°F (22°C) lower than in sunbelt communities such as Miami, Florida. (RK Reinstein/The Image Works) (Wesley Bocxe/Photo Researchers)

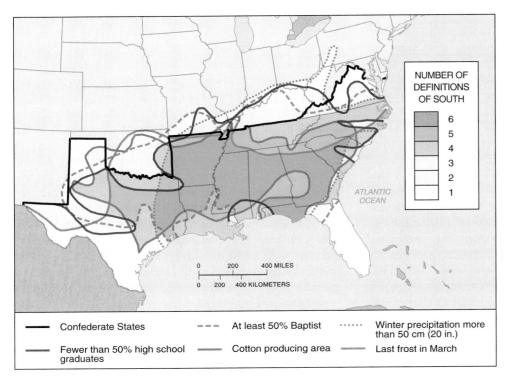

Figure 1–16 Defining the South as a vernacular region. The South is popularly distinguished as a distinct vernacular region within the United States according to a number of factors, such as mild climate, propensity for growing cotton, and importance of the Baptist Church. When several factors are mapped together, the South does in fact stand out from the rest of the country, but the region's boundaries become muddled. Should Kentucky and south Florida be excluded altogether? Where in Texas and Virginia should the region's western and northern boundaries be drawn?

respondent's list, Iowa. The magazine decided to define the Midwest as extending between Ohio and the Dakotas.

A useful way to identify perceptual regions is to get people to draw mental maps, which are internal representations of a portion of Earth's surface. A mental map depicts what an individual knows about a place, containing personal impressions of what is in a place and where places are located. A student and a professor are likely to have different mental maps of a college campus, based on differences in where they work, live, and eat, and a senior is likely to have a more detailed and "accurate" map than a first-year student.

Mental maps of perceptual regions can play a critical role in organizing daily life. For example, students at one university were shown a map of their campus divided into squares. They were asked to indicate in which squares they felt safe walking alone at 10:30 P.M. When combined, the responses portrayed a campus divided into regions that were widely regarded as safe and regions that were widely regarded as dangerous. Such studies can also determine whether perceptions of safety are uniform among groups of students or vary by age, gender, and ethnicity.

Regional Integration

A region gains uniqueness, not from possessing a single human or environmental characteristic, but from a combination of them. Not content merely to identify these characteristics, geographers seek relationships among them. Geographers recognize that, in the real world, characteristics are integrated.

For example, geographers divide the world into formal regions that are more (or relatively) developed economically and those that are less developed (or developing). The regions of more developed countries (MDCs), such as Europe, North America, and Japan, are located primarily in the northern latitudes, while the regions of less developed countries (LDCs) are concentrated in the southern latitudes. This north–south regional split underlies many of the world's social and economic problems.

Various shared characteristics—such as per capita income, literacy rates, televisions per capita, and hospital beds per capita—distinguish more developed from less developed regions. Geographers demonstrate that the distribution of one characteristic of development is associated with others. As with other formal regions, geographers also caution that the division between MDCs and LDCs hides diversity within each region. Not everybody is rich in an MDC or poor in an LDC; the common characteristics represent averages rather than universal conditions.

The geographer's job is to sort out the associations among various social characteristics, each of which is uniquely distributed across Earth's surface. For example, geographers conclude that political unrest in the Middle East, Eastern Europe, and other areas derives in large measure from the fact that the distributions of important features, such as ethnicity and resources, do not match the political boundaries of individual countries.

In some cases, geographers can build models to prove that the distribution of one characteristic causes the distribution of another. For instance, we shall see in Chapter 2 that differences among regions in the rate of population growth are caused primarily by differences in the crude birth rates. However, geographers often hedge their bets: They recognize that one characteristic must be associated across Earth's surface with others, even if the relationship cannot be modeled precisely. Geographers may have difficulty in constructing exact models of cause and effect, because they must integrate many cultural and physical features to explain a region's distinctiveness.

Integrating Cancer Information. Recognizing that the distributions of various cultural and physical characteristics are integrated helps us understand social problems at a variety of scales, from local to global. At a national scale, the percentage of people who die each year from cancer differs among regions within the United States. The mid-Atlantic region has the highest level, with Maryland ranked first among the 50 states, followed by Delaware. The rate in Washington, DC, which is adjacent to Maryland, is higher than in any state.

Why does Maryland have the highest cancer rate among the 50 states? Mapping the distribution of cancer among Maryland's major subdivisions (23 counties plus the independent city of Baltimore), as well as the District of Columbia, shows important regional differences even within one of the smallest states in the country. The cancer rate in Baltimore City is more than 50 percent higher than in westernmost Garrett County (Figure 1–17).

The map of cancer rates by county in Maryland does not communicate useful information to someone who knows little about the regions of the state. By integrating other spatial information about people, activities, and environments, we can begin to see factors that may be associated with regional differences in cancer.

We can divide the state's counties into two regions: those that comprise part of the populous Washington–Baltimore metropolitan area, and those that do not. (See Chapter 13 for details on how counties are classified as metropolitan or nonmetropolitan.) This division is not yet helpful in explaining the distribution of cancer, because we can find counties with high and low rates within both the metropolitan and nonmetropolitan regions. However, if we further subdivide the Washington–Baltimore metropolitan area into two smaller regions a pattern emerges: The highest cancer rates are in the cities of Baltimore and Washington, whereas the suburban counties surrounding the two cities have lower rates.

Once we recognize that the cities have higher cancer rates than the suburbs, we can integrate that information

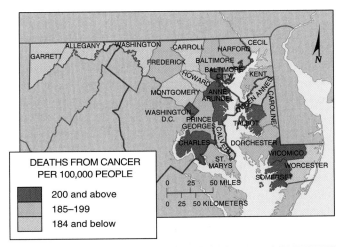

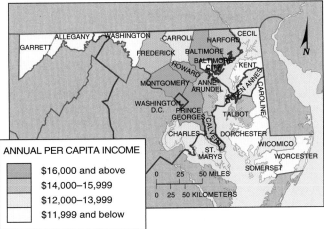

Figure 1–17 Regional integration. Maryland has the highest incidence of cancer in the United States, although the rate varies widely within the state. Maryland can be divided into two regions—the Washington–Baltimore metropolitan area in the center of the state and the nonmetropolitan area in the west and east. Within the metropolitan area, rates are higher in the two large cities of Washington and Baltimore and lower in the suburbs. Within the nonmetropolitan area, rates are higher in the east and lower in the west.

with a variety of other characteristics. People in Baltimore City and Washington, DC, are more likely than suburbanites to have low incomes and low levels of education. As a result of these characteristics, people living in the cities may be less aware of the risks associated with activities such as smoking and consuming alcohol and less able to afford medical care to minimize the risk of dying from cancer.

We can also identify two regions within Maryland's nonmetropolitan area: the west, where cancer rates are relatively low, and the east, where rates are relatively high. Income and education do not explain the difference, as was the case within the metropolitan area, because levels are comparable in the two nonmetropolitan regions, and are lower than in the metropolitan area. Instead, we must attempt to integrate other economic and environmental factors into our explanation.

People living in counties on the Chesapeake Bay's Eastern Shore may be especially exposed to cancer-causing chemicals, because relatively high percentages are engaged in fishing and farming compared to people living in the mountainous western counties. The nearby Chesapeake Bay is one of the nation's principal sources of shellfish, and many Eastern Shore residents work in seafood-processing industries. But the Chesapeake Bay also suffers from runoff of chemicals from Eastern Shore farms, which make heavy use of pesticides, as well as discharges of waste from factories, for the most part located in the metropolitan counties on the western side of the bay. Prevailing winds also carry pollutants eastward from industries in the metropolitan areas.

Key Issue 3

Why Are Different Places Similar?

- Spatial Interaction: Interdependence among places
- Diffusion: Movement between places

Although accepting that each place or region on Earth is unique, geographers recognize that human activities are rarely confined to one location. This section discusses two basic concepts—spatial interaction and diffusion—that help geographers understand why two regions can display similar characteristics.

People living in a particular region display some form of spatial interaction, which is interdependence with people in other regions established through the movement of people, ideas, and objects between the regions. Geography studies the ways that spatial interaction occurs, especially through the movement between regions of cultural values and artifacts, as well as of money, goods, and other economic activities. The people of one region may have considerable or limited spatial interaction with those in other regions, but in the modern world no group lives in complete and total isolation.

Because spatial interaction requires some sort of movement of people, objects, and ideas across the barrier of space, geographers are interested in diffusion, which is the process of spreading something from one place to another over time. Geographers study how diffusion of something occurred, where the diffusing item originated and reached, and how the diffusion altered the places of origin and destination.

Spatial Interaction: Interdependence Among Places

Human geographers are especially interested in spatial interaction because of the increasing importance of **globalization**, which is an action or process that involves the entire world and results in making something worldwide in scope. For spatial interaction to occur, two areas must be connected to each other through a network. Historically, regions were linked primarily by transportation networks, but electronic communications networks have become increasingly important means of spatial interaction.

Globalization of Economy and Culture

Globalization means that the world is "shrinking"—not literally in size, of course, but in the ability of a person, object, or idea to interact with a person, object, or idea in another place. People are plugged into a global economy and culture, producing a world that is more uniform, integrated, and interdependent.

Economic Interaction. A few people living in very remote regions of the world may be able to provide all of their daily necessities. The crop grown or product manufactured in a particular place may be influenced by the distinctive features and assets of the place. But most economic activities undertaken in one region are influenced by interaction with decision-makers located elsewhere. The choice of crop is influenced by demand and prices set in markets elsewhere. The factory is located to facilitate bringing in raw materials and shipping out products to the markets.

Globalization of the economy has been led primarily by transnational corporations, sometimes called multinational corporations. A **transnational corporation** conducts research, operates factories, and sells products in many countries, not just where its headquarters and principal shareholders are located.

Most transnational corporations have their headquarters in one of three regions—North America (especially the United States), Western Europe (especially the United Kingdom, Germany, and France), and Japan. Transnational corporations also locate most of their factories and markets within these three regions, but the United Nations reported that in 1994 transnational corporations employed 61 million people in the core regions and 12 million elsewhere.

An increasing percentage of investment reaches Latin America and Africa, as well as other countries in Asia (Figure 1–18). U.S. transnational corporations are more likely than Western European or Japanese transnationals to invest in Latin America, whereas Western European transnationals are more likely to invest in Eastern Europe and Africa, and Japanese transnationals are more likely to invest in Asia.

In the 1980s and 1990s, governments in the three regions where transnational corporations are based changed tax codes and regulations that had hindered international operations of private firms. Other countries where transnational corporations wished to invest changed laws and bureaucratic procedures that had effectively prevented transnational corporations from operating within their borders.

Historically, people and companies had difficulty moving even small sums of money from one country to another. International transfer of money involved a cumbersome set of procedures, and funds could be frozen for several weeks until all of the paperwork had cleared. Most governments prohibited the removal of large sums of money—and in the case of Communist countries, no

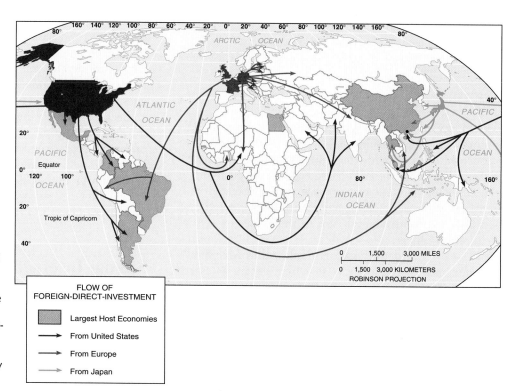

Figure 1–18 Majors flows of foreign investment. Most transnational companies invest in the three core regions of North America, Western Europe, and Japan. Outside the core regions, the largest amount of investment by transnational corporations is in Latin America (especially by transnationals based in the United States) and in Asia (especially by Japanese transnationals).

FLOW OF
FOREIGN-DIRECT-INVESTMENT

Largest Host Economies
From United States
From Europe
From Japan

money at all could be removed without government approval.

Modern communications and transportation systems provide the technical means to easily move money—as well as materials, products, technology, and other economic assets—around the world. Thanks to the electronic superhighway, companies can now organize economic activities across vast distances.

Banks, corporations, and other financial institutions are able to operate worldwide in part because the major centers where decisions are made that affect the global economy—New York, London, and Tokyo—are located in different time zones. When Tokyo's stock market closes, at 3 P.M., it is 6 A.M. in London, only 3 hours before the opening of the day's trading there. The stock market opens in New York at 9:30 A.M., while London's is still open. When the market closes in New York at 4 P.M., it is 6 A.M. the next morning in Tokyo, only 3 hours before the opening of the market there the next day. As a result, investors can react immediately to changes in the value of gold, the rate of exchange between the dollar and the yen, and other constantly changing elements of the global economy. (Refer to the map of time zones in the Globalization and Local Diversity box.)

Cultural Interaction. Geographers observe that increasingly uniform cultural preferences produce uniform "global" landscapes of material artifacts and of cultural values. Fast-food restaurants, service stations, and retail chains deliberately create a visual appearance that varies among locations as little as possible, so that customers know what to expect regardless of where in the world they happen to be. Houses built on the edge of one urban area will look very much like houses built on the edge of urban areas in other regions.

The globalization of material culture is based primarily on interaction by people in LDCs with lifestyles and products from MDCs, especially the United States. Regardless of cultural traditions, people around the world aspire to drive an automobile, watch television, and own a house. The survival of a culture's distinctive beliefs, forms, and traits is threatened by interaction with such social customs as wearing jeans and Nike shoes, consuming Coca-Cola and McDonald's hamburgers, and other preferences in food, clothing, shelter, and leisure activities.

Underlying the uniform cultural landscape is globalization of cultural beliefs and forms, especially religion and language. Africans, in particular, have moved away from traditional religions and have adopted Christianity or Islam, religions shared with hundreds of millions of people throughout the world. Increased interaction requires a form of common communication, and increasingly the English language is playing that role. People still speak thousands of different languages, traditionally a people's most distinctive social form, but English has become increasingly important as the language of international communication. More than three-fourths of college-age Europeans have learned to speak English, and three-fourths of the material on the Internet is in English.

Networks

Interaction takes place through networks, which are chains of communication that connect places. Today, ideas that originate in one area diffuse rapidly to other areas owing to our sophisticated communications and transportation networks. As a result of this rapid diffusion, interaction in the contemporary world is complex. People in more than one region may improve and modify an idea at the same time but in different ways.

A well-known example of a network in the United States is the television network (ABC, CBS, FOX, NBC, PBS). Each comprises a chain of stations around the country simultaneously broadcasting the same program, such as a football game. Transportation systems also form networks that connect places to each other. Airlines in the United States, for example, have adopted distinctive networks known as "hub-and-spokes." Under the "hub-and-spokes" system, airlines fly planes from a large number of places into one hub airport within a short period of time and then a short time later send the planes to another set of places. In principle, travelers originating in relatively small towns can reach a wide variety of destinations by changing planes at the hub airport (Figure 1–19).

Typically, the farther away one group is from another, the less likely the two groups are to interact. Contact diminishes with increasing distance and eventually disappears. This trailing-off phenomenon is called **distance decay**.

Interaction among groups can be retarded by barriers. These can be physical, such as oceans and deserts, or cultural, such as language and legal systems. We regard the landscape as part of our inheritance from the past. As a result, we may be reluctant to modify it unless we are under heavy pressure to do so. A major change in the landscape may reflect an upheaval in a people's culture.

Enhanced Communications

Cultural groups in different regions of the world can share beliefs, forms, and traits because of enhanced communications. We know more about what is happening elsewhere in the world, and we know sooner. Distant places seem less remote, more accessible to us. Thanks to technological improvements, we are able to reach into the everyday lives of people in far-off places (Figure 1–20). Geographers apply the term **time–space compression** to describe the reduction in the time it takes to diffuse something to a distant place.

In the past, most forms of interaction among cultural groups required the physical movement of settlers,

Figure 1–19 Delta Airlines' network. Delta, like other major U.S. airlines, has configured its route network in a system known as "hub and spokes." Lines connect each airport to the city to which it sends the most nonstop flights. Most flights originate or end at one of the company's hubs, especially Atlanta and Cincinnati. Delta's largest hub, Atlanta Hartsfield International Airport, has several parallel concourses connected to each other by an underground train. (Ron Sherman/Tony Stone Images)

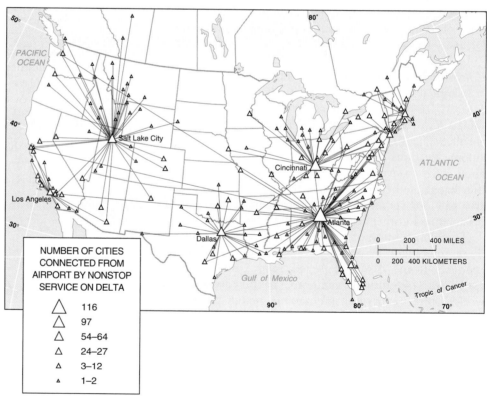

explorers, and plunderers from one location to another. As recently as 1800 A.D., people traveled in the same ways and at about the same speeds as in 1800 B.C.—they either were carried by an animal, took a sailboat, or walked.

Today, travel by motor vehicle or airplane is much quicker. But we do not even need to travel to know about another place. We can transmit images and mes-

sages from one part of the world to another at the touch of a button. We can communicate instantly with people in distant places through computers and telecommunications, and we can see instantly people in distant places on television. These and other forms of communications have made it possible for people in different places to be aware of the same cultural beliefs, forms, and traits.

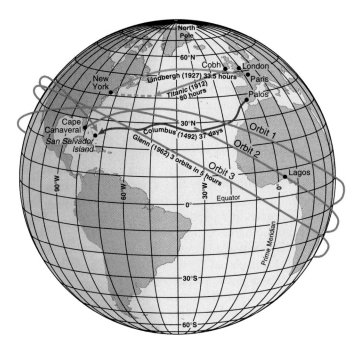

Figure 1–20 Time–space compression. Transportation improvements have "shrunk" the world. In 1492, Christopher Columbus took 37 days (nearly 900 hours) to sail across the Atlantic Ocean from the Canary Islands to San Salvador Island. In 1912, the *Titanic* was scheduled to sail from Queenstown (now Cobh) Ireland to New York in about 5 days, although two-thirds of the way across, after 80 hours at sea, it hit an iceberg and sank. In 1927, Charles Lindbergh was the first person to fly nonstop across the Atlantic, taking 33.5 hours to go from New York to Paris. In 1962, John Glenn, the first American to orbit in space, crossed above the Atlantic in about a quarter hour and circled the globe three times in 5 hours.

Diffusion: Movement Between Places

Diffusion is the process by which a characteristic spreads from one place to another over time. Something originates at a hearth or node and diffuses from there to other locations. Geographers document the location of nodes and the processes by which diffusion carries things elsewhere over time.

The region from which an innovation originates is called a **hearth**. How does a hearth emerge? A cultural group must be willing to try something new and be able to allocate resources to nurture the innovation. To develop a hearth, a group of people must also have the technical ability to achieve the desired idea and the economic structures, such as financial institutions, to facilitate implementation of the innovation.

As discussed in subsequent chapters, geographers can trace the dominant cultural, political, and economic features of contemporary United States and Canada primarily to hearths in Europe and the Middle East. However, other areas of the world also contain important hearths. In some cases an idea, such as an agricultural practice, may originate independently in more than one hearth. In

other cases, hearths may emerge in two regions because two cultural groups modify a shared concept in two different ways.

Types of Diffusion

For a person, object, or idea to have interaction with persons, objects, or ideas in other regions, diffusion must occur. Geographers observe two basic types of diffusion: relocation and expansion.

Relocation Diffusion. The spread of an idea through physical movement of people from one place to another is termed **relocation diffusion**. We shall see in Chapter 3 that people migrate for a variety of political, economic, and environmental reasons. When they move, they carry with them their culture, including language, religion, and ethnicity. The most commonly spoken languages in North and South America are Spanish, English, French, and Portuguese, primarily because several hundred years ago, Europeans who spoke those languages comprised the largest number of migrants. Thus, these languages spread through relocation diffusion. We will examine the diffusion of languages, religions, and ethnicity in Chapters 5 through 7.

The process of relocation diffusion helps us understand the distribution of acquired immunodeficiency syndrome (AIDS) within the United States. During the early 1980s, New York, California, and Florida were the nodes of origin for the disease within the United States (Figure 1–21). Half of the 50 states had no reported cases, whereas New York City, with only 3 percent of the nation's population, contained more than one-fourth of the AIDS cases. In the neighboring state of New Jersey, AIDS cases dropped with increasing distance from New York City.

During the 1980s and early 1990s, new AIDS cases diffused to every state, although California, Florida, and New York remained the focal points. These three states, plus Texas, accounted for half of the nation's new AIDS cases in the peak year of 1993, and the District of Columbia had the largest rate.

At a national scale, the diffusion of AIDS in the United States through relocation halted after 1993. The number of new AIDS cases dropped by one-fourth in just 2 years. Between 1993 and 1995, the rate declined by more than one-half in Texas, one-third in California, and one-fifth in Florida. However, within the United States local-scale diffusion had not yet stopped. The rate of new AIDS cases dropped one-fourth in New York but remained virtually unchanged in the states of Connecticut and New Jersey where many ex-New Yorkers now live. Similarly, the rate dropped one-fifth in the District of Columbia but remained unchanged in Virginia and actually rose in Maryland, the two states to which many Washingtonians have moved.

The AIDS Memorial Quilt on display in Washington, DC. The quilt was assembled as a memorial to people who have died of AIDS. By increasing awareness of AIDS, the creators of the quilt hope to slow diffusion of the disease. (Mark Phillips/Photo Researchers, Inc.)

Relocation diffusion can explain the rapid rise in the number of AIDS cases in the United States during the 1980s and early 1990s but not the rapid decline beginning in the mid-1990s. Instead, the decline resulted from the rapid diffusion of preventive methods and medicines such as AZT. The rapid spread of these innovations is an example of expansion diffusion rather than relocation diffusion.

Expansion and Related Types of Diffusion. **Expansion diffusion** is the spread of a feature from one place to another in a snowballing process. This expansion may result from one of three processes:

- Hierarchical diffusion
- Contagious diffusion
- Stimulus diffusion

Hierarchical diffusion is the spread of an idea from persons or nodes of authority or power to other persons or places. Hierarchical diffusion may result from the spread of ideas from political leaders, socially elite people, or other important persons to others in the commu-

nity. Innovations may also originate in a particular node or place of power, such as a large urban center, and diffuse later to isolated rural areas. Hip-hop or rap music is an example of an innovation that diffused from low-income African-Americans rather than from socially elite people, but it originated in urban areas.

Contagious diffusion is the rapid, widespread diffusion of a characteristic throughout the population. As the term implies, this form of diffusion is analogous to the spread of a contagious disease, such as influenza. Contagious diffusion spreads like a wave among fans in a stadium, without regard for hierarchy and without requiring permanent relocation of people. The rapid adoption throughout the United States of AIDS prevention methods and new medicines is an example of contagious diffusion. Ideas placed on the World Wide Web spread through contagious diffusion, because web surfers throughout the world have access to the same material simultaneously—and quickly.

Stimulus diffusion is the spread of an underlying principle, even though a characteristic itself apparently fails to diffuse. For example, early desktop computer sales in the United States divided about evenly between Macintosh Apple and IBM-compatible DOS systems. By the 1990s, Apple sales had fallen far behind IBM-compatibles in the United States, and the company had limited presence in rapidly expanding overseas markets. But principles pioneered by Apple, notably making selections by pointing a mouse at an icon rather than typing a string of words, diffused through a succession of IBM-compatible Windows systems.

Expansion diffusion occurs much more rapidly in the contemporary world than in the past. Modern methods of communications, such as computers, facsimile machines, and electronic mail systems have encouraged more rapid hierarchical diffusion than in the past. The Internet, especially the World Wide Web, has encouraged more rapid contagious diffusion. All the new technologies support the possibility of stimulus diffusion. Diffusion from one place to another can be instanta-neous in time, even if the physical distance between two places—as measured in kilometers or miles—is large.

Diffusion of Economy

In a global economy, transportation and communications systems have been organized to rapidly diffuse raw materials, goods, services, and capital from nodes of origin to other regions. Every area of the world plays some role intertwined with the roles played by other regions. Companies and workers that in the past were largely unaffected by events elsewhere in the world now share a single economic world with other companies and workers. The fate of an auto worker in Detroit is tied to investment decisions made in Mexico City, Seoul, Stuttgart, and Tokyo.

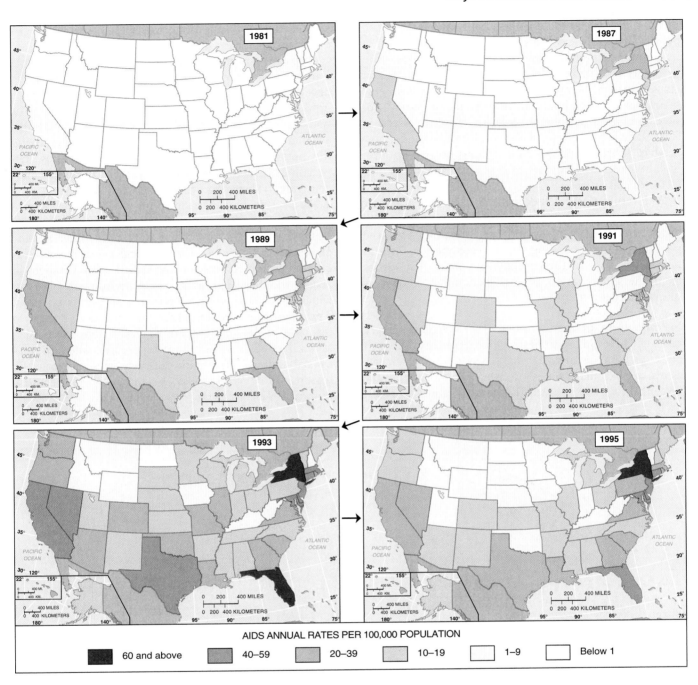

Figure 1–21 Diffusion of AIDS. Acquired immunodeficiency syndrome (AIDS) diffused across the United States from nodes in New York, California, and Florida. In 1981, virtually all people with AIDS were found in these three nodes. During the 1980s, the number of cases increased everywhere, but the incidence remained highest in the three original nodes. During the 1990s, the number of cases declined relatively rapidly in the original nodes.

Investment Flows from Three Core Regions. The global economy is increasingly centered on three core or hearth regions of North America, Western Europe, and Japan. These three regions have a large percentage of the world's advanced technology, capital to invest in new activities, and wealth to purchase goods and services. From "command centers" in the three major world cities of New York, London, and Tokyo, key decision-makers employ modern telecommunications to send out orders to factories, shops, and research centers around the world, an example of hierarchical diffusion.

Meanwhile, "nonessential" employees of the companies can be relocated to lower-cost offices outside the major financial centers. For example, Fila maintains headquarters in Italy but has moved 90 percent of its production of sportswear to Asian countries. Mitsubishi's corporate offices are in Japan, but all of its VCRs are produced in other Asian countries.

Countries in Africa, Asia, and Latin America contain three-fourths of the world's population, and nearly all of its population growth, but they find themselves on a periphery, or outer edge, of global investment that arrives through hierarchical diffusion of decisions made by transnational corporations through hierarchical diffusion. People in peripheral regions, who once toiled in isolated farm fields to produce food for their family, now produce crops for sale in core regions, or give up farm life altogether and migrate to cities in search of jobs in factories and offices.

As a result, the global economy has produced greater disparities than in the past between the levels of wealth and well-being enjoyed by people in the core and in the periphery. The increasing gap in economic conditions between regions in the core and periphery that results from the globalization of the economy is known as **uneven development**.

Specialization in the Location of Production. Every place in the world is part of the global economy, but each place plays a distinctive role, based on its particular assets. A place may be near valuable minerals, or it may be inhabited by especially well-educated workers.

Transnational corporations assess the particular economic assets of each place.

A particular place may be especially suitable for a transnational corporation to conduct research, to develop new engineering systems, to extract raw materials, to produce parts, to store finished products, to sell them, or to manage operations. In a global economy, transnational corporations remain competitive by correctly identifying the optimal location for each of these activities. Especially suitable places may be clustered in one country or region or dispersed around the world.

As a result, globalization of the economy has heightened economic differences among places. Factories are closed in some locations and opened in others. Some places become centers for technical research, whereas others become centers for low-skilled tasks. Changes in production have led to a spatial division of labor, in which a region's workers specialize in particular tasks. Transnationals decide where to produce things in response to characteristics of the local labor force, such as level of skills, prevailing wage rates, and attitudes toward unions. Transnationals may close factories in regions with high wage rates and strong labor unions (Figure 1–22).

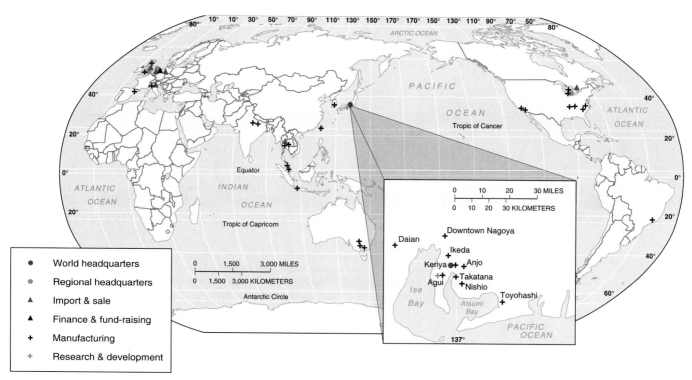

Figure 1–22 Globalization of economy. Denso, a transnational corporation that makes parts for cars, such as heaters and air conditioners, has its world headquarters, research labs, and eight factories in its "home town" of Nagoya, Japan. Regional headquarters are located in the world's two other core regions—North America and Western Europe—the company's main overseas markets. A financial center is located in the Netherlands. Factories and sales centers are located in a number of more developed and less developed countries.

Diffusion of Culture

The world contains only a handful of individuals who lead such isolated and sheltered lives that they have never watched a television set, used a telephone, or been in a motor vehicle. Even these extremely isolated and sheltered people are at least aware of the existence of these important elements by which global culture is communicated.

Unequal Access to Cultural Elements. Many people take for granted the ability to watch events in distant places by television, speak to others in distant places by telephone, and travel to far-off places by motor vehicle. An increasing number of the world's population regard access to these communications systems as novelties, perhaps recently experienced for the first time. But for some people, access to these elements is a distant aspiration. Knowledge of these communications systems is global but ability to purchase them is not.

Access to television, telephones, motor vehicles, and other means of communicating culture is restricted by an uneven division of wealth in the world. In some regions, possession of these objects is widespread, but in other regions few people have enough wealth to buy them. Even within regions, access to cultural elements may be restricted because of uneven distribution of wealth, or because of discrimination against women or minority groups.

Maintaining Local Traditions. As more people become aware of elements of global culture and aspire to possess them, local cultural beliefs, forms, and traits are threatened with extinction. Yet despite the globalization of culture, differences among places not only persist, but in many places local cultural traditions are flourishing. Global standardization of products does not mean that everyone wants the same cultural products. And even if the same message or image is transmitted simultaneously around the world, people in different places may not derive the same meanings from exposure to the same cultural elements.

The communications revolution that promotes globalization of culture also permits preservation of cultural diversity. Television, for example, is no longer restricted to a handful of channels displaying one set of cultural values. With the distribution of programming through cable and satellite systems, people have a choice of hundreds of programs rather than a handful. The proliferation in programming means that people in English-speaking countries now have the opportunity to watch programs in other languages, such as Spanish in the United States, Welsh in the United Kingdom, or Gaelic in Ireland.

Local cultural traditions may be transmitted around the world. Chinese, Ethiopian, Greek, Italian, Mexican, and Thai restaurants might coexist side-by-side along a single street in Chicago, London, or Toronto.

With the globalization of communications, people in two distant places can watch the same television program. At the same time, with the fragmentation of the broadcasting market, two people in the same house can watch different programs. Groups of people on every continent may aspire to wear jeans, but jeans buyers might live with someone who prefers khakis. In a global culture, companies can target groups of consumers with similar tastes in different parts of the world.

When two groups interact, the more dynamic and powerful culture is likely to dominate the weaker one. The modification of one culture as a result of contact with a more powerful one is called **acculturation**.

One of two things might happen to the weaker culture through acculturation. First, the weaker culture may be obliterated. For example, most immigrants to the United States quickly lost touch with most of the cultural characteristics of their former home and adopted the cultural traits of their new community. Second, the weaker culture might be transformed into a new culture, in which a new set of characteristics may coexist with older ones. New patterns emerge through the integration of the two cultures, but elements of the older culture remain.

Strong determination on the part of a group to retain its cultural traditions in the face of globalization of culture can lead to intolerance of people who display other beliefs, social forms, and material traits. Political disputes, unrest, and wars erupt in places such as Southeast Europe, East Africa, and the Middle East, where different cultural groups have been unable to share the same space peacefully (see Chapter 7).

Cargo Cult

The diffusion of cultural elements from Europe and North America may transform rather than destroy local cultural beliefs. For example, people on the Pacific island of Tana, part of the country of Vanuatu, worship Prince Phillip, the husband of Britain's Queen Elizabeth, as a god. According to local customary belief, Prince Phillip is a messiah who grew up on Tana Island, and Queen Elizabeth broke with her great council of chiefs to marry him. Prince Phillip was added to the collection of local gods a few years ago when he visited the country, formerly the British colony of New Hebrides. His advance person apparently distributed photographs, which the local people regard as holy icons.

The introduction of Prince Phillip as a god in Vanuatu is an example of a **cargo cult**. A cargo cult is a belief that the arrival of a ship or airplane in a locality has spiritual meaning. Several hundred years ago, some American Indians regarded Europeans who arrived on ships as gods.

Belief in a cargo cult persists in Papua New Guinea and other Pacific Ocean islands because American ships and airplanes brought new technology and equipment during World War II. Some residents believe that if they remain faithful, the planes will return with vast wealth. People in Papua New Guinea prepare large wooden planes as female sirens to lure male planes and their cargo from the sky.

Tension Between Globalization and Local Diversity

Globalization has not destroyed the uniqueness of an individual place's culture, economy, and environment. Human geographers understand that many contemporary social problems result from a tension between forces promoting global culture and economy on the one hand and preservation of local economic autonomy and cultural traditions on the other hand.

Culturally, people residing in different places are displaying fewer differences and more similarities in their cultural preferences, made possible through enhanced communications. But while consumers in different places express increasingly similar cultural preference, they do not share the same access to them. And the desire of some people to retain their traditional cultural elements, in the face of increased globalization of cultural preferences, has led to political conflict and market fragmentation in some regions.

Economically, large corporations are able to operate throughout the world and move money instantaneously from one place to another. Global investment decisions diffuse primarily from the three more developed core regions of North America, Western Europe, and Japan. But globalization of the economy has led to more specialization at the local level.

■ Summary

Here again are the key issues for Chapter 1:

1. **How do geographers describe where things are?** The most fundamental concept in geography is location, which is the position on Earth's surface that something occupies. Collections of things form distributions, or regular arrangements across Earth. Geographers use maps to display the location of objects and to extract information about places. Early geographers drew maps of Earth's surface based on exploration and observation, and during the past century geographic inquiry has become more scientific.

2. **Why is each place unique?** To understand why every place in the world is unique, geographers identify the distinctive culture of a group of people, and identify regions of the world distinguished by distinctive combinations of cultural as well as economic and environmental features.

3. **Why are different places similar?** Geographers study the interactions of groups of people and human activities across space, and they identify processes by which people and ideas diffuse from one location to another over time.

Case Study Revisited

Why Is More Than One Location Named Miami?

The case study that opened this chapter asked a "where" question—where is Miami located?—and answered the question using the concept of location. To conclude the chapter, we return to the opening case study to ask one of geography's basic "why" questions: Why do two places a thousand miles apart have the same name? Earlier in this chapter, we saw that two basic concepts—spatial interaction and diffusion—are especially important in explaining why different places are similar. These concepts help to explain the existence of places named Miami in both Ohio and Florida.

The name *Miami* originated with the Miami people, a Native American tribe belonging to the Algonquin family. When European explorers first encountered them, the Miami people lived in northeastern Wisconsin's Door County peninsula, near Sturgeon Bay. The word "Miami" means "people on the peninsula," a reference to the tribe's Door County homeland.

In the late seventeenth century, the Miamis migrated southward and settled along the St. Joseph River (today in southwestern Michigan) and the Wabash River (northeastern

Indiana). Pushed out of the southern Lake Michigan area by other tribes, the Miamis then migrated eastward to present-day Ohio, where they were living when the territory became part of the United States. Early nineteenth-century settlers from the East Coast retained the name Miami on the landscape to identify rivers, towns, and a university. Maumee, a river in northern Ohio, is a variant spelling of Miami.

Settlements called Miami were established in several other states through relocation diffusion by migrants from Ohio or by members of the Miami tribe. For example, most of the Miamis were forced to migrate from Ohio to eastern Kansas between 1832 and 1840. The present-day town of Miami, Kansas, is located near the land that the government gave to the tribe. In 1867, the tribe relocated with the Illinois Indians farther westward to Oklahoma, then known as Indian Territory. In 1891, Miami chief Thomas P. Richardville and Colonel W. C. Lykins established a town in the northeastern corner of Oklahoma that they named Miami. In the late nineteenth century, settlers from Ohio established a settlement in Gila County, Arizona, also named Miami. Settlements called Miami were likewise established in Missouri, New Mexico, Texas, and West Virginia.

In Florida, the name Miami did not derive from relocation diffusion of Miami Indians. Instead, it came from a Spaniard, Hernando d'Escalante Fontañeda, the first recorded European to live near present-day Miami, Florida. Following a shipwreck, Fontañeda became a prisoner of the Tequesta Indians, a branch of the Calusas people, between 1545 and 1562. In his 1575 memoir, Fontañeda used the Calusa Indian word *mayaimi* to describe Lake Okeechobee, as well as a river that flowed from the lake to the Atlantic Ocean. He translated the word *mayaimi* as "very large."

About 1830, Richard R. Fitzgerald established a plantation at the future site of the city of Miami, Florida. He named the plantation Miami, the first time that the Ohio spelling rather than the Spanish/Calusa version was used in Florida. Born in Columbia, South Carolina, Fitzgerald may have read the Ohio spelling Miami, but he had no known spatial interaction with the Miami in Ohio.

The U.S. Army took over the Miami plantation from the late 1830s until the early 1850s and renamed it Fort Dallas. After its abandonment as a fort, the site was sparsely inhabited until the 1890s. Henry M. Flagler started to build shops and hotels on the site in March 1896 and extended his Florida East Coast Railroad to the settlement one month later. By July 1896, the new settlement had 1,500 inhabitants and was incorporated as a city. Given a choice in naming the new settlement, the voters selected the former plantation name Miami instead of Flagler or Fort Dallas. The presence of several former Ohio residents in the new settlement may have influenced the vote, perhaps an example of stimulus diffusion.

Thus, the name Miami originated with Native Americans in northeastern Wisconsin and diffused to other locations in the United States as a result of the tribe's migration. The name Miami originated independently in Florida, the sort of coincidence that occurs when groups of people lack spatial interaction with each other. When south Florida became part of the United States, the process of stimulus diffusion produced a modified spelling in conformance with the northern Indian tribe.

■ Key Terms

Acculturation The modification of a culture as a result of contact with a more powerful culture.

Baseline An east-west line designated in the Land Ordinance of 1785 to facilitate the surveying and numbering of townships.

Cargo cult A belief that the arrival of a ship or airplane in a locality has spiritual meaning.

Cartography The science of making maps.

Concentration The spread of something over a given area.

Contagious diffusion The rapid, widespread diffusion of a feature or trend throughout a population.

Cultural ecology Geographic approach that emphasizes human–environment relationships.

Culture The body of customary beliefs, social forms, and material traits that together constitute a group of people's distinct tradition.

Density The frequency with which something exists within a given unit of area.

Diffusion The process of spread of a feature or trend from one place to another over time.

Distance decay The diminishing in importance and eventual disappearance of a phenomenon with increasing distance from its origin.

Distribution The arrangement of something across Earth's surface.

Environmental determinism A nineteenth- and early twentieth-century approach to the study of geography that argued that the general laws sought by human geographers could be found in the physical sciences. Geography was therefore the study of how the physical environment caused human activities.

Expansion diffusion The spread of a feature or trend among people from one area to another in a snowballing process.

Formal region (or uniform or homogeneous region) An area in which everyone shares in one or more distinctive characteristics.

Functional region (or nodal region) An area organized around a node or focal point.

GIS A computer system that stores, organizes, analyzes, and displays geographic data.

Globalization Actions or processes that involve the entire world and result in making something worldwide in scope.

Greenwich Mean Time The time in that time zone encompassing the prime meridian, or 0° longitude.

Hearth The region from which innovative ideas originate.

Hierarchical diffusion The spread of a feature or trend from one key person or node of authority or power to other persons or places.

International Date Line An arc that for the most part follows 180° longitude, although it deviates in several places to avoid dividing land areas. When you cross the International Date Line heading east (toward America), the clock moves back 24 hours, or one entire day. When you go west (toward Asia), the calendar moves ahead one day.

Land Ordinance of 1785 A law that divided much of the United States into a system of townships to facilitate the sale of land to settlers.

Latitude The numbering system used to indicate the location of parallels drawn on a globe and measuring distance north and south of the equator (0°).

Location The position of anything on Earth's surface.

Longitude The numbering system used to indicate the location of meridians drawn on a globe and measuring distance east and west of the prime meridian (0°).

Map A two-dimensional, or flat, representation of Earth's surface or a portion of it.

Meridian An arc drawn on a map between the North and South poles.

Parallel A circle drawn around the globe parallel to the equator and at right angles to the meridians.

Pattern The geometric or regular arrangement of something in a study area.

Polder Land created by the Dutch by draining water from an area.

Possibilism The theory that the physical environment may set limits on human actions, but people have the ability to adjust to the physical environment and choose a course of action from many alternatives.

Prime meridian The meridian, designated as 0° longitude, which passes through the Royal Observatory at Greenwich, England.

Principal meridian A north-south line designated in the Land Ordinance of 1785 to facilitate the surveying and numbering of townships.

Projection The system used to transfer locations from Earth's surface to a flat map.

Region An area distinguished by a unique combination of trends or features.

Regional studies (or cultural landscape) An approach to geography that emphasizes the relationships among social and physical phenomena in a particular study area.

Relocation diffusion The spread of a feature or trend through bodily movement of people from one place to another.

Remote sensing The acquisition of data about Earth's surface from a satellite orbiting the planet or other long-distance methods.

Resource A substance that has value or usefulness.

Scale The relationship between the size of an object on a map and the size of the actual feature on Earth's surface.

Section A square normally 1 mile long and 1 mile wide under the Land Ordinance of 1785; townships were normally divided into 36 sections.

Site The physical character of a place.

Situation The location of a place relative to other places.

Spatial interaction The interdependence between people in two areas through the movement of people, ideas, and objects between the two areas.

Stimulus diffusion The spread of an underlying principle, even though a specific characteristic is rejected.

Time-space compression The reduction in the time it takes to diffuse something to a distant place, as a result of improved communications and transportation systems.

Toponym The name given to a portion of Earth's surface.

Township A square normally 6 miles on a side. The Land Ordinance of 1785 divided much of the United States into a series of townships.

Transnational corporation A company that conducts research, operates factories, and sells products in many countries, not just where its headquarters or shareholders are located.

Uneven development The increasing gap in economic conditions between core and peripheral regions as a result of the globalization of the economy.

Vernacular region (or perceptual region) An area that people believe to exist as part of their cultural identity.

■ Thinking Geographically

1. Cartography is not simply a technical exercise in penmanship and coloring; nor are decisions confined to scale and projection. Mapping is a politically sensitive undertaking. Look at how maps in this book distinguish between the territories of Israel and its neighbors, the locations of borders in South Asia and the Arabian peninsula, and the relationship of China and Taiwan. Are there other logical ways to draw boundaries and distinguish among territories in these regions? What might they be?

2. Imagine that a transportation device (perhaps the one in *Star Trek*) would enable all humans to travel instantaneously to any location on Earth's surface. What impact would that invention have on the distribution of peoples and activities across Earth?

3. When earthquakes, hurricanes, or other environmental disasters strike, humans tend to "blame" nature and see themselves as innocent victims of a harsh and cruel nature. To what extent do environmental hazards stem from un-

predictable nature, and to what extent do they originate from human actions? Should victims blame nature, other humans, or themselves for the disaster? Why?

4. The construction of dams is a particularly prominent example of human–environment interaction in regions throughout the world. Turkey is building the Ataturk Dam on the Euphrates River, a move opposed by Syria and Iraq, the two downstream countries. Egypt, which operates the Aswan Dam on the Nile River, has blocked loans to Ethiopia that could be used to divert the source of the Nile. Some Russians oppose construction of the Gorskaya Dam in the Gulf of Finland near St. Petersburg. Similarly, the Balbina Dam on the Uatruma River, a tributary of the Amazon, has generated considerable opposition in Brazil. Why do governments push the construction of dams so forcefully, and why do others oppose their construction so passionately?

5. Geographic concepts are supposed to help explain contemporary issues. Are there any stories in your newspaper to which geographic concepts can be applied to help understand the issues? Discuss.

■ Further Readings

Agnew, John; David N. Livingstone, and Alisdair Rogers, eds. *Human Geography: An Essential Anthology*. Oxford: Blackwell, 1996.

Andriot, John L., ed. *Township Atlas of the United States*. McLean, VA: Andriot Associates, 1979.

Ayers, Edward L., Patricia Nelson Limerick, Stephen Nussbaum, and Peter S. Onuf. *All Over the Map: Rethinking American Regions*. Baltimore: The Johns Hopkins University Press, 1996.

Azaryahu, Maoz. "German Reunification and the Politics of Street Names: The Case of East Berlin." *Political Geography* 16 (1997): 479–94.

Baer, Anne. "Not Enough Water to Go Round?" *International Social Science Journal* 48 (1996): 277–92.

Bennett, John W. *The Ecological Transition: Cultural Anthropology and Human Adaptation*. Oxford: Pergamon Press, 1976.

Blaut, J. M. "Diffusionism: A Uniformitarian Critique." *Annals of the Association of American Geographers* 77 (1987): 48–62.

Bodman, Andrew R. "Weavers of Influence: The Structure of Contemporary Geographic Research." *Transactions of British Geographers New Series* 16 (1991): 21–37.

Brown, Lawrence A. *Innovation Diffusion: A New Perspective*. London: Methuen, 1981.

Brunn, Stanley D. "Sunbelt USA." *Focus* 36 (1986): 34–35.

Carlson, Helen S. *Nevada Place Names: A Geographical Dictionary*. Reno: University of Nevada Press, 1974.

Claval, Paul. "The Region as a Geographical, Economic and Cultural Concept." *International Social Science Journal* 39 (1987): 159–72.

Cohen, Saul B., and Nurit Kliot. "Place-Names in Israel's Ideological Struggle over the Administered Territories." *Annals of the Association of American Geographers* 82 (1992): 653–80.

Constandse, A. K. *Planning and Creation of an Environment*. Lelystad, The Netherlands: Rijksdienst voor de IJsselmeerpolders, 1976.

Cutter, Susan L. "Societal Responses to Environmental Hazards." *International Social Science Journal* 48 (1996): 525–36.

Domosh, Mona. "Geography and Gender: The Personal and the Political." *Progress in Human Geography* 21 (1997): 81–87.

Golledge, Reginald G. "Geographical Theories." *International Social Science Journal* 48 (1996): 461–76.

Eldridge, J. Douglas, and John Paul Jones III. "Warped Space: A Geography of Distance Decay." *Professional Geographer* 43 (1991): 500–11.

Entrikin, J. Nicholas, and Stanley D. Brunn, eds. *Reflections on Richard Hartshorne's "The Nature of Geography."* Washington, DC: Association of American Geographers, 1989.

Escolar, Marcelo. "Exploration, Cartography and the Modernization of State Power." *International Social Science Journal* 49 (1997): 55–76.

Escott, Paul D., and David R. Goldfield, eds. *The South for New Southerners*. Chapel Hill: University of North Carolina Press, 1991.

Espenshade, Edward B., Jr., ed. *Goode's World Atlas*, 19th ed. Chicago: Rand McNally, 1995.

Forman, R. T. T., and M. Godron. *Landscape Ecology*. New York: John Wiley, 1986.

Freeman, Donald B. "The Importance of Being First: Preemption by Early Adopters of Farming Innovations in Kenya." *Annals of the Association of American Geographers* 75 (1985): 1–16.

Gaile, Gary L., and Cort J. Willmott, eds. *Geography in America*. New York: Merrill, 1989.

Gardner, Lytt I., Jr., et al. "Spatial Diffusion of the Human Immunodeficiency Virus Infection Epidemic in the United States, 1985–87." *Annals of the Association of American Geographers* 79 (1989): 25–43.

Geography Education Standards Project. *Geography for Life: National Geography Standards*. Washington, DC: National Geographic Research & Exploration, 1994.

Goliber, Thomas J. "Sub-Saharan Africa: Population Pressures on Development. World Population in Transition." *Population Bulletin*, 40 (1). Washington, DC: Population Reference Bureau, 1985.

Gould, Peter R. "Space, Time and the Human Being." *International Social Science Journal* 48 (1996): 449–60.

_____. *The Geographer at Work*. Boston: Routledge and Kegan Paul, 1985.

Gritzner, Charles F., Jr. "The Scope of Cultural Geography." *Journal of Geography* 65 (1966): 4–11.

Gross, Jonathan L., and Steve Rayner. *Measuring Culture*. New York: Columbia University Press, 1985.

Hägerstrand, Torsten. *Innovation Diffusion as a Spatial Process*. Chicago: University of Chicago Press, 1967.

Haggett, Peter. *The Geographer's Art*. Cambridge, MA: Blackwell, 1990.

Hamm, Bernd, and Martin Lutsch. "Sunbelt v. Frostbelt: A Case for Convergence Theory?" *International Social Science Journal* 39 (1987): 199–214.

Hartshorne, Richard. *The Nature of Geography*. Lancaster, PA: Association of American Geographers, 1939.

Jackson, John Brinckerhoff. *American Space*. New York: W. W. Norton & Co., 1972.

James, Preston E. *All Possible Worlds: A History of Geographical Ideas*. New York: Bobbs-Merrill, 1972.

Janelle, Donald G. "Central Place Development in a Time-Space Framework." *Professional Geographer* 20 (1968): 5–10.

_____, ed. *Geographical Snapshots of North America*. New York: Guilford Press, 1992.

Johnson, Hildegard B. *Order Upon the Land*. New York: Oxford University Press, 1976.

Johnston, R. J. "A Place for Everything and Everything in Its Place." *Transactions of British Geographers New Series* 16 (1991): 131–47.

_____. *Philosophy and Human Geography*, 2d ed. London: Edward Arnold, 1986.

_____, ed. *The Dictionary of Human Geography*, 3d ed. Oxford: Blackwell, 1994.

Kramer, A. *Hawaii, Ostmikronesien und Samoa*. Stuttgart: Strecker and Schroder, 1902.

Leighly, John, ed. *Land and Life: A Selection from the Writings of Carl Ortwin Sauer*. Berkeley: University of California Press, 1963.

Lewis, David. *Voyaging Stars*. New York: W.W. Norton & Co., 1978.

Macgill, Sally M. "Environmental Questions and Human Geography." *International Social Science Journal* 38 (1986): 357–76.

Matless, David, "New Material? Work in Cultural and Social Geography, 1995." *Progress in Human Geography* 20, (1996): 379–91.

Meinig, D. W., ed. *The Interpretation of Ordinary Landscapes*. New York: Oxford University Press, 1979.

Mikesell, Marvin W. "Tradition and Innovation in Cultural Geography." *Annals of the Association of American Geographers* 68 (1978): 1–16.

Mitchell, Don. "There's No Such Thing as Culture: Towards a Reconceptualization of the Idea of Culture in Geography." *Transactions of the Institute of British Geographers New Series* 20 (1995): 102–16.

Monmonier, Mark. *How to Lie with Maps*. Chicago: University of Chicago Press, 1991.

Noronha, Valerian T., and Michael F. Goodchild. "Modeling Interregional Interaction: Implications for Defining Functional Regions." *Annals of the Association of American Geography* 82 (1992): 86–102.

Norton, William. *Explorations in the Understanding of Landscape: A Cultural Geography*. Westport, CT: Greenwood Press, 1989.

Peet, Richard. "The Social Origins of Environmental Determinism." *Annals of the Association of American Geographers* 75 (1985): 309–33.

Penning-Rowsell, Edmund C., and David Lowenthal, eds. *Landscape, Meanings and Values*. London: Allen and Unwin, 1986.

Pinkerton, James P. "Enviromanticism." *Foreign Affairs*. 76 (1997): 2–7.

Reed, Michael, ed. *Discovering Past Landscapes*. London: Croom Helm, 1986.

Rice, Bradley R. "Searching for the Sunbelt." *American Demographics* 3 (1981): 22–23.

Roberts, Neil. "The Human Transformation of the Earth's Surface." *International Social Science Journal* 48 (1996): 493–510.

Rowntree, Lester B., and Margaret W. Conkey. "Symbolism and the Cultural Landscape." *Annals of the Association of American Geographers* 70 (1980): 459–74.

Salter, Christopher L. *The Cultural Landscape*. Belmont, CA: Duxbury Press, 1971.

_____. "What Can I Do with Geography?" *Professional Geographer* 35 (1983): 266–73.

Santos, Milton. "Geography in the Late Twentieth Century: New Roles for a Theoretical Discipline." *International Social Science Journal* 36 (1984): 657–72.

Sauer, Carl O. "Morphology of Landscape." *University of California Publications in Geography* 2 (1925): 19–54.

Sawers, Larry, and William K. Tabb, eds. *Sunbelt/Snowbelt: Urban Development and Regional Restructuring*. New York: Oxford University Press, 1984.

Shannon, Gary W., and Gerald F. Pyle. "The Origin and Diffusion of AIDS: A View from Medical Geography." *Annals of the Association of American Geographers* 79 (1989): 1–24.

_____, and Rashid L. Bashshur. *The Geography of AIDS*. New York: Guilford Press, 1991.

Shortridge, James R. *The Middle West: Its Meaning in American Culture*. Lawrence: University of Kansas Press, 1989.

Smith, Stanley K., and Christopher McCarty. "Demographic Effects of Natural Disasters: A Case Study of Hurricane Andrew." *Demography* 33 (1996) 265–75.

Solot, Michael. "Carl Sauer and Cultural Evolution." *Annals of the Association of American Geographers* 76 (1986): 508–20.

Tikunov, Vladimir S. "The Information Revolution in Geography." *International Social Science Journal* 48 (1996): 477–92.

Tuan, Yi-Fu. "Cultural Pluralism and Technology." *Geographical Review* 79 (1989): 269–79.

Wagner, Philip L., and Marvin W. Mikesell, eds. *Readings in Cultural Geography*. Chicago: University of Chicago Press, 1962.

Zimmerer, Karl S. "Human Geography and the 'New Ecology': The Prospect and Promise of Integration." *Annals of the Association of American Geographers* 84 (1994): 108–25.

Also consult the following journals: *Annals of the Association of American Geographers*; *Area*; *Canadian Geographer (Géographe canadien)*; *Focus*; *Geografiska Annaler Series B Human Geography*; *Geographical Analysis*; *Geographical Review*; *Geography*; *Journal of Geography*; *Professional Geographer*; *Progress in Human Geography*; *Transactions of the Institute of British Geographers*.

CHAPTER 2

Population

How many brothers and sisters do you have? How many brothers and sisters did your parents or grandparents have? Did they have more, fewer, or the same number of siblings as yourself? How many children do you have, or intend to have? Is that figure larger, smaller, or the same number as your parents and grandparents had?

The typical family in a more developed country (MDC) today contains fewer people than in the past, and the number of children is declining. In much of North America and Europe, a majority of people have the same number or fewer siblings than their parents and grandparents. And the number of children your generation has, or will have, appears to be fewer on average, although only the future can reveal the actual trend.

In other regions of the world, the number of children per household tends to be much higher than in the MDCs. The ability of less developed countries (LDCs) to provide food, clothing, and shelter for their people is severely hampered by the continued rapid growth of their population.

A study of population is the basis for understanding a wide variety of issues in human geography. To study the challenge of increasing the food supply, reducing pollution, and encouraging economic growth, geographers must ask where and why a region's population is distributed as it is. Therefore, our study of human geography begins with a study of population.

Key Issues

1. Where is the world's population distributed?

2. Where has the world's population increased?

3. Why is population increasing at different rates in different countries?

4. Why might the world face an overpopulation problem?

Jakarta, Indonesia (Chuck O'Rear/Woodfin Camp)

Case Study

Population Growth in India

The Phatak family lives in a village of 600 inhabitants in India. At age 40, Indira Phatak has been pregnant five times. Four of her children have survived; they are aged 5 to 18.

When the two Phatak daughters marry a few years from now, how many children will each of them bear? The Indian government hopes that they will choose to have fewer children than their mother. About 28 million babies will be born this year in India, and the country's population is growing by 18 million annually. Unless attitudes and behavior drastically change in the next few years, India's population—currently 1 billion—could exceed 5 billion a century from now.

Three-fourths of Indians live in rural settlements that have fewer than 5,000 inhabitants. For many of these people, children are an economic asset, because they help perform chores on the farm and are expected to provide for their parents in their old age. The high percentage of children who will die before they reach working age also encourages large families. One out of every 13 infants in India dies within one year of birth, and 160,000 women die annually during pregnancy and childbirth.

In recent years, India has made significant progress in diffusing modern agricultural practices, building new industry, and developing natural resources, all of which have increased national wealth. However, in a country with a rapidly expanding population, much of the newly created wealth must be used to provide food, housing, and other basic services for the additional people. With more than one-third of the population under the age of 15, the government must build schools, hospitals, and day-care centers. Therefore, the growing wealth is going primarily to provide a reasonable standard of living for an expanding population. Further, will employment be available to these 350 million children when they are old enough to work?

The study of population is critically important for three reasons:

- More people are alive at this time—6 billion— than at any point in Earth's long history.
- The world's population increased at a faster rate during the second half of the twentieth century than ever before in history.
- Virtually all global population growth is concentrated in less developed countries.

These facts lend urgency to the task of understanding the diversity of population problems in the world today.

The scientific study of population characteristics is **demography**. Demographers look statistically at how people are distributed spatially and by age, gender, occupation, fertility, health, and so on.

As introduced in Chapter 1, geographers ask where and why questions. As we begin our study of the major topics in human geography, note the wording of the four key issues that organize the material in this chapter. The first two issues ask "where" questions, the second two "why" questions.

Geographers study population problems by first describing *where* people are found across Earth's surface. The locations of Earth's 6 billion people form regular distributions. The second key issue looks at another "where" question, this time where population is growing. The chapter then turns to explaining *why* population is growing at different rates in different regions. As with other geographic themes, local variations in population growth stem from a combination of diverse cultural conditions in various regions and the diffusion around the world of a process by which population growth rates change.

The final issue explains why geographers consider local differences in growth rates to be important. Some demographers predict that the world may become overburdened with too many people in the future. They ask whether the world's population will exceed the capacity of Earth to provide food, space, and resources for the people. Geographers who specialize in demography cannot offer a simple yes or no answer. However, geography's focus on answering the where and why questions helps to explain the global population problem and to suggest solutions.

From the perspective of *globalization*, geographers argue that the world's so-called **overpopulation** problem is not simply a matter of the total number of people on Earth, but the relationship between the number of people and the availability of resources. Problems result when an area's population exceeds the capacity of the environment to support them at an acceptable standard of living.

From a *local diversity* perspective, geographers find that overpopulation is a threat in some regions of the world but not in others. The capacity of Earth as a whole

to support human life may be high, but some regions have a favorable balance between people and available resources, while others do not. Further, the regions with the most people are not necessarily the same as the regions with an unfavorable balance between population and resources.

Key Issue 1

Where Is the World's Population Distributed?

- Population concentrations
- Sparsely populated regions
- Population density

Human beings are not distributed uniformly across Earth's surface. We can understand how population is distributed by examining two basic properties, concentration and density. Geographers identify regions of Earth's surface where population is clustered and regions where it is sparse. We also construct several density measures to help geographers explain the relationship between the number of people and available resources.

Population Concentrations

Approximately three-fourths of the world's population live on only 5 percent of Earth's surface. The balance of Earth's surface consists of oceans (about 71 percent) and less intensively inhabited land.

The world's population is clustered in five regions: East Asia, South Asia, Southeast Asia, Western Europe, and Eastern North America (Figure 2–1). These five regions display some similarities. Most of their people live near an ocean, or near a river with easy access to an ocean, rather than in the interior of major landmasses. In fact, approximately two-thirds of the world's population live within 500 kilometers (300 miles) of an ocean, and 80 percent live within 800 kilometers (500 miles). The five population clusters occupy generally low-lying areas, with fertile soil and temperate climate. The regions all are located in the Northern Hemisphere between 10° and 55° north latitude, with the exception of part of the Southeast Asia concentration. Despite these similarities, we can see significant differences in the pattern of occupancy of the land in the five concentrations.

East Asia

One-fourth of the world's people live in East Asia, the largest cluster of inhabitants. The region, bordering the Pacific Ocean, includes eastern China, the islands of Japan, the Korean peninsula, and the island of Taiwan.

Five-sixths of the people in this concentration live in the People's Republic of China, the world's most populous country. China is the world's third largest country in

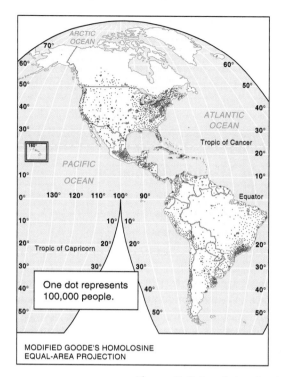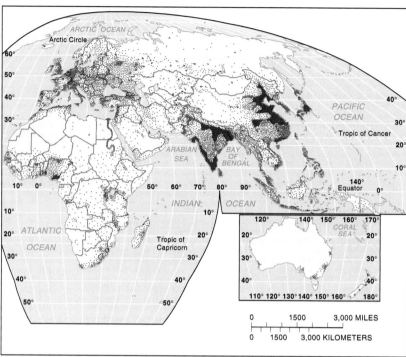

Figure 2–1 Population distribution. People are not distributed uniformly across Earth's surface. Most people live near an ocean or a river.

land area, but much of its interior is sparsely inhabited mountains and deserts. The Chinese population is clustered near the Pacific Coast and in several fertile river valleys that extend inland, such as the Huang and the Yangtze. Although China has 11 cities with more than 2 million inhabitants, three-fourths of the people live in rural areas where they work as farmers.

In Japan and South Korea, population is not distributed uniformly either. More than one-third of the people live in three large metropolitan areas—Tokyo and Osaka in Japan and Seoul in South Korea—that cover less than 3 percent of the two countries' land area. In sharp contrast to China, more than three-fourths of the Japanese and Koreans live in urban areas and work at industrial or service jobs.

South Asia

The second largest concentration of people, more than one-fifth, is in South Asia, which includes India, Pakistan, Bangladesh, and the island of Sri Lanka. India, the world's second most populous country, contains more than three-fourths of the South Asia population concentration.

The most important concentration of people within South Asia lives along a 1,500-kilometer (900-mile) corridor from Lahore, Pakistan, through India and Bangladesh to the Bay of Bengal. Much of this area's population is concentrated along the plains of the Indus and Ganges rivers. Population is also heavily concentrated near India's two long coastlines—the Arabian Sea to the west and the Bay of Bengal to the east.

Jakarta, Indonesia. Rapid population growth in less developed countries taxes the ability of services, such as transportation. As a freight train passes through a low-income neighborhood of Jakarta, a city of more than 11 million inhabitants, children "hitch" a ride on the outside of a freight train. (Chuck O'Rear/Woodfin Camp)

Like the Chinese, most people in South Asia are farmers living in rural areas. The region contains 14 cities with more than 2 million inhabitants, but only one-fourth of the total population lives in an urban area.

Southeast Asia

A third important Asian population cluster, and the world's fourth largest (after Europe, described next), is in Southeast Asia. One-half billion people live in Southeast Asia, mostly on a series of islands that lie between the Indian and Pacific oceans. These islands include Java, Sumatra, Borneo, Papua New Guinea, and the Philippines. The largest concentration is on the island of Java, inhabited by more than 100 million people.

Indonesia, which consists of 13,677 islands, including Java, is the world's fourth most populous country. Several islands that belong to the Philippines contain high population concentrations, and population is also clustered along several river valleys and deltas at the southeastern tip of the Asian mainland, known as Indochina. Like China and South Asia, the Southeast Asia concentration is characterized by a high percentage of people working as farmers in rural areas.

The three Asian population concentrations together comprise over half of the world's total population, but together they live on less than 10 percent of Earth's land area. The same held true 2,000 years ago, when approximately half of the world's population was found in these same regions.

Europe

Combining the populations of Western Europe, Eastern Europe, and the European portion of Russia forms the world's third largest population cluster, one-eighth of the world's people. The region includes four dozen countries, ranging from Monaco, with 1 square kilometer (0.7 square miles) and a population of 30,000, and San Marino, with 62 square kilometers (24 square miles) and a population of 20,000, to Russia, the world's largest country in land area when its Asian part is included.

In contrast with the three Asian concentrations, three-fourths of Europe's inhabitants live in cities, and less than 20 percent are farmers. A dense network of road and rail lines links settlements. The highest concentrations in Europe are near the coalfields of England, Germany, and Belgium, historically the major source of energy for industry.

Although the region's temperate climate permits cultivation of a variety of crops, Europeans do not produce enough food for themselves. Instead, they import food and other resources from elsewhere in the world. The search for additional resources was a major incentive for Europeans to explore and colonize other parts of the world during the previous six centuries. Today, Europeans turn many of these resources into manufactured products.

Eastern North America

The largest population concentration in the Western Hemisphere is in the northeastern United States and southeastern Canada. This cluster extends along the Atlantic Coast from Boston to Newport News, Virginia, and westward along the Great Lakes to Chicago. About 2 percent of the world's people live in the area. Like the Europeans, most Americans are urban dwellers; less than 5 percent are farmers.

The clustering of the world's population can be displayed on a cartogram, which depicts the size of countries according to a statistic other than land area, as is the case with most maps. In this case, the size of countries varies according to population (Figure 2–2). The shapes of several large or populous countries, including Brazil, Canada, China, Indonesia, Russia, and the United States, have been exaggerated to show the regions within the countries where most of the population is clustered.

When compared to a more typical equal-area map, such as Figure 2–1, the population cartogram displays the major population clusters of Europe and East, South, and Southeast Asia as much larger and Africa and the Western Hemisphere as much smaller. As you look at maps of population growth and other topics in this and subsequent chapters, pay special attention to Asia and Europe, because global patterns are heavily influenced by conditions in these regions, where two-thirds of the world's people live.

Sparsely Populated Regions

Human beings avoid clustering in certain physical environments. Relatively few people live in regions that are too dry, too wet, too cold, or too mountainous for activities such as agriculture. The portion of Earth's surface occupied by permanent human settlement is called the **ecumene**.

The areas of Earth that humans consider too harsh for occupancy have diminished over time, while the ecumene has increased (Figure 2–3). Seven thousand years ago, humans occupied only a small percentage of Earth's land area, primarily in the Middle East, Eastern Europe, and East Asia. Even 500 years ago, much of North America and Asia lay outside the ecumene.

Dry Lands

Areas too dry for farming cover approximately 20 percent of Earth's land surface. The two largest desert regions in the world lie in the Northern Hemisphere between 15° and 50° north latitude and in the Southern Hemisphere between 20° and 50° south latitude. The largest desert region, extending from North Africa to Southwest and Central Asia, is known by several names, including the Sahara, Arabian, Thar, Takla Makan, and Gobi deserts. A smaller desert region, in the Southern Hemisphere, comprises much of Australia. Earth's desert regions are

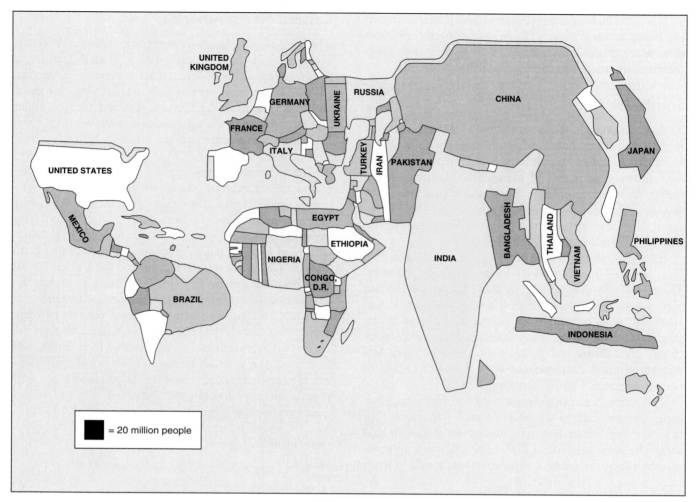

Figure 2–2 Population cartogram. Countries are displayed by size of population rather than land area. Countries named on the cartogram have at least 50 million inhabitants.

shown on Figure 1–12. Regions where desert conditions are advancing appear in Figure 14–16.

Deserts generally lack sufficient water to grow crops that could feed a large population, although some people survive there by raising animals, such as camels, that are adapted to the climate. By constructing irrigation systems, people can grow crops in some parts of the desert. Although dry lands are generally inhospitable to intensive agriculture, they may contain natural resources useful to people—notably, much of the world's oil reserves. The increasing demand for these resources has led to a growth in settlements in or near deserts.

Wet Lands

Lands that receive very high levels of precipitation may also be inhospitable for human occupation. These lands are located primarily near the equator between 20° north and south latitude in the interiors of South America, Central Africa, and Southeast Asia. Rainfall averages more than 1.25 meters (50 inches) per year, with most areas receiving more than 2.25 meters (90 inches) per year. The combination of rain and heat rapidly depletes nutrients from the soil, thus hindering agriculture.

Precipitation may be concentrated into specific times of the year or spread throughout the year. In seasonally wet lands, such as those in Southeast Asia, enough food can be grown to support a large population (see the rice production map, Figure 10–6).

Cold Lands

Much of the land near the North and South poles is perpetually covered with ice, or the ground is permanently frozen (permafrost). The polar regions receive less precipitation than some Central Asian deserts, but over thousands of years the small annual snowfall has accumulated into thick ice. Consequently, the polar regions are unsuitable for planting crops, few animals can survive the extreme cold, and few human beings live there.

High Lands

Finally, relatively few people live at high elevations. The highest mountains in the world are steep, snow-covered, and sparsely settled. For example, approximately half of Switzerland's land is more than 1,000 meters (3,300 feet) above sea level, and only 5 percent of the country's people live there.

We can find some significant exceptions, especially in Latin America and Africa. People may prefer to occupy higher lands if temperatures and precipitation are uncomfortably high at lower elevations. In fact, Mexico City, one of the world's largest cities, is located at an elevation of 2,243 meters (7,360 feet).

Population Density

Density, defined in Chapter 1 as the number of people occupying an area of land, can be computed in several ways, including arithmetic density, physiological density, and agricultural density. These measures of density help geographers to describe the distribution of people in comparison to available resources.

Arithmetic Density

Geographers most frequently use **arithmetic density**, which is the total number of people divided by total land area. (This measure also is called *population density*.) Geographers rely on the arithmetic density to compare conditions in different countries because the two pieces of information needed to calculate the measure—total population and total land area—are easy to obtain.

For example, to compute the arithmetic or population density for the United States, we can divide the population (approximately 270 million people) by the land area (approximately 9.2 million square kilometers, or 3.7 million square miles). The result shows that the United States has an arithmetic density of 28 persons per square kilometer (73 persons per square mile). By comparison, the arithmetic density is much higher in South Asia. In Bangladesh, it is approximately 849 persons per square kilometer (2,220 persons per square mile) and 306 (792) in India. On the other hand, the arithmetic density is only 3 persons per square kilometer (8 persons per square mile) in Canada and 2 (6) in Australia (Figure 2–4).

Arithmetic density varies even more within individual countries. In the United States, for example, New York County (Manhattan Island) has a population density of approximately 21,000 persons per square kilometer (55,000 persons per square mile), whereas Loving County, Texas, has a population density of approximately 0.08 persons per square kilometer (0.2 per square mile). In Egypt, the arithmetic density is only 65 persons per square kilometer (168 persons per square mile) overall, but it is 2,100 persons per square kilometer (5,600 persons per square mile) in the delta and valley of the Nile River.

Arithmetic density enables geographers to make approximate comparisons of the number of people trying to live on a given piece of land in different regions of the world. Thus, arithmetic density answers the "where" question. However, to explain *why* people are not uniformly distributed across Earth's surface, other density measures are more useful.

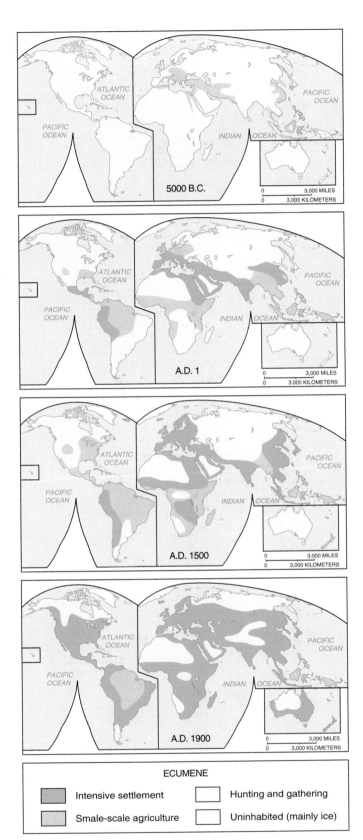

Figure 2–3 Ecumene. The portion of Earth occupied by permanent human settlement—the ecumene—has expanded from the Middle East and East Asia to encompass most of the world's land area.

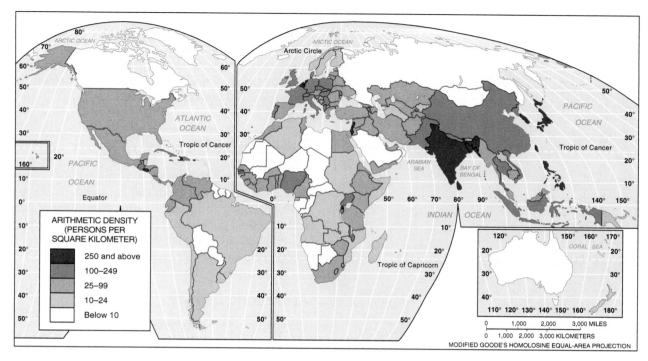

Figure 2–4 Arithmetic density. Arithmetic, or population, density is the total number of people divided by the total land area. The highest population densities are found in Asia, Europe, and Central America, while the lowest are in North and South America and Australia.

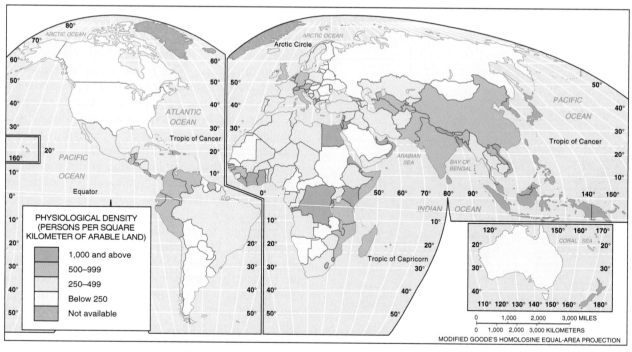

Figure 2–5 Physiological density. Physiological density is the number of people per unit area of arable land, which is land suitable for agriculture. Physiological density is a better measure than arithmetic density of the relationship between population and the availability of resources in a society.

Physiological Density

A more meaningful population measure is afforded by looking at the number of people per area of a certain type of land in a region. Land suited for agriculture is called *arable land*. In a region, the number of people supported by a unit area of arable land is called the **physiological density** (Figure 2–5). For example, in the United States, the physiological density is 140 persons per square kilometer (363 per square mile) of arable land. This contrasts sharply with Egypt, which has 2,167 persons per square kilometer (5,612 per square mile) of arable land. This large difference in physiological densities demonstrates that crops grown on a hectare of land in Egypt must feed far more people than in the United States.

The higher the physiological density, the greater is the pressure that people may place on the land to produce enough food. Physiological density provides insights into the relationship between the size of a population and the availability of resources in a region.

Comparing physiological and arithmetic densities helps geographers to understand the capacity of the land to yield enough food for the needs of people. In Egypt, for example, the large difference between the physiological density (2,167 people per square kilometer of arable land) and arithmetic density (65 persons per square kilometer over the entire country) indicates that most of the country's land is unsuitable for intensive agriculture. In fact, all but 5 percent of the Egyptian people live in the Nile River valley and delta, because it is the only area in the country that receives enough moisture (by irrigation from the river) to allow intensive cultivation of crops (Table 2–1).

Agricultural Density

Two countries can have similar physiological densities, but they may produce significantly different amounts of food because of different economic conditions. **Agricultural density** is the ratio of the number of farmers to the amount of arable land. This density measure helps account for economic differences. For example, the United States has an extremely low agricultural density (4 farmers per square kilometer of arable land), whereas Egypt has a very high density (737 farmers per square kilometer of arable land). MDCs have lower agricultural densities because technology and finance allow a few people to farm extensive land areas and feed many people. This frees most of the MDC population to work in factories, offices, or shops, rather than in the fields.

To understand the relationship between population and resources in a country, geographers examine its phys-

Cairo, Egypt. Population is not distributed uniformly within Egypt. Egyptians are highly clustered in the delta and valley of the Nile River, including Cairo. The remainder of the country consists of sparsely inhabited desert lands. (Dilip Mehta/Contact Press Images)

iological and agricultural densities together. As Table 2–1 shows, the physiological densities of both Egypt and the Netherlands are high, but the Dutch have a much lower agricultural density than the Egyptians. Geographers conclude that both the Dutch and Egyptians put heavy pressure on the land to produce food, but the more efficient Dutch agricultural system requires many fewer farmers than does the Egyptian system.

Similarly, the Netherlands has a much higher physiological density than does India, but a much lower agricultural density. This difference demonstrates that, compared with India, the Dutch have extremely limited arable land to meet the needs of their population. (Recall

Table 2–1 Measures of Density In Selected Countries, Expressed as Population Per Square Kilometer					
	Arithmetic Density	Physiological Density	Agricultural Density	Percent Farmers	Percent Arable
Canada	3	33	1	4	9
United States	28	140	4	3	20
Egypt	65	2,167	737	34	3
India	306	556	361	65	55
Japan	333	2,562	179	7	13
Netherlands	379	1,450	58	4	26
Bangladesh	849	1,267	824	65	67
United Kingdom	240	828	17	2	29

Physiological and agricultural density in Egypt. With most of the country desert, arable land farmed intensively, with very high numbers of persons and farmers per area of arable land, Egypt is under pressure to produce food from a limited amount of suitable land. (Sherif Sonbol/UNEP-Select/The Image Works)

from Chapter 1 how the Dutch have built dikes and created polders, areas of land made usable by draining water from them.) However, the highly efficient Dutch farmers can generate a large food supply from a limited resource.

Key Issue 2

Where Has the World's Population Increased?

- Natural increase
- Fertility
- Mortality

After identifying where people are distributed across Earth's surface, we can describe the locations where the numbers of people are increasing. Population increases rapidly in places where many more people are born than die, increases slowly in places where the number of births exceeds the number of deaths by only a small margin, and declines in places where deaths outnumber births. The population of a place also increases when people move in and decreases when people move out; this element of population change, called migration, is discussed in Chapter 3.

Geographers most frequently measure population change in a country or the world as a whole through three measures: crude birth rate, crude death rate, and natural increase rate.

- The **crude birth rate** (CBR) is the total number of live births in a year for every 1,000 people alive in the society. A crude birth rate of 20 means that for every 1,000 people in a country, 20 babies are born over a 1-year period.
- The **crude death rate** (CDR) is the total number of deaths in a year for every 1,000 people alive in the society. Comparable to the crude birth rate, the crude death rate is expressed as the annual number of deaths per 1,000 population.
- The **natural increase rate** (NIR) is the percentage by which a population grows in a year. It is computed by subtracting CBR − CDR, after first converting the two measures from numbers per 1,000 to percentages (numbers per 100). Thus, if the CBR is 20 and the CDR is 5 (both per 1,000), then the NIR is 15 per 1,000, or 1.5 percent. The term *natural* means that a country's growth rate excludes migration.

The word *crude* means that we are concerned with society as a whole, rather than a refined look at particular individuals or groups. In communities with an unusually large number of people of a certain age—such as a college town—we may study separate birth rates for women of each age. These numbers are *age-specific* rather than crude birth rates. In communities with large numbers of elderly people, demographers may compute separate death rates for males and females or for each group.

Natural Increase

During the 1990s, the world natural increase rate was 1.5, meaning that world population grew each year by 1.5 percent. For many things in life, such as an examination grade, the difference between 1 percent and 2 percent may not be important. However, for population the difference is critical.

The rate of natural increase affects the **doubling time**, which is the number of years needed to double a population, assuming a constant rate of natural increase. At the current NIR of 1.5 percent per year, world population would double in about 50 years. Should the current NIR continue through the twenty-first century, global population in the year 2100 would reach 24 billion.

During the 1960s and 1970s, when global NIR was 2.0 percent per year, the world's doubling time was 35 years. Had the 2.0 percent rate continued through the 1980s and 1990s instead of declining to 1.5 percent, Earth's population today would be one-half billion higher than it is, and a 2.0 percent NIR through the twenty-first century would have produced a total population of 48 billion in 2100. On the other hand, should the natural increase rate immediately decline to 1.0, doubling time would stretch out to 70 years, and world population in 2100 would be only 16 billion.

Very small changes in the natural increase rate dramatically affect the size of the population, because the base population from which we derive the percentage is so high. For example, when we multiply the natural increase rate of 1.5 by the current global population base of 6 billion, the result is an addition of 90 million people per year. If the natural increase rate immediately dropped to 1.0, then the annual growth in the world's population would drop to 60 million.

Although the world's NIR is lower today than its all-time peak of 2.0 percent per year during the 1960s and 1970s, the number of people added to the world annually is greater today than ever before. The number of people added to the world each year during the 1960s and 1970s—between 60 million and 80 million—was less than today, because the population base was much lower then. As the base continues to grow in the twenty-first century, a change of only one-tenth of 1 percent would produce very large swings in population growth.

The distribution of natural increase rates shows very large regional differences (Figure 2–6). The NIR exceeds 3.0 percent in a number of countries in central Africa, the Middle East, and Central America. At the other extreme, the natural increase rate is 0 percent or even negative in much of Europe, meaning that in the absence of immigrants, population actually is declining.

Not only is the world adding more people annually than ever before, virtually all the growth is concentrated in poorer countries. Over the past three decades, approximately 54 percent of the world's population growth has been in Asia, 15 percent each in sub-Saharan Africa and the Middle East, and 10 percent in Latin America. Europe and North America each account for only 3 percent of global population growth (Figure 2–7). Regional differences in natural increase rates mean that virtually all the world's additional people live in the countries that are least able to maintain them. To explain these differences in growth rates, geographers point to regional differences in fertility and mortality rates.

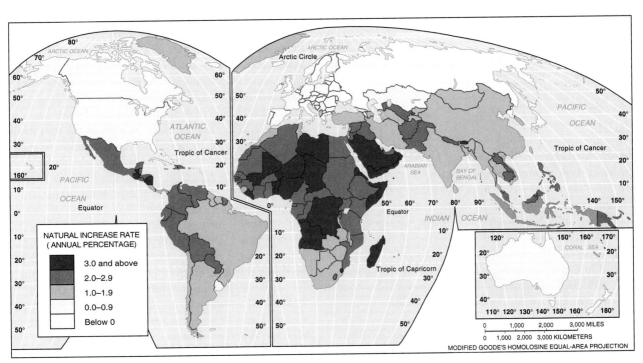

Figure 2–6 Natural increase rate. The natural increase rate is the percentage by which the population of a country grows in a year. World average is currently about 1.5 percent. The countries with the highest natural increase rates are concentrated in Africa and Southwest Asia.

Percent World Population 2000

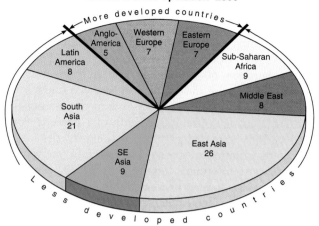

Percent World Population Growth 1960–2000

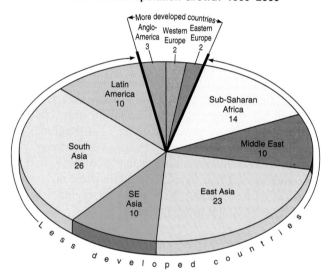

Figure 2–7 Population distribution by world region. About 81 percent of the world's population is in less developed countries, but about 93 percent of the population growth during the past four decades has occurred in these countries.

Fertility

The world map of crude birth rates (Figure 2–8) mirrors the distribution of natural increase rates (compare with Figure 2–6). As was the case with natural increase rates, the highest crude birth rates are in Africa and the lowest are in Europe and North America. Most African countries have a CBR over 40, and some exceed 50. Crude birth rates over 30 are common in Asia and Central America. On the other hand, crude birth rates are about 10 in Europe as a whole and below 10 in several Eastern and Southern European countries. The CBR is also about 10 in Japan.

Geographers also use the **total fertility rate (TFR)** to measure the number of births in a society. The TFR is the average number of children a woman will have throughout her childbearing years (roughly ages 15 through 49). To compute the TFR, scientists must assume that a woman reaching a particular age in the

future will be just as likely to have a child as are women of that age today. Therefore, the crude birth rate provides a picture of a society as a whole in a given year, whereas the total fertility rate attempts to predict the future behavior of individual women in a world of rapid cultural change.

The total fertility rate for the world as a whole is 3, and, again, the figures vary between more and less developed countries. The TFR exceeds 6 in many countries of sub-Saharan Africa and the Middle East, compared to less than 2 in Western and Eastern Europe (Figure 2–9).

Mortality

Two useful measures of mortality in addition to the crude death rate already discussed are the infant mortality rate and life expectancy. The **infant mortality rate (IMR)** is the annual number of deaths of infants under one year of age, compared with total live births. As was the case with the CBR and CDR, the IMR is usually expressed as the number of deaths among infants per 1,000 births, rather than as a percentage (per 100).

The global distribution of infant mortality rates follows the pattern that by now has become familiar. The highest rates are in the poorer countries of sub-Saharan Africa, the Middle East, and South and Southeast Asia, whereas the lowest rates are in Western Europe. Infant mortality rates exceed 100 in some LDCs, meaning that more than 10 percent of all babies die before reaching their first birthday. Infant mortality rates are about 5 in Western Europe (Figure 2–10).

In general, the IMR reflects a country's health care system. We find lower infant mortality rates in countries with well-trained doctors and nurses and large supplies of hospitals and medicine. Ironically, although the United States is well-endowed with medical facilities, it suffers from somewhat higher infant mortality rates than Canada and many European countries. African-Americans and other minorities in the United States have infant mortality rates that are twice as high as the national average, comparable to levels in Latin America and Asia. Some health experts attribute this to the fact that many poor people in the United States, especially minorities, cannot afford good health care for their infants.

Life expectancy at birth measures the average number of years a newborn infant can expect to live at current mortality levels. Like every other mortality and fertility rate discussed thus far, life expectancy is most favorable in the wealthy countries of Western Europe and least favorable in the poor countries of sub-Saharan Africa. Babies born today can expect to live only into their forties in many African countries and into their late seventies in Western Europe and North America (Figure 2–11).

Natural increase, crude birth, total fertility, infant mortality, life expectancy—the descriptions have become repetitious because their distributions follow similar patterns. Western Europe has the lowest rates of natural

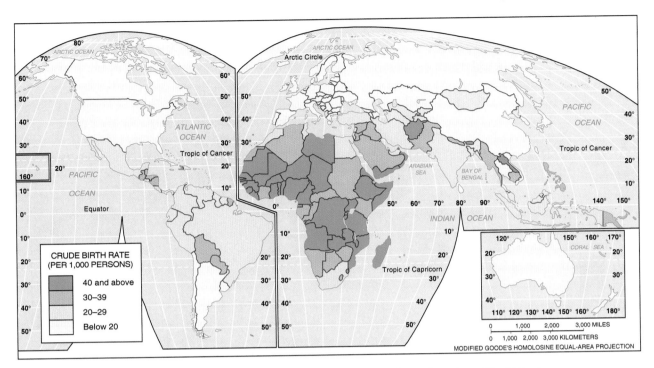

Figure 2–8 Crude birth rate (CBR). The crude birth rate is the total number of live births in a year for every 1,000 people alive in the society. The global distribution of crude birth rates parallels that of natural increase rates. Again, the highest crude birth rates are found in sub-Saharan Africa and the Middle East, whereas the lowest are in Europe.

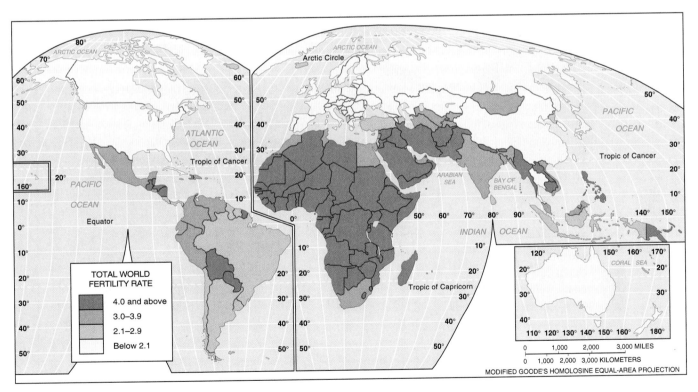

Figure 2–9 Total fertility rate (TFR). Total fertility rate is the number of children a woman will have throughout her childbearing years. Again, the highest rates are in sub-Saharan Africa and the Middle East, the lowest in Europe.

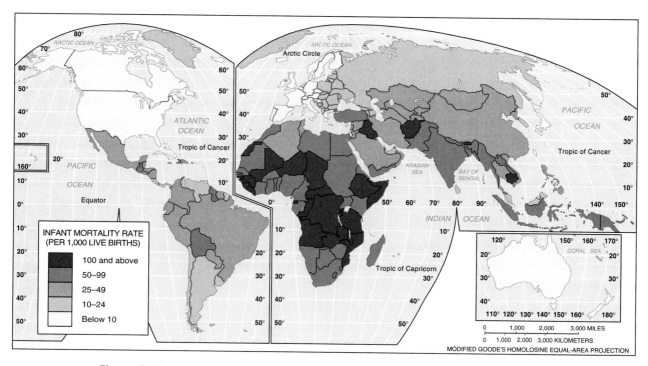

Figure 2–10 Infant mortality rate (IMR). The infant mortality rate is the number of deaths of infants under age 1 per 1,000 live births in a year. European and North American countries generally have infant mortality rates of under 10 per 1,000, whereas rates of more than 100 per 1,000 are common in Africa.

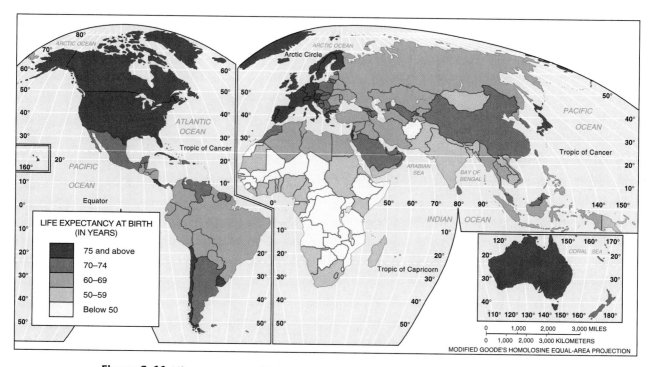

Figure 2–11 Life expectancy at birth. Life expectancy at birth is the average number of years a newborn infant can expect to live. Babies born this year are expected to live until their mid-sixties. Life expectancy for babies, however, ranges from the low forties in several African countries to the late seventies in much of Europe, Australia, North America, and Japan.

increase, crude birth, total fertility, and infant mortality, and the highest average life expectancies. The highest rates of natural increase, crude birth, total fertility, and infant mortality, and the lowest average life expectancies are in sub-Saharan Africa, followed by other less developed regions.

Our final map of global distributions of demographic variables—crude death rates—does not follow the familiar pattern. Consistent with the other demographic characteristics, the highest crude death rates are in Africa. But perhaps unexpectedly, the lowest crude death rates are in poorer countries of Latin America, Asia, and the Middle East, rather than in wealthy countries of North America and Western Europe (Figure 2–12).

Overall, the combined crude death rate for all LDCs is virtually the same—if not slightly lower some years—as the rate for all MDCs, whereas the combined crude birth rate for all LDCs is still substantially higher than for all MDCs. Furthermore, the variation between the world's highest and lowest crude death rates is much lower than the variation in crude birth rates. The difference between the countries with the highest and lowest crude death rates is only about 20, whereas crude birth rates for individual countries range from less than 10 to over 50, a spread of more than 40.

Why does Sweden, one of the world's wealthiest countries, have a higher crude death rate than Thailand, one of the poorest? Why does the United States, with its extensive system of hospitals and physicians, have a higher crude death rate than Costa Rica, Mexico, or Panama? The answer is that the populations of different countries are at various stages in an important process known as the demographic transition, upon which we focus in the third key issue of the chapter.

Key Issue 3

Why Is Population Increasing at Different Rates in Different Countries?

- The demographic transition
- Population pyramids
- Countries in different stages of demographic transition
- Demographic transition and world population growth

All countries have experienced some changes in natural increase, fertility, and mortality rates, but at different times and at different rates. While rates vary among countries, a similar process of change in a society's population, known as the **demographic transition**, is operating. Because of diverse local cultural and economic conditions, the demographic transition diffuses to indi-

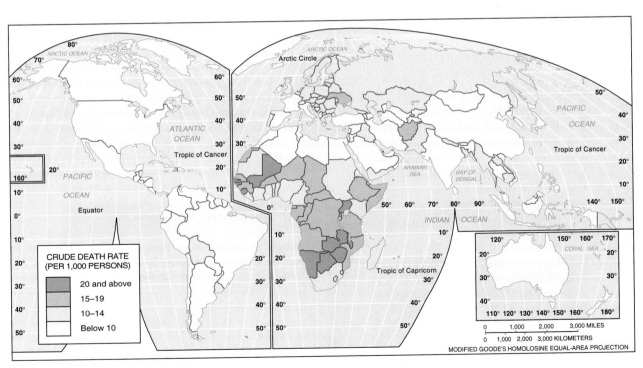

Figure 2–12 Crude death rate (CDR). Crude death rate is the total number of deaths in a year for every 1,000 people alive in the society. The global pattern of crude death rates varies from those for the other demographic variables already mapped in this chapter. First, while Europe has the lowest natural increase, crude birth, and infant mortality rates, it has relatively high crude death rates. Second, the variance between the highest and lowest crude death rates is much lower than was the case for the crude birth rates. The concept of the demographic transition helps to explain the distinctive distribution of crude death rates.

Figure 2–13 Demographic transition. The demographic transition consists of four stages: *Stage 1*—very high birth and death rates produce virtually no long-term natural increase. *Stage 2*—rapidly declining death rates combined with very high birth rates produce very high natural increase. *Stage 3*—birth rates rapidly decline, while death rates continue to decline; natural increase rates begin to moderate. *Stage 4*—very low birth and death rates produce virtually no long-term natural increase, and possibly a decrease.

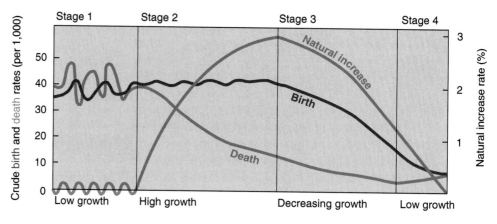

vidual countries at different rates and produces local variations in natural increase, fertility, and mortality.

The Demographic Transition

The demographic transition is a process with several stages, and every country is in one of them. The process has a beginning, middle, and end, and—barring a catastrophe such as a nuclear war—it is irreversible. Once a country moves from one stage of the process to the next, it does not revert to an earlier stage. The four stages are shown in Figure 2–13.

Stage 1: Low Growth

For most of human kind's several hundred-thousand-year occupancy of Earth, they were in stage 1 of the demographic transition. At times, the species would multiply in some regions and decline in others, but it remained sparse overall. Crude birth and death rates varied considerably from one year to the next, but over the long term they were roughly comparable, at very high levels. As a result, the natural increase rate was essentially zero, and Earth's population was unchanged, at perhaps one-half million.

During most of this period, people depended on hunting and gathering for food (see Chapter 10). When food was easily obtained the population increased, but it declined when people were unable to locate enough animals or vegetation.

About the year 8000 B.C., the world's population began to grow by several thousand per year. Between 8000 B.C. and A.D. 1750, Earth's human population increased from approximately 5 million to 800 million (Table 2–2). The burst of population growth around 8000 B.C. was caused by the **agricultural revolution**, which was the time when human beings first domesticated plants and animals and no longer relied entirely on hunting and gathering. By growing plants and raising animals, human beings created larger and more stable sources of food, so more people could survive.

Table 2–2 World Population and Growth Rates			
Date	Estimated Population	Percent Average Yearly Growth in Prior Period	Number of Years in Which Population Doubles at Current Growth Rate
400,000 B.C.	500,000	—	
8000 B.C.	5,000,000	0.001	59,007
A.D. 1	300,000,000	0.05	1,354
1750	791,000,000	0.06	1,250
1800	978,000,000	0.43	163
1850	1,262,000,000	0.51	136
1900	1,650,000,000	0.54	129
1950	2,517,000,000	0.85	82
1997	5,852,000,000	1.80	47

Despite the agricultural revolution, the human population remained in stage 1 of the demographic transition, because food supplies were still unpredictable. Farmers prospered during abundant harvests and the population expanded, but when unfavorable climatic conditions resulted in low food production the crude death rate would soar. Wars and diseases also took their toll in stage 1 societies.

Most of human history was spent in stage 1 of the demographic transition, but today no such country remains there. Every nation has moved on to at least stage 2 of the demographic transition and with that transition has experienced profound changes in population.

Stage 2: High Growth

For nearly 10,000 years after the agricultural revolution, world population grew at a modest pace. After around A.D. 1750, the world's population suddenly began to grow ten times faster than in the past. The average annual increase jumped from about 0.05 percent (one-twentieth of 1 percent) to 0.5 percent (one-half of 1 percent). World population grew by about 5 million in 1800, compared to about one-half million in 1750.

The sudden burst of population growth occurred because in the late eighteenth and early nineteenth centuries several countries moved into stage 2 of the demographic transition. In stage 2 of the demographic transition, the crude death rate suddenly plummets, while the crude birth rate remains roughly the same as in stage 1. Because the difference between the crude birth rate and crude death rate is suddenly very high, the natural increase rate is also very high, and population grows rapidly.

Some demographers divide stage 2 of the demographic transition into two parts. The first part is the period of accelerating population growth. During the second part the growth rate begins to slow, although the gap between births and deaths remains high.

Countries entered stage 2 of the demographic transition after 1750 as a result of the **industrial revolution**, which began in England in the late eighteenth century and spread to the European continent and North America during the nineteenth century. The industrial revolution was a conjunction of major improvements in industrial technology (invention of the steam engine, mass production, powered transportation) that transformed the process of manufacturing goods and delivering them to market (see Chapter 11). The result of this transformation was an unprecedented level of wealth, some of which was used to make communities healthier places to live.

New machines helped farmers increase agricultural production and feed the rapidly growing population. More efficient agriculture freed people to work in factories, producing other goods and generating enough food for the industrial workers.

The wealth produced by the industrial revolution was also used to improve sanitation and personal hygiene. Sewer systems were installed in cities, and food and water supplies were protected against contamination. As a result of these public improvements, people were healthier and lived longer.

Whereas countries in Europe and North America entered stage 2 of the demographic transition about 1800, stage 2 diffused to countries in Africa, Asia, and Latin America much more recently, in most cases after 1950. With the diffusion of stage 2 of the demographic transition, world population has grown by 1.8 percent per year since 1950, compared to 0.5 percent per year during the nineteenth century. The world is currently adding 90 million people annually, compared to about 8 million annually in 1900.

Countries in Africa, Asia, and Latin America moved into stage 2 of the demographic transition in the late twentieth century for a different reason than was the case for Europe and North America two hundred years earlier. The recent push of countries into stage 2 has been caused by the **medical revolution**. Medical technology invented in Europe and North America has diffused to the less developed countries of Africa, Asia, and Latin America.

Improved medical practices suddenly eliminated many of the traditional causes of death in LDCs and enabled more people to have longer and healthier lives. Penicillin, vaccines, and insecticides effectively and inexpensively controlled many infectious diseases, such as malaria, smallpox, and tuberculosis. Current mortality rates in Africa, Asia, and Latin America are only one-fourth the levels of the late 1940s.

Stage 3: Moderate Growth

A country moves from stage 2 to stage 3 of the demographic transition when the crude birth rate begins to drop sharply. The crude death rate continues to fall in stage 3 but at a much slower rate than in stage 2. Consequently, the population continues to grow because the crude birth rate is still greater than the crude death rate. However, the rate of natural increase is more modest in stage 3 countries than in stage 2 because the gap between the crude birth and death rates narrows.

European and North American countries generally moved from stage 2 to stage 3 of the demographic transition during the first half of the twentieth century. Some countries in Africa, Asia, and Latin America have moved to stage 3 in recent years, while others remain in stage 2.

The sudden drop in the crude birth rate during stage 3 occurs for different reasons than the rapid decline of the crude death rate during stage 2. The crude death rate declined in stage 2 following introduction of new technology into the society, but the crude birth rate declines in stage 3 because of changes in social customs.

A society enters stage 3 of the demographic transition when people choose to have fewer children. The decision is partly a delayed reaction to a decline in mortality, especially the infant mortality rate. In stage 1 societies, the survival of any one infant could not be confidently predicted, and families typically had a large number of babies to improve the chances of some surviving to adulthood. Medical practices introduced in stage 2 societies greatly improved the probability of an infant surviving, but many years elapsed before families reacted by conceiving fewer babies.

Economic changes in stage 3 societies also induce people to have fewer offspring. People in stage 3 societies are more likely to live in cities rather than the countryside and to work in offices, shops, or factories rather than on farms. Farmers often consider a large family to be an asset because children can do some of the chores. In contrast, children living in cities are generally not economic assets to their parents, because they are prohibited from working in most types of urban jobs. In addition, urban homes are relatively small and may not have space to accommodate large families.

Stage 4: Low Growth

A country reaches Stage 4 of the demographic transition when the crude birth rate declines to the point where it equals the crude death rate, and the natural increase rate approaches zero. This condition is called **zero population growth (ZPG)**, a term often applied to stage 4 countries.

Zero population growth may occur when the crude birth rate is still slightly higher than the crude death rate, because some females die before reaching childbearing years, and the number of females in their childbearing years can vary. To account for these discrepancies, demographers more precisely define zero population growth as the total fertility rate (TFR) that results in a lack of change in the total population over a long term. A TFR of approximately 2.1 produces ZPG, although a country that receives many immigrants may need a lower total fertility rate to achieve ZPG.

Countries in stage 4 of the demographic transition can be identified on the map of total fertility rates (Figure 2–9). Most European countries have reached stage 4 of the demographic transition, because they have TFRs well below the ZPG replacement level of 2.1. Also note in Figure 2–6 that most European countries have NIRs near zero, or even negative. The United States has not completely moved into stage 4 because of its ethnic diversity. The total fertility rate is still above the ZPG level among African-Americans and Hispanic-Americans, whereas fertility among Americans of European descent is comparable to the stage 4 levels found in Europe.

Social customs again explain the movement from one stage of the demographic transition to the next. Increasingly, women in stage 4 societies enter the labor force rather than remain at home as full-time homemakers. When most families lived on farms, employment and child rearing were conducted at the same place, but in urban societies most parents must leave the home to work in an office, shop, or factory. An employed parent must arrange for someone to take care of preschool children during working hours.

Changes in lifestyle also encourage smaller families. People who have access to a wider variety of birth-control methods are more likely to use some of them. With increased income and leisure time, more people participate in entertainment and recreation activities that may not be suitable for young children, such as attending cultural events, traveling overseas, going to bars, and eating at upscale restaurants.

Several Eastern European countries, including Bulgaria, Hungary, and Russia, have negative natural increase rates, meaning that the number of deaths exceed the number of births. Eastern Europe's relatively high death rates and low birth rates are a legacy of a half-century of Communist rule. Higher death rates have resulted from inadequate pollution controls (see Chapter 14), while lower birth rates have resulted from very strong family-planning programs and deep-seated pessimism about having children in an uncertain world. As memories of the Communist era fade, Eastern Europeans may display birth and death rates more comparable to those in Western Europe.

A country that has passed through all four stages of the demographic transition has in some ways completed a cycle—from little or no natural increase in stage 1, to little or no natural increase in stage 4. Two crucial demographic differences underlie this process, however. First, at the beginning of the demographic transition, the crude birth and death rates are high—35 to 40 per 1,000—while at the end of the process the rates are very low, approximately 10 per 1,000. Second, the total population of the country is much higher in stage 4 than in stage 1.

The Demographic Transition in England

England provides a good case study of the long-term impact of the demographic transition, for several reasons. England has reached stage 4, and at least fragmentary information on its population is available for the past 1,000 years. Further, unlike the United States and many other countries, England has not changed its boundaries, nor has it been affected by migration of enough people to affect national trends.

Stage 1: Low Growth Until 1750. In 1066, when the Normans invaded England, the country's population was approximately 1 million. Seven hundred years later, the population was only 6 million, and the country was still in stage 1 of the demographic transition (Figure 2–14).

During that 700-year period, population rose in some years and fell in others. For example, England's population declined from 4 million in the year 1250 to 2 million

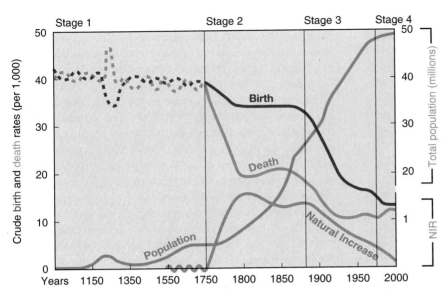

Figure 2–14 Demographic transition for England. Demographers must estimate birth and death rates prior to 1750, because precise records are not available. Church parish records of births, baptisms, marriages, and burials help in making estimates. England entered stage 2 of the demographic transition in the mid-eighteenth century, stage 3 in the late nineteenth century, and stage 4 in the mid-twentieth century.

a century later, after the Black Death (bubonic plague) and famines swept the country. Crude birth and death rates averaged more than 35 per 1,000 but varied considerably from one year to the next. As recently as the 1740s, the crude death rate skyrocketed following a series of bad harvests.

Stage 2: High Growth (1750–1880). In 1750, the crude birth and death rates in England were both 40 per 1,000. In 1800, the crude birth rate remained very high at 34, but the crude death rate had plummeted to 20. This 50-year period marked the start of the industrial revolution in England. New production techniques increased the nation's food supply and generated money that was spent on improvements in public health.

England remained in stage 2 of the demographic transition for about 125 years. During that period the population rose from 6 million to 30 million, an average annual natural increase rate of 1.4 percent.

Stage 3: Moderate Growth (1880–early 1970s). Crude birth and death rates changed little in England during most of the nineteenth century. In 1880 the crude birth rate was 33 per 1,000 and the crude death rate 19, in both cases only 1 per 1,000 lower than in 1800. After 1880, England entered stage 3 of the demographic transition. The crude death rate continued to fall somewhat over the next century, from 19 per 1,000 in 1880 to 12 in 1970. However, the crude birth rate declined sharply, from 33 per 1,000 in 1880 to 18 by 1930 and 15 in 1970. The population increased between 1880 and 1970 from 26 million to 49 million, about 0.7 percent per year.

Stage 4: Low Growth (Early 1970s–present). Since the early 1970s, England has been in stage 4 of the demographic transition. The population has increased only 1 million since 1970, an average natural increase rate of

0.1. The crude death rate has consistently rested at 12 per 1,000 since the 1970s, while the crude birth rate has varied between 12 and 14. The crude birth rate increases slightly in some years because the number of women in their childbearing years is greater, not because of decisions by women to have more children. Total fertility rate has been well below the 2.1 needed for replacement, and the main cause of population growth has been immigration from former colonies.

When England began to progress through the demographic transition around 1750, the country had 6 million people, crude birth and death rates of 40 per 1,000, and a record of little population growth over the previous 700 years. For the past quarter-century, England has been in another period of little population growth. The difference is that the crude birth and death rates are now around 12 rather than 40, and the country has 50 million inhabitants instead of 6 million.

Population Pyramids

As you might expect, the stage of demographic transition in which a country exists gives it a distinctive population structure. A country's population is influenced by the demographic transition in two ways: the percentage of the population in each age group, and the distribution of males and females.

We can display the distribution of a country's population by age and gender groups on a bar graph called a **population pyramid**. A population pyramid normally shows the percentage of the total population in 5-year age groups, with the youngest group (0 to 4 years old) at the base of the pyramid and the oldest group at the top. The length of the bar represents the percentage of the total population contained in that group. By convention, males are usually shown on the left side of the pyramid and females on the right.

The shape of a pyramid is determined primarily by the crude birth rate in the community. A country in stage 2 of the demographic transition, with a high crude birth rate, has a relatively large number of young children, making the base of the population pyramid very broad. On the other hand, a country in stage 4, with a relatively large number of older people, has a graph with a wider top that looks more like a rectangle than a pyramid.

Age Distribution

The age structure of a population is extremely important in understanding similarities and differences among countries. The most important factor is the **dependency ratio**, which is the number of people who are too young or too old to work, compared to the number of people in their productive years. The larger the percentage of dependents, the greater is the financial burden on those who are working to support those who cannot.

To compare the dependency ratios of different countries, we can divide the population into three age groups: 0 to 14, 15 to 64, and 65 and older. People who are 0-14 and 65-plus normally are classified as dependents. Approximately one-half of all people living in countries in stage 2 of the demographic transition are dependents, compared to only one-third in stage 4 countries. Consequently, the dependency ratio is nearly 1:1 in stage 2 countries, whereas in stage 4 countries the ratio is 1:2 (1 dependent for every two workers). Young dependents outnumber elderly ones by ten to one in stage 2 countries, but the numbers of young and elderly dependents are roughly equal in stage 4 countries.

In nearly every African country, and in many Asian and Latin American countries, more than 40 percent of the people are under age 15. This high percentage follows from the high crude birth rates in these regions. In contrast, in European and North American countries, which are at or near stage 4 of the demographic transition, the percentage of children under 15 is only about 20 percent (Figure 2–15).

In Africa, Asia, and Latin America, the large percentage of children strains the ability of poorer countries to provide needed services, such as schools, hospitals, and day-care centers. When children reach the age of leaving school, jobs must be found for them, but the government must continue to allocate scarce resources to meet the needs of the still-growing number of young people.

As countries pass through the stages of the demographic transition, the percentage of elderly people increases. The higher percentage partly reflects the lower percentage of young people produced by declining crude birth rates. Older people also benefit in stage 4 countries from improved medical care and higher incomes. People over age 65 exceed 15 percent of the population in several European countries, such as Denmark, Sweden, the

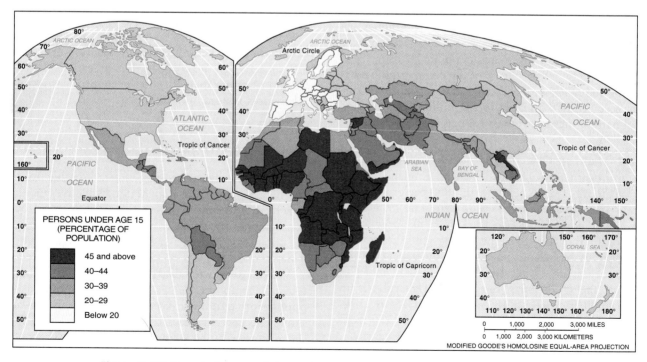

Figure 2–15 Percent of the population under age 15. Approximately one-third of the world's inhabitants are under age 15, but the percentage varies from over 40 percent in most African countries to less than 20 percent in many European countries. A map of the percentage of people over age 65 would show a reverse pattern, with the highest percentages in Europe and the lowest in sub-Saharan Africa and the Middle East.

(*Left*) The percentage of children under age 15 is especially high in more remote areas of less developed countries, such as the Amazon Basin of northwestern Peru. This family lives along the Rio Javari, which separates Peru from Brazil. (*Right*) In Italy, a country in stage 4 of the demographic transition, people over age 65 account for 16 percent of the population. (Alison Wright/The Image Works, K. Harrison/The Image Works)

United Kingdom, and Germany, compared to less than 5 percent in most African countries. In some cities in Florida, more than one-third of the residents are over age 65.

Older people must receive adequate levels of income and medical care after they retire from their jobs. The "graying" of the population places a burden on European and North American governments to meet these needs. More than one-fourth of all government expenditures in the United States, Canada, Japan, and many European countries go to social security, health care, and other programs for the older population. Because of the larger percentage of older people, countries in stage 3 or 4 of the demographic transition, such as the United States and Sweden, have higher crude death rates than do stage 2 countries.

Sex Ratio

The number of males per hundred females in the population is the **sex ratio**. It varies among countries, depending on birth and death rates. In general, slightly more males than females are born, but males have higher death rates. In Europe and North America, the ratio of men to women is about 95:100 (that is, 95 men for each 100 women). In the rest of the world, the ratio is 102:100.

In the United States, males under 15 exceed females 105:100. Women start outnumbering men about age 30, and they comprise 60 percent of the population over age 65. In poorer countries, high mortality rates during childbirth partly explain the lower percentage of women. The difference also relates to the age structure, because poorer countries have a larger percentage of young people—where males generally outnumber females—and a lower percentage of older people—where females are much more numerous.

Societies with a high rate of immigration typically have more males than females because males are more likely to undertake long-distance migration. Frontier areas and boom towns typically have more men than women. The rapidly growing state of Alaska, for example, has 111 men for every 100 women. On the other hand, retirement communities have relatively high percentages of women, because they have longer life expectancies than do men.

The shape of a community's population pyramid tells a lot about its distinctive character, especially compared to other places. In Figure 2–16, compare the shapes of the overall U.S. population pyramid with those for Cedar Rapids, Detroit, Honolulu, and Laredo. Cedar Rapids and Honolulu have relatively flat pyramids; Detroit and Laredo have relatively broad-based ones.

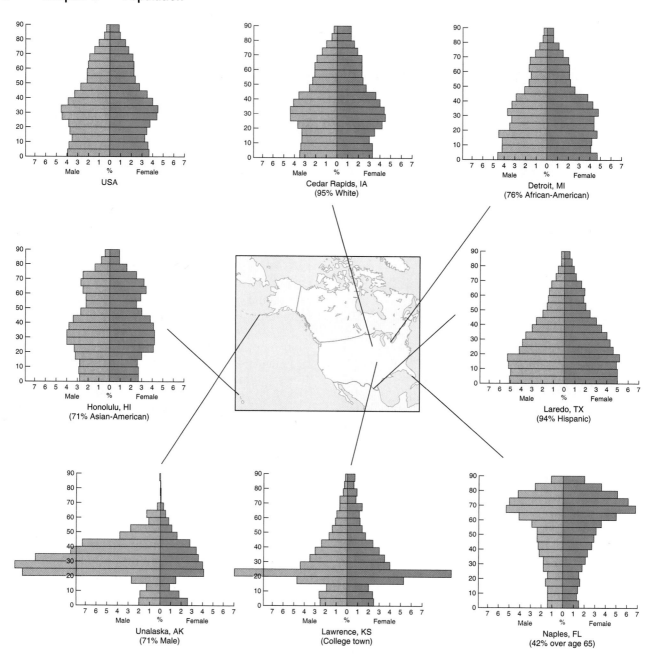

Figure 2–16 Population pyramids for the United States and selected communities. Detroit and Laredo have broader pyramids than Cedar Rapids and Honolulu, indicating higher percentages of young people and higher fertility rates. Unalaska has a high percentage of males, because it contains an isolated military base. Lawrence has a high percentage of people in their twenties because it is the home of the University of Kansas. Naples has a high percentage of elderly people, especially women, so its pyramid is upside down.

The different shapes result from differences in the ethnic composition of the four cities. Detroit and Laredo have relatively broad-based pyramids, because birth rates are relatively high among African-Americans and Hispanic-Americans, who form the majority in these two cities. On the other hand, birth rates are relatively low among the Asian-American and European-descended communities, the majorities in Honolulu and Cedar Rapids, respectively.

The population pyramid for Naples, where 42 percent of the people are over age 65, resembles an upside-down pyramid. Unalaska, a small town with a military base, has an exceptionally high percentage of males, whereas Naples, with a large percentage of elderly people, has substantially more females than males, because females have longer life expectancies. Cities with large universities, such as Lawrence, have an exceptionally high percentage of people in their twenties. See the Globalization and Local Diversity box for information about collecting statistics about a country's population.

Countries in Different Stages of Demographic Transition

Countries display distinctive population characteristics depending on their stage in the demographic transition. No country today remains in stage 1 of the demographic transition, but it is instructive to compare countries in each of the other three stages. Let us look at three case studies of countries in stages 2, 3, and 4.

Cape Verde: Stage 2 (High Growth)

Cape Verde, a collection of 12 small islands in the Atlantic Ocean off the coast of West Africa, moved from stage 1 to stage 2 about 1950. Cape Verde was a colony of Portugal until becoming independent in 1975, and the Portuguese administrators left better records of births and deaths than is typical for a colony in stage 1.

During the first half of the twentieth century, Cape Verde's population declined, from 147,000 in 1900 to 137,000 in 1949. Crude birth rates were generally in the forties and crude death rates in the twenties. The large gap between births and deaths most years produced a high natural increase rate typical of stage 2 of the demographic transition, yet Cape Verde remained in stage 1 until 1950.

Cape Verde remained in stage 1 because several severe famines dramatically disrupted the typical patterns of birth, death, and natural increase. For example, famine made Cape Verde's crude death rate rocket to 74 per 1,000 in 1941 and 101 in 1942. Because fewer babies were conceived at the height of the famine in 1942, the crude birth rate fell in 1943 to only 22. Population also declined during periods of famine because survivors

migrated to other countries. Wide fluctuations in the crude birth and death rates from one year to the next, depending on economic and environmental conditions, are typical of stage 1 countries.

This long-term pattern of demographic uncertainty suddenly ended in 1950, and Cape Verde quickly moved into stage 2 of the demographic transition. During the half-century since entering stage 2, the population of Cape Verde has nearly tripled, to approximately 400,000, and natural increase has averaged nearly 3.0 percent per year since 1950.

Cape Verde moved into stage 2 when an antimalarial campaign was launched. The crude death rate dropped by more than one-third between 1949 and 1950, from 27 to 17 per 1,000. It further declined during the 1950s and 1960s to about 10 per 1,000. Since the early 1970s, the crude death rate has changed very little, under 10 most years, although a drought in 1971 and a famine in 1986 temporarily lifted the rate above 10 (Figure 2–17).

Meanwhile, as is typical of stage 2 countries, Cape Verde's crude birth rate has remained relatively high and still fluctuates wildly. The crude birth rate increased in the early 1950s to a maximum of 53 per 1,000 in 1954, declined during the 1960s and 1970s to below 30, increased during the 1980s back above 40, and declined in the late 1990s to below 30 again.

The wild fluctuations in Cape Verde's crude birth rate are a legacy of the severe famine during the 1940s. Birth rates were lower during the 1960s because Cape Verde had relatively few women in their twenties, the prime childbearing years. Women in their twenties during the 1960s would have been born during the 1940s, when the famine kept birth rates very low. Similarly, the decline in the birth rates during the 1990s reflects the small number of women in prime childbearing years, who would have been born during the 1960s and 1970s. Conversely,

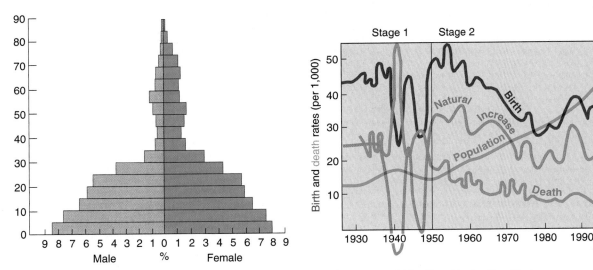

Figure 2–17 Demographic transition and population pyramid for Cape Verde. Cape Verde entered stage 2 of the demographic transition in approximately 1950, as indicated by the large gap between birth and death rates since then. As is typical of countries in stage 2 of the demographic transition, Cape Verde has a population pyramid with a very wide base.

Globalization and Local Diversity

Counting Our Numbers

The most important source of knowledge about *where* people live is the **census**. In the United States, Article 1, Section 2 of the Constitution requires that a census be taken every 10 years, beginning 3 years after the first meeting of Congress in 1787. In accordance with the Constitution, the U.S. census has taken place in every year ending in zero since 1790. Canada, the United Kingdom, and a number of other countries once ruled by the British take the census every 10 years, in years ending in the numeral one. The French average 7 years between censuses, but the government decrees exactly when each will be done.

The reasons *why* governments take a census are varied. First, the census provides information about the diversity of conditions in which people live at various locations in the country. The census collects information about the number of inhabitants in each dwelling, their ages, their relationships to each other, their levels of education, and their income. Questions are also asked about the conditions of the dwellings, including number of rooms, rent, age, and availability of heat and hot water.

Obtaining an accurate census is critical in most countries because a portion of the national budget is turned over to localities in proportion to their population. Further, congressional, parliamentary, and local government districts are adjusted in many countries following a new census, to promote equal representation. Places shown in the census to be gaining population may be entitled to more money and representatives, whereas places with declining populations may lose funds and representatives.

Rulers counted their subjects to assess taxes and raise an army as far back as the fourth century B.C. in such places as present-day Korea and Pakistan. Sweden was the first country to take a comprehensive, modern census, in 1745. People in many African and Asian territories were counted during the nineteenth century by their European colonial rulers. After gaining independence in the twentieth century, taking an accurate census was a priority for many countries. In general, though, we must estimate the past population of the world from fragmentary information, such as reports compiled in ancient Rome and China and parish church records of baptisms, marriages, and burials.

Successfully completing the census shows the *local diversity* with which people live in various countries. Countries face two types of obstacles.

First, actually counting everyone is difficult. Nigeria orders shops and factories to close and millions of people to remain at home for 3 days. Turkey declares an all-day curfew on census day, with everyone, including foreigners and travelers, required to remain indoors between 7:00 A.M. and 5:00 P.M.

Success depends on the cleverness and resourcefulness of census takers. In less developed countries, census takers must communicate with people who live in isolated locations that lack modern services, such as electricity, telephones, and mail delivery. Surveyors in India must count millions of people who have no permanent address and live on sidewalks. South African census takers threw block parties and set up tables with food to find homeless people. Sometimes a census brings surprises: A census worker in Rio de Janeiro, Brazil, opened an icebox in a butcher shop and discovered that it provided access to an otherwise unknown street where people lived.

Even in the United States, with its relatively high degree of literacy and few isolated communities, not everyone is counted. Many undocumented aliens fail to complete a census form because they fear that the government will find and deport them. Other people simply ignore the census, despite the fact that noncompliance is illegal. Some people do not participate in the census because they are unable to read the forms.

Officials of large U.S. cities claim that their populations have been undercounted because they contain high concentrations of people who do not participate in the census. As a result, the cities may be receiving less funding

---

the higher birth rates in the 1950s and 1980s resulted from a larger number of women in childbearing years.

The decline in the crude birth rate during the late 1990s does not yet signal that Cape Verde has moved to stage 3 of the demographic transition. For Cape Verde to enter stage 3, birth rates must continue to decline even in years when the number of women in prime childbearing years is greater.

Chile: Stage 3 (Moderate Growth)

Chile provides an example of a country outside Europe and North America that has reached stage 3 of the demographic transition, but is likely to take some time before continuing to stage 4. Chile has changed from a predominantly rural society based on agriculture to an urban society, in which most people now work in factories, offices, and shops. However, many Chileans still prefer to have large families.

Like most countries outside Europe and North America, Chile entered the twentieth century still in stage 1 of the demographic transition. Population had grown modestly during the nineteenth century, at a natural increase rate of less than 1 percent per year. However, much of Chile's population growth—as in other countries in the Western Hemisphere—resulted from European immigration.

One of India's 1.5 million census takers questions a family in a poor section of Delhi, as the government tries to find out how many people actually live in the country. (Reuters/Corbis-Bettmann)

and fewer state and federal representatives than they should.

A more fundamental obstacle is the strong opposition of many people to conducting a census at all. Arguing that the census invaded their privacy, opponents managed to delay Germany's 1981 census for 6 years, and when the census was finally conducted Germans were urged to boycott it and fill in incorrect answers. Nigeria was unable to conduct a census for 28 years because several ethnic groups feared that their people would be undercounted, and therefore would be entitled to fewer positions and lower funds

from the government. Nigeria's 1973 census was halted following reports that some census takers had been beaten or kidnapped and others bribed to inflate figures.

We have not taken the ultimate step in **globalization** of the census, for a simultaneous census of the entire planet has never been held. But the collection of national censuses is a prominent symbol of the fact that most people do not live in complete isolation but must be counted. Unless we periodically ask questions, our understanding of people is based on speculation or old data, rather than on current facts.

Chile's crude death rate declined sharply in the 1930s, moving the country into stage 2 of the demographic transition. As elsewhere in Latin America, Chile's crude death rate was lowered by the infusion of medical technology from MDCs such as the United States, bringing under control such diseases as smallpox, malaria, and dysentery. During the 1940s and 1950s, Chile's rate of natural increase exceeded 2 percent per year, and the crude death rate dropped from the mid-30s to less than 15 (Figure 2–18).

Chile has been in stage 3 of the demographic transition since about 1960. The crude death rate declined further during the 1960s and 1970s to less than 10, while the

crude birth rate dropped sharply, from about 35 in the early 1960s to about 20 by the late 1970s. However, Chile has failed to make further progress over the past quarter-century in reducing the gap between births and deaths. The natural increase rate has remained around 1.5 percent annually since the 1960s.

Chile moved into stage 3 of the demographic transition primarily because of a vigorous government family-planning policy, initiated in 1966. Reduced income and high unemployment at that time also induced couples to postpone marriage and delay childbearing.

While Chile's natural increase rate is lower today than in the 1950s, the country is unlikely to move into

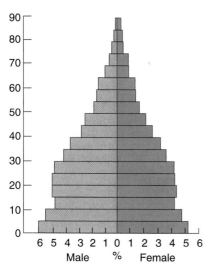

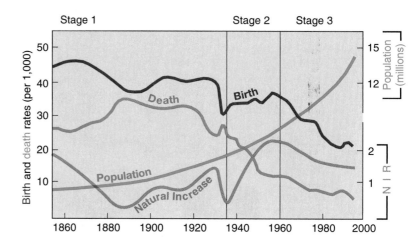

Figure 2–18 Demographic transition and population pyramid for Chile. Chile entered stage 2 of the demographic transition in the 1930s, when death rates declined sharply, and stage 3 in the 1960s, when birth rates declined sharply. Since the mid-1980s, however, birth rates have no longer declined, and Chile's natural increase rate has remained over 1.5.

stage 4 of the demographic transition in the near future. By 1979, Chile's government reversed its policy and renounced support for family planning. The government policy was that population growth could help promote national security and economic development. Further reduction in the crude birth rate is also hindered by the fact that most Chileans belong to the Roman Catholic Church, which opposes the use of what it calls artificial birth-control techniques.

Denmark: Stage 4 (Low Growth)

Denmark, like several other Northern and Western European countries, has reached stage 4 of the demographic transition. Denmark's history is similar to that of England's, already discussed. The country entered stage 2 of the demographic transition in the nineteenth century, when the crude death rate began its permanent decline. The crude birth rate then dropped in the late nineteenth century, and the country moved into stage 3 (Figure 2–19).

Since the 1970s, the crude birth and the crude death rates have been roughly equal, about 12 per 1,000. The country has reached zero population growth, and the population is unlikely to increase from the current level of just over 5 million.

Denmark's population pyramid shows the impact of the demographic transition. Instead of a classic pyramid shape, Denmark has a column, demonstrating that the percentages of young and elderly people are nearly the same. With further medical advances, the number of elderly people may actually exceed the number of young people in a few years. Denmark's crude death rate has actually increased somewhat in recent years because of the increasing number of elderly people. The CDR is unlikely to decline unless another medical revolution,

such as a cure for cancer, keeps older elderly people alive much longer.

Demographic Transition and World Population Growth

Having used case studies to see the patterns of the demographic transition in individual countries, we now are ready to take a global view. Why is worldwide population increasing rapidly today? Because few countries are in the two stages of the demographic transition that have low population growth—no country remains in stage 1, and few have reached stage 4. The overwhelming majority of countries are either in stage 2 or stage 3 of the demographic transition—stages with rapid population growth—and only a few are likely to reach stage 4 in the near future.

The four-stage demographic transition is characterized by two big breaks with the past. The first break—the sudden drop in the death rate that comes from technological innovation—has been accomplished everywhere. The second break—the sudden drop in the birth rate that comes from changing social customs—has yet to be achieved in many countries. If most countries in Europe and North America have reached—or at least are approaching—stage 4 of the demographic transition, why aren't countries elsewhere in the world? The answer is that fundamental problems prevent other countries from replicating the experience in Europe and North America.

The first demographic change—the sudden decline in crude death rate—occurred for different reasons in the past. The nineteenth-century decline in the CDR in Europe and North America took place in conjunction with the industrial revolution. The unprecedented level of wealth generated by the industrial revolution was used

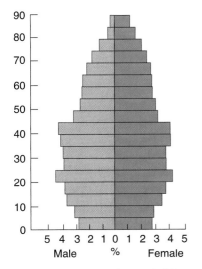

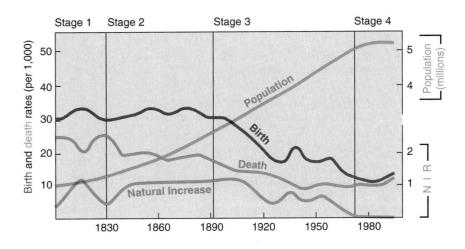

Figure 2–19 Demographic transition and population pyramid for Denmark. Denmark has been in stage 4 of the demographic transition and has experienced virtually no change in total population since the 1970s. The population pyramid is much straighter than those of Cape Verde and Chile, a reflection of the relatively large percentage of elderly people and small percentage of children.

in part to stimulate research by European and North American scientists into the causes and cures for diseases. These studies ultimately led to medical advances, such as pasteurization, X rays, penicillin, and insecticides.

In contrast, the sudden drop in the CDR in Africa, Asia, and Latin America in the twentieth century was accomplished by different means, and with less internal effort by local citizens. For example, the crude death rate on the island of Sri Lanka (then known as Ceylon) plummeted 43 percent between 1946 and 1947. The most important reason for the sharp drop was the use of the insecticide DDT to control the mosquitoes that spread malaria.

European and North American countries invented and manufactured the DDT and trained the experts to supervise its use. The spraying of Sri Lankans' houses and other medical services, which cost only $2 per person per year, were paid for primarily by international organizations.

Thus, Sri Lanka's CDR was reduced by nearly one-half in a single year with no change in the country's economy or culture. Medical technology was injected from Europe and North America instead of arising within the country as part of an economic revolution. This pattern has been repeated throughout Africa, Asia, and Latin America.

Having caused the first break with the past through diffusion of medical technology, European and North American countries now urge other countries to complete the second break with the past, the reduction in the birth rate. However, reducing the crude birth rate is more difficult. A decline in the crude death rate can be induced through introduction of new technology by outsiders, but the CDR will drop only when people decide for themselves to have fewer children.

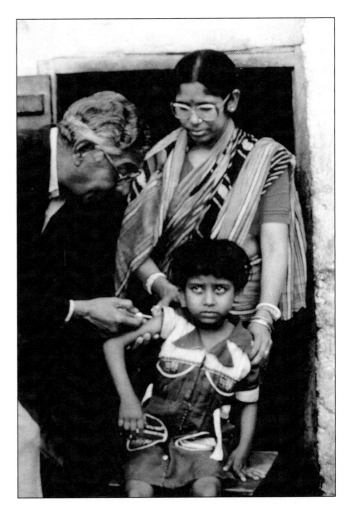

Inoculations reduce the crude death rates in LDCs and move the countries into stage 2 of the demographic transition. (Steve Maines/Stock Boston)

Many in Africa, Asia, and Latin America may be unprepared for this second break with the past; also, they are being urged to move through the demographic transition rapidly. In Europe and North America, stage 2 of the demographic transition lasted for approximately 100 years. During that time, global population increased by about 1 billion. If stage 2 of the demographic revolution in Africa, Asia, and Latin America also lasts for 100 years—from about 1950 to 2050—15 billion people will be added to the world during that time.

Key Issue 4

Why Might the World Face an Overpopulation Problem?

- Malthus on overpopulation
- Debate over how to reduce natural increase

Why does global population growth matter? In view of the current size of Earth's population, and the natural increase rate, will there soon be too many of us? Will continued population growth lead to global starvation, war, and lower quality of life?

Geographers are particularly well suited to address these questions, because answers require understanding both human behavior and the physical environment. Further, geographers observe that diverse local cultural and environmental conditions may produce different answers in different places.

Malthus on Overpopulation

English economist Thomas Malthus (1766–1834) was one of the first to argue that the world's rate of population increase was far outrunning the development of food supplies. Malthus's views remain influential today.

Population Growth vs. Food Supply

In his *An Essay on the Principle of Population*, published in 1798, Malthus claimed that population was growing much more rapidly than Earth's food supply, because population increased geometrically, while food supply increased arithmetically. According to Malthus, these growth rates would produce the following relationships between people and food in the future:

- Today: 1 person, 1 unit of food
- 25 years from now: 2 persons, 2 units of food
- 50 years from now: 4 persons, 3 units of food
- 75 years from now: 8 persons, 4 units of food
- 100 years from now: 16 persons, 5 units of food

Malthus wrote several decades after England had become the first country to enter stage 2 of the demographic transition, in association with the industrial revolution.

He concluded that population growth would press against available resources in every country, unless "moral restraint" produced lower crude birth rates or unless disease, famine, war, or other disasters produced higher crude death rates.

Neo-Malthusians. Contemporary geographers and other analysts are taking another look at Malthus's theory, because of the unprecedented rate of natural increase in LDCs. In Malthus's time, only a few relatively wealthy countries had entered stage 2 of the demographic transition, characterized by rapid population increase. He failed to anticipate that relatively poor countries would have the most rapid population growth, because of transfer of medical technology (but not wealth) from MDCs.

Analysts such as Robert Kaplan and Thomas Fraser Homer-Dixon have broadened Malthus's theory to encompass a wide variety of resources, rather than only food. They paint a frightening picture of a world in which billions of people are engaged in a desperate search for food and fuel. According to their "neo-Malthusian" argument, wars and civil violence will increase in the coming years because of scarcities of food, as well as other resources, such as clean air, suitable farmland, and fuel.

Many LDCs have expanded their food production significantly in recent years, but they have more poor people than ever before. For example, income in East African countries rose during the past two decades by approximately 2 percent per year above inflation, but the population grew by approximately 3 percent per year. Because population growth outpaced economic development, all the economic growth was absorbed simply in accommodating the additional population. Despite this economic growth, the average East African is worse off today than a decade or two ago.

Malthus's Critics

Malthus's theory has been severely criticized from a variety of perspectives. The Marxist theorist Friedrich Engels dismissed Malthus's arithmetic as an artifact of capitalism. Engels argued that the world possessed sufficient resources to eliminate global hunger and poverty, if only these resources were shared equally. Under capitalism, workers do not have enough food because they do not control the production and distribution of food and are not paid sufficient wages to purchase it.

Contemporary analysts such as Esther Boserup and Julian Simon criticize Malthus's theory that population growth produces problems. To the contrary, a larger population could stimulate economic growth, and as a result production of more food. Population growth could generate more customers and more good ideas for improving technology.

Many geographers consider Malthusian beliefs unrealistically pessimistic, because they are based on a belief that the world's supply of resources is fixed rather than expanding. According to the principles of possibilism discussed in Chapter 1, our well-being is influenced by conditions in the physical environment, but humans have some ability to choose courses of action that can expand the supply of food and other resources.

On a global scale, conditions during the past half-century have not supported Malthus's theory. Even though the human population has grown at its most rapid rate ever, world food production has consistently grown at a faster rate than the natural increase rate since 1950, according to geographer Vaclav Smil (Figure 2–20). Smil has shown that Malthus was fairly close to the mark on food production, but much too pessimistic on population growth.

Food production increased during the last half of the twentieth century somewhat more rapidly than Malthus predicted. Better growing techniques, higher-yielding seeds, and cultivation of more land all contributed to the expansion in food supply (see Chapter 14). Many people in the world cannot afford to buy food, or do not have access to sources of food, but these are problems of distribution of wealth, rather than insufficient global production of food as Malthus theorized.

Malthus's model expected world population to quadruple between 1950 and 2000, from 2.5 billion to 10 billion people, but world population actually grew during the period to only 6 billion. Malthus did not foresee critical cultural, economic, and technological changes that

would induce societies sooner or later to move into stages 3 and 4 of the demographic transition.

Debate Over How to Reduce Natural Increase

Although the Malthus theory seems unduly pessimistic at a global scale, geographers recognize the diversity of conditions among regions of the world. Although the world as a whole may not be in danger of "running out" of food, some regions with rapid population growth do face shortages of food.

Consequently, most people hope for further reductions in the natural increase rate. For countries currently in stage 2 of the demographic transition—high growth—natural increase can be reduced in only two ways:

- Return to stage 1 by raising the crude death rate up to the level of the crude birth rate.
- Move to stages 3 and 4 by lowering the crude birth rate to the level of the crude death rate.

Few people wish to see the first alternative realized, leaving the second alternative as the only humane choice. However, experts sharply disagree on the best way to move countries into stages 3 and 4.

Higher Death Rates

A return to stage 1 by increasing the crude death rate is dismaying, but we must explore it. An increase in the CDR could halt the growth of the human population. For example, if the population increases faster than the expansion of the food supply, widespread famine could result. Some argue that sending food to starving Africans is well-intentioned but a mistake, because a higher death rate now may prevent even greater mass starvation in the future. Other natural causes could drastically cut the population; millions could die in wars—especially nuclear—or from a natural disaster, such as a major earthquake.

The crude death rate may also rise from the spread of disease. One-third of all deaths among children in LDCs (other than China) are from diarrhea that results from poor sanitation and resulting infections. Another one-third of child deaths result from six infectious diseases: polio, measles, diphtheria, tetanus, whooping cough, and tuberculosis. These diseases have been virtually eliminated in MDCs through immunization, improved nutrition, and hygiene. However, only a minority of children in developing countries have been immunized against these diseases, and water supplies remain unsafe in many places. Even where programs have been implemented to fight these preventable diseases, they have not always been successful because of a lack of qualified medical staff.

The diffusion of AIDS (acquired immunodeficiency syndrome) in both MDCs and LDCs could produce a rise in the CDR. The distribution of AIDS within the

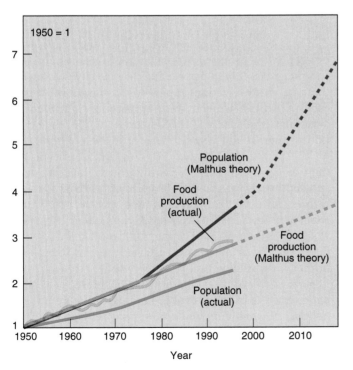

Figure 2–20 Malthus's theory compared to actual world food production and population, 1950–1997. Malthus expected population to grow more rapidly than food production. In reality, during the second half of the twentieth century—when world population grew at its most rapid rate ever—food production actually expanded even more rapidly.

United States was discussed in Chapter 1 (see Figure 1–21). Although information is not as easy to obtain, the incidence of AIDS is more extensive in Africa than in MDCs, and its diffusion across that continent has not been stopped.

Lower Birth Rates

Few people wish to see population growth curbed through an increase in the death rate. The only demographic alternative is to reduce the birth rate. Analysts and public health officials debate over best means to achieve lower birth rates. One alternative emphasizes reliance on economic development, the other on distribution of contraceptives.

Economic Development Alternative. One approach to lowering birth rates emphasizes the importance of improving local economic conditions. A wealthier community has more money to spend on education and health care programs that would promote lower birth rates.

According to this approach, if more women are able to attend school, and to remain in school longer, they are more likely to learn employment skills, and gain more economic control over their lives. With better education, women would better understand their reproductive rights, make more informed reproductive choices, and select more effective methods of contraception.

With improved health care programs, infant mortality rates would decline through such programs as improved prenatal care, counseling about sexually transmitted diseases, and child immunization. With the survival of more infants assured, women would be more likely to choose to make more effective use of contraceptives to limit the number of children.

Distribution of Contraceptives. The other approach to lowering birth rates emphasizes the importance of rapidly diffusing modern contraceptive methods. Economic development may promote lower birth rates in the long run, but the world cannot wait around for that alternative to take effect. Putting resources into family-planning programs can reduce birth rates much more rapidly.

In less developed countries, demand for contraceptive devices is greater than the available supply. Therefore, the most effective way to increase their use is to distribute more of them, cheaply and quickly. According to this approach, contraceptives are the best method for lowering the birth rate.

Bangladesh is an example of a country that has seen little improvement in the wealth and literacy of its people, but the percentage using contraceptives rose from 6 percent in 1974 to 45 percent in 1993, compared to 70 percent in more developed countries. Similar growth has occurred in other LDCs, including Colombia, Morocco, and Thailand. Rapid growth in the acceptance of family planning is evidence that, in the modern world, ideas can diffuse rapidly, even to places where people

"Family Planning helps us live better," proclaims a billboard in Huacho, Peru. The government hopes that open communication between couples will reduce the crude bith rate. (Richard Lord/ The Image Works)

have limited access to education and modern communications.

The percentage of women using contraceptives is especially low in Africa, so the alternative of distributing contraceptives could have an especially strong impact there. A United Nations report found that only 12 percent of African women were employing contraceptive devices in the late 1980s, compared to 55 percent in Latin America and over 70 percent in Asia. The reason for this is partly economics, religion, and education, but there is another reason. Very high birth rates in Africa and southwestern Asia reflect the relatively low status of women. In societies where women receive less formal education and hold fewer legal rights than do men, women regard having a large number of children as a measure of their high status, and men regard it as a sign of their own virility.

Regardless of which alternative is more successful, many oppose birth-control programs for religious and political reasons. Adherents of several religions, including Roman Catholics, fundamentalist Protestants, Muslims, and Hindus, have religious convictions that prevent them from using some or all birth-control devices. Opposition is strong within the United States to terminating pregnancy by abortion, and the U.S. government has at times withheld aid to countries and family-planning organizations that advise abortion, even when such advice is only a small part of the overall aid program.

Analysts agree that the most effective means of reducing births would employ both alternatives. But LDC governments and international family-planning organizations have limited funds to promote lower birth rates, so they must set priorities and make choices for allocating scarce funds.

■ Summary

Overpopulation—too many people for the available resources—has already hit regions of Africa and threatens other countries in Asia and Latin America. The world as a whole does not face overpopulation immediately, but current trends must be reversed to prevent a future crisis.

Geographers caution that the number of people living in a region is not by itself an indication of overpopulation. Some densely populated regions are not overpopulated, while some sparsely inhabited areas are. Instead, overpopulation is a relationship between the size of the population and a region's level of resources. The capacity of the land to support life derives partly from characteristics of the natural environment and partly from human actions to modify the environment through agriculture, industry, and exploitation of raw materials.

The track toward overpopulation already may be irreversible in Africa. Rapid population growth has led to the overuse of land. As the land declines in quality, more effort is needed to yield the same amount of crops. This extends the working day of women, who have the primary responsibility for growing food for their families. Women then regard having another child as a means of securing additional help in growing food.

Should the current world rate of natural increase continue for several decades, global population would far exceed even the most optimistic estimates of world food and energy capacities. In another thousand years, there would be less than 1 square foot of land per person in the world, including deserts, mountains, and ice caps.

These projections are not intended as *predictions*. They are offered to illustrate the significance of current growth rates and to demonstrate the need to modify current trends. The challenge is to lower the current rate of population growth before the negative consequences of a large population pose insoluble social and economic problems.

We cannot completely explain the overpopulation problem until we see how people in different regions earn a living and modify the environment. However, we can reach some conclusions by briefly reviewing the key issues raised at the beginning of this chapter.

1. **Where is the world's population distributed?** Global population is concentrated in a few places. Human beings tend to avoid those parts of Earth's surface that they consider to be too wet, too dry, too cold, or too mountainous. The capacity of Earth to support a much larger population depends heavily on people's ability to use sparsely settled lands more effectively.

2. **Where has the world's population increased?** Virtually all the world's natural increase is concentrated in the relatively poor countries of Africa, Asia, and Latin America. In contrast, most European and North American countries now have low population growth rates, and some are experiencing population declines. The difference in natural increase between MDCs and LDCs is attributable to differences in crude birth rates rather than in crude death rates.

Overpopulation in Mali. A region can be sparsely inhabited yet overpopulation if it has rapid population growth and limited resources, as is the case in Mali. (Steve McCurry/Magnum Photos)

3. **Why is population increasing at different rates in different countries?** The demographic transition is a change in a country's population. A country moves from a condition of high birth and death rates, with little population growth, to a condition of low birth and death rates, with low population growth. During this process, the total population increases enormously, because the death rate declines some years before the birth rate does. The MDCs of Europe and North America have reached or neared the end of the demographic transition. African, Asian, and Latin American countries are at the stages of the demographic transition characterized by rapid population growth, in which death rates have declined sharply, but birth rates remain relatively high.

4. **Why might the world face an overpopulation problem?** The rate at which global population has grown in the last three decades is unprecedented in history. A dramatic decline in the death rate has produced the increase. With death rates controlled, for the first time in history the most critical factor determining the size of the world's population is the birth rate. Scientists agree that the current rate of natural increase must be reduced, but they disagree on the appropriate methods for achieving this goal.

Case Study Revisited

India Versus China

The world's two most populous countries, China and India, will heavily influence future prospects for global overpopulation. These two countries—together encompassing more than one-third of the world's population—have adopted different policies to control population growth. In the absence of strong family-planning programs, India adds about 6 million more people each year than does China. Current projections show that India would surpass China as the world's most populous country by 2050.

India's Population Policies

India, like most countries in Africa, Asia, and Latin America, remained in stage 1 of the demographic transition until the late 1940s. During the first half of the twentieth century, population increased modestly—less than 1 percent per year—and even decreased in some years because of malaria, famines, plagues, and cholera epidemics. For example, more than 16 million Indians—approximately 5 percent of the population—died of influenza in 1918 and 1919, and the population at the 1921 census was lower than that ten years earlier.

Immediately following independence from England in 1947, India's death rate declined sharply, to 20 per 1,000 by 1951, while the crude birth rate remained about 40. Consequently, the natural increase rate jumped to 2 percent per year. The demographic pattern has not changed much in India during the past 40 years. Birth and death rates have both drifted a few points lower since the 1950s, but the natural increase has consistently remained around 2 percent per year. In the half-century since independence, India's population has grown by more than one-half billion.

The government of India has launched various programs to encourage family planning, but none have been very successful. In 1952, India became the first country to embark on a national family-planning program. The government has established clinics and distributes information about alternative methods of birth control. Birth-control devices are distributed free or at subsidized prices. Abortions, legalized in 1972, have been performed at a rate of several million per year. Altogether, the government spends several hundred million dollars annually on various family-planning programs.

India's most controversial family-planning program was the establishment of camps in 1971 to perform sterilizations, surgical procedures by which people were made incapable of reproduction. A sterilized person was entitled to a payment, which has been adjusted several times but generally has been equivalent to the average monthly income in India. At the height of the program, in 1976, 8.3 million sterilizations were performed during a 6 month period, mostly on women.

The birth-control drive declined in India after 1976. Widespread opposition to the sterilization program grew in the country, because people feared that they would be forcibly sterilized. The prime minister, Indira Gandhi, was defeated in 1977, and the new government emphasized the voluntary nature of birth-control programs. The term "family planning," which the Indian people associated with the forced sterilization policy, was replaced by "family welfare" to indicate that compulsory birth control programs had been terminated. Although Mrs. Gandhi served again as prime minister from 1980 until she was assassinated in 1984, she did not emphasize family planning because of the opposition during her previous administration.

A government-sponsored family-planning program continues, but it emphasizes education, including advertisements on national radio and television networks and information distributed through local health centers. Given the cultural diversity of the Indian people, the national campaign has had only limited success. The dominant form of birth control continues to be sterilization of women, many of whom have already borne several children, rather than vasectomies of men. Effective methods have not been devised to induce recently married couples to have fewer children.

China's Population Policies

In contrast with India, China has made substantial progress in reducing its rate of growth. Between the 1950s and 1990s, the natural increase rate was halved, from 2 to 1 percent per year. The government of the People's Republic of China has acted forcefully to reduce the number of children. The core of the government's policy is to limit families to one child.

Couples receive financial subsidies, a long maternity leave, better housing, and (in rural areas) more land if they agree to

The government of China aggressively promotes a one-child policy on public billboards. (United Nations)

have just one child. The government prohibits marriage for men until they are 22 and women until they are 20. To discourage births further, people receive free contraceptives, abortions, and sterilizations. A family with more than one child must pay a fine, amounting to 5 or 10 percent of its income for ten years, and job promotions may be denied. Some officials in rural villages maintain records of women's menstrual cycles to assure that no unplanned babies are born.

Another factor is female infanticide. If limited to one child, most Chinese families prefer to have a boy, in part because of cultural tradition and in part because a boy is regarded as stronger and better able to take care of aging parents. The one-child policy encouraged the killing of baby girls. In our American culture, such a practice is abhorrent, but one of geography's great lessons is that other cultures are very, very different from our own, including their fundamental values. Because of international criticism, the Chinese government has relaxed enforcement of the one-child rule.

China is likely to maintain a much lower natural increase rate than India into the twenty-first century. Following years of intensive educational programs, as well as coercion, the Chinese people have accepted to a greater degree than the Indian people the benefits of family planning. As China moves closer to a market economy, especially in rural areas, women increasingly recognize that having fewer children opens greater opportunities to obtain a job and earn more money.

■ Key Terms

Agricultural density The ratio of the number of farmers to the total amount of land suitable for agriculture.

Agricultural revolution The time when human beings first domesticated plants and animals and no longer relied entirely on hunting and gathering. See Chapter 10.

Arithmetic density The total number of people divided by the total land area.

Census A complete enumeration of a population.

Crude birth rate (CBR) The total number of live births in a year for every 1,000 people alive in the society.

Crude death rate (CDR) The total number of deaths in a year for every 1,000 people alive in the society.

Demographic transition The process of change in a society's population from a condition of high crude birth and death rates and low rate of natural increase to a condition of low crude birth and death rates, low rate of natural increase, and a higher total population.

Demography The scientific study of population characteristics.

Dependency ratio The number of people under the age of 15 and over age 64, compared to the number of people active in the labor force.

Doubling time The number of years needed to double a population, assuming a constant rate of natural increase.

Ecumene The portion of Earth's surface occupied by permanent human settlement.

Industrial revolution A series of improvements in industrial technology that transformed the process of manufacturing goods. See Chapter 11.

Infant mortality rate (IMR) The total number of deaths in a year among infants under 1 year old for every 1,000 live births in a society.

Life expectancy The average number of years an individual can be expected to live, given current social, economic, and medical conditions. Life expectancy at birth is the average number of years a newborn infant can expect to live.

Medical revolution Medical technology invented in Europe and North America that is diffused to the poorer countries of Latin America, Asia, and Africa. Improved medical practices have eliminated many of the traditional causes of death in poorer countries and enabled more people to live longer and healthier lives.

Natural increase rate (NIR) The percentage growth of a population in a year, computed as the crude birth rate minus the crude death rate.

Overpopulation The number of people in an area exceeds the capacity of the environment to support life at a decent standard of living.

Physiological density The number of people per unit of area of arable land, which is land suitable for agriculture.

Population pyramid A bar graph representing the distribution of population by age and sex.

Sex ratio The number of males per 100 females in the population.

Total fertility rate (TFR) The average number of children a woman will have throughout her childbearing years.

Zero population growth (ZPG) A decline of the total fertility rate to the point where the natural increase rate equals zero.

■ Thinking Geographically

1. The current method of counting a country's population by requiring every household to complete a census form once every ten years has been severely criticized as inaccurate. The census allegedly fails to count people who cannot read the form, or who do not wish to be found. This undercounting produces a geographic bias, because people who are missed are more likely to live in inner cities, remote rural areas, or communities that attract a relatively high number of recent immigrants. Given the availability of reliable statistical tests, should the current method of trying to count 100 percent of the population be replaced by a survey of a carefully drawn sample of the population, as is done with political polling and consumer preferences? Why or why not?

2. Scientists disagree about the effects of high density on human behavior. Some laboratory tests have shown that rats display evidence of increased aggressiveness, competition, and violence when very large numbers of them are placed in a box. Is there any evidence that very high density causes humans to behave especially aggressively or violently? Discuss.

3. Paul and Anne Ehrlich argue in *The Population Explosion*

(1990) that a baby born in an MDC such as the United States poses a graver threat to global overpopulation than a baby born in an LDC. The reason is that people in MDCs place much higher demands on the world's supply of energy, food, and other limited resources. Do you agree with this view? Why?

4. The baby-boom generation—people born between 1946 and 1964—totals nearly one-third of the U.S. population. (They are the bulge in the U.S. population pyramid, Figure 2–16.) Baby boomers have received more education than their parents, and women are more likely to enter the labor force. They have delayed marriage and parenthood and have fewer children compared to their parents. They are more likely to divorce, to bear children while unmarried, and to cohabit. As they grow older, what impact will baby boomers have on the American population in the twenty-first century?

5. What policies should governments in MDCs pursue to reduce global population growth? If an MDC provides funds and advice to promote family planning, does it gain the right to tell developing countries how to spend the funds and how to use the expertise? Explain your answer.

■ Further Readings

Ashford, Lori S. "New Perspectives on Population: Lessons from Cairo." *Population Bulletin* 50 (1). Washington, DC: Population Reference Bureau, 1995.

Beaujeu-Garnier, Jacqueline. *Geography of Population*, 2d ed. London: Longman, 1978.

Bender, William H. "How Much Food Will We Need in the 21st Century?" *Environment* 39 (1997): 6–11.

——, and Margaret Smith. "Population, Food, and Nutrition." *Population Bulletin* 51 (4). Washington, DC: Population Reference Bureau, 1997.

Bennett, D. Gordon. *World Population Problems: An Introduction to Population Geography*. Champaign, IL: Park Press, 1984.

Bongaarts, John, and Susan Cotts Watkins. "Social Interactions and Contemporary Fertility Transitions." *Population and Development Review* 22 (1996): 639–82.

Bouvier, Leon F., and Carol J. DeVita. "The Baby Boom—Entering Midlife." *Population Bulletin* 46 (3). Washington, DC: Population Reference Bureau, 1991.

Brass, William. "Demographic Data Analysis in Less Developed Countries, 1946–1996." *Population Studies* 50 (1996): 451–68.

Brown, Lester R., and Jodi L. Jacobson. "Our Demographically Divided World." *Worldwatch Paper* 74. Washington, DC: Worldwatch Institute, December 1986.

Caldwell, J.C. "Demography and Social Science." *Population Studies* 50 (1996): 305–34.

Carr-Saunders, A. B. *World Population: Past Growth and Present Trends*. New York: Oxford University Press, 1936.

Chesnais, Jean-Claude. "Fertility, Family, and Social Policy in Contemporary Western Europe." *Population and Development Review* 22 (1996): 729–40.

Clarke, John I. *Population Geography*, 2d ed. Oxford and New York: Pergamon Press, 1972.

——. *Geography and Population: Approaches and Applications*. Oxford and New York: Pergamon Press, 1984.

Cleland, John. "Demographic Data Collections in Less Developed Countries, 1946–1996." *Population Studies* 50 (1996): 433–50.

Coale, Ansley, and James Trussell. J "The Development and Use of Demographic Models." *Population Studies* 50 (1996): 469–84.

Coleman, David, and Roger Schofield, eds. *The State of Population Theory: Forward from Malthus*. Oxford and New York: Basil Blackwell, 1986.

Demko, George; George Schnell, and Harold Rose. *Population Geography: A Reader*. New York: McGraw-Hill, 1970.

Donaldson, Peter J., and Amy Ong Tsui. "The International Family Planning Movement." *Population Bulletin* 45 (3). Washington, DC: Population Reference Bureau, 1990.

Ehrlich, Paul, and Anne Ehrlich. *The Population Explosion*. New York: Simon & Schuster, 1990.

Freedman, Ronald. "Family Planning Programs in the Third World." *Annals of the American Academy of Political and Social Science* 510 (1990): 33–43.

Goliber, Thomas J. "Africa's Expanding Population: Old Problems, New Policies." *Population Bulletin* 44 (3). Washington, DC: Population Reference Bureau, 1989.

Gould, W. T. S., and R. Lawton, eds. *Planning for Population Change*. Totowa, NJ: Barnes and Noble Books, 1986.

Haub, Carl. "Population Change in the Former Soviet Republics." *Population Bulletin* 49 (4). Washington, DC: Population Reference Bureau, 1994.

Hobcraft, John. "Fertility in England and Wales: A Fifty-Year Perspective." *Population Studies* 50 (1996): 485–524.

Homer-Dixon, Thomas F. *Environmental Scarcity and Global Security*. Ephrata, PA: Science Press, 1993.

Jacobsen, Judith. "Promoting Population Stabilization: Incentives for Small Families." *Worldwatch Paper* 54. Washington, DC: Worldwatch Institute, June 1983.

Hodgson, Dennis, and Susan Cotts Watkins. "Feminists and Neo-Malthusians: Past and Present Alliances." *Population and Development Review* 23 (1997):469–524.

Jisen, Ma. "1.2 Billion: Retrospect and Prospect of Population in China." *International Social Science Journal* 48 (1996): 261–68.

Keyfitz, Nathan. "Population Growth, Development and the Environment." *Population Studies* 50 (1996): 335–60.

Kirk, Dudley. "Demographic Transition Theory." *Population Studies* 50 (1996): 361–88.

Lutz, Wolfgang. "The Future of World Population." *Population Bulletin* 49 (1). Washington, DC: Population Reference Bureau, 1994.

Malthus, Thomas. *An Essay on the Principles of Population*. 1978 Reprint; London: Royal Economic Society, 1926 (first published 1798).

McFalls, Joseph A., Jr. "Population: A Lively Introduction." *Population Bulletin* 46 (2). Washington, DC: Population Reference Bureau, 2nd ed., 1995.

Menken, Jane, ed. *World Population and U.S. Policy*. New York: W. W. Norton & Co. 1986.

Merrick, Thomas W. "World Population in Transition." *Population Bulletin* 41 (2). Washington, DC: Population Reference Bureau, 1986.

Mosley, W. Henry, and Peter Cowley. "The Challenge of World Health." *Population Bulletin* 46 (4). Washington, DC: Population Reference Bureau, 1991.

Nash, Alan. "Population Geography." *Progress in Human Geography* 20 (1996): 203–14.

Peters, Gary L., and Robert P. Larkin. *Population Geography: Problems, Concepts, and Prospects*, 3d ed. Dubuque, IA: Kendall-Hunt, 1989.

Preston, Samuel H. "Population Studies of Mortality." *Population Studies* 50 (1996): 525–36.

Robert, Godfrey. *Population Policy, Contemporary Issues*. New York: Praeger, 1990.

Saito, Osamu. "Historical Demography: Achievements and Prospects." *Population Studies* 50 (1996): 537–53.

Scientific American. *The Human Population*. San Francisco: W. H. Freeman, 1974.

Simon, Julian. *The Ultimate Resource*. Princeton, NJ: Princeton University Press, 1981.

Smil, Vaclav. "How Many People Can the Earth Feed?" *Population and Development Review* 20 (1994): 255–92.

Thompson, Warren S. *Population Problems*, 4th ed. New York: McGraw-Hill, 1953.

Tien, H. Yuan, with Zhang Tianlu, Ping Yu, Li Jingneng, and Liang Zhongtang. "China's Demographic Dilemmas." *Population Bulletin* 47 (1). Washington, DC: Population Reference Bureau, 1992.

Trewartha, Glenn T. *A Geography of Population*. New York: John Wiley, 1969.

_____ . *The Less Developed Realm: A Geography of Its Population*. New York: John Wiley, 1972.

United Nations. *Demographic Yearbook*. New York: United Nations, published annually.

_____ . *Statistical Yearbook*. New York: United Nations, published annually.

United States Department of Commerce, Bureau of the Census. *Statistical Abstract of the United States*. Washington, DC: Government Printing Office, published annually.

Van De Kaa, D.J. "Anchored Narratives: The Story and Findings of Half a Century of Research into the Determinants of Fertility." *Population Studies* 50 (1996): 389–432.

_____ . "Europe's Second Demographic Transition." *Population Bulletin* 42 (1). Washington, DC: Population Reference Bureau, 1987.

Visaria, Leela, and Pravin Visaria. "India's Population in Transition." *Population Bulletin* 50 (3). Washington, DC: Population Reference Bureau, 1995.

Weeks, John R. *Population: An Introduction to Concepts and Issues*, 4th ed. Belmont, CA: Wadsworth, 1989.

World Bank. *World Bank Development Report*. New York: Oxford University Press, published annually.

Wrigley, E. A. *Population and History*. New York: World University Library, 1969.

Zelinsky, Wilbur. "A Bibliographic Guide to Population Geography." Research Paper 80. Chicago: University of Chicago Department of Geography, 1962.

_____ . *A Prologue to Population Geography*. Englewood Cliffs, NJ: Prentice-Hall, 1966.

_____ , Leszek A. Kosinski, and R. Mansell Prothero, eds. *Geography and a Crowding World*. New York: Oxford University Press, 1970.

Also consult the following journals: *American Demographics*; *Demography*; *Intercom*; *Population*; *Population and Development Review*; *Population Bulletin*; *Population Studies*. In addition, the Population Reference Bureau publishes a World Population Data Sheet every year.

CHAPTER 3

Migration

Refer back to Figure 2–1 (world population distribution) for a moment. If a series of such maps were created for several points in time and shown one after the other like frames in a movie, the pattern would change constantly. Some regions would suddenly increase in population, while other regions would decline. In part, these changes result from regional differences in crude birth and death rates. Most of the changes, however, result from the movement of people from some places to others. Where do people migrate from? Where do they migrate to?

How many times has your family moved? In the United States, the average family moves once every 5 years. Was your last move traumatic or exciting? The loss of old friends and familiar settings can hurt, but the experiences awaiting you at a new location can be stimulating. Think about the multitude of Americans—maybe including yourself—who have migrated from other countries. Imagine the feelings of people migrating from another country when they arrive in a new land without a job, friends, or—for many—the ability to speak the local language.

Why would people make a perilous journey across thousands of kilometers of ocean? Why did the pioneers cross the Great Plains, the Rocky Mountains, or the Mojave Desert to reach the American West? Why do people continue to migrate by the millions today? The hazards that many migrants have faced are a measure of the strong lure of new locations and the desperate conditions in their former homelands. Most people migrate in search of three objectives: economic opportunity, cultural freedom, and environmental comfort. This chapter studies the reasons why people migrate.

Key Issues

1. Why do people migrate?
2. Why has U.S. immigration changed?
3. Why do migrants face obstacles?
4. Why do people migrate within a country?

Flood victims, Dhaka, Bangladesh. (Brace Brander/Photo Researchers, Inc.)

Case Study

Migrating to and from Ireland

Maighread Rutledge, a resident of Cobh, Ireland, in her mid-30s, is an unusual member of her family. The youngest of four sisters, Maighread has never moved away from Cobh. One of her sisters moved to England, one to the United States, and one to Australia.

Jeff Robinson left Ireland, as did 13 of 15 other engineering majors from Dublin's Trinity College class of 1986. In the class of 1987, Stephen Kilroy moved away, as did 22 of 23 other engineering majors.

Robinson, Kilroy, and the three older Rutledge sisters joined a long tradition of leaving Ireland. Since 1820, nearly 5 million Irish have moved to the United States, millions more to other countries. Poor economic conditions pushed people out of Ireland, and better job prospects lured them to other countries.

During the 1840s, people started leaving Ireland in large numbers because a blight (fungus) destroyed most of the potato crop, their major food source, and produced mass starvation. Millions died during the famine, and many of the survivors left in search of better economic conditions. The famine was made worse by the fact that most of the land was owned by people living in England rather than in Ireland (see Chapter 6). These absentee landowners did not encourage the Irish who were working the land to change their farming practices. By the end of the famine, Ireland's population was reduced by one-half.

During the 1980s and 1990s, economic factors again induced people to leave Ireland. Of the 3.5 million citizens of Ireland, more than one-half million live outside the country, mostly in Europe. About 5,000 people each year officially move from Ireland to the United States, and an even larger number are estimated to be moving illegally.

Most of the people who have left Ireland, such as the Rutledge sisters, Robinson, and Kilroy, are young and well-educated. In a country where one-third of the young people are unemployed, a high school or college diploma is a ticket out of Ireland. More than one-fourth of Ireland's working-age population moved to other countries in the 1980s.

In the late 1990s, the situation suddenly changed. The three Rutledge sisters all moved back to Ireland, as did Robinson and Kilroy, lured by improvements in the Irish economy. The economy is one of the fastest growing in Europe, unemployment has dropped by one-half, and average income has tripled in a decade.

The decision to move to another country is difficult for many people, torn between cultural ties to their homeland and better economic opportunities elsewhere. In a booming economy, staying home is a choice; in a depressed economy, staying home is a sacrifice.

In Chapter 1 we defined diffusion as a process by which a characteristic spreads from one area to another, and relocation diffusion as the spread through the bodily movement of people from one place to another. The subject of this chapter is a specific type of relocation diffusion called **migration**, which is a permanent move to a new location. Migrants permanently change their place of residence—where they sleep, store their possessions, and receive legal documents. Geographers document *where* people migrate from, and where they migrate to.

The flow of migration is always two-way. Given two locations A and B, some people migrate from A to B, while at the same time others migrate from B to A. **Emigration** is migration *from* a location; **immigration** is migration *to* a location. The difference between the number of immigrants and the number of emigrants is the **net migration**. If the number of immigrants exceeds the emigrants, the net migration is positive, and the region has *net in-migration*. If the number of emigrants exceeds the immigrants, the net migration is negative, and the region has *net out-migration*.

Migration is a form of **mobility**, which is a more general term covering all types of movements from one place to another. People display mobility in a variety of ways, such as by journeying every weekday from their homes to places of work or education and once a week to shops, places of worship, or recreation areas. These types of short-term, repetitive, or cyclical movements that recur on a regular basis, such as daily, monthly, or annually, are called **circulation**. College students display another form of mobility—*seasonal* mobility—by moving to a dormitory each fall and returning home the following spring.

Geographers are especially interested in *why* people migrate, even though migration occurs much less frequently than other forms of mobility, because it produces profound changes for individuals and entire cultures. A permanent move to a new location disrupts traditional cultural ties and economic patterns. At the same time, when people migrate they take with them to their new home their language, religion, ethnicity, and other cultural traits, as well as their methods of farming and other economic practices.

Modern transportation systems, such as motor vehicles, airplanes, and railroads, make relocation diffusion more feasible than in the past, when people had to rely on walking, animal power, or slow ships. However, thanks to modern communications systems, relocation diffusion is no longer essential for transmittal of ideas from one place to another. Culture and economy can diffuse rapidly around the world through forms of expansion diffusion.

If people can participate in the *globalization* of culture and economy regardless of place of residence, why do they still migrate in large numbers? The answer is that location is still important to an individual's cultural identity and economic prospects. Within a global economy, an individual's ability to earn a living varies by location. Within a global culture, people migrate to escape from domination by other cultural groups or to be reunited with others of similar culture. Migration of people with similar cultural values creates pockets of *local diversity*.

Although migration is a form of relocation diffusion, reasons for migrating can be gained from expansion diffusion. Someone may migrate and send back a message that gives others the idea of migrating. For example, many Europeans migrated to the United States in the nineteenth century because very favorable reports from early migrants led them to believe that the streets of American cities were paved with gold.

Key Issue 1

Why Do People Migrate?

- Reasons for migrating
- Distance of migration
- Characteristics of migrants

Geography has no comprehensive theory of migration, but a nineteenth-century outline of 11 migration "laws" written by E.G. Ravenstein is the basis for contemporary geographic migration studies. To understand where and why migration occurs, we can organize Ravenstein's "laws" into three main elements of migration: the reasons why migrants move, the distance they typically move, and their characteristics.

Reasons for Migrating

- Most people migrate for economic reasons.
- Cultural and environmental factors also induce migration, although not as frequently as economic factors.

People decide to migrate because of push factors and pull factors. A **push factor** induces people to move out of their present location, whereas a **pull factor** induces people to move into a new location. As migration for most people is a major step not taken lightly, both push and pull factors typically play a role. To migrate, people view their current place of residence so negatively that they feel pushed away, and another place so attractive that they feel pulled toward it.

We can identify three major kinds of push and pull factors: economic, cultural, and environmental. Usually one of the three factors emerges as most important, although as will be discussed later in this chapter, ranking the relative importance of the three factors can be difficult, even controversial.

Economic Push and Pull Factors

Most people migrate for economic reasons, as was the case in Ireland discussed at the beginning of the chapter. People think about emigrating from places that have few

Famine, a migration push factor. During the 1840s potato famine, many people migrated from Ireland to the United States. (Bettmann-Corbis)

job opportunities, and they immigrate to places where the jobs seem to be available. Because of economic restructuring, job prospects often vary from one country to another and within regions of the same country.

An area that has valuable natural resources, such as petroleum or uranium, may attract miners and engineers. A new industry may lure factory workers, technicians, and scientists. Construction workers, restaurant employees, and public service officials may move to areas where rapid population growth stimulates demand for additional services and facilities.

The United States and Canada have been especially prominent destinations for economic migrants. Many European immigrants to North America in the nineteenth century truly expected to find streets paved with gold. While not literally so gilded, the United States and Canada did offer Europeans prospects for economic advancement. This same perception of economic plenty now lures people from Latin America and Asia.

The relative attractiveness of a region can shift with economic change, as the introductory example from Ireland demonstrated. Similarly, Scotland has attracted migrants in recent years, after decades of net out-migration. Following the discovery of petroleum in the North Sea off the coast of northeast Scotland, thousands of people have been lured to jobs in the drilling or refining of petroleum or in supporting businesses.

Cultural Push and Pull Factors

Cultural factors can be especially compelling push factors, forcing people to emigrate from a country. Forced international migration has historically occurred for two main cultural reasons: slavery and political instability. Millions of people were shipped to other countries as slaves or as prisoners, especially from Africa to the Western Hemisphere, during the eighteenth and early nineteenth centuries.

Large groups of people were no longer forced to migrate as slaves in the twentieth century, but forced international migration increased because of political instability resulting from cultural diversity. Boundaries of newly independent states often have been drawn to segregate two ethnic groups. Because at least some intermingling among ethnicities inevitably occurs, members of an ethnic group caught on the "wrong" side of a boundary may be forced to migrate to the other side. Wars have also forced large-scale migration of ethnic groups in the twentieth century, especially in Europe and Africa. Forced migration of ethnicities is discussed in more detail in Chapter 7.

According to the United Nations, **refugees** are people who have been forced to migrate from their home country and cannot return for fear of persecution because of their race, religion, nationality, membership in a social group, or political opinion. In 1997, the U.S. Committee for Refugees, a nonprofit organization independent of the U.S. government, counted more than 15 million refugees in need of protection or assistance offered by international organizations, such as the Red Cross and the United Nations (Figure 3–1).

Refugees have no home until another country agrees to allow them in, or improving conditions make possible a return to their former home. In the interim, they must camp out in tents, board in shelters, or lie down by the side of a road. Refugees returned to their former home or given permission to live permanently in other countries are not included in the U.S. Committee's count of 15 million.

Political conditions can also operate as pull factors, especially the lure of freedom. People are attracted to democratic countries that encourage individual choice in education, career, and place of residence. This pull factor is particularly difficult to disentangle from a push factor, because the pull of democracy is normally accompanied by the push from a totalitarian country.

After Communists gained control of Eastern Europe in the late 1940s, many people in that region were pulled toward the democracies in Western Europe and North America. After permitting some emigration to the West, the Communist governments in Eastern Europe clamped down, for fear of losing their most able workers. The most dramatic symbol of restricted emigration was the Berlin Wall, which the Communists built to prevent emigration from Communist-controlled East Berlin into democratic West Berlin (see Chapter 8).

Refugees from civil war in Rwanda during the mid-1990s migrated to Congo Democratic Republic. (Mariella Furrer/SABA Press Photos, Inc.)

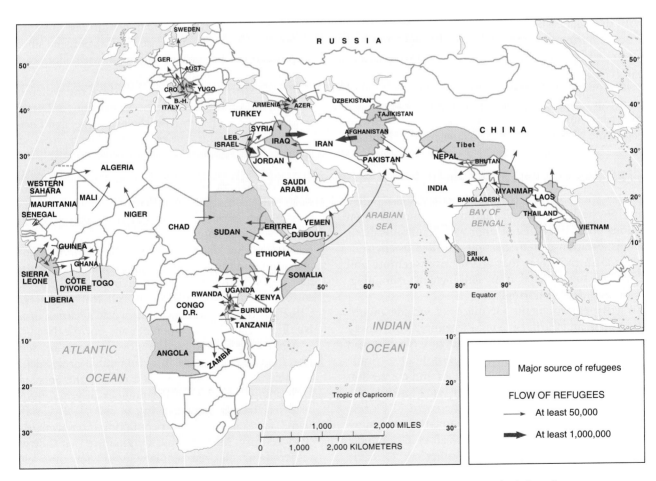

Figure 3–1 Major sources and destinations of refugees. A refugee is a person who is forced to migrate from a country, usually because of political reasons. The U.S. Committee for Refugees estimated that in 1997 there were 15 million refugees in the world, although the organization warns that the figure is difficult to determine because many refugees are not documented.

East Berliners cross the Berlin Wall at the Invalidenstrasse check point, following a shopping trip to West Berlin, November 12, 1989, one day after the Berlin Wall was opened for unrestricted travel between East and West Berlin. Within a year, the Berlin Wall was demolished, and the two parts of Berlin were united. (Reuters/Juergen Schwarz/Archive Photos)

With the election of democratic governments in Eastern Europe during the 1990s, Western Europe's political pull has disappeared as a migration factor. Eastern Europeans now can visit where they wish, although few have the money to pay for travel-related expenses beyond a round-trip bus ticket. However, Western Europe pulls an increasing number of migrants from Eastern Europe for economic reasons, as discussed later in this chapter.

Environmental Push and Pull Factors

People also migrate for environmental reasons, pulled toward physically attractive regions and pushed from hazardous ones. In an age of improved communications and transportation systems, people can live in environmentally attractive areas that are relatively remote and still not feel too isolated from employment, shopping, and entertainment opportunities.

Attractive environments for migrants include mountains, seasides, and warm climates. Proximity to the Rocky Mountains lures Americans to the state of Colorado, and the Alps pull French people to eastern France. Some migrants are shocked to find polluted air and congestion in these areas. The southern coast of England, the Mediterranean coast of France, and the coasts of Florida attract migrants, especially retirees, who enjoy swimming and lying on the beach. One-third of all elderly people who migrate from one U.S. state to another select Florida as their destination. Regions with warm winters, such as southern Spain and the southwestern United States, attract migrants from harsher climates.

Those with bronchitis, asthma, tuberculosis, and allergies have been pulled to Arizona by the dry desert climate. Ironically, the large number of migrants has modified Arizona's environmental conditions. The pollen count in Tucson has increased 3,500 percent since the 1940s, and the percentage of people with allergies there is now twice the national average.

Local experts attribute two-thirds of the pollen count in Tucson to three types of vegetation imported by migrants: the mulberry tree, the olive tree, and Bermuda grass. Some communities have banned these three species. The mulberry tree dies after 30 years, but the olive tree—an attractive species in Arizona because it is drought-resistant—can live for 500 years. Bermuda grass sinks deep roots and is difficult to eradicate. Arizona's recent experience shows that migration may no longer be the answer for people with allergies.

Migrants are also pushed from their homes by adverse physical conditions. Water—either too much or too little—poses the most common environmental threat. According to a study by Ian Burton, Robert Kates, and Gilbert White, 40 percent of the world's natural disasters are flood-related and 20 percent are storm-related.

Many people are forced to move by water-related disasters because they live in a vulnerable area, such as a floodplain. The **floodplain** of a river is the area subject to flooding during a specific number of years, based on historical trends. People living in the "100-year floodplain," for example, can expect flooding on average once every century. Many people are unaware that they live in a floodplain, and even people who do know often choose to live there anyway. In the United States, families living in floodplains may not be eligible for government insurance to help rebuild after flood damage.

A lack of water pushes others from their land. Hundreds of thousands have been forced to move from the Sahel region of northern Africa because of drought conditions. The people of the Sahel have traditionally been pastoral nomads, a form of agriculture adapted to dry lands but effective only at low population densities (see Chapter 10). The capacity of the Sahel to sustain human life—never very high—has declined recently because of population growth and several years of unusually low rainfall. Consequently, many of these nomads have been forced to move into cities and rural camps, where they survive on food donated by the government and international relief organizations.

In the United States, people were pushed from their land by severe drought as recently as the 1930s. Portions of Oklahoma and surrounding states became known as the Dust Bowl, following several years of limited rainfall. Strong, dry winds blew across the plains and buried farms under several feet of dust. Thousands of families abandoned their farms and migrated to California, where they were called "Okies." The plight of the Okies was graphically portrayed by John Steinbeck in his novel *The Grapes of Wrath* (1939).

Distance of Migration

Ravenstein's theories made two main points about the distance that migrants travel to their home:

- Most migrants relocate a short distance and remain within the same country.
- Long-distance migrants to other countries head for major centers of economic activity.

Internal Migration

International migration is permanent movement from one country to another, whereas **internal migration** is permanent movement within the same country. Consistent with the principle of distance-decay presented in Chapter 1, the farther away a place is located the less likely that people will migrate to it. Thus, international migrants are much less numerous than internal migrants.

Most people find migration within a country less traumatic than international migration because they find familiar language, foods, broadcasts, literature, music, and other social customs after they move. Moves within a country also generally involve much shorter distances than those in international migration. However, internal migration can involve long-distance moves in large countries, such as the United States and Russia.

Internal migration can be divided into two types: **Interregional migration** is movement from one region of a country to another, while **intraregional migration** is movement within one region. Historically, the main type of interregional migration has been from rural to urban areas in search of jobs. In recent years, some developed countries have seen migration from urban to environmentally attractive rural areas. The main type of intraregional migration has been within urban areas, from older cities to newer suburbs.

International Migration

International migration is further divided into two types: forced and voluntary. **Voluntary migration** implies that the migrant has *chosen* to move for economic improvement, whereas **forced migration** means that the migrant has been *compelled* to move by cultural factors. Economic push and pull factors usually induce voluntary migration, while cultural factors normally compel forced migration. In one sense, migrants may also feel compelled by pressure inside themselves to migrate for economic reasons,

U.S. 1930s Dust Bowl. This family broke down on the road between Dallas and Austin, Texas, August 1936. (Corbis-Bettmann)

such as search for food or jobs, but they have not been explicitly compelled to migrate by the violent actions of other people.

Geographer Wilbur Zelinsky has identified a **migration transition**, which consists of changes in a society comparable to those in the demographic transition. The migration transition is a change in the migration pattern in a society that results from the social and economic changes that also produce the demographic transition.

According to the migration transition, international migration is primarily a phenomenon of countries in stage 2 of the demographic transition, whereas internal migration is more important in stages 3 and 4. A society in stage 1 of the demographic transition—characterized by high birth and death rates and a low natural increase rate—is unlikely to migrate permanently to a new location, although it does have high daily or seasonal mobility in search of food.

In stage 2 of the demographic transition—when the natural increase rate goes up rapidly as a result of a sharp decline in the crude death rate—international migration becomes important, as well as interregional migration from one country's rural areas to its cities. Like the sudden decline in the crude death rate, migration patterns in stage 2 societies are a consequence of technological change. Improvement in agricultural practices reduces the number of people needed in rural areas, whereas jobs in factories attract migrants to the cities in another region of the same country, or in another country.

Crude birth rates begin to decline in stages 3 and 4 of the demographic transition as a result of social changes—people deciding to have fewer children. According to migration transition theory, societies in stages 3 and 4 are the destinations of the international migrants leaving the stage 2 countries in search of economic opportunities. The principal form of internal migration within countries in stages 3 and 4 of the demographic transition is intraregional, from cities to surrounding suburbs.

Characteristics of Migrants

Ravenstein noted distinctive gender and family-status patterns in his migration theories:

- Most long-distance migrants are male.
- Most long-distance migrants are adult individuals rather than families with children.

Gender of Migrants

A century ago, Ravenstein theorized that males were more likely than females to migrate long distances to other countries, because searching for work was the main reason for international migration, and males were much more likely than females to be employed. This held true for U.S. immigrants: During the nineteenth and much of the twentieth centuries, about 54 percent

were male. But in the 1990s the gender pattern reversed, and women now constitute about 54 percent of U.S. immigrants.

Mexicans who come to the United States without proper immigration documents, currently the largest group of U.S. immigrants, show similar gender changes. As recently as the late 1980s, males constituted 85 percent of the Mexican migrants arriving in the United States without proper documents, according to U.S. census and immigration service estimates. But in the 1990s, women accounted for about half of the undocumented immigrants from Mexico.

The increased female migration to the United States partly reflects the changing role of women in Mexican society: In the past, rural Mexican women were obliged to marry at a young age and to remain in the village to care for children. Now, some Mexican women are migrating to the United States to join husbands or brothers already in the United States, but most are seeking jobs. At the same time, women also feel increased pressure to get a job in the United States because of poor economic conditions in Mexico.

Family Status of Migrants

Ravenstein also believed that most long-distance migrants were young adults seeking work, rather than children or elderly people. For the most part, this pattern continues for the United States. About 55 percent of immigrants are between the ages of 18 and 44, compared to about 45 percent of the entire U.S. population. Immigrants are less likely to be elderly; only 5 percent of immigrants are over age 65, compared to 13 percent of the entire U.S. population.

But an increasing percentage of U.S. immigrants are children, and the percentage of immigrants under age 17 is now about the same as the percentage in the total U.S. population. With the increasing number of women migrating to the United States, more children are coming with their mother.

Recent immigrants to the United States have attended school for fewer years and are less likely to have high school diplomas than American citizens. The typical undocumented Mexican immigrant has attended school for 4 years, less than the average American, but a year more than the average Mexican.

For the most part, the origin of Mexican immigrants to the United States matches the expectations of the migration transition and distance-decay theories. With Mexico in stage 2 of the demographic transition, more than three-fourths of migrants are from rural areas. The destination of choice within the United States is overwhelmingly states that border Mexico, with California receiving more than half, Texas another fifth, and other southwestern states most of the remainder.

But most immigrants originate not from Mexico's northern states, but from interior states far from the U.S. border, as the distance-decay theory would suggest. The four leading sources of Mexican migrants are the states of Guanajuato, Jalisco, Chihuahua, and Zacatecas, and only Chichuahua is on the U.S. border. Residents of Mexico's border states are less likely to migrate to the United States, because jobs are relatively plentiful there, (as discussed in Chapter 11), a result of increased economic integration with the United States.

Most illegal Mexican immigrants have jobs in their home village but migrate to the United States to earn more money. The largest number work in agriculture, picking fruits and vegetables, although some work in clothing factories. Even those who work long hours for a few dollars a day as farm laborers or factory workers prefer to earn relatively low wages by American standards than to live in poverty at home.

Most undocumented residents have no difficulty finding jobs in the United States. Some employers like to hire immigrants who do not have visas that permit them to work in the United States, because they can pay lower wages and do not have to provide health care, retirement plans, and other benefits. Unsatisfactory or troublesome workers can be fired and threatened with deportation.

Because farm work is seasonal, the flow of immigrants varies throughout the year. The greatest number of Mexicans head north to the United States in the autumn and return home in the spring. The money brought back by seasonal migrants is the primary source of income for many Mexican villages (and, of course, that money is removed from the U.S. economy). Shops give credit to the villagers through the winter until the men return in the spring with dollars. During the winter, these villages may be inhabited almost entirely by women and children.

Key Issue 2

Why Has U.S. Immigration Changed?

- European immigration to the United States
- Recent immigration from less developed regions

Most people choose to migrate from one country to another because they hope for economic advancement. In general, people emigrate from countries where they have limited prospects for earning a living, and they migrate to countries where they believe that economic opportunities await them.

The United States plays a special role in the study of international migration. The world's third most populous country is inhabited overwhelmingly by direct descendants of immigrants. About 60 million people have migrated to the United States since 1820, including about 20 million currently alive. Immigrants make up a higher percentage of a few other much less populous countries, such as Israel and Canada, but the United States has been the destination for a much larger number of international migrants than any other country in recorded history.

The United States has had two main eras in immigration. The first era began in the mid-nineteenth century and culminated in the early twentieth century. The second era began in the 1970s and continues today. The two eras drew migrants from different regions. During the first era, more than 90 percent of the immigrants came from Europe, whereas during the second era more than three-fourths are coming from Latin America and Asia.

While the origins vary, the reasons for migrating have remained essentially the same: rapid population growth limited prospects for economic advancement at home. Europeans left when their countries entered stage 2 of the demographic transition in the nineteenth century, and Latin Americans and Asians began to leave in large numbers in recent years after their countries entered stage 2. But Europeans arriving in the United States in the nineteenth century found a very different country than Latin Americans and Asians who have recently arrived.

European Immigration to the United States

In the 500-plus years since Christopher Columbus sailed from Spain to the Western Hemisphere, about 60 million Europeans have migrated to other continents. For 36 million of them, the destination was the United States. The remainder went primarily to the temperate climates of Canada, Australia, New Zealand, southern Africa, and southern South America, where farming methods used in Europe could be most easily transplanted.

Waves of European Immigration

For European migrants, the United States offered the greatest opportunity for economic success. Early migrants extolled the virtues of the United States to friends and relatives back in Europe, which encouraged still others to come. The lure of the United States was summarized in the following popular nineteenth-century song. Steel magnate and philanthropist Andrew Carnegie remarked that this song had inspired his father to come to America:

To the west, to the west, to the land of the free
Where mighty Missouri rolls down to the sea;
Where a man is a man if he's willing to toil,
And the humblest may gather the fruits of the soil.
Where children are blessings and he who hath most
Has aid for his fortune and riches to boast.
Where the young may exult and the aged may rest,
Away, far away, to the land of the west.
Away, far away, let us hope for the best
And build up a home in the land of the west.

The total flow of European migrants to the United States and the number from individual countries have varied from year to year. Overall, Germany has sent the largest number of immigrants to the United States,

7.1 million. Other major European sources include Italy, 5.4 million; United Kingdom, 5.2 million; Ireland, 4.8 million; and Russia and the former Soviet Union, 3.5 million. About one-fourth of Americans trace their ancestry to German immigrants, and one-eighth each to Irish and English immigrants.

Note that frequent boundary changes in Europe make precise national counts impossible. For example, most Poles migrated to the United States at a time when Poland did not exist as an independent country. Therefore, most were counted as immigrants from Germany, Russia, or Austria.

First Peak of European Immigration. From the first permanent English settlers to arrive at the Virginia colony's Jamestown, in 1607, until 1840, a steady stream of Europeans migrated to the American colonies (and after 1776 to the newly independent United States of America). Although early migrants included some Dutch, Swedes, French, Germans, German-Swiss, Spanish, and Portuguese, 90 percent of U.S. immigrants prior to 1840 came from Great Britain. Perhaps 1 million Europeans migrated to the American colonies prior to independence, and another million from the late 1700s until 1840.

During the 1840s and 1850s, the level of immigration to the United States surged (Figure 3–2). More than 4 million people migrated to the United States during those two decades, more than twice as many as in the previous 250 years combined. Immigration jumped from approximately 20,000 per year during the first 50 years of independence to over one-quarter million in the peak immigration years of the 1840s and 1850s.

More than 90 percent of all U.S. immigrants during the 1840s and 1850s came from Northern and Western Europe, including two-fifths from Ireland and another one-third from Germany. At first, desperate economic push factors compelled the Irish and Germans to cross the Atlantic. Germans migrated to escape from political unrest, as well as from poor economic conditions.

Second Peak of European Immigration. U.S. immigration declined somewhat during the 1860s as a result of the Civil War (1861–1865). But it began to climb again in the 1870s. A second peak was reached during the

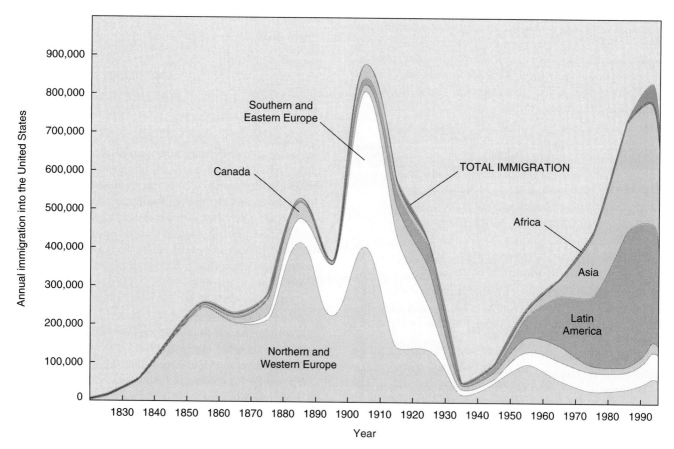

Figure 3–2 Migration to the United States by region of origin. Europeans comprised more than 90 percent of the immigrants to the United States during the nineteenth century and even as recently as the early 1960s continued to account for more than 50 percent. Since the 1960s, Latin America has replaced Europe as the dominant source of immigrants to the United States, although Asia was the largest source during the early 1980s.

1880s, when more than one-half million people annually immigrated to the United States.

Again, more than three-fourths of the immigrants during the late 1800s came from Northern and Western Europe. Germans accounted for one-third and the Irish still constituted a large percentage. However, other countries in Northern and Western Europe sent increasing numbers of migrants, especially the Scandinavian countries of Norway and Sweden. The industrial revolution had diffused to these countries, and population was growing rapidly, as a result of entering stage 2 of the demographic transition (rapidly declining crude death rates). Most of the people who could not find land to farm at home—such as those whose older siblings had inherited their parents' farm—migrated to the cities. But some decided to migrate to other countries in search of farmland or jobs in foreign cities.

Third Peak of European Immigration. Economic problems in the United States discouraged immigration during the early 1890s, but by the end of the decade the level reached a third peak. Nearly a million people each year immigrated during the first 15 years of the twentieth century. The record year was 1907, with 1.3 million immigrants.

During the third peak, more than 90 percent of the immigrants were European. But instead of coming from Great Britain, Ireland, and Germany, most came from countries that previously had sent few people. Nearly one-fourth each came from Italy, Russia, and Austria-Hungary. (Austria-Hungary encompassed portions of present-day Austria, Bosnia-Herzegovina, Croatia, Czech Republic, Hungary, Italy, Poland, Romania, Slovakia, Slovenia, and Ukraine.)

Immigrants came from Southern and Eastern Europe in the early twentieth century for the same reason that Northern and Western Europeans had migrated in the previous century. The shift in the primary source of immigrants coincided with the diffusion of the industrial revolution from Northern and Western Europe to Southern and Eastern Europe. The population of these countries grew rapidly as a result of improved technology and health care. For many, the option of migrating to the United States proved irresistible.

According to the 1910 U.S. census, taken at the peak of immigration, 12.9 million U.S. residents were either born in a foreign country or had at least one foreign-born parent. This amounted to 13.9 percent of the country's total population of 92.2 million. These recent immigrants comprised more than 20 percent of the population in the Northeast, across a northern tier between Michigan and Montana, and along the Pacific Coast.

Impact of European Migration

The era of massive European migration to the United States ended with the start of World War I in 1914, be-

This Italian family immigrated through Ellis Island, in New York harbor, in 1905. (Lewis W. Hine/Corbis/Bettmann)

cause the war involved the most important source countries, such as Austria-Hungary, Germany, and Russia, as well as the United States. The level of European emigration has steadily declined since that time. Europeans accounted for one-third of all U.S. immigrants in the 1960s and only 10 percent since 1980.

Europe's Demographic Transition. Rapid population growth in Europe fueled emigration, especially after 1800. Application of new technologies spawned by the industrial revolution—such as public health, medicine, and food—produced a rapid decline in the death rate and pushed much of Europe into stage 2 of the demographic transition (high growth rate). As the population increased, many Europeans found limited opportunities for economic advancement.

To promote more efficient agriculture, some European governments forced the consolidation of several small farms into larger units. Historically, family farms often had to be divided among a great number of relatives, and the average farm was becoming too small to be profitable. In England, this consolidation policy was known as the "enclosure movement." The enclosure movement forced millions of people to emigrate from rural areas. Displaced farmers could choose between working in factories in the large cities or migrating to the United States or another country where farmland was plentiful.

For several hundred years, the United States was Europe's safety valve. When Europe's population began to increase rapidly because of the industrial revolution, migration to the United States drained off some of the growth. As a result, people remaining in Europe enjoyed more of the economic and social benefits from the industrial revolution.

Most European countries now have very low natural increase rates (stage 3 or 4 of the demographic transition) and economies capable of meeting the needs of their people. Countries such as Germany, Italy, and Ireland, which once sent several hundred thousand people annually to the United States, now send only a few thousand. The safety valve is no longer needed.

Diffusion of European Culture. The emigration of 60 million Europeans has profoundly changed world culture. As do all migrants, Europeans brought their cultural heritage to their new homes. Because of migration, Indo-European languages now are spoken by half of the world's people (as discussed in the next chapter), and Europe's most prevalent religion, Christianity, has the world's largest numbers of adherents. European art, music, literature, philosophy, and ethics have also diffused throughout the world.

Regions that were sparsely inhabited prior to European immigration, such as North America and Australia, have become closely integrated into Europe's cultural traditions. Distinctive European political structures and economic systems have diffused to these regions.

However, Europeans also planted the seeds of conflict by migrating to regions that have large indigenous populations, especially in Africa and Asia. Europeans frequently imposed political domination on existing populations and injected their cultural values with little regard for local traditions. Economies in Africa and Asia became based on extracting resources for export to Europe rather than on using those resources to build local industry.

In more tropical climates, especially in Latin America and Asia, European migrants established plantations that grew cotton, rice, sugar, and tobacco for sale back in Europe. Europeans owned most of the plantations, but relatively few worked on them. Instead, most of the many workers were native Asians or Latin Americans, or were slaves from Africa.

Many of today's conflicts in former European colonies result from past practices by European immigrants, such as drawing arbitrary boundary lines and discriminating among different local ethnic groups.

Recent Immigration from Less Developed Regions

Immigration to the United States dropped sharply in the 1930s and 1940s, during the Great Depression and World War II. During the 1930s, only 50,000 immigrants a year arrived in the United States, and the number of emigrants leaving the United States actually exceeded the number of immigrants by one-fourth. The number of immigrants to the United States steadily increased during the 1950s, 1960s, and 1970s, then surged during the 1980s and 1990s to historically high levels.

Immigration from Asia and Latin America

Latin America and Asia have provided most of the recent U.S. immigrants, but patterns have varied between the two regions. Immigration from Asia has consistently increased, whereas immigration from Latin America has fluctuated.

Immigration from Asia. During the nineteenth and first half of the twentieth centuries, only 1 million Asians migrated to the United States, nearly all from China, Turkey, and Japan. During the last quarter of the twentieth century, about 7 million Asians arrived in the United States, and annual immigration from Asia increased from 40,000 in the 1960s to 150,000 in the 1970s, 200,000 in the 1980s, and 300,000 in the 1990s. Asia was the leading source of immigrants between the late 1970s and late 1980s until overtaken by Latin America.

Over the years, the largest number of Asian immigrants have come from China (including Taiwan), followed by India, not surprising considering that these are the world's two most populous countries, and both are in stage 2 of the demographic transition. But during the 1980s and 1990s, the three leading sources of U.S. immigrants from Asia were the Philippines, Vietnam, and South Korea, which together accounted for more than one-half of all Asian immigrants (Figure 3–3).

Asians also comprise more than 40 percent of Canadian immigrants, but compared to the United States, Canada receives a much higher percentage of Europeans and a lower percentage of Latin Americans. Canada, however, takes in 50 percent more immigrants per capita than does the United States.

Immigration from Latin America. About 2 million Latin Americans migrated to the United States between 1820 and 1960, about 10 million between 1960 and 2000. Annual immigration from Latin America increased from 60,000 in the 1950s to 130,000 in the 1960s, 180,000 in the 1970s, and 350,000 in the 1980s, and more than 1 million in 1990 and 1991, before declining to 250,000 later in the 1990s.

Mexico passed Germany during the 1980s as the country that has sent to the United States the most immigrants ever, currently more than 8 million, and increasing by 100,000 each year. About 4 million of the 20 million immigrants currently alive in the United States were born in Mexico. The Dominican Republic is currently the second leading source of immigrants from Latin America, followed by Jamaica, and Haiti. El Salvador, Colombia, Peru, Guyana, Ecuador, and Guatemala are other leading sources in Latin America (Figure 3–4).

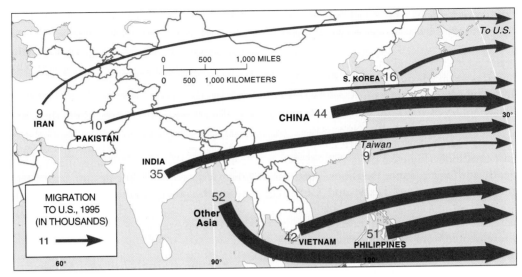

Figure 3–3 Migration to the United States from Asia. The largest numbers of Asians in the late 1990s came from the Philippines and Vietnam, followed by China, India, and South Korea.

The unusually large number of immigrants from Mexico and other Latin American countries in 1990 and 1991 resulted from the 1986 Immigration Reform and Control Act, which issued visas to several hundred thousand who had entered the United States in previous years without legal documents. Counting those legalized under the 1986 act, the United States in 1991 admitted more immigrants—1.8 million—than any other year in history, with 1990 the second highest ever, at 1.5 million.

The pattern of immigration to the United States has changed from predominantly European to Asian and Latin American, although the reason for immigration remains the same. People are pushed by poor conditions at home and lured by economic opportunity and social advancement in the United States. Europeans came in the nineteenth century because they saw the United States as a place to escape from the pressures of land shortage and rapid population increase. Similar motives exist today for people in Asia and Latin America. Several Caribbean countries in stage 2 of the demographic transition are transferring the equivalent of most of their annual natural increase in population to the United States.

While the motives for moving to the United States are similar, the country has changed over time. Unfortunately for the people in less developed countries, the United States is no longer a sparsely settled, economically booming country with a large supply of unclaimed land. In 1912, New Mexico and Arizona were admitted as the forty-seventh and forty-eighth states. Thus, for the first time in its history, all the contiguous territory of this country was a "united" state (other than the District of Columbia). This symbolic closing of the frontier meant to many Americans that the country no longer had the space to accommodate an unlimited number of immigrants.

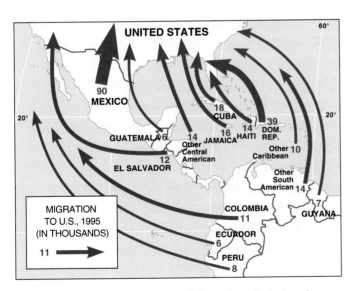

Figure 3–4 Migration to the United States from Latin America. Mexico has been the largest source of immigrants to the United States for many years. The Dominican Republic, Cuba, Jamaica, Haiti, El Salvador, and Colombia are other Latin American countries that sent at least 10,000 immigrants annually to the United States during the late 1990s.

Destination of Immigrants within the United States

Recent immigrants are not distributed uniformly through the United States. One-fourth are clustered in California, one-fourth in New York and New Jersey, one-fourth in Florida, Texas, and Illinois, and one-fourth in the other 44 states. Coastal states were once the main entry points for immigrants, because most arrived by ship. Today, nearly all arrive by motor vehicle or airplane. California and Texas are the two most popular states for entry of motor vehicles from Mexico, and these six states have the country's busiest airports for international arrivals.

Individual states attract immigrants from different countries. Immigrants from Mexico head for California, Texas, or Illinois, while immigrants from Caribbean island countries head for New York or Florida. Chinese and Indians immigrate primarily to New York or California, other Asians to California. Eastern Europeans choose New York or Illinois (Figure 3–5).

Proximity clearly influences some decisions, such as Mexicans preferring California or Texas, and Cubans Florida. But proximity is not a factor in Poles heading for Illinois or Iranians for California. Immigrants cluster in communities where people from the same country previously settled. **Chain migration** is the migration of people to a specific location because relatives or members of the same nationality previously migrated there.

Undocumented Immigration to the United States

Legal immigration to the United States has reached the highest level since the early twentieth century, yet the number of people who wish to migrate to the United States is much higher than the quotas permit. Many people who cannot legally enter the United States are now immigrating illegally. Those who do so are entering without proper documents, and thus are called **undocumented immigrants**.

No one knows how many people immigrate to the United States without proper documents. The U.S. Immigration and Naturalization Service (INS) places the figure at more than 3 million, although others are as high as 20 million. The INS estimates that about 1 million of the 3 million illegal immigrants it claims to be in the United States are from Mexico, 300,000 from El Salvador, and 100,000 each from Guatemala, Canada, Poland, the Philippines, and Haiti.

The INS apprehends more than a million persons annually trying to cross the southern U.S. border, more than 95 percent of whom are Mexican. People enter or remain in the United States illegally primarily because they wish to work but do not have permission to do so from the government. Foreigners who fail to receive work visas have two choices if they still wish to work in the United States:

- Approximately half of the undocumented residents legally enter the country as students or tourists and then remain after they are supposed to leave;
- The other half simply slip across the border without showing a passport and visa to a border guard.

Once in the United States, undocumented immigrants can become "documented" by purchasing forged documents for as little as $25, including a birth certificate, alien registration card, and social security number.

What happens to the minority of illegal immigrants who are caught? To save time and money, the U.S. Immigration and Naturalization Service escorts most of them out of the country. However, the overwhelming majority simply retrace their steps and recross the border (see Globalization and Local Diversity box).

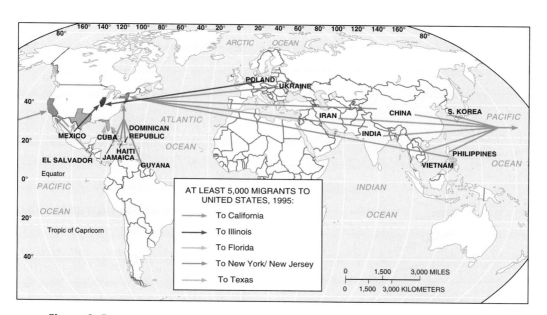

Figure 3–5 Destination of immigrants by U.S. states. California receives about one-fourth of all immigrants, with the largest numbers from Mexico, the Philippines, Vietnam, and China. New York and New Jersey receive another one-fourth of immigrants, especially from the Dominican Republic. A large number of Cubans go to Florida, Mexicans to Texas, and Mexicans and Poles to Illinois.

Globalization and Local Diversity

Intervening Obstacles

Where migrants go is not always their desired destination. The reason **why** is that they may be blocked by an **intervening obstacle**, which is an environmental or cultural feature that hinders migration.

In the past, intervening obstacles were primarily environmental. Before the invention of modern transportation such as railroads and motor vehicles, people migrated across landmasses by horse or on foot. Such migration was frequently difficult because of hostile features in the physical environment like mountains and deserts. For example, many migrants lured to California during the nineteenth century by the economic pull factor of the Gold Rush failed to reach their destination because they could not cross such intervening obstacles as the Great Plains, the Rocky Mountains, or desert country.

Bodies of water long have been important intervening obstacles. The Atlantic Ocean proved a particularly significant intervening obstacle for most European immigrants to North America. Tens of millions of Europeans spent their life savings for the right to cross the rough and dangerous Atlantic in the hold of a ship shared with hundreds of other immigrants.

In the late nineteenth and early twentieth centuries, some Eastern Europeans who booked passage on ships to North America never made it. An unscrupulous ship owner would sail the boat through the Baltic Sea and North Sea and land at Liverpool or some other British port. Told that they had reached America, the passengers—none of whom could speak English—paid for a transatlantic journey of 7,000 kilometers (4,400 miles) but received a voyage of 1,300 kilometers (800 miles) to an undesired destination.

Transportation improvements that have promoted **globalization**, such as motor vehicles and airplanes, have diminished the importance of environmental features as intervening obstacles. However, today's migrant faces intervening obstacles created by **local diversity** in government and politics. A migrant needs a passport to legally emigrate from a country and a visa to legally immigrate to a new country.

The 1986 Immigration Reform and Control Act tried to reduce the flow of illegal immigrants to the United States. Under the law, aliens who could prove that they had lived in the United States continuously between 1982 and 1987 could become permanent resident aliens and apply for U.S. citizenship after 5 years. Seasonal agricultural workers could also qualify for permanent residence and citizenship. However, only 1.3 million agricultural workers and 1.8 million others applied for permanent residence, far fewer than government officials estimated would take advantage of the program. Other undocumented residents apparently feared that if their applications were rejected, they would be deported.

At the same time, the law discouraged further illegal immigration by making it harder for recent immigrants to get jobs without proper documentation. An employer must verify that a newly hired worker can legally work in the United States and may be fined or imprisoned for hiring an undocumented worker.

Crossing the U.S.–Mexican Border. Crossing the U.S.–Mexican border illegally is not difficult. Guards heavily patrol the official border crossings, most of which are located in urban areas such as El Paso, Texas, and San Diego, California, or along highways. However, the border is 3,600 kilometers (2,000 miles) long. It runs through sparsely inhabited regions and is guarded by only a handful of agents. A barbed wire fence runs along the border itself but is broken in many places.

The typical illegal immigrant from Mexico may have more difficulty reaching the U.S. border than actually crossing it. One documented case is illustrated in Figure 3–6. A group of illegal immigrants started from Ahuacatlan, a village of 1,000 in the state of Querétaro. The group first took a bus from the village to the state capital of Querétaro City, followed by another bus for the 1,800-kilometer (1,100-mile) trip to Sonoita, a town near the U.S. border. At Sonoita, the Mexicans hired a driver to transport them to a deserted border location, where they crossed on foot to Pia Oik, Arizona.

Once inside the United States, the group contacted a smuggler, known as a *coyote*, or sometimes a *pollero* ("one who sells chickens for a living"). The smuggler took them by car to Phoenix, approximately 250 kilometers (150 miles) away. A U.S. Border Patrol agent arrested them in Phoenix and took them to Tucson for processing and then across the border to Nogales. From Nogales, they took a bus 110 kilometers (70 miles) to Santa Ana and a second one 260 kilometers (160 miles) back to Sonoita. Repeating their earlier moves to cross the border at Pia Oik, Arizona, they eventually reached Phoenix once more, where they remained. The entire journey cost several hundred dollars.

Illegal immigration presents a dilemma for the United States. On the one hand, allowing illegal immigrants to stay could encourage more to come, thus threatening the U.S. unemployment rate. On the other hand, most undocumented residents take very low-paying jobs that most U.S. citizens will not accept.

Figure 3–6 Crossing the U.S.-Mexican border. The documented route of one group of undocumented immigrants from Mexico to the United States began in Ahuacatlan (1), a village of 1,000 inhabitants in Querétaro State. The immigrants took a bus to Querétaro (2) and another bus to Sonoita (3), hired a driver to take them to a remote location on the border. Looking at the detailed right-hand map, they crossed the United States border on foot near Pia Oik, Arizona (4), and paid a driver to take them to Phoenix (5). Arrested in Phoenix by the Border Patrol, they were driven to Tucson (6) and then to the Mexican border at Nogales (7), where they boarded buses to Santa Ana (8) and back to Sonoita. They then repeated the same route back to Phoenix (9), where they found work.

Why Do Migrants Face Obstacles?

- Immigration policies in other countries
- Cultural problems living in other countries

The principal obstacle traditionally faced by migrants to other countries was the long, arduous, and expensive passage over land or by sea. Think of the cramped and unsanitary conditions endured by nineteenth-century immigrants to the United States who had to sail across the Atlantic or Pacific Ocean in tiny ships.

Today, motor vehicles and airplanes bring most immigrants speedily and reasonably comfortably to the United States and other countries. The major obstacles faced by most immigrants now begin only after they arrive at their desired destination. Immigrants face two major difficulties—gaining permission to enter a new country in the first place and hostile attitudes of citizens once they have entered the new country.

Immigration Policies in Other Countries

Countries to which immigrants wish to migrate have adopted two policies to control the arrival of foreigners seeking work. The United States uses a quota system to limit the number of foreign citizens who can migrate permanently to the country and obtain work. Other major recipients of immigrants, especially in Western Europe and the Middle East, permit guest workers to work temporarily but not stay permanently.

A fence erected in 1997 along the international border between the United States and Mexico at Nogales, Arizona, was designed to be more attractive than the barbed-wire fence that previously demarcated the border. The fourteen-foot-high fence has openings that permit people on the two sides of the border to see each other and to talk. (Monica Almeida/The New York Times)

U.S. Quota Laws

The era of unrestricted immigration to the United States ended when Congress passed the Quota Act in 1921 and the National Origins Act in 1924. These laws established **quotas**, or maximum limits on the number of people who could immigrate to the United States from each country during a 1-year period. The quota was simple: For each country that had native-born persons living in the United States, 2 percent of their number (based on the 1910 census) could immigrate each year. This limited the number of immigrants from the Eastern Hemisphere to 150,000 per year, virtually all of whom had to be from Europe. The system continued with minor modifications until the 1960s.

Quota laws were designed to assure that most immigrants to the United States continued to be Europeans. Although Asians never accounted for more than 5 percent of immigrants during the late nineteenth and early twentieth centuries, many Americans nevertheless were alarmed at the prospect of millions of Asians flooding into the country, especially to states along the Pacific Coast.

Following passage of the Immigration Act of 1965, quotas for individual countries were eliminated in 1968, and replaced with hemisphere quotas. The annual number of U.S. immigrants was restricted to 170,000 from the Eastern Hemisphere and 120,000 from the Western Hemisphere. In 1978, the hemisphere quotas were replaced by a global quota of 290,000, including a maximum of 20,000 per country. The Immigration Act of 1990 raised the total to 714,000 per year for fiscal years 1992, 1993, and 1994, and 675,000 beginning in fiscal year 1995. (The U.S. federal fiscal year runs from October 1 each year to the following September 30.)

Preference System Under Quotas. Because the number of applicants for admission to the United States far exceeds the quotas, Congress has set preferences. Under the 1980 Refugee Act, refugees are a special case: The quota does not apply to them and they are admitted if they are judged genuine refugees.

Beginning in 1995, some 480,000 of the 675,000 annual visas have been issued to relatives of people already living in the United States, in nearly all cases spouses or unmarried children. Skilled workers and exceptionally talented professionals have received 140,000 visas. The remaining 55,000 immigrants have been admitted by lottery under a diversity category.

Because most immigrants to the United States now come from Latin America and Asia, the 55,000 visas awarded by lottery go primarily to people from other regions. Quotas for lottery visas have been set at 43 percent from Europe, 37 percent from Africa, 13 percent from Asia, 4 percent the Western Hemisphere, and 1 percent from Oceania. Immigrants from one country cannot receive more than 7 percent of all diversity category visas.

Asians have made especially good use of the priorities set by the U.S. quota laws. Many well-educated Asians enter the United States under the preference for skilled workers. Once admitted, they can bring in relatives under the family-reunification provisions of the quota. Eventually, these immigrants can bring in a wider range of other relatives from Asia, through a process of chain migration.

Some of today's immigrants to the United States and Canada are poor people pushed from their homes by economic desperation, but most are young, well-educated people lured to economically growing countries. Scientists, researchers, doctors, and other professionals migrate to countries where they can make better use of their abilities. After earning PhDs, young scholars find more teaching positions available at American universities than at home. Other countries charge that by giving preference to skilled workers, U.S. immigration policy now contributes to a **brain drain**, which is a large-scale emigration by talented people.

British Policy. The United Kingdom also severely restricts the ability of foreigners to obtain work permits. However, British policy is complicated by the legacy of the country's former worldwide empire. When some of the United Kingdom's former colonies were granted independence, residents there could choose between remaining British citizens and becoming citizens of the new country. Millions of former colonials in India, Ireland, Pakistan, and the West Indies retained their British citizenship and eventually moved to the United Kingdom. However, spouses and other family members who are citizens of the new countries do not have the right to come to Britain.

Temporary Migration for Work

Migration in search of work is not confined to the United States, of course. But lack of economic opportunity and immigration restrictions imposed by governments reduce the volume on other continents.

People unable to migrate permanently to a new country for employment opportunities may be allowed to migrate temporarily. Prominent forms of temporary work migrants include guest workers in Europe and the Middle East and historically time-contract workers in Asia.

Guest Workers. Citizens of poor countries who obtain jobs in Western Europe and the Middle East are known as **guest workers**. In Europe, guest workers are protected by minimum-wage laws, labor union contracts, and other support programs.

Guest workers serve a useful role in Western European and Middle Eastern countries because they take low-status and low-skilled jobs that local residents won't accept. In cities such as Berlin, Brussels, Paris, and

Zurich, guest workers provide essential services, such as driving buses, collecting garbage, repairing streets, and washing dishes. In the oil-producing countries of the Middle East, guest workers from poorer Middle East countries perform many of the dirty and dangerous functions in the oil fields.

While relatively low-paid by European standards, guest workers earn far more than they would at home. The economy of the guest worker's native country also gains from the arrangement. By letting their people work elsewhere, poorer countries reduce their own unemployment problem. Guest workers also help their native country by sending a large percentage of their earnings back home to their families. The injection of foreign currency then stimulates the local economy.

Origin and Destination of Guest Workers. Guest workers exceed 10 percent of the population in Switzerland and Luxembourg, 9 percent in Germany, and 6 percent in France. Two-thirds of the workers in Middle Eastern petroleum-exporting states such as Kuwait, Qatar, Saudi Arabia, and United Arab Emirates are foreign.

Most guest workers in Europe come from Southern and Eastern Europe, North Africa, the Middle East, and Asia. Distinctive migration routes have emerged among the exporting and importing countries. Italy and Turkey send the largest number of guest workers to Northern Europe, especially to Germany as a result of government agreements. Many guest workers in France come from former French colonies in North Africa—Algeria, Morocco, and Tunisia. Switzerland attracts a large number of Italians; Luxembourg receives primarily Portuguese (Figure 3–7). Migration to Northern Europe from

Eastern Europe has also increased since the fall of communism, as Eastern Europeans leave home in search of jobs elsewhere.

The petroleum-exporting countries of the Middle East attract guest workers primarily from poorer Middle Eastern countries and from Asia. One-fourth of the labor force in Jordan, Lebanon, Syria, and Yemen migrate to petroleum-exporting states to seek employment. India, Pakistan, Thailand, and South Korea also send several million guest workers to the Middle East.

Time-Contract Workers

Millions of Asians migrated in the nineteenth century as time-contract laborers, recruited for a fixed period to work in mines or on plantations. When their contracts expired, many would settle permanently in the new country. Indians went as time-contract workers to Burma (Myanmar), Malaysia, British Guiana (present-day Guyana in South America), eastern and southern Africa, and the islands of Fiji, Mauritius, and Trinidad. Japanese and Filipinos went to Hawaii, and Japanese also went to Brazil. Chinese worked on the U.S. West Coast and helped build the first railroad to span the United States, completed in 1869.

More than 29 million ethnic Chinese currently live permanently in other countries, for the most part in Asia. Chinese comprise three-fourths of the population in Singapore, one-third in Malaysia, and one-tenth in Thailand. Most migrants were from southeastern China. Migration patterns vary among ethnic groups of Chinese. Chiu Chownese migrate to Cambodia, Laos, and Singapore; Hakka to Indonesia, Malaysia, and Thailand; and Hokkien to Indonesia and the Philippines (Figure 3–8).

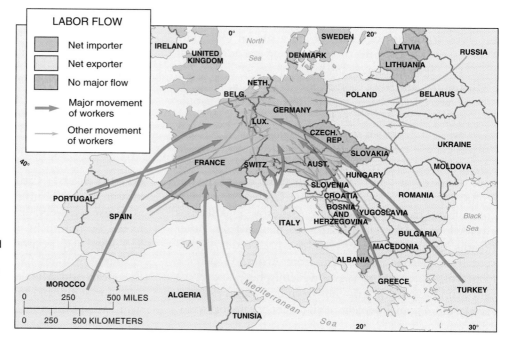

Figure 3–7 Guest workers in Europe. Guest workers emigrate primarily from Southern Europe and North Africa to work in the more developed countries of Northern and Western Europe. Guest workers follow distinctive migration routes. The selected country may be a former colonial ruler, have a similar language, or have an agreement with the exporting country.

Guest workers from North Africa in Paris. Guest workers take low-status and low-skilled jobs in France and other Western European countries that local residents don't want. (M. Jacob/The Image Works)

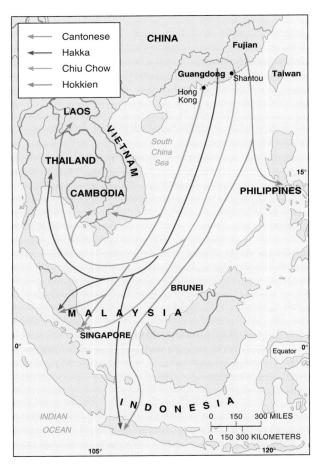

Figure 3–8 Emigration from China. Various ethnic Chinese peoples have distinctive streams of migration to other Asian countries. Most migrate to communities where other members of the same ethnic group have already established businesses. Most emigrate from Guangdong and Hokkien (Fujian) provinces.

In recent years, people have immigrated illegally in Asia to find work in other countries. Estimates of illegal foreign workers in Taiwan range from 20,000 to 70,000. Most are Filipinos, Thais, and Malaysians who are attracted by employment in textile manufacturing, construction, and other industries. These immigrants accept half the pay demanded by Taiwanese, for the level is much higher than what they are likely to get at home, if they could even find employment.

Distinguishing Between Economic Migrants and Refugees

It is sometimes difficult to distinguish between migrants seeking economic opportunities and refugees fleeing from the persecution of an undemocratic government. Distinguishing between the two reasons has been especially difficult for emigrants from Cuba, Haiti, and Vietnam.

The distinction between economic migrants and refugees is important, because the United States, Canada, and Western European countries treat the two groups differently. Economic migrants are generally not admitted to these countries unless they possess special skills or have a close relative already there, and even then they must compete with similar applicants from other countries. However, refugees receive special priority in admission to other countries.

Emigrants from Cuba. Since the 1959 revolution that brought the Communist government of Fidel Castro to power, the U.S. government has regarded emigrants

Haitian boat people. Haitians attempted to migrate to the United States in the 1980s and again in the 1990s in overcrowded boats that were not seaworthy. U.S. officials claiming that Haitians were migrating for economic rather than political reasons, prevented some of the boats from reaching the United States. (Randy Taylor/Sygma)

from Cuba as political refugees. Under Castro's leadership, the Cuban government took control of privately owned banks, factories, and farms, and political opponents of the government were jailed. The U.S. government has prevented companies from buying and selling in Cuba, and Cuba has been excluded from cooperative organizations of Western Hemisphere countries.

In the years immediately following the revolution, more than 600,000 Cubans were admitted to the United States. The largest number settled in southern Florida, where they have become prominent in the region's economy and politics.

A second flood of Cuban emigrants reached the United States in 1980, when Fidel Castro suddenly decided to permit political prisoners, criminals, and mental patients to leave the country. More than 125,000 Cubans left within a few weeks to seek political asylum in the United States, a migration stream that became known as the "Mariel boatlift," named for the port from which the Cubans were allowed to embark.

To reach the United States, most crossed the 200-kilometer (125-mile) Straits of Florida in small boats, many of which were unseaworthy and capsized. When they learned about Castro's new policy, many Cubans already living in Florida sailed from the United States to Cuba, found their relatives, and returned to Florida with them.

U.S. officials were unprepared for the sudden influx of Cuban immigrants. Most Cubans were processed at Key West, Florida, and transferred to camps. Officials identified families or social service agencies willing to sponsor the refugees. Sponsors were expected to provide food and shelter and help the people secure jobs. Most refugees quickly found sponsors, but several thousand who did not lived in army camps and temporary settlements. Approximately 1,000 inhabited Miami's Orange Bowl stadium until the start of the football season, when they were transferred to tents pitched under Interstate 95 in downtown Miami.

Beginning in 1987, the United States agreed to permit 20,000 Cubans per year to migrate to the United States. Cuba also agreed to the return of 2,500 criminals or mental patients who had come in the 1980 Mariel boatlift.

Emigrants from Haiti. Shortly after the 1980 Mariel boatlift from Cuba, several thousand Haitians also sailed in small vessels for the United States. Under the dictatorship of Francois (Papa Doc) Duvalier (1957–1971) and his son Jean-Claude (Baby Doc) Duvalier (1971–1986), the Haitian government persecuted its political opponents at least as harshly as the Cuban. But the U.S. government drew a distinction between the governments of the two neighboring Caribbean countries, because Castro was an ally of the Soviet Union.

Claiming that they had migrated for economic advancement rather than political asylum, U.S. immigration officials would not let the Haitian boat people stay in the United States. However, the Haitians brought a lawsuit against the U.S. government, arguing that if the Cubans were admitted, they should be, too. The government settled the case by agreeing to admit the Haitians.

After a 1991 coup that replaced Haiti's elected president Jean-Bertrand Aristide with military leaders, thousands of Haitians fled their country. In boats that often were overcrowded and unseaworthy, they headed for the U.S. Guantánamo Bay naval base in southeastern Cuba, about 160 kilometers (100 miles) across the Windward Passage from Haiti. Although situated on Cuba, Guantánamo Bay naval base has been controlled by the United States for years.

Once safely ashore at Guantánamo, the Haitians could apply as refugees for migration to the U.S. mainland. Similarly, Haitians picked up by the U.S. Coast Guard from boats drifting in the Windward Passage were eligible to claim political asylum in the United States. The U.S. Immigration and Naturalization Service recognized the claim of political persecution made by many of the Haitians, but the U.S. State Department decided that most left Haiti for economic rather than political reasons.

The United States invaded Haiti in 1994 to reinstate Aristide as president, and a year later a United Nations peacekeeping force assured democratic elections. Although political persecution has subsided, many Haitians still try to migrate to the United States, reinforcing the view that economic factors may always have been important in emigration from the Western Hemisphere's poorest country.

Emigrants from Vietnam. In 1975, the long Vietnam War ended when Communist-controlled North Vietnam captured South Vietnam's capital city of Saigon (since renamed Ho Chi Minh City). The United States, which had supported the government of South Vietnam, evacuated from Saigon several thousand people who had been closely identified with the American position during the war, and who were therefore vulnerable to persecution after the Communist victory.

Thousands of other pro-U.S. South Vietnamese who were not politically prominent enough to get space on an American evacuation helicopter tried to leave by boat. Fleeing overland to neighboring Cambodia, China, and Laos was unattractive because of Communist domination or political unrest in those countries. The so-called ***boat people*** drifted into the South China Sea, hoping they would be saved by the U.S. Navy.

U.S. naval officers wished to save the boat people, but hesitated because of U.S. law. Once taken on board, the boat people would technically be on U.S. territory and could apply for admission to the United States as refugees. This would be unfair to the large numbers of people elsewhere in the world, as well as those still in Vietnam, who had been waiting a long time for the U.S. government to consider their claims for admission as refugees. Consequently, some boat people were not allowed to board U.S. vessels.

A second surge of Vietnamese boat people began in the late 1980s. Their most popular destinations were Malaysia, Hong Kong, and Thailand, with smaller numbers sailing to Indonesia, the Philippines, and Singapore (Figure 3–9). As memories of the Vietnam War faded, officials in these countries no longer considered boat people as refugees, except for a handful who could prove they had been victims of specific incidents of political persecution. Thailand, in particular, pushed Vietnamese boats back out to sea, even though some of them capsized and many boat people drowned.

According to an international agreement, most of the Vietnamese boat people who were judged refugees were transferred to other places, especially the United States, Canada, Australia, and France. However, the majority of the boat people, who were considered economic migrants, were placed in detention camps surrounded by barbed wire and patrolled by armed soldiers. The United Nations helped to fund the camps and monitor living conditions in them until 1996, when the camps were closed and the remaining boat people were sent back to Vietnam.

Altogether, about one-half million Vietnamese have reached the United States since the end of the Vietnam War, another 1 million in other countries. Vietnam remains a major source of immigrants to the United States, but the pull of economic opportunity in the United States is a greater incentive than the push of political persecution.

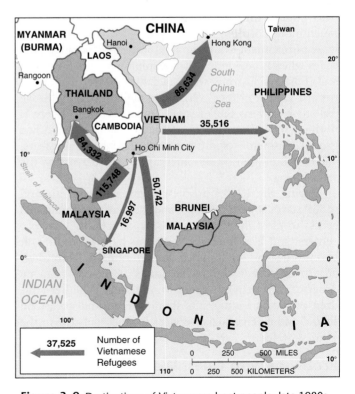

Figure 3–9 Destinations of Vietnamese boat people, late 1980s and early 1990s. During the 1970s, Vietnamese boat people were regarded as political refugees following the end of a long war. In recent years, neighboring countries have severely restricted the number of Vietnamese permitted to stay. Other countries have argued that the boat people can no longer make legitimate claims to be refugees from a war that ended back in 1975.

Cultural Problems Living in Other Countries

For many immigrants, admission to another country does not end their problems. Citizens of the host country may dislike the newcomers' cultural differences. More significantly, politicians exploit immigrants as scapegoats for local economic problems.

U.S. Attitudes Toward Immigrants

Americans have always regarded new arrivals with suspicion, but tempered their dislike during the nineteenth century because immigrants helped to settle the frontier and extend U.S. control across the continent. European immigrants converted the forests and prairies of the vast North American interior into productive farms. By the early twentieth century, though, most Americans believed that the frontier had closed. When the U.S. frontier closed, the gates to the country partially closed as well.

Opposition to immigration intensified when the majority of immigrants ceased to come from Northern and Western Europe. German and Irish immigrants in the nineteenth century suffered some prejudice from so-called native Americans, who had in reality arrived only a few years earlier from Britain. However, Italians, Russians, Poles, and other Southern and Eastern Europeans who poured into the United States about 1900 faced much more hostility.

A government study in 1911 reflected popular attitudes when it concluded that immigrants from Southern and Eastern Europe were racially inferior, "inclined toward violent crime," resisted assimilation, and "drove old-stock citizens out of some lines of work." (There is nothing new about racism, prejudice, fear of unknown groups, suspicion of different cultures, economic fears, and anti-immigration sentiment. Only the players on the stage change.)

During the 1990s, hostile citizens in California and other states voted to deny undocumented immigrants access to most public services, such as schools, day-care centers, and health clinics. The laws have been difficult to enforce and of dubious constitutionality, but their enactment reflects the unwillingness on the part of many Americans to help out needy immigrants. Whether children of recent immigrants should be entitled to attend school and receive social services is much debated in the United States.

Problems with Guest Workers

In Europe, many guest workers suffer from poor social conditions. The guest worker is typically a young man who arrives alone in a city. He has little money for food, housing, or entertainment, because his primary objective is to send home as much money as possible. He is likely to use any surplus money for a railway ticket home for the weekend.

Far from his family and friends, the guest worker can lead a lonely life. His isolation may be heightened by unfamiliarity with the host country's language and distinctive cultural activities. Many guest workers pass their leisure time at the local railway station. There they can buy native-language newspapers, mingle with other guest workers, and meet people who have just arrived by train from home.

Both guest workers and their host countries regard the arrangement as temporary. In reality, however, many guest workers remain indefinitely, especially if they are joined by other family members. Some guest workers apply their savings to starting a grocery store, restaurant, or other small shop. These businesses can fill a need in European cities by remaining open on weekends and evenings when most locally owned establishments are closed.

In the Middle East, petroleum-exporting countries fear that the increasing numbers of guest workers will spark political unrest and abandonment of traditional Islamic customs. After the 1991 Gulf War, Kuwaiti officials expelled hundreds of thousands of Palestinian guest workers who had sympathized with Iraq's invasion of Kuwait in 1990. To minimize long-term stays, other host countries in the Middle East force migrants to return home if they wish to marry and prevent them from returning once they have wives and children.

As a result of lower economic growth rates, Middle Eastern and Western European countries have reduced the number of guest workers in recent years. Several Western European governments pay guest workers to return home, but some of these countries have their own unemployment problems, and sometimes refuse to take back their own nationals.

Many Western Europeans dislike the guest workers and oppose government programs to improve their living conditions. Political parties that support restrictions on immigration have gained support in France, Germany, and other European countries, and attacks by local citizens on immigrants have increased.

Migration by Asians nearly a century ago is producing contemporary problems in several countries. For example, between 1879 and 1920, the British brought Indians as indentured laborers to the Fiji Islands in the South Pacific. Today, Fiji includes slightly more Indians than native Fijians. For many decades, Fiji was a model of how two culturally diverse groups could live together peacefully under a democratically elected government. Indians controlled most of the country's businesses, while Fijians dominated the government and army. However, after an Indian party won the elections in 1987, rioting broke out between the two groups, and Fijian army officers seized temporary control of the government. A new constitution in 1990 ensured that Fijians would hold a majority of seats in the parliament.

The argument of anti-immigrant politicians is seductive to many voters in Western Europe, as well as the United States: If all of the immigrants were thrown out of the country then the unemployment rate would drop, and if all of the immigrants were cut off from public programs then taxes would drop. In an economically integrated world, such arguments have little scientific basis, and in a culturally diverse world these arguments have racist overtones.

Key Issue 4

Why Do People Migrate Within a Country?

- Migration between regions of a country
- Migration within one region

Internal migration for most people is less disruptive than international migration. International migration involves movement to a country with different cultural traditions, such as language and religion. Even migration among culturally similar countries can be disorienting in less profound ways, such as major sports and popular television programs.

Two main types of internal migration are interregional and intraregional. The principal type of interregional migration is between rural and urban areas, while the main type of intraregional migration is from older cities to suburbs.

Migration Between Regions of a Country

In the United States, interregional migration was more prevalent in the past, when most people were farmers. Lack of farmland pushed many people from the more densely settled regions of the country and lured them to the frontier, where land was abundant. Today, most people move to a new region for a better job, although many also move for noneconomic reasons.

Migration Between Regions Within the United States

The most famous example of large-scale internal migration is the opening of the American West. Two hundred years ago, the United States consisted of a collection of settlements concentrated on the Atlantic Coast. Through mass interregional migration, the interior of the continent was settled and developed.

Changing Center of Population. The U.S. Census Bureau computes the country's population center at the time of each census. The population center is the average location of everyone in the country, the "center of population gravity." If the United States were a flat plane placed on top of a pin, and each individual weighed the same, the population center would be the point where the population distribution causes the flat plane to balance on the pin.

The changing location of the population center graphically demonstrates the march of the American people across the North American continent over the past 200 years. When the first U.S. census was taken in 1790, the population center was located in Chesapeake Bay, east of Baltimore, Maryland. Throughout the colonial period, the population center remained roughly in the same place. This location reflects the fact that virtually all settlements were near the Atlantic Coast (Figure 3–10).

Few colonists ventured far from coastal locations because they depended on shipping links with Europe to receive products and to export raw materials. Settlement in the interior was also hindered by an intervening obstacle, the Appalachian Mountains. The Appalachians blocked Western development because of their steep slopes, thick forests, and few gaps that allowed easy passage. Hostile indigenous residents, commonly called Indians, also retarded western settlement.

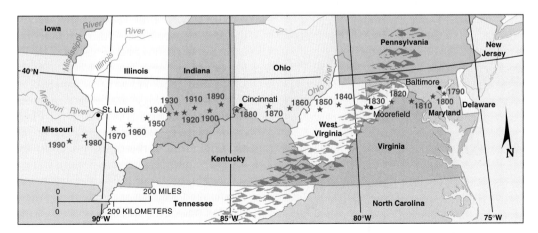

Figure 3–10 Changing center of population in the United States. The center has consistently shifted westward, although the rate of movement has varied in different eras. In 1980, the center of population shifted west of the Mississippi River for the first time. In 1990, the center was the farthest south ever, as well as the farthest west.

Early Settlement in the Interior. Settlement of the interior began after 1790. By 1830 the center of population moved west of Moorefield, West Virginia. Encouraged by the opportunity to obtain a large amount of land at a low price, people moved into river valleys and fertile level lowlands as far west as the Mississippi River.

Transportation improvements helped open the interior in the early 1800s, especially the building of canals. Most important was the Erie Canal, which enabled people to travel inexpensively by boat between New York City and the Great Lakes. When the Erie Canal opened in 1825, the fare from New York to Detroit was only $10, yet traffic was so heavy on the canal that tolls paid for construction costs within 9 years. Between 1816 and 1840, the network of new canals dug in the United States totaled 5,352 kilometers (3,326 miles). The diffusion of steam-powered boats further speeded water travel.

After 1830, the U.S. population center moved west more rapidly. By 1880, it was just west of Cincinnati, Ohio. The population center moved 11 kilometers (7 miles) per year during that period, compared to only 7 kilometers (4 miles) per year during the previous 40 years.

What accounts for the more rapid westward shift between 1830 and 1880? The primary reason was that most western pioneers at the time passed through the interior of the country and headed for California. For nearly 40 years, the continuous westward advance of settlement stopped at the 98th meridian as migrants jumped directly to California. (The 98th meridian runs north-south through the eastern Dakotas, Nebraska, Kansas, Oklahoma, and Texas.)

Large numbers of migrants passed through the interior without stopping in part because they were pulled to California, especially by the Gold Rush beginning in the late 1840s. At the same time, the interior of the country confronted early settlers with a physical environment that was unsuited to familiar agricultural practices.

Early nineteenth-century Americans preferred to start farms in forested areas that receive 100 centimeters (40 inches) or more precipitation a year. They cut down the trees and used the wood to build homes, barns, and fences. But when they crossed west of the 98th meridian, pioneers found few trees. Instead, they saw vast rolling grasslands that average less than 50 centimeters (20 inches) of precipitation annually.

Without the technology to overcome this dry climate, lack of trees, and tough grassland sod, early explorers such as Zebulon Pike declared the region unfit for farming. Maps at the time labeled the region from the Dakotas through Nebraska, Kansas, and Oklahoma to Texas as "the Great American Desert." Ironically, with today's agricultural practices, the region west of the 98th meridian to the Rocky Mountains, which we call the Great Plains, is one of the world's richest farming areas.

Settlement of the Great Plains. After 1880, the U.S. population center continued to migrate westward at a much slower pace. Between 1880 and 1950, the center moved approximately 5 kilometers (3 miles) per year, less than half the rate of the previous half century. The rate slowed in part because large-scale migration to the East Coast from Europe offset some of the migration from the East Coast to the U.S. West.

The westward movement of the U.S. population center also slowed after 1880 because people began to fill in the area between the 98th meridian and California that earlier generations had bypassed. The Dakota Territory, for example, grew from 14,000 inhabitants in 1870 to 135,000 in 1880 and 539,000 by 1890. Advances in agricultural technology in the late nineteenth century enabled people to cultivate the Great Plains. Farmers used barbed wire to reduce dependence on wood fencing, the steel plow to cut the thick sod, and windmills and well-drilling equipment to pump more water.

Beginning in the 1840s, the expansion of the railroads encouraged western settlement. By the 1880s, an extensive rail network permitted settlers on the Great Plains to transport their products to the large concentrations of customers in East Coast cities. The railroad companies also promoted western settlement by selling land to farmers. Companies that built the railroad lines received

Hazards traveling west in the nineteenth century. Pioneers were ill-prepared to cross the high mountains and extensive deserts of the U.S. west. (North Wind)

large land grants from the federal government, not just narrow right-of-way strips to lay tracks. The railroad companies in turn financed construction of their lines by selling small parcels of the adjacent land to farmers. Rail companies established offices in major East Coast and European cities to sell land.

Recent Growth of the South.   Since 1950, the population center has moved west faster, at 10 kilometers (6 miles) per year. In 1980, it was located near DeSoto, Missouri, southwest of St. Louis. The site was significant: For the first time in U.S. history, population center had crossed the Mississippi River. By 1990, the center had migrated farther westward.

The recent movement of the population center also shows a second trend, namely movement southward. In 1790, the center was at 39°16′30″ north latitude. By 1920, it had moved only slightly southward to 39°10′21″. Beginning in the 1920s, the center moved southward, at first slowly, but since 1950 at 4 kilometers (2 miles) per year. By 1990, the center had reached 37°52′20″ north latitude.

The population center drifted southward because of net migration into the southern and western states. Between 1980 and 1988, for example, about 3 million people moved from the Midwest to the West, whereas about 2 million moved from the West to the Midwest (Figure 3–11). Net migration from the Midwest to the South and from the Northeast to the South were both about 1.7 million during the period; in other words, 1.7 million more people moved to the South from each of the two northern areas than moved from the South to the Northeast and Midwest.

Interregional Migration to the South and West.   Why are Americans emigrating from the North and East and immigrating to the South and West? More than half move primarily for job opportunities. New jobs created each year since 1960 have averaged about 3 percent in the United States as a whole, 5 percent in the South and West, but only 1 percent in the Northeast and Midwest.

People also migrate to the South and West for environmental reasons. Americans commonly refer to the South and West as the "sunbelt," because of its more temperate climate, while the Northeast and Midwest are labeled the "rustbelt," because of the regions' dependency on declining steel and other manufacturing industries (as well as the ability of the regions' climate to rust out cars relatively quickly). As people gain more leisure time, they are lured to the sunbelt for outdoor recreation throughout the year.

The growth in population and employment of the South and West and the decline of the Northeast and Midwest have aggravated interregional antagonism. Some people in the Northeast and Midwest believe that southern and western states have stolen industries from them. The fact is, however, that although some industries have relocated from the Northeast and Midwest, most industrial growth in the South and West comes from newly established companies.

To some extent, the regional difference in economic growth reduces a historical imbalance, because traditionally people in the Northeast have enjoyed higher incomes than residents of the South. If the average income of a U.S. family is arbitrarily designated as 100, then the average income in 1990 was 116 for people in the Northeast and 88 for people in the South. By comparison, as recently as 1929, average income was 138 in the Northeast and only 51 in the South. However, the gap between the Northeast and the South has widened since 1970, when the average was 107 in the Northeast and 90

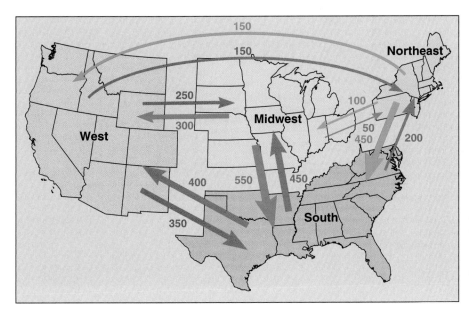

Figure 3–11 U.S. interregional migration. Figures show average annual migration (in thousands), 1980s and 1990s. The largest flows during the period were from the Midwest and the Northeast to the South. Note that large numbers of people migrated out of the South during the period, but net migration was strongly directed into the South.

in the South; the increase results from recent problems faced by southern industries (see Chapter 11).

Net migration of African-Americans has followed a different pattern. A century ago, most African-Americans lived in the South, because their ancestors had been forced to migrate to the region from Africa. During the twentieth century, large numbers of African-Americans migrated from the South to take jobs in the large cities of the Northeast, Midwest, and West (see Chapter 7).

Migration Between Regions in Other Countries

As in the United States, long-distance interregional migration has been an important means of opening new regions for economic development in other large countries. Incentives have been used to stimulate migration to other regions.

Russia.   Interregional migration was important in developing the former Soviet Union. Soviet policy encouraged factory construction near raw materials rather than near existing population concentrations (see Chapter 11). Not enough workers lived nearby to fill all the jobs at the mines, factories, and construction sites established in these remote resource-rich regions. To build up an adequate labor force, the Soviet government had to stimulate interregional migration.

Soviet officials were especially eager to develop Russia's Far North, which included much of Siberia, because it is rich in natural resources—fossil fuels, minerals, and forests. The Far North encompassed 45 percent of the Soviet Union's land area but contained

less than 2 percent of its people. Earlier in this century, the Soviet government had forced people to migrate to the Far North to construct and operate steel mills, hydro-electric power stations, mines, and other enterprises. In later years, the Soviet government reduced the use of forced migration and instead provided incentives, including higher wages, more paid holidays, and earlier retirement, to induce voluntary migration to the Far North.

However, the incentives failed to pull as many migrants to the Far North as Soviet officials desired. People were reluctant because of the region's harsh climate and remoteness from population clusters. Each year, as many as half of the people in the Far North migrated back to other regions of the country and had to be replaced by other immigrants, especially young males willing to work in the region for a short period. One method the Soviet government used was to send a brigade of young volunteers, known as *Komsomol*, during school vacations to help construct projects. An example is the Baikal-Amur Railroad, which runs for 3,145 kilometers (1,955 miles) from Taishet to Sovetskaia Gavan.

The collapse of the Soviet Union ended policies that encouraged interregional migration. In the transition to a market-based economy, Russian government officials no longer dictate "optimal" locations for factories.

Brazil.   Another large country, Brazil, has encouraged interregional migration. Most Brazilians live in a string of large cities near the Atlantic Coast, including Recife, Salvador, Rio de Janeiro, São Paulo, and Porto Alegre. São Paulo and Rio de Janeiro have become two of the world's largest cities. In contrast, Brazil's tropical interior is very sparsely inhabited.

Brasília. Brazil's capital was moved here in 1960. Since then, thousands of migrants have arrived in search of jobs. Many live in poor-quality housing on the edge of the city, a stark contrast to the carefully planned central area. (Tony Stone Images)

To increase the attractiveness of the interior, in 1960 the government moved its capital from Rio to a newly built city called Brasília, situated 1,000 kilometers (600 miles) from the Atlantic Coast. From above, Brasília's design resembles an airplane, with government buildings located at the center of the city and housing arranged along the "wings."

At first, Brasília's population grew slowly, because government workers and foreign embassy officials resented the forced move from Rio, one of the world's most animated cities. In recent years, thousands of people have migrated to Brasília in search of jobs. In a country with rapid population growth, many people will migrate where they think they can find employment. Many of these workers could not afford housing in Brasília and were living instead in hastily erected shacks on the outskirts.

Indonesia. Since 1969, the Indonesian government has paid for the migration of more than 5 million people, primarily from the island of Java, where nearly two-thirds of its people live, to less populated islands. Under the government program, families receive 2 hectares (5 acres) of land, materials to build a house, seeds and pesticides, and food to tide them over until the crops are ready.

The number of participants has declined in recent years, primarily because of environmental concerns. Some families moved to land that could not support intensive agriculture, while others disrupted the habitats of indigenous peoples. The program siphoned off only a very small percentage of Java's population growth during the past quarter-century.

Europe. The pattern of interregional migration throughout Western Europe is reflected in differences in per capita income and unemployment in different regions. The regions with net immigration are also the ones with the highest per capita incomes (Figure 3–12).

Even countries that occupy relatively small land areas have important interregional migration trends. People in Italy migrate from the south, known as the Mezzogiorno, to the north in search of job opportunities. Compared to the Mezzogiorno, Italy's north benefits from relatively rich agricultural land and a strong industrial base. The Mezzogiorno comprises 40 percent of Italy's land area and contains 35 percent of the population, but only 24 percent of the national income. Per capita income is nearly twice as high in the north as in the south, and unemployment rates are less than 5 percent in the north, compared to more than 20 percent in the south.

Similarly, people in the United Kingdom are migrating because of regional differences in job opportunities, although the pattern is the opposite of Italy's: Economic growth is in the south, while the north is declining. The northern regions of the United Kingdom were the first in the world to enter the industrial revolution in the eighteenth century. Today, many of the region's industries

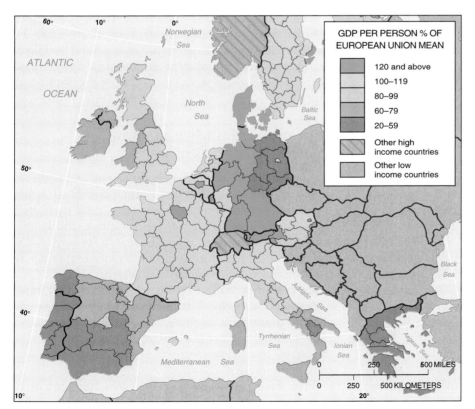

Figure 3–12 Per capita gross domestic product as a percentage of the European Union average. Sharp differences in wealth exist among regions of some European countries. Within Italy, for example, wealth is twice as high in the north as in the south. For the European Union as a whole, incomes are higher in regions near the center and lower in regions that are relatively remote from the center.

are no longer competitive in the global economy. On the other hand, industries in the south and east—especially the region around London—are relatively healthy.

Regional differences in economic conditions within European countries may become greater with increased integration of the Continent's economy. Regions closer to European markets, such as the south of Britain and the north of Italy, may hold a competitive advantage over more peripheral regions.

India. A number of governments limit the ability of people to migrate from one region to another. For example, Indians require a permit to migrate—or even to visit—the State of Assam in the northeastern part of the country. The restrictions, which date from the British colonial era, are designed to protect the ethnic identity of Assamese by limiting the ability of outsiders to compete for jobs and purchase land. Because Assam is situated on the border with Bangladesh, the restrictions also limit international migration.

Migration Within One Region

While interregional migration attracts considerable attention, far more people move within the same region, which is *intraregional* migration. Since 1800, the most prominent type of intraregional migration in the world has been from rural to urban areas. Less than 5 percent of the world's people lived in urban areas in 1800, compared to nearly half today.

Migration from Rural to Urban Areas

Urbanization began in the 1800s in the countries of Europe and North America that were undergoing rapid industrial development. The percentage of people living in urban areas in the United States, for example, increased from 5 percent in 1800 to 50 percent in 1920. Today, approximately three-fourths of the people in the United States and other more developed countries live in urban areas.

Migration from rural to urban areas has skyrocketed in recent years in the less developed countries of Africa, Asia, and Latin America. Studies conducted in a variety of less developed countries show that migration from rural areas accounts for nearly half of the population increase in urban areas, while the natural increase (excess of births over deaths) accounts for the remainder. Worldwide, more than 20 million people are estimated to migrate each year from rural to urban areas.

Migration to one of the world's largest cities, São Paulo, Brazil, has reached 300,000 people per year. Many of these migrants cannot find housing in the city and must live in squatter settlements, known in Brazil as *favelas*. The favelas may lack electricity, running water, and paved streets (see Chapter 13).

Like interregional migrants, most people who move from rural to urban areas seek economic advancement. They are pushed from rural areas by declining opportunities in agriculture and are pulled to the cities by the prospect of work in factories or in service industries.

Migration from Urban to Suburban Areas

In more developed countries, most intraregional migration is from central cities out to the suburbs. Annual net migration from cities to suburbs exceeds 1 million people in the United States; comparable rates of suburbanization are found in Canada, the United Kingdom, and other Western European countries. The population of most central cities has declined in North America and Western Europe, while suburbs have grown rapidly.

The major reason for the large-scale migration to the suburbs is not related to employment, as was the case with other forms of migration. For most people, migration to suburbs does not coincide with changing jobs. Instead, people are pulled by a suburban lifestyle.

Suburbs offer the opportunity to live in a detached house rather than an apartment, surrounded by a private yard where children can play safely. A garage or driveway on the property guarantees space to park automobiles at no charge. Suburban schools tend to be more modern, better equipped, and safer than those in cities. Automobiles and trains enable people to live in suburbs, yet have access to jobs, shops, and recreation facilities throughout the urban area (see Chapter 13).

As a result of suburbanization, the territory occupied by urban areas has rapidly expanded (see Chapter 13). To accommodate suburban growth, farms on the periphery of urban areas are converted to housing developments, where new roads, sewers, and other services must be built.

Migration from Metropolitan to Nonmetropolitan Areas

During the 1970s, the more developed countries of North America and Western Europe witnessed a new trend. For the first time in U.S. history, rural areas grew more rapidly than did urban areas. Canada, the United Kingdom, and several other European countries displayed similar patterns. Net migration from urban to rural areas is called **counterurbanization**.

Counterurbanization results in part from very rapid expansion of suburbs. The boundary where suburbs end and the countryside begins cannot be precisely defined. However, most counterurbanization represents genuine migration from cities and suburbs to small towns and rural communities.

Like suburbanization, people move from urban to rural areas for lifestyle reasons. People are lured to rural areas by the prospect of swapping the frantic pace of urban life for the opportunity to live on a farm where they can own horses or grow vegetables. However, most people who move to farms do not earn their living from agriculture. Instead, they work in nearby factories, small-town shops, or other services.

With modern communications and transportation systems, no location in a more developed country is truly isolated, either economically or socially. Computers enable us to work anywhere and still have access to an international network. We can obtain money at any time from a conveniently located electronic transfer machine rather than by going to a bank building. We can select clothing from a mail-order catalogue, place the order by telephone, pay by credit card, and have the desired items delivered within a few days. We can follow the fortunes of our favorite baseball teams on television anywhere in the country, thanks to satellite dishes.

Many migrants from urban to rural areas are retired people who are attracted by access to leisure activities, such as fishing and hiking. Retirement communities—in reality, small towns restricted to older people, typically over age 50—appeal to retired people who like to participate in recreation activities. In France, some elderly people migrate from Paris to the rural village where they were born, whereas others are attracted to the mild climate in the south of the country along the Mediterranean coast.

Counterurbanization has stopped in the United States since the early 1980s, because job opportunities have declined in rural areas. Many factories that located in rural areas during the 1970s are no longer competitive in a rapidly changing global economy. Industries that located in rural areas to take advantage of the lower costs of doing business are being undersold by Asian competitors who have even lower production costs. Surviving industries in rural parts of the United States and other more developed countries have had to become more efficient, often by eliminating jobs.

The rural economy has also been hurt by poor agricultural conditions. The price of farm products has declined, and many farmers have gone bankrupt. Although farmers constitute a small percentage of the labor force, they play an important role in the economy of rural areas. For example, the typical farmer borrows large sums of money from local banks and buys expensive equipment from local stores.

Future migration trends are unpredictable in more developed countries, because future economic conditions are difficult to forecast. Have these countries reached long-term equilibrium, in which approximately three-fourths of the people live in urban areas and one-fourth in rural areas? Will counterurbanization resume in the future, because people prefer to live in rural areas? Is the decline of the rural economy reversible?

■ Summary

Migration will play an increasing role in determining population growth of countries in stages 3 and 4 of the demographic transition. In the United States, the crude birth rate is approximately 16, the crude death rate is approximately 9, and the natural increase rate is approximately 0.7 percent per year. These rates translate into approximately 3.9 million births and 2.3 million deaths annually, and a natural increase of 1.6 million people each year. However, the annual population increase in the United States is actually more than 2.5 million. The difference between actual growth and natural increase is due to net in-migration.

In a couple of decades, the crude birth and crude death rates will be roughly equal in the United States. At that time, virtually all population growth will be attributable to net in-migration rather than to natural increase. Here again are the key issues we have raised about migration:

1. **Why do people migrate?** We can group the reasons into push and pull factors. People feel compelled (pushed) to emigrate from a location for political, economic, and environmental reasons. Similarly, people are induced (pulled) to immigrate because of the political, economic, or environmental attractiveness of a new location. We can also distinguish between international and internal migration.

2. **Why has U.S. immigration changed?** The principal sources of immigrants to the United States have been countries in stage 2 of the demographic transition. In the nineteenth century, the largest stream of international migration came from Europe to the United States. Today, Latin Americans and Asians comprise the largest group trying to migrate to the United States for economic opportunities.

3. **Why do migrants face obstacles?** Migrants have difficulty getting permission to enter other countries, and they face hostility from local citizens once they arrive. Immigration laws restrict the number who can legally enter the United States. In Europe and the Middle East, guest workers migrate temporarily to perform menial jobs.

4. **Why do people migrate within a country?** We can distinguish between interregional and intraregional migration within the same country. Historically, interregional migration was especially important in settling the frontier of large countries such as the United States, Russia, and Brazil. Today, interregional migration persists because of differences among regions of a country in economic conditions, climate, and other environmental factors. The most important intraregional migration trends are from rural to urban areas within less developed countries and from cities to suburbs within more developed countries.

Case Study Revisited

Give Me Your Tired, Your Poor . . .

The most famous symbol of migration in the world is surely the Statue of Liberty. Its inscription, written by Emma Lazarus, includes the famous words, "Give me your tired, your poor, your huddled masses yearning to breathe free." The statue stands at the mouth of New York Harbor, near Ellis Island, which was for many years the initial landing and processing point for millions of immigrants from Europe. The Statue of Liberty was the first landmark seen by many European immigrants when they sailed into the United States.

European countries like Ireland no longer supply most of the migrants to the United States, and as the introductory case study pointed out, Irish people are being lured back to Ireland by an improving economy. Desire to migrate to the United States has shifted to less developed countries in Asia and Latin America.

Latin American countries in stage 2 of the demographic transition export a large percentage of their population growth to the United States. For example, Jamaica, has a population of 2.6 million, a crude birth rate of 22 (per 1,000), a crude death rate of 6, and a natural increase rate of 1.6 percent per year (calculated as crude birth rate minus crude death rate). Otherwise stated, in one year Jamaica has approximately 57,000 births, approximately 16,000 deaths, and therefore a natural increase of approximately 41,000. However, 17,000 Jamaicans have migrated to the United States annually in recent years. As a result of this emigration, Jamaica's annual population increase is actually only about 24,000, or 0.9 percent, a bit more than half of the natural increase rate.

Jamaicans claim that the large-scale emigration subsidizes the U.S. economy, because most of the migrants are nurses, teachers, doctors, and other professionals who have been trained at the expense of the Jamaican government. However, many of these migrants send their savings earned in the United States back to Jamaica to help out relatives who remain in the country. In addition, Jamaicans who have immigrated to the United States often return to Jamaica for visits, taking money and goods back with them. More importantly, they bring back to Jamaica the cultural values acquired through living in a more developed society.

Residents of the United States in the nineteenth and early twentieth centuries did not greet European immigrants with open arms. However, immigrants were given the opportunity to enter the country and make new lives. Some of them were successful. Even recently arrived undocumented immigrants stand a good chance of success, if given the chance.

For many people, the only way to enter the United States or Canada is illegally. In the latter part of the twentieth century, the tradition of universal ability to migrate to North America no longer exists. Paradoxically, in an era when human beings have invented easy means of long-distance transport, the right of free migration has been replaced by human barriers. The United States no longer asks for immigration of the world's tired, poor, huddled masses yearning to be free.

■ Key Terms

Brain drain Large-scale emigration by talented people.

Chain migration Migration of people to a specific location because relatives or members of the same nationality previously migrated there.

Circulation Short-term, repetitive, or cyclical movements that recur on a regular basis.

Counterurbanization Net migration from urban to rural areas in more developed countries.

Emigration Migration *from* a location.

Floodplain The area subject to flooding during a given number of years according to historical trends.

Forced migration Permanent movement compelled usually by cultural factors.

Guest workers Workers who migrate to the more developed countries of Northern and Western Europe, usually from Southern and Eastern Europe or from North Africa, in search of higher-paying jobs.

Immigration Migration *to* a new location.

Internal migration Permanent movement within a particular country.

International migration Permanent movement from one country to another.

Interregional migration Permanent movement from one region of a country to another.

Intervening obstacle An environmental or cultural feature of the landscape that hinders migration.

Intraregional migration Permanent movement within one region of a country.

Migration Form of relocation diffusion involving permanent move to a new location.

Migration transition Change in the migration pattern in a society that results from industrialization, population growth, and other social and economic changes that also produce the demographic transition.

Mobility All types of movement from one location to another.

Net migration The difference between the level of immigration and the level of emigration.

Pull factors Factors that induce people to move to a new location.

Push factors Factors that induce people to leave old residences.

Quota In reference to migration, a law that places maximum limits on the number of people who can immigrate to a country each year.

Refugees People who are forced to migrate from their home country and cannot return for fear of persecution because of their race, religion, nationality, membership in a social group, or political opinion.

Undocumented immigrants People who enter a country without proper documents.

Voluntary migration Permanent movement undertaken by choice.

■ Thinking Geographically

1. Should preference for immigrating to the United States and Canada be given to individuals with special job skills, or should priority be given to reunification of family members? Should quotas be raised to meet increasing demand for both types of immigrants? Why or why not?

2. What is the impact of large-scale emigration on the places from which migrants depart? On balance, do these places suffer because of the loss of young, upwardly mobile workers, or do these places benefit from the draining away of surplus labor? In the communities from which migrants depart, is the quality of life improved overall through reduced pressures on local resources, or is it damaged overall through the deterioration of social structures and institutions? Explain.

3. According to the concept of chain migration, current migrants tend to follow the paths of relatives and friends who have moved earlier. Can you find evidence of chain migration in your community? Does chain migration apply primarily to the relocation of people from one community in a less developed country to one community in a more developed country, or is chain migration more applicable to movement within a more developed country? Explain.

4. Which demographic characteristics (such as rates of natural increase, crude birth, and crude death) prevail in the three regions with the largest numbers of refugees—the Horn of Africa, Afghanistan, and the Middle East? Is large-scale forced migration alleviating or exacerbating population growth in these regions? Explain.

5. At the same time that some people are migrating from less developed countries to more developed countries in search of employment, transnational corporations have relocated some low-skilled jobs to less developed countries to take advantage of low wage rates. Should less developed countries care whether their surplus workers emigrate or remain as employees of foreign companies? Why?

■ Further Readings

Appleyard, Reginald, ed. *International Migration Today. Vol. I: Trends and Prospects.* Paris: UNESCO, 1988.

Bennett, D. Gordon, and Ole Gade. *Geographic Perspectives on Migration Behavior: A Bibliographic Survey.* University of North Carolina Studies in Geography, No. 12. Chapel Hill: University of North Carolina Press, 1979.

Berry, Brian J. L., and Lester Silverman, eds. *Population Redistribution and Public Policy.* Washington, DC: National Academy of Sciences, 1978.

Bigger, Jeanne C. "The Sunning of America: Migration to the Sunbelt." *Population Bulletin* 34 (3). Washington, DC: Population Reference Bureau, 1979.

Bouvier, Leon F. "Immigration and Its Impact on U.S. Population Size." *Population Bulletin* 36. Washington, DC: Population Reference Bureau, 1981.

———, and Robert W. Gardner. "Immigration to the U.S.: The Unfinished Story." *Population Bulletin* 41 (4). Washington, DC: Population Reference Bureau, 1986.

Brown, Lawrence A., and Victoria A. Lawson. "Migration in Third World Settings, Uneven Development, and Conventional Modeling: A Case Study of Costa Rica." *Annals of the Association of American Geographers* 75 (1985): 29–47.

Brown, Larry A., and R. L. Sanders. "Toward a Development Paradigm of Migration: With Particular Reference to Third World Settings." In *Migration Decision Making: Multidisciplinary Approaches to Micro-level Studies in Developed and Developing Countries,* edited by G. F. DeJong and R. W. Gardner. New York: Pergamon, 1981.

Burton, Ian; Robert W. Kates, and Gilbert F. White. *The Environment as Hazard.* New York: Oxford University Press, 1978.

Cadwallader, M. *Migration and Residential Mobility: Macro and Micro Approaches.* Madison: University of Wisconsin Press, 1992.

Champion, A. G., ed. *Counterurbanization: The Changing Pace and Nature of Population Deconcentration.* London: Edward Arnold, 1989.

Chant, Sylvia, ed. *Gender and Migration in Developing Countries.* London: Belhaven Press, 1992.

Clark, Gordon L. *Interregional Migration, National Policy and Social Justice.* Totowa, NJ: Rowman and Allanheld, 1983.

Clark, W. A. V. *Human Migration.* Beverly Hills, CA: Sage Publications, 1986.

———, and James E. Burt. "The Impact of Workplace on

Residential Relocation." *Annals of the Association of American Geographers* 70 (1980): 59–67.

Clark, W. A. V., and Eric G. Moore, eds. *Residential Mobility and Public Policy*. Beverly Hills, CA: Sage Publications, 1980.

Cohen, Robin, ed. *Theories of Migration*. Lyme, NH: Edward Elgar, 1996.

Durand, Jorge; William Kandel, and Douglas S. Massey. "International Migration and Development in Mexican Communities." *Demography* 33 (1996) 249–64.

du Toit, Brian M., and Helen I. Safa, eds. *Migration and Development*. The Hague, Netherlands: Mouton, 1975.

Findlay, A.M., F.L.N. Li, A.J. Jowett, and R. Skeldon. "Skilled International Migration and the Global City: A Study of Expatriates in Hong Kong." *Transactions of the Institute of British Geographers New Series* 21 (1996): 49–61.

Frey, William H. "Immigrants and Native Migrant Magnets." *American Demographics* 18 (1996): 36–41.

_____. "Immigration, Domestic Migration, and Demographic Balkanization in America: New Evidence for the 1990s." *Population and Development Review* 22 (1996): 741–64.

_____. "Migration and Metropolitan Decline in Developed Countries: A Comparative Study." *Population and Development Review* 14 (1988): 599–628.

Gober, Patricia. "Americans on the Move." *Population Bulletin* 48 (3). Washington, DC: Population Reference Bureau, 1993.

Grigg, D. B. "E.G. Ravenstein and the 'Laws of Migration'." *Journal of Historical Geography* 3 (1977): 41–54.

Hyndman, Jennifer. "Border Crossings." *Antipode* 29 (1997): 149–76.

Jackson, J. D., ed. *Migration*. London: Cambridge University Press, 1969.

Jensen, Leif. *The New Immigration: Implications for Poverty*. Westport, CT: Greenwood Press, 1989.

Jones, Huw; Nicholas Ford, James Caird, and William Berry. "Counter-urbanization in Societal Context: Long-Distance Migration to the Highlands and Islands of Scotland." *Professional Geographer* 36 (1984): 437–43.

Jones, Richard C. "Immigration Reform and Migrant Flows: Compositional and Spatial Changes in Mexican Migration after the Immigration Reform Act of 1968." *Annals of the Association of American Geographers* 85 (1995): 715–30.

_____. "Undocumented Migration from Mexico: Some Geographical Questions." *Annals of the Association of American Geographers* 72 (1982): 77–78.

Kidron, Michael, and Ronald Segal. *The State of the World Atlas*, 5th ed. New York: Penguin, 1995.

King, Russell. "The Geopolitics of International Migration in Europe: An Introduction." *Transactions of the Institute of British Geographers New Series* 21 (1996): 62–63.

Kontuly, Thomas, and Roland Vogelsang. "Explanations for the Intensification of Counterurbanization in the Federal Republic of Germany." *Professional Geographer* 40 (1988): 42–53.

Kosinski, Leszek A., and R. Mansell Prothero. *People on the Move*. London: Methuen, 1975.

Kritz, Mary M.; Charles B. Keely, and Silvano M. Tomasi. *Global Trends in Migration: Theory and Research on International Population Movements*. New York: Center for Migration Studies, 1981.

Lee, Everett. "A Theory of Migration." *Demography* 3 (1966): 47–57.

Liu, Xiao-Feng, and Glen Norcliffe. "Closed Windows, Open Doors: Geopolitics and Post-1949 Mainland Chinese Immigration to Canada." *The Canadian Geographer* 40 (1996): 306–18.

Martin, Philip, and Elizabeth Midgley. "Immigration to the United States: Journey to an Uncertain Destination." *Population Bulletin* 49 (2). Washington, DC: Population Reference Bureau, 1994.

Martin, Philip, and Jonas Widgren. "International Migration: A Global Challenge" *Population Bulletin* 51 (1). Washington, DC: Population Reference Bureau, 1996.

McHugh, Kevin E., and Robert C. Mings. "The Circle of Migration: Attachment to Place in Aging." *Annals of the Association of American Geographers* 86 (1996): 530–50.

McNeill, William, and Ruth S. Adams. *Human Migration: Patterns and Policies*. Bloomington: Indiana University Press, 1978.

Morrison, Peter A. *Population Movements: Their Form and Functions in Urbanization and Development*. Liège, Belgium: Ordina Editions for International Union for the Scientific Study of Population, 1983.

Nam, Charles B.; William J. Serow, and David F. Sly. *International Handbook on Internal Migration*. Westport, CT: Greenwood Press, 1990.

Newland, Kathleen. "International Migration: The Search for Work." *Worldwatch Paper* 33. Washington, DC: Worldwatch Institute, November 1979.

_____. "Refugees: The New International Politics of Displacement." *Worldwatch Paper* 43. Washington, DC: Worldwatch Institute, March 1981.

Ogeden, P.E. *Migration and Geographical Change*. Cambridge: Cambridge University Press, 1984.

Organisation for Economic Co-operation and Development. *Migration: The Demographic Aspects*. Paris: OECD, 1991.

Pandit, Kavita. "Cohort and Period Effects in U.S. Migration: How Demographic and Economic Cycles Influence the Migration Schedule." *Annals of the Association of American Geographers* 87 (1997): 439–50.

Papademetrion, Demetrios G. "International Migration in a Changing World." *International Social Science Journal* 36 (1984): 409–24.

Plane, David A. "Age-Composition Change and the Geographical Dynamics of Interregional Migration in the U.S." *Annals of the Association of American Geographers* 82 (1992): 64–85.

_____, and Peter A. Rogerson. "Tracking the Baby Boom, the Baby Bust, and the Echo Generations: How Age Composition Regulates U.S. Migration." *Professional Geographer* 43 (1991): 416–30.

Ravenstein, Ernest George. "The Laws of Migration." *Journal of the Royal Statistical Society* 48 (1885): 167–227.

Robinson, Vaughan, ed. *Geography and Migration*. Lyme, NH: Edward Elgar, 1996.

Rogerson, Peter A. "Changes in U.S. National Mobility Levels." *Professional Geographer* 39 (1987): 344–50.

Rogge, John R., ed. *Refugees: A Third World Dilemma*. Totowa, NJ: Rowman & Littlefield, 1987.

Roseman, Curtis C. *Changing Migration Patterns within the United States*. Washington, DC: Association of American Geographers, 1977.

_____. "Migration as a Spatial and Temporal Process." *Annals of the Association of American Geographers* 61 (1971): 589–98.

Sanders, Alvin J., and Larry Long. "New Sunbelt Migration Patterns." *American Demographics* 9 (1987): 38–41.

Simon, Rita J., and Caroline B. Brettell, eds. *International Migration: The Female Experience*. Totowa, NJ: Rowman & Allanheld, 1986.

Stephenson, George M. *A History of American Immigration*. New York: Russell and Russell, 1964.

Stillwell, J., and P. Congdon. *Migration Models: Macro and Micro Approaches*. London: Belhaven Press, 1991.

Svart, Larry M. "Environmental Preference Migration: A Review." *Geographical Review* 66 (1976): 314–30.

Tabbarah, Riad. "Prospects of International Migration." *International Social Science Journal* 36 (1984): 425–40.

Vernez, Georges, and David Ronfeldt. "The Current Situation in Mexican Immigration." *Science* 251 (1991): 1189–93.

Waldorf, B. "The Internal Dynamic of International Migration Systems." *Environment and Planning A* 28 (1996): 631–50.

White, Paul E., and Robert I. Woods, eds. *The Geographic Impact on Migration*. London and New York: Longman, 1980.

Williams, James D., and Andrew J. Sofranko. "Why People Move." *American Demographics* 3 (1981): 30–31.

Wolpert, Julian. "Behavioral Aspects of the Decision to Migrate." *Papers, Regional Science Association* 15 (1965): 159–69.

Zelinsky, Wilbur. "The Hypothesis of the Mobility Transition." *Geographical Review* 61 (1971): 219–49.

CHAPTER 4

Folk and Popular Culture

What did you do today? Presumably, your first activity was to get out of bed—for some of us the most difficult task of the day. Shortly thereafter, you got dressed. What did you wear? That depended on both the weather (shorts or sweater) and the day's activities (suit or T-shirt).

After work or school you returned home (ranch house, apartment, or dorm room). You then ate dinner (pizza or salad). After studying or finishing some work, you may now have some free time during the evening for leisure activities (watching television, listening to music, or playing or watching sports).

This narrative may not precisely describe you, but you can recognize the day of a "typical" North American. However, the routine described and the choices mentioned in parentheses do not accurately reflect the practices of many people elsewhere in the world. People living in other locations often have extremely different social customs. Geographers ask why such differences exist and how social customs are related to the cultural landscape.

As you watch television in your single-family dwelling, wearing jeans and munching on a pizza, consider the impact if people from rural Botswana or Papua New Guinea were suddenly placed in the room. Despite such striking differences in social customs across the landscape, you might be surprised to find that your visitors are familiar with some of your customs, as Earth becomes more and more a "global village." Your visitors might be attracted within a short period of time to change their customs.

Key Issues

1. Where do folk and popular cultures originate and diffuse?

2. Why is folk culture clustered?

3. Why is popular culture widely distributed?

4. Why does globalization of popular culture cause problems?

Loading Coca Cola in Zamboanga, Philippines. (David H. Wells/The Image Works)

Case Study

The Aboriginal Artists of Australia at Lincoln Center

The Aboriginal Artists of Australia, a group of aborigines living in the isolated Australian interior, visited New York a few years ago and danced at the Lincoln Center for the Performing Arts. Their series of dances, handed down from their ancestors, reflected their customs and local landscape.

The aboriginal dancers challenged their New York audience to understand the meaning of their movements and music. Aborigines consider such dances an essential social custom, reflecting their daily experiences and activities, such as the need for rain or the behavior of particular animals. At best, the New York audience could recognize that the dances were meaningful to the aborigines. But understanding was inevitably limited by the lack of a comparable role for dance in Western customs.

The geographic contrast between the aboriginal dancers and the New York theater audience was heightened by differing attitudes toward the physical environment. The aboriginal dancers respond to specific landscape features and environmental conditions in their Australian homeland. In contrast, New York's Lincoln Center is not a product of an isolated and unique set of social customs. Nothing at Lincoln Center is "indigenous to the unique conditions of the site"—not the arrangement of structures, the building materials, the variety of performances, nor the performers' places of origin. Lincoln Center reflects the diffusion of social customs across a large portion of Earth's surface. Lincoln Center exemplifies how regional differences in social and physical characteristics become less important in the distribution of cultural activities, through interaction and integration.

In Chapter 1, *culture* was shown to combine three things: values, material artifacts, and political institutions. Geographers are interested in all three components of the definition of culture—where these various elements of culture are found in the world and reasons why the observed distributions occur.

This chapter deals with the material artifacts of culture, the visible objects that a group possesses and leaves behind for the future. Chapters 5, 6, and 7 examine three important components of a group's beliefs and values, including language, religion, and ethnicity. Chapter 8 concludes the emphasis on the cultural elements of human geography by looking at the political institutions that maintain values and protect their artifacts.

Culture follows logically from the discussion of migration in Chapter 3. Two locations have similar cultural beliefs, objects, and institutions because people bring along their culture when they migrate. Differences emerge when two groups have limited interaction.

This chapter examines two facets of material culture. First is material culture deriving from survival activities of everyone's daily life—food, clothing, and shelter. Each cultural group provides these in its own way. Second is culture involving leisure activities—the arts and recreation. Each cultural group has its own definition of meaningful art and stimulating recreation.

Habit, custom, and culture: We must clearly distinguish among these three terms. A **habit** is a repetitive act that a particular *individual* performs, such as wearing jeans to class every day. A **custom** is a repetitive act of a *group*, performed to the extent that it becomes *characteristic* of the group—most American university students wear jeans to class every day. Unlike custom, habit does not imply that the act has been adopted by most of the society's population. A custom is therefore a habit that a group of people has widely adopted.

A collection of social customs produces a group's material culture—jeans typically represent American informality and a badge of youth. In this chapter, *custom* may be used to denote a specific element of material culture, such as wearing jeans, whereas *culture* refers to a group's entire collection of customs.

Material culture falls into two basic categories: folk and popular. **Folk culture** is traditionally practiced primarily by small, homogeneous groups living in isolated rural areas—such as wearing a sarong (a loose skirt made of a long strip of cloth wrapped around the body) in Malaysia or a sari (a long cloth draped so that one end forms a skirt and the other a head or shoulder covering) in India. **Popular culture** is found in large, heterogeneous societies that share certain habits despite differences in other personal characteristics—such as wearing jeans.

Geographers focus on two aspects of *where* folk and popular cultures are located. First is spatial distribution, for each cultural activity, like wearing jeans, has its own. Geographers study a particular social custom's origin, its diffusion, and integration with other social characteristics. Second, geographers study the relation between material culture and the physical environment. Each cultural group takes particular elements from the environment into its culture, and in turn constructs landscapes (what geographers call "built environments") that modify nature in distinctive ways.

Geographers observe that popular culture has a more widespread distribution than folk culture. The reason *why* the distributions are different is interaction, or lack of it. A group develops distinctive customs from experiencing local social and physical conditions, in isolation from other groups. Even groups living in proximity may generate a variety of folk customs in a limited geographic area, because of limited communication. Landscapes dominated by a collection of folk customs change relatively little over time.

In contrast, popular culture is based on rapid simultaneous global interaction through communications sytems, transportation networks, and other modern technology. Rapid diffusion facilitates frequent changes in popular customs. Thus, folk culture is more likely to vary from place to place at a given time, whereas popular culture is more likely to vary from time to time at a given place.

In Earth's *globalization*, popular culture is becoming more dominant, threatening the survival of unique folk cultures. These folk customs—along with language, religion, and ethnicity—provide a unique identity to each group of people who occupy a specific portion of Earth's surface. The disappearance of local folk customs reduces *local diversity* in the world and the intellectual stimulation that arises from differences in background.

The dominance of popular culture can also threaten the quality of the environment. Folk culture derived from local natural elements may be more sensitive to the protection and enhancement of their environment. Popular culture is less likely to reflect the diversity of local physical conditions and is more likely to modify the environment in accordance with global values.

Key Issue 1

Where Do Folk and Popular Cultures Originate and Diffuse?

- Origin of folk and popular cultures
- Diffusion of folk and popular cultures

Each social custom has a unique spatial distribution, but in general, distribution is more extensive for popular culture than for folk culture. Two basic factors help explain the spatial differences between popular and folk cultures: the process of origin and the pattern of diffusion.

Origin of Folk and Popular Cultures

A social custom originates at a hearth, a center of innovation. Folk customs often have anonymous hearths, originating from anonymous sources, at unknown dates, through unidentified originators. They may also have multiple hearths, originating independently in isolated locations.

In contrast to folk customs, popular culture is most often a product of the economically more developed countries, especially in North America, Western Europe, and Japan. Popular music and fast food are good examples. They arise from a combination of advances in industrial technology and increased leisure time. Industrial technology permits the uniform reproduction of objects in large quantities (CDs, stylish clothing, pizzas). Many of these objects help people enjoy leisure time, which has increased as a result of the widespread change in the labor force from predominantly agricultural work to predominantly jobs in service and manufacturing.

Origin of Folk Music

Music exemplifies the differences in the origins of folk and popular culture. According to a Chinese legend, music was invented in 2697 B.C., when the Emperor Huang Ti sent Ling Lun to cut bamboo poles that would produce a sound matching the call of the phoenix bird. But in reality, folk songs are usually composed anonymously and transmitted orally. A song may be modified from one generation to the next as conditions change, but the content is most often derived from events in daily life that are familiar to the majority of the people.

Folk songs tell a story or convey information about daily activities such as farming, life-cycle events (birth, death, and marriage), or mysterious events such as storms and earthquakes. In Vietnam, where most people are subsistence farmers, information about agricultural technology is conveyed through folk songs. For example, the following folk song provides advice about the difference between seeds planted in the summer and winter:

> *Ma chiêm ba tháng không già*
> *Ma mùa tháng ruỡi ất la `không non**

This song can be translated as follows:

> *While seedlings for the summer crop are not old when they are three months of age,*
>
> *Seedlings for the winter crop are certainly not young when they are one-and-a-half months old.*

The song hardly sounds lyrical to a Western ear. But when English-language folk songs appear in cold print, similar themes emerge, even if the specific information conveyed about the environment differs.

Folk customs may have multiple origins owing to non-communication among groups in different places. U.S. country music provides a recent example of the process by which folk customs originate independently at multiple hearths. Country music originated in the Upper South, a belt from western Virginia to western Texas. Within the Upper South, geographer George Carney identified four major hearths of country music during the late nineteenth and early twentieth centuries: southern Appalachia, central Tennessee and Kentucky, the Ozark and Ouachita uplands of western Arkansas and eastern Oklahoma, and north-central Texas (Figure 4–1). Carney documented these hearths on the basis of the birthplaces of performers and other individuals active in the field.

---

*From John Blacking and Joann W. Kealiinohomoku, eds., *The Performing Arts: Music and Dance* (The Hague: Mouton, 1979), 144. Reprinted by permission of the publisher.

Folk singing in Nepal. The song tells of a queen who must bid goodbye to family and friends. (Jon Burbank/The Image Works)

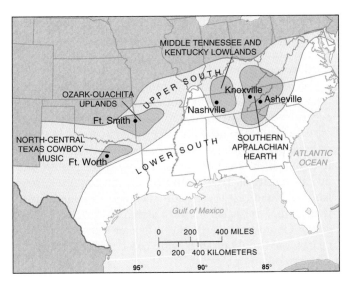

Figure 4–1 Origin of country music. Country music in the United States has four major hearths, or regions of origin. These include southern Appalachia, central Tennessee and Kentucky, the Ozark Plateau and Ouachita mountains of western Arkansas and eastern Oklahoma, and north-central Texas.

Origin of Popular Music

In contrast to folk music, popular music is written by specific individuals for the purpose of being sold to a large number of people. It displays a high degree of technical skill and is frequently capable of being performed only in a studio with electronic equipment.

Popular music as we know it today originated around 1900. At that time, the main popular musical entertainment in the United States and Western Europe was the variety show, called the *music hall* in the United Kingdom and *vaudeville* in the United States. To provide songs for music halls and vaudeville, a music industry developed in New York, along 28th Street between Fifth Avenue and Broadway, a district that became known as Tin Pan Alley (Figure 4–2). The name derived from the sound of pianos being furiously pounded by people called song pluggers, who were demonstrating tunes to publishers.

Tin Pan Alley was home to song writers, music publishers, orchestrators, and arrangers. Companies in Tin Pan Alley originally tried to sell as many printed song sheets as possible, although sales of recordings ultimately became the most important measure of success. The location of Tin Pan Alley later moved uptown to Broadway and 32nd Street, and then along Broadway between 42nd and 50th streets. After World War II, Tin Pan Alley disappeared, as recorded music became more important than printed song sheets.

The diffusion of American popular music worldwide began in earnest during World War II, when the Armed Forces Radio Network broadcast music to American soldiers and to citizens of countries where American forces were stationed or fighting. English became the international language for popular music. Today popular musicians in Japan, Poland, Russia, and other countries often

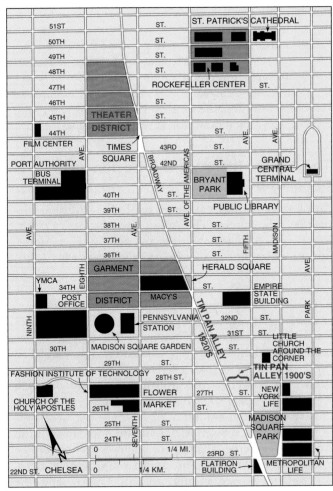

Figure 4–2 Tin Pan Alley. At the beginning of this century, most writers and publishers of popular music were clustered in New York City in a few buildings along 28th Street between Fifth Avenue and Broadway, which became known as Tin Pan Alley. Tin Pan Alley relocated north to 32nd Street and Broadway, and then along Broadway between 42nd and 50th streets. Tin Pan Alley is no longer a node of popular music, but performing arts are still clustered in New York City. The Theater District, near 45th Street and Broadway, contains the country's largest concentration of theaters featuring live plays and shows. Lincoln Center for the Performing Arts, a place for music and theater productions, is located near 63rd Street and Broadway.

write and perform in English, even though few people in their audiences understand the language.

Diffusion of Folk and Popular Cultures

The broadcasting of American popular music on Armed Forces radio illustrates the difference in diffusion of folk and popular cultures. The spread of popular culture typically follows the process of hierarchical diffusion from hearths or nodes of innovation. In the United States, prominent nodes of innovation for popular culture include Hollywood, California, for the film industry and Madison Avenue in New York City for advertising agencies. Popular culture diffuses rapidly and extensively through the use of modern communications and transportation.

In contrast, folk culture is transmitted from one location to another more slowly and on a smaller scale, primarily through migration rather than electronic communication. The spread of folk culture is an example of relocation diffusion, the spread of a characteristic through migration. We will now look at several examples.

The Amish: Relocation Diffusion of Folk Culture

Amish customs illustrate how relocation diffusion distributes folk culture. The Amish have distinctive clothing, farming, religious practices, and other customs. They leave a unique pattern on landscapes where they settle. Shunning mechanical and electrical power, the Amish still travel by horse and buggy and continue to use hand tools for farming.

Although the Amish population in the United States numbers only about 70,000, a mere 0.03 percent of the total population, Amish folk culture remains visible on the landscape in at least 17 states. The distribution of Amish folk culture across a major portion of the U.S. landscape is explained by examining the diffusion of their culture through migration.

In the 1600s, a Swiss Mennonite bishop names Jakob Ammann gathered a group of followers who became known as the Amish. The Amish originated in Bern, Switzerland; Alsace in northeastern France; and the Palatinate region of southwestern Germany. They migrated to other portions of northwestern Europe in the 1700s, primarily for religious freedom. In Europe, the Amish did not develop distinctive language, clothing, or farming practices and gradually merged with various Mennonite church groups.

Several hundred Amish families migrated to North America in two waves. The first group, primarily from Bern and the Palatinate, settled in Pennsylvania in the early 1700s, enticed by William Penn's offer of low-priced land. Because of lower land prices, the second group, from Alsace, settled in Ohio, Illinois, Iowa, and Ontario in the early 1800s. From these core areas, groups of Amish migrated to other locations where inexpensive land was available.

Living in rural and frontier settlements relatively isolated from other groups, Amish communities retained their traditional customs, even as other European immigrants to the United States adopted new ones. We can observe Amish customs on the landscape in such diverse areas as southeastern Pennsylvania, northeastern Ohio, and east-central Iowa (Figure 4–3). These communities are relatively isolated from each other but share cultural traditions distinct from those of other Americans.

Amish folk culture continues to diffuse slowly through interregional migration within the United States. In recent years, a number of Amish families have sold their farms in Lancaster County, Pennsylvania—the oldest and at one time largest Amish community in the United States—and migrated to Christian and Todd counties in southwestern Kentucky.

According to Amish tradition, every son is given a farm when he is an adult, but land suitable for farming is expensive and hard to find in Lancaster County because of its proximity to growing metropolitan areas. With the average price of farmland in southwestern Kentucky less than one-fifth that in Lancaster County, an Amish family can sell its farm in Pennsylvania and acquire enough land in Kentucky to provide adequate farmland for all their sons. Amish families are also migrating from Lancaster County to escape the influx of tourists from the nearby metropolitan areas, who come to gawk at the distinctive folk culture.

Sports: Hierarchical Diffusion of Popular Culture

In contrast with the diffusion of folk customs, organized sport provides an example of how popular culture is diffused. Many sports originated as isolated folk customs and were diffused like other folk culture, through the migration of individuals. The contemporary diffusion of organized sports, however, displays the characteristics of popular culture.

Folk Culture Origin of Soccer. Soccer is the world's most popular sport (it is called football outside North America). Its origin is obscure, although the earliest documented contest took place in England in the eleventh century. According to football historians, after the Danish invasion of England between 1018 and 1042, workers excavating a building site encountered a Danish soldier's head, which they began to kick. "Kick the Dane's head" was imitated by boys, one of whom got the idea of using an inflated cow bladder.

Early football games resembled mob scenes. A large number of people from two villages would gather to kick the ball. The winning side was the one that kicked the ball into the center of the rival village. In the twelfth century, the game—by then commonly called football—was confined to smaller vacant areas, and the rules became standardized. Because football disrupted village life, King Henry II banned the game from England in the late twelfth century. It was not legalized again until 1603 by King James I. At this point, football was an English folk custom, rather than a global popular custom.

Globalization of Soccer. The transformation of football from an English folk custom to global popular culture began in the 1800s. Football and other recreation clubs were founded in Britain, frequently by churches, to provide factory workers with organized recreation during leisure hours. Sport became a subject that was taught in school.

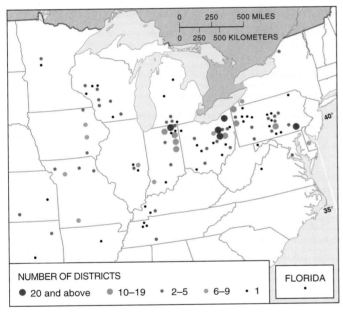

Figure 4–3 Amish settlements are distributed throughout the northeastern United States. According to William Crowley, who documented this distribution, the number of church districts within a settlement indicates the relative number of Amish in the community. In the photograph, an Amish father and son near Oakland, Maryland, are bringing hay in from the field. (Vanessa Vick/Photo Researchers, Inc.)

NUMBER OF DISTRICTS
● 20 and above ● 10–19 ● 2–5 ● 6–9 · 1

FLORIDA

Increasing leisure time permitted people not only to view sporting events but to participate in them. With higher incomes, spectators paid to see first-class events. To meet public demand, football clubs began to hire professional players. Several British football clubs formed an association in 1863 to standardize the rules and to organize professional leagues. Organization of the sport into a formal structure in Great Britain marks the transition of football from folk to popular culture.

The word "soccer" originated after 1863, when supporters of the game formed the Football Association. "Association" was shortened to "assoc," which ultimately became twisted around into the word "soc-cer." The terms "soccer" and "association football" also helped to distinguish the game from rugby football, which permits both kicking and carrying of the ball. Rugby originated in 1823, when a football player at Rugby College picked up the ball and ran with it.

Beginning in the late 1800s, the British exported association football around the world, first to continental Europe and then to other countries. Football was first played in continental Europe in the late 1870s by Dutch students who had been in Britain. The game was diffused to other countries through contact with English players. For example, football went to Spain via English engineers working in Bilbao in 1893 and was quickly adopted by local miners. British citizens further diffused the game throughout the worldwide British Empire. In the twentieth century, soccer, like other sports, has been further diffused by new communication systems, especially radio and television.

Soccer diffused to Russia in a curious manner. The English manager of a textile factory near Moscow organized a team at the factory in 1887 and advertised in London for workers who could play football. After the Russian Revolution in 1917, both the factory and its football team were absorbed into the Soviet Electric Trade Union. The team, renamed the Moscow Dynamo, became the country's most famous, although the official history of Soviet football never acknowledged its English origin.

Although soccer was also exported to the United States, it never gained the popularity it won in Europe and Latin America. The first college football game played in the United States, between Princeton and Rutgers in 1869, was really soccer, and officials of several colleges met 4 years later to adopt football rules consistent with those of British soccer. But Harvard's representatives successfully argued for adoption of rugby rules instead. Rugby was so thoroughly modified by U.S. colleges that an entirely new game—American football—emerged. Similar modifications of football were undertaken in other English-speaking countries, including Canada, Australia, and Ireland. This complex tale of diffusion is typical of many popular customs.

Other Sports. Each country has its own preferred sports (see Globalization and Local Diversity box for a discussion of lacrosse, a sport played in Canada and a few Eastern U.S. cities, especially Baltimore and New York). Cricket is popular primarily in Britain and former British colonies. Ice hockey prevails, logically, in colder climates, especially in Canada, Northern Europe, and Russia. The most popular sports in China are martial arts, known as *wushu*, including archery, fencing, wrestling, and boxing. Baseball, once confined to North America, became popular in Japan after it was introduced by American soldiers who occupied the country after World War II.

Despite the diversity in distribution of sports across Earth's surface and the anonymous origin of some games, organized spectator sports today are part of popular culture. The common element in professional sports is the willingness of people throughout the world to pay for the privilege of viewing, in person or on TV, events played by professional athletes. Competition for the World Cup in soccer is clear evidence of the global diffusion of sports. National soccer teams worldwide compete every 4 years, including in France in 1998. Thanks to television, more spectators view the final match than any other event in history.

Why Is Folk Culture Clustered?

- Isolation promotes cultural diversity
- Influence of physical environment

Folk culture typically has unknown or multiple origins among groups living in relative isolation. Folk culture diffuses slowly to other locations through the process of migration. A combination of physical and cultural factors influences the distinctive distributions of folk culture.

Isolation Promotes Cultural Diversity

A group's unique folk customs develop through centuries of relative isolation from customs practiced by other cultural groups. As a result, folk customs observed at a point in time vary widely from one place to another, even among nearby places.

Himalayan Art

In a study of artistic customs in the Himalaya Mountains, geographers P. Karan and Cotton Mather demonstrate that distinctive views of the physical environment emerge among neighboring cultural groups that are isolated. The study area, a narrow corridor of 2,500 kilometers (1,500 miles) in the Himalaya Mountains of Bhutan, Nepal, northern India, and southern Tibet (China), contains four religious groups: Tibetan Buddhists in the north, Hindus in the south, Muslims in the west, and Southeast Asian animists in the east (Figure 4–4). Despite their spatial proximity, limited interaction among these groups produces distinctive folk customs.

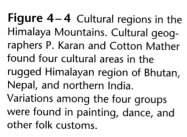

Figure 4–4 Cultural regions in the Himalaya Mountains. Cultural geographers P. Karan and Cotton Mather found four cultural areas in the rugged Himalayan region of Bhutan, Nepal, and northern India. Variations among the four groups were found in painting, dance, and other folk customs.

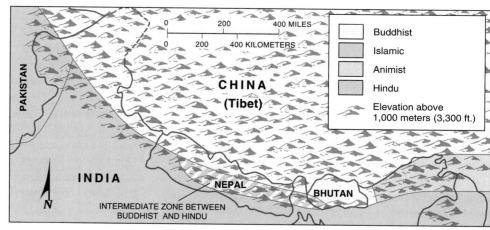

Globalization and Local Diversity

Revival of Lacrosse, a Folk Custom

The distribution of sports is a curious mixture of folk and popular cultures. Nothing seems more local than **where** particular sports are played and where specific teams are supported. At the same time, what could be a better example of **globalization** than such sports events as the Olympics and World Cup? The reason **why** sports events have become part of a global culture owes a lot to television, which can broadcast the events simultaneously to billions of people around the world.

Fans who migrate to new regions experience the local diversity of sports. Coverage of a favorite team or sport will be minimal at a new location. However, **local diversity** in sports can extend beyond the area of a team's coverage to become an important component of a group's cultural identity. One such example is the revival of lacrosse as part of the cultural identity of Native Americans in the United States and Canada.

Lacrosse has fostered cultural identity among the Iroquois Confederation of Six Nations (Cayugas, Mohawks, Oneidas, Onondagas, Senecas, and Tuscaroras) who live in the northeastern United States and southeastern Canada. As early as 1636, European explorers observed the Iroquois playing lacrosse, known in their language as "guhchigwaha," which means "bump hips." European colonists in Canada picked up the game from the Iroquois and diffused it to a handful of U.S. communities, especially in Maryland, upstate New York, and Long Island. The name "lacrosse" derived from the French words *la crosse*, for a bishop's crosier or staff, which has a similar shape to the lacrosse stick.

In recent years, the International Lacrosse Federation has invited the Iroquois nation to participate in the Lacrosse World Championships, along with teams from Australia, Canada, England, and the United States. Although the Iroquois have not won, they have had the satisfaction of hearing their national anthem played and seeing their flag fly alongside those of the other participants.

A final element of local diversity in sports: Many people care passionately about sports, whereas many others are indifferent or hostile to all sports.

An Iroquois defender (yellow shirt) tries to stop an Australian attacker in the 1992 Lacrosse World Championships (Porter Gifford/Gamma-Liaison, Inc.)

Subjects of paintings by each group reveal how their folk culture mirrors their religions and individual views of their environment:

- Tibetan Buddhists in the northern region paint idealized divine figures, such as monks and saints. Some of these figures are depicted as bizarre or terrifying, perhaps reflecting the inhospitable environment.
- Hindus in the southern region create scenes from everyday life and familiar local scenes. Their paintings sometimes portray a deity in a domestic scene and frequently represent the region's violent and extreme climatic conditions.
- Paintings in the Islamic western portion show the region's beautiful plants and flowers, because the Muslim faith prohibits displaying animate objects in art. In contrast with the paintings from the Buddhist and Hindu regions, Muslims do not depict harsh climatic conditions.
- Animist groups from Myanmar (Burma) and elsewhere in Southeast Asia, who have migrated to the eastern region of the study area, paint symbols and designs that derive from their religion rather than from the local environment.

The distribution of artistic subjects in the Himalayas shows how folk customs are influenced by cultural institutions like religion, and by environmental processes such as climate, landforms, and vegetation. These groups display similar uniqueness in their dance, music, architecture, and crafts.

Influence of the Physical Environment

Recall from Chapter 1 that environmental determinists theorized how processes in the environment cause social customs. This may sound reasonable on the surface, but most contemporary geographers reject the idea. Many examples exist of peoples who live in similar environments, but who adopt different social customs. Conversely, many examples exist of peoples who live under different environmental conditions, but who adopt similar social customs. Of course, people respond to their environment, but the environment is only one of several controls over social customs.

Customs such as provision of food, clothing, and shelter are clearly influenced by the prevailing climate, soil, and vegetation. For example, residents of arctic climates may wear fur-lined boots, which protect against the cold, and snowshoes to walk on soft, deep snow without sinking in. On the other hand, people living in warm and humid climates may not need any footwear if heavy rainfall and time spent in water discourage such use. The custom in the Netherlands of wearing wooden shoes may appear quaint, but it actually derives from environmental

conditions. Dutch farmers wear the wooden shoes, which are waterproof, as they work in fields that often are extremely wet because much of the Netherlands is below sea level.

Environmental conditions can limit the variety of human actions anywhere, but folk societies are particularly responsive to the environment because of their low level of technology and the prevailing agricultural economy. People living in folk cultures are likely to be farmers growing their own food, using hand tools and animal power.

Yet, folk culture may ignore the environment. Not all arctic residents wear snowshoes, nor do all people in wet temperate climates wear wooden shoes. Geographers observe that broad differences in folk culture arise in part from physical conditions, and that these conditions produce varied customs.

Two necessities of daily life—food and shelter—demonstrate the influence of cultural values and the environment on development of unique folk culture. Different folk societies prefer different foods and styles of house construction.

Distinctive Food Preferences

Folk food habits derive from the environment. According to nineteenth-century geographer Vidal de la Blache, "Among the connections that tie [people] to a certain environment, one of the most tenacious is food supply; clothing and weapons are more subject to modification than the dietary regime, which experience has shown to be best suited to human needs in a given climate."

Humans eat mostly plants and animals—living things that spring from the soil and water of a region. Inhabitants of a region must consider the soil, climate, terrain, vegetation, and other characteristics of the environment in deciding to produce particular foods. For example, rice demands a milder, moist climate, while wheat thrives in colder, drier regions.

People adapt their food preferences to conditions in the environment. A good example is soybeans, which are an excellent source of protein and are widely grown in Asia. In the raw state they are toxic and indigestible. Lengthy cooking renders them edible, but fuel is scarce in Asia. Asians have adapted to this environmental dilemma by deriving foods from soybeans that do not require extensive cooking. These include bean sprouts (germinated seeds), soy sauce (fermented soybeans), and bean curd (steamed soybeans).

In Europe, traditional preferences for quick-frying foods in Italy resulted in part from fuel shortages. In Northern Europe, an abundant wood supply encouraged the slow stewing and roasting of foods over fires, which also provided home heat in the colder climate.

Food Diversity in Transylvania. Food customs are inevitably affected by the availability of products, but people do not simply eat what is available in their particular environment. Food habits are strongly influenced by cultural traditions. What is eaten establishes one's social, religious, and ethnic memberships. The surest way to identify a family's ethnic origins is to look in its kitchen.

In Transylvania, currently part of Romania, food preferences distinguish among groups who long have lived in close proximity. A century ago, before killings and emigrations during the World War II era, Transylvania contained about 4 million Hungarians, 4 million Romanians, 500,000 to 600,000 Saxons, 50,000 to 75,000 Jews, 20,000 to 25,000 Armenians, and several thousand Szeklers. The Saxons and Szeklers were German peoples who migrated to Transylvania in the ninth century. The Hungarians conquered Transylvania in 1003 and ruled it with few interruptions until losing it to Romania after World War I. Most Jews came to the region with the Hungarians. Most of the Armenians migrated to Transylvania in the 1600s to escape the Muslim-controlled Ottoman Empire to the southeast.

Soup, the food consumed by poorer people, shows the distinctive traditions of the neighboring cultural groups in Transylvania. Romanians made sour bran soups from cracked wheat, corn, brown bread, and cherry tree twigs. Saxons instead simmered fatty pork in water, added sauerkraut or vinegar, and often used fruits. Jews preferred soups made from beets and sorrel (a leafy vegetable), rather than from meat. Armenians made soup based on churut (curdled milk) and ground vegetables. Hungarians added smoked bacon to the soup and thickened it with flour and onion fried in lard. Szeklers—who adopted many Jewish dietary practices, including avoidance of pork products—substituted smoked goose or other poultry for the bacon in the Hungarian recipes.

Distinctive food preferences among groups from Transylvania have continued, even after many migrated to the United States. Long after dress, manners, and speech have become indistinguishable from those of the majority, old food habits often continue as the last vestige of traditional folk customs.

Food Attractions and Taboos. According to many folk customs, everything in nature carries a signature, or distinctive characteristic, based on its appearance and natural properties. Consequently, people may desire or avoid certain foods in response to perceived beneficial or harmful natural traits.

Certain foods are eaten because their natural properties are perceived to enhance qualities considered desirable by the society, such as strength, fierceness, or lovemaking ability. The Abipone Indians of Paraguay eat jaguars and bulls to make them strong, brave, and swift. The mandrake, a plant native to Mediterranean climates, was thought to enhance an individual's lovemaking abilities. The smell of the plant's orange-colored berries is attractive, but the mandrake's association with sexual prowess comes primarily from the appearance of the root, which is thick, fleshy, and forked, suggesting a man's torso. In parts of Africa and the Middle East, the

mandrake's root is administered as a drug, and several references to its powers are found in the Bible.

People refuse to eat particular plants or animals that are thought to embody negative forces in the environment. Such a restriction on behavior imposed by social custom is a **taboo**. Other social customs, such as sexual practices, carry prohibitions, but taboos are especially strong in the area of food.

The Ainus in Japan avoid eating otters, because they are believed to be forgetful animals, and consuming them could cause loss of memory. Europeans blamed the potato, the first edible plant they had encountered that grew from tubers rather than seeds, for a variety of problems during the seventeenth and eighteenth centuries, including typhoid, tuberculosis, and famine. Initially, Europeans also resisted eating the potato because it resembled human deformities caused by leprosy.

Before becoming pregnant, Mbum Kpau women of Chad do not eat chicken or goat. Abstaining from consumption of these animals is thought to help escape pain in childbirth and to prevent birth of an abnormal child. During pregnancy, Mbum Kpau women avoid meat from antelopes with twisted horns, which could cause them to bear deformed offspring. In the Trobriand Islands off the eastern tip of Papua New Guinea, couples are prohibited from eating meals together before marriage, while premarital sexual relations are an accepted feature of social life.

Cassava (or manioc) roots are sources of starch, flour, and tapioca. This girl on Sumba Island, Indonesia, is pounding cassava. (Margie Politzer/Photo Researchers, Inc.)

Some folk cultures may establish food taboos because of concern for the natural environment. These taboos may help to protect endangered animals or to conserve scarce natural resources. For example, to preserve scarce animal species, only a few high-ranking people in some tropical regions are permitted to hunt, whereas the majority cultivate crops. However, most food-avoidance customs arise from cultural values.

Relatively well-known taboos against consumption of certain foods can be found in the Bible. The ancient Hebrews were prohibited from eating a wide variety of foods, including animals that do not chew their cud or that have cloven feet, and fish lacking fins or scales. These taboos arose partially from concern for the environment by the Hebrews, who lived as pastoral nomads in lands bordering the eastern Mediterranean. The pig, for example, is prohibited in part because it is more suited to sedentary farming than pastoral nomadism, and in part because its meat spoils relatively quickly in hot climates, such as the Mediterranean.

Similarly, Muslims embrace the taboo against pork, because pigs are unsuited for the dry lands of the Arabian Peninsula (Figure 4–5). Pigs would compete with humans for food and water, without offering compensating benefits, such as being able to pull a plow, carry loads, or provide milk and wool. Widespread raising of pigs would be an ecological disaster in Islam's hearth.

Hindu taboos against consuming cows can also be explained partly for environmental reasons. Cows are the source of oxen (castrated male bovine), the traditional choice for pulling plows as well as carts. A large supply of oxen must be maintained in India, because every field has to be plowed at approximately the same time, when the monsoon rains arrive. Religious sanctions have kept India's cow population large as a form of insurance against the loss of oxen and increasing population.

But the taboo against consumption of meat among many people, including Muslims, Hindus, and Jews, cannot be explained primarily by environment factors. Social values must influence the choice of diet, because people in similar climates and with similar levels of income consume different foods. The biblical food taboos were established in part to set the Hebrew people apart from others. That Christians ignore the biblical food injunctions reflects their desire to distinguish themselves from Jews. Furthermore, as a universalizing religion, Christianity was less tied to taboos that originated in the Middle East.

Food taboos are significant even in countries dominated by popular culture, such as the United States. Americans avoid eating insects, despite their nutritional value. In Colombia, on the other hand, movie theaters sell roasted leaf-cutter abdomens rather than popcorn, and giant water bugs are a delicacy in Thailand and Myanmar (Burma). Mixing insects with rice provides lysine, an amino acid that is often deficient in the diet of people in developing countries where rice is the staple food. The aversion of most Americans to eating insects is

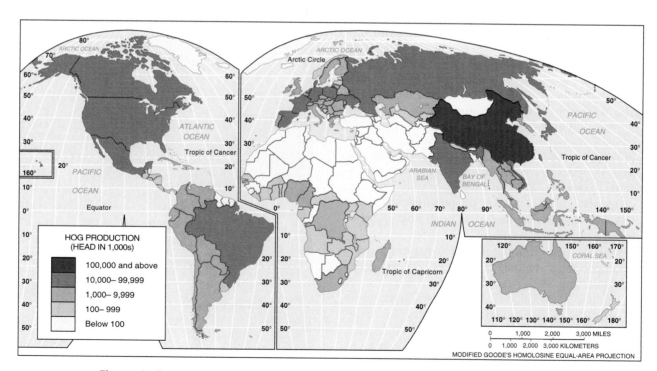

Figure 4–5 Annual hog production. The number of hogs produced in different parts of the world is influenced to a considerable extent by religious taboos against consuming pork. Hog production is virtually nonexistent in predominantly Muslim regions, such as northern Africa and southwestern Asia, while the level is high in predominantly Buddhist China and predominantly Christian countries.

amusing, because many foods, such as canned mushrooms and tomato paste, contain insects even though that fact is not commonly recognized.

Folk Housing

French geographer Jean Brunhes, a major contributor to the cultural landscape tradition, views the house as being among the essential facts of human geography. It is a product of both cultural tradition and natural conditions. American cultural geographer Fred Kniffen considered the house to be a good reflection of cultural heritage, current fashion, functional needs, and the impact of environment.

Distinctive Building Materials. The type of building materials used to construct folk houses is influenced partly by the resources available in the environment. The two most common building materials in the world are wood and brick, although stone, grass, sod, and skins are also used. If available, wood is generally preferred for house construction because it is easy to build with it. In the past, pioneers who settled in forested regions built log cabins for themselves.

Today, people in more developed societies buy lumber that has been cut by machine into the needed shapes. Cut lumber is used to erect a frame, and sheets or strips of wood are attached for the floors, ceilings, and roof. Shingles, stucco, vinyl, aluminum or other materials may be placed on the exterior for insulation or decoration.

Some societies have limited access to forests, and use alternative materials. In relatively hot, dry climates—such as the U.S. Southwest, Mexico, northern China, and parts of the Middle East—bricks are made by baking wet mud in the Sun. Stone is used to build houses in parts of Europe and South America and as decoration on the outside of brick or wood houses in other countries.

The choice of building materials is influenced both by social factors and by what is available from the environment. If the desired material is not locally available, then it must be imported. For example, migrants sometimes paved streets and built houses in their new location with the stone ballast placed in the hold of the ship that transported them. Building materials may be available but be more expensive than alternatives. For example, to save money (as well as trees) most new homes in the United States have interior walls made of drywall (filled with gypsum, a widely available mineral) rather than wood.

Distinctive House Form and Orientation. Social groups may share building materials, but the distinctive form of their houses may result from customary beliefs or environmental factors. In addition, the orientation of the houses on their plots of land can vary.

The form of houses in some societies might reflect religious values. For example, houses may have sacred walls or corners. The east wall of a house is considered sacred in Fiji, as is the northwest wall in parts of China.

Sacred walls or corners are also noted in parts of the Middle East, India, and Africa.

In Madagascar, religious considerations influence the use of each part of the house, and even furniture arrangement. The main door is on the west, considered the most important direction, while the northeast corner is the most sacred. The north wall is for honoring ancestors; in addition, important guests enter a room from the north and are seated against the north wall. The bed is placed against the east wall of the house, with the head facing north.

Beliefs govern the arrangement of household activities in a variety of Southeast Asian societies. In the south-central part of the island of Java, the front door always faces south, the direction of the South Sea Goddess, who holds the key to Earth.

Figure 4–6 (left) shows an interesting housing custom in northern Laos, where the Lao people arrange beds perpendicular to the center ridgepole of the house. Because the head is considered high and noble and the feet low and vulgar, people sleep so that their heads will be opposite their neighbor's heads and their feet opposite their neighbor's feet. The principal exception to this arrangement: A child who builds a house next door to the parents sleeps with his or her head toward the parents' feet as a sign of obeying the customary hierarchy.

Although they speak similar Southeast Asian languages and adhere to Buddhism, the Lao do not orient their houses in the same manner as the Yuan and Shan peoples in nearby northern Thailand (Figure 4–6, right). The Yuan and Shan ignore the position of neighbors and all sleep with their heads toward the east, which Buddhists consider the most auspicious direction. Staircases must not face west, the least auspicious direction, the direction of death and evil spirits.

Housing and Environment. The form of housing is related in part to environmental as well as social conditions. The construction of a pitched roof is important in wet or snowy climates to facilitate runoff and to reduce the weight of accumulated snow. Windows may face south in temperate climates to take advantage of the Sun's heat and light. In hot climates, on the other hand, window openings may be smaller to protect the interior from the full heat of the Sun.

Even in areas that share similar climates and available building materials, folk housing can vary owing to minor differences in environmental features. For example, R. W. McColl compared house types in four villages situated in the dry lands of northern and western China. All use similar building materials, including adobe and timber from the desert poplar tree, and they share a similar objective: protection from the extremes of temperatures, from very hot summer days to subfreezing winter nights.

Despite their similarities, the houses in these four Chinese villages have individual designs. Houses have second-floor open-air patios in Kashgar, small open

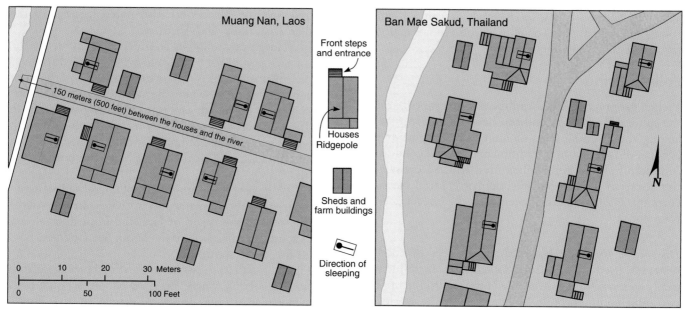

Figure 4–6 (left) Houses of Lao people in northern Laos. The fronts of Lao houses, such as those in the village of Muang Nan, Laos, face one another across a path, and the backs face each other at the rear. Their ridgepoles (the centerline of the roof) are set perpendicular to the path, but parallel to a stream if one is nearby. Inside adjacent houses, people sleep in the orientation shown, so neighbors are head-to-head or feet-to-feet. (right) Houses of Yuan and Shan peoples in northern Thailand. In the village of Ban Mae Sakud, Thailand, the houses are not set in a straight line because of a belief that evil spirits move in straight lines. Ridgepoles parallel the path, and the heads of all sleeping persons point eastward.

courtyards in Turpan, large private courtyards in Yinchuan, and sloped roofs in Dunhuang. McColl attributed the differences to local cultural preferences (Figure 4–7).

U.S. Folk House Forms

Older houses in the United States display local folk culture traditions. When families migrated westward in the 1700s and 1800s, they cut trees to clear fields for planting, and used the wood to build a house, barn, and fences. The style of pioneer homes reflected whatever upscale style was prevailing at the place on the East Coast from which they migrated. In contrast, houses built in the United States during the past half-century display popular culture influences.

Fred Kniffen identified three major hearths or nodes of folk house forms in the United States: New England, Middle Atlantic, and Lower Chesapeake. Migrants carried house types from New England northward to upper New England and westward across the southern Great Lakes region; from the Middle Atlantic westward across the Ohio Valley and southwestward along the Appalachian trails; and from the lower Chesapeake southward along the Atlantic Coast (Figure 4–8).

New England Houses. Four major house types were popular in New England at various times during the eighteenth and early nineteenth centuries. They are described here and shown in Figure 4–9, along with a map of how they diffused.

- *Saltbox.* During the early 1700s, the house preferred by successful New Englanders typically consisted of two full stories plus an attic, with four rooms per story (two in the front and two in the rear), organized around a massive central chimney. In some cases, a one-story addition was placed on the rear of the house, and the roof had a longer slope in the rear than in the front. This type of house became known as the saltbox, because it resembled an old-fashioned salt container.

- *Two Chimney.* After 1750, prosperous New Englanders slightly modified their idealized house type to give it a more formal, symmetrical appearance. A central hall was added, and the central chimney was replaced with two, set at opposite ends.

- *Cape Cod.* During the late 1700s and early 1800s, smaller houses became popular, especially among people who could not afford very large saltboxes. These smaller dwellings, later known as Cape Cod houses, had similar floor plans to the saltboxes, but had one full story rather than two.

- *Front Gable and Wing.* During the 1800s, New Englanders began to build houses turned 90 degrees so that the ridge and gable (the triangular area of the wall near the roof) faced the

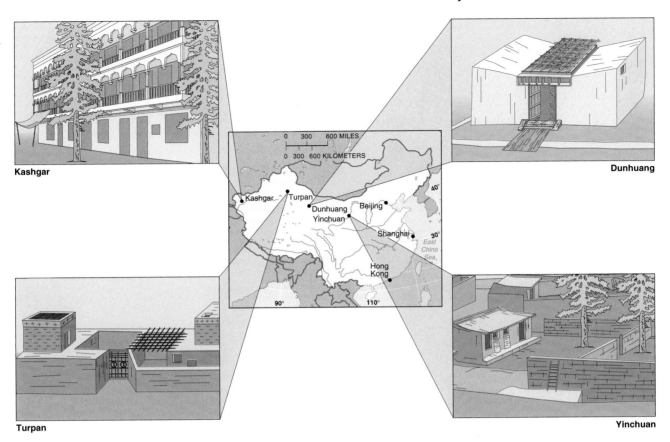

Kashgar

Dunhuang

Turpan

Yinchuan

Figure 4–7 House types in four communities of western China.

(upper left) Kashgar houses have second-floor open-air patios, where the residents can catch evening breezes. Poplar and fruit trees can be planted around the houses, because the village has a river that is constantly flowing rather than seasonal, as is the case in much of China's dry lands. These deciduous trees provide shade in the summer and openings for sunlight in the winter.

(lower left) Turpan houses have small, open courtyards for social gatherings. Turpan is situated in a deep valley with relatively little open land, because much of the space is allocated to drying raisins. Second-story patios, which would use even less land, are avoided, because the village is subject to strong winds.

(lower right) Yinchuan houses are built around large, open-air courtyards, which contain tall trees to provide shade. Most residents are Muslims, who regard courtyards as private spaces to be screened from outsiders. The adobe bricks are square or cubic rather than rectangular, as is the case in the other villages, though R. W. McColl found no reason for this distinctive custom.

(upper right) Dunhuang houses are characterized by walled central courtyards, covered by an open-lattice grape arbor. The cover allows for the free movement of air but provides shade from the especially intense direct summer heat and light. Rather than the flat roofs characteristic of dry lands, houses in Dunhuang have sloped roofs, typical of wetter climates, so that rainfall can run off. The practice is apparently influenced by Dunhuang's relative proximity to the population centers of eastern China, where sloped roofs predominate.

front and rear rather than the sides. One or two side wings were sometimes added, with their gable at 90 degrees to the main structure. This arrangement gave the houses a more classical look.

When settlers from New England migrated westward, they took their house type with them. The New England house type can be found throughout the Great Lakes region as far west as Wisconsin, because this area was settled primarily by migrants from New England. As the house preferred by New Englanders changed over time, the predominant form found on the landscape varies, based on the date of initial settlement.

Middle Atlantic "I"-houses. The major house type in the Middle Atlantic region was known as the "I"-house, typically two full stories in height, with gables to the sides. The "I"-house resembled the letter "I"—it was only one room deep and at least two rooms wide. The two rooms could be of equal or different size, and connected by a central hall, a direct interior door, or only outside doors. The house could contain one central stair-

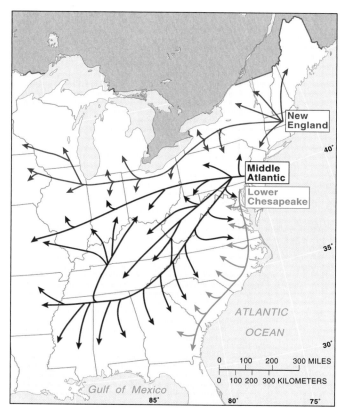

Figure 4–8 Source areas of U.S. house types. According to Fred Kniffen, house types in the United States originated in three main source areas and diffused westward along different paths. These paths coincided with predominant routes taken by migrants from the East Coast toward the interior of the country.

case or two, one at either end, affording more privacy to the two second-floor bedrooms. One fireplace could be set at the center of the house or one at each end. "I"-houses could be built from brick, stones, timber, or logs, depending on local availability of materials.

The "I"-house became the most extensive style of construction in much of the eastern half of the United States, especially in the Ohio Valley and Appalachia. Settlers built "I"-houses in much of the Midwest because most of them had migrated from the Middle Atlantic region.

Lower Chesapeake Houses. The Lower Chesapeake style of house typically comprised one story with a steep roof and chimneys at either end. These houses spread from the Chesapeake Bay–Tidewater, Virginia, area along the southeast coast. As was the case with the Middle Atlantic "I"-house, the form of housing that evolved along the southeast coast typically was only one room deep. In wet areas, houses in the coastal southeast were often raised on piers or a brick foundation.

Today, such distinctions are relatively difficult to observe in the United States. The style of housing does not display the same degree of regional distinctiveness because rapid communication and transportation systems provide people throughout the country with knowledge of alternative styles. Furthermore, most people do not

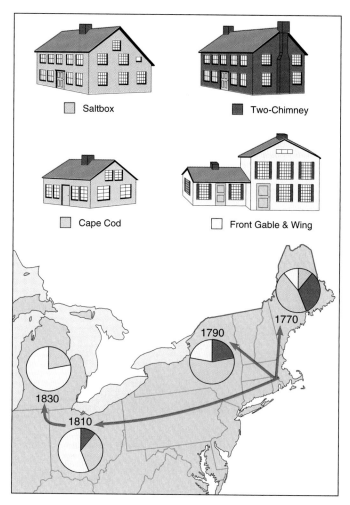

Figure 4–9 Diffusion of New England house types. Fred Kniffen suggests that these four major house types were popular in New England at various times during the eighteenth and early nineteenth centuries. As settlers migrated, they carried memories of familiar house types with them and built similar structures on the frontier. Thus, New Englanders were most likely to build houses like the yellow one when they began to migrate to upstate New York in the 1790s, because that was the predominant house type they knew. During the 1800s, when New Englanders began to migrate farther westward to Ohio and Michigan, they built the type of house typical in New England at that time, shown here in yellow.

build the houses in which they live. Instead, houses are usually mass-produced by construction companies.

Key Issue 3

Why Is Popular Culture Widely Distributed?

- Diffusion of popular housing, clothing, and food
- Role of television in diffusing popular culture

Popular culture varies more in time than in place. Like folk culture, it may originate in one location, within the context of a particular society and environment. But, in contrast to folk culture, it diffuses rapidly across Earth to

Exposure to modern technology does not necessarily change the traditional role of women in many societies. In Kyoto, Japan, a geisha girl, who is trained to provide entertainment for men, arranges appointments on her way to the restaurant where she entertains her male clients. (Paul Chesley)

locations with a variety of physical conditions. Rapid diffusion depends on a group of people having a sufficiently high level of economic development to acquire the material possessions associated with popular culture.

Diffusion of Popular Housing, Clothing, and Food

Some regional differences in food, clothing, and shelter persist in more developed countries, but differences are much less than in the past. Go to any recently built neighborhood on the outskirts of an American city from Portland, Maine, to Portland, Oregon: The houses look the same, the people wear jeans, and the same chain delivers pizza.

Popular Housing Styles

Housing built in the United States since the 1940s demonstrates how popular customs vary more in time than in place. In contrast with folk housing characteristic of the early 1800s, newer housing in the United States has been built to reflect rapidly changing fashion concerning the most suitable house form.

Houses show the influence of shapes, materials, detailing, and other features of architectural style in vogue at any one point in time. In the years immediately after World War II, which ended in 1945, most U.S. houses were built in a *modern style*. Since the 1960s, styles that architects call *neo-eclectic* have predominated (Figure 4–10).

Modern House Styles (1945–1960). Specific types of modern-style houses were popular at different times. In the late 1940s and early 1950s, the dominant type was known as *minimal traditional*, reminiscent of Tudor-style houses popular in the 1920s and 1930s. Minimal traditional houses were usually one-story, with a dominant front gable and few decorative details. They were small, modest houses designed to house young families and veterans returning from World War II.

The *ranch* house replaced minimal traditional as the dominant style of housing in the 1950s and into the 1960s. The ranch house was one-story, with the long side parallel to the street. With all the rooms on one level rather than two or three, the ranch house took up a larger lot and encouraged the sprawl of urban areas (see Chapter 13).

The *split-level* house was a popular variant of the ranch house between the 1950s and 1970s. The lower level of the typical split-level house contained the garage and the newly invented "family" room, where the television set was placed. The kitchen and formal living and dining rooms were placed on the intermediate level, with the bedrooms on the top level above the family room and garage.

The *contemporary* style was an especially popular choice between the 1950s and 1970s for architect-designed houses. These houses frequently had flat or low-pitched roofs. The *shed* style, popular in the late 1960s, was characterized by high-pitched shed roofs, giving the house the appearance of a series of geometric forms.

Neo-eclectic House Styles (Since 1960). In the late 1960s, *neo-eclectic* styles became popular, and by the 1970s had surpassed modern styles in vogue. The first popular neo-eclectic style was the *mansard* in the late 1960s and early 1970s. The shingle-covered second-story walls sloped slightly inward and merged into the roof line.

The *neo-Tudor* style, popular in the 1970s, was characterized by dominant, steep-pitched front-facing gables and half-timbered detailing. The *neo-French* style also appeared in the early 1970s, and by the early 1980s was the most fashionable style for new houses. It featured dormer windows, usually with rounded tops, and high-hipped roofs. The *neo-colonial* style, an adaptation of English colonial houses, has been continuously popular since the 1950s but never dominant. Inside many neo-eclectic houses, a large central "great room" has replaced

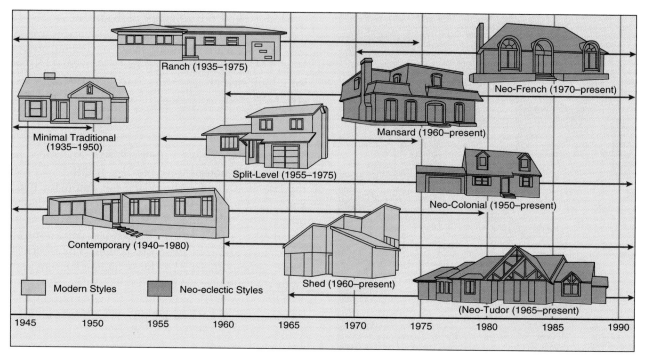

Ranch (1935–1975)

Neo-French (1970–present)

Minimal Traditional (1935–1950)

Mansard (1960–present)

Split-Level (1955–1975)

Neo-Colonial (1950–present)

Contemporary (1940–1980)

Modern Styles   Neo-eclectic Styles

Shed (1960–present)

(Neo-Tudor (1965–present)

1945 1950 1955 1960 1965 1970 1975 1980 1985 1990

Figure 4–10 U.S. house types 1945–1990. The dominant type of house construction in the United States was *minimal traditional* during the late 1940s and early 1950s, followed by *ranch* houses during the late 1950s and 1960s. The *split-level* was a popular variant of the ranch between the 1950s and 1970s, while the *contemporary* style was popular for architect-designed houses during the same period. The *shed* style was widely built in the late 1960s. *Neo-eclectic* styles, beginning with the *mansard*, were in vogue during the late 1960s. The *neo-Tudor* was popular in the 1970s and the *neo-French* in the 1980s. The *neocolonial* style has been widely built since the 1950s, but never dominated popular architecture.

separate family and living rooms, which were located in different wings or floors of ranch and split-level houses.

Regional differences in the predominant type of house do persist to some extent in the United States, as you can see in Figure 4–11. According to geographers Jakle, Bastian, and Meyer, small towns in the southeastern United States were more likely to contain ranch houses. In northeastern small towns, the most numerous form was the so-called double-pile, which was two rooms wide and two rooms deep. Northeastern houses were larger, more likely to be painted white, and have garages, whereas southeastern houses were smaller, more likely to be painted beige or brown, and have carports. Differences in roofs, porches, and building materials also distinguish northeastern and southeastern houses.

However, differences in housing among U.S. communities derive largely from differences in the time period in which the houses were built. The ranch house was more common in the Southeast than in the Northeast primarily because the Southeast grew much more rapidly during the 1950s and 1960s, the period when the ranch house was especially popular. A housing development built in one region will resemble more closely developments built at the same time elsewhere in the country than will developments built in the same region at other points in time.

Rapid Diffusion of Clothing Styles

Individual clothing habits reveal how popular culture can be distributed across the landscape with little regard for distinctive physical features. Such habits reflect availability of income, as well as social forms such as job characteristics.

In the more developed countries of North America and Western Europe, clothing habits generally reflect occupations rather than particular environments. A lawyer or business executive, for example, tends to wear a dark suit, light shirt or blouse, and necktie or scarf, whereas a factory worker wears jeans and a work shirt. A lawyer in California is more likely to dress like a lawyer in New York than like a steelworker in California.

A second influence on clothing in MDCs is higher income. Women's clothes, in particular, change in fashion from one year to the next. The color, shape, and design of dresses change to imitate pieces created by clothing designers. For social purposes, people with sufficient income may update their wardrobe frequently with the latest fashions.

Improved communications have permitted the rapid diffusion of clothing styles from one region of Earth to another. Original designs for women's dresses, created in Paris, Milan, London, or New York, are reproduced in

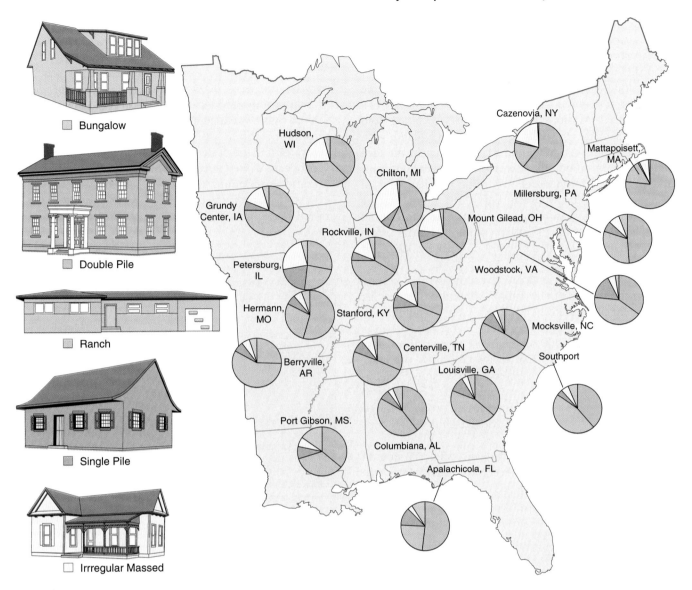

Figure 4–11 Regional differences in house types. Jakle, Bastian, and Meyer allocated the single-family housing in 20 small towns in the eastern United States into five groups: *bungalow*, *double-pile*, *irregularly massed*, *ranch*, and *single-pile*. Ranch houses were more common in the southeastern towns, while double-pile predominated in northeastern areas.

large quantities at factories in Asia and sold for relatively low prices at North American and European chain stores. Speed is essential in manufacturing copies of designer dresses because fashion tastes change quickly.

Until recently, a year could elapse from the time an original dress was displayed to the time that inexpensive reproductions were available in the stores. Now, the time lag is less than 6 weeks because of the diffusion of facsimile machines, computers, and satellites. Sketches, patterns, and specifications are sent instantly from European fashion centers to American corporate headquarters, and then on to Asian factories. Buyers from the major retail chains can view the fashions on large, high-definition televisions linked by satellite networks.

The globalization of clothing styles has involved increasing awareness by North Americans and Euro-

peans of the variety of folk costumes around the world. Increased travel and the diffusion of television have exposed people in MDCs to other forms of dress, just as people in other parts of the world have come into contact with Western dress. The poncho from South America, the dashiki of the Yoruba people of Nigeria, and the Aleut parka have been adopted by people elsewhere in the world. The continued use of folk costumes in some parts of the globe may persist not because of distinctive environmental conditions or traditional cultural values but to preserve past memories or to attract tourists.

Jeans. An important symbol of the diffusion of Western popular culture is jeans, which became a prized possession for young people throughout the world. In the late 1960s, jeans acquired an image of youthful independence in the United States, as young people adopted a style of clothing previously associated with low-status manual laborers and farmers.

Locally made denim trousers are available throughout Europe and Asia for under $10, but "genuine" jeans made by Levi Strauss, priced at $50 to $100, are preferred as a status symbol. Millions of second-hand Levis are sold each year in Asia, especially in Japan and Thailand, with most priced between $100 and $1,000. Even in the face of globalization of popular culture such as wearing jeans, some local variation persists: According to sellers of used jeans, Asians especially prefer Levi's 501 model with a button fly rather than a zipper. And within the United States, the button fly is more common on the West Coast, whereas Easterners prefer the zipper fly because it doesn't let in cold air.

Jeans became an obsession and a status symbol among youth in the former Soviet Union, when the Communist government prevented their import. Gangs would attack people to steal their American-made jeans, and authentic jeans would sell for $400 on the black market. Ironically, jeans were brought into the Soviet Union by the elite, including diplomats, bureaucrats, and business executives—essentially those who were permitted to travel to the West. These citizens obtained scarce products in the West and resold them inside the Soviet Union for a considerable profit.

The scarcity of high-quality jeans was just one of many consumer problems that were important motives in the dismantling of Communist governments in Eastern Europe around 1990. Eastern Europeans, who were aware of Western fashions and products—thanks to television—could not obtain them, because government-controlled industries were inefficient and geared to producing tanks rather than consumer-oriented goods (see Chapter 11).

With the end of communism, jeans are now imported freely into Russia. Levi Strauss opened a store in the center of Moscow that sells jeans for about $50, about a week's wage for a typical Russian. In an integrated global economy, prominent symbols of popular culture have diffused around the world. Access to these products is now limited primarily by lack of money rather than government regulation.

Popular Food Customs

Popular culture flourishes where people in a society have sufficient income to acquire the tangible elements of popular culture and the leisure time to make use of them. People in a country with a more developed economy are likely to have the income, time, and inclination to facilitate greater adoption of popular culture.

Alcohol and Fresh Produce. Consumption of large quantities of alcoholic beverages and snack foods are characteristic of the food customs of popular societies. Nonetheless, the amounts of alcohol and snacks con-

sumed, as well as preferences for particular types, vary by region within more developed countries, such as the United States.

Americans choose particular beverages or snacks in part on the basis of preference for what is produced, grown, or imported locally. Bourbon consumption in the United States is concentrated in the Upper South, where most of it is produced. Rum consumption is heavily concentrated on the East Coast, where it arrives from the Caribbean, while Canadian whiskey is preferred in communities contiguous to Canada (Figure 4–12). Southerners may prefer pork rinds because more hogs are raised there, and Northerners may prefer popcorn and

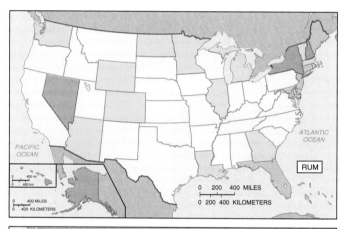

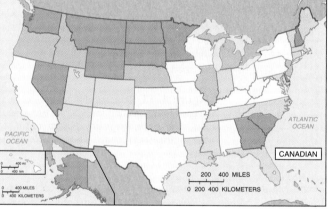

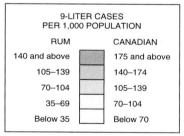

9-LITER CASES PER 1,000 POPULATION		
RUM		CANADIAN
140 and above		175 and above
105–139		140–174
70–104		105–139
35–69		70–104
Below 35		Below 70

Figure 4–12 Per capita consumption of rum (top) and Canadian whiskey (bottom). Rum has a higher incidence of consumption in East Coast states, where most of it is imported. States that have a high incidence of consumption of Canadian whiskey are located in the north, along the Canadian border. Preference for Canadian whiskey has apparently diffused southward from Canada into the United States.

potato chips because more corn and potatoes are grown there (Figure 4–13).

However, cultural backgrounds also affect the amount and types of alcohol and snack foods consumed. Alcohol consumption relates partially to religious backgrounds and partially to income and advertising. Baptists and Mormons, for example, drink less than do adherents of other denominations. Because Baptists are concentrated in the Southeast and Mormons in Utah, these regions have relatively low consumption rates. Nevada has a high rate because of the heavy concentration of gambling and other resort activities there. Texans may prefer tortilla chips because of the large number of Hispanic-Americans there, and Westerners may prefer multigrain chips because of greater concern for the nutrition content of snack foods.

Geographers cannot explain all the regional variations in food preferences. Why do urban residents prefer Scotch, and New Englanders consume nuts? Why is per capita consumption of snack food one–third higher in the Midwest than in the west? Why does consumption of gin and vodka show little spatial variation within the United States?

In general, though, consumption of alcohol and snack foods is part of popular culture primarily dependent on two factors—high income and national advertising. Variations within the United States are much less significant than differences between the United States and less developed countries of Africa and Asia.

Wine Production. The spatial distribution of wine production demonstrates that the environment plays a role in the distribution of popular as well as folk food customs. The distinctive character of a wine derives from a unique combination of soil, climate, and other physical characteristics at the place where the grapes are grown.

Vineyards are best cultivated in temperate climates of moderately cold, rainy winters and fairly long, hot summers. Hot, sunny weather is necessary in the summer for the fruit to mature properly, while winter is the preferred season for rain, because plant diseases that cause the fruit to rot are more active in hot, humid weather. Vineyards are planted on hillsides, if possible, to maximize exposure to sunlight and to facilitate drainage. A site near a lake or river is also desirable because water can temper extremes of temperature.

Grapes can be grown in a variety of soils, but the best wine tends to be produced from grapes grown in soil that is coarse-grained and well drained—a soil not necessarily very fertile for other crops. For example, the soil is generally sandy and gravelly in the Bordeaux wine region, chalky in Champagne country, and of a slate composition in the Moselle Valley. The distinctive character of each region's wine is especially influenced by the unique combination of trace elements, such as boron, manganese, and zinc, in the rock or soil. In large quantities, these elements could destroy the plants, but in small quantities they lend a unique taste to the grapes.

Because of the unique product created by the distinctive soil and climate characteristics, the world's finest wines are most frequently identified by their place of origin. Wines may be labeled with the region, town, district, or specific *estate*. A wine expert can determine the precise origin of a wine just by tasting because of the unique taste imparted to the grapes by the specific soil composition of each estate. (Similarly, a coffee expert can tell precisely where the beans were grown.)

The year of the harvest is also indicated on finer wines because specific weather conditions each year affect the quality and quantity of the harvest. Wines may also be

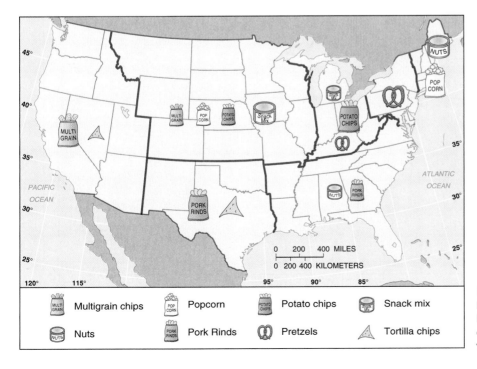

Figure 4–13 Preferences for snack foods. Preferences vary among U.S. regions. The large symbols are placed where per capita consumption is highest, the small symbols where consumption is second highest.

identified by the variety of grape used rather than the location of the vineyard. Less expensive wines might contain a blend of grapes from a variety of estates and years.

Although grapes can be grown in a wide variety of locations, wine distribution is based principally on cultural values, both historical and contemporary. Wine is made today primarily in locations that have a tradition of excellence in making it and people who like to drink it and can afford to purchase it.

The social custom of wine production in much of France and Italy extends back at least to the Roman Empire. Wine consumption declined after the Fall of Rome, and many vineyards were destroyed. Monasteries preserved the wine-making tradition in medieval Europe, for both sustenance and ritual. Wine consumption has become extremely popular again in Europe in recent centuries, as well as in the Western Hemisphere, which was colonized by Europeans. Vineyards are now typically owned by private individuals and corporations rather than religious organizations.

Wine production is discouraged in regions of the world dominated by religions other than Christianity (Figure 4–14). Hindus and Muslims in particular avoid alcoholic beverages. Thus, wine production is limited in the Middle East (other than Israel) and southern Asia primarily because of cultural values, especially religion. The distribution of wine production shows that the diffusion of popular customs depends less on the distinctive environment of a location than on the presence of beliefs, institutions, and material traits conducive to accepting those customs.

Role of Television in Diffusing Popular Culture

Watching television is an especially significant popular custom for two reasons. First, it is the most popular leisure activity in more developed countries throughout the world. Second, television is the most important mechanism by which knowledge of popular culture, such as professional sports, is rapidly diffused across Earth.

Diffusion of Television

Inventors in a number of countries, including the United States, the United Kingdom, France, Germany, Japan, and the Soviet Union, simultaneously contributed to the development of television. The U.S. public first saw television in the 1930s, but its diffusion was blocked for a number of years when broadcasting was curtailed or suspended entirely during World War II. In 1945, for example, there were only 10,000 television receivers in the United States. By 1949, the number rapidly increased to 1 million. It was 10 million in 1951, and 50 million in 1959. By the mid-1950s, three-fourths of all U.S. homes had TV receivers.

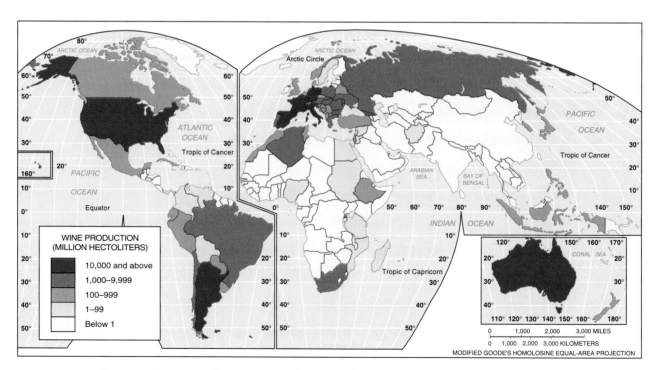

Figure 4–14 Annual wine production. The distribution of wine production is influenced in part by the physical environment, and in part by social customs. Most grapes used for wine are grown near the Mediterranean Sea or in areas of similar climate. Income, preferences, and other social customs also influence the distribution of wine consumption, as seen in the lower production levels of predominantly Muslim countries south of the Mediterranean.

Watching TV is an increasingly popular activity in less developed countries, although many people must share a television. This village in Niger lacks electricity, so the residents have gathered to watch a soccer match on a solar-powered set. (John Chaisson/Gamma/Liaison, Inc.)

During the early 1950s, television sets were being sold in only 20 countries, and more than 85 percent of the world's 37 million sets were in the United States. The United Kingdom had 9 percent of the world's television sets, the Soviet Union and Canada 2 percent each, and the rest of the world (primarily Cuba, Mexico, France, and Brazil) the remaining 2 percent. The United States still possessed more than half of the world's 98 million television receivers as late as the early 1960s, and the number of countries where sets were available increased to 62.

During the 1960s, the number of countries where people possessed television sets increased to 91, and the United States had one-third of the world's 229 million sets. By the early 1990s, more than 180 countries had 900 million television sets, with less than one-fourth in the United States (Figure 4–15).

Currently, the level of television service falls into four categories. The first category consists of countries where nearly every household owns a TV set. This category includes the more developed countries of North America and Europe as well as Australia, New Zealand, and Japan. A second category consists of countries in which ownership of a television is common but by no means universal. These are primarily Latin American countries and the poorer European states, such as Yugoslavia and Portugal.

The third category consists of countries in which television exists but has not yet been widely diffused to the population as a whole because of the high cost of receivers. This category includes some countries in Africa, Asia, and Latin America. Finally, about 30 countries, most of which are in Africa and Asia, have very few television sets. Some of these countries do not have operating TV stations, although programs from neighboring countries may be received.

Government Control of Television

In the United States, most television stations are owned by private corporations, which receive licenses from the government to operate at specific frequencies (channels). The company makes a profit by selling air time for advertisements. Some stations, however, are owned by local governments or other nonprofit organizations and are devoted to educational or noncommercial programs.

The U.S. pattern of private commercial stations is found in other Western Hemisphere countries but is rare elsewhere in the world. In most countries, the government either directly operates the stations or appoints an autonomous board of directors to manage them. Most governments control TV stations to minimize the likelihood that programs hostile to current policies will be broadcast—in other words, they are censored.

Operating costs are typically paid by the national government from tax revenues, although some government-controlled stations do sell air time to private advertisers. These advertisements are typically scheduled in extended blocks of time between programs, rather than in the midst of programs, as in the United States. The British Broadcasting Corporation (BBC) accepts no advertising and obtains revenue from the sale of licenses, required of all TV receiver owners; the fee is higher for color televisions than for black and white sets. A number of Western European countries have transferred some government-controlled television stations to private companies.

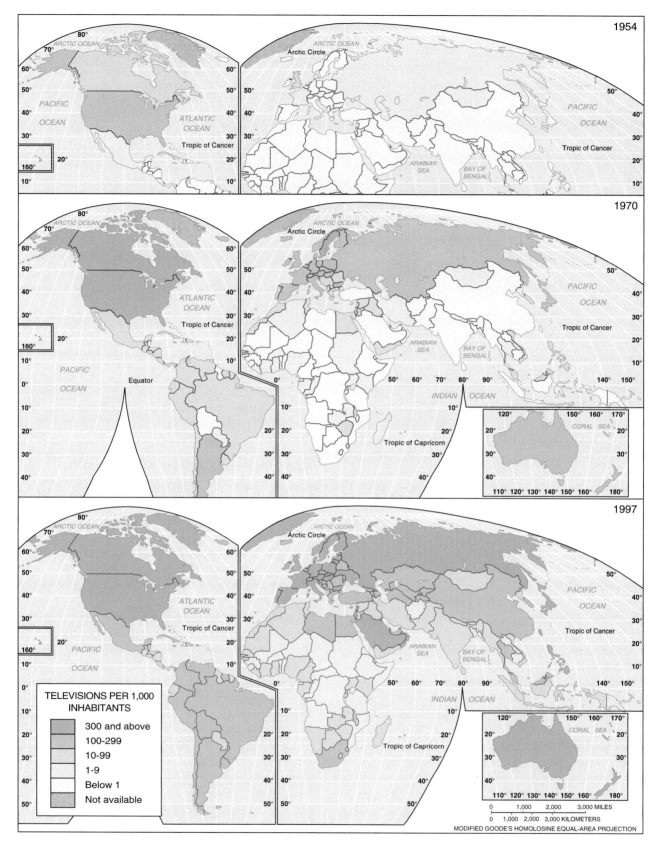

Figure 4–15 Televisions per 1,000 inhabitants, 1954, 1970, and 1997. Television has diffused from North America and Europe to other regions of the world, but the number of TV sets per capita still varies considerably among countries. The number of TV sets per capita is also an important indicator of a society's level of development, as shown in Chapter 9.

Reduced Government Control. In the past, many governments viewed television as an important tool for fostering cultural integration; television could extol the exploits of the leaders or the accomplishments of the political system. People turned on their TV sets and watched what the government wanted them to see. Because television signals weaken with distance, and are strong out to roughly 100 kilometers (60 miles), few people could receive television broadcasts from other countries. George Orwell's novel *1984*, published in 1949, anticipated that television—then in its infancy—would play a major role in the ability of a totalitarian government to control people's daily lives.

In recent years, changing technology—especially the diffusion of small satellite dishes—has made television a force for political change rather than stability. Satellite dishes enable people to choose from a wide variety of programs, produced in other countries, not just the local government-controlled station.

A number of governments in Asia have tried to prevent consumers from obtaining satellite dishes. The Chinese government banned private ownership of satellite dishes by its citizens, although foreigners and fancy hotels were allowed to keep them. The government of Singapore banned ownership of satellite dishes, yet it encourages satellite services, including MTV and HBO, to locate their Asian headquarters in the country. The government of Saudi Arabia ordered 150,000 satellite dishes dismantled, claiming that they were "un-Islamic."

Governments have had little success in shutting down satellite technology. Despite the threat of heavy fines, several hundred thousand Chinese still own satellite dishes. Consumers can outwit the government, because the small size of satellite dishes makes them easy to smuggle into the country and erect out of sight, perhaps behind a brick wall or under a canvas tarpaulin. A dish may be expensive by local standards—twice the annual salary of a typical Chinese, for example—but several neighbors can share the cost and hook up all of their TV sets to it.

The diffusion of small satellite dishes hastened the collapse of Communist governments in Eastern Europe during the late 1980s. For the first time, Eastern Europeans living beyond the signal range of Western broadcast stations could watch TV programs from Western Europe and North America. Eastern European countries have allocated some of their channels to such foreign broadcasters as CNN and MTV, because after many years under Communist control, citizens still do not trust the accuracy of locally produced television programs.

Satellite dishes represent only one assault on government control of the flow of information. Facsimile machines, portable video recorders, and cellular telephones have also put chinks in government censorship.

Why Does Globalization of Popular Culture Cause Problems?

- Threat to folk culture
- Environmental impact of popular culture

The international diffusion of popular culture has led to two problems, both of which can be understood from geographic perspectives. First, the diffusion of popular culture may threaten the survival of traditional folk culture in many countries. Second, popular culture may be less responsive to the diversity of local environments, and consequently may generate adverse environmental impacts.

Threat to Folk Culture

Many countries fear the loss of folk culture. They fear the disappearance of folk culture may be symbolic of the loss of traditional values in society. And they are concerned that the diffusion of popular culture from more developed countries can lead to dominance of Western perspectives.

Loss of Traditional Values

One example of the symbolic importance of folk culture is clothing. In African and Asian countries today, there is a contrast between the clothes of rural farm workers and of urban business and government leaders. Adoption of a more developed society's types of clothing is part of a process of imitation and replication of foreign symbols of success. Leaders of African and Asian countries have traveled to MDCs and experienced the sense of social status attached to clothes, such as men's business suits. Adoption of clothing customs from more developed countries has become a symbol of authority and leadership at home. The Western business suit has been accepted as the uniform for business executives and bureaucrats around the world.

Wearing clothes typical of MDCs is controversial in some Middle Eastern countries. Some political leaders in the region choose to wear Western business suits as a sign that they are trying to forge closer links with the United States and Western European countries. Fundamentalist Muslims oppose the widespread adoption of Western clothes, especially by women living in cities, as well as other social customs and attitudes typical of more developed countries. Women are urged to abandon skirts and blouses in favor of the traditional black *chador*, a combination head covering and veil.

In its 1997 presidential election, Iran was presented with a sharp contrast between Ali Akbar Nateq-Nouri, who favored banning Western popular culture not in accordance with strict Muslim practices, and a more

A diplomat from Niger, dressed in traditional clothing, meets two government officials in Senegal, dressed in Western-style business suits. (Owen Franken/Stock Boston)

moderate candidate, Mohammad Khatami, who favored more tolerance of Western cultural influences. Religious and military leaders supported Nateq-Nouri, but young people overwhelmingly supported Khatami. Said one 21-year-old woman, "I want Khatami to win because I want to continue wearing my blue jeans." Khatami won.

Change in Traditional Role of Women. The global diffusion of popular culture threatens the subservience of women to men that is embedded in many folk customs. Women were traditionally relegated to performing household chores, such as cooking and cleaning, and to bearing and raising large numbers of children. Those women who worked outside the home were likely to be obtaining food for the family, either through agricultural work or by trading handicrafts.

Advancement of women was limited by low levels of education and high rates of victimization from violence, often inflicted by husbands. The concepts of legal equality and availability of economic and social opportunities outside the home have become widely accepted in more developed countries, even where women in reality continue to suffer from discriminatory practices.

However, contact with popular culture also has brought negative impacts for women in less developed countries, such as an increase in prostitution. Hundreds of thousands of men from more developed countries, such as Japan and Northern Europe (especially Norway, Germany, and the Netherlands) purchase tours from travel agencies that include airfare, hotels, and the use of a predetermined number of women. The principal destinations of these "sex tours" include the Philippines, Thailand, South Korea, and to a lesser extent Indonesia and Sri Lanka. International prostitution is encouraged in these countries as a major source of foreign currency.

Through this form of global interaction, popular culture may regard women as essentially equal at home, but as objects that money can buy in foreign folk societies.

Threat of Foreign Media Imperialism

Less developed countries fear the incursion of popular culture for other reasons. Leaders of some LDCs consider the dominance of popular customs by MDCs as a threat to their independence. The threat is posed primarily by the media, especially news-gathering organizations and television.

Three MDCs—the United States, the United Kingdom, and Japan—dominate the television industry in LDCs. The Japanese operate primarily in South Asia and East Asia, selling their electronic equipment. British companies have invested directly in management and programming for television in Africa. U.S. corporations own or provide technical advice to many Latin American stations. These three countries are also the major exporters of programs. For example, only 6 percent of all television programs in Japan are foreign-made, compared to 83 percent in Uganda and 66 percent in Ecuador.

Leaders of many LDCs view the spread of television as a new method of economic and cultural imperialism on the part of the more developed countries, especially the United States. American television, like other media, presents characteristically American beliefs and social forms, such as upward social mobility, relative freedom for women, glorification of youth, and stylized violence. These attractive themes may conflict with and drive out traditional social customs.

For some Asian governments, MTV has become a metaphor for all that is bad about Western culture. To avoid offending Asian governments, the largest satellite broadcaster in Asia, Star TV (owned by Rupert Murdoch, who also owns the Fox television network), does not carry MTV in most markets. Star TV resumed showing MTV in India only after it agreed to allow the government-owned channel to censor unacceptable videos.

Satellite broadcasters, which are owned by companies based in developed countries, try to be sensitive to local cultural preferences. The Turner Broadcasting System, which operates an all-cartoon channel in Indonesia, does not show cartoons featuring Porky Pig, because most Indonesians are Muslim and avoid pork products. Entertainment programs emphasize family values and avoid controversial cultural, economic, and political issues.

Western Control of News Media. Less developed countries fear the effects of the news-gathering capability of the media even more than their entertainment function. The diffusion of information to newspapers around the world is dominated by the Associated Press (AP) and Reuters, which are owned by American and British companies, respectively.

The process of gathering news worldwide is expensive, and most newspapers and broadcasters are unable to afford their own correspondents. Instead, they buy the right to use the dispatches of one or more of the main news organizations. The AP transmits most news photographs and provides radio stations around the world with reports from its correspondents. Similarly, two joint British-American organizations, Visnews Ltd. and Worldwide Television News Corporation (WTN), supply most of the world's television news video.

The news media in most LDCs are dominated by the government, which typically runs the radio and TV service as well as the domestic news-gathering agency. Newspapers may be owned by the government, a political party, or a private individual, but in any event they are dependent on the government news-gathering organization for information. Sufficient funds are not available to establish a private news service.

Many African and Asian government officials criticize the Western concept of freedom of the press. They argue that the American news organizations reflect American values and do not provide a balanced, accurate view of other countries. U.S. news-gathering organizations are more interested in covering earthquakes, hurricanes, or other sensational disasters than more meaningful but less visual and dramatic domestic stories, such as birth-control programs, health care innovations, or construction of new roads.

Nevertheless, according to a study by the British International Institute of Communications, television newscasts throughout the world allocated the vast majority of time to domestic stories. On the same night, these were the first stories on the most widely watched nationwide newscasts:

- Brazil: traffic jam in Rio de Janeiro.
- India: the birthday of the assassinated former prime minister, Indira Gandhi.
- Japan: sumo wrestling results.
- Kuwait: the day's activities of the ruling sheik.
- Thailand: the increasing cost of eggs.

Veteran travelers and journalists invariably pack a portable shortwave radio when they visit other countries. In many regions of the world, the only reliable and unbiased news accounts come from the British Broadcasting Corporation (BBC) World Service shortwave radio newscasts. During the civil war in Iran in 1979, the television news anchor, who was not allowed to report the government's deteriorating position, urged citizens to

listen to the BBC World Service to learn the most accurate local news. During the 1990s, the BBC World Service again provided the most reliable information to local residents during the Persian Gulf War, the Somalian relief operation, and the Bosnian civil war.

Environmental Impact of Popular Culture

Popular culture is less likely than folk culture to be distributed with consideration for physical features. The spatial organization of popular culture reflects the distribution of social and economic features. In a global economy and culture, popular culture appears increasingly uniform.

Modifying Nature

Popular culture can significantly modify or control the environment. It may be imposed on the environment, rather than springing forth from it, as with many folk customs. For many popular customs, the environment is something to be modified to enhance participation in a leisure activity or to promote the sale of a product. Even if the resulting built environment looks "natural," it is actually the deliberate creation of people in pursuit of popular social customs.

Diffusion of Golf. Golf courses, because of their large size (80 hectares, or 200 acres), provide a prominent example of imposing popular culture on the environment. A surge in U.S. golf popularity has spawned construction of roughly 200 courses annually since the late 1980s. Geographer John Rooney attributes this to increased income and leisure time, especially among recently retired older people and younger people with flexible working hours.

According to Rooney, the provision of golf courses is not uniform across the United States. Although perceived as a warm-weather sport, the number of golf courses per person is actually greatest in north-central states, from Kansas to North Dakota, as well as the northeastern states abutting the Great Lakes, from Wisconsin to upstate New York (Figure 4–16). People in these regions have a long tradition of playing golf, and social clubs with golf courses are important institutions in the fabric of the regions' popular customs.

In contrast, access to golf courses is more limited in the South, in California, and in the heavily urbanized Middle Atlantic region between New York City and Washington, DC. Rapid population growth in the South and West and lack of land on which to build in the Middle Atlantic region have reduced the number of courses per capita. However, selected southern and western areas, such as coastal South Carolina, southern Florida, and central Arizona, have high concentrations of golf courses as a result of the arrival of large numbers of

Figure 4–16 The 50 best-served and worst-served metropolitan areas in terms of the number of golf holes per capita. In the north-central states, people have a long tradition of playing golf, even if it is confined to summer months. The ratio is less favorable for golfers in the large urban areas of the East Coast, as well as in the rapidly growing areas of the South and West.

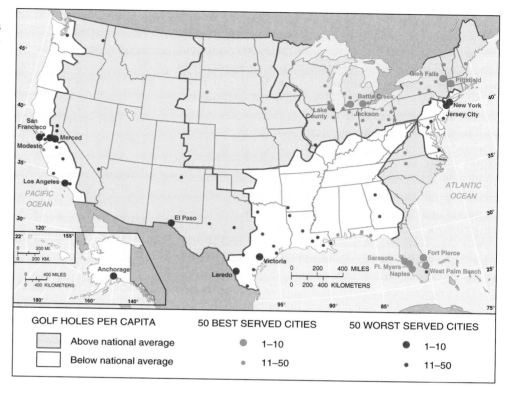

GOLF HOLES PER CAPITA	50 BEST SERVED CITIES	50 WORST SERVED CITIES
Above national average	● 1–10	● 1–10
Below national average	· 11–50	· 11–50

golf-playing northerners, either as vacationers or permanent residents.

Golf courses are designed partially in response to local physical conditions. Grass species are selected to thrive in the local climate and still be suitable for the needs of greens, fairways, and roughs. Existing trees and native vegetation are retained if possible (few fairways in Michigan are lined by palms). Yet, like other popular customs, golf courses remake the environment: creating or flattening hills, cutting grass or letting it grow tall, carting in or digging up sand for traps, and draining or expanding bodies of water to create hazards.

Uniform Landscapes

The distribution of popular culture around the world tends to produce more uniform landscapes. The spatial expression of a popular custom in one location will be similar to another. In fact, promoters of popular culture want a uniform appearance, to generate "product recognition" and greater consumption.

Fast-food Restaurants. The diffusion of fast-food restaurants is a good example of such uniformity. Such restaurants are usually organized as franchises. A franchise is a company's agreement with business people in a local area to market that company's product. The franchise agreement lets the local outlet use the company's name, symbols, trademarks, methods, and architectural styles. To both local residents and travelers, the buildings are immediately recognizable as part of a national or multinational company. A uniform sign is prominently displayed.

Tokyo McDonald's. U.S. fast-food chains have diffused to other countries, including Japan. Corporate logos enable customers to instantly identify the establishment regardless of whether they know the language. (Greg Davis/The Stock Market)

Much of the attraction of fast-food restaurants comes from the convenience of the product and the use of the building as a low-cost socializing location for teenagers or families with young children. At the same time, the success of fast-food restaurants depends on large-scale mobility: People who travel or move to another city immediately recognize a familiar place. Newcomers to a particular place know what to expect in the restaurant, because the establishment does not reflect strange and unfamiliar local customs that could be uncomfortable.

Fast-food restaurants originally developed to attract people who arrived by car. The buildings generally were brightly colored, even gaudy, to attract motorists. Recently built fast-food restaurants are more subdued, with brick facades, pseudo-antique fixtures, and other stylistic details. To facilitate reuse of the structure in case the restaurant fails, company signs often are free-standing, rather than integrated into the building design.

Uniformity in the appearance of the landscape is promoted by a wide variety of other popular structures in North America, such as gas stations, supermarkets, and motels. These structures are designed so that both local residents and visitors immediately recognize the purpose of the building, even if not the name of the company.

Global Diffusion of Uniform Landscapes. Physical expression of uniformity in popular culture has diffused from North America to other parts of the world. American motels and fast-food chains have opened in other countries. These establishments appeal to North American travelers, yet most customers are local residents who wish to sample American customs they have seen on television.

Diffusion of popular culture across Earth is not confined to products that originate in North America. With faster communications and transportation, customs from any place on Earth can rapidly diffuse elsewhere. Japanese automobiles and electronics, for example, have diffused in recent years to the rest of the world, including North America. Until the 1970s, automobiles produced in North America, Europe, and Japan differed substantially in appearance and size, but in recent years, styling has become more uniform, largely because of consumer preference around the world for Japanese cars. Automakers such as General Motors, Ford, Toyota, and Honda now manufacture similar models in North and South America, Europe, and Asia, instead of separately designed models for each continent.

Negative Environmental Impact

The diffusion of some popular customs can adversely impact environmental quality in two ways: depletion of scarce natural resources and pollution of the landscape.

Increased Demand for Natural Resources. Diffusion of some popular customs increases demand for raw materials, such as minerals and other substances found beneath Earth's surface. The depletion of resources used to produce energy, especially petroleum, is discussed in Chapter 14.

Popular culture may demand a large supply of certain animals, resulting in depletion, or even extinction, of some species. For example, some animals are killed for their skins, which can be shaped into fashionable clothing and sold to people living thousands of kilometers from the animals' habitat. The skins of the mink, lynx,

Route 66. When it connected Chicago and Los Angeles, Route 66 was once a well-known symbol of an especially prominent element of U.S. popular culture—the freedom to drive a car across the country's wide open spaces. Most of Route 66 has been replaced by interstate highways, and the remaining stretches are often cluttered by unattractive strip development, dominated by large signs for national motel, gasoline, and restaurant chains. (Bryan F. Peterson/The Stock Market)

jaguar, kangaroo, and whale have been heavily consumed for various articles of clothing, to the point that the survival of these species is endangered. This unbalances ecological systems of which the animals are members.

Folk culture may also encourage the use of animal skins, but the demand is usually smaller than for popular culture. Nonetheless, overhunting by Native Americans resulted in the near extinction of the buffalo.

Increased demand for some products can strain the capacity of the environment. An important example is increased meat consumption. This has not caused extinction of cattle and poultry; we simply raise more. But animal consumption is an inefficient way for people to acquire calories—90 percent less efficient than if people simply ate grain directly.

To produce 1 kilogram (2.2 pounds) of beef sold in the supermarket, nearly 10 kilograms (22 pounds) of grain are consumed by the animal. For every kilogram of chicken, nearly 3 kilograms (6.6 pounds) of grain are consumed by the fowl. This grain could be fed to people directly, bypassing the inefficient meat step. With a large percentage of the world's population undernourished, some question this inefficient use of grain to feed animals for eventual human consumption.

Pollution.   Popular culture also can pollute the environment. The environment can accept and assimilate some level of waste from human activities. But popular culture generates a high volume of waste—solids, liquids, and gases—that must be absorbed into the environment. Although waste is discharged in all three forms, the most visible is solid waste—cans, bottles, old cars, paper, and plastics. These products are often discarded rather than recycled. With more people adopting popular customs worldwide, this problem grows.

Folk culture, like popular culture, can also cause environmental damage, especially when natural processes are ignored. A widespread belief exists that indigenous peoples of the Western Hemisphere practiced more "natural," ecologically sensitive agriculture before the arrival of Columbus and other Europeans. Geographers increasingly question this. In reality, pre-Columbian folk customs included burning grasslands for planting and hunting, cutting extensive forests, and overhunting of some species. Very high rates of soil erosion have been documented in Central America from the practice of folk culture.

The more developed societies that produce endless supplies for popular culture have created the technological capacity both to create large-scale environmental damage and to control it. However, a commitment of time and money must be made to control the damage. Adverse environmental impact of popular culture is further examined in Chapter 14.

◼ Summary

Material culture can be divided into two types, folk and popular. Folk culture most often exists among small, homogeneous groups living in relative isolation at a low level of economic development. Popular culture is characteristic of societies with good communications and transportation, which enable rapid diffusion of uniform concepts. Geographers are concerned with several aspects of folk and popular culture.

Geographers study an array of thousands of social customs with distinctive spatial distributions. Groups display preferences in providing material needs such as food, clothing, and shelter, and in leisure activities such as performing arts and recreation. Examining where various social customs are practiced helps us to understand the extent of cultural diversity in the world.

Folk culture is especially interesting to geographers, because its distribution is relatively clustered, and its preservation can be seen as enhancing diversity in the world. Popular culture is important too, because it derives from the high levels of material wealth characteristic of societies that are economically developed. As societies seek to improve their economic level, they may abandon traditional folk culture and embrace popular culture associated with more developed countries.

Underlying the patterns of material culture are differences in the way people relate to their environment. Material culture contributes to the modification of the environment, and in turn nature influences the cultural values of an individual or a group.

Geographers, then, classify culture into popular and folk, based on differences in the ways the environment is modified and meaning is derived from environmental conditions.

Popular culture makes relatively extensive modifications of the environment, given society's greater technological means and inclination to do so. Here again are the key issues concerning folk and popular culture:

1. **Where do folk and popular cultures originate and diffuse?** Because of distinctive processes of origin and diffusion, folk culture has different distribution patterns than does popular culture. Folk culture is more likely to have an anonymous origin and to diffuse slowly through migration, whereas popular culture is more likely to be invented and diffused rapidly with the use of modern communications.

2. **Why is folk culture clustered?** Unique regions of folk culture arise because of lack of interaction among groups, even those living nearby. Folk culture is more likely to be influenced by the local environment.

3. **Why is popular culture widely distributed?** Popular culture diffuses rapidly across Earth, facilitated by modern communications, especially television. Differences in popular culture are more likely to be observed in one place at different points in time than among different places at one point in time.

4. **Why does globalization of popular culture cause problems?** Geographers observe two kinds of problems from diffusion of popular culture across the landscape. First, popular culture—generally originating in Western MDCs—may cause elimination of some folk culture. Second, popular culture may adversely affect the environment.

Case Study Revisited

The Aboriginal Artists Return to Australia

The Aboriginal Artists of Australia and their audience in New York's Lincoln Center highlight the contrast between folk culture—rooted in the uniqueness of an isolated landscape—and popular culture, which imposes uniform standards on the landscape. Will the aboriginal dancers maintain their traditions? Or will they be enticed by the consumer goods characteristic of popular customs, such as televisions and cars? What from the United States did they take back with them to Australia?

Many Aboriginals were not given the choice of maintaining their traditional folk customs or becoming part of popular culture. Between 1910 and 1970, the Australian government forcibly removed nearly 100,000 Aboriginal children from their families. Selected children usually had a white father or grandfather. Children with lighter skins were adopted by white families, while darker skinned ones were placed in orphanages. Because those with the darkest skins were not included in the program, mothers tried to hold on to their light-skinned babies by rubbing charcoal on them.

The Australian government removed Aboriginal youngsters from their homes in the belief that growing up in the country's dominant white society would be in the best interest of the children. Aboriginals would soon die out from a low fertility rate, and with them their folk culture, including the use of 400 languages.

The Aboriginal removal program has been terminated, but the number of Aboriginals in Australia is now less than one-half million, less than 3 percent of the national population. The folk culture of the remaining Aboriginals will now be preserved through groups such as the Australian Artists rather than obliterated. But the experience of the Aboriginals demonstrates how frail the preservation of folk culture can be in the face of popular culture.

Aborigine, near Kimberley, Australia, awaits *corroboree*, a nocturnal "celebration of important events." (Paul Chesley/Tony Stone)

◼ Key Terms

Custom The frequent repetition of an act, to the extent that it becomes characteristic of the group of people performing the act.

Folk culture Culture traditionally practiced by a small, homogeneous, rural group living in relative isolation from other groups.

Habit A repetitive act performed by a particular individual.

Popular culture Culture found in a large, heterogeneous society that shares certain habits despite differences in other personal characteristics.

Taboo A restriction on behavior imposed by social custom.

◼ Thinking Geographically

1. Should geographers regard culture and social customs as meaningful generalizations about a group of people, or should they concentrate instead on understanding how specific individuals interact with the physical environment? Why?

2. In what ways might gender affect the distribution of social customs in a community?

3. Are there examples of groups, either in more developed countries or in less developed countries, that have successfully resisted the diffusion of popular customs? Describe

such a group and tell how it has succeeded in preserving its culture.

4. Which elements of the physical environment are emphasized in the portrayal of various places on television?

5. Which images of social customs do countries depict in campaigns to promote tourism? To what extent do these images reflect local social customs realistically?

■ Further Readings

Adams, Robert L., and John F. Rooney. "American Golf Courses: A Regional Analysis of Supply." *Sport Place International* 3 (1989): 2–17.

Bale, John. *Sport and Place: A Geography of Sport in England, Scotland, and Wales.* Lincoln: University of Nebraska Press, 1983.

_____. *Sports Geography.* London: E. & F. N. Spon, 1989.

Ballas, Donald J., and Margaret J. King. "Cultural Geography and Popular Culture: Proposal for a Creative Merger." *Journal of Cultural Geography* 2 (1981): 154–63.

Bennett, Merril K. *The World's Foods.* New York: Harper and Bros., 1954.

Bigsby, C. W. E., ed. *Superculture: American Popular Culture and Europe.* Bowling Green, OH: Bowling Green Popular Press, 1975.

Blacking, John, and Joann W. Kealiinohomoku. *The Performing Arts: Music and Dance.* The Hague: Mouton, 1979.

Bourdier, Jean-Paul, and Nezar Alsayyad. *Dwellings, Settlements, and Tradition.* Lanham, MD: University Press of America, 1989.

Bull, Adrian. *The Economics of Travel and Tourism.* Melbourne, Australia: Pitman, 1991.

Carlson, Alvar W. "The Contributions of Cultural Geographers to the Study of Popular Culture." *Journal of Popular Culture* 11 (Spring 1978): 830–31.

Carney, George O. "Bluegrass Grows All Around: The Spatial Dimensions of a Country Music Style." *Journal of Geography* 73 (1974): 34–55.

_____, ed. *Fast Food, Stock Cars, and Rock-n-Roll: Space and Place in American Pop Culture.* Lanham, MD: Rowman & Littlefield, 1996.

_____. *The Sounds of People and Places: A Geography of American Folk and Popular Music.* Lanham, MD: Rowman & Littlefield, 1994.

Chakravarti, A. K. "Regional Preference for Foods: Some Aspects of Food Habit Patterns in India." *Canadian Geographer* 18 (Winter 1974): 395–410.

Chubb, Michael, and Holly R. Chubb. *One Third of Our Time?* New York: John Wiley, 1981.

Crowley, William K. "Old Order Amish Settlement: Diffusion and Growth." *Annals of the Association of American Geographers* 68 (June 1978): 249–65.

DeBlij, Harm J. *A Geography of Viticulture.* Miami: University of Miami Geographical Society, 1981.

Denevan, William E. "The Pristine Myth: The Landscape of the Americas in 1492." *Annals of the Association of American Geographers* 82 (1992): 367–85.

Farb, Peter, and George Armelagos. *Consuming Passions: The Anthropology of Eating.* Boston: Houghton Mifflin, 1980.

Fusch, Richard, and Larry Ford. "Architecture and the Geography of the American City." *Geographical Review* 73 (1983): 324–39.

Hart, John Fraser, and John T. Morgan. "Mobile Homes." *Journal of Cultural Geography* 15, no. 2 (Spring 1995): 35–54.

Jakle, John A. "Roadside Restaurants and Place-Product-Packaging." *Journal of Cultural Geography* 3 (1982): 76–93.

_____, Robert W. Bastian, and Douglas K. Meyer. *Common Houses in America's Small Towns.* Athens: The University of Georgia Press, 1989.

Jakle, John A., and Richard L. Mattson. "The Evolution of a Commercial Strip." *Journal of Cultural Geography* 1 (1981): 12–25.

Karan, Pradyumna P., and Cotton Mather. "Art and Geography: Patterns in the Himalayas." *Annals of the Association of American Geographers* 66 (1976): 487–515.

Kniffen, Fred B. "Folk-Housing: Key to Diffusion." *Annals of the Association of American Geographers* 55 (1965): 549–77.

Lamme, Ary J., III, ed. *North American Culture.* Vol. 1. Stillwater, OK: Society for the North American Cultural Survey, 1984.

Lewis, Peirce F.; Yi-Fu Tuan, and David Lowenthal. *Visual Blight in America.* Washington, DC: Association of American Geographers, 1973.

Leyshon, Andrew; David Matless, and George Revill. "The Place of Music." *Transactions of the Institute of British Geographers New Series* 20 (1995): 423–33.

Lomax, Alan. *The Folk Songs of North America.* Garden City, NY: Doubleday, 1960.

Lomax, John A., and Alan Lomax. *American Ballads and Folk Songs.* New York: Dover, 1994.

Lornell, Christopher, and W. Theodore Mealor, Jr. "Traditions and Research Opportunities in Folk Geography." *Professional Geographer* 35 (1983): 51–56.

McAlester, Virginia, and Lee McAlester. *A Field Guide to American Houses.* New York: Alfred A. Knopf, 1984.

McColl, Robert W. "By Their Dwellings Shall We Know Them: Home and Setting Among China's Inner Asian Ethnic Groups." *Focus* 39 (1989): 1–6.

Nash, Peter H., and George O. Carney. "The Seven Themes of Music Geography." *Canadian Geographer* 40, (1996): 69–74.

Pounds, Norman J.G. *Hearth and Home: A History of Material Culture.* Bloomington: Indiana University Press, 1989.

Rapoport, Amos. *House Form and Culture.* Englewood Cliffs, NJ: Prentice-Hall, 1969.

Rooney, John F., Jr. *A Geography of American Sport.* Reading, MA: Addison-Wesley, 1974.

_____, and Paul L. Butt. "Beer, Bourbon and Boone's Farm: A Geographical Examination of Alcoholic Drink in the United States." *Journal of Popular Culture* 11 (1978): 832–56.

Rooney, John F., Jr.; Wilbur Zelinsky, and Dean R. Louder, eds. *This Remarkable Continent: An Atlas of United States and Canadian Society and Culture.* College Station: Texas A & M University Press for the Society for the North American Cultural Survey, 1982.

Rose, Damaris; Susan Hanson, and Geraldine Pratt. "Gender, Work and Space." *Antipode* 28, (1996): 338–42.

Rowe, Peter G. *Making a Middle Landscape.* Cambridge, MA: MIT Press, 1991.

Rubin, Barbara. "A Chronology of Architecture in Los Angeles." *Annals of the Association of American Geographers* 69 (1979): 339–61.

Shortridge, Barbara G., and James R. Shortridge. "Consumption of Fresh Produce in the Metropolitan United States." *Geographical Review* 79 (1989): 79–98.

———. "Cultural Geography of American Foodways: An Annotated Bibliography." *Journal of Cultural Geography* 15, no. 2 (1995): 79–108.

Skaburskis, Andrejs. "Gender Differences in Housing Demand." *Urban Studies* 34, (1997): 275–320.

Szalai, Alexander, ed. *The Use of Time.* The Hague: Mouton, 1972.

Tunstall, Jeremy. *The Media Are American.* New York: Columbia University Press, 1977.

Van Doren, Carlton S.; George B. Priddle, and John E. Lewis. *Land and Leisure: Concepts and Methods in Outdoor Recreation,* 2d ed. Chicago: Maaroufa Press, 1979.

van Elteren, Mel. "Conceptualizing the Impact of US Popular Culture Globally." *Journal of Popular Culture* 30, (1996): 47–90.

Walker, Richard. "For Better or Worcester: Reflections on Gender, Work and Space." *Antipode* 28, (1996): 329–337.

Zelinsky, Wilbur. "North America's Vernacular Regions." *Annals of the Association of American Geographers* 70 (1980): 1–16.

Also consult the following journals: *International Folk Music Council Journal*; *Journal of American Culture*; *Journal of American Folklore*; *Journal of American Studies*; *Journal of Cultural Geography*; *Journal of Leisure Research*; *Journal of Popular Culture*; *Journal of Sport History*; *Landscape*; *Leisure Science*.

CHAPTER 5

Language

How many languages do you speak? If you are Dutch, you are likely to be able to speak at least four languages. All schoolchildren in the Netherlands are required to learn at least Dutch, English, German, and one other language, usually French or Russian.

For those of you who do not happen to be Dutch, the number is probably a bit lower. In fact, most people in the United States know only English. Only one-sixth of U.S. high school students are currently studying a foreign language. In contrast, five-sixths of European high school students learn a second language, and one-fourth learn at least two foreign languages.

Even in other English-speaking countries, foreign languages are studied more frequently than in the United States. For example, 60 percent of British and 70 percent of Irish high school students learn French.

Earth's heterogeneous collection of languages is one of its most obvious examples of cultural diversity. Estimates of distinct languages in the world range from 2,000 to 4,000. Twelve languages, including English, are spoken by at least 100 million people. Several of these are relatively familiar to North Americans (Spanish, Portuguese, Russian, German, and French), and several less familiar (Mandarin, Hindi, Bengali, Arabic, Japanese, and Malay-Indonesian).

Including the 12 large ones, altogether only about 100 languages are spoken by at least 5 million people, another 115 by between 1 million and 5 million people. The remaining several thousand languages are spoken by fewer than 1 million people. The distribution of some of these languages is easy for geographers to document, while others—especially in Africa and Asia—are difficult, if not impossible.

Key Issues

1. Where are different languages distributed?

2. Why do people in different locations speak similar languages?

3. Why is one language spoken differently in different locations?

4. Why do people preserve local languages?

Two women sitting on a bench and talking in Vilnius, Lithuania's Lukishkiv Square. (Jeff Greenbero/Peter Arnold, Inc.)

Case Study

French and Spanish in the United States and Canada

The Tremblay family lives in a suburb of Montréal, Québec. The parents and two young children speak French at home, work, school, and shops. The Lopez family—also two parents and two children—live in San Antonio, Texas, and speak Spanish in their household.

The Tremblay and Lopez families share a common condition: They live in countries with an English-speaking majority, but English is not their native language. The French-speaking inhabitants of Canada and the Spanish-speaking residents of the United States continue to speak their languages, although English dominates the political, economic, and cultural life of their countries. The two families use languages other than English because they believe that language is important in retaining and enhancing their cultural heritage. At the same time, both families recognize that knowledge of English is essential for career advancement and economic success.

French is one of Canada's two official languages, along with English. French speakers comprise one-fourth of the country's population, most of whom are clustered in Québec, where they comprise more than three-fourths of the province's speakers. Colonized by the French in the seventeenth century, Québec was captured by the British in 1763 and in 1867 became one of the provinces in the Confederation of Canada.

In the United States, Spanish has become an increasingly important language in recent years because of large-scale immigration from Latin America. In some communities, public notices, government documents, and advertisements are printed in Spanish. Several hundred Spanish-language newspapers and radio and television stations operate in the United States, especially in southern Florida, the Southwest, and large northern cities, where most of the 17 million Spanish-speaking people live.

These examples—French-speaking residents of Canada and Spanish-speaking residents of the United States—illustrate the *where* and *why* questions that concern geographers who study languages. Where are different languages spoken? English, French, Spanish, and other languages are spoken in distinct locations around the world, and geographers can document the distribution of this important element of cultural identity. Why in some cases are two different languages spoken in two locations, whereas in other cases the same language is spoken in two locations? The geography of language displays especially clearly this book's overall theme of interplay between forces of globalization and local diversity.

Language is a system of communication through speech, a collection of sounds that a group of people understands to have the same meaning. Many languages also have a **literary tradition**, or a system of written communication. However, hundreds of spoken languages lack a literary tradition. The lack of written record makes it difficult to document the distribution of many languages.

Language is part of culture, which, as shown in Chapter 1, has two main meanings—people's values and their tangible artifacts. This chapter and the next two discuss the three traits that best distinguish cultural values: language, religion, and ethnicity. Chapter 4 looked at the material objects of culture. We start our study of the geographic elements of cultural values with language in part because it is the means through which other cultural values, such as religion and ethnicity, are communicated.

Consistent with this book's where and why approach, this chapter first looks at *where* different languages are used, and how these languages can be logically grouped. The second and third key issues examine *why* languages have distinctive distributions. The study of language follows logically from migration, because the contemporary distribution of languages around the world results largely from past migrations of peoples.

Language is like luggage: People carry it with them when they move from place to place. They incorporate new words into their own language when they reach new places, and they contribute words brought with them to the existing language at the new location. Geographers look at the similarities among languages to understand the diffusion and interaction of people around the world.

The final section of the chapter discusses contradictory trends of *globalization* and local diversity in language. On the one hand, English has achieved an unprecedented global dominance, because people around the world are learning it to participate in a global economy and culture. On the other hand, people are trying to preserve *local diversity* in language, because language is one of the basic elements of cultural identity and a major feature of a region's uniqueness. Language is a source of pride to a people, a symbol of cultural unity. As a culture develops, language is both a cause of that development, and a consequence.

Key Issue 1

Where Are Different Languages Distributed?

- Language families
- Indo-European languages
- Other language families

This section describes where different languages are found around the world. Although several thousand languages are spoken, they can be organized logically into a small number of language families. The larger language families in turn can be further divided into language branches and language groups.

Language Families

A **language family** is a collection of languages related through a common ancestor that existed long before recorded history. Figure 5–1 shows the world's language families:

- About 50 percent of all people speak a language in the *Indo-European family*. English is one example.
- About 20 percent speak a language in the *Sino-Tibetan family*. Examples include the languages spoken by most Chinese people.
- About 5 percent each speak a language in one of these four families:
  - *Austronesian* (in Southeast Asia)
  - *Afro-Asiatic* (in the Middle East)
  - *Niger-Congo* (in Africa)
  - *Dravidian* (in India).
- The remaining 10 percent of the world's population speak languages belonging to a number of smaller families.

Within a language family, a **language branch** is a collection of languages related through a common ancestor that existed several thousand years ago. Differences are not as extensive or as old as with language families, and archaeological evidence can confirm that the branches derived from the same family. For example, the Indo-European family has eight branches—Germanic, Romance, Balto-Slavic, Indo-Iranian, Greek, Albanian, Armenian, and Celtic. English is part of the Germanic branch of the Indo-European family.

A **language group** is a collection of languages within a branch that share a common origin in the relatively recent past and display relatively few differences in grammar and vocabulary. English is a language in the West Germanic group of the Germanic branch of the Indo-European family. Although they sound very different, English and German are both languages in the West Germanic group because they are structurally similar and have many words in common.

Figure 5–2 attempts to depict differences among language families, branches, and groups. Language families form the trunks of the trees, while individual languages are displayed as leaves. The larger the trunks and leaves, the greater the number of speakers of those families and languages. Some trunks divide into several branches, which logically represent language branches. The branches representing Germanic, Balto-Slavic, and Indo-Iranian in Figure 5–2 divide a second time into language groups.

Figure 5–1 Language families. Most language can be classified into one of a handful of language families. The pie chart shows the percentage of people who speak a language from each major family. You can see that Indo-European and Sino-Tibetan languages dominate, with Indo-European spoken by about 50 percent of Earth's people, and Sino-Tibetan spoken by about 20 percent. The map colors show the distribution of each family. Note especially the worldwide span of Indo-European languages, but the relatively narrow diffusion of Sino-Tibetan tongues. Languages that have more than 100 million speakers are identified on the map.

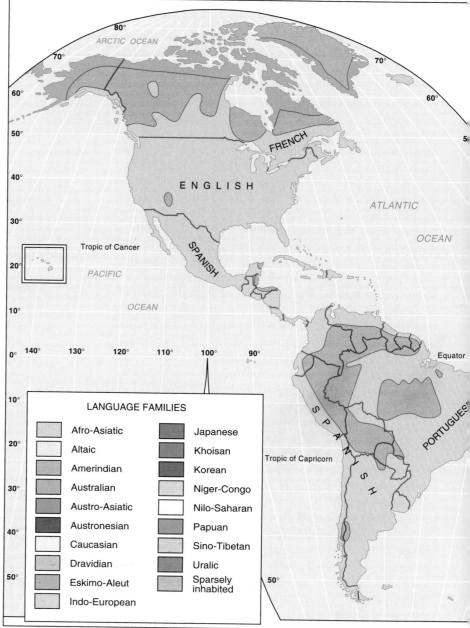

LANGUAGE FAMILIES

Afro-Asiatic	Japanese
Altaic	Khoisan
Amerindian	Korean
Australian	Niger-Congo
Austro-Asiatic	Nilo-Saharan
Austronesian	Papuan
Caucasian	Sino-Tibetan
Dravidian	Uralic
Eskimo-Aleut	Sparsely inhabited
Indo-European	

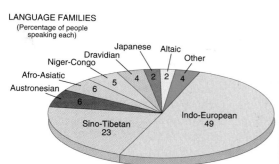

LANGUAGE FAMILIES
(Percentage of people speaking each)

Japanese · Altaic
Dravidian
Niger-Congo
Afro-Asiatic · Other
Austronesian
Sino-Tibetan 23
Indo-European 49
6 · 6 · 5 · 4 · 2 · 2 · 4

Figure 5–2 displays each language family as a separate tree at ground level, because differences among families predate recorded history. Linguists speculate that language families were joined together as a handful of superfamilies tens of thousands of years ago. Superfamilies are shown as roots below the surface, because their existence is highly controversial and speculative.

A **dialect** is a regional variety of a language distinguished by vocabulary, spelling, and pronunciation. Just as languages evolve from a common ancestor, so do several dialects derive from one language. British and American English are examples of different dialects of English. Generally, speakers of one dialect can understand speakers of another dialect of the same language. Figure 5–2 would become much too complex if dialects were included. The leaves would need to be drawn like oak leaves, with marginal teeth.

Countries designate at least one language as their **official language**, which is the one used by the government for laws, reports, and public objects, such as road signs, money, and stamps. A country with more than one official language may require all public documents to be in all languages. Logically, an official language would be understood by most if not all of the country's citizens, but some countries that were once British colonies designate English as an official language, even though few of their citizens can speak it.

Indo-European Languages

The world's most extensively spoken language family, by a wide margin, is Indo-European. More than 3 billion people can speak an Indo-European language, although an Indo-European language is the first language for

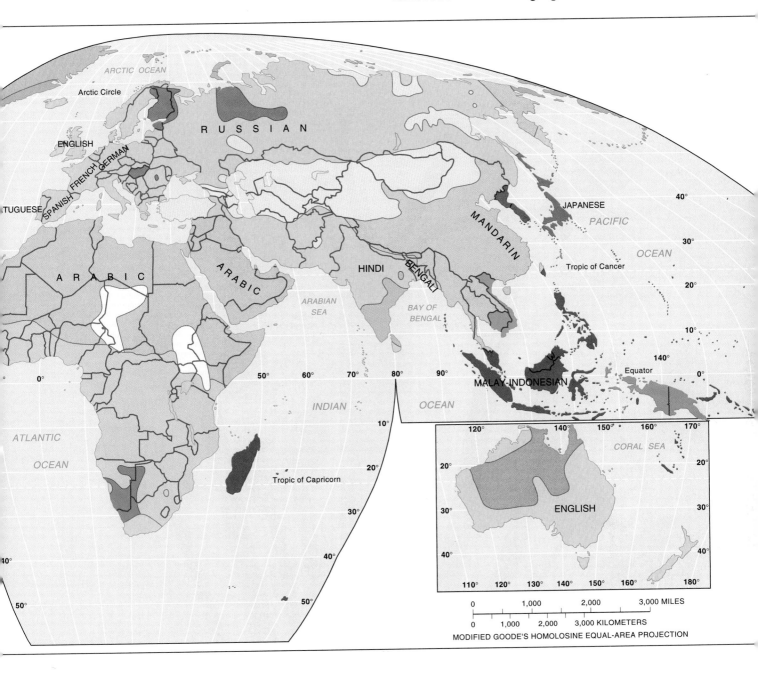

MODIFIED GOODE'S HOMOLOSINE EQUAL-AREA PROJECTION

2.5 billion people and the second language for the other 0.5 billion.

Four of Indo-Europeans's branches—Indo-Iranian, Romance, Germanic, and Balto-Slavic—are spoken by large numbers of people. Indo-Iranian languages are clustered in South Asia, Romance languages in southwestern Europe and Latin America, Germanic languages in northwestern Europe and North America, and Balto-Slavic languages in Eastern Europe. The four less extensively used Indo-European language branches are Albanian, Armenian, Greek, and Celtic.

The numbers of speakers of the Indo-European branches and major languages are included in Figure 5–2. Celtic is not shown in Figure 5–2 because it has fewer than 5 million speakers. Figure 5–3 shows the location of the eight branches in Europe and western Asia.

Germanic Language Branch

German may seem a difficult language for many English speakers to learn, but the two languages are actually closely related. Both belong to the Germanic language branch of Indo-European. The Germanic branch can be divided into the West Germanic and North Germanic groups (Figure 5–4).

West Germanic Group: English and German. West Germanic is the group within the Germanic branch of Indo-European to which English belongs. West Germanic is further divided into High Germanic and Low Germanic subgroups, so named because they are found in high and low elevations within present-day Germany. High German, spoken in the southern mountains of Germany, is the basis for the modern standard German language.

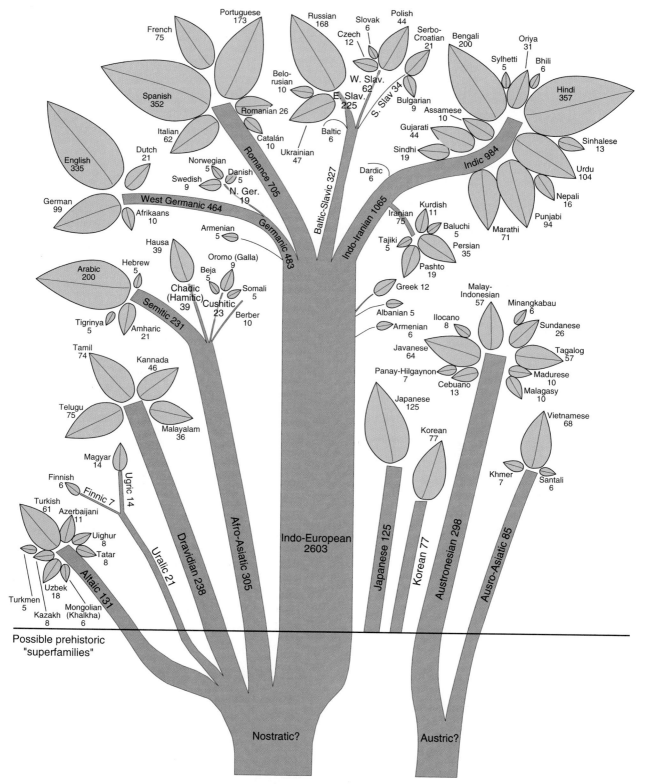

Figure 5–2 Language family tree. Language families are divided into branches and groups. Shown here are language families and individual languages that have more than 5 million speakers. Numbers on the tree are in millions of native speakers. Native speakers are people for whom the language is their first language. The totals exclude those who use the languages as second languages. Below ground level, the language tree's "roots" are shown. However, the theory that several language families had common origins tens of thousands of years ago is a highly controversial speculation advocated by some linguists and rejected by others.

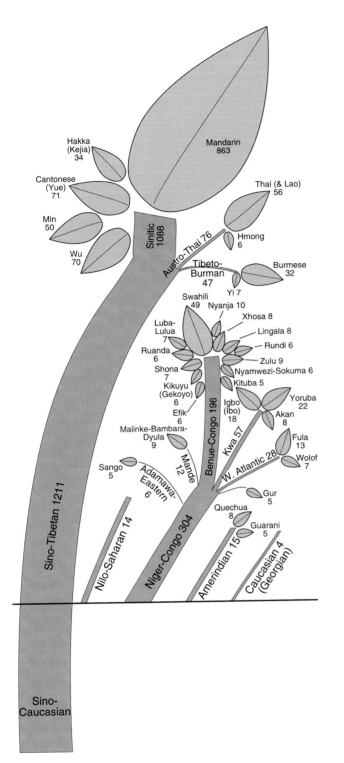

North Germanic Group: Scandinavian Languages.
The Germanic language branch also includes North Germanic languages, spoken in Scandinavia. The four Scandinavian languages—Swedish, Danish, Norwegian, and Icelandic—all derive from Old Norse, which was the principal language spoken throughout Scandinavia before A.D. 1000. Four distinct languages emerged after that time because of migration and the political organization of the region into four independent and isolated countries.

Romance Language Branch

The Romance language branch evolved from the Latin language spoken by the Romans 2,000 years ago, giving the branch its name, as discussed later in this chapter. The four most widely used contemporary Romance languages are Spanish, Portuguese, French, and Italian. Spanish and French are two of the six official languages of the United Nations.

The European regions in which these four languages are spoken correspond somewhat to the boundaries of the modern states of Spain, Portugal, France, and Italy (Figure 5–5). Rugged mountains serve as boundaries among these four countries. France is separated from Italy by the Alps and from Spain by the Pyrenees, while several mountain ranges mark the border between Spain and Portugal. Physical boundaries such as mountains are strong intervening obstacles, creating barriers to communication between people living on opposite sides (see Chapter 3).

The fifth most important Romance language, Romanian, is the principal language of Romania and Moldova. It is separated from the other Romance-speaking European countries by Slavic-speaking peoples.

Indo-Iranian Language Branch of Indo-European

The branch of the Indo-European language family with the most speakers is Indo-Iranian. This branch includes more than 100 individual languages, spoken by more than 1 billion people. The branch can be divided into an eastern group (Indic) and a western group (Iranian).

Indic (Eastern) Group of Indo-Iranian Language Branch. The most widely used languages in India, as well as in the neighboring countries of Pakistan and Bangladesh, belong to the Indo-European language family, and more specifically to the Indic group of the Indo-Iranian branch of Indo-European.

Approximately one-third of Indians, mostly in the north, use an Indic language called Hindi. Hindi is spoken many different ways—and therefore could be regarded as a collection of many individual languages—but there is only one official way to write the language, using

English is classified in the Low Germanic subgroup of the West Germanic group. Other Low Germanic languages include Dutch, which is spoken in the Netherlands, as well as Flemish, which is generally considered a dialect of Dutch spoken in northern Belgium. Afrikaans, a language of South Africa, is similar to Dutch, because Dutch settlers migrated to South Africa 300 years ago. Frisian is spoken by a few residents in northeastern Netherlands. A dialect of German spoken in the northern lowlands of Germany is also classified as Low Germanic.

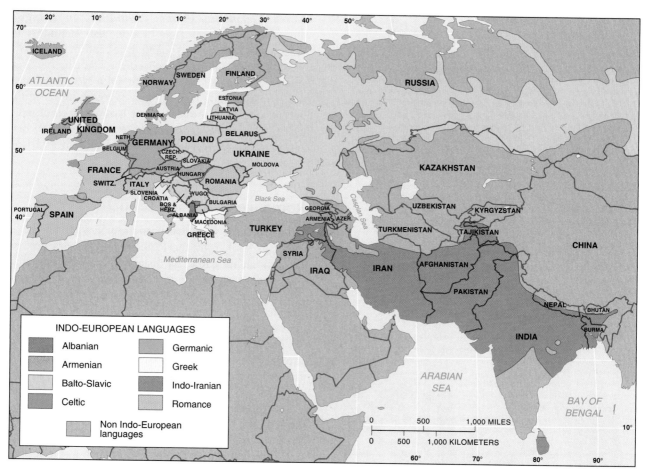

Figure 5–3 Branches of Indo-European language family. Most Europeans speak languages from the Indo-European language family. In Europe, the three most important branches are Germanic (north and west), Romance (south and west), and Slavic (east). The fourth major branch, Indo-Iranian, clustered in southern and western Asia, has over 1 billion speakers, the greatest number of any Indo-European branch.

a script called Devanagari, which has been used in India since the seventh century A.D. (For example, the word for "Sun" is written in Hindi as सूरज , pronounced *surag*.) Local differences arose in the spoken forms of Hindi but not in the written form, because until recently few speakers of that language could read or write it.

Pakistan's principal language, Urdu, is spoken very much like Hindi but is written with the Arabic alphabet, a legacy of the fact that most Pakistanis are Muslims, and their holiest book (the Quran) is written in Arabic. The basis of both languages is Hindustani, a form of the language used in communication among different groups of people in much of India for many centuries. Hindi, originally a variety of Hindustani spoken in the area of New Delhi, grew into a national language in the nineteenth century when the British encouraged its use in government. Collectively, Indic languages constitute the world's second largest language group.

One of the main elements of cultural diversity among the nearly 1 billion residents of India is language. Figure 5–6 shows that India has four important language families: Indo-European (the yellowish area dominating the north), Dravidian (south), Sino-Tibetan (northeast), and Austro-Asiatic (central and eastern highlands).

After India became an independent state in 1947, Hindi was proposed as the official language, but Dravidian speakers from southern India strongly objected. Therefore, India's constitution as amended recognizes 18 official languages, including 13 Indo-European (Assamese, Bengali, Gujarati, Hindi, Kashmiri, Konkani, Marathi, Nepali, Oriya, Punjabi, Sanskrit, Sindhi, and Urdu), four Dravidian languages (Kannada, Malayalam, Tamil, and Telugu) and one Sino–Tibetan language (Manipuri). More than 90 percent of the population speak at least one of these 14 languages, but as many as 10 million Indians use other languages. Bengali is the most important language in Bangladesh.

As the language of India's former colonial ruler, English has an "associate" status, even though only 1 percent of the Indian population can speak it. However,

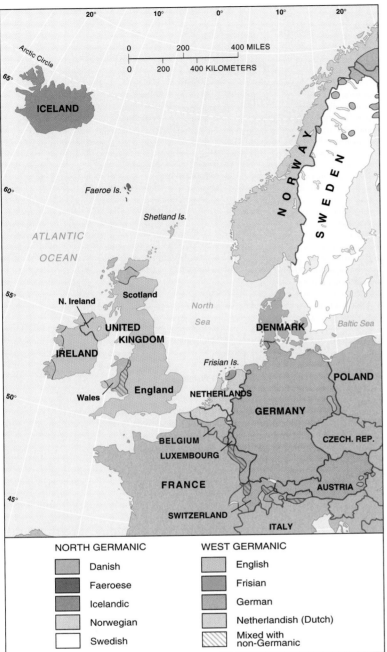

Figure 5–4 Germanic branch of Indo-European language family. Germanic languages predominate in Northern and Western Europe. The main North Germanic languages include Swedish, Danish, Norwegian, and Icelandic. The main West Germanic languages are English and German, with Dutch spoken in the Netherlands and northern Belgium. Two less widely used Germanic languages are Faeroese, spoken by inhabitants of the Faeroe Islands (part of Denmark), and Frisian, used in the northeastern Netherlands.

speakers of two different Indian languages who wish to communicate with each other sometimes are forced to turn to English as a common language.

Iranian (Western) Group of Indo-Iranian Language Branch. Indo-Iranian languages are also spoken in Iran and neighboring countries in southwestern Asia. These form a separate group from Indic within the Indo-Iranian branch of the Indo-European family.

The major Iranian group languages include Persian (sometimes called Farsi) in Iran, Pathan in eastern Afghanistan and western Pakistan, and Kurdish, used by the Kurds of western Iran, northern Iraq, and eastern

Turkey. These languages are written in the Arabic alphabet.

Balto-Slavic Language Branch of Indo-European

The other Indo-European language branch with large numbers of speakers is Balto-Slavic. Slavic was once a single language, but differences developed in the seventh century A.D. when several groups of Slavs migrated from Asia to different areas of Eastern Europe and thereafter lived in isolation from one other. As a result, this branch can be divided into East,

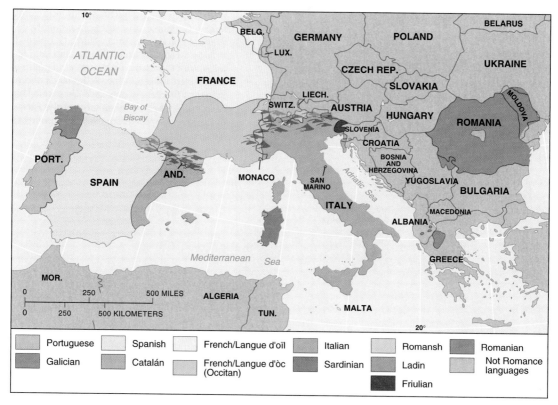

Figure 5–5 Romance branch of Indo-European language family. Romance includes three of the world's 12 most widely spoken languages (Spanish, Portuguese, and French) plus two other widely spoken tongues (Italian and Romanian). The map also shows boundaries among some dialects of Spanish and French. Catalán is a dialect of Spanish and the official language of Andorra. French dialects include Occitan (langue d'oïl) and langue d'òc. Rhaeto-Romanic languages include Romansh, Ladin, and Friulian.

West, and South Slavic groups as well as a Baltic group. Figure 7–20 shows the widespread area populated with Balto-Slavic speakers.

East Slavic and Baltic Groups of Balto-Slavic Language Branch. The most widely used Slavic languages are the eastern ones, primarily Russian, which is spoken by more than 80 percent of Russian people. Russian is one of the six official languages of the United Nations.

The importance of Russian increased with the Soviet Union's rise to power after World War II ended in 1945. Soviet officials forced native speakers of other languages to learn Russian as a way of fostering cultural unity among the country's diverse peoples. In Eastern European countries that were dominated politically and economically by the Soviet Union, Russian was taught as the second language. With the demise of the Soviet Union, the newly independent republics adopted official languages other than Russian, although Russian remains the language for communications among officials in the countries that were formerly part of the Soviet Union.

After Russian, Ukrainian and Belorusian (sometimes written Byelorussian) are the two most important East Slavic languages, official languages in Ukraine and Belarus. Ukraine is a Slavic word meaning "border," and Belo- is translated as "white." The presence of so many non-

Russian speakers was a measure of cultural diversity in the Soviet Union, and the desire to use languages other than Russian was a major drive in its breakup a decade ago.

The two principal Baltic languages are Latvian and Lithuanian, official languages of Latvia and Lithuania. Estonian, the official language of Estonia, is a Uralic language unrelated to the Indo-European family.

West and South Slavic Groups of Balto-Slavic Language Branch. The most-spoken West Slavic language is Polish, followed by Czech and Slovak. The latter two are quite similar, and speakers of one can understand the other. The government of the former state of Czechoslovakia tried to balance the use of the two languages, even though the country contained twice as many Czechs as Slovaks. For example, the announcers on televised sports events used one of the languages during the first half and switched to the other for the second half. These balancing measures were effective in promoting national unity during the Communist era, but in 1993, some four years after the fall of communism, Slovakia split from the Czech Republic. Slovaks rekindled their long-suppressed resentment of perceived dominance of the national culture by the Czech ethnic group.

The two most important South Slavic languages are Serbo-Croatian and Bulgarian. Although Serbs and Croats

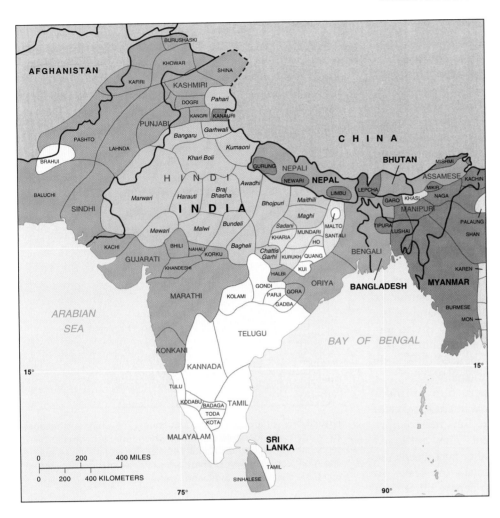

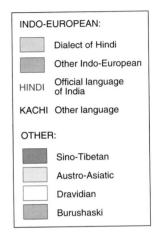

Figure 5–6 Languages and language families in South Asia. The region has four main language families: Indo-European, Dravidian, Sino-Tibetan, and Austro-Asiatic. More than 90 percent of the people of India speak at least one of the country's 14 official languages, written in red on the map.

speak the same language, they use different alphabets: Croatian is written in the Roman alphabet (what you are reading now), whereas Serbian is written in Cyrillic (for example, *Yugoslavia* written in Serbian is ЈУГОСЛАВИЈА). Slovene is the official language of Slovenia, while Macedonian is used in the former Yugoslav republic of Macedonia.

In general, differences among all Slavic languages are relatively small. A Czech, for example, can understand most of what is said or written in Slovak and could become fluent without much difficulty. However, because language is a major element in a people's cultural identity, relatively small differences among Slavic as well as other languages are being preserved and even accentuated in recent independence movements.

Since Bosnia and Croatia broke away from Serb-dominated Yugoslavia in the early 1990s, regional differences within Serbo-Croatian have increased. Bosnian Muslims have introduced Arabic words used in their religion, and Croats have replaced words regarded as having a Serbian origin with words considered to be purely Croatian. For example, the Serbo-Croatian word for martyr or hero—*junak*—has been changed to *heroj* by Croats and *shahid* by Bosnian Muslims. The term *Serbo-Croatian* now offends

Bosnians and Croatians, because it recalls when they were part of the same country as Serbs. In the future, after a generation of isolation and hostility among Bosnians, Croats, and Serbs, the languages spoken by the three groups may be sufficiently different to justify their classification as distinct languages.

Other Language Families

Half the people in the world speak an Indo-European language. The second largest family is Sino-Tibetan, spoken by nearly one-fourth of the world. Other major language families include Afro-Asiatic, Altaic, Austronesian, Japanese, and Niger-Congo. Refer to Figure 5–1 to see the distribution of these language families, and Figure 5–2 to see the number of people who speak each of them.

Sino-Tibetan Language Family

The Sino-Tibetan family encompasses languages spoken in the People's Republic of China—the world's most populous state at over 1 billion—as well as several smaller countries in Southeast Asia. The languages of China generally belong to the Sinitic branch of the Sino-

Tibetan family. Austro-Thai and Tibetan-Burman are two smaller branches of the family.

Sinitic Branch. There is no single Chinese language. Rather, the most important is Mandarin (or, as the Chinese call it, *pu tong hua*—common speech). Spoken by approximately three-fourths of the Chinese people, Mandarin is by a wide margin the most used language in the world. Once the language of emperors in Beijing, Mandarin is now the official language of both the People's Republic of China and Taiwan, as well as one of the six official languages of the United Nations.

Four other Sinitic branch languages are spoken by tens of millions of people in China, mostly in the southern and eastern parts of the country—Cantonese (also known as Yue), Wu, Min, and Hakka (also known as Kejia). However, the Chinese government is imposing Mandarin countrywide. The relatively small number of languages in China (compared to India, for example) is a source of national strength and unity. Unity is also fostered by a consistent written form for all Chinese languages. Although the words are pronounced differently in each language, they are written the same way.

You already know the general structure of Indo-European quite well, because you are a fluent speaker of at least one Indo-European language. But the structure of Chinese languages is quite different. They are based on 420 one-syllable words. This number far exceeds the possible one-syllable sounds that humans can make, so Chinese languages use each sound to denote more than one thing. The sound *shi*, for example, may mean "lion," "corpse," "house," "poetry," "ten," "swear," or "die." The sound *jian* has more than 20 meanings, including "to see." The listener must infer the meaning from the context in the sentence and the tone of voice the speaker uses.

In addition, two one-syllable words can be combined into two syllables, forming a new word. For example, the two-syllable word "Shanghai" is a combination of words that mean "above" and "sea." *Kan jian*—a combination of the words for "look" and "see," which would be redundant in English—clarifies that "to see" is the intended meaning for the multiple meanings of *jian*.

The other distinctive characteristic of the Chinese languages is the method of writing (Figure 5–7). The Chinese languages are written with a collection of thousands of characters. Some of the characters represent sounds pronounced in speaking, as in English. However, most are **ideograms**, which represent ideas or concepts, not specific pronunciations. The system is intricate and mature, having developed over 4,000 years.

The main language problem for the Chinese is the difficulty in learning to write, owing to the large number of characters. The Chinese government reports that 16 percent of the population over age 16 is unable to read or write more than a few characters.

Austro-Thai and Tibeto-Burman branches of Sino-Tibetan family. In addition to the Chinese languages

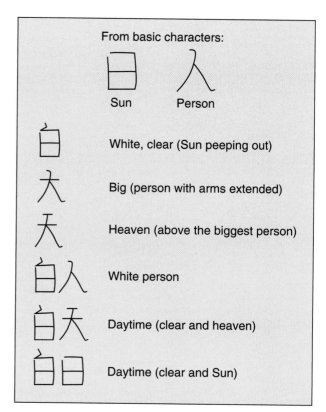

Figure 5–7 Chinese language ideograms. The Chinese languages are written with ideograms, most of which represent ideas or concepts rather than sounds. About 240 key characters may be built into more complex words. These are examples of words built from two basic characters—"Sun" and "person."

included in the Sinitic branch, the Sino-Tibetan family includes two smaller branches, Austro-Thai and Tibeto-Burman. The major language of the Austro-Thai branch is Thai, used in Laos, Thailand, and parts of Vietnam. Lao, a dialect of Thai used in Laos, is classified by some linguists as a separate language. Burmese, the principal language of the Tibeto-Burman branch, is used in Myanmar (Burma).

Other East and Southeast Asian Language Families

To some Western observers, the written languages of the large East Asian population concentrations may be difficult to distinguish because they are written with such unfamiliar characters, and their sound has a general similarity. However, Japanese and Korean both form distinctive language families. If you look at their distribution in Figure 5–1, you can see a physical reason for their independent development: Japan is isolated because it is an island country, and Korea is isolated to some extent because it is a peninsular state.

Japanese. Chinese cultural traits have diffused into Japanese society, including the original form of writing the Japanese language. But the structures of the two languages differ. Japanese is written in part with Chinese ideograms, but it also uses two systems of phonetic symbols, like Western languages, used either in place of the ideograms or alongside them. Foreign terms may be written with one of these sets of phonetic symbols.

Japanese Internet users have developed distinctive symbols that reflect cultural traditions. For example, ^.^ represents a girl smiling, because it is impolite for a girl to bear her teeth in a grin.

Korean. Korean is usually classified as a separate language family, although it may be related to the Altaic languages of Central Asia, or to Japanese. However, in contrast to Sino-Tibetan languages and Japanese, Korean is written not with ideograms but in a system known as *hankul* (also called *hangul* and *onmun*). In this system, each letter represents a sound, as in Western languages. More than half of the Korean vocabulary derives from Chinese words. In fact, Chinese and Japanese words are the principal sources for creating new words to describe new technology and concepts.

Austro-Asiatic. Austro-Asiatic, spoken by about 1 percent of the world's population, is based in Southeast Asia. Vietnamese, the most-spoken tongue of the Austro-Asiatic language family, is written with our familiar Roman alphabet, with the addition of a large number of diacritical marks above the vowels. The Vietnamese alphabet was devised in the seventh century by Roman Catholic missionaries.

Afro-Asiatic Language Family

The Afro-Asiatic—once referred to as the Semito-Hamitic—language family includes Arabic and Hebrew, as well as a number of languages spoken primarily in northern Africa and southwestern Asia. The world's fourth largest language family, Afro-Asiatic's international significance transcends the number of speakers because its languages were used to write the holiest books of three major world religions, the Judeo-Christian Bible and the Islamic Quran.

Arabic is the major Afro-Asiatic language, an official language in two dozen countries of North Africa and southwestern Asia, from Morocco to the Arabian Peninsula. Besides the 200 million native speakers of Arabic, a large percentage of the world's Muslims have at least some knowledge of Arabic because the Quran (Koran) was written in that language in the seventh century A.D. Although a number of dialects exist in Arabic, a standard Arabic has developed because of the influence of the Quran, newspapers, and radio. The United Nations added Arabic as its sixth official language in the General Assembly in 1973 and in the Security Council in 1982.

Altaic and Uralic Language Families

The Altaic and Uralic language families were once thought to be linked as one family because the two display similar word formation, grammatical endings, and other structural elements. Recent studies, however, point to geographically distinct origins of the two families. The Altaic languages are thought to have originated in the

steppes bordering the Qilian Shan and Altai mountains between Tibet and China. Linguists do not know whether one group originally spoke a single Altaic language or whether the language originated through a mixture of several others, which merged through interaction and acculturation of different peoples living in the steppes.

Altaic Languages. The Altaic languages are spoken across an 8,000-kilometer (5,000-mile) band of Asia between Turkey on the west and Mongolia and China on the east. Turkish, by far the most widely used Altaic language, was once written with Arabic letters. But in 1928, the Turkish government, led by Kemal Ataturk, ordered that the language be written with the Roman alphabet instead. Ataturk believed that switching to Roman letters would help modernize the economy and culture of Turkey through increased communications with European countries.

Other Altaic languages with at least 1 million speakers include Azerbaijani, Bashkir, Chuvash, Kazakh, Kyrgyz, Mongolian, Tatar, Turkmen, Uighur, and Uzbek. When the Soviet Union governed most of the Altaic-speaking region, use of Altaic languages was suppressed to create a homogeneous national culture. One element of Soviet policy was to force everyone to write with the Russian Cyrillic alphabet, although some Altaic languages traditionally employed Arabic letters. Most speakers of Altaic

This sign in Georgia for the Terek River is in Georgian (Caucasian language family), Russian, and English. The use of Russian is a remnant from Georgia's former status as a republic in the Soviet Union, where Russian was the most important language. (Peter Arnold)

languages are Muslims and are familiar with Arabic letters because Islamic holy books are written in Arabic.

With the dissolution of the Soviet Union in the early 1990s, Altaic languages became official in several newly independent countries, including Azerbaijan, Kazakhstan, Kyrgyzstan, Turkmenistan, and Uzbekistan. People in these countries may no longer be forced to learn Russian and write with Cyrillic letters. But unrest continues among speakers of Altaic languages, because enthusiasm for restoring languages long discouraged by the Soviet Union threatens the rights of minorities in these countries to speak other languages that are not officially recognized.

Problems also persist because the boundaries of the countries do not coincide with the regions in which the speakers of the various languages are clustered. The speakers of one Altaic language may find themselves divided among several countries, while the speakers of other Altaic languages—such as Bashkir, Chuvash, Tatar, and Uighur—do not control the governments of independent states.

Uralic Languages. Every European country is dominated by Indo-European speakers, except for three: Estonia, Finland, and Hungary (refer to Figure 5–3). The Estonians, Finns, and Hungarians speak languages that belong to the Uralic family. Uralic languages are traceable back to a common language, Proto-Uralic, first used 7,000 years ago by people living in the Ural Mountains of present-day Russia, north of the Kurgan homeland.

Migrants carried the Uralic languages to Europe. One branch moved north along the Volga River and then either turned westward toward Estonia and Finland or eastward into Siberia. The second branch moved southward and then westward to present-day Hungary. These Uralic-speaking migrants carved out homelands for themselves in the midst of Germanic- and Slavic-speaking peoples and retained their language as a major element of cultural identity.

African Language Families

No one knows the precise number of languages spoken in Africa, and scholars disagree on classifying the known ones into families. Nearly 1,000 distinct languages and several thousand named dialects have been documented. Figure 5–8 shows the broad view of African language families, and Figure 5–9 of Nigeria hints at the complex pattern of multiple tongues. This great number of languages results from at least 5,000 years of minimal inter-

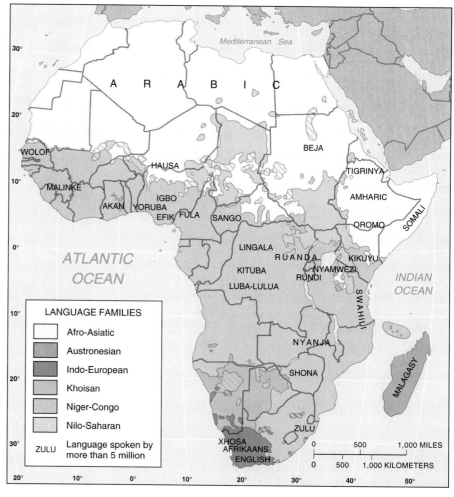

Figure 5–8 Africa's language families. Nearly 1,000 languages have been identified in Africa, and experts do not agree on how to classify them into families, especially languages in central Africa. On the large island of Madagascar, the language is unrelated to other African languages. Madagascar's Austronesian language is from a language family spoken across a wide area of the South Pacific (see Figure 5–1). This wide diffusion indicates that early speakers of Austronesian on the island of Madagascar must have migrated long distances. Languages with more than 5 million speakers are named on the map.

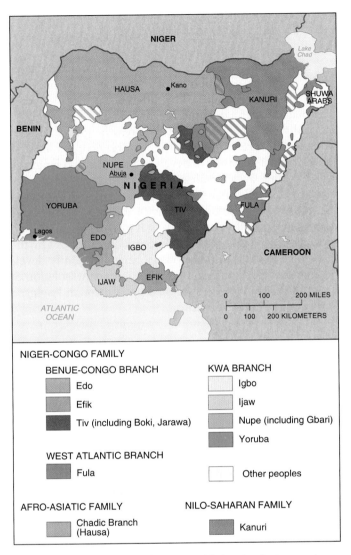

Figure 5–9 Nigeria's main languages. National unity is severely strained by the lack of a common language that a large percentage of the population can understand. To encourage unity among the disparate cultural groups, Nigeria has moved the national capital from Lagos, in the Yoruba-speaking southwest, to Abuja in the country's center. This central and "neutral" location was selected to avoid existing concentrations of the major rival cultural groups.

NIGER-CONGO FAMILY

BENUE-CONGO BRANCH
- Edo
- Efik
- Tiv (including Boki, Jarawa)

WEST ATLANTIC BRANCH
- Fula

AFRO-ASIATIC FAMILY
- Chadic Branch (Hausa)

KWA BRANCH
- Igbo
- Ijaw
- Nupe (including Gbari)
- Yoruba

- Other peoples

NILO-SAHARAN FAMILY
- Kanuri

action among the thousands of cultural groups inhabiting the African continent. Each group developed its own language, religion, and other cultural traditions in isolation from other groups.

Documenting African languages is a formidable task, because most lack a written tradition, and only ten are spoken by more than 10 million people. In the 1800s, European missionaries and colonial officers began to record African languages using the Roman or Arabic alphabet. Twentieth-century researchers continue to add newly discovered languages to the African list. They have found no evidence that any have become extinct.

In northern Africa, the language pattern is relatively clear where Afro-Asiatic languages dominate (Figure 5–8). Arabic dominates, although in a variety of dialects.

Other Afro-Asiatic languages spoken by more than 5 million Africans include Amharic, Oromo, and Somali in the Horn of Africa and Hausa in northern Nigeria. In sub-Saharan Africa, however, languages grow far more complex.

Niger-Congo Language Family. More than 95 percent of the people in sub-Saharan Africa speak languages of the Niger-Congo family, which includes six branches with many hard-to-classify languages. The remaining 5 percent speak languages of the Khoisan or Nilo-Saharan families. In addition, several million South Africans speak Indo-European languages, either English or Afrikaans, a Germanic Dutch-like language reflecting South Africa's Dutch colonial history.

The largest branch of the Niger-Congo family is the Benue-Congo branch, and its most important language is Swahili. Although it is the official language only of Tanzania, Swahili is spoken in much of eastern Africa. Swahili originally developed through interaction among African groups and Arab traders, so its vocabulary has strong Arabic influences. Also, Swahili is one of the few African languages with an extensive literature.

Nilo-Saharan Language Family. Nilo-Saharan languages are spoken by a few million people in north-central Africa, immediately north of the Niger-Congo language region. Divisions within the Nilo-Saharan family exemplify the problem of classifying African languages. Despite fewer speakers, the Nilo-Saharan family is divided into six branches: Chari-Nile, Fur, Koma, Maba, Saharan, and Songhai. The Chari-Nile branch (East Africa from Egypt to Tanzania) can be subdivided into four groups: Berta, Central Sudanic, East Sudanic, and Kunama. The Central Sudanic group in turn comprises ten subgroups. Therefore, the total number of speakers of each individual Nilo-Saharan language is extremely small.

Khoisan Language Family. The third important language family of sub-Saharan Africa—Khoisan—is concentrated in the southwest. A distinctive characteristic of the Khoisan languages is the use of clicking sounds. Upon hearing this, whites in southern Africa derisively and onomatopoeically named the most important Khoisan language Hottentot.

Austronesian Language Family. About 6 percent of the world's people speak an Austronesian language, once known as the Malay-Polynesian family. The most frequently used Austronesian language is Malay-Indonesian, the most important language of Indonesia, the world's fourth most populous country.

The map of world languages (Figure 5–1) shows a striking oddity with Madagascar, the large island off the east coast of Africa. The people of Madagascar speak

Malagasy, which belongs to the Austronesian family, even though the island is separated by 3,000 kilometers (1,900 miles) from any other Austronesian-speaking country, such as Indonesia. This is certainly strong evidence of migration to Madagascar from the South Pacific. Malayo-Polynesian people apparently sailed in small boats across the Indian Ocean to reach Madagascar approximately 2,000 years ago.

Nigeria: Conflict Among Speakers of Different Languages. Africa's most populous country, Nigeria, displays problems that can arise from the presence of many speakers of many languages. More than 200 distinct languages are spoken in Nigeria. In the north, Hausa, an Afro-Asiatic language, is spoken by approximately one-fourth of the population, mostly Hausa and Fulani peoples. In the southeast, Igbo is the most common language, followed by Efik and Ijaw. In the southwest, Yoruba is the most important language, followed by Edo (Figure 5–9).

Nigeria's principal problem as a country is that none of its 200-plus indigenous languages has widespread use. Reflecting its colonial history, English is spoken by 2 percent of Nigerians. In fact, English is the official language, which has the considerable advantage of being intelligible to governments of other countries.

Groups living in different regions of Nigeria have often battled. The southern Ibos attempted to secede from Nigeria during the 1960s, and northerners have repeatedly claimed that the Yorubas discriminate against them. To reduce these regional tensions, the government has moved the capital from Lagos in the Yoruba-dominated southwest to Abuja in the center of Nigeria.

Nigeria reflects the problems that can arise when great cultural diversity—and therefore, language diversity—is packed into a relatively small region. Nigeria also illustrates the importance of language in identifying distinct cultural groups at a local scale. Speakers of one language are unlikely to understand any of the others in the same family, let alone languages from other families.

Key Issue 2

Why Do People in Different Locations Speak Similar Languages?

- Origin and diffusion of English
- Origin and diffusion of Romance languages
- Origin and diffusion of Indo-European
- Isolated languages

The global distribution of languages results from a combination of two geographic processes—interaction and isolation. People in two locations speak the same language because of migration from one of the locations to another. If the two groups have little contact with each other after the migration, the language spoken by each will begin to differ. After a long period without contact, the two groups will speak languages that are so different they are classified as separate languages.

The interplay between interaction and isolation helps to explain the distribution of individual languages and entire language families. The difference is that individual languages emerged in the recent past as a result of historically documented events, whereas language families emerged several thousand years before recorded history.

For example, English developed as a distinct language in England as a result of migration and subsequent isolation of Germans 1,500 years ago and Normans 1,000 years ago. Similarly, individual Romance languages developed 2,000 years ago as a result of migration and isolation of Romans to other parts of Europe. On the other hand, the Indo-European language family developed as a result of migration and subsequent isolation of people that can only be reconstructed through linguistic and archaeological theories.

Origin and Diffusion of English

English is spoken fluently by one-half billion people, more than any language except for Mandarin. Whereas nearly all Mandarin speakers are clustered in one country, China, English speakers are distributed around the world. English is an official language in 42 countries, more than any other language, and is spoken by a significant percentage of people in a number of other countries (Figure 5–10). Two billion people—one-third of the world—live in a country where English is an official language, even if they cannot speak it.

The origin and diffusion of English speakers, discussed in this section of the book, serves as a case study for understanding the geographic processes by which any language is distributed around the world. A language originates at a particular place and diffuses to other locations through the migration of its speakers.

English Colonies

The contemporary distribution of English speakers around the world exists because the people of England migrated with their language when they established colonies during the past four centuries. Compare Figure 5–10 to Figure 8–4, which shows the location of former British colonies. English is an official language in most of the former British colonies.

English first diffused west from England to North America in the seventeenth century. The first English colonies were built in North America, beginning with Jamestown, Virginia, in 1607, and Plymouth, Massachusetts, in 1620. After England defeated France in a battle to dominate the North American colonies during the eighteenth century, the position of English as the principal language of North America was assured, even after the United States and Canada became independent countries.

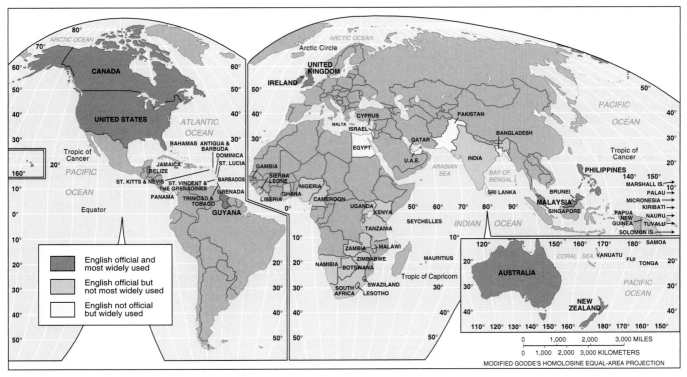

Figure 5–10 English-speaking countries. English is the official language in 42 countries, although in 26 of these it is not the most widely used language. English is also understood by a significant number of people in several other countries that were once British colonies.

Similarly, the British took control of Ireland in the seventeenth century, South Asia in the mid-eighteenth century, the South Pacific in the late eighteenth and early nineteenth centuries, and southern Africa in the late nineteenth century. In each case, English became an official language, even if only the colonial rulers and a handful of elite local residents could speak it.

More recently, the United States has been responsible for diffusing English to several places, most notably the Philippines, which Spain ceded to the United States in 1899 a year after losing the Spanish-American War. After gaining full independence in 1946, the Philippines retained English as one of its official languages, along with Filipino (Tagalog).

Origin of English in England

The global distribution of English may be a function primarily of migration from England since the seventeenth century, but that does not explain how English came to be the principal language of the British Isles in the first place, nor why English is classified as a Germanic language.

The British Isles had been inhabited for thousands of years, but we know nothing of their early languages, until tribes called the Celts arrived around 2000 B.C., speaking languages we call Celtic. Then, around A.D. 450, tribes from mainland Europe invaded, pushing the Celts into the remote northern and western parts of Britain, including Cornwall and the highlands of Scotland and Wales.

German Invasion. The invading tribes were the Angles, Jutes, and Saxons. All three were Germanic tribes, the Jutes from northern Denmark, the Angles from southern Denmark, and the Saxons from northwestern Germany (Figure 5–11). Today, English people and others who trace their cultural heritage back to England are often called Anglo-Saxons, after the two larger tribes.

The name *England* comes from *Angles' land*. In Old English, *Angles* was spelled *Engles*, and the Angles' language was known as *englisc*. They came from a corner, or *angle*, of Germany known as Schleswig-Holstein.

Modern English has evolved primarily from the language spoken by the Angles, Jutes, and Saxons when they migrated to England fifteen hundred years ago. The three tribes who brought the beginnings of English to the British Isles came from present-day Denmark and Germany, where they shared a language similar to that of other peoples in the region. Because the people who migrated to England came from the northern lowlands rather than the southern highlands of Germany, English is classified as a Low Germanic language, within the West Germanic group.

At some time in history, all Germanic people spoke a common language, but that time predates written records. The common origin of English with other Germanic languages can be reconstructed by analyzing language differences that emerged after Germanic groups migrated to separate territories and lived in isolation from each other, allowing their languages to continue evolving independently.

Figure 5–11 Invasions of England. The first speakers of the language that became known as English were tribes that lived in present-day Germany and Denmark. They invaded England in the fifth century A.D. The Jutes settled primarily in southeastern England, the Saxons in the south and west, and the Angles in the north, eventually giving the country its name: Angles' Land, or England. From this original spatial separation, the first major regional differences in English dialect developed, as Figure 5–14 shows. Invasions by Vikings in the ninth century and Normans in the eleventh century brought new words to the language spoken in the British Isles. The Normans were the last successful invaders of England. (Source: From Albert C. Baugh and Thomas Cable, *A History of the English Language*, 3d ed., © 1978, p. 47. Reprinted by permission of Prentice Hall, Englewood Cliffs, NJ.)

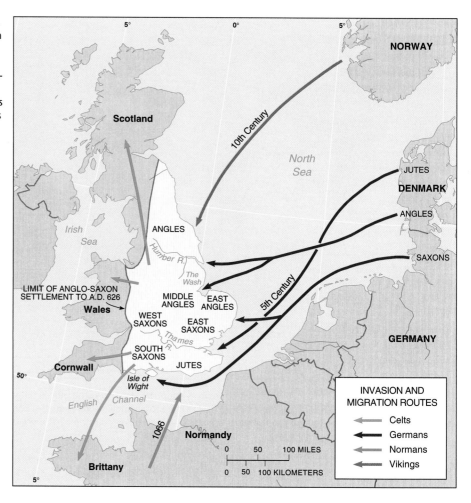

Other peoples subsequently invaded England and added their languages to the basic English. Vikings from present-day Norway landed on the northeast coast of England in the ninth century. Although defeated in their effort to conquer the islands, many Vikings remained in the country to enrich the language with new words.

Norman Invasion. English is a good bit different from German today primarily because England was conquered by the Normans in 1066. The Normans, who came from present-day Normandy in France, spoke French, which they established as England's official language for the next 150 years. The leaders of England, including the royal family, nobles, judges, and clergy, therefore spoke French. However, the majority of the people, who had little education, did not know French, so they continued to speak English to each other.

In 1204, during the reign of King John, England lost control of Normandy and entered a long period of conflict with France. As a result, fewer people in England wished to speak French, and English again became the country's unchallenged dominant language. Recognizing that nearly everyone in England was speaking English, Parliament enacted the Statute of Pleading in 1362 to change the official language of court business from French to English.

During the 300-year period that French was the official language of England, the Germanic language used by the common people and the French used by the leaders mingled to form a new language. Modern English owes its simpler, straightforward words, such as *sky, horse, man,* and *woman,* to its Germanic roots, and fancy, more elegant words, such as *celestial, equestrian, masculine,* and *feminine* to its French invaders.

Origin and Diffusion of Romance Languages

The Romance languages, including Spanish, Portuguese, French, Italian, and Romanian, are part of the same branch, because they all developed from Latin, the "Romans' language" (refer to Figure 5–5). The rise in importance of the city of Rome 2,000 years ago brought a diffusion of its Latin language.

Role of Roman Empire

At its height in the second century A.D., the Roman Empire extended from the Atlantic Ocean on the west to the Black Sea on the east and encompassed all lands bordering the Mediterranean Sea (the empire's boundary is shown in Figure 6–5). As the conquering Roman armies occupied the provinces of this vast empire, they brought the Latin language with them. In the process, the languages spoken by the natives of the provinces were either extinguished or suppressed in favor of the language of the conquerors.

Even during the period of the Roman Empire, Latin varied to some extent from one province to another. The empire grew over a period of several hundred years, so the Latin used in each province was based on that spoken by the Roman army at the time of occupation. The Latin spoken in each province also integrated words from the language formerly spoken in the area.

The Latin that people in the provinces learned was not the standard literary form but a spoken form, known as **Vulgar Latin**, from the Latin word referring to "the masses" of the populace. Vulgar Latin was introduced to the provinces by the soldiers stationed throughout the empire. For example, the literary term for "horse" was *equus*, from which English has derived such words as "equine" and "equestrian." However, the Vulgar term, used by the common people, was *caballus*, from which are derived the modern terms for "horse" in Italian (*cavallor*), Spanish (*caballo*), Portuguese (*cavalo*), French (*cheval*), and Romanian (*cal*).

Collapse of Roman Empire

Following the collapse of the Roman Empire in the fifth century, communication among the former provinces declined, creating still greater regional variation in spoken Latin. By the eighth century, regions of the former empire had been isolated from each other long enough for distinct languages to evolve.

Latin persisted in parts of the former empire. People in some areas reverted to former languages, while others adopted the languages of conquering groups from the north and east, which spoke Germanic and Slavic.

In the past, when migrants were unable to communicate with speakers of the same language back home, major differences emerged between the languages spoken in the old and new locations, leading to the emergence of distinct, separate languages. This was the case with the migration of Latin speakers 2,000 years ago.

Origin and Diffusion of Indo-European

If Germanic, Romance, Balto-Slavic, and Indo-Iranian languages are all part of the same Indo-European language family, then they must be descended from a single common ancestral language. Unfortunately, the existence of a single ancestor—which can be called Proto-Indo-European—cannot be proven with certainty, because it would have existed thousands of years before the invention of writing or recorded history.

Evidence of Common Origin of Indo-European Languages

The evidence that Proto-Indo-European once existed is "internal," derived from the physical attributes of words themselves in various Indo-European languages. For example, the words for some animals and trees in modern Indo-European languages have common roots, including *beech*, *oak*, *bear*, *deer*, *pheasant*, and *bee*. Because all Indo-European languages share these similar words, linguists believe the words must represent things experienced in the daily lives of the original Proto-Indo-European speakers.

In contrast, words for other features, such as *elephant*, *camel*, *rice*, and *bamboo*, have different roots in the various Indo-European languages. Such words therefore cannot be traced back to a common Proto-Indo-European ancestor, and must have been added later, after the root language split into many branches.

Interestingly, individual Indo-European languages share common root words for *winter* and *snow*, but not for *ocean*. Therefore, linguists conclude that original Proto-Indo-European speakers probably lived in a cold climate, or one that had a winter season, but did not come in contact with oceans.

Two Theories of Origin and Diffusion

Linguists and anthropologists generally accept that Proto-Indo-European must have existed, but they disagree on when and where the language originated and the process and routes by which it diffused. The debate over place of origin and paths of diffusion is significant, because one theory argues that language diffused primarily through warfare and conquest, while the other theory argues that the diffusion resulted from peaceful sharing of food.

Theory 1: Kurgan Origin. So where did Indo-European originate? One influential hypothesis, espoused by Marija Gimbutas, is that the first Proto-Indo-European speakers were the Kurgan people, whose homeland was in the steppes near the border between present-day Russia and Kazakhstan. The earliest archaeological evidence of the Kurgans dates to around 4300 B.C.

The Kurgans were nomadic herders. Among the first to domesticate horses and cattle, they migrated in search of grasslands for their animals. This took them westward through Europe, eastward to Siberia, and southeastward to Iran and South Asia. Between 3500 and 2500 B.C., Kurgan warriors, using their domesticated horses as weapons, conquered much of Europe and South Asia (Figure 5–12).

Theory 2: Origin in Anatolia. Not surprisingly, scholars disagree on where and when the first speakers of Proto-Indo-European lived. Archaeologist Colin Renfrew argues that they lived 2,000 years before the Kur-

Figure 5–12 Origin and diffusion of Indo-European (Kurgan hearth theory). The Kurgan homeland was north of the Caspian Sea, near the present-day border between Russia and Kazakhstan. According to this theory, the Kurgans may have infiltrated into Eastern Europe beginning around 4000 B.C. and into central Europe and southwestern Asia beginning around 2500 B.C.

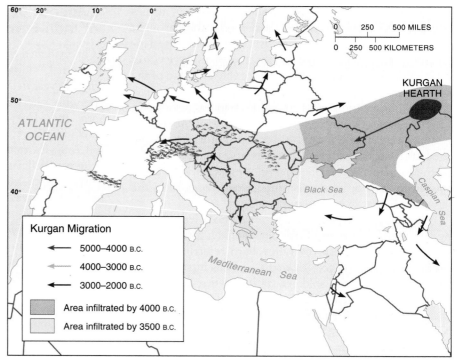

gans, in eastern Anatolia, part of present-day Turkey (Figure 5–13). Renfrew believes they diffused from Anatolia westward to Greece (the origin of the Greek language branch), and from Greece westward toward Italy, Sicily, Corsica, the Mediterranean coast of France, Spain, and Portugal (the origin of the Romance language branch). From the Mediterranean coast, the speakers migrated northward toward central and northern France and on to the British Isles (perhaps the origin of the Celtic language branch).

Renfrew believes that Indo-European also diffused northward from Greece toward the Danube River (Ro-

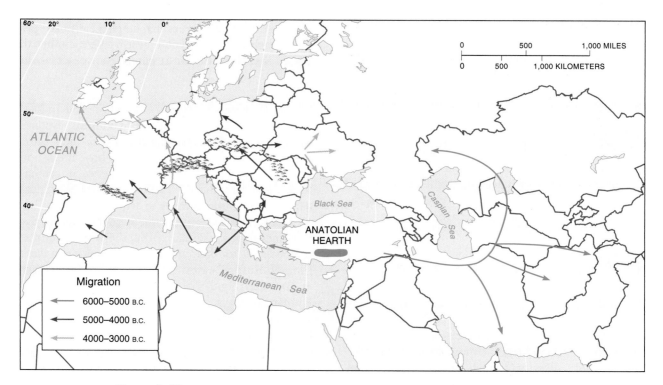

Figure 5–13 Origin and diffusion of Indo-European (Anatolian hearth theory). Indo-European may have originated in present-day Turkey 2,000 years before the Kurgans. According to this theory, the language diffused along with agricultural innovations west into Europe and east into Asia.

mania) and westward to central Europe. From there, the language diffused northward toward the Baltic Sea (the origin of the Germanic language branch) and eastward toward the Dnestr River near Ukraine (the origin of the Slavic language branch). From the Dnestr River, speakers migrated eastward to the Dnepr River (the homeland of the Kurgans).

The Indo-Iranian branch of the Indo-European language family originated either directly through migration from Anatolia along the south shores of the Black and Caspian seas by way of Iran and Pakistan, or indirectly by way of Russia north of the Black and Caspian seas.

Renfrew argues that Indo-European diffused into Europe and South Asia along with agricultural practices, rather than by military conquest. The language triumphed because its speakers became more numerous and prosperous by growing their own food instead of relying on hunting. Regardless of how Indo-European diffused, communication was poor among different peoples, whether warriors or farmers. After many generations of complete isolation, individual groups evolved increasingly distinct languages.

Isolated Languages

An **isolated language** is a language unrelated to any other and therefore not attached to any language family. Similarities and differences between languages—our main form of communication—are a measure of the degree of interaction among groups of people. The diffusion of Indo-European languages demonstrates that a common ancestor dominated much of Europe before recorded history. Similarly, the diffusion of Indo-European languages to the Western Hemisphere is a result of conquests by Indo-European speakers in more recent times. On the other hand, isolated languages arise through lack of interaction with speakers of other languages.

A Pre-Indo-European Survivor: Basque

The best example of an isolated language in Europe is Basque, apparently the only language currently spoken in Europe that survives from the period before the arrival of Indo-European speakers. No attempt to link Basque to the common origin of the other European languages has been successful. Basque was probably once spoken over a wider area but was abandoned where its speakers came in contact with Indo-Europeans.

Basque is spoken by 1 million people in the Pyrenees Mountains of northern Spain and southwestern France (refer to Figure 5–3). Basque's lack of connection to other languages reflects the isolation of the Basque people in their mountainous homeland. This isolation has helped the Basque people preserve their language in the face of the wide diffusion of Indo-European languages.

An Unchanging Language: Icelandic

Unlike Basque, Icelandic is related to other languages, in the North Germanic group of the Germanic branch of the Indo-European family. Icelandic's significance is that over the past thousand years it has changed less than any other in the Germanic branch.

As was the case with England, people in Iceland speak a Germanic language because their ancestors migrated to the island from the east, in this case from Norway. Norwegian settlers colonized Iceland in A.D. 874.

When an ethnic group migrates to a new location, it takes along the language spoken in the former home.

Basque, an isolated language. Basque is probably the only modern European language that predates Indo-European. Basques separatists seeking independence from Spain hold a rally in San Sebastian in northern Spain. (Eric Vandeville/Gamma-Liaison, Inc.)

The language spoken by most migrants—such as the Germanic invaders of England—changes in part through interaction with speakers of other languages. But in the case of Iceland, the Norwegian immigrants had little contact with speakers of other languages when they arrived in Iceland, nor did they have contact with speakers of their language back in Norway. After centuries of interaction with other Scandinavians, Norwegian and other North Germanic languages had adopted new words and pronunciation, while the isolated people of Iceland had less opportunity to learn new words and no reason to change their language.

Key Issue 3

Why Is One Language Spoken Differently in Different Locations?

- Dialects of English
- Romance-language dialects

We have seen how a language—be it English during the past several centuries or Proto-Indo-European thousands of years ago—diffuses through migration of the language's speakers from one place to another. After people arrive at a new location, over a period of time their language becomes different from the one spoken by the people at the former location. Differences develop through isolation from other speakers of the same language, as well as by contact with speakers of other languages.

When migrants are able to maintain some communication with other speakers of the language, differences large enough to cause the emergence of an entirely new language are unlikely. Instead, various dialects of a language may develop. This was the case with the migration of English speakers to North America several hundred years ago. Geographers are especially interested in differences in dialects, because they reflect distinctive features of the environments in which groups live.

Dialects of English

"If you use proper English, you're regarded as a freak; why can't the English learn to speak?" asked Professor Henry Higgins in the Broadway musical *My Fair Lady*. He was referring to the Cockney-speaking Eliza Doolittle, who pronounced "rain" like "rine" and dropped the /h/ sound from the beginning of words like "happy." Eliza Doolittle's speech illustrates that English, like other languages, has a wide variety of dialects that use different pronunciations, spellings, and meanings for particular words.

Because of its large number of speakers and widespread distribution, English has an especially large number of dialects. North Americans are well aware that they speak English differently from the British, not to mention Indians and other South Asians, or Australians and

In *My Fair Lady*, the 1950s Broadway and Hollywood musical by Alan J. Lerner and Frederick Loewe, based on George Bernard Shaw's play *Pygmalion*, language expert Professor Henry Higgins (played by Rex Harrison) encounters Eliza Doolittle, a Cockney from the poor East End of London (played in the movie by Audrey Hepburn), selling flowers in front of London's Covent Garden Opera House. Higgins accepts a wager from a friend that he can transform Doolittle into an upper-class woman primarily by teaching her to speak with the accent used by upper-class Britons. (Warner Bros./Photofest)

other South Pacific people. Further, English varies by regions within individual countries. In both the United States and England, northerners sound different from southerners.

In a language with multiple dialects, one dialect may be recognized as the **standard language**, which is a dialect that is well-established and widely recognized as the most acceptable for government, business, education, and mass communication. One particular dialect of English, the one associated with upper-class Britons living in the London area, is recognized in much of the English-speaking world as the standard form of British speech. This speech, known as **British Received Pronunciation (BRP)**, is well known, because it is commonly used by politicians, broadcasters, and actors. Why don't Americans or, for that matter, other British people speak that way?

Dialects in England

As you have seen, English originated with three invading groups from Northern Europe, who settled in different parts of Britain: the Angles in the north, the Jutes in the southeast, and the Saxons in the south and west. The language each spoke was the basis of distinct regional dialects of Old English: Kentish in the southeast, West Saxon in the southwest, Mercian in the center of the island, and Northumbrian in the north (Figure 5–14, left).

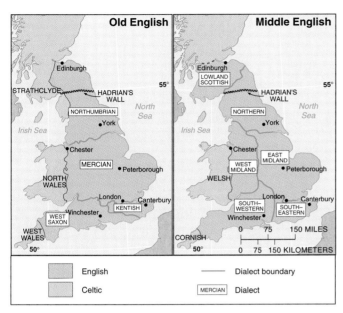

Figure 5–14 (Left) Old English dialects, before the Norman invasion of A.D. 1066. (Right) Middle English dialects (1150–1500). Comparing the maps, you can see that important dialects of Middle English corresponded closely to those of Old English. The Old English Northumbrian dialect, spoken by the Angles, split into Scottish and Northern dialects. The Old English Mercian dialect, spoken by the Saxons, divided into East Midland and West Midland, and the West Saxon dialect became known as the Southwestern dialect. The Old English Kentish dialect, spoken by the Jutes, extended considerably in area and became known as the Southeastern dialect. (Source: From Albert C. Baugh and Thomas Cable, *A History of the English Language*, 3d ed., © 1978, p. 53. Reprinted by permission of Prentice Hall, Englewood Cliffs, NJ.)

Following the Norman invasion of 1066, French replaced English as the language of the government and aristocracy. By the time English again became the country's dominant language, five major regional dialects had emerged: Northern, East Midland, West Midland, Southwestern, and Southeastern or Kentish. The boundaries of these five regional dialects roughly paralleled the pattern before the Norman invasion (compare Figure 5–14, left and right). However, after 150 years of living in isolation in rural settlements under the control of a French-speaking government, people spoke English differently in virtually every county of England.

From this large collection of local dialects, one eventually emerged as the standard language for writing and speech throughout England: the dialect used by upper-class residents in the capital city of London and the two important university cities of Cambridge and Oxford. The diffusion of the dialect spoken in London and the university cities was first encouraged by the introduction of the printing press to England in 1476. Grammar books and dictionaries printed in the eighteenth century established rules for spelling and grammar that were based on the London dialect. These frequently arbitrary rules were then taught in schools throughout the country.

Current Dialect Differences in England. Despite the current dominance of British Received Pronunciation, strong regional differences persist in English dialects spoken in the United Kingdom, especially in rural areas. Although several dozen dialects are identifiable, they can be grouped into three main ones: Northern, Midland, and Southern. People in the south of England pronounce words like "grass" and "path" with an /*ah*/ sound, whereas people in the Midlands and North use a short /*a*/, as do most people in the United States. People in the

The British call sausages *bangers*, which they like to eat with mashed potatoes, called *mash*. This bangers and mash stand is in the marketplace at Norwich, England. (Lee Snider/The Image Works)

Midlands and North pronounce "butter" and "Sunday" with the /oo/ sound of words like "boot." Northerners pronounce "ground" and "pound" like "grund" and "pund," with the /uh/ sound similar to the word "punt" in U.S. football.

Further, distinctive southwestern and southeastern accents occur within the Southern dialect. People in the southwest, for example, pronounce "thatch" and "thing" with the /th/ sound of "then," rather than "thin." "Fresh" and "eggs" have an /ai/ sound. Southeasterners pronounce the /a/ in "apple" and "cat" like the short /e/ in "bet." Local dialects can be further distinguished, and some words have distinctive pronunciations and meanings in each county of the United Kingdom.

Children's words particularly show strong regional differences. For example, words used by youngsters in games vary in different regions of the country and in specific communities within particular regions (Figure 5–15).

English in North America

The English language was brought to the North American continent by colonists from England who settled along the Atlantic Coast beginning in the seventeenth century. The early colonists naturally spoke the language used in England at the time and established seventeenth-century English as the dominant form of European speech in colonial America. Later immigrants from other countries found English already implanted here. Although they made significant contributions to American English, they became acculturated into a society that already spoke English. Therefore, the earliest colonists were most responsible for the dominant language patterns that exist today in the English-speaking part of the Western Hemisphere.

Differences Between British and American English. Why is the English language in the United States so different from that in England? As is so often the case with languages, the answer is isolation. Separated by the Atlantic Ocean, English in the United States and England evolved independently during the eighteenth and nineteenth centuries, with little influence on one another. Few residents of one country could visit the other, and the means to transmit the human voice over long distances would not become available until the twentieth century.

U.S. English differs from that of England in three significant ways: vocabulary, spelling, and pronunciation. The vocabulary is different largely because settlers in America encountered many new objects and experiences. The new continent contained physical features, such as large forests and mountains, that had to be given new names. New animals were encountered, including the *moose*, *raccoon*, and *chipmunk*, all of which were given names borrowed from Native Americans. Indigenous American "Indians" also enriched American English with names for objects such as *canoe*, *moccasin*, and *squash*.

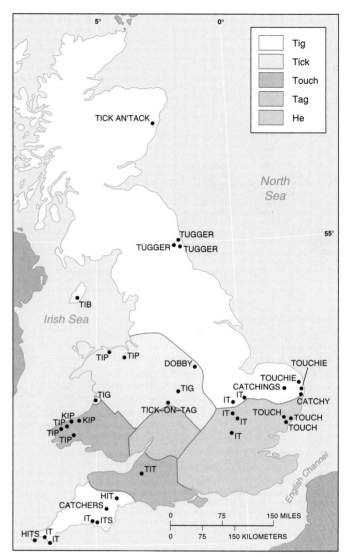

Figure 5–15 Contemporary dialects in Great Britain. Regional differences in vocabulary continue to exist in Great Britain despite the small size of the country and the diffusion of standard language through television, radio, and print. One example is the word that children use in a game of tag to signal that they have touched another participant. The map also shows that distinctive words indicating "touch" can exist in individual communities as well. (Adapted from Iona and Peter Opie, *Children's Games in Street and Playground*. London: Claredon Press, 1969.)

As new inventions appeared, they acquired different names on either side of the Atlantic. For example, the elevator is called a *lift* in England, and the flashlight is known as a *torch*. The British call the hood of a car the *bonnet* and the trunk the *boot*.

Spelling diverged from the British standard because of a strong national feeling in the United States for an independent identity. Noah Webster, the creator of the first comprehensive American dictionary and grammar books, was not just a documenter of usage; he had an agenda. Webster was determined to develop a uniquely American dialect of English. He either ignored or was unaware of recently created rules of grammar and spelling developed in England.

Webster argued that spelling and grammar reforms would help establish a national language, reduce cultural dependence on England, and inspire national pride. The spelling differences between British and American English, such as the elimination of the "u" from the British spelling of words like "honour" and "colour" and the substitution of "s" for "c" in "defence," are due primarily to the diffusion of Webster's ideas inside the United States.

Differences in pronunciation between British and U.S. speakers are immediately recognizable. Again, geographic concepts help explain the reason for the differences. From the time of their arrival in North America, colonists began to pronounce words differently from the British. Such divergence is normal, for interaction between the two groups was largely confined to exchange of letters and other printed matter rather than direct speech.

One prominent difference between British and U.S. English is the pronunciation of the letters "a" and "r." Such words as "fast," "path," and "half" are pronounced in England like the /ah/ in "father" rather than the /a/ in "man." The British also eliminate the letter "r" from pronunciation except before vowels. Thus, "lord" in British pronunciation sounds like "laud." Further, Americans pronounce unaccented syllables with more clarity. The words "secretary" and "necessary" have four syllables in American English but only three in British ("secret'ry" and "necess'ry").

Surprisingly, pronunciation has changed more in England than in the United States. The letters "a" and "r" are pronounced in the United States the way they used to be pronounced in Britain, specifically in the seventeenth century when the first colonists arrived.

A single dialect of Southern English did not emerge as the British national standard until the late eighteenth century, after the American colonies had declared independence and were politically as well as physically isolated from England. Thus, people in the United States do not speak "proper" English because when the colonists left England, "proper" English was not what it is today. Furthermore, few colonists were drawn from the English upper classes.

Dialects in the United States

Major differences in U.S. dialects originated because of differences in dialects among the original settlers. The English dialect spoken by the first colonists, who arrived in the seventeenth century, determined the future speech patterns for their communities, because later immigrants adopted the language used in their new homes when they arrived. The language may have been modified somewhat by the new arrivals, but the distinctive elements brought over by the original settlers continued to dominate.

Settlement in the East. The original American settlements stretched along the Atlantic Coast in 13 separate colonies. The settlements can be grouped into three areas: New England, Middle Atlantic, and Southeastern. Massachusetts and the other New England colonies were established and inhabited almost entirely by settlers from England. Two-thirds of the New England colonists were Puritans from East Anglia in southeastern England, and only a few came from the north of England.

The nucleus of the southeastern colonies was Virginia, where the first permanent settlement by the English in North America was established at Jamestown in 1607. About half of the southeastern settlers came from southeast England, although they represented a diversity of social-class backgrounds, including deported prisoners, indentured servants, and political and religious refugees. The English dialects now spoken in the U.S. Southeast and New England are easily recognizable. Current distinctions result from the establishment of independent and isolated colonies in the seventeenth century.

The immigrants to the Middle Atlantic colonies were more diverse. The early settlers of Pennsylvania were predominantly Quakers from the north of England. Scots and Irish also went to Pennsylvania, as well as to New Jersey and Delaware. In addition, the Middle Atlantic colonies attracted many German, Dutch, and Swedish immigrants who learned their English from the English-speaking settlers in the area. The dialect spoken in the Middle Atlantic colonies thus differed significantly from those spoken farther north and south, because most of the settlers came from the north rather than the south of England or from other countries.

Current Dialect Differences in the East.   Today, major dialect differences within the United States continue to exist, primarily on the East Coast, although some distinctions can be found elsewhere in the country. The different dialects have been documented through the study of particular words. Every word that is not used nationally has some geographic extent within the country and therefore has boundaries. Such a word-usage boundary, known as an **isogloss**, can be constructed for each word. These isoglosses are determined by collecting data directly from people, particularly natives of rural areas. They are shown pictures to identify or are given sentences to complete with a particular word. While every word has a unique isogloss, boundary lines of different words coalesce in some locations to form regions.

Two important isoglosses separate the eastern United States into three major dialect regions, known as Northern, Midland, and Southern. The northern boundary runs across Pennsylvania, while the southern one runs along the Appalachian Mountains (Figure 5–16).

Some words are commonly used within one of the three major dialect areas, but rarely in the other two. In most instances, these words relate to rural life, food, and objects from daily activities. Language differences tend to be greater in rural areas than in cities, because farmers are relatively isolated from interaction with people from other dialect regions.

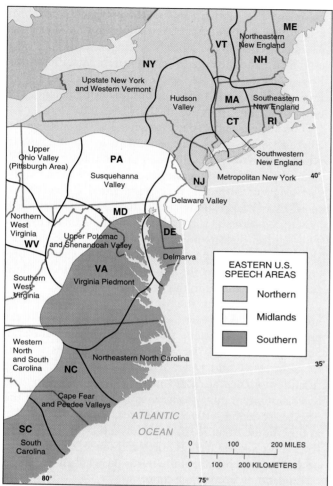

Figure 5–16 Dialects in eastern United States. The most comprehensive classification of dialects in the United States was made by Hans Kurath in 1949. He found the greatest diversity of dialects in the eastern part of the country, especially in vocabulary used on farms. Kurath divided the eastern United States into three major dialect regions—Northern, Midlands, and Southern—each of which contained a number of important subareas.

For example, a container commonly used on farms is known as a "pail" in the north and a "bucket" in the Midlands and South. A small stream is known as a "brook" in the North, a "run" in the Midlands, and a "branch" in the South. The term "run" was apparently used in the north of England and Scotland, which was the area of origin for many Middle Atlantic settlers but for few New England or Southern settlers.

Phrases for some farm activities, such as calling cows from pasture, show particularly sharp differences among the three regional dialects. New England farmers call cows with "Boss!" or "Bossie!", sometimes preceded by "Co" or "Come." In the Midlands, the preferred call is "Sook!" or sometimes "Sookie!" or "Sook cow!" The choice in the South is "Co-wench!" or its alternative forms, "Co-inch!" and "Co-ee!"

Many words that were once regionally distinctive are now national in distribution. Mass media, especially television and radio, influence the adoption of the same words throughout the country. For example, a "frying pan" was once commonly called a "spider" in New England and a "skillet" in the Middle Atlantic area.

Pronunciation Differences. Regional pronunciation differences are more familiar to us than word differences, although it is harder to draw precise isoglosses for them. Pronunciations that distinguish the Southern dialect include making such words as "half" and "mine" into two syllables ("ha-af" and "mi-yen"), pronouncing "poor" as "po-ur," and pronouncing "Tuesday" and "due" with a /y/ sound ("Tyuesday" and "dyue").

The New England accent is well known for dropping the /r/ sound, so that "heart" and "lark" are pronounced "hot" and "lock." Also, "ear" and "care" are pronounced with /ah/ substituted for the /r/ endings. This characteristic dropping of the /r/ sound is shared with speakers from the south of England and reflects the place of origin of most New England colonists. It also reflects the relatively high degree of contact between the two groups. Residents of Boston, New England's main port city, maintained especially close ties to the important ports of southern England such as London, Plymouth, and Bristol. Compared to other colonists, New Englanders received more exposure to changes in pronunciation that occurred in Britain during the eighteenth century.

The New England and southern accents sound odd to the majority of Americans because the standard pronunciation throughout the American West comes from the Middle Atlantic states rather than the New England and Southern regions. This pattern occurred because the Middle Atlantic states provided most of the western settlers.

The diffusion of particular English dialects into the middle and western parts of the United States is a result of the westward movement of colonists from the three dialect regions of the East. The area of the Midwest south of the Ohio River was settled first by colonists from Virginia and the other southern areas. The Middle Atlantic colonies sent most of the early settlers north of the Ohio River, although some New Englanders moved to the Great Lakes area.

As more of the West was opened to settlement during the nineteenth century, people migrated from all parts of the East Coast. The California gold rush attracted people from throughout the East, many of whom subsequently moved to other parts of the West. The mobility of Americans has been a major reason for the relatively uniform language that exists throughout much of the West.

Romance Language Dialects

In other languages, the dialect spoken by upper-class residents of the capital city also has emerged as the standard language. For example, the Parisian dialect became the standard form of French, and the Madridian dialect became the standard form of Spanish. As was the case with language families and individual languages, the dominance of one dialect over others within a country is a measure of the relative political strength of the speak-

Globalization and Local Diversity

How Many Romance Languages Exist?

The distribution of **where** different Romance-branch languages are spoken shows the difficulty in trying to establish the number of distinct languages in the world. The reason **why** sorting out languages is difficult stems from the wide variety of possible criteria for identifying distinct languages. Three Romance languages—Spanish, Portuguese, and French—are widely spoken around the world, a measure of the **globalization** of European culture. But the large number of other languages that can be counted as distinct Romance languages reflects on the role of language in promoting **local diversity**.

If official languages are counted, we add four to the three mentioned above: Italian, Romanian, Romansh, and Catalán. Italian and Romanian are official languages of Italy and Romania, respectively. Romansh is one of the four official languages of Switzerland, although it is spoken by only 25,000 people. Catalán, a Spanish dialect, is the official language of Andorra, a tiny country of approximately 50,000 inhabitants situated in the Pyrenees Mountains between Spain and France. Catalán is also spoken by another 10 million people, mostly around the city of Barcelona. A third Romance language, Sardinian—a mixture of Italian, Spanish, and Arabic—once was the official language of the Mediterranean island of Sardinia (refer to Figure 5–5).

In addition to these official languages, several unofficial Romance languages have individual literary traditions. In Italy, Ladin (not Latin) is spoken by 20,000 people living in the South Tyrol, while Friulian is spoken by 500,000 in the northeast. Ladin and Friulian (along with the official Romansh) are dialects of Rhaeto-Romanic. A Romance tongue called Ladino—a mixture of Spanish, Greek, Turkish, and Hebrew—is spoken by 140,000 Sephardic Jews, most of whom now live in Israel. None of these languages have an official status in any country, although all are used in literature.

Difficulties arise in determining whether two languages are distinct or merely two dialects of the same language. Occitan is a dialect of French, although it has a separate literary tradition. Moldovan (or Moldavian) is the official language of Moldova, but is generally classified as a dialect of Romanian, and Flemish, the official language of northern Belgium, is generally considered a dialect of Dutch. Galician, spoken in northwestern Spain, is generally classified as a dialect of Portuguese The task of distinguishing individual languages from dialects is difficult, because many residents of these regions choose to regard their languages as distinct.

Romance languages spoken in some former colonies can also be classified as separate languages because they differ substantially from the original introduced by European colonizers. Examples include French Creole in Haiti, Papiamento (Creolized Spanish) in Netherlands Antilles (West Indies), and Portuguese Creole in the Cape Verde Islands off the African coast. A **creole** or **creolized language** is defined as a language that results from the mixing of the colonizer's language with the indigenous language of the people being dominated.

A creolized language forms when the colonized group adopts the language of the dominant group but makes some changes, such as simplifying the grammar and adding words from their former language. The word *creole* derives from a word in several Romance languages for a slave who is born in the master's house.

---

ers of the various dialects (see Globalization and Local Diversity box).

Dialects in France

The dialect of the Île-de-France region, known as *Francien*, became the standard form of French because the region included Paris, which became the capital and largest city of the country. *Francien* French became the country's official language in the sixteenth century, and local dialects tended to disappear as a result of the capital's long-time dominance over French political, economic, and social life.

The most important surviving dialect difference within France is between the north and the south (refer to Figure 5–5). The northern dialect is known as *langue d'oïl* and the southern as *langue d'òc*. It is worth exploring these names, for they provide insight into how languages evolve. These terms derive from different ways in which the word for "yes" was said.

One Roman term for "yes" was *hoc illud est*, meaning "that is so." In the south, the phrase was shortened to *hoc*, or *òc*, because the /h/ sound was generally dropped, just as we drop it on the word *honor* today. Northerners shortened the phrase to *o-il* after the first sound in the first two words of the phrase, again with the initial /h/ suppressed. If the two syllables of *o-il* are spoken very rapidly, they are combined into a sound like the English word "wheel." Eventually, the final consonant was eliminated, as in many French words, giving a sound for "yes" like the English "we," spelled in French *oui*.

A province where the southern dialect is spoken in southwestern France is known as Languedoc. The southern French dialect itself is now sometimes called Occitan, derived from the French region of Aquitaine, which in French has a similar pronunciation to Occitan. About 3 million people in southeastern France speak a form of Occitan known as Provençal.

Worldwide Diffusion of Spanish and Portuguese

Spain, like France, contained many dialects during the Middle Ages. One dialect, known as Castilian, arose during the ninth century in Old Castile, located in the north-central part of the country. The dialect spread southward over the next several hundred years as independent kingdoms were unified into one large country. Spain grew to its approximate present boundaries in the fifteenth century, when the Kingdom of Castile and Léon merged with the Kingdom of Aragón. At that time, Castilian became the official language for the entire country. Regional dialects, such as Aragón, Navarre, Léon, Asturias, and Santander, survived only in secluded rural areas. The official language of Spain is now called Spanish, although the term "Castilian" is still used in Latin America.

Spanish and Portuguese have achieved worldwide importance because of the colonial activities of their European speakers. Approximately 90 percent of the speakers of these two languages live outside Europe, mainly in Central and South America. Spanish is the official language of 18 Latin American states, while Portuguese is spoken in Brazil, which has as many people as all the other South American countries combined and 15 times more than Portugal itself.

These two Romance languages were diffused to the Americas by Spanish and Portuguese explorers. The division of Central and South America into Portuguese- and Spanish-speaking regions is the result of a 1493 decision by Pope Alexander VI to give the western portion of the New World to Spain and the eastern part to Portugal. The Treaty of Tordesillas, signed a year later, carried out the papal decision.

The Portuguese and Spanish languages spoken in the Western Hemisphere differ somewhat from their European versions, as is the case with English. Brazil, Portugal, and several Portuguese-speaking countries in Africa agreed in 1994 to standardize the way their common language is written. Many people in Portugal are upset that the new standard language more closely resembles the Brazilian version, which eliminates most of the accent marks—such as tildes (São Paulo), cedillas (Alcobaça), circumflexes (Estância), and hyphens—and the agreement recognizes as standard thousands of words that Brazilians have added to the language.

The standardization of Portuguese is a reflection of the level of interaction that is possible in the modern world between groups of people who live tens of thousands of kilometers apart. Books and television programs produced in one country diffuse rapidly to other countries where the same language is used.

Key Issue 4

Why Do People Preserve Local Languages?

- Preserving language diversity
- Global dominance of English

The distribution of a language is a measure of the fate of an ethnic group. English has been diffused around the world from a small island in northwestern Europe because of the cultural dominance of England and the United States over other territory on Earth's surface. On the other hand, Icelandic has remained a little-used language because of the isolation of the Icelandic people.

As in other cultural traits, language displays the two competing geographic trends of globalization and local diversity. On the one hand, English has become the principal language of communication and interaction for the entire world. At the same time, local languages endangered by the global dominance of English are being protected and preserved.

Preserving Language Diversity

Thousands of languages are **extinct languages**, once in use—even in the recent past—but no longer spoken or read in daily activities by anyone in the world. One example is Gothic, widely spoken by people in Eastern and Northern Europe in the third century A.D. Not only is Gothic extinct, but so is the entire language group to which it belonged, the East Germanic group of the Germanic branch of Indo-European.

The last speakers of Gothic lived in the Crimea in Russia in the sixteenth century. The language died because the descendants of the Goths were converted to other languages through processes of integration, such as political dominance and cultural preference. For example, many Gothic people switched to speaking the Latin language after their conversion to Christianity.

Today, endangered languages are being preserved. The European Union has established the European Bureau of Lesser Used Languages, based in Dublin, Ireland, to provide financial support for the preservation of about two dozen languages, especially languages belonging to the Celtic branch of Indo-European.

Hebrew: Reviving Extinct Languages

Hebrew is a rare case of an extinct language that has been revived. Most of the Bible's Old Testament was written in Hebrew (a small part of it was written in another Afro-Asiatic language, Aramaic). A language of daily activity in biblical times, Hebrew diminished in use in the fourth century B.C. and was thereafter retained only for Jewish religious services. At the time of Jesus, people in present-day Israel generally spoke Aramaic, which in turn was replaced by Arabic.

When Israel was established as an independent country in 1948, Hebrew became one of the new country's two official languages, along with Arabic. Hebrew was chosen because the Jewish population of Israel consisted of refugees and migrants from many countries who spoke many languages. Because Hebrew was still used in Jewish prayers, no other language could so symbolically unify the disparate cultural groups in the new country.

The task of reviving Hebrew as a living language was formidable. Words had to be created for thousands of objects and inventions unknown in biblical times, such as telephones, cars, and electricity. The effort was initiated by Eliezer Ben-Yehuda, who lived in Palestine before the creation of the state of Israel and who refused to speak any language other than Hebrew. Ben-Yehuda is credited with the invention of 4,000 new Hebrew words—related when possible to ancient ones—and the creation of the first modern Hebrew dictionary.

Celtic: Preserving Endangered Languages

The Celtic branch of Indo-European is of particular interest to English speakers because it was the major language in the British Isles before the Germanic Angles, Jutes, and Saxons invaded. Two thousand years ago, Celtic languages were spoken in much of present-day Germany, France, and northern Italy, as well as in the British Isles. Today, Celtic languages survive only in remoter parts of Scotland, Wales, and Ireland, and on the Brittany peninsula of France.

Celtic Groups.   The Celtic language branch is divided into Goidelic (Gaelic) and Brythonic groups. Two Goidelic languages survive: Irish Gaelic and Scottish Gaelic. Irish Gaelic and English are the Republic of Ireland's two official languages, but only 75,000 people speak Irish Gaelic exclusively. In Scotland, fewer than

80,000 of the people (2 percent) speak Scottish Gaelic. An extensive body of literature exists in Gaelic languages, including the Robert Burns poem *Auld Lang Syne* ("old long since"), the basis for the popular New Year's Eve song. Gaelic was carried from Ireland to Scotland about fifteen hundred years ago.

Over time, speakers of Brythonic (also called Cymric or Britannic) fled westward to Wales, southwestward to Cornwall, or southward across the English Channel to the Brittany peninsula of France. Wales—the name derived from the Germanic invaders' word for *foreign*—was conquered by the English in 1283. However, Welsh remained dominant in Wales until the nineteenth century, when many English speakers migrated there to work in coal mines and factories. An estimated one-fourth of the people in Wales still use Welsh as their primary language, although all but a handful know English as well. In some isolated communities in the northwest, especially in the county of Gwynedd, as many as 80 percent of the people speak Welsh.

Cornish became extinct in 1777, with the death of the language's last known native speaker, Dolly Pentreath, who lived in Mousehole (pronounced "muzzle"). Before Pentreath died, an English historian recorded as much of her speech as possible, so that future generations could study the Cornish language. One of her last utterances was later translated as "I will not speak English . . . you ugly, black toad!"

In Brittany—like Cornwall, an isolated peninsula that juts out into the Atlantic Ocean—50,000 people still speak Breton. Breton differs from the other Celtic languages in that it has more French words.

Road signs in Eire (Republic of Ireland) are written in both English and Goidelic (Gaelic). The English versions of the names are displayed underneath the Goidelic. This road sign, in Ballyvaughan, County Clare, shows the way to Galway, known in Goidelic as Gallimh. (C.E. Nagele/FPG International)

The survival of any language depends on the political and military strength of its speakers. The Celtic languages declined because the Celts lost most of the territory they once controlled to speakers of other languages. In the 1300s, the Irish were forbidden to speak their own language in the presence of their English masters. By the nineteenth century, Irish children were required to wear "tally sticks" around their necks at school. The teacher carved a notch in the stick every day the child used an Irish word, and at the end of the day meted out punishment based on the number of tallies. Parents encouraged their children to learn English so that they could compete for jobs. Most remaining Celtic speakers also know the language of their English or French conquerors.

Revival of Celtic Languages.   Recent efforts have prevented the disappearance of Celtic languages. In Wales, the *Cymdeithas yr Iaith Gymraeg* (Welsh Language Society) has been instrumental in preserving the language. Britain's 1988 Education Act made Welsh language training a compulsory subject in all schools in Wales, and Welsh history and music have been added to the curriculum. All local governments and utility companies are now obliged to provide services in Welsh. Welsh-language road signs have been posted throughout Wales, and the British Broadcasting Corporation (BBC) produces Welsh-language television and radio programs.

The number of people fluent in Irish Gaelic has grown in recent years, as well, especially among younger people. Irish singers, including many rock groups (although not U2), have begun to record and perform in Gaelic. An Irish-language TV station began broadcasting in 1996. The revival is being led by young Irish living in other countries who wish to distinguish themselves from the English (in much the same way that Canadians traveling abroad often make efforts to distinguish themselves from U.S. citizens).

A couple of hundred people have now become fluent in the formerly extinct Cornish language, which was revived in the 1920s. Cornish is taught in grade schools and adult evening courses, and is used in some church services; some banks accept checks written in Cornish. However, a dispute has erupted over the proper way to spell Cornish words. Some prefer to revive the confusing, illogical medieval spellings, whereas others, including the Cornish Language Board, advocate spelling words phonetically. When officials in Camborne erected a welcome sign with the name of the town spelled "Kammbronn," traditionalists were outraged, because the medieval spelling was "Cambron." They argued that "Kammbronn" looked too "German," a harsh insult because it recalled both the successful invasion by Germanic people fifteen hundred years ago and the failed attempt by the Nazis in 1940.

The long-term decline of languages such as Celtic provides an excellent example of the precarious struggle for survival that many languages experience. Faced with the diffusion of alternatives used by people with greater political and economic strength, speakers of Celtic and other languages must make sacrifices to preserve their cultural identity.

Multilingual States

Difficulties can arise at the boundary between two languages. Note on Figure 5–3, the map of Indo-European language branches, that the boundary between the Romance and Germanic branches runs through the middle of two small European countries, Belgium and Switzerland. Belgium has had more difficulty than Switzerland in reconciling the interests of the different language speakers.

Belgium.   Southern Belgians (known as Walloons) speak French, whereas northern Belgians (known as Flemings) speak a dialect of the Germanic language of Dutch, called Flemish. The language boundary sharply divides the country into two regions. Antagonism between the Flemings and Walloons is aggravated by economic and political differences. Historically, the Walloons dominated Belgium's economy and politics, and French was the official state language (Figure 5–17).

In response to pressure from Flemish speakers, Belgium was divided into two independent regions, Flanders and Wallonia. Each elects an assembly that controls cultural affairs, public health, road construction, and urban development in its region. The national government turns over approximately 15 percent of its tax revenues to pay for the regional governments.

Motorists in Belgium clearly see the language boundary on expressways. Heading north, the highway signs suddenly change from French to Flemish at the boundary between Wallonia and Flanders. Brussels, the capital city, is an exception. Although located in Flanders, Brussels is officially bilingual and signs are in both French and Flemish. As an example, some stations on the subway map of Brussels are identified by two names, one French and one Flemish (For instance, Porte de Hal and Halle Poort—see Figure 13–24).

Belgium had difficulty fixing a precise boundary between Flemish and French speakers, because people living near the boundary may actually use the language spoken on the other side. During the late 1980s, this problem jailed one town's mayor and collapsed the national government. The town is named *Voeren* in Flemish and *Fourons* in French. Jose Happart, its mayor, refused to speak Flemish, which is required by national law because the town is in Flanders. Happart had been elected on a platform of returning the town to French Wallonia, from which it had been transferred in 1963, when the national government tried to clear up the language boundary. After refusing to be tested on his knowledge of Dutch, Happart (who in fact knew Dutch) was jailed and removed from office. In protest, French-speaking members quit the coalition governing the country, forcing the Belgian prime minister to resign.

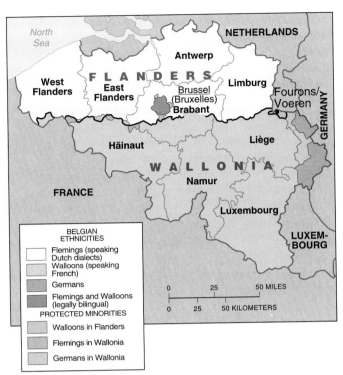

Figure 5–17 Languages in Belgium. Belgians are sharply divided by their language differences. Flemings in the north speak Flemish, a Dutch dialect. Walloons in the south speak French. The two groups have had difficulty sharing national power. As a result, considerable power has been transferred to two regional assemblies, one each for Flanders and Wallonia. In Brussels, the capital of Belgium, signs, such as this one for *The European Bookstore*, are in both French and Flemish. (D. Greco/The Image Works)

Switzerland. In contrast, Switzerland peacefully exists with multiple languages. The key is a very decentralized government, in which local authorities hold most of the power, and decisions are frequently made by voter referenda. Switzerland has four official languages: German (used by 65 percent of the population), French (18 percent), Italian (12 percent), and Romansh (1 percent). Swiss voters made Romansh an official language in a 1938 referendum, despite the small percentage who use the language (Figure 5–18).

Dominance of English

One of the most fundamental needs in a global society is a common language for communication. Increasingly in the modern world, the language of international communication is English.

A Polish airline pilot who flies over France speaks to the traffic controller on the ground in English. When well-educated speakers of two different languages wish to communicate with each other in countries such as India or Nigeria, they frequently use English. Most information on the World Wide Web is in English.

English: An Example of a Lingua Franca

A language of international communication, such as English, is known as a **lingua franca**. To facilitate trade, speakers of two different languages would create a lingua franca by mixing elements of the two language into a simple common language. The term, which means *language of the Franks*, was originally applied by Arab traders during the Middle Ages to describe the language they

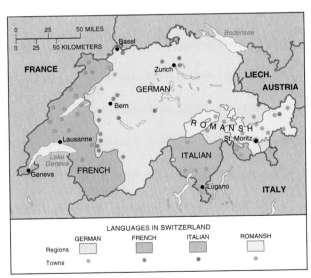

Figure 5–18 Languages in Switzerland. Switzerland lives peacefully with four official languages, including Romansh, which is used by only 1 percent of the population. Although the country can be divided into four main linguistic regions as shown, people living in individual communities, especially in the mountains, may use a language other than the prevailing local one. The Swiss, relatively tolerant of speakers of other languages, have institutionalized cultural diversity by creating a form of government that places considerable power in small communities.

English is the lingua franca in places around the world where tourists are likely to visit, such as this shrine in Macau. (Macduff Everton/The Image Works)

used to communicate with Europeans, whom they called *Franks*.

A group that learns English or another lingua franca may learn a simplified form, called a **pidgin language**. To communicate with speakers of another language, two groups construct a pidgin language by learning a few of the grammar rules and words of a lingua franca, while mixing in some elements of their own languages. A pidgin language has no native speakers—it is always spoken in addition to one's native language.

Other than English, modern lingua franca languages include Swahili in East Africa, Hindustani in South Asia, and Russian in the former Soviet Union. A number of African and Asian countries that became independent in the twentieth century adopted English or Swahili as an official language for government business, as well as for commerce, even if the majority of the people couldn't speak it.

The rapid growth in importance of English is reflected in the percentage of students learning English as a second language in school. Among European Union countries, 83 percent of high school students are learning English, including more than 90 percent of students in Denmark, Germany, the Netherlands, and Spain. Seventy percent of Europeans between ages 18 and 24 speak English.

Around the world, approximately 200 million people speak English fluently as a second language, and an unknown number have some working knowledge of the lan-

guage. Foreign students increasingly seek admission to universities in countries that teach in English rather than in German, French, or Russian. Students around the world want to learn in English as the most effective way to work in a global economy and participate in a global culture.

Expansion Diffusion of English

In the past, a lingua franca achieved widespread distribution through migration and conquest. Two thousand years ago, use of Latin spread through Europe along with the Roman Empire, and in recent centuries use of English spread around the world primarily through the British Empire. In contrast, the current growth in use of English around the world is a result not of military conquest, nor of migration by English-speaking people.

Rather, the current growth in the use of English is an example of expansion diffusion, the spread of a trait through the snowballing effect of an idea, rather than through the relocation of people. Expansion diffusion has occurred in two ways with English. First, English is changing through diffusion of new vocabulary, spelling, and pronunciation. Second, English words are fusing with other languages.

New Vocabulary, Spelling, and Pronunciation.   For a language to remain vibrant, new words and usage must always be coined to deal with new situations. Unlike most examples of expansion diffusion, though, recent changes in English have percolated up from common usage and ethnic dialects rather than directed down to the masses by elite people.

Some African-Americans speak a dialect of English heavily influenced by the group's distinctive heritage of forced migration from Africa during the eighteenth century to be slaves in the southern colonies. African-American slaves preserved a distinctive dialect in part to communicate in a code not understood by their white masters. Black dialect words such as "gumbo" and "jazz" have long since diffused into the standard English language.

In the twentieth century, many African-Americans migrated from the South to the large cities in the Northeast and Midwest (see Chapter 7). Living in racially segregated neighborhoods within northern cities, and attending segregated schools, many blacks preserved their distinctive dialect. That dialect has been termed **Ebonics**, a combination of *ebony* and *phonics*.

The American Speech, Language and Hearing Association has classified Ebonics as a distinct dialect, with a recognized vocabulary, grammar, and word meaning. Among the distinctive elements of Ebonics are the use of double negatives, such as *I'm not going there no more*, and *be* instead of *is* in such sentences as *She be at home*.

Use of Ebonics is controversial within the African-American community. On the one hand, some regard it as substandard, a measure of poor education, and an obstacle to success in the United States. Others see Ebonics as a means for preserving a distinctive element of

African-American culture and an effective way to teach African-Americans who otherwise perform poorly in school. The Oakland, California, school board voted to recognize Ebonics as a second language in 1996, but rescinded the vote after protests from many African-Americans, as well as whites.

Diffusion to Other Languages

English words have become increasingly integrated into other languages. The Japanese, for example, refer to *beisboru* (baseball), *naifu* (knife), and *sutoroberi keki* (strawberry cake). French speakers regard the invasion of English words with alarm, but Spanish speakers find the mixing of the two languages stimulating.

Franglais. Traditionally, language has been an especially important source of national pride and identity in France. The French are particularly upset with the increasing worldwide domination of English, especially the invasion of their language by English words and the substitution of English for French as the most important language of international communications.

French is an official language in 26 countries and for hundreds of years served as the lingua franca for international diplomats. Many French are upset that English words like *cowboy*, *hamburger*, *jeans*, and *T-shirt* were allowed to diffuse into the French language and destroy the language's purity. The widespread use of English

in the French language is called **franglais**, a combination of *français* and *anglais*, the French words for *French* and *English*.

Since 1635, the French Academy has been the supreme arbiter of the French language. In modern times, it has promoted the use of French terms in France, such as *stationnement* rather than *parking*, and *fin du semaine* rather than *le weekend*. *logiciel* instead of *software*. In 1994, however, France's highest court ruled that most of the country's laws banning franglais were illegal.

Protection of the French language is even more extreme in Québec, which is completely surrounded by English-speaking provinces and U.S. states (Figure 5–19). Québécois are committed to preserving their distinctive French-language culture, and to do so they may secede from Canada.

Spanglish. English is diffusing into the Spanish language spoken by 17 million Hispanics in the United States, a process called **Spanglish**, a combination of *Spanish* and *English*. In Miami's large Cuban-American community, Spanglish is sometimes called Cubonics, a combination of Cuban and phonetics.

As with franglais, Spanglish involves converting English words to Spanish forms. Some of the changes modify the spelling of English words to conform to Spanish preferences and pronunciations, such as dropping final consonants and replacing *v* with *b*. For example,

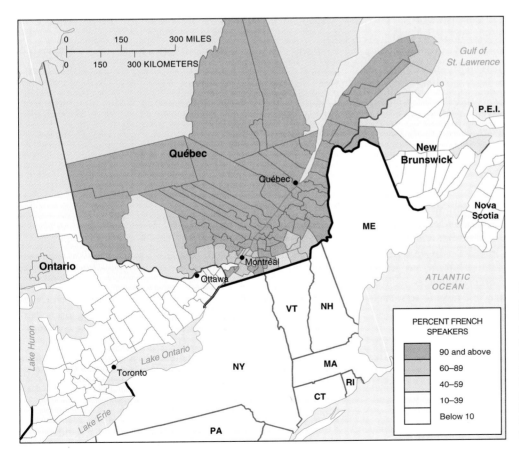

Figure 5–19 English/French language boundary in Canada. More than 80 percent of Québec's residents speak French, compared to approximately 6 percent for the rest of Canada. The boundary between Canada's French- and English-speaking regions is not precise; mixed areas exist along the borders with New Brunswick, Newfoundland, Ontario, and the United States.

(Left) French dominates in Québec City, although signs are bilingual where many English-speaking tourists venture, such as Dufferin Terrace, which offers a spectacular view of Québec's Old Lower Town and the St. Lawrence River. (R. Sidney/The Image Works) (Right) Spanish dominates along Calle Ocho (Eighth Street), the main street of Little Havana, west of downtown Miami, Florida. Many Cubans immigrated to Miami after Fidel Castro gained power in Cuba in 1959. (Tony Savino/The Image Works)

shorts (pants) becomes *chores*, and *vacuum cleaner* becomes *bacuncliner*. In other cases, awkward Spanish words or phrases are dropped in favor of English words. For example, *parquin* is used rather than *estacionamiento* for *parking*, and *taipear* is used instead of *escribir a máquina* for *to type*.

Spanglish is a richer integration of English with Spanish, than the mere borrowing of English words. New words have been invented in Spanglish that do not exist in English, but would be useful if they did. For example, *bipiar* is a verb derived from the English *beeper* that means to *beep someone on a pager*, and *i-meiliar* is a verb that means to *E-mail someone*. Spanglish also mixes English and Spanish words in the same phrase. For example, a magazine article is titled *When he says me voy . . . what does he really mean?* (*me voy* means *I'm leaving*).

Spanglish has become especially widespread in popular culture, such as song lyrics, television, and magazines aimed at young Hispanic women, but it has also been adopted by writers of serious literature. Inevitably, critics charge that Spanglish is a substitute for rigorously learning the rules of standard English and Spanish. And Spanglish has not been promoted for use in schools, as has Ebonics.

Rather than a threat to existing languages, Spanglish is generally regarded as an enriching of both English and Spanish by adopting the best elements of each—English's ability to invent new words and Spanish's ability to convey nuances of emotion. Many Hispanic-Americans like being able to say *Hablo un mix de los dos languages*.

■ Summary

The emergence of the Internet as an important means of communication has further strengthened the dominance of English. Because five-sixths of the material on the Internet is in English, knowledge of English is essential for Internet users around the world. Most E-mail systems and interactive Internet programs do not accept accent marks used in other languages such as French. Languages that are not written in Arabic letters, such as Japanese and Russian, are extremely cumbersome if not impossible to write on the Web.

The dominance of English as an international language has facilitated the diffusion of popular culture and science and the growth of international trade. However, people who forsake their native language must weigh the benefits of using English against the cost of losing a fundamental element of local cultural identity.

People in smaller countries need to learn English to participate more fully in a global economy and culture. All children learn English in the schools of countries such as the Netherlands and Sweden to facilitate international communication. This may seem culturally unfair, but obviously it is more likely that several million Dutch people will learn English, than that several hundred million English speakers around the world will learn Dutch.

In view of the global dominance of English, many U.S. citizens do not recognize the importance of learning other languages. (Does your own college or university have a foreign-language requirement for graduation?) However, one of the best ways to learn about the beliefs, traits, and values of people living in other regions is to learn their language. The lack of effort by Americans to learn other languages is a source of resentment among people elsewhere in the world, especially when Americans visit or work in other countries.

The inability to speak other languages is also a handicap for Americans who try to conduct international business. Successful entry into new overseas markets requires knowledge of local culture, and officials who can speak the local language are better able to obtain important information. Japanese businesses that wish to expand in the United States send English-speaking officials, but American businesses that wish to sell products to the Japanese are rarely able to send a Japanese-speaking employee.

Here again are the key issues raised by the geography of languages:

1. **Where are different languages distributed?** English is a language in the West Germanic group of the Germanic branch of the Indo-European language family, a collection of languages used by half the world's population. Other branches of Indo-European with large numbers of speakers are Romance, Indo-Iranian, and Balto-Slavic. The language family with the second-most speakers is Sino-Tibetan, used by 20 percent of the world's population.

2. **Why do people in different locations speak similar languages?** English is a Germanic language, because Germanic tribes invaded England fifteen hundred years ago and brought their language with them. Similarly, all Indo-European languages can be traced to a common ancestor, Proto-Indo-European. Individual languages developed from this single root through migration, followed by the isolation of one group from others who formerly spoke the same language.

3. **Why is one language spoken differently in different locations?** Speakers of English as well as other languages use a wide variety of dialects. Differences in vocabulary, spelling, and pronunciation emerge primarily because speakers of one language, especially in rural areas, are isolated from other speakers of the same language. Geographers can document the boundaries (isoglosses) that separate different dialect regions within countries, such as those that exist in the United States and the United Kingdom.

4. **Why do people in different locations speak English differently?** English has become the most important language for international communication in popular arts, science, and business. In the face of the global dominance of a lingua franca such as English, less widely used languages can face extinction, but recent efforts have been made to preserve and revive local languages, because of the importance of language as an element of cultural identity.

Case Study Revisited

The Future of French and Spanish in Anglo-America

The French-speaking people of Canada and the Spanish-speaking people of the United States both live on a continent dominated by English speakers. However, future prospects for these two languages in North America are different.

French Canada. Until recently, Québec was one of Canada's poorest and least developed provinces. Its economic and political activities were dominated by an English-speaking minority, and the province suffered from cultural isolation and lack of French-speaking leaders. In recent years, Québec has strengthened its links to France. When French President Charles de Gaulle visited Québec in 1967, he encouraged the development of an independent Québec by shouting in his speech, *"Vive le Québec libre!"* ("Long live free Québec!").

During the 1970s, the Québec government made the use of French mandatory in many daily activities. Alarmed at these pro-French policies, more than 100,000 English speakers and dozens of major corporations moved from Montréal, Québec's largest city, to English-speaking Toronto, Ontario. Many Québécois favored total separation of the province from Canada as the only way to preserve their cultural heritage. Voters in Québec have rejected separation from Canada, but by a slim majority.

In recent years, Québec has further restricted the use of languages other than French. Québec's Commission de Toponyme is renaming towns, rivers, and mountains that have names with English-language origins. The word *Stop* has been replaced by *Arrêt* on the red octagonal road signs, even though *Stop* is used throughout the world, even in France and other French-speaking countries. French must be the predominant language on all commercial signs, and the legislature passed a law banning non-French outdoor signs altogether (ruled unconstitutional by the Canadian Supreme Court).

Whether or not Québec remains part of Canada, people who wish to be integrated into the English-speaking culture and economy of North America are emigrating from Québec. Since 1969, some 500,000 people have immigrated to Québec—mostly from Italy, Greece, and Portugal—but more than 300,000 have simultaneously emigrated from Québec to English-speaking Canadian provinces.

Hispanic America. Linguistic unity is an apparent feature of the United States, a nation of immigrants who learn English to become Americans. However, the diversity of languages in the United States is greater than first appears. According to the 1990 census, 17 million speak Spanish at home, a 50 percent increase in a decade. The United States also contains more than a million speakers each of French, German, Italian, and Chinese, and more than a half-million each of Tagalog, Polish, Korean, and Vietnamese. Altogether, 32 million people in the United States speak a language other than English at home, 14 percent of the population over 5 years old.

In reaction against the increasing use of Spanish in the United States, about 20 states and a number of localities have laws making English the official language. Some courts have judged these laws to be unconstitutional restrictions on free speech. The U.S. Congress has debated enacting similar legislation. For a state like Montana, the law is symbolic, because it has few non-English speakers. But for states like California, Texas, and Florida, with large Hispanic populations, the debate affects access to jobs, education, and social services.

Americans have also debated whether schools should offer bilingual education. Some people want Spanish-speaking children to be educated in Spanish, because they think that children will learn more effectively if taught in their native language, and that this will also preserve their own cultural heritage. Others argue that learning in Spanish creates a handicap for people in the United States when they look for jobs, virtually all of which require knowledge of English. Bilingual education has also been hampered by the lack of teachers able to speak two languages and by the high cost of hiring added personnel and purchasing additional materials.

Promoting the use of English symbolizes that language is the chief cultural bond in the United States in an otherwise heterogeneous society. With the growing dominance of the English language in the global economy and culture, knowledge of English is important for people around the world, not just inside the United States. At the same time, the increasing use of other languages in the United States itself is a reminder of the importance that groups place on preserving cultural identity, and the central role that language plays in maintaining that identity.

■ Key Terms

British Received Pronunciation (BRP) The dialect of English associated with upper-class Britons living in the London area and now considered standard in the United Kingdom.

Creole or creolized language A language that results from the mixing of a colonizer's language with the indigenous language of the people being dominated.

Dialect A regional variety of a language distinguished by vocabulary, spelling, and pronunciation.

Ebonics Dialect spoken by some African-Americans.

Extinct language A language that was once used by people in daily activities but is no longer used.

Franglais A term used by the French for English words that have entered the French language, a combination of *français* and *anglais*, the French words for "French" and "English," respectively.

Ideograms The system of writing used in China and other East Asian countries in which each symbol represents an idea or a concept rather than a specific sound, as is the case with letters in English.

Isogloss A boundary that separates regions in which different language usages predominate.

Isolated language A language that is unrelated to any other languages and therefore not attached to any language family.

Language A system of communication through the use of speech, a collection of sounds understood by a group of people to have the same meaning.

Language branch A collection of languages related through a common ancestor that existed several thousand years ago. Differences are not as extensive or as old as with language families, and archaeological evidence can confirm that the branches derived from the same family.

Language family A collection of languages related to each other through a common ancestor long before recorded history.

Language group A collection of languages within a branch that share a common origin in the relatively recent past and display relatively few differences in grammar and vocabulary.

Lingua franca A language mutually understood and commonly used in trade by people who have different native languages.

Literary tradition A language that is written as well as spoken.

Official language The language adopted for use by the government for the conduct of business and publication of documents.

Pidgin language A form of speech that adopts a simplified grammar and limited vocabulary of a lingua franca, used for communications among speakers of two different languages.

Spanglish Combination of Spanish and English, spoken by Hispanic-Americans.

Standard language The form of a language used for official government business, education, and mass communications.

Vulgar Latin A form of Latin used in daily conversation by ancient Romans, as opposed to the standard dialect, which was used for official documents.

■ Thinking Geographically

1. Twenty U.S. states have passed laws mandating English as the language of all government functions. In 1990, Arizona's law making English the official language was ruled an unconstitutional violation of free speech. Should the use of English be encouraged in the United States to foster cultural integration, or should bilingualism be encouraged to foster cultural diversity? Why?

2. Does the province of Québec possess the resources, economy, political institutions, and social structures to be a viable, healthy country? What would be the impact of Québec's independence on the remainder of Canada, on the United States, and on France?

3. How is American English different from British English as a result of contributions by African-Americans and immigrants who speak languages other than English?

4. The southern portion of Belgium (Wallonia) suffers from higher rates of unemployment, industrial decline, and other economic problems, compared to Flanders in the

north. How do differences in language exacerbate Belgium's regional economic differences?

5. Many countries now receive Cable News Network (CNN) broadcasts that originate in the United States, but even English-speaking viewers in other countries have difficulty understanding some American English. A recent business program on CNN created a stir outside the United States when it reported that McDonald's was a major IRA con-tributor. Viewers in the United Kingdom thought that the American hamburger chain was financing the purchase of weapons by the Irish Republican Army, which sometimes resorts to violence in its attempt to achieve the unification of Ireland. However, McDonald's, in fact, was contributing to Individual Retirement Accounts for its employees. Can you think of other examples where the use of a word could cause a British–American misunderstanding?

■ Further Readings

Aitchison, J. W., and H. Carter. "The Welsh Language in Cardiff: A Quiet Revolution." *Transactions of the Institute of British Geographers*, New Series 12 (1987): 482–92.

Allen, Harold B. *The Linguistic Atlas of the Upper Midwest.* 3 vols. Minneapolis: University of Minnesota Press, 1973–1976.

Baugh, Albert C., and Thomas Cable. *A History of the English Language*, 3d ed. Englewood Cliffs, NJ: Prentice-Hall, 1978.

Bryson, Bill. *Made in America.* London: Martin Secker & Warburg, 1994.

Cardona, George; Henry M. Hoeningswald, and Alfred Senn, eds. *Indo-European and Indo-Europeans.* Philadelphia: University of Pennsylvania Press, 1970.

Cartwright, Don, "The Expansion of French Language Rights in Ontario, 1968–1993: The Uses of Territoriality in a Policy of Gradualism." *The Canadian Geographer* 40 (1996): 238–57.

Doran, Charles F., "Will Canada Unravel?" *Foreign Affairs* 75 (1996): 97–109.

Delgado de Carvalho, C. M. "The Geography of Languages." In *Readings in Cultural Geography*, ed. by Philip L. Wagner and Marvin W. Mikesell. Chicago: University of Chicago Press, 1962.

Dugdale, J. S. *The Linguistic Map of Europe.* London: Hutchinson University Library, 1969.

Gade, Daniel W. "Foreign Languages and American Geography." *Professional Geographer* 35 (1983): 261–65.

Gamkrelidze, Thomas V., and V.V. Ivanov. *The Indo-European Language and the Indo-Europeans.* 2 vols. The Hague: Mouton, 1990.

Greenberg, Joseph H. *Studies in African Language Classification.* Bloomington: Indiana University Press, 1963.

_____. *Language in the Americas.* Palo Alto, CA: Stanford University Press, 1987.

Hughes, Arthur, and Peter Trudgill. *English Accents and Dialects.* Birkenhead, UK: Edward Arnold, 1979.

Hymes, Dell H. *Language in Culture and Society.* New York: Harper and Row, 1964.

Kaplan, David H. "Population and Politics in a Plural Society: The Changing Geography of Canada's Linguistic Groups." *Annals of the Association of American Geographers* 84 (1994): 46–67.

_____. "Two Nations in Search of a State: Canada's Ambivalent Spatial Identities." *Annals of the Association of American Geographers* 84 (1994): 585–606.

Katzner, Kenneth. *The Languages of the World.* New York: Funk and Wagnalls, 1975.

Kirk, John M.; Stewart Sanderson, and J.D.A. Widdowson, eds. *Studies in Linguistic Geography: The Dialects of English in Britain and Ireland.* London: Croom Helm, 1985.

Krantz, Grover S. *Geographical Development of European Languages.* New York: Peter Lang, 1988.

Kurath, Hans. *Word Geography of the Eastern United States.* Ann Arbor: University of Michigan Press, 1949.

Laird, Charlton. *Language in America.* New York and Cleveland: World Publishing, 1970.

Lind, Ivan. "Geography and Place Names." In *Readings in Cultural Geography*, ed. by Philip L. Wagner and Marvin W. Mikesell. Chicago: University of Chicago Press, 1962.

Luckmann, Thomas. "Language in Society." *International Social Science Journal* 36 (1984): 5–20.

McCrum, Robert; William Cran, and Robert McNeil. *The Story of English.* New York: Viking, 1986.

Meillet, Antoine, and Marcel Cohen. *Les langues du monde.* Paris: Centre National de la Recherche Scientifique, 1952.

Muller, Siegfried H. *The World's Living Languages.* New York: Frederick Ungar, 1964.

Opie, Iona, and Peter Opie. *Children's Games in Street and Playground.* London: Clarendon Press, 1969.

Ramanujan, A. K., and Colin Masica. "A Phonological Typology of the Indian Linguistic Area." In *Current Trends in Linguistics*, ed. by Thomas A. Sebeok. Vol. 5. The Hague: Mouton, 1969.

Renfrew, Colin. *Archaeology and Language.* Cambridge: Cambridge University Press, 1988.

Sopher, David E., ed. *An Exploration of India: Geographical Perspectives on Society and Culture.* Ithaca, NY: Cornell University Press, 1980.

Thomas, Peter. "Belgium's North-South Divide and the Walloon Regional Problem." *Geography* 75 (1990): 36–50.

Trudgill, Peter. "Linguistic Geography and Geographical Linguistics." *Progress in Geography* 7 (1975): 227–52.

Wagner, Philip L. "Remarks on the Geography of Language." *Geographical Review* 48 (1958): 86–97.

Wakelin, Martyn F. *English Dialects.* London: Athlone Press, 1972.

Williams, Colin H., ed. *Language in Geographic Context.* Clevedon, UK: Multilingual Matters, 1988.

Wixman, Ronald. *Language Aspects of Ethnic Patterns and Processes in the North Caucasus.* Chicago: University of Chicago Department of Geography, 1980.

Zelinsky, Wilbur. "Generic Terms in the Place Names of the Northeastern United States." In *Readings in Cultural Geography*, ed. by Philip L. Wagner and Marvin W. Mikesell. Chicago: University of Chicago Press, 1962.

CHAPTER

Religion

And He shall judge between the nations,
And shall decide for many peoples;
And they shall beat their swords into
  ploughshares,
And their spears into pruning-hooks:
Nation shall not lift up sword against
  nation,
Neither shall they learn war any more.
                              Isaiah 2:4

This passage from the Bible, the holiest book of Christianity and Judaism, is one of the most eloquent pleas for peace among the nations of the world. For many religious people, especially in the Western Hemisphere and Europe, Isaiah evokes a highly attractive image of the ideal future landscape.

Islam's holiest book, the Quran (sometimes spelled Koran), also evokes powerful images of a peaceful landscape:

> *He it is who sends down water from the sky, whence ye have drink, and whence the trees grow whereby ye feed your flocks.*

> *He makes the corn to grow, and the olives, and the palms, and the grapes, and some of every fruit; verily, in that is a sign unto a people who reflect.*
>
> *Sûrah (Chapter) of the Bee XVI.9*

Most religious people pray for peace, but religious groups may not share the same vision of how peace will be achieved. Geographers see that the process by which one religion diffuses across the landscape may conflict with the distribution of others. Geographers are concerned with the regional distribution of different religions and the resulting potential for conflict.

Geographers also observe that religions are derived in part from elements of the physical environment, and that religions, in turn, modify the landscape. As evidence of this, note the rich agricultural images in the passages just quoted from the Bible and the Quran.

Key Issues

1. Where are religions distributed?
2. Why do religions have different distributions?
3. Why and how do religions organize space?
4. Why do territorial conflicts arise among religious groups?

Mountain Light Photography, Inc.

191

Case Study

The Dalai Lama vs. the People's Republic of China

The Dalai Lama, the spiritual leader of Tibetan Buddhists, is as important to that religion as the Pope is to Roman Catholics. Traditionally, the Dalai Lama—which translates as *oceanic teacher*—was not only the spiritual leader of Tibetan Buddhism, but also the head of the government of Tibet. The photograph on the first page of this chapter shows the Dalai Lama's former palace in Tibet's capital Lhasa, situated in the Himalaya Mountains.

China, which had ruled Tibet from 1720 until its independence in 1911, invaded the rugged, isolated country in 1950, turned it into a province named Xizang in 1951, and installed a communist government in Tibet in 1953. After crushing a rebellion in 1959, China executed or imprisoned tens of thousands and forced another 100,000, including the Dalai Lama, to emigrate. Buddhist temples were closed and demolished, and religious artifacts and scriptures were destroyed.

Why did the Chinese try to dismantle the religious institutions of a poor, remote country? At issue was the fact that the presence of strong religious feelings among the Tibetan people conflicted with the aims of the Chinese government. The conflict between traditional Buddhism and the Chinese government is one of many examples of the impact of religion. However, in the modern world of global economics and culture, local religious belief continues to play a strong role in people's lives.

Religion interests geographers because it is essential for understanding how humans occupy Earth. As always, human geographers start by asking *where* and *why*. The predominant religion varies among regions of the world, as well as among regions within North America. Geographers document *where* various religions are located in the world and offer explanations for why some religions have widespread distributions while others are highly clustered.

To understand *why* some religions have more widespread distributions than others, geographers must look at differences among practices of various faiths. Geographers, though, are not theologians, so they stay focused on those elements of religions that are geographically significant. Utilizing basic geographic concepts, geographers study the distinctive place of origin of religions, the extent of diffusion of religions from their places of origin, the processes by which religions diffused to other locations, and the religious practices and beliefs that lead some religions to have more widespread distributions.

Geographers find the tension between *globalization* and *local diversity* especially acute in religion for a number of reasons:

- People care deeply about their religion and draw from religion their core values and beliefs, an essential element of the definition of culture.
- Some religions are actually *designed* to appeal to people throughout the world, whereas other religions are designed to appeal primarily to people in geographically limited areas.
- Religious values are important in understanding not only how people identify themselves, as was the case with language, but also the meaningful ways that they organize the landscape.
- Most (though not all) religions require exclusive adherence, so adopting a global religion usually requires turning away from a traditional local religion, whereas people can learn a globally important language like English and at the same time still speak the language of their local culture.
- Like language, migrants take their religion with them to new locations, but while migrants typically learn the language of the new location, they retain their religion.

This chapter starts by describing the distribution of major religions, then in the second section explains why some religions have diffused widely whereas others have not. As a major facet of culture, religion leaves a strong imprint on the physical environment, discussed in the third section of the chapter. Religion, like other cultural characteristics, can be a source of pride and a means of identification with a distinct culture. Unfortunately, intense identification with one religion can lead adherents into conflict with followers of other religions, discussed in the fourth key issue of the chapter.

Key Issue 1

Where Are Religions Distributed?

- Universalizing religions
- Ethnic religions

Only a few religions can claim the adherence of large numbers of people. Each of these faiths has a distinctive distribution across Earth's surface (Figure 6–1).

Geographers distinguish two types of religions: universalizing and ethnic. **Universalizing religions** attempt to be global, to appeal to all people, wherever they may live in the world, not just to those of one culture or location. An **ethnic religion** appeals primarily to one group of people living in one place. About 58 percent of the world's population adhere to a universalizing religion, 23 percent to an ethnic religion, and 19 percent to no religion. This section examines the world's three main universalizing religions and some representative ethnic religions.

Universalizing Religions

The three main universalizing religions are Christianity, Islam, and Buddhism. Each of the three is divided into branches, denominations, and sects. A **branch** is a large and fundamental division within a religion. A **denomination** is a division of a branch that unites a number of local congregations in a single legal and administrative body. A **sect** is a relatively small group that has broken away from an established denomination. Two universalizing religions with smaller numbers of adherents are discussed in the Globalization and Local Diversity box.

Christianity

Christianity has about 2 billion adherents, far more than any other world religion, and has the most widespread distribution. It is the predominant religion in North America, South America, Europe, and Australia, and countries with a Christian majority exist in Africa and Asia as well.

Branches of Christianity. Christianity has three major branches: Roman Catholic, Protestant, and Eastern Orthodox. Roman Catholics comprise approximately 50 percent of the world's Christians, Protestants 24 percent, and Eastern Orthodox 11 percent. The remaining 14 percent include Catholics other than Roman and followers of a variety of African, Asian, and Latin American Christian churches.

Within Europe, Roman Catholicism is the dominant Christian branch in the southwest and east, Protestantism in the northwest, and Eastern Orthodoxy in the east and southeast (Figure 6–2). The regions of Roman Catholic and Protestant majorities frequently have sharp boundaries, even when they run through the middle of

Figure 6–1 World distribution of religions. More than two-thirds of Earth's people adhere to one of four religions:

- *Christianity*: 35 percent, especially in Europe and the Western Hemisphere
- *Islam*: 19 percent, especially in northern Africa and Southwest and Southeast Asia
- *Hinduism*: 14 percent, virtually all in India
- *Buddhism*: 6 percent, especially in East and Southeast Asia.

About 6 percent adhere to one of a number of ethnic religions based in China, Japan, and Korea. Three percent belong to another ethnic religion. The remaining 19 percent are nonreligious or atheists. In China and sub-Saharan Africa, many people profess adherence to both an ethnic and a universalizing religion. The small pie charts on the map show the overall proportion of the world's religions in each world region. The large pie chart below shows the worldwide percentage of people adhering to the various religions.

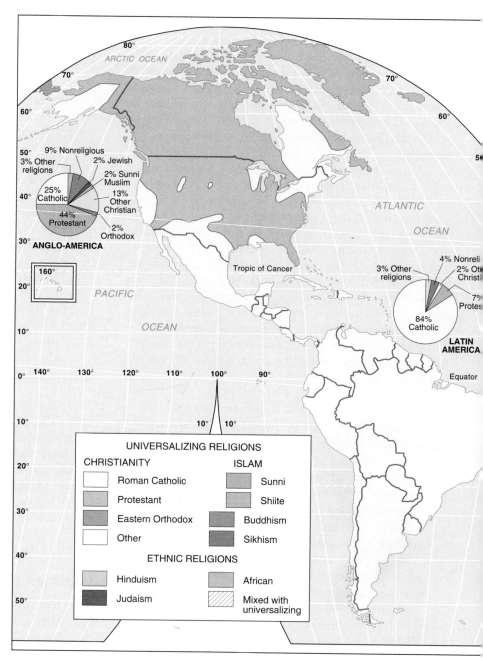

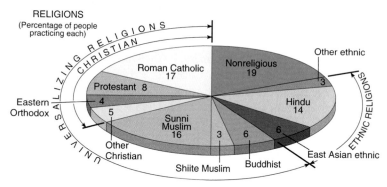

RELIGIONS
(Percentage of people practicing each)

countries. For example, the Netherlands and Switzerland have approximately equal percentages of Roman Catholics and Protestants, but the Roman Catholic populations are concentrated in the south of these countries and the Protestant populations in the north.

The Eastern Orthodox branch of Christianity is a collection of 14 self-governing churches in Eastern Europe and the Middle East. More than 40 percent of all Eastern Orthodox Christians belong to one of these 14, the Russian Orthodox Church. Christianity came to Russia

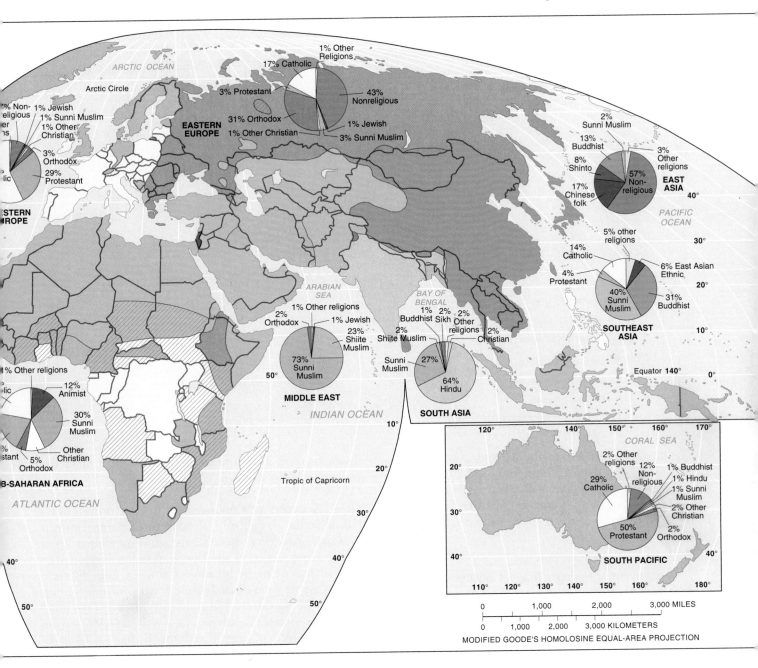

MODIFIED GOODE'S HOMOLOSINE EQUAL-AREA PROJECTION

in the tenth century, and the Russian Orthodox Church was established in the sixteenth century.

Nine of the other 13 self-governing churches were established in the nineteenth or twentieth century. The largest of these nine, the Romanian church, includes 20 percent of all Eastern Orthodox Christians. The Bulgarian, Greek, and Serbian Orthodox churches have approximately 10 percent each. The other five recently established Orthodox churches—Albania, Cyprus, Georgia, Poland, and Sinai—combined have about 2 percent of all Eastern Orthodox Christians.

The remaining four of the 14 Eastern Orthodox churches—Constantinople, Alexandria, Antioch, and Jerusalem—trace their origins to the earliest days of Christianity. They have a combined membership of about 3 percent of all Eastern Orthodox Christians.

Christianity in the Western Hemisphere. The overwhelming percentage of people living in the Western Hemisphere are Christian, including 84 percent of North Americans and 96 percent of Latin Americans. About 7 percent in North America and 3 percent in Latin America belong to other religions. The remaining 9 percent in North America and 1 percent in Latin America are not religious.

A fairly sharp boundary exists within the Western Hemisphere in the predominant branches of Christianity. Roman Catholics comprise 88 percent of Christians in Latin America, compared with 30 percent in North America. Within North America, Roman Catholics are clustered in the southwestern and northeastern United States and the Canadian province of Québec (Figure 6–3).

Figure 6–2 Branches of Christianity in Europe. In the United Kingdom, Germany, and Scandinavia, the majority adhere to a Protestant denomination. In Eastern and Southeastern Europe, Eastern Orthodoxy dominates. Roman Catholicism is dominant in Southern, Central, and Southwestern Europe.

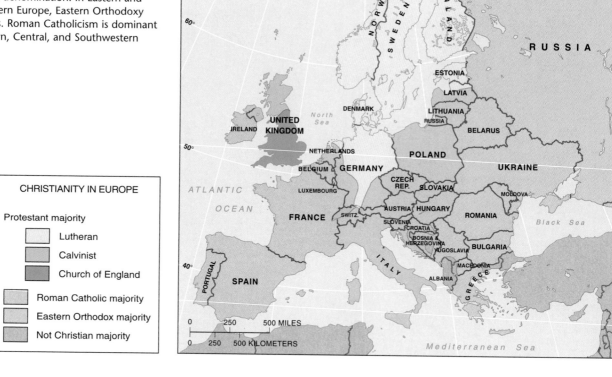

The three largest Protestant denominations in the United States are Baptist, Methodist, and Pentecostal, followed by Lutheran, Latter-Day Saint, Presbyterian, and Episcopal. The percentages of Americans who belong to the largest churches are as follows:

14 percent a Baptist church

  6 percent a Southern Baptist Convention church

  4 percent a National Baptist Convention U.S.A. or America church

  4 percent another Baptist church

  5 percent a Methodist church

  3 percent a United Methodist church

  2 percent an African Methodist Episcopal or Episcopal Zion church

  4 percent a Pentecostal church

  2 percent a Church of God in Christ

  1 percent one of the Assemblies of God churches

  1 percent another Pentecostal church

  3 percent a Lutheran church

  2 percent a Latter-Day Saint church

  2 percent a Presbyterian church

  1 percent an Episcopal church

Membership in some Protestant churches varies by region of the United States. Baptists, for example, are highly clustered in the southeast, Lutherans in the upper midwest, and Latter-Day Saints in Utah. Other Christian denominations are more evenly distributed around the country.

Other Branches of Christianity. Several other Christian churches developed independent of the three main branches. Many of these Christian communities were isolated from others at an early point in the development of Christianity, partly because of differences in doctrine and partly as a result of Islamic control of intervening territory in Southwest Asia and North Africa. Two small Christian churches survive in northeast Africa: the Coptic Church of Egypt and the Ethiopian Church. The Ethiopian Church, with perhaps 10 million adherents, split from the Egyptian Coptic Church in 1948, although it traces its roots to the fourth century, when two shipwrecked Christians, who were taken as slaves, ultimately converted the Ethiopian king to Christianity.

The Armenian Church originated in Antioch, Syria, and was important in diffusing Christianity to South and East Asia between the seventh and thirteenth centuries. The church's few present-day adherents are concentrated in Lebanon and Armenia, as well as northeastern Turkey and western Azerbaijan. Despite the small number of adherents, the Armenian Church, like other small sects, plays a significant role in regional conflicts. For example, Armenian Christians have fought for the independence of Nagorno-Karabakh, a portion of Azerbaijan, because Nagorno-Karabakh is predominantly Armenian, whereas the remainder of Azerbaijan is overwhelmingly Shiite Muslim (see Chapter 8).

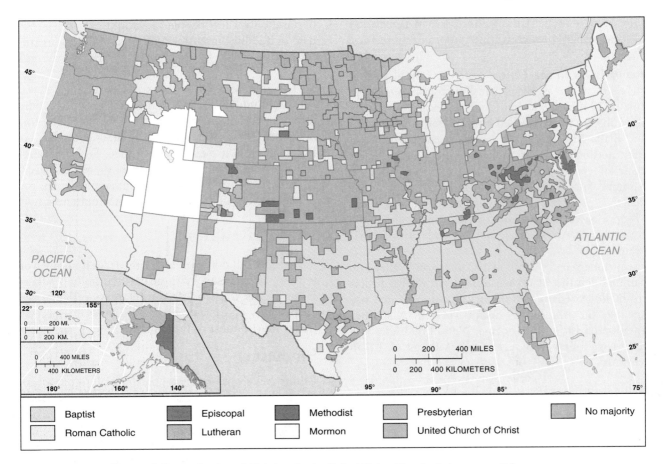

Figure 6–3 Distribution of Christians in the United States. The shaded areas are U.S. counties in which more than 50 percent of church membership is concentrated in either Roman Catholicism or one Protestant denomination. Baptists are concentrated in the Southeast, Lutherans in the Upper Midwest, Mormons in Utah and contiguous states, and Roman Catholics in the Northeast and Southwest. The distinctive distribution of religious groups within the United States results from patterns of migration, especially from Europe in the nineteenth century and from Latin America in recent years.

The Maronites are another example of a small Christian sect that plays a disproportionately prominent role in political unrest. They are clustered in Lebanon, which has suffered through a long civil war fought among religious groups (see Chapter 7).

Islam

Islam, the religion of more than 1 billion people, dominates a region from North Africa to Central Asia, from Morocco to Pakistan (Figure 6–1). The two most important concentrations of Muslims outside this region are in Bangladesh and Indonesia. The word *Islam* in Arabic means *submission to the will of God*, and it has a similar root to the Arabic word for *peace*. An adherent of the religion of Islam is known as a Muslim, which in Arabic means *one who surrenders to God*.

The core of Islamic belief is represented by five pillars of faith:

1. There is no god worthy of worship except the one God, the source of all creation, and Muhammad is the messenger of God.

2. Five times daily, a Muslim prays, facing the city of Makkah (Mecca), as a direct link to God.

3. A Muslim gives generously to charity, as an act of purification and growth.

4. A Muslim fasts during the month of Ramadan, as an act of self-purification.

5. If physically and financially able, a Muslim makes a pilgrimage to Makkah.

Branches of Islam. Islam is divided into two important branches: *Sunni* (from the Arabic word for *orthodox*) and *Shiite* (from the Arabic word for *sectarian* and sometimes written *Shia* in English). Sunnis comprise 83 percent of Muslims and are the largest branch in most Muslim countries (see Figure 6–1).

Sixteen percent of Muslims are Shiites, clustered in a handful of countries. About 70 percent of all Shiites live in Iran, where they constitute about 90 percent of the country's population, and another 15 percent live in Iraq, where they are twice as numerous as Sunnis. Shiites also outnumber Sunnis in Azerbaijan, Lebanon, and

Bahrain. Most of the remaining Shiites live in Yemen and Afghanistan, where they constitute important minorities.

Nation of Islam in the United States. The Nation of Islam, also known as Black Muslims, was founded in Detroit in 1930 and was led for more than 40 years by Elijah Muhammad, who called himself "the messenger of Allah." Black Muslims lived austerely and advocated a separate autonomous nation within the United States for their adherents. During the 1960s, tension between Muhammad and a Black Muslim minister Malcolm X divided the sect. After a pilgrimage to Makkah in 1963, Malcolm X converted to orthodox Islam and founded the Organization of Afro-American Unity. He was assassinated in 1965.

Since Muhammad's death in 1975, his son Wallace D. Muhammad led the Black Muslims closer to the principles of orthodox Islam, and the organization's name was changed to the American Muslim Mission. A splinter group adopted the original name, Nation of Islam, and continues to follow the separatist teachings of Elijah Muhammad.

Buddhism

Buddhism, the third of the world's major universalizing religions, has over 300 million adherents, especially in China and Southeast Asia (refer to Figure 6–1). The foundation of Buddhism is represented by these concepts, known as the Four Noble Truths:

1. All living beings must endure suffering.
2. Suffering, which is caused by a desire to live, leads to reincarnation (repeated rebirth in new bodies or forms of life).
3. The goal of all existence is to escape from suffering and the endless cycle of reincarnation into Nirvana (a state of complete redemption), which is achieved through mental and moral self-purification.
4. Nirvana is attained through an Eightfold Path, which includes rightness of belief, resolve, speech, action, livelihood, effort, thought, and meditation.

Branches of Buddhism. Like the other two universalizing religions, Buddhism split into more than one branch, as followers disagreed on interpreting statements by the founder Siddhartha Gautama. The two main branches are Theravada and Mahayana.

Theravada Buddhism is most prevalent in Southeast Asia, especially Cambodia, Laos, Myanmar, Sri Lanka, and Thailand. Mahayana Buddhism predominates farther north in Asia, including China, Japan, Korea, Mongolia, and Tibet. Mahayana Buddhism is divided into at least six distinct denominations. Because the Mahayana denominations do not occupy distinct geographical areas, geographers cannot map their spatial distribution.

Buddhism currently has more than 300 million adherents, although an accurate count is especially difficult,

because only a few people in Buddhist countries participate in Buddhist institutions. Religious functions are performed primarily by monks rather than by the general public.

The number of Buddhists is also difficult to count, because Buddhism, although a universalizing religion, differs in significant respects from the Western concept of a formal religious system. Someone can be both a Buddhist and a believer in other Eastern religions, whereas Christianity and Islam both require exclusive adherence. Most Buddhists in China and Japan, in particular, believe in an ethnic religion simultaneously.

Ethnic Religions

The ethnic religion with by far the largest number of followers is Hinduism. With nearly 800 million adherents, Hinduism is the world's third largest religion, behind Christianity and Islam. Ethnic religions in Asia and Africa comprise most of the remainder.

Hinduism

Ethnic religions typically have much more clustered distributions than do universalizing religions. Although Hinduism is the world's third largest religion, 97 percent of Hindus are concentrated in one country—India—2 percent are in the neighboring country of Nepal, and the remaining 1 percent are dispersed around the world. Hindus comprise 83 percent of the population of India and 90 percent of Nepal, and a small minority in every other country.

Hinduism adheres to the belief that there is more than one path to reach God. Because people start from different backgrounds and experiences, the appropriate form of worship for any two individuals may not be the same. Hinduism does not have a central authority or a single holy book, so that each individual selects suitable rituals. If one person practices Hinduism in a particular way, other Hindus will not think that the individual has made a mistake or strayed from orthodox doctrine.

The average Hindu has allegiance to a particular god or concept within a broad range of possibilities. The three approaches that have the largest number of followers are probably Sivaism, Vaishnavism, and Shaktism. Although various deities and approaches are supported throughout India, some geographic concentration exists: Siva and Shakti in the north, Shakti and Vishnu in the east, Vishnu in the west, and Siva, along with some Vishnu, in the south. However, holy places for Siva and Vishnu are dispersed throughout India.

Other Ethnic Religions

Several hundred million people practice ethnic religions in East Asia, especially in China and Japan. The coexistence of Buddhism with these ethnic religions in East Asia differs from the Western concept of exclusive religious belief. Confucianism and Daoism (sometimes spelled Taoism) are often distinguished as separate ethnic religions in China, but many Chinese consider themselves both Buddhists and either Confucian, Daoist, or some other Chinese ethnic religion.

Buddhism does not compete for adherents with Confucianism, Daoism, and other ethnic religions in China, because many Chinese accept the teachings of both universalizing and ethnic religions. Such commingling of diverse philosophies is not totally foreign to Americans. The tenets of Christianity or Judaism, the wisdom of the ancient Greek philosophers, and the ideals of the Declaration of Independence can all be held dear without doing grave injustice to the others.

Confucianism. Confucius (551–479 B.C.) was a philosopher and teacher in the Chinese province of Lu. His sayings, which were recorded by his students, emphasized the importance of the ancient Chinese tradition of *li*, which can be translated roughly as *propriety* or *correct behavior*. Confucianism is an ethnic religion because of its especially strong rooting in traditional values of special importance to Chinese people.

Confucianism prescribed a series of ethical principles for the orderly conduct of daily life in China, such as following traditions, fulfilling obligations, and treating others with sympathy and respect. These rules applied to China's rulers, as well as to their subjects.

Daoism (Taoism). Lao-Zi (604–531? B.C., also spelled Lao Tse), a contemporary of Confucius, organized Daoism. Although a government administrator by profession, Lao-Zi's writings emphasized the mystical and magical aspects of life rather than the importance of public service, as had Confucius.

Daoists seek *dao* (or *tao*), which means the *way* or *path*. A virtuous person draws power (*de* or *te*) from being absorbed in *dao*. *Dao* cannot be comprehended by reason and knowledge, because not everything is knowable. Because the universe is not ultimately subject to rational analysis, myths and legends develop to explain events. Only by avoidance of daily activities and introspection can a person live in harmony with the principles that underlie and govern the universe.

Daoism split into many sects, some acting like secret societies, and followers embraced elements of magic. The religion was officially banned by the Communists after they took control of China in 1949, but it is still practiced in China, and it is legal in Taiwan.

Shintoism. Since ancient times, Shintoism has been the distinctive ethnic religion of Japan. Ancient Shintoists considered forces of nature to be divine, especially the Sun and Moon, as well as rivers, trees, rocks, mountains, and certain animals. The religion was transmitted from one generation to the next orally until the fifth century A.D., when the introduction of Chinese writing facilitated the recording of ancient rituals and prayers. Gradually, deceased emperors and other ancestors became more important deities for Shintoists than natural features.

Under the reign of the Emperor Meiji (1868–1912), Shintoism became the official state religion, and the emperor was regarded as divine. Shintoism therefore was as much a political cult as a religion, and in a cultural sense all Japanese were Shintoists. After defeating Japan in World War II, the victorious Allies ordered Emperor Hirohito to renounce his divinity in a speech to the Japanese people, although he was allowed to retain ceremonial powers.

Shintoism still thrives in Japan, although no longer as the official state religion. Prayers are recited to show reverance for ancestors, and pilgrimages are made to shrines believed to house deities.

Judaism. About 6 million Jews live in the United States, 4 million in Israel, 2 million in former Soviet Union republics, especially Russia, Belarus, Lithuania, and Ukraine, and 2 million elsewhere. Within the United States, Jews are heavily concentrated in the large cities, including one-third in the New York area alone. Jews constitute a majority in Israel, where for the first time since the biblical era an independent state has had a Jewish majority.

The number of Jews living in the former Soviet Union has declined rapidly since the late 1980s, when emigration laws were liberalized. For many years, their religious practices were strongly discouraged, but few were allowed to emigrate. The number of synagogues in the Soviet Union declined from about 400 in 1960 to 62 by 1975. Conditions for Jews remaining in Eastern Europe have worsened since the fall of communism, because they have been blamed by some for recent economic problems.

Judaism plays a more substantial role in Western civilization than its number of adherents would suggest, because two of the three main universalizing religions—Christianity and Islam—find some of their roots in Judaism. Jesus was born a Jew, and Muhammad traced his ancestry to Abraham.

Judaism is an ethnic religion based in the lands bordering the eastern end of the Mediterranean Sea, called Canaan in the Bible, Palestine by the Romans, and the state of Israel since 1948. About four thousand years ago, Abraham, considered the patriarch or father of Judaism, migrated from present-day Iraq to Canaan, along a route known as the Fertile Crescent (see discussion of the Fertile Crescent in Chapter 8 and Figure 8–3). The Old Testament recounts the ancient history of the Jewish people.

Fundamental to Judaism was belief in one all-powerful God. It was the first recorded religion to espouse **monotheism**, belief that there is only one God. Judaism offered a sharp contrast to the **polytheist** practice of neighboring people, who worshipped a collection of gods. Jews considered themselves the "chosen" people, because God had selected them to live according to His ethical and moral principles, such as the Ten Commandments.

The name *Judaism* derives from *Judah*, one of the patriarch Jacob's 12 sons; *Israel* is another biblical name for Jacob. Descendants of ten of Jacob's sons, plus two of his grandsons, constituted the 12 tribes of Hebrews who emigrated from Egypt in the Exodus narrative. Each received a portion of Canaan. Judah is one of the surviving tribes of Hebrews; ten of the tribes were considered lost after they were conquered and forced to migrate to Assyria in 721 B.C.

Ethnic African Religions. About 10 percent of Africans follow traditional ethnic religions, sometimes called **animism**. Animists believe that such inanimate objects as plants and stones or such natural events as thunderstorms and earthquakes are "animated," or have discrete spirits and conscious life. Relatively little is known about African religions because few holy books or other written documents have come down from ancestors. Religious rituals are passed from one generation to the next by word of mouth.

African animist religions are apparently based on monotheistic concepts, although below the supreme god there is a hierarchy of divinities. These divinities may be assistants to god or personifications of natural phenomena, such as trees or rivers.

As recently as 1980, some 200 million Africans—half the population of the continent—were classified as animists. Some atlases and textbooks persist in classifying Africa as predominantly animist, even though the actual percentage is small and declining. Followers of traditional African religions now constitute a clear majority in Mozambique and four small West African countries, including Benin, Guinea-Bissau, Liberia, and Togo. About half the population is animist in Angola, Botswana, Cameroon, Congo Republic, Madagascar, and Zimbabwe.

The rapid decline in animists in Africa has been caused by increases in the numbers of Christians and Muslims. Africa is now nearly 50 percent Christian—split about evenly among Roman Catholic, Protestant, and other—and another 40 percent are Muslims. The growth in the two universalizing religions at the expense of ethnic religions reflects fundamental geographical differences between the two types of religions, discussed in the next key issue.

Key Issue 2

Why Do Religions Have Different Distributions?

- Origin of religions
- Diffusion of religions
- Holy places
- The calendar

We can identify several major geographical differences between universalizing and ethnic religions. These dif-ferences include the locations where the religions originated, the processes by which they diffused from their place of origin to other regions, the types of places that are considered holy, the calendar dates identified as important holidays, and attitudes toward modifying the physical environment.

Origin of Religions

Universalizing religions have precise places of origin, based on events in the life of a man. Ethnic religions have unknown or unclear origins, not tied to single historical individuals.

Origin of Universalizing Religions

Each of the three universalizing religions can be traced to the actions and teachings of a man who lived since the start of recorded history. The beginnings of Buddhism go back about 2,500 years, Christianity 2,000 years, and Islam 1,500 years. Specific events also led to the division of the universalizing religions into branches.

Origin of Christianity. Christianity was founded upon the teachings of Jesus, who was born in Bethlehem between 8 and 4 B.C. and died on a cross in Jerusalem about A.D. 30. Raised as a Jew, Jesus gathered a small band of disciples and preached the coming of the Kingdom of God. The four Gospels of the Christian Bible—Matthew, Mark, Luke, and John—documented miracles and extraordinary deeds that Jesus performed. He was referred to as *Christ*, from the Greek word for the Hebrew word *messiah*, which means *anointed*.

In the third year of his mission, he was betrayed to the authorities by one of his companions, Judas Iscariot. After sharing the Last Supper (the Jewish Passover seder) with his disciples in Jerusalem, Jesus was arrested and put to death as an agitator. On the third day after his death, his tomb was found empty. Christians believe that Jesus died to atone for human sins, that he was raised from the dead by God, and that his Resurrection from the dead provides people with hope for salvation.

Roman Catholics accept the teachings of the Bible, as well as the interpretation of those teachings by the Church hierarchy, headed by the Pope. According to Roman Catholic belief, God conveys His grace directly to humanity through seven sacraments, including Baptism, Confirmation, Penance, Anointing the sick, Matrimony, Holy Orders, and the Eucharist (the partaking of bread and wine that repeats the actions of Jesus at the Last Supper). Roman Catholics believe that the Eucharist literally and miraculously become the body and blood of Jesus while keeping only the appearances of bread and wine, an act known as transubstantiation.

Eastern Orthodoxy comprises the faith and practices of a collection of churches that arose in the Eastern part of the Roman Empire. The split between the Roman and Eastern churches dates to the fifth century, as a result of rivalry between the Pope of Rome and the Patriarchy of Constantinople, which was especially intense after the collapse of the Roman Empire. The split between the two churches became final in 1054, when Pope Leo IX condemned the Patriarch of Constantinople. Eastern Orthodox Christians accepted the seven sacraments but rejected doctrines that the Roman Catholic Church had added since the eighth century.

Protestantism originated with the principles of the Reformation in the sixteenth century. The Reformation movement is regarded as beginning when Martin Luther posted 95 theses on the door of the church at Wittenberg on October 31, 1517. According to Luther, individuals had primary responsibility for achieving personal salvation through direct communication with God. Grace is achieved through faith rather than through sacraments performed by the Church.

Origin of Islam. Islam traces its origin to the same narrative as Judaism and Christianity. All three religions consider Adam to have been the first man and Abraham to have been one of his descendants. According to legend, Abraham married Sarah, who did not bear children. Polygamy being a custom of the culture, Abraham then married Hagar, who bore a son, Ishmael. However, Sarah's fortunes changed, and she bore a son, Isaac. Sarah then successfully prevailed upon Abraham to banish Hagar and Ishmael.

Jews and Christians trace their story through Abraham's original wife and son, Sarah and Isaac. Muslims trace their story through his second wife and son, Hagar and Ishmael. After their banishment, Ishmael and Hagar wandered through the Arabian desert, eventually reaching Makkah (spelled Mecca on many English-language maps), in present-day Saudi Arabia. Centuries later, one of Ishmael's descendants, Muhammad, became the Prophet of Islam.

Muhammad was born in Makkah about A.D. 570. At age 40, while engaged in a meditative retreat, Muhamad received his first revelation from God through the Angel Gabriel. The Quran, the holiest book in Islam, is a record of God's words, as revealed to the Prophet Muhammad through Gabriel. Arabic is the lingua franca, or language of communication, within the Muslim world, since it is the language in which the Quran is written.

As he began to preach the truth that God had revealed to him, Muhammad suffered persecution, and in 622 he was commanded by God to emigrate. His migration from Makkah to the city of Yathrib—an event known as the *Hijra* (from the Arabic word for *migration*, sometimes spelled *hegira*)—marks the beginning of the Muslim calendar. Yathrib was subsequently renamed Madina, Arabic for *the City of the Prophet*. After several years, Muhammad and his followers returned to Makkah and established Islam as the city's religion. By Muhammad's death in 632 at about age 63, Islam had diffused to most of present-day Saudi Arabia.

Differences between the two main branches—Shiites and Sunnis—go back to the earliest days of Islam, and basically reflect disagreement over the line of succession in Islamic leadership. Muhammad had no surviving son, and no follower of comparable leadership ability. His successor was Abu Bakr (573–634), an early supporter from Makkah, who became known as *caliph* ("successor of the prophet"). The next two caliphs, Umar (634–644) and Uthman (644–656), expanded the territory under Muslim influence to Egypt and Persia.

Uthman was a member of a powerful Makkah clan that had initially opposed Muhammad before the clan's conversion to Islam. More zealous Muslims criticized Uthman for seeking compromises with other formerly pagan families in Makkah. Uthman's opponents found a leader in Ali (600?–661), a cousin and son-in-law of Muhammad, and thus Muhammad's nearest male heir. When Uthman was murdered in 656, Ali became caliph, although 5 years later he, too, was assassinated.

Ali's descendants claim leadership of Islam, and Shiites support this claim. But Shiites disagree among themselves about the precise line of succession from Ali to modern times. They acknowledge that the chain of leadership was broken, but they dispute the date and events surrounding the disruption. During the 1970s, both the shah (king) of Iran and an ayatollah (religious scholar) named Khomeini claimed to be the divinely appointed interpreter of Islam for the Shiites. The allegiance of the Iranian Shiites switched from the shah to the ayatollah largely because the ayatollah made a more convincing case that he was more faithfully adhering to the rigid laws laid down by Muhammad in the Quran.

Origin of Buddhism. The founder of Buddhism, Siddhartha Gautama, was born about 563 B.C. in Lumbinī, in present-day Nepal, near the border with India, about 160 kilometers (100 miles) from Vāranāsi (Benares). The son of a lord, he led a privileged existence sheltered from life's hardships. Gautama had a beautiful wife, palaces, and servants.

According to Buddhist legend, Gautama's life changed after a series of four trips. He encountered a decrepit old man on the first trip, a disease-ridden man on the second trip, and a corpse on the third trip. After witnessing these scenes of pain and suffering, Gautama began to feel he could no longer enjoy his life of comfort and security. Then, on a fourth trip, Gautama saw a monk, who taught him about withdrawal from the world.

At age 29, Gautama left his palace one night and lived in a forest for the next 6 years, thinking and experimenting with forms of meditation. Gautama emerged as the *Buddha*, the "awakened or enlightened one," and spent 45 years preaching his views across India. In the process, he trained monks, established orders, and preached to the public.

Theravada is the older of the two main branches of Buddhism. The word means "the way of the elders," indicating the Theravada Buddhists' belief that they are closer to Buddha's original approach. Theravadists believe that Buddhism is a full-time occupation, so to become a good Buddhist, one must renounce worldly goods and become a monk.

Mahayana split from Theravada Buddhism about 2,000 years ago. *Mahayana* is translated as "the bigger ferry" or "raft," and Mahayanists call Theravada Buddhism by the name *Hinayana*, or "the little raft." Mahayanists claim that their approach to Buddhism can help more people because it is less demanding and all-encompassing. While the Theravadists emphasize Buddha's life of self-help and years of solitary introspection, Mahayanists emphasize Buddha's later years of teaching and helping others. The Theravadists cite Buddha's wisdom, the Mahayanists his compassion.

Origin of Hinduism, an Ethnic Religion

Unlike the three universalizing religions, Hinduism did not originate with a specific founder. The word *Hinduism* is simply a term for *the religious system of India*. While the origins of Christianity, Islam, and Buddhism are recorded in the relatively recent past, Hinduism existed prior to recorded history. The earliest surviving Hindu documents were written around 1500 B.C., although archaeological explorations have unearthed older objects relating to the religion. Aryan tribes from Central Asia invaded India about 1400 B.C. and brought with them

Indo-European languages, as discussed in Chapter 5. In addition to their language, the Aryans brought their religion.

The Aryans first settled in the area now called the Punjab in northwestern India, and later migrated east to the Ganges River valley, as far as Bengal. Centuries of intermingling with the Dravidians already living in the area modified their religious beliefs.

Diffusion of Religions

The three universalizing religions diffused from specific hearths, or places of origin, to other regions of the world. In contrast, ethnic religions typically remain clustered in one location.

Diffusion of Universalizing Religions

The hearths where each of the three universalizing religions originated are based on the events in the lives of the three key individuals (Figure 6–4). All three hearths are in Asia (Christianity and Islam in Southwest Asia, Buddhism in South Asia). Followers transmitted the messages preached in the hearths to people elsewhere, diffusing them across Earth's surface along distinctive paths, as shown in Figure 6–4. Today, these three universalizing religions together have several billion adherents distributed across wide areas of the world.

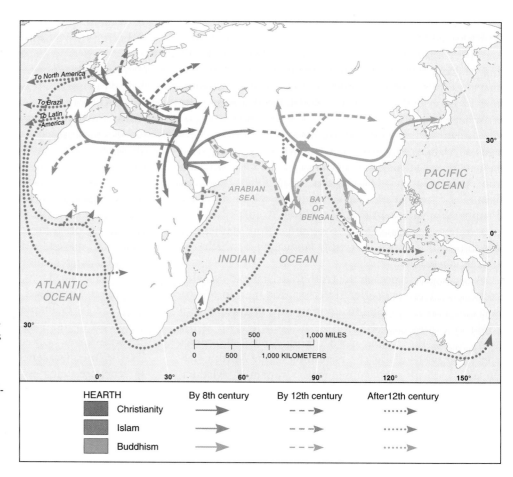

Figure 6–4 Diffusion of universalizing religions. Buddhism's hearth is in present-day Nepal and northern India, Christianity's in present-day Israel, and Islam's in present-day Saudi Arabia. Buddhism diffused primarily east toward East and Southeast Asia, Christianity west toward Europe, and Islam west toward northern Africa and east toward southwestern Asia.

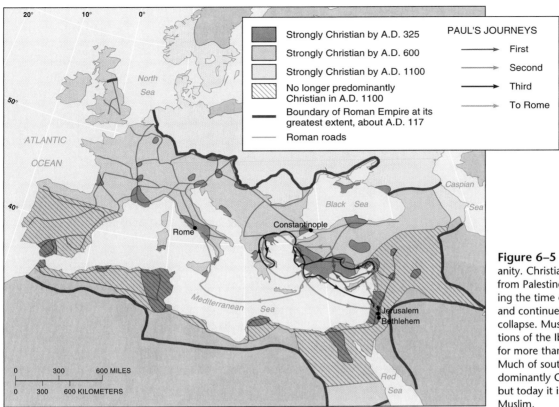

Figure 6–5 Diffusion of Christianity. Christianity began to diffuse from Palestine through Europe during the time of the Roman Empire, and continued after the empire's collapse. Muslims controlled portions of the Iberian Peninsula (Spain) for more than 700 years, until 1492. Much of southwestern Asia was predominantly Christian at one time, but today it is predominantly Muslim.

Origin and Diffusion. How did Christianity become the world's most practiced and most widely distributed religion? Christianity's diffusion has been rather clearly recorded since Jesus first set forth its tenets in the Roman province of Palestine. Consequently, geographers can examine its diffusion by reconstructing patterns of communications, interaction, and migration.

In Chapter 1, we identified two processes of diffusion—relocation (diffusion through migration) and expansion (diffusion through a snowballing effect)—and within expansion diffusion we distinguished between hierarchical (diffusion through key leaders) and contagious (widespread diffusion). Christianity diffused through a combination of all of these forms of diffusion.

Christianity first diffused from its hearth in Palestine through relocation diffusion. **Missionaries**—individuals who help to transmit a universalizing religion through relocation diffusion—carried the teachings of Jesus along the Roman Empire's protected sea routes and excellent road network to people in other locations. Paul of Tarsus, a disciple of Jesus, traveled especially extensively through the Roman Empire as a missionary. The outline of the empire, and spread of Christianity, are shown in Figure 6–5.

People in commercial towns and military settlements that were directly linked by the communications network received the message first from Paul and other missionaries. But Christianity spread widely within the Roman Empire through contagious diffusion—daily contact between believers in the towns and nonbelievers in the surrounding countryside. **Pagan**, the word for a follower of a polytheistic religion in ancient times, derives from the Latin word for *countryside*.

The dominance of Christianity throughout the Roman Empire was assured during the fourth century through hierarchical diffusion—acceptance of the religion by the empire's key elite figure, the emperor. Emperor Constantine encouraged the spread of Christianity by embracing it in A.D. 313, and Emperor Theodosius proclaimed it the empire's official religion in 380. In subsequent centuries, Christianity further diffused into Eastern Europe through conversion of kings or other elite figures.

Migration and missionary activity by Europeans since the year 1500 has extended Christianity to other regions of the world, as shown in Figure 6–1. Through permanent resettlement of Europeans, Christianity became the dominant religion in North and South America, Australia, and New Zealand. Christianity's dominance was further achieved by conversion of indigenous populations and by intermarriage. In recent decades, Christianity has further diffused to Africa, where it is now the most widely practiced religion.

Latin Americans are predominantly Roman Catholic because their territory was colonized by the Spanish and Portuguese, who brought with them to the Western Hemisphere their religion as well as their languages. Canada (except Québec) and the United States have Protestant majorities because their early colonists came primarily from Protestant England.

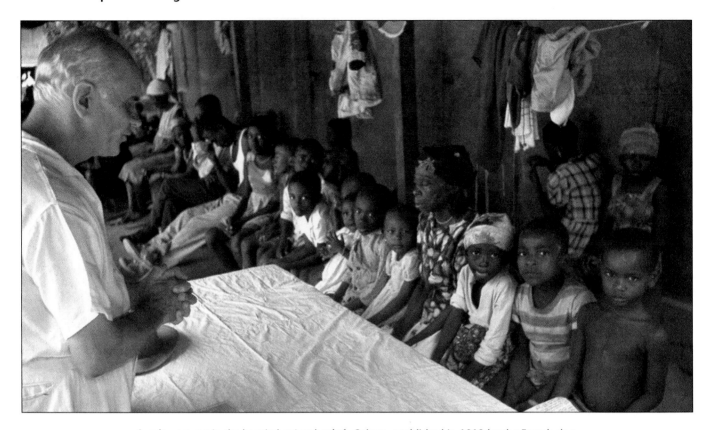

Sunday prayers in the hospital at Lambaréné, Gabon, established in 1913 by the French doctor and theologian Albert Schweitzer (1875–1965). For most of his life, Dr. Schweitzer resided in Lambaréné, where he taught modern medical practices and Christian ethics. He was awarded the Nobel Peace Prize in 1952 for his work in Gabon. (Tom Pix/Peter Arnold, Inc.)

Some regions and localities within the United States and Canada are predominantly Roman Catholic because of immigration from Roman Catholic countries (refer to Figure 6–3). New England and large midwestern cities such as Cleveland, Chicago, Detroit, and Milwaukee have concentrations of Roman Catholics because of immigration from Ireland, Italy, and Eastern Europe, especially in the late nineteenth and early twentieth centuries. Immigration from Mexico and other Latin American countries has concentrated Roman Catholics in the Southwest, while French settlement from the seventeenth century, as well as recent immigration, has produced a predominantly Roman Catholic Québec.

Similarly, geographers trace the distribution of Protestant denominations within the United States to the fact that migrants came from different parts of Europe, especially during the nineteenth century. Followers of the Church of Jesus Christ of Latter-Day Saints, popularly known as Mormons, settled at Fayette, New York, but after the death of their founder, Joseph Smith, the group moved several times in search of religious freedom. Eventually, under the leadership of Brigham Young, they migrated to the sparsely inhabited Salt Lake Valley in the present-day state of Utah.

Diffusion of Islam.
Muhammad's successors organized followers into armies that extended the region of Muslim control over an extensive area of Africa, Asia, and Europe. Within a century of Muhammad's death, Muslim armies conquered Palestine, the Persian Empire, and much of India, resulting in the conversion of many non-Arabs to Islam, often through intermarriage. To the west, Muslims captured North Africa, crossed the Strait of Gibraltar, and retained part of Western Europe, particularly much of present-day Spain, until 1492 (Figure 6–6). During the same century that the Christians regained all of Western Europe, Muslims took control of much of southeastern Europe and Turkey.

As was the case with Christianity, Islam, as a universalizing religion, diffused well beyond its hearth in Southwest Asia through relocation diffusion of missionaries to portions of sub-Saharan Africa and Southeast Asia. Although it is spatially isolated from the Islamic core region in Southwest Asia, Indonesia, the world's fourth most populous country, is predominantly Muslim, because Arab traders brought the religion there in the thirteenth century.

Diffusion of Buddhism.
Buddhism did not diffuse rapidly from its point of origin in northeastern India. Most responsible for the spread of Buddhism was Asoka, emperor of the Magadhan Empire from about 273 to 232 B.C. The Magadhan Empire formed the nucleus of several

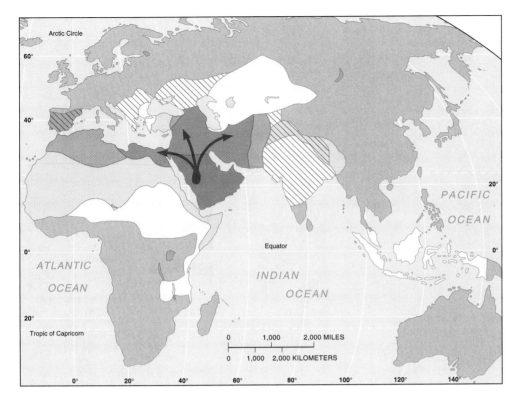

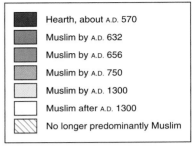

Figure 6–6 Diffusion of Islam. Islam diffused rapidly from its point of origin in present-day Saudi Arabia. Within 200 years, Islamic armies controlled much of North Africa, southwestern Europe, and southwestern Asia. Subsequently, Islam became the predominant religion as far east as Indonesia.

Map legend:
- Hearth, about A.D. 570
- Muslim by A.D. 632
- Muslim by A.D. 656
- Muslim by A.D. 750
- Muslim by A.D. 1300
- Muslim after A.D. 1300
- No longer predominantly Muslim

powerful kingdoms in South Asia between the sixth century B.C. and the eighth century A.D. About 257 B.C., at the height of the Magadhan Empire's power, Asoka became a Buddhist and thereafter attempted to put into practice Buddha's social principles.

A council organized by Asoka at Pataliputra decided to send missionaries to territories neighboring the Magadhan Empire. Emperor Asoka's son, Mahinda, led a mission to the island of Ceylon (now Sri Lanka), where the king and his subjects were converted to Buddhism. As a result, Sri Lanka is the country that claims the longest continuous tradition of practicing Buddhism. Missionaries were also sent in the third century B.C. to Kashmir, the Himalayas, Burma (Myanmar), and elsewhere in India.

In the first century A.D., merchants along the trading routes from northeastern India introduced Buddhism to China. Many Chinese were receptive to the ideas brought by Buddhist missionaries, and Buddhist texts were translated into Chinese languages. Chinese rulers allowed their people to become Buddhist monks during the fourth century A.D., and in the following centuries, Buddhism turned into a genuinely Chinese religion. Buddhism further diffused from China to Korea in the fourth century and from Korea to Japan two centuries later. During the same era, Buddhism lost its original base of support in India (Figure 6–7).

Lack of Diffusion of Ethnic Religions

Most ethnic religions have limited if any diffusion. These religions lack missionaries who are devoted to converting people from other religions. Thus the diffusion of universalizing religions, especially Christianity and Islam, typically comes at the expense of ethnic religions.

Mingling of Ethnic and Universalizing Religions. Universalizing religions may supplant ethnic religions, or mingle with them. In some African countries, Christian practices are similar to those in their former European colonial masters. Equatorial Guinea, a former Spanish colony, is mostly Roman Catholic, whereas Namibia, a former German colony, is heavily Lutheran.

Elsewhere, traditional African religious ideas and practices have been merged with Christianity. For example, African rituals may give relative prominence to the worship of ancestors. Anglican bishops have decided that an African man who has more than one wife can become a Christian, as long as he does not add to the number of wives by further marriages. Desire for a merger of traditional practices with Christianity has led to the formation of several thousand churches in Africa not affiliated with established churches elsewhere in the world.

In East Asia, Buddhism is the universalizing religion that has most mingled with ethnic religions, such as Shintoism in Japan. Shintoists first resisted Buddhism, when it first diffused to Japan from Korea in the ninth century A.D. Later, Shintoists embraced Buddhism and amalgamated elements of the two religions. Buddhist priests took over most of the Shinto shrines, but Buddhist deities came to be regarded by the Japanese as Shintoist deities instead.

Figure 6–7 Diffusion of Buddhism. In contrast to the other large universalizing religions, Buddhism diffused slowly from its core in northeastern India. Buddhism was not well established in China until 800 years after Buddha's death.

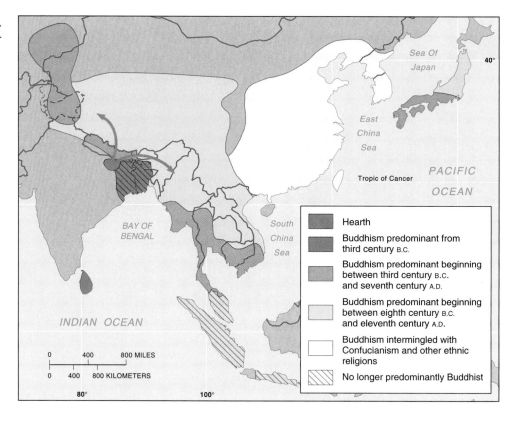

The current situation in Japan offers a strong caution to anyone attempting to document the number of adherents of any religion. Although Japan is a wealthy country with excellent record-keeping, the number of Shintos in the country is currently estimated at anywhere from 3 million to 110 million—and these extreme figures come from the same authoritative source, the *Encyclopaedia Britannica Yearbook*. About 90 percent of Japanese say they are Shintos and about 75 percent say they are Buddhists, so clearly most Japanese profess adherence to both religions. And it is impossible to gauge the strength of adherence to the two religions—is one Japanese more Buddhist than Shinto, while another is more Shinto than Buddhist?

Mapping the distribution of support for Shintoism and Buddhism in Japan shows the extent of the overlap of the two religions. Nearly every locality displays either above-average support for both religions or below-average support for both religions, and only a handful are above average for one of the religions and below average for the other (Figure 6–8).

Ethnic religions can diffuse if adherents migrate to new locations for economic reasons and are not forced to adopt a strongly entrenched universalizing religion. For example, the 1 million inhabitants of Mauritius include 52 percent Hindu, 28 percent Christian, and 17 percent Muslim. The religious diversity is a function of the country's history of immigration.

A 2,040 square-kilometer (788 square-mile) island located in the Indian Ocean 800 kilometers (500 miles) east of Madagascar, Mauritius was uninhabited until 1638, so it had no traditional ethnic religion. That year, Dutch settlers arrived to plant sugarcane and naturally brought their religion—Christianity—with them. France gained control in 1721 and imported African slaves to work on the sugarcane plantations. Then the British took over 1810 and brought workers from India. Mauritius became independent in 1992. Hinduism on Mauritius traces back to the Indian immigrants, Islam to the African immigrants, and Christianity to the European immigrants.

Judaism, an Exception.   The spatial distribution of Jews differs from that of other ethnic religions, because Judaism is practiced in many countries not just its place of origin. Only since the creation of the state of Israel in 1948 has a significant percentage of the world's Jews lived in their Eastern Mediterranean homeland.

Most Jews have not lived in the Eastern Mediterranean since A.D. 70, when the Romans forced them to disperse throughout the world, an action known as the *diaspora*, from the Greek word for *dispersion*. The Romans forced the diaspora after crushing an attempt by the Jews to rebel against Roman rule.

Most Jews migrated from the Eastern Mediterranean to Europe, although some went to North Africa and Asia. Having been exiled from the home of their ethnic religion, Jews lived among other nationalities, retaining separate religious practices but adopting other cultural characteristics of the host country, such as language.

Other nationalities often persecuted the Jews living in their midst. Historically, the Jews of many European

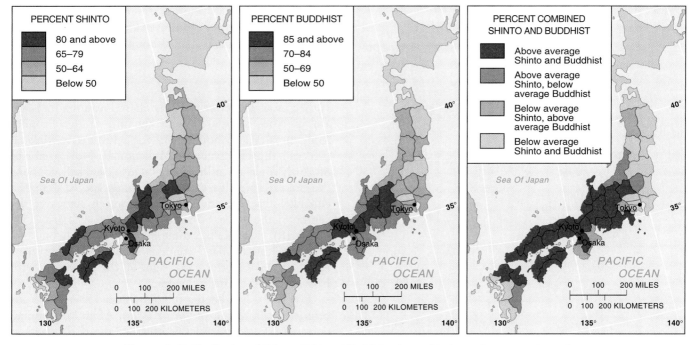

Figure 6–8 Distribution of Shintos (left) and Buddhists (center) in Japan. In some regions of the country, more than two-thirds of the people are Buddhists and more than two-thirds are Shintoists (right). This is possible because many people adhere simultaneously to both religions. In most places, people are either more likely than average to be both Shinto and Buddhist or less likely than average to be both.

countries were forced to live in a **ghetto**, defined as a city neighborhood set up by law to be inhabited only by Jews. The term *ghetto* originated during the sixteenth century in Venice, Italy, as a reference to the city's foundry or metal-casting district, where Jews were forced to live. Ghettos were frequently surrounded by walls, and the gates were locked at night to prevent escape.

Beginning in the 1930s, but especially during World War II (1939–1945), the Nazis systematically rounded up a large percentage of European Jews, transported them to concentration camps, and exterminated them. About 4 million Jews died in the camps, and 2 million in other ways. Many of the survivors migrated to Israel. Today, about 10 percent of the world's 14 million Jews live in Europe, compared to 90 percent a century ago.

Holy Places

Religions may elevate particular places to a holy position. However, universalizing and ethnic religions differ on the types of places that are considered holy. An ethnic religion typically has a less widespread distribution than a universalizing one in part because its holy places derive from the distinctive physical environment of its hearth, such as mountains, rivers, or rock formations. A universalizing religion endows with holiness cities and other places associated with the founder's life. Its holy places do not necessarily have to be near each other, nor do they need to be related to any particular physical environment.

Making a **pilgrimage** to these holy places—a journey for religious purposes to a place considered sacred—is incorporated into the rituals of some universalizing and ethnic religions. Hindus and Muslims are especially encouraged to make pilgrimages to visit holy places in accordance with recommended itineraries, and Shintoists are encouraged to visit holy places in Japan.

Holy Places in Universalizing Religions

Buddhism and Islam are the universalizing religions that place the most emphasis on identifying shrines. Places are holy because they are the locations of important events in the life of Buddha or Muhammad.

Buddhist Shrines. Eight places are holy to Buddhists because they were the locations of important events in Buddha's life. The four most important of the eight places are concentrated in a small area of northeastern India and southern Nepal (Figure 6–9). Most important is Lumbinī, in southern Nepal, where Buddha was born, around 563 B.C. Many sanctuaries and monuments were built there, but all are in ruins today.

The second great event in Buddha's life occurred at Bodh Gayā, 250 kilometers (150 miles) southeast of his birthplace, where Buddha reached perfect wisdom. A temple has stood near the site since the third century B.C., and part of the surrounding railing built in the first century A.D. still stands. Because Buddha reached perfect

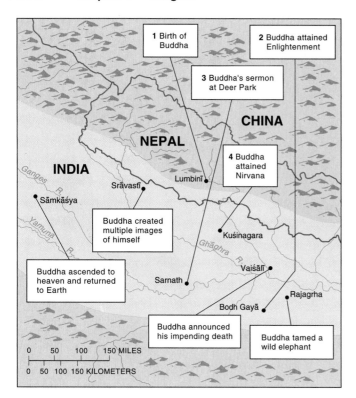

Figure 6–9 Holy places in Buddhism. Most are clustered in northeastern India and southern Nepal, because they were the locations of important events in Buddha's life. Most of the sites are in ruins today.

enlightenment while sitting under a bo tree, that tree has become a holy object as well. To honor Buddha, the bo tree has been diffused to other Buddhist countries, such as China and Japan.

The third important location is Deer Park in Sarnath, where Buddha gave his first sermon. The Dhamek pagoda at Sarnath, built in the third century B.C., is probably the oldest surviving structure in India. Nearby is an important library of Buddhist literature, including many works removed from Tibet when Tibet's Buddhist leader, the Dalai Lama, went into exile.

The fourth holy place is Kuśinagara, where Buddha died at age 80 and passed into Nirvana, a state of peaceful extinction. Temples built at the site are currently in ruins.

Four other sites in northeastern India are particularly sacred because they were the locations of Buddha's principal miracles. At Śrāvastī, Buddha performed his greatest miracle. Before an assembled audience of competing religious leaders, Buddha created multiple images of himself and visited heaven. Śrāvastī became an active center of Buddhism, and one of the most important monasteries was established there.

At the second miracle site, Sāmkāśya, Buddha is said to have ascended to heaven, preached to his mother, and returned to Earth. The third site, Rajagrha, is holy because Buddha tamed a wild elephant there, and shortly after Buddha's death, it became the site of the first Buddhist Council. Vaiśālī, the fourth location, is

the site of Buddha's announcement of his impending death and the second Buddhist Council. All four miracle sites are in ruins today, although excavation activity is under way.

Holy Places in Islam.   The holiest locations in Islam are in cities associated with the life of the Prophet Muhammad. The holiest city for Muslims is Makkah (Mecca), the birthplace of Muhammad. Now a city of more than a half million inhabitants, Makkah contains the holiest object in the Islamic landscape, al–Ka'ba, a cubelike structure encased in silk, which stands at the center of the Great Mosque, al-Haram al-Sharīf (Figure 6–10). The Ka'ba, thought to have been built by Abraham and Ishmael, contains a black stone given to Abraham by Gabriel as a sign of the covenant with Ishmael and the Muslim people.

The Ka'ba had been a religious shrine in Makkah for centuries before the origin of Islam. After Muhammad defeated the local people, he captured the Ka'ba, cleared it of idols, and rededicated it to the all-powerful Allah (God). The al-Haram mosque also contains the well of Zamzam, considered to have the same water source as that used by Ishmael and Hagar when they were wandering in the desert after their exile from Canaan.

The second most holy geographic location in Islam is Madinah (Medina), approximately 350 kilometers (220 miles) north of Makkah. Muhammad received his first support from the people of Madinah and became the city's chief administrator. Muhammad's tomb is at Madinah, inside Islam's second mosque.

Every healthy Muslim who has adequate financial resources is expected to undertake a pilgrimage, called a *hajj*, to Makkah (Mecca). Regardless of nationality and economic background, all pilgrims dress alike in plain white robes to emphasize common loyalty to Islam and the equality of people in the eyes of Allah. A precise set of rituals is practiced, culminating in a visit to the Ka'ba. The word "mecca" now has a general meaning in the English language as a goal sought or a center of activity.

The *hajj* attracts 1 million Muslims a year to Makkah from countries other than Saudi Arabia. Roughly 40 percent each come from the Middle East and northern Africa, with the largest numbers from Nigeria, Turkey, and Yemen. Asian countries are responsible for most of the remaining 20 percent. Although Indonesia is the world's most populous Muslim country, it does not send the largest number of pilgrims to Makkah because of the relatively long travel distance.

Holy Places in Ethnic Religions

One of the principal reasons that ethnic religions are highly clustered is that they are closely tied to the physical geography of a particular place. Pilgrimages are undertaken to view these physical features.

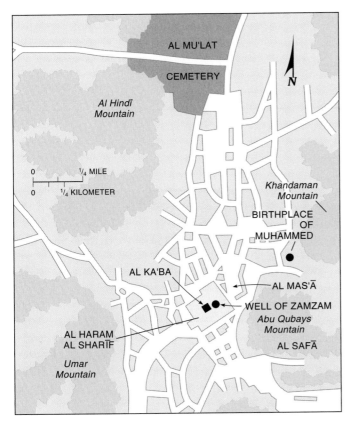

Figure 6–10 Makkah (Mecca), in Saudi Arabia, is the holiest city for Muslims, because Muhammad was born there. Thousands of Muslims make a pilgrimage to Makkah each year and gather at al-Haram al-Sharif, a mosque in the center of the city. The black, cubelike structure in the center of the mosque, called al–Ka'ba, once had been a shrine to tribal idols until Muhammad rededicated it to Allah. Muslims believe that Abraham and Ishmael originally built the Ka'ba. (Mohamed Lounes/Gamma-Liaison, Inc.)

Holy Places in Hinduism. As an ethnic religion of India, Hinduism is closely tied to the physical geography of India. According to a survey conducted by the geographer Surinder Bhardwaj, the natural features most likely to rank among the holiest shrines in India are riverbanks or coastlines.

Hindus consider a pilgrimage, known as a *tirtha*, to be an act of purification. Although not a substitute for meditation, the pilgrimage is an important act in achieving redemption. Hindu holy places are organized into a hierarchy. Particularly sacred places attract Hindus from all over India, despite the relatively remote locations of some, while less important shrines attract primarily local pilgrims (Figure 6–11).

Because Hinduism has no central authority, the relative importance of shrines is established by tradition, not by doctrine. For example, many Hindus make long-distance pilgrimages to Mt. Kailās, located at the source of the Ganges in the Himalayas, which is holy because Siva lives there. At the same time, other mountains may attract only local pilgrims. Throughout India, local residents may consider a nearby mountain to be holy if Siva is thought to have visited it at one time.

Hindus believe that they achieve purification by bathing in holy rivers. The Ganges is the holiest river in India, because it is supposed to spring forth from the hair of Siva, one of the main deities. Indians come from all over the country to Hardwār, the most popular location for bathing in the Ganges.

The remoteness of holy places from population clusters once meant that making a pilgrimage required major commitments of time and money as well as undergoing considerable physical hardship. However, recent improve-

209

Figure 6–11 Hierarchy of Hindu holy places. Some places are important to Hindus all over India and are visited frequently, while others have importance only to nearby residents. The map also shows that holy places for particular deities are somewhat clustered in different regions of the country—Shakti in the east, Vishnu in the west, and Siva in the north and south.

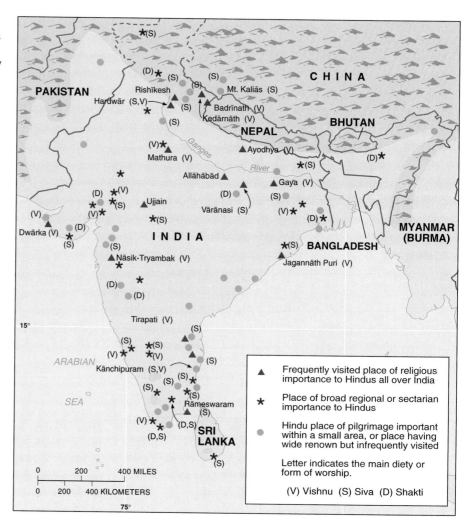

Bathing in the Ganges River. The river attracts Hindu pilgrims from all over India, because they believe that the Ganges springs from the hair of Siva, one of the main deities. Hindus achieve purification by bathing in the Ganges, and bodies of the dead are washed with water from it before being cremated. (Richard Vogel/Gamma-Liaison, Inc.)

ments in transportation have increased the accessibility of shrines. Hindus can now reach holy places in the Himalaya Mountains by bus or car, and Muslims from all over the world can reach Makkah by airplane.

Cosmogony in Ethnic Religions. Ethnic religions differ from universalizing religions in their understanding of relationships between human beings and nature. These differences derive from distinctive concepts of **cosmogony**, which is a set of religious beliefs concerning the origin of the universe. A variety of events in the physical environment are more likely to be incorporated into the principles of an ethnic religion. These events range from familiar and predictable to unexpected disasters.

For example, Chinese ethnic religions, such as Confucianism and Daoism, believe that the universe is made up of two forces, yin and yang, which exist in everything. The yin force is associated with earth, darkness, female, cold, depth, passivity, and death. The yang force is associated with heaven, light, male, heat, height, activity, and life. Yin and yang forces interact with each other to achieve balance and harmony, but they are in a constant state of change. An imbalance results in disorder and chaos. The principle of yin and yang applies to the creation and transformation of all natural features.

The universalizing religions that originated in Southwest Asia, notably Christianity and Islam, consider that God created the universe, including Earth's physical environment and human beings. A religious person can serve God by cultivating the land, draining wetlands, clearing forests, building new settlements, and otherwise making productive use of natural features that God created. As the very creator of Earth itself, God is more powerful than any force of nature, and if in conflict the laws of God take precedence over laws of nature.

Christian and Islamic cosmogony do differ in some respects. For example, Christians believe that Earth was given by God to humanity to finish the task of creation. Obeying the all-supreme power of God meant independence from the tyranny of natural forces. Muslims regard humans as representatives of God on Earth, capable of reflecting the attributes of God in their deeds, such as growing food or other hard work to improve the land. But humans are not partners with God, who alone was responsible for Earth's creation.

In the name of God, some people have sought mastery over nature, not merely independence from it. Large-scale development of remaining wilderness is advocated by some religious people as a way to serve God. To those who follow this approach, failure to make full and complete use of Earth's natural resources is considered a violation of biblical teachings.

Christians are more likely to consider floods, droughts, and other natural disasters to be preventable and may take steps to overcome the problem by modifying the environment. However, some Christians regard natural disasters as punishment for human sins.

Adherents of ethnic religions do not attempt to transform the environment to the same extent. To animists, for example, God's powers are mystical, and only a few people on Earth can harness these powers for medical or other purposes. God can be placated, however, through prayer and sacrifice. Environmental hazards may be accepted as normal and unavoidable.

The Calendar

Universalizing and ethnic religions have different approaches to the calendar. An ethnic religion typically has a more clustered distribution than a universalizing religion, in part because its holidays are based on the distinctive physical geography of the homeland. In universalizing religions, major holidays relate to events in the life of the founder, rather than to the changing seasons of one particular place.

The Calendar in Ethnic Religions

A prominent feature of ethnic religions is celebration of the seasons—the calendar's annual cycle of variation in climatic conditions. Knowledge of the calendar is critical to successful agriculture, whether for sedentary crop farmers or nomadic animal herders. The seasonal variations of temperature and precipitation help farmers select the appropriate times for planting and harvesting and make the best choice of crops.

Rituals are performed to pray for favorable environmental conditions or to give thanks for past success. The major religious events of the Bontok people of the Philippines, for example, revolve around the agricultural calendar. Sacred moments, known as *obaya*, include the times when the rice field is initially prepared, when the seeds are planted, when the seedlings are transplanted, when the harvest is begun, and when the harvest is complete.

The Jewish Calendar. Judaism is classified as an ethnic, rather than a universalizing, religion in part because its major holidays are based on events in the agricultural calendar of the religion's homeland in present-day Israel. In that Mediterranean agricultural region, grain crops generally are planted in autumn, which is a time of hope and worry over whether the winter's rainfall will be sufficient. The two holiest days in the Jewish calendar, Rosh Hashanah (New Year) and Yom Kippur (Day of Atonement), come in the autumn.

The other three most important holidays in Judaism originally related even more closely to the agricultural cycle. *Sukkot* celebrates the final gathering of fruits for the year, and prayers, especially for rain, are offered to bring success in the upcoming agricultural year. *Pesach* (Passover) derives from traditional agricultural practices in which farmers offered God the first fruits of the new spring harvest and herders sacrificed a young animal at the time when cows began to calve. *Shavuot* (Feast of Weeks) comes at the end of the grain harvest.

These three agricultural holidays later gained importance because they also commemorated events in the Exodus of the Jews from Egypt, as recounted in the Old Testament. Pesach recalled the liberation of the Jews from slavery in Egypt and the miracle of their successful flight under the leadership of Moses. Sukkot derived from the Hebrew word for the *booths* or *temporary shelters* occupied by Jews during their wandering in the wilderness for 40 years after fleeing Egypt. Shavuot was considered the date during the wandering when Moses received the Ten Commandments from God. The reinterpretation of natural holidays in the light of historical events has been especially important for Jews in the United States, Western Europe, and other regions who are unfamiliar with the agricultural calendar of the Middle East.

In daily business, North Americans use the solar calendar of 12 months, each containing 30 or 31 days, taking up the astronomical slack with 28 or 29 days in February. But Israel—the only country where Jews are in the majority—uses a lunar rather than a solar calendar. The Moon has a mystical quality because of its variation from one day to the next. From its fullest disk, the Moon becomes smaller and disappears altogether ("new Moon") before reappearing and expanding to a full Moon again. The appearance of the new Moon marks the new month in Judaism and Islam, and is a holiday for both religions.

The lunar month is only about 29 days long, so a lunar year of about 350 days quickly becomes out-of-step with the agricultural seasons. The Jewish calendar solves the problem by adding an extra month seven out of every 19 years, so that its principal holidays are celebrated in the same season every year.

The Solstice. The **solstice** has special significance in some ethnic religions. A major holiday in some pagan religions is the winter solstice, December 21 or 22 in the Northern Hemisphere and June 21 or 22 in the Southern Hemisphere. The winter solstice is the shortest day and longest night of the year, when the Sun appears lowest in the sky and appears to stand still (*solstice* comes from the Latin *to stand still*). Stonehenge, a collection of rocks erected in southwestern England, probably by the Druids, is a prominent remnant of a pagan structure apparently aligned so the Sun rises between two stones on the solstice.

If you stand at the western facade of the U.S. Capitol in Washington, at exactly noon on the summer solstice (June 21 or 22 in the Northern Hemisphere), and look down Pennsylvania Avenue, the Sun is directly over the center of the avenue. Similarly, at the winter solstice the Sun is directly aligned with the view from the Capitol down Maryland Avenue. Will archaeologists of the distant future think we erected the Capitol Building and aligned the streets as a religious ritual? Did the planner of Washington, Pierre L'Enfant, create the pattern accidentally or deliberately, and if so, why?

The Calendar in Universalizing Religions

The principal purpose of the holidays in universalizing religions is to commemorate events in the founder's life. Christians in particular associate their holidays with seasonal variations in the calendar, but climate and the agricultural cycle are not central to the liturgy and rituals.

Christian Holidays. Christians commemorate the resurrection of Jesus on Easter, observed on the first Sunday

Stonehenge is a collection of enormous stones (weighing 100 tons each), erected about 4,000 years ago, probably for religious ceremonies. The stones are arranged so that the sun is centered between two of them at astronomically significant dates and times, such as sunrise on the summer solstice. (Robert Hallman/Tony Stone Images)

after the first full Moon following the spring equinox in late March. But not all Christians observe Easter on the same day, because Protestant and Roman Catholic branches calculate the date on the Gregorian calendar, but Eastern Orthodox churches use the Julian calendar.

Christians may relate Easter to the agricultural cycle, but that relationship differs with where they live. In Southern Europe, Easter is a joyous time of harvest. Northern Europe and North America do not have a major Christian holiday at harvest time, which would be placed in the fall. Instead, Easter in Northern Europe and North America is a time of anxiety over planting new crops, as well as a celebration of spring's arrival after a harsh winter. In the United States and Canada, Thanksgiving has been endowed with Christian prayers to play the role of harvest festival.

Most Northern Europeans and North Americans associate Christmas, the birthday of Jesus, with winter conditions, such as low temperatures, snow cover, and absence of vegetation except needleleaf evergreens. But for Christians in the Southern Hemisphere, December 25 is the height of the summer, with warm days and abundant sunlight.

Muslim Calendar. Islam, like Judaism, uses a lunar calendar. While the Jewish calendar inserts an extra month every few years to match the agricultural and solar calendars, Islam as a universalizing religion retains a strict lunar calendar. In a 30-year cycle, the Islamic calendar has 19 years with 354 days and 11 years with 355 days.

As a result of using a lunar calendar, Muslim holidays arrive in different seasons from generation to generation. For example, during the holy month of Ramadan, Muslims fast during daylight every day and try to make a pilgrimage to the holy city of Makkah. At the moment, the start of Ramadan is occurring in the Northern Hemisphere autumn—for example, November 27, 2000, on the western Gregorian calendar. In A.D. 1990, Ramadan fell in March, and in A.D. 2010, Ramadan will be in August. Because Ramadan occurs at different times of the solar year in different generations, the number of hours of the daily fast varies widely, because the amount of daylight varies by season and by location on Earth's surface.

Observance of Ramadan can be a hardship by interfering with critical agricultural activities, depending on the season. However, as a universalizing religion with more than 1 billion adherents worldwide, Islam is practiced in various climates and latitudes. If Ramadan were fixed at the same time of the Middle East's agricultural year, Muslims in various places of the world would need to make different adjustments to observe Ramadan.

Buddhist Holidays. All Buddhists celebrate as major holidays Buddha's birth, Enlightenment, and death. However, Buddhists do not all observe them on the same days. Japanese Buddhists celebrate Buddha's birth on April 8, his Enlightenment on December 8, and his death on February 15, whereas Theravadist Buddhists observe all three events on the same day, usually in April.

Key Issue 3

Why and How Do Religions Organize Space?

- Places of worship
- Sacred space
- Administration of space

Geographers study the major impact on the landscape made by all religions, regardless of whether they are universalizing or ethnic. In large cities and small villages around the world, regardless of the region's prevailing religion, the tallest most elaborate buildings are often religious structures.

The distribution of religious elements on the landscape reflects the importance of religion in people's values. The impact of religion on the landscape is particularly profound, for many religious people believe that their life on Earth ought to be spent in service to God.

Places of Worship

Church, basilica, mosque, temple, pagoda, synagogue. These familiar names identify places of worship in various faiths. Sacred structures are physical "anchors" of religion. All major religions have structures, but the functions of the buildings influence the arrangement of the structures across the landscape. They may house shrines, or be places where people assemble for worship. Some religions require a relatively large number of elaborate structures, whereas others have more modest needs.

Christian Churches

The Christian landscape is dominated by a high density of churches. The word *church* derives from a Greek terms meaning *lord*, *master*, and *power*. *Church* also refers to a gathering of believers, as well as the building where the gathering occurs.

The church building plays a more critical role in Christianity than in other religions, in part because the structure is an expression of religious principles, an environment in the image of God. The church is also more prominent in Christianity because attendance at a collective service of worship is considered extremely important.

The prominence of churches on the landscape also stems from their style of construction and location. Traditionally in some communities, the church was the largest and tallest building and was placed at an important square or other prominent location. Although such characteristics may no longer apply in large cities, they are frequently still true for small towns and neighborhoods within cities.

Underlying the large number and size of Christian churches is their considerable expense. Because of the importance of a place of worship, Christians have contributed much wealth to the construction and maintenance of churches. A wealthier congregation may build an elaborate structure designed by an architect to provide an environment compatible with the religious doctrine and ritual. Over the centuries, the most prominent architects have been commissioned to create religious structures, such as those designed by Christopher Wren in London during the late seventeenth century.

Church Architecture. Early churches were modeled after Roman buildings for public assembly, known as *basilicas*. The basilica was a rectangular building divided by two rows of columns that formed a central nave (hall) and two side aisles. At the western end of the church stood a semicircular apse, in front of which was the altar where the priest conducted the service. Later, the apse was placed on the eastern wall. The raised altar, symbolizing the hill of Calvary, where Jesus was crucified, facilitated the reenactment at every service of Christ's sacrifice. Churches built during the Gothic period, between the twelfth and fourteenth centuries, had a floor plan in the form of the cross.

Since Christianity split into many denominations, no single style of church construction has dominated. Churches reflect both the cultural values of the denomination and the region's architectural heritage. Eastern Orthodox churches, for example, follow an architectural style that developed in the Byzantine Empire during the fifth century. Byzantine-style Eastern Orthodox churches tend to be highly ornate, topped by prominent domes. Many Protestant churches in North America, on the other hand, are simple, with little ornamentation. This austerity is a reflection of the Protestant conception of a church as an assembly hall for the congregation.

Availability of building materials also influences church appearance. In the United States, early churches were most frequently built of wood in the Northeast, brick in the Southeast, and adobe in the Southwest. Stucco and stone predominated in Latin America. This diversity reflected differences in the most common building materials found by early settlers.

Places of Worship in Other Religions

Religious buildings are highly visible and important features of the landscapes in regions dominated by religions other than Christianity. But unlike Christianity, other major religions do not consider their important buildings a sanctified place of worship.

Muslim Mosques. Muslims consider the mosque as a space for community assembly. In contrast to a church, however, a *mosque* is not viewed as a sanctified place but rather as a location for the community to gather together for worship. Mosques are found primarily in larger cities of the Muslim world, while simple structures may serve as places of prayer in rural villages.

The mosque is organized around a central courtyard—traditionally open air, although it may be enclosed in harsher climates. The pulpit is placed at the end of the courtyard facing Makkah, the direction toward which all Muslims pray. Surrounding the courtyard is a cloister used for schools and nonreligious activities. A distinctive feature of the mosque is the *minaret*, a tower where a man known as a *muzzan* summons people to worship (Figure 6–10 shows two minarets).

Hindu Temples. Sacred structures for collective worship are relatively unimportant in Asian ethnic and universalizing religions. Instead, important religious functions are more likely to take place at home within the family. Temples are built to house shrines for particular gods rather than for congregational worship.

The Hindu temple serves as a home to one or more gods, although a particular god may have more than one temple. Wealthy individuals or groups usually maintain local temples. Size and frequency of temples are determined by local preferences and commitment of resources rather than standards imposed by religious doctrine.

The typical Hindu temple contains a small, dimly lit interior room where a symbolic artifact or some other image of the god rests. The remainder of the temple may be devoted to space for ritual processions. Because congregational worship is not part of Hinduism, the temple does not need a large closed interior space filled with seats. The site of the temple, usually demarcated by a wall, may also contain a structure for a caretaker and a pool for ritual baths.

Buddhist and Shintoist Pagodas. The pagoda is a prominent and visually attractive element of the Buddhist and Shintoist landscapes. Frequently elaborate and delicate in appearance, pagodas typically include tall, many-sided towers arranged in a series of tiers, balconies, and slanting roofs.

Pagodas contain relics that Buddhists believe to be a portion of Buddha's body or clothing. After Buddha's death, his followers scrambled to obtain these relics. As part of the process of diffusing the religion, Buddhists carried these relics to other countries and built pagodas for them. Pagodas are not designed for congregational worship. Individual prayer or meditation is more likely to be undertaken at an adjacent temple, a remote monastery, or in a home.

Sacred Space

The impact of religion is clearly seen in the arrangement of human activities on the landscape at several scales, from relatively small parcels of land to entire communi-

The Great Buddha, Kamakura, Japan. The 13-meter (42-foot) tall bronze statue was cast in 1252. (Raga/The Stock Market)

ties. How each religion distributes its elements on the landscape depends on its beliefs. The most significant religious land uses are for burial of the dead and religious settlements.

Disposing of the Dead

A prominent example of religiously inspired arrangement of land at a smaller scale is burial practices. Climate, topography, and religious doctrine combine to create differences in practices to shelter the dead.

Burial. Christians, Muslims, and Jews usually bury their dead in a specially designated area called a *cemetery*. The Christian burial practice can be traced to the early years of the religion. In ancient Rome, underground passages known as *catacombs* were used to bury early Christians (and to protect the faithful when the religion was still illegal).

After Christianity became legal, Christians buried their dead in the yard around the church. As these burial places became overcrowded, separate burial grounds had to be established outside the city walls. Public health and sanitation considerations in the nineteenth century led to public management of many cemeteries. However, some cemeteries are still operated by religious organizations.

The remains of the dead are customarily aligned in some traditional direction. Some Christians bury the dead with the feet toward Jerusalem so that they may meet Christ there on the Day of Judgment. The Mandan Indians of the North American Plains placed the dead on scaffolds with the feet to the southeast, the direction the spirits were said to take to reach the Heart River, the place where the ancestors used to live. The face is often aligned toward the west, the direction where the Sun "dies" in its daily setting.

Cemeteries may consume significant space in a community, increasing the competition for scarce space. In congested urban areas, Christians and Muslims have traditionally used cemeteries as public open space. Before the widespread development of public parks in the nineteenth century, cemeteries were frequently the only green space in rapidly growing cities. Cemeteries are still used as parks in Muslim countries, where the idea faces less opposition than in Christian societies.

Traditional burial practices in China have put pressure on agricultural land. By burying dead relatives, rural residents have removed as much as 10 percent of the land from productive agriculture. The government in China has ordered the practice discontinued, even encouraging farmers to plow over old burial mounds. Cremation is encouraged instead.

Other Methods of Disposing of Bodies. Not all faiths bury their dead. Hindus generally practice cremation rather than burial. The body is washed with water from the Ganges River and then burned with a slow fire on a funeral pyre. Burial is reserved for children, ascetics, and people with certain diseases. Cremation is considered an act of purification, although it tends to strain India's wood supply.

Cremation was the principal form of disposing of bodies in Europe before Christianity. Outside of India, it is still practiced in parts of Southeast Asia, possibly because of Hindu influence.

Motivation for cremation may have originated from unwillingness on the part of nomads to leave their dead behind, possibly because of fear that the body could be attacked by wild beasts or evil spirits, or even return to life. Cremation could also free the soul from the body for departure to the afterworld and provide warmth and comfort for the soul as it embarked on the journey to the afterworld.

To strip away unclean portions of the body, Parsis (Zoroastrians) expose the dead to scavenging birds and animals (see Globalization and Local Diversity box). The ancient Zoroastrians did not want the body to contaminate the sacred elements of fire, earth, or water. Tibetan Buddhists also practice exposure for some dead, with cremation reserved for the most exalted priests.

Disposal of bodies at sea is used in some parts of Micronesia, but the practice is much less common than in the past. The bodies of lower-class people would be flung into the sea, whereas elites could be set adrift on a raft or boat. Water burial was regarded as a safeguard against being contaminated by the dead.

Cremation is the most common form of disposal of bodies in India. In middle class families, bodies are more likely to be cremated in an electric oven at a crematorium. A poor person may be cremated in an open fire, such as this one within sight of the Taj Mahal. High-ranking officials and strong believers in traditional religious practices may also be cremated on an outdoor fire. (Raghu Rai/Magnum Photos, Inc.)

Religious Settlements

Buildings for worship and burial places are smaller-scale manifestations of religion on the landscape, but there are larger-scale examples: entire settlements. Most human settlements serve an economic purpose (Chapter 12), but some are established primarily for religious reasons.

A utopian settlement is an ideal community built around a religious way of life. By 1858, some 130 different utopian settlements had begun in the United States in conformance with a group's distinctive religious beliefs. Examples include Oneida, New York; Ephrata, Pennsylvania; Nauvoo, Illinois; and New Harmony, Indiana. Buildings were sited and economic activities organized to integrate religious principles into all aspects of daily life.

An early utopian settlement in the United States was Bethlehem, Pennsylvania, founded in 1741 by Moravians, Christians who had emigrated from the present-day Czech Republic. The culmination of the utopian movement in the United States was the construction of Salt Lake City by the Mormons, beginning in 1848. The layout of Salt Lake City is based on a plan of the city of Zion given to the church elders in 1833 by the Mormon prophet Joseph Smith. The city has a regular grid pattern, unusually broad boulevards, and church-related buildings situated at strategic points.

Most utopian communities declined in importance or disappeared altogether. Some utopian communities disappeared because the inhabitants were celibate and could not attract immigrants, whereas in other cases residents moved away in search of better economic conditions. The utopian communities that have not been demolished are now inhabited by people who are not members of the original religious sect, although a few have been preserved as museums.

Although most colonial settlements were not planned primarily for religious purposes, religious principles affected many of the designs. Most early New England settlers were members of a Puritan Protestant denomination. The Puritans generally migrated together from England and preferred to live near each other in clustered settlements rather than on dispersed, isolated farms. Reflecting the importance of religion in their lives, New England settlers placed the church at the most prominent location in the center of the settlement, usually adjacent to a public open space, known as a *common* because it was for common use by everyone.

Religious Place-Names

Roman Catholic immigrants frequently have given religious place-names, or toponyms, to their settlements in the New World, particularly in Québec and the U.S. Southwest. Québec's boundaries with Ontario and the United States clearly illustrate the difference between toponyms selected by Roman Catholic and Protestant settlers. Religious place-names are common in Québec but rare in the two neighbors (Figure 6–12).

Administration of Space

Followers of a universalizing religion must be connected so as to assure communication and consistency of doctrine. The method of interaction varies among universalizing religions, branches, and denominations. Ethnic religions tend not to have organized, central authorities.

Hierarchical Religions

A **hierarchical religion** has a well-defined geographic structure and organizes territory into local administrative

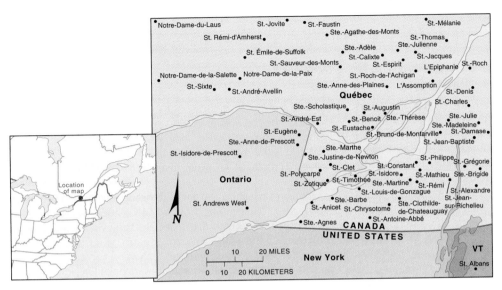

Figure 6–12 Place-names near Québec's boundaries with Ontario and New York State shows the impact of religion on the landscape. In Québec, a province with a predominantly Roman Catholic population, a large number of settlements are named for saints, whereas relatively few religious toponyms are found in predominantly Protestant Ontario and New York.

units. Roman Catholicism provides a good example of a hierarchical religion.

Roman Catholic Hierarchy. The Roman Catholic Church has organized much of Earth's inhabited land into an administrative structure, ultimately accountable to the Pope in Rome. Here is the top hierarchy of Roman Catholicism:

- The *Pope* (he is also the bishop of the Diocese of Rome).

- Reporting to the Pope are *archbishops*. Each heads a *province*, which is a group of several dioceses. (The archbishop also is bishop of one diocese within the province, and some distinguished archbishops are elevated to the rank of *cardinal*.)

- Reporting to each archbishop are *bishops*. Each administers a **diocese**, of which there are several thousand. The diocese is the basic unit of geographic organization in the Roman Catholic Church. The bishop's headquarters, called a "see," is typically the largest city in the diocese.

Saint Peter's Church, in Vatican City, within Rome, Italy, is the principal and largest church of the Roman Catholic world, built mainly in the sixteenth century. St. Peter's dome is 123 meters (404 feet) tall, one-third larger than the U.S. Capitol dome. Architects included Donato Bramante (1444–1514), Raphael (1483–1520), and Michelangelo (1475–1564). The elliptical forecourt was designed by Giovanni Lorenzo Bernini (1598–1680) in the seventeenth century. (Dan Budnik/Woodfin Camp)

Globalization and Local Diversity

Sikhism and Zarathustrianism: Rise and Fall of Universalizing Religions

Christianity, Islam, and Buddhism are by far the three largest universalizing religions, but they are not the only ones. Geographers are interested in smaller universalizing religions because they can provide manageable case studies of changing spatial processes. Sikhism is a recently established universalizing religion that has expanded to 20 million adherents around the world. Zarathustrianism (also known as Zoroastrianism), which predates all three of the large universalizing religions, has declined to less than 200,000 adherents.

Sikhism: A Newer Universalizing Religion

As with other universalizing religions, the place *where* Sikhism originated, as well as the originator, can be documented. The first guru (religious teacher or enlightener) of Sikhism was Nanak (A.D. 1469–1538), who lived in a village near the city of Lahore, in present-day Pakistan. Guru Nanak traveled widely through South Asia preaching his new faith, and many people became his *Sikhs*, which is the Hindi word for *disciples*.

God was revealed to Guru Nanak as The One Supreme Being, or Creator, who rules the universe by divine will. Thus, like other universalizing religions, Sikhism is monotheistic. People can influence their destiny under divine will by taking individual responsibility for their deeds and actions on Earth. Only God is perfect, but through heartfelt adoration, devotion, and surrender to the one God, people have the capacity for continual improvement and movement toward perfection.

Nine other gurus succeeded Guru Nanak. Arjan, the fifth guru, compiled and edited in A.D. 1604 the *Guru Granth Sahib* (the Holy Granth of Enlightenment), which became the book of Sikh holy scriptures. Sikhs point out that the their scripture is the only one that a religion's founders compiled while they were still alive. Arjan also built Sikhism's most holy structure, the Darbar Sahib, or Golden Temple, at Amritsar, in the far west of present-day India.

Sikhism's most important ceremony, introduced by the tenth guru Gobind Singh (A.D. 1666–1708), is the Amrit (or Baptism), in which Sikhs declare they will uphold the principles of the faith. Sikhism, though, does not require the performing of special rituals. Salvation can be attained by responsibly performing activities of daily life, including hard work, sharing food with others, giving to charity, providing selfless service to the community, and devotion to family.

Guru Gobind Singh declared that after his death instead of an eleventh guru, Sikhism's highest spiritual authority would instead be the holy scriptures of the Guru Granth Sahib. A major holiday in Sikhism is the day when the Holy Granth was installed as the religion's spiritual guide. Typical of a universalizing religion, Sikhism celebrates as its other main holidays the births and deaths of the ten gurus. Other holidays recall important dates in Sikh history, such as when prominent Sikhs died fighting for the religion. By commemorating historical events rather than glorifying the physical geography of India, Sikhism promoted an essential element of *globalization*, the ability of a universalizing religion to transfer to other places rather than remain rooted in the physical geography of one place, as is the case with the ethnic religion of Hinduism.

Sikhism did not diffuse as much as Christianity and Islam. Most of the 20 million Sikhs are still clustered in the region known as the Punjab, where the religion originated.

• A diocese in turn is spatially divided into *parishes*, each headed by a *priest*.

The area and population of parishes and dioceses vary according to historical factors and the distribution of Roman Catholics across Earth's surface (for example, the United States is shown in Figure 6–13). In parts of Southern and Western Europe, the overwhelming majority of the dense population is Roman Catholic. Consequently, the density of parishes is high. A typical parish may encompass only a few square kilometers and fewer than 1,000 people.

At the other extreme, Latin American parishes may encompass several hundred square kilometers and 5,000 people. The more dispersed Latin American distribution is attributable partly to a lower population density than in Europe. Because Roman Catholicism is a hierarchical religion, individual parishes must work closely with centrally located officials concerning rituals and procedures. If Latin America followed the European model of small parishes, many would be too remote for the priest to communicate with others in the hierarchy. The less intensive network of Roman Catholic institutions also results in part from colonial traditions, for both Portuguese and Spanish rulers discouraged parish development in Latin America.

The Roman Catholic population is growing rapidly in the U.S. Southwest and suburbs of some large North American and European cities. Some of these areas have a low density of parishes and dioceses compared to the population, so the Church must adjust its territorial organization. New local administrative units can be created, although funds to provide the desired number of churches, schools, and other religious structures might be scarce. Conversely, the Roman Catholic population is declining in inner cities and rural areas. Maintaining ser-

But Sikhism did survive in its place of origin, and it has expanded in recent years to other locations. Others live elsewhere in India. One-half million each are in Western Europe, mostly in the United Kingdom, and in North America, and several thousand are in East Africa.

The reason *why* Sikhism survived in South Asia is comparable to the fate of other universalizing religions, although on a smaller scale. The tenth guru, Gobind Singh, shaped Sikhs into a powerful army, noted as strong, disciplined warriors. Under subsequent leaders, called maharajas (princes), rather than gurus, the Sikhs fought with the Muslims to gain control of the region known as the Punjab where their religion originated. In 1802, an independent Sikh state was established in the Punjab that lasted until 1849, when it was taken over by the British. Sikhs enjoyed a privileged position in South Asia under British rule, and they fought in the British army.

When India and Pakistan became independent countries in 1947, the Punjab was divided between the two rather than made a separate independent state. Preferring to live in Hindu-dominated India rather than Muslim-dominated Pakistan, 2.5 million Sikhs moved from Pakistan's West Punjab region to East Punjab in India.

As part of India, Sikhs have maintained aspirations for an independent country. Militant Sikhs used the Golden Temple at Amritsar as a base for launching attacks in support of greater autonomy for the Punjab. In 1984, the Indian army attacked the Golden Temple at Amritsar and killed approximately a thousand Sikhs defending the temple. In retaliation later that year, India's Prime Minister Indira Gandhi was assassinated in 1984 by two of her guards, who were Sikhs.

Living in India and other countries, Sikhs have retained elements of **local diversity**. The tenth guru, Gobind Singh, introduced the practice of men wearing turbans on their heads and never cutting their beards or hair. Wearing a uniform gave Sikhs a disciplined outlook and a sense of unity of purpose. Now, because many young men are shaving their beards, cutting their hair, and discarding turbans, traditional Sikhs fear that the religion will lose its cultural uniqueness.

Zarathustrianism: A Contracting Universalizing Religion

Zarathustrianism is a good example of territorial contraction, the opposite process to the diffusion that is typical of universalizing religions. The religion's founder, Zarathustra (628?–?551 B.C.), lived in the northeastern part of Persia (present-day Iran). Challenging the superstitions and sacrifices in the local ethnic religion, Zarathustra (Zoroaster is the Greek form of the Persian name) preached his message of one God. After the ruler Vishtaspa and his family accepted the religion and became missionaries, Zarathustrianism diffused rapidly. Cyrus the Great (600?-529 B.C.), the first ruler to unify the entire territory of present-day Iran into the Persian Empire, further diffused the religion. For nearly 1,000 years, Zarathustrianism was the religion of the Persian Empire, until the seventh century A.D., when it was conquered by the Muslims. A few Zarathustris fled from Persia to India, where their descendants are known as Parsis, but most converted to Islam.

vices in these areas is expensive, but the process of combining parishes and closing schools is very difficult.

Latter-Day Saints. Among other Christian religions, Latter-Day Saints (Mormons) exercise strong organization of the landscape. The territory occupied by Mormons, primarily Utah and portions of surrounding states, is organized into *wards*, with populations of approximately 750 each. Several wards are combined into a *stake* of approximately 5,000 population. The highest authority in the Church—the board and president—frequently redraws ward and stake boundaries in rapidly growing areas to reflect the ideal population standards.

Locally Autonomous Religions

Some universalizing religions are highly **autonomous religions**, or self-sufficient, and interaction among communities is confined to little more than loose cooperation and shared ideas. Islam and some Protestant denominations are good examples.

Local Autonomy in Islam. Among the three large universalizing religions, Islam provides the most local autonomy. Like other locally autonomous religions, Islam has neither a religious hierarchy nor a formal territorial organization. A mosque is a place for public ceremony, and a leader calls the faithful to prayer, but everyone is expected to participate equally in the rituals and is encouraged to pray privately.

In the absence of a hierarchy, the only formal organization of territory in Islam is through the coincidence of religious territory with secular states. Governments in some predominantly Islamic countries include in their bureaucracy people who administer Islamic institutions. These administrators interpret Islamic law and run welfare programs.

Figure 6–13 Roman Catholic hierarchy in the United States. The Roman Catholic Church divides the United States into provinces, each headed by an archbishop. Provinces are subdivided into dioceses, each headed by a bishop. The archbishop of a province also serves as the bishop of a diocese. Dioceses that are headed by archbishops are called archdioceses.

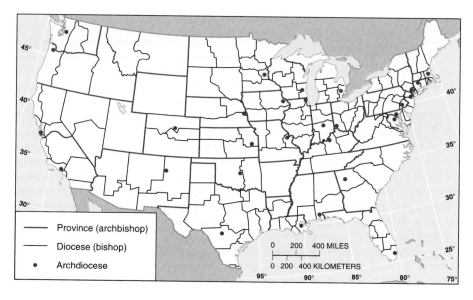

Strong unity within the Islamic world is maintained by a relatively high degree of communication and migration, such as the pilgrimage to Makkah. In addition, uniformity is fostered by Islamic doctrine, which offers more explicit commands than other religions.

Protestant   Denominations. Protestant Christian denominations vary in geographic structure from extremely autonomous to somewhat hierarchical. Extremely autonomous denominations such as Baptists and United Church of Christ are organized into self-governing congregations. Each congregation establishes the precise form of worship and selects the leadership.

Presbyterian churches represent an intermediate degree of autonomy. Individual churches are united in a *presbytery*, several of which in turn are governed by a *synod*, with a *general assembly* as ultimate authority over all churches. Each Presbyterian church is governed by an elected board of directors with lay members. The Episcopalian, Lutheran, and most Methodist churches have hierarchical structures, somewhat comparable to the Roman Catholic Church.

Ethnic Religions. Judaism and Hinduism also have no centralized structure of religious control. To conduct a full service, Judaism merely requires the presence of ten adult males. (Females count in some Jewish communities.) Hinduism is even more autonomous, because worship is usually done alone or with others in the household. Hindus share ideas primarily through undertaking pilgrimages and reading traditional writings.

Key Issue 4

Why Do Territorial Conflicts Arise Among Religious Groups?

- Religion vs. government policies
- Religion vs. religion

The twentieth century was a century of global conflict—two world wars during the first half of the century, the Cold War between supporters of democracy and communism during the second half. With the end of the Cold War in the late twentieth century, the threat of global conflict has receded, but local conflicts have increased in areas of cultural diversity, as will be discussed in Chapters 7 and 8.

The element of cultural diversity that has led to conflict in many localities is religion. Contributing to more intense religious conflict has been a resurgence of religious **fundamentalism**, which is a literal interpretation and a strict and intense adherence to basic principles of a religion (or a religious branch, denomination, or sect). In a world increasingly dominated by a global culture and economy, religious fundamentalism is one of the most important ways that a group maintains a distinctive cultural identity.

As this section demonstrates, the attempt by very intense adherents of one religion to organize Earth's surface can conflict with the spatial expression of other religious or nonreligious ideas. A group convinced that their religious view is *the* correct one may spatially intrude upon the territory controlled by other religious groups. In addition, religious groups oppose government policies seen as promoting social change conflicting with traditional religious values. We will look at examples of each type of conflict.

Religion vs. Government Policies

The role of religion in organizing Earth's surface has diminished in some societies, owing to political and economic change. Islam has been particularly affected by a perceived conflict between religious values and modernization of the economy. Hinduism also has been forced to react to new nonreligious ideas from the West. Buddhism, Christianity, and Islam have all been challenged by Communist governments that diminish the importance of religion in society. Yet, in recent years, religious principles have become increasingly important in the political organization of countries, especially where a branch of Christianity or Islam is the prevailing religion.

Religion vs. Social Change

In less developed countries, participation in the global economy and culture can expose local residents to values and beliefs originating in more developed countries of North America and Western Europe. North Americans and Western Europeans may not view economic development as incompatible with religious values, but many religious adherents in less developed countries do, especially where Christianity is not the predominant religion.

Hinduism vs. Social Equality.
Hinduism has been strongly challenged since the 1800s, when British colonial administrators introduced their social and moral concepts to India. The most vulnerable aspect of the Hindu religion was its rigid **caste** system, which was the class or distinct hereditary order into which a Hindu was assigned according to religious law. In Hinduism, because everyone was different, it was natural that each individual should belong to a particular caste or position in the social order.

The caste system apparently originated around 1500 B.C. when Aryans invaded India from the west. The Aryans divided themselves into four castes, which developed strong differences in social and economic position: Brahmans, the priests and top administrators; Kshatriyas or warriors; Vaisyas or merchants; and Shudras, or agricultural workers and artisans. The Shudras occupied a distinctly lower status than the other three castes. Over the centuries, these original castes split into thousands of subcastes.

The type of Hinduism practiced will depend in part on the individual's caste. A high-caste Brahman may practice a form of Hinduism based on knowledge of relatively obscure historical texts. At the other end of the caste system, a low-caste illiterate in a rural village may perform religious rituals without a highly developed set of written explanations for them.

Below the four castes were the outcasts, or untouchables, who did the work considered too dirty for other castes. In theory, the untouchables were descended from the indigenous people who dwelt in India prior to the Aryan conquest. Until recently, social relations among the five groups were limited, and the rights of non-Brahmans, especially untouchables, were restricted.

British administrators and Christian missionaries pointed out the shortcomings of the caste system, such as neglect of the untouchables' health and economic problems. The rigid caste system has been considerably relaxed in recent years. The Indian government legally abolished the untouchable caste, and the people formerly in that caste now have equal rights with other Indians.

Religion vs. Communism

Organized religion was challenged in the twentieth century by the rise of communism in Eastern Europe and Asia. The three religions most affected were Eastern Orthodox Christianity, Islam, and Buddhism.

Eastern Orthodox Christianity and Islam vs. the Soviet Union.
In 1721, Czar Peter the Great made the Russian Orthodox Church a part of the Russian government. The patriarch of the Russian Orthodox Church was replaced by a 12-member committee, known as the Holy Synod, nominated by the czar.

Following the 1917 Bolshevik revolution, which overthrew the czar, the Communist government of the Soviet Union pursued antireligious programs. Karl Marx had called religion "the opium of the people," a view shared by V. I. Lenin and other early Communist leaders. Marxism became the official doctrine of the Soviet Union, so religious doctrine was a potential threat to the success of the revolution. People's religious beliefs could not be destroyed overnight, but the role of organized religion in Soviet life could be reduced, and was.

In 1918, the Soviet government eliminated the official church-state connection that Peter the Great had forged. All Church buildings and property were nationalized and could be used only with local government permission. The Orthodox religion retained adherents in the Soviet Union, especially among the elderly, but younger people

St. Basil's Cathedral, Red Square, Moscow (Michael Justice/The Image Works)

generally had little contact with the church beyond attending a service perhaps once a year. With religious organizations prevented from conducting social and cultural work, religion dwindled in daily life.

The end of Communist rule in the late twentieth century has brought a religious revival in Eastern Europe, especially where Roman Catholicism is the most prevalent branch of Christianity, including Croatia, the Czech Republic, Hungary, Lithuania, Poland, Slovakia, and Slovenia. Property confiscated by the Communist governments has reverted to Church ownership, and attendance at church services has increased.

In Central Asian countries that were former parts of the Soviet Union—Kazakhstan, Kyrgyzstan, Tajikistan, Turkmenistan, and Uzbekistan—most people are Muslims. These newly independent countries are struggling to determine the extent to which laws should be rewritten to conform to Islamic custom rather than to the secular tradition inherited from the Soviet Union.

Buddhism vs. Southeast Asian Countries.   In Southeast Asia, Buddhists were hurt by the long Vietnam War, waged between the French and, later, by Americans, on one side and Communist groups on the other. Neither antagonist was particularly sympathetic to Buddhists. U.S. air raids in Laos and Cambodia destroyed many Buddhist shrines, while others were vandalized by Vietnamese and by the Khmer Rouge Cambodian Communists. On a number of occasions, Buddhists immolated

(burned) themselves to protest policies of the South Vietnamese government.

The current Communist governments in Southeast Asia have discouraged religious activities and permitted monuments to decay, most notably the Angkor Wat complex in Cambodia, considered one of the world's most beautiful Buddhist structures. In any event, these countries do not have the funds necessary to restore the structures.

Religion vs. Religion

Refer back to the map of world religions, (Figure 6–1), near the beginning of this chapter. Conflicts are most likely to occur where colors change, indicating a boundary between two religious groups. Two long-standing conflicts involving religious groups are in the Middle East and Northern Ireland.

Religious Wars in the Middle East

Conflict in the Middle East is among the world's longest-standing and most intractable. Jews, Christians, and Muslims have fought for 2,000 years to control the same small strip of land in the Eastern Mediterranean. All three religions have especially strong attachments to the city of Jerusalem. To some extent, the hostility among Christians, Muslims, and Jews in the Middle East stems

Jews pray in front of the Western Wall in Jerusalem (right). The wall is the only remaining portion of the Second Temple, which was destroyed in A.D. 70. For hundreds of years, Jews were allowed to visit the site only once a year. Jews are still restricted from visiting the rest of the temple site because it is occupied by structures holy to Muslims, including the Dome of the Rock (left), where Muhammad is thought to have ascended to heaven. (Harold Glaser/Gamma-Liaison, Inc.,left; Brent Petersen/The Stock Market, right)

from their similar heritage. All three groups trace their origins to Abraham in the Old Testament narrative, but the religions diverged in ways that have made it difficult for them to share the same territory.

As an ethnic religion, Judaism makes a special claim to the territory it calls the Promised Land. The major events in the development of Judaism took place there, and the religion's customs and rituals acquired meaning from the agricultural life of the ancient Hebrew tribe.

Jerusalem is especially holy to Jews because it was the location of the Temple, their center of worship in ancient times. The First Temple, built by King Solomon in the tenth century B.C., was destroyed by the Babylonians in 586 B.C. A few years later, when the Persian Empire, led by Cyrus the Great, gained control of Jerusalem, Jews were allowed to build the Second Temple. After the Romans gained control of the area, which they called the province of Palestine, they dispersed the Jews from Palestine and only a handful were permitted to live in the region until the twentieth century. The Romans destroyed the Jewish Second Temple in A.D. 70.

Most inhabitants of Palestine accepted Christianity, especially after the religion was officially adopted by the Roman Empire. Christians consider Palestine the Holy Land and Jerusalem the Holy City because the major events in Jesus' life, death, and Resurrection were concentrated there.

Muslims regard Jerusalem as their third holiest city, after Makkah and Madinah. The most important Muslim structure in Jerusalem is the mosque at the Dome of the Rock, built in 691. The rock is thought to be the place from which Muhammad ascended to heaven. Christians and Jews regard the rock as the altar on which Abraham prepared to sacrifice his son, Isaac, although Muslims believe that Abraham prepared to sacrifice his other son, Ishmael in Makkah.

Immediately to the south is the al-Aqsa Mosque, built on the site of the Second Temple. The only remaining portion of the Second Temple, is the Western Wall, called the Wailing Wall by Christians and Muslims, because for many years Jews were allowed to visit the site only once a year to lament the destruction of their temple (Figure 6–14).

Christians vs. Muslims. In the seventh century, Muslims, now also called Arabs, because they came from the Arabian peninsula, captured most of the Middle East, including Palestine and Jerusalem. The Arab army diffused the Arabic language across the Middle East and converted most of the people from Christianity to Islam.

The Arab army moved west across North Africa and invaded Europe at Gibraltar in 710. It conquered most of the Iberian Peninsula, crossed the Pyrenees Mountains a few years later, and for a time occupied much of present-

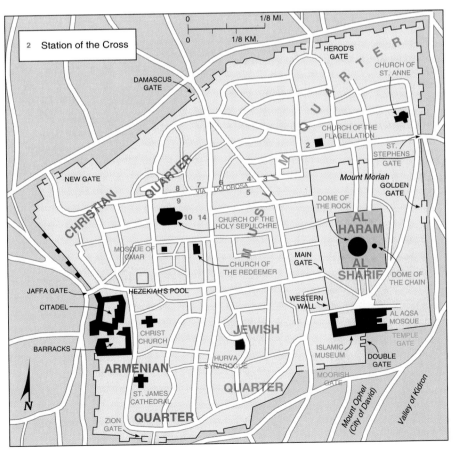

Figure 6–14 Jerusalem. The Old City of Jerusalem contains holy places for three religions. The flattened hill on the eastern side of the Old City is the site of two structures holy to Muslims, the Dome of the Rock and the al-Aqsa Mosque, both of which were built on the site of ancient Jewish temples. The west side of the Old City contains the most important Christian shrines, including the Church of the Holy Sepulchre, where Jesus is thought to have been buried.

day France (see Figure 6–6). Its initial advance in Europe was halted by the Franks (a West Germanic people), led by Charles Martel, at Poitiers, France, in 732. The Arab army made further gains in Europe in subsequent years and continued to control portions of present-day Spain until 1492, but Martel's victory ensured that Christianity rather than Islam would be Europe's dominant religion.

To the east, the Arab army captured Eastern Orthodox Christianity's most important city, Constantinople (present-day Istanbul in Turkey), in 1453 and advanced a few years later into Southeast Europe, as far north as present-day Bosnia and Herzegovina. The current civil war in that country is a legacy of the fifteenth-century Muslim invasion.

To recapture the Holy Land from its Muslim conquerors, European Christians launched a series of military campaigns, known as Crusades, over a 150-year period. Crusaders captured Jerusalem from the Muslims in 1099 during the First Crusade, lost it in 1187 (which led to the Third Crusade), regained it in 1229 as part of a treaty ending the Sixth Crusade, and lost it again in 1244.

Jews vs. Muslims. The Muslim Ottoman Empire controlled Palestine for most of the two centuries between 1516 and 1917. Upon the empire's defeat in World War I, Great Britain took over Palestine under a mandate from the League of Nations, and later from the United Nations. For a few years, the British allowed some Jews to return to Palestine, but immigration was restricted

again during the 1930s in response to intense pressure by Arabs in the region.

As violence initiated by both Jewish and Muslim settlers escalated after World War II, the British announced their intention to withdraw from Palestine. The United Nations voted to partition Palestine into two independent states, one Jewish and one Muslim. Jerusalem was to be an international city, open to all religions, and run by the United Nations.

When the British withdrew in 1948, Jews declared an independent state of Israel within the boundaries prescribed by the U.N. resolution. The next day, its neighboring Arab Muslim states declared war. The combatants signed an armistice in 1949 that divided control of Jerusalem. The Old City of Jerusalem, which contained the famous religious shrines, became part of the Muslim country of Jordan. The newer, western portion of Jerusalem became part of Israel, but Jews were still not allowed to visit the historic shrines in the Old City.

Israel won three more wars with its neighbors in 1956, 1967, and 1973. During the 1967 Six-Day War, Israel captured the entire city of Jerusalem and removed the barriers that had prevented Jews from visiting and living in the Old City of Jerusalem. Israel also captured four other territories in that war. More than three decades after the Six-Day War, Israel has withdrawn from some of the territory, while the status of others has still not been settled (see Chapter 7 for more information on the current situation).

Figure 6–15 Distribution of Protestants in Ireland, 1911. Long a colony of England, Ireland became a self-governing dominion within the British Empire in 1921. In 1937, it became a completely independent country, but 26 districts in the north of Ireland chose to remain part of the United Kingdom. The Republic of Ireland today is more than 95 percent Roman Catholic, whereas Northern Ireland has a Protestant majority. The boundary between Roman Catholics and Protestants does not coincide precisely with the international border, so Northern Ireland includes some communities that are predominantly Roman Catholic. This is the root of a religious conflict that continues today. (below) Women march in protest after release of British soldier Lee Clegg and arrests following riots. (Boyes Kelvin/Gamma-Liaison, Inc.)

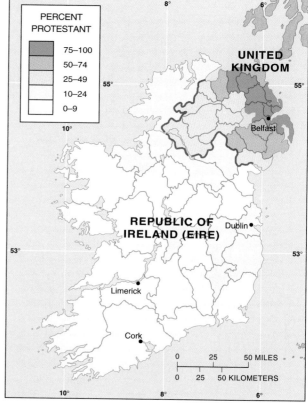

The ultimate obstacle to comprehensive peace in the Middle East is the status of Jerusalem. Israel allows Muslims unlimited access to the religion's holy structures in Jerusalem and some control over them. But as long as any one religion—Jewish, Muslim, or Christian—maintains exclusive political control over Jerusalem, the other religious groups will not be satisfied.

Religious Wars in Ireland

The most troublesome religious boundary in Western Europe lies on the island of Eire (Ireland). The Republic of Ireland, which occupies five-sixths of the island, is 93 percent Roman Catholic, but the island's northern one-sixth, which is part of the United Kingdom rather than Ireland, is about 58 percent Protestant and 42 percent Roman Catholic (Figure 6–15).

The entire island was an English colony for many centuries and was made part of the United Kingdom in 1801. Agitation for independence from Britain increased in Ireland during the nineteenth century, especially after poor economic conditions and famine in the 1840s led to mass emigration, as described in Chapter 3. Following a succession of bloody confrontations, Ireland became a self-governing dominion within the British Empire in 1921. Complete independence was declared in 1937, and a republic was created in 1949.

When most of Ireland became independent, a majority in six northern counties voted to remain in the United Kingdom. Protestants, who comprised the majority in Northern Ireland, preferred to be part of the predominantly Protestant United Kingdom rather than join the predominantly Roman Catholic Republic of Ireland.

Roman Catholics in Northern Ireland have been victimized by discriminatory practices, such as exclusion from higher-paying jobs and better schools. Demonstrations by Roman Catholics protesting discrimination began in 1968. Since then, more than 3,000 have been killed in Northern Ireland—both Protestants and Roman Catholics—in a never-ending cycle of demonstrations and protests.

A small number of Roman Catholics in both Northern Ireland and the Republic of Ireland joined the Irish Republican Army (IRA), a militant organization dedicated to achieving Irish national unity by whatever means available, including violence. Similarly, a scattering of Protestants created extremist organizations to fight the IRA, including the Ulster Defense Force (UDF). While the overwhelming majority of Northern Ireland's Roman Catholics and Protestants are willing to live peacefully with the other religious group, extremists disrupt daily life for everyone. As long as most Protestants are firmly committed to remaining in the United Kingdom and most Roman Catholics are equally committed to union with the Republic of Ireland, peaceful settlement appears difficult. Peace agreements negotiated in 1998 may permit more links between Roman Catholics in the two parts of Ireland, while protecting the political power of Northern Ireland's Protestants.

■ Summary

North Americans pride themselves on tolerance of religious diversity. Most North Americans are Christian, but they practice Christianity in many ways, including Roman Catholicism, many denominations of Protestantism, and other Christian faiths. North America is also home to about 6 million each Jews and Muslims and about 1 million each Buddhists and Hindus. The freedom to establish a religion is a protected right.

The religious landscape looks different outside North America. One-third of the world's people are Christian, but that leaves two-thirds who are not. Around the world, people care deeply about their religion and are willing to fight other religious groups and governments to protect their right to worship as they choose.

Almost all religions preach a doctrine of peace and love, yet religion has been at the center of conflicts throughout history. For geographers, religion represents a critical factor in explaining cultural differences among locations, as well as interrelationships between the environment and culture. Given the importance of religion to people everywhere, geographers are sensitive to the importance of accurately understanding global similarities and local diversity among religions.

The key issues of this chapter demonstrate the impact of religion on the cultural landscape. Here again are the key issues for Chapter 6:

1. **Where are religions distributed?** The world has three large universalizing religions: Christianity, Islam, and Buddhism, each of which is divided into branches and denominations. Hinduism is the largest ethnic religion.

2. **Why do religions have different distributions?** A universalizing religion has a known origin and clear patterns of diffusion, whereas ethnic religions typically have unknown origins and little diffusion. Holy places and holidays in a universalizing religion are related to events in the life of its founder or prophet, and related to the local physical geography in an ethnic religion. Some religions encourage pilgrimages to holy places.

3. **Why and how do religions organize space?** Some religions have elaborate places of worship. Religions affect the landscape in other ways: religious communities are built, religious toponyms mark the landscape, and extensive tracts are reserved for burying the dead. Some but not all universalizing religions organize their territory into a rigid administrative structure to disseminate religious doctrine.

4. **Why do territorial conflicts arise among religious groups?** With Earth's surface dominated by four large religions, expansion of the territory occupied by one religion may reduce the territory of another. In addition, religions must compete for control of territory with nonreligious ideas, notably communism and economic modernization.

Case Study Revisited

Future of Buddhism in Tibet

When the Dalai Lama dies, Tibetan Buddhists believe that his spirit enters the body of a child. In 1937, a group of priests located and recognized a 2-year-old child named Tenzin Gyatso as the fourteenth Dalai Lama, the incarnation of the deceased thirteenth Dalai Lama, Bodhisattva Avalokiteshvara. The child was brought to Lhasa in 1939 when he was 4, and enthroned a year later. Priests trained the young Dalai Lama to assume leadership and sent him to college when he was 16.

Daily life in Tibet was traditionally dominated by Buddhist rites. As recently as the 1950s, one-fourth of all males were monks, and polygamy was encouraged among other males to produce enough children to prevent the population from declining.

After taking control of Tibet in 1950, the Chinese Communists sought to reduce the domination of Buddhist monks in the country's daily life by destroying monasteries and temples.

Farmers were required to join agricultural communes unsuitable for their nomadic style of raising livestock, especially yaks.

In recent years, the Chinese have built new roads and power plants to help raise the low standard of living in Tibet. The Chinese argue that they have brought modern conveniences to Tibet, including paved roads, hospitals, schools, and agricultural practices. Some monasteries have been rebuilt, but no new monks are being trained. Tibet has been given a small degree of autonomy to operate local government.

The Dalai Lama has become an articulate spokesperson for religious freedom, and in 1989 he was awarded the world's most prestigious award for peace, the Nobel Prize. Despite the efforts of the Dalai Lama and other Buddhists, though, when the current generation of priests dies, many Buddhist traditions in Tibet may be lost forever.

■ Key Terms

Animism Belief that objects, such as plants and stones, or natural events, like thunderstorms and earthquakes, have a discrete spirit and conscious life.

Autonomous religion A religion that does not have a central authority but shares ideas and cooperates informally.

Branch A large and fundamental division within a religion.

Caste The class or distinct hereditary order into which a Hindu is assigned according to religious law.

Cosmogony A set of religious beliefs concerning the origin of the universe.

Denomination A division of a branch that unites a number of local congregations in a single legal and administrative body.

Diocese The basic unit of geographic organization in the Roman Catholic Church.

Ethnic religion A religion with a relatively concentrated spatial distribution whose principles are likely to be based on the physical characteristics of the particular location in which its adherents are concentrated.

Fundamentalism Literal interpretation and strict adherence to basic principles of a religion (or a religious branch, denomination, or sect).

Ghetto During the Middle Ages, a neighborhood in a city set up by law to be inhabited only by Jews; now used to denote a section of a city in which members of any minority group live because of social, legal, or economic pressure.

Hierarchical religion A religion in which a central authority exercises a high degree of control.

Missionary An individual who helps to diffuse a universalizing religion.

Monotheism The doctrine or belief of the existence of only one god.

Pagan A follower of a polytheistic religion in ancient times.

Pilgrimage A journey to a place considered sacred for religious purposes.

Polytheism Belief in or worship of more than one god.

Sect A relatively small group that has broken away from an established denomination.

Solstice Time when the Sun is farthest from the equator.

Universalizing religion A religion that attempts to appeal to all people, not just those living in a particular location.

■ Thinking Geographically

1. Sharp differences in demographic characteristics, such as natural increase, crude birth, and migration rates, can be seen among Jews, Christians, and Muslims in the Middle East and between Roman Catholics and Protestants in Northern Ireland. How might demographic differences affect future relationships among the groups in these two regions?

2. People carry their religious beliefs with them when they migrate. Over time, change occurs in the regions from which most U.S. immigrants originate, and in the U.S. regions where they settle. How has the distribution of U.S. religious groups been affected by these changes?

3. To what extent have increased interest in religion and ability to practice religious rites served as forces for unification in Eastern Europe and the countries that formerly were part of the Soviet Union? Has the growing role of religion in the region fostered political instability? Explain.

4. Why does Islam seem strange and threatening to some

people in predominantly Christian countries? To what extent is this attitude shaped by knowledge of the teachings of Muhammad and the Quran, and to what extent is it based on lack of knowledge of the religion?

5. Some Christians believe that they should be prepared to carry the word of God and the teachings of Jesus Christ to

people who have not been exposed to them, at any time and at any place. Are evangelical activities equally likely to occur at any time and at any place, or are some places more suited than others? Why?

■ Further Readings

Al Faruqi, Isma'il R., and Lois Lamaya' Al Faruqi. *The Cultural Atlas of Islam*. New York: Macmillan, 1986.

Al Faruqi, Isma'il R., and David E. Sopher. *Historical Atlas of the Religions of the World*. New York: Macmillan, 1974.

Archer, John Clark, and Carl E. Purinton. *Faiths Men Live By*. 2d ed. New York: Ronald Press, 1958.

Bapat, P. V., ed. *2500 Years of Buddhism*. Delhi: Government of India Ministry of Information and Broadcasting, 1959.

Barraclough, Geoffrey, ed. *The Times' Concise Atlas of World History*. Maplewood, NJ: Hammond, Inc., 1982.

Barrett, David B., ed. *World Christian Encyclopedia*. Oxford: Oxford University Press, 1982.

Berger, Arthur; Paul Badham, Austin H. Kutscher, Joyce Berger, Michael Perry, and John Beloff, eds. *Perspectives on Death and Dying: Cross-Cultural and Multi-Disciplinary Views*. Philadelphia: The Charles Press, 1989.

Bhardwaj, Surinder M. *Hindu Places of Pilgrimage in India*. Berkeley: University of California Press, 1973.

Boal, Frederick W., and J. Neville H. Douglas, eds. *Integration and Division: Geographical Perspectives on the Northern Ireland Problem*. London and New York: Academic Press, 1982.

Cooper, Adrian. "New Directions in the Geography of Religion." *Area* 24 (1992): 123–29.

Curry-Roper, Janet M. "Contemporary Christian Eschatologies and their Relation to Environmental Stewardship." *Professional Geographer* 42 (1990): 157–69.

Fickeler, Paul. "Fundamental Questions in the Geography of Religions." In *Readings in Cultural Geography*, ed. by Philip L. Wagner and Marvin W. Mikesell. Chicago: University of Chicago Press, 1962.

Francaviglia, Richard V. *The Mormon Landscape*. New York: AMS Press, 1978.

Gaustad, E. S. *Historical Atlas of Religion in America*. New York: Harper and Row, 1962.

Hardon, John A. *Religions of the World*. 2 vols. Garden City, NY: Image Books, 1968.

Heatwole, Charles A. "Exploring the Geography of America's Religious Denominations: A Presbyterian Example." *Journal of Geography* 76 (1977): 99–104.

_____. "Sectarian Ideology and Church Architecture." *Geographical Review* 79 (1989): 63–78.

Hiller, Carl E. *Caves to Cathedrals: Architecture of the World's Great Religions*. Boston: Little, Brown, 1974.

Jackson, Richard H. "Mormon Perception and Settlement." *Annals of the Association of American Geographers* 68 (1978): 317–34.

_____, and Roger Henrie. "Perception of Sacred Space." *Journal of Cultural Geography* 3 (1983): 94–107.

Kay, Jeanne. "Human Dominion over Nature in the Hebrew Bible." *Annals of the Association of American Geographers* 79 (1989): 214–32.

Kong, L. "Geography and Religion: Trends and Prospects." *Progress in Human Geography* 14: 355–71.

Levine, Gregory J. "On the Geography of Religion." *Transactions of the Institute of British Geographers*, New Series 11, No. 4 (1987): 428–40.

Ling, Trevor. *A History of Religion East and West*. London: Macmillan, 1968.

Mazrui, Ali A. "Islamic and Western Values." *Foreign Affairs* 72 (1997): 118–32.

Meinig, Donald W. "The Mormon Culture Region: Strategies and Patterns in the Geography of the American West, 1847–1964." *Annals of the Association of American Geographers* 55 (965): 191–220.

Nolan, Mary Lee, and Sidney Nolan. *Christian Pilgrimage in Modern Western Europe*. Chapel Hill: University of North Carolina Press, 1989.

Quinn, Bernard; Herman Anderson, Martin Bradley, Paul Geotting, and Peggy Shriver. *Churches and Church Membership in the U.S.* Atlanta: Glenmary Research Center, 1982.

Sampat, Payal. "The Ganges: Myth and Reality." *Worldwatch* 9 (1996): 24–32.

Short, Ernest. *A History of Religious Architecture*. New York: W. W. Norton & Co., 1951.

Shortridge, James R. "Patterns of Religion in the United States." *Geographical Review* 66 (1976): 420–34.

Smith, Huston. *The Religions of Man*. New York: Harper and Bros., 1958.

Sopher, David E. *The Geography of Religions*. Englewood Cliffs, NJ: Prentice-Hall, 1967.

_____. "Geography and Religions." *Progress in Human Geography* 5 (1981): 510–24.

Stump, Roger W. "Regional Variations in Denominational Switching Among White Protestants." *Professional Geographer* 39 (November 1987): 438–49.

Thompson, Jan, and Mel Thompson. *The R. E. Atlas: World Religions in Maps and Notes*. London: Edward W. Arnold, 1986.

Topping, Gary. "Religion in the West." *Journal of American Culture* 3 (1980): 330–50.

Viorst, Milton, "The Muslims of France." *Foreign Affairs* 75 (1996): 78–96.

Zelinsky, Wilbur. "An Approach to the Religious Geography of the United States: Patterns of Church Membership in 1952." *Annals of the Association of American Geographers* 51 (1961): 139–67.

CHAPTER 7

Ethnicity

Few humans live in total isolation. The overwhelming majority are members of groups of people with whom they share important attributes. If you are a citizen of the United States of America, you are identified as an American, which is a type of nationality.

Many Americans further identify themselves as belonging to an ethnicity, a group with whom they share cultural background. One-fifth of Americans identify their ethnicity as African-American or Hispanic. Other Americans identify with ethnicities tracing back to Europe or Asia.

Ethnicity is a source of pride to people, a link to the experiences of ancestors and to cultural traditions, such as food and music preferences. The ethnic group to which one belongs has important measurable differences, such as average income, life expectancy, and infant mortality rate. Ethnicity also matters in places with a history of discrimination by one ethnic group against another.

The significance of ethnic diversity is controversial in the United States:

- To what extent does discrimination persist against minority ethnicities, especially African-Americans and Hispanics?
- Should preferences be given to minority ethnicities to correct past patterns of discrimination?
- To what extent should the distinct cultural identity of ethnicities be encouraged or protected?

Key Issues

1. Where are ethnicities distributed?
2. Where have ethnicities been transformed into nationalities?
3. Why do ethnicities clash?
4. Why does ethnic cleansing occur?

Lower East Side of New York, 1890s, home to hundreds of thousands of recent immigrants from Eastern Europe. (North Wind Pictures)

Case Study

Ethnic Conflict in Rwanda

Samuel Ntawiniga, Helene Mukabutera, and their five children had a comfortable life in Rwanda, a small central African country about the size of Maryland. The family lived in the capital Kigali, in a three-bedroom house with a modern kitchen and a television with videocassette recorder.

The Ntawiniga family's comfortable life was shattered in April 1994, when the presidents of Rwanda and neighboring Burundi died in a mysterious plane explosion. The plane crash destroyed a peace accord being negotiated at the time between Rwanda's two ethnicities, Hutu and Tutsi.

The next day, soldiers broke into the Ntawiniga house and took their money, after threatening to shoot them. Family members living elsewhere in the city were killed that night. The next day, the family was ordered to leave their house or risk being shot. They walked for a week, until they reached the border town of Cyangugu, 130 kilometers (80 miles) away.

The Ntawinigas are Hutu, the ethnicity of 85 percent of the people of Rwanda when the killing started in 1994. It was Hutu soldiers who forced the Ntawinigas to flee, because Samuel and Helene looked like members of a different ethnicity, Tutsi, and the Hutus feared a Tutsi uprising after the president died. Both Samuel and Helene's mothers were Tutsi, but they were both considered Hutu because their fathers were Hutu.

After the Tutsi rebels defeated the Hutu army and gained control of Rwanda, the Ntawinigas were forced to flee again, this time to a refugee camp at Nyarushishi. They joined millions of other Rwandans—Hutu and Tutsi—huddling together under blue plastic tarpaulins in a vain attempt to stay warm and dry. Many have died in the camps from tuberculosis, pneumonia, malaria, and dysentery.

In Rwanda, Hutus were forced to migrate to neighboring countries after Tutsis won the civil war in 1994. (Steve Lehman/SABA Press Photos, Inc.)

Ethnicity is identity with a group of people who share the cultural traditions of a particular homeland or hearth. Ethnicity comes from the Greek word *ethnikos*, which means *national*.

Geographers are interested in *where* ethnicities are distributed, like other elements of culture. An ethnic group is tied to a particular place, because members of the group—or their ancestors—were born and raised there. The cultural traits displayed by an ethnicity derives from particular conditions and practices in the group's homeland.

The reason *why* ethnicities have distinctive traits should by now be familiar. Like other cultural elements, ethnic identity derives from the interplay of migration and isolation from other groups.

Ethnicity is an especially important cultural element of *local diversity* because our ethnic identity is immutable. We can deny or suppress our ethnicity, but we cannot choose to change it the way we can choose to speak a different language or practice a different religion. If our parents come from two ethnic groups or our grandparents from four, our ethnic identity may be extremely diluted, but it never completely disappears.

The study of ethnicity lacks the tension between preservation of local diversity and *globalization* observed in other cultural elements. Despite efforts to preserve local languages, it is not farfetched to envision a world where virtually all educated people speak English. And universalizing religions continue to gain adherents around the world. But no ethnicity is attempting or even aspiring to achieve global dominance, although ethnic groups are fighting with each other to control specific areas of the world.

Ethnicity is especially important to geographers, because in the face of globalization trends in culture and economy, ethnicity stands as the strongest bulwark for the preservation of local diversity. Even if globalization engulfs language, religion, and other cultural elements, the diversity of ethnic identity will remain.

Key Issue 1

Where Are Ethnicities Distributed?

- Distribution of ethnicities in the United States
- Differentiating ethnicity and race

An ethnicity may be clustered in specific areas within a country, or the area it inhabits may match closely the boundaries of a country. This section of the chapter examines the clustering of ethnicities within countries, and the next key issue looks at ethnicities at the national scale.

Distribution of Ethnicities in the United States

The two most numerous ethnicities in the United States are African-Americans, about 12 percent, and Hispanics

or Latinos, about 9 percent. In addition, about 3 percent are Asian-American and 1 percent American Indian.

Clusterings of Ethnicities

Within a country, clustering of ethnicities can occur at two scales. Ethnic groups may live in particular regions of the country, and they may live in particular neighborhoods within cities.

Regional Concentrations of Ethnicities. At a regional scale, ethnicities have distinctive distributions within the United States. African-Americans are clustered in the Southeast, Hispanics in the Southwest, Asian-Americans in the West, and American Indians in the Southwest and Plains states.

African-Americans comprise at least one-fourth of the population in Alabama, Georgia, Louisiana, and South Carolina, and more than one-third in Mississippi (Figure 7–1). At the other extreme, nine states have fewer than 1 percent African-Americans, including the upper New England states of Maine, New Hampshire, and Vermont, as well as the Plains states of Idaho, Montana, North Dakota, South Dakota, Utah, and Wyoming.

About 9 percent of Americans identify themselves as being of Hispanic or Latino ethnicity. *Hispanic* or *Hispanic-American* is a term that the U.S. government chose in 1973 to describe the group, because it was an inoffensive label that could be applied to all people from Spanish-speaking countries. Some Americans of Latin-American descent have adopted the term *Latino* instead. A 1995 U.S. Census Bureau survey found that 58 percent of Americans of Latin-American descent preferred the term *Hispanic* and 12 percent *Latino*.

Most Hispanics identify with a more specific ethnic or national origin. The largest number of Hispanics, about 64 percent, come from Mexico, and are sometimes called Chicanos (males) or Chicanas (females). Originally the term was considered insulting, but beginning in the

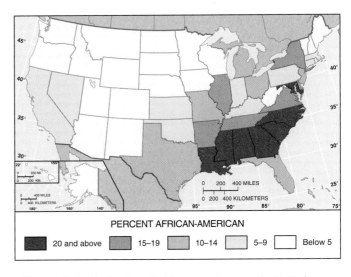

PERCENT AFRICAN-AMERICAN

| 20 and above | 15–19 | 10–14 | 5–9 | Below 5 |

Figure 7–1 Distribution of African-Americans in the United States. The highest percentages of African-Americans are in the rural South and in northern cities.

1960s Mexican-American youths in Los Angeles began to call themselves Chicanos and Chicanas with pride. Puerto Ricans comprise the second-largest group of Hispanics, about 11 percent, followed by Cubans, about 4 percent.

Within the United States, Hispanics are heavily clustered in the four southwestern states of Arizona, California, New Mexico, and Texas, where they constitute more than one-fourth of the total population. More than half of all Hispanics are in these four states, and nearly three-fourths are in these four states plus Florida and New York (Figure 7–2). Eleven states have fewer than 1 percent Hispanic, mostly in the Southeast.

About 3 percent of the U.S. population are Asian-Americans. Chinese account for about 25 percent of Asian-Americans, Filipinos 20 percent, and Japanese, Asian Indians, and Vietnamese 12 percent each. The largest concentration of Asian-Americans is in Hawaii, where they comprise 62 percent of the population. California, with 9.6 percent, is home to one-third of all Asian-Americans (Figure 7–3).

American Indians and Alaska Natives make up about 1 percent of the U.S. population. Within the 48 continental United States, American Indians are most numerous in the Southwest and the Plains states.

Concentration of Ethnicities in Cities. African-Americans are highly clustered within cities. About one-fourth of all Americans live in cities, whereas more than half of African-Americans live in cities.

The contrast is greater at the state level. For example, African-Americans comprise 76 percent of the population in the city of Detroit and only 6 percent in the rest of Michigan. Otherwise stated, Detroit contains 11 percent of Michigan's total population, but 60 percent of the state's African-American population. Similarly, Chicago is 39 percent African-American, compared to 7 percent

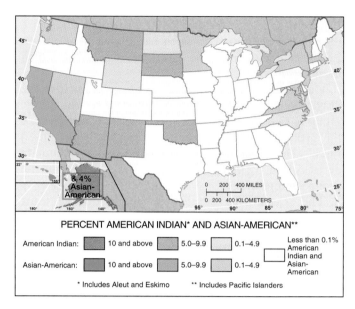

Figure 7–3 Distribution of Asian-Americans and American Indians in the United States. The highest percentages of American Indians are in Alaska and Plains states. The highest percentages of Asian-Americans are in Hawaii and California.

in the rest of Illinois. Chicago has 24 percent of Illinois' total population and 64 percent of the state's African-Americans.

The distribution of Hispanics is similar to that of African-Americans in large northern cities. For example, New York City is 24 percent Hispanic, compared to 4 percent in the rest of New York State, and New York City contains 41 percent of the state's total population and 81 percent of its Hispanics.

In the states with the largest Hispanic populations—California and Texas—the distribution is mixed. In California, Hispanics comprise 40 percent of Los Angeles's population, but the percentage of Hispanics in California's other large cities is less than or about equal to the overall state average. In Texas, El Paso and San Antonio—the two large cities closest to the Mexican border—contain 67 and 56 percent Hispanic, respectively, but the state's other large cities have percentages below or about equal to the state's average.

The clustering of ethnicities is especially pronounced at the scale of neighborhoods within cities. In the early twentieth century, Chicago, Cleveland, Detroit, and other midwestern cities attracted ethnic groups primarily from Southern and Eastern Europe to work in the rapidly growing steel, automotive, and related industries. For example, in 1910, when Detroit's auto production was exploding, three-fourths of the city's residents were immigrants and children of immigrants. Southern and Eastern European ethnic groups clustered in newly constructed neighborhoods that were often named for their predominant ethnicities, such as Detroit's Greektown and Poletown (Figure 7–4).

During the twentieth century, the children and grandchildren of European immigrants moved out of most of the original inner-city neighborhoods. For descendants

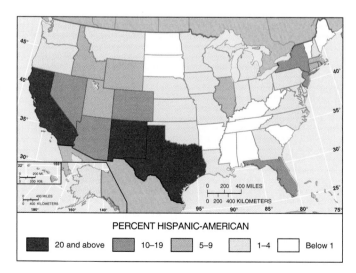

Figure 7–2 Distribution of Hispanic-Americans in the United States. The highest percentages are in the Southwest, near the Mexican border, and in northern cities.

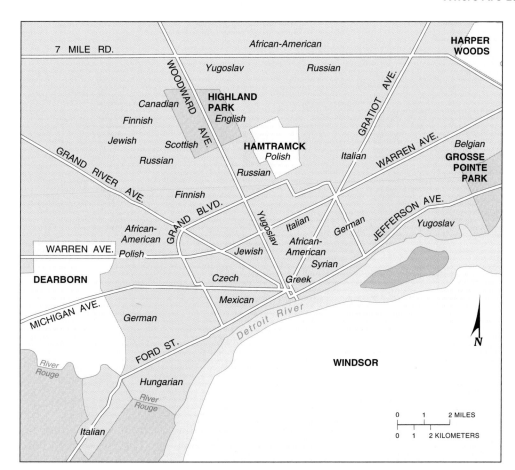

Figure 7–4 Distribution of ethnicities in Detroit, 1910. The most numerous ethnicities in Detroit in 1910 were immigrants from Poland, followed by immigrants from Germany and Russia.

of European immigrants, ethnic identity is more likely to be retained through religion, food, and other cultural traditions, rather than through location of residence. A visible remnant of early twentieth-century European ethnic neighborhoods is the clustering of restaurants in such areas as Little Italy and Greektown.

During the twentieth century, ethnic concentrations in U.S. cities have increasingly consisted of African-Americans who migrated from the South, or immigrants from Latin America and Asia. In cities such as Detroit, African-Americans now comprise the majority and live in neighborhoods originally built by European ethnic groups. Chicago has extensive African-American neighborhoods on the South and West sides of the city, but the city also contains a mix of neighborhoods inhabited by European, Latin American, and Asian ethnicities (Figure 7–5).

In Los Angeles, which contains large percentages of African-Americans, Hispanics, and Asian-Americans, the major ethnic groups are clustered in different areas (Figure 7–6). African-Americans are located in south-central Los Angeles and Hispanics in the east. Asian-Americans are located to the south and west contiguous to the African-American and Hispanic areas.

The proximity of Asian-American ethnic groups to African-Americans proved volatile in Los Angeles in 1992. After white police officers were acquitted—despite videotape evidence—of beating an African-American (Rodney King), unrest broke out in African-American

south side neighborhoods. Many of the stores that were looted or burned were owned by Asian-Americans.

African-American Migration Patterns

The clustering of ethnicities within the United States is partly a function of the same process that help geographers to explain the regular distribution of other cultural factors, such as language and religion—namely migration. The migration patterns of African-Americans have been especially distinctive. Three major migration flows have shaped their current distribution within the United States:

- immigration from Africa to the American colonies in the eighteenth century;
- immigration from the U.S. South to northern cities during the first half of the twentieth century;
- immigration from inner-city ghettos to other urban neighborhoods in the second half of the twentieth century.

Forced Migration from Africa. Most African-Americans are descended from Africans forced to migrate to the Western Hemisphere as slaves. Slavery is a system whereby one person owns another person like a piece of property and can force that slave to work for the owner's benefit.

The first Africans brought to the American colonies as slaves arrived at Jamestown, Virginia, on a Dutch ship in

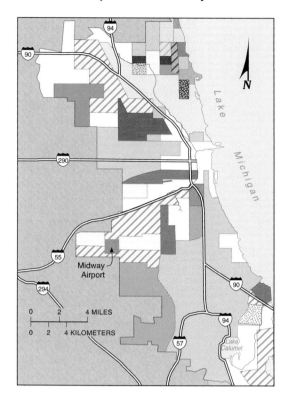

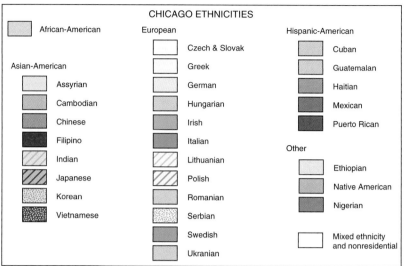

CHICAGO ETHNICITIES

African-American		European		Hispanic-American	
			Czech & Slovak		Cuban
Asian-American			Greek		Guatemalan
	Assyrian		German		Haitian
	Cambodian		Hungarian		Mexican
	Chinese		Irish		Puerto Rican
	Filipino		Italian		
	Indian		Lithuanian	Other	
	Japanese		Polish		Ethiopian
	Korean		Romanian		Native American
	Vietnamese		Serbian		Nigerian
			Swedish		
			Ukranian		Mixed ethnicity and nonresidential

Figure 7–5 Distribution of ethnicities in Chicago. African-Americans occupy extensive areas on the South and West sides. Hispanic-Americans are clustered in several neighborhoods on the West Side. European ethnic groups are located to the northwest, southwest, and far South Side. Asian ethnic groups are clustered in the far North Side.

1619. During the eighteenth century, the British shipped about 400,000 Africans to the 13 colonies that later formed the United States. In 1808, the United States banned bringing in additional Africans as slaves, but an estimated 250,000 were illegally imported during the next half-century.

Slavery was widespread during the time of the Roman Empire, about 2,000 years ago. During the Middle Ages, slavery was replaced in Europe by a feudal system, in which laborers working the land (known as serfs) were bound to the land and not free to migrate elsewhere. Serfs had to turn over a portion of their crops to the lord and provide other services as demanded by the lord.

Although slavery was rare in Europe, Europeans were responsible for diffusing the practice to the Western Hemisphere. This large-scale slave trade was a response to a shortage of labor in the sparsely inhabited Americas. Europeans who owned large plantations in the Americas turned to African slaves as a cheap and abundant source of labor.

At the height of the slave trade between 1710 and 1810, at least 10 million Africans were uprooted from their homes and sent on European ships to the Western Hemisphere for sale in the slave market. During that period, the British and Portuguese each shipped about 2 million slaves to the Western Hemisphere, with most of the British slaves going to Caribbean islands, and the Portuguese ships to Brazil.

The forced migration began when people living along the east and west coasts of Africa, taking advantage of their superior weapons, captured members of other groups living farther inland and sold the captives to Europeans. Europeans in turn shipped the captured Afri-

cans to the Americas, selling them as slaves either on consignment or through auctions. The Spanish and Portuguese first participated in the slave trade in the early sixteenth century, and the British, Dutch, and French joined in during the next century.

Different European countries operated in various regions of Africa, each sending slaves to different destinations in the Americas (Figure 7–7, right). The Portuguese shipped slaves primarily from their principal African colonies—Angola and Mozambique—to their major American colony, Brazil. Other European countries took slaves primarily from a coastal strip of West Africa between Liberia and the Congo, 4,000 kilometers (2,500 miles) long and 160 kilometers (100 miles) wide. The majority of these slaves went to Caribbean islands, and most of the remainder to Central and South America. Fewer than 5 percent of the slaves ended up in the United States.

At the height of the eighteenth century slave demand, a number of European countries adopted the **triangular slave trade**, an efficient triangular trading pattern (Figure 7–7, left). Ships left Europe for Africa with cloth and other trade goods, used to buy the slaves. They then transported slaves and gold from Africa to the Western Hemisphere, primarily to the Caribbean islands. To complete the triangle, the same ships then carried sugar and molasses from the Caribbean on their return trip to Europe. Some ships added another step, making a rectangular trading pattern, in which molasses was carried from the Caribbean to the North American colonies and rum from the colonies to Europe.

The large-scale forced migration of Africans obviously caused them unimaginable hardship, separating families

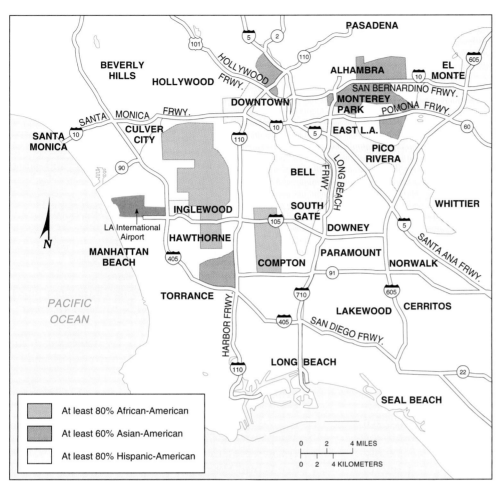

Figure 7–6 Distribution of ethnicities in Los Angeles. African-Americans are clustered to the south of downtown Los Angeles, and Hispanic to the east. Asian-American neighborhoods are contiguous to the African-American and Hispanic areas.

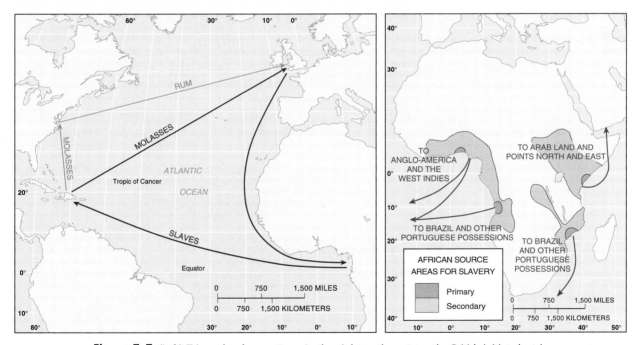

Figure 7–7 (Left) Triangular slave pattern. In the eighteenth century, the British initiated a triangular slave trading pattern. Cloth, iron bars, and other goods were carried by ship from Britain to Africa to buy slaves. The same ships transported slaves from Africa to the Caribbean islands. The ships then completed the triangle by returning to Britain with molasses to make rum. Sometimes the ships formed a rectangular pattern by carrying the molasses from the Caribbean islands to the North American colonies, where the rum was distilled and shipped to Britain. (Right) The British and other European powers obtained slaves primarily from a narrow strip along the west coast of Africa, from Liberia to Angola. In the early days of colonization, Europeans secured territory along the Atlantic Coast and rarely ventured more than 160 kilometers (100 miles) into the interior of the continent.

and destroying villages. Traders generally seized the stronger and younger villagers, who could be sold as slaves for the highest price. The Africans were packed onto ships at extremely high density, kept in chains, and provided with minimal food and sanitary facilities. Approximately one-fourth died crossing the Atlantic.

In the 13 colonies that later formed the United States, most of the large plantations in need of labor were located in the South, primarily those growing cotton, as well as tobacco. Consequently, nearly all blacks shipped to the 13 colonies ended up in the Southeast.

Attitudes toward slavery dominated U.S. politics during the nineteenth century. During the early 1800s, when new states were carved out of western territory, anti-slavery northeastern states and pro-slavery southeastern states bitterly debated whether to permit slavery in the new states. The Civil War (1861–1865) was fought to prevent 11 pro-slavery southern states from seceding from the Union. In 1863, during the Civil War, Abraham Lincoln issued the Emancipation Proclamation, freeing the slaves in the 11 Confederate states. The Thirteenth Amendment to the Constitution, adopted 8 months after the South surrendered, outlawed slavery.

Freed as slaves, most African-Americans remained in the rural South during the late nineteenth century working as sharecroppers. A **sharecropper** works fields rented from a landowner and pays the rent by turning over to the landowner a share of the crops. To obtain seed, tools, food, and living quarters, a sharecropper gets a line of credit from the landowner and repays the debt with yet more crops. The sharecropper system burdened poor African-Americans with high interest rates and heavy debts. Instead of growing food that they could eat, share-croppers were forced by landowners to plant extensive areas of crops such as cotton that could be sold for cash.

Immigration to the North.
Sharecropping declined in the early twentieth century, as the introduction of farm machinery and decline in land devoted to cotton reduced demand for labor. At the same time sharecroppers were being pushed off the farms, they were being pulled to the prospect of jobs in the booming industrial cities of the North.

African-Americans migrated out of the South along several clearly defined channels (Figure 7–8). Most traveled by bus and car along the major two-lane long-distance U.S. roads that had been paved and signposted in the early decades of the twentieth century and since replaced by interstate highways.

- From the Carolinas and other South Atlantic states north to Baltimore, Philadelphia, New York, and other northeastern cities, along U.S. Route 1 (parallel to present-day I-95).
- From Alabama and eastern Tennessee north to either Detroit, along U.S. Route 25 (present-day I-75), or Cleveland, along U.S. route 21 (present-day I-77).
- From Mississippi and western Tennessee north to St. Louis and Chicago along U.S. routes 61 and 66 (present-day I-55).
- From Texas west to California, along U.S. routes 80 and 90 (present-day I-10 and I-20).

Southern African-Americans migrated north and west in two main waves, the first in the 1910s and 1920s before and after World War I and the second in the 1940s and 1950s before and after World War II. The world wars stimulated expansion of factories in the 1910s and 1940s to produce war material, while the demands of the armed forces created shortages of factory workers. After the wars, during the 1920s and 1950s, factories produced steel, motor vehicles, and other goods demanded in civilian society.

For example, in 1910 only 5,741 of Detroit's 465,766 inhabitants were African-American. With the expansion

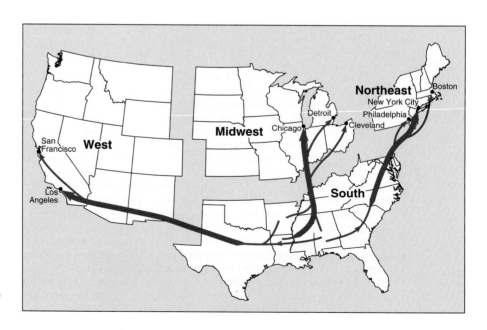

Figure 7–8 African-American twentieth-century migration within the United States. Migration followed distinctive channels, including from the Carolinas to the Northeast, from Alabama and Mississippi to the Midwest, and from Texas to California.

of the auto industry during the 1910s and 1920s, the African-American population increased to 120,000 in 1930, 300,000 in 1950, and 500,000 in 1960.

Expansion of the Ghetto. When they reached the big cities, African-American immigrants clustered in the one or two neighborhoods where the small numbers who had arrived in the nineteenth century were already living. These areas became known as ghettos, after the term for neighborhoods in which Jews were forced to live in the Middle Ages (see Chapter 6). A half-million African-Americans jammed in Chicago's 3-square-mile (8-square-kilometer) South Side ghetto.

In 1950, most of Baltimore's quarter-million African-Americans lived in a 1-square-mile (3-square-kilometer) neighborhood northwest of downtown. The remainder were clustered east of downtown or in a large isolated housing project on the south side built for black war-time workers in port industries (Figure 7–9).

Densities in the ghettos were high, with 100,000 inhabitants per square mile (40,000 per square kilometer) common. Contrast that density with the current level found in typical American suburbs of 10 inhabitants per square mile (4 per square kilometer). Because of the shortage of housing in the ghettos, families were forced to live in one room. Many dwellings lacked bathrooms, kitchens, hot water, and heat.

During the 1950s and 1960s, African-Americans moved from the tight ghettos into immediately adjacent neighborhoods. In Chicago, African-Americans pushed south from the old South Side neighborhood at the rate of 1 square mile (2.5 square kilometers) per year. In Baltimore, the West Side African-American area expanded from 1 square mile (2.5 square kilometers) in 1950 to 10 square miles (25 square kilometers) in 1970, and a 2-square-mile (5-square-kilometer) area on the East Side became virtually all African-American. The expansion of the ghetto followed major avenues to the northwest and northeast.

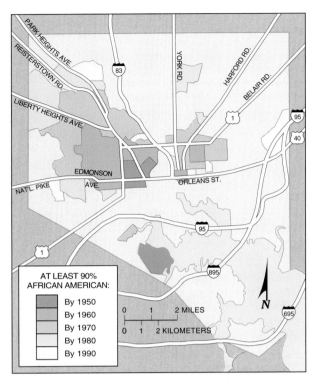

Figure 7–9 Expansion of African-American ghetto in Baltimore, Maryland. In 1950, most African-Americans in Baltimore lived in a small area northwest of downtown. During the 1950s and 1960s, the African-American area expanded to the northwest, along major radial roads, and a second node opened on the east side. The south-side African-American area was an isolated public housing complex built for wartime workers in the nearby port industries.

Differentiating Ethnicity and Race

Ethnicity is distinct from **race**, which is identity with a group of people who share a biological ancestor. Race comes from a middle-French word for *generation.*

Race and ethnicity are often confused. In the United States, consider three prominent ethnic groups—Asian-Americans, African-Americans, and Hispanic-Americans.

This African-American family arrived in Chicago after migrating from Mississippi, about 1910. (Stock Montage, Inc.)

During the first half of the twentieth century, African-Americans were jammed into deteriorated inner-city neighborhoods. In the absence of back yards or nearby playgrounds, children play in a narrow alleyway of an inner-city neighborhood in Washington, DC, about 1940. (Library of Congress)

All three ethnicities display distinct cultural traditions that originate at particular hearths, but the three are regarded in different ways:

- Asian is recognized as a distinct race by the U.S. Bureau of the Census, so Asian as a race and Asian-American as an ethnicity encompass basically the same group of people. However, the Asian-American ethnicity lumps together people with ties to many countries in Asia.
- African-American and black are different groups. Most black Americans are descended from African immigrants and therefore also belong to an African-American ethnicity. Some American blacks, however, trace their cultural heritage to regions other than Africa, including Latin America, Asia, or Pacific islands.
- Hispanic or Latino is not considered a race, so on the census form members of the Hispanic or Latino ethnicity select any race they wish—white, black, or other.

The traits that characterize race are those that can be transmitted genetically from parents to children. For example, lactose intolerance is relatively common among persons of Asian, African, and Eastern European descent and extremely rare among persons of Northern European descent. Nearly everyone is born with the ability to produce lactase, which enables infants to digest the large amount of lactose in milk. Lactase production typically slackens during childhood, leaving some with difficulty in absorbing a large amount of lactose as adults. A large percentage of persons of Northern European descent have a genetic mutation that results in lifelong production of lactase.

Biological features of all humans, such as skin color, hair type and color, blood traits, and shape of body, head, and facial features, were once thought to be scientifically classifiable into a handful of world races. At best, however, biological features are so highly variable among

members of a race that any pre-judged classification is meaningless. Perhaps many tens or hundreds of thousands of years ago, early "humans" (however they emerged as a distinct species) lived in such isolation of other early "humans" that they were truly distinct genetically. But the degree of isolation needed to keep biological features distinct genetically vanished when the first human crossed a river or climbed a hill.

At worst, biological classification by race is the basis for **racism**, which is the belief that race is the primary determinant of human traits and capacities and that racial differences produce an inherent superiority of a particular race. A **racist** is a person who subscribes to the beliefs of racism.

Ethnicity is important to geographers because its characteristics derive from the distinctive features of particular places on Earth. In contrast, contemporary geographers reject the entire biological basis of classifying humans into a handful of races because these features are not rooted in specific places. Geographers stress the heterogeneity of the human population, and an examination of biological differences among people does not explain why people live as they do.

However, one feature of race does matter to geographers—the color of skin. The distribution of persons of color matters to geographers because it is the most fundamental basis by which people in many societies sort out where they reside, attend school, recreate, and perform many other activities of daily life.

The term *African-American* identifies a group with an extensive cultural tradition, whereas the term *black* in principle denotes nothing more than a dark skin. Because many Americans make judgments about the values and behavior of others simply by observing skin color, black is substituted for African-American in daily language.

Race in the United States

Every 10 years, the U.S. Bureau of the Census asks people to classify themselves according to the race with which they most closely identify. Americans are asked to identify themselves by checking the box next to one of the following 6 races: white, black or African–American, Asian, American Indian or Alaska Native, Native Hawaiian or other Pacific Islander, or other race. The 2000 census permits people to check more than one box.

In 1990, about 80 percent of Americans checked that they were white, 12 percent black, and 3 percent Asian, 1 percent American Indian or Alaska Native, 0.1 percent Native Hawaiian or other Pacific Islander, and 4 percent other race.

In explaining spatial regularities, geographers look for patterns of spatial interaction. A distinctive feature of race relations in the United States has been the strong discouragement of spatial interaction, in the past through legal means, today through cultural preferences or discrimination.

"Separate But Equal" Doctrine. In 1896, the U.S. Supreme Court upheld a Louisiana law that required black and white passengers to ride in separate railway cars. In *Plessy v. Ferguson*, the Supreme Court stated that Louisiana's law was constitutional because it provided separate, *but equal*, treatment of blacks and whites, and equality did not mean that whites had to mix socially with blacks.

Once the Supreme Court permitted "separate but equal" treatment of the races, southern states enacted a comprehensive set of laws to segregate blacks from whites as much as possible. These were called "Jim Crow" laws, named for a nineteenth century song-and-dance act that depicted blacks offensively. Blacks had to sit in the back of buses, and shops, restaurants, and hotels could serve only whites. Separate schools were established for blacks and whites. After all, white southerners argued, the bus got blacks sitting in the rear to the destination at the same time as the whites in the front, some commercial establishments served only blacks, and all of the schools had teachers and classrooms.

Throughout the country, not just in the South, house deeds contained restrictive covenants that prevented the owners from selling to blacks, as well as to Roman Catholics or Jews in some places. Restrictive covenants kept blacks from moving into an all-white neighborhood. And because schools, especially at the elementary level, were located to serve individual neighborhoods, most were segregated in practice, even if not legally mandated.

"White Flight." Segregation laws were eliminated during the 1950s and 1960s. The landmark Supreme Court decision, *Brown v. Board of Education of Topeka, Kansas*, in 1954, found that separate schools for blacks and whites was unconstitutional, because no matter how equivalent the facilities, racial separation branded minority children as inferior, and therefore was inherently unequal. A year later the Supreme Court further ruled that schools had to be desegregated "with all deliberate speed."

Rather than integrate, whites fled. The expansion of the black ghettos in American cities was made possible by "white flight," the emigration of whites from an area in anticipation of blacks immigrating into the area. Detroit provides a clear example. During the 1950s, black immigration into Detroit from the South subsided, but as legal barriers to integration crumbled whites began to emigrate out of Detroit. Detroit's white population dropped by nearly 1 million between 1950 and 1970, and by another one-half million between 1970 and 1990. While whites fled, Detroit's black population continued to grow, but at a more modest rate, as a result of natural increase.

Until the 1960s in the U.S. South, whites and blacks had to use separate drinking fountains, as well as separate restrooms, bus seats, hotel rooms, and other public facilities (Corbis-Bettmann)

Figure 7–10 Homelands in South Africa. As part of its apartheid system, the government of South Africa designated ten homelands, expecting that ultimately every black would become a citizen of one of them. South Africa declared four of these homelands to be independent states, but no other country recognized the action. With the end of apartheid and the election of a black majority government, the homelands were abolished, and South Africa was reorganized into nine provinces.

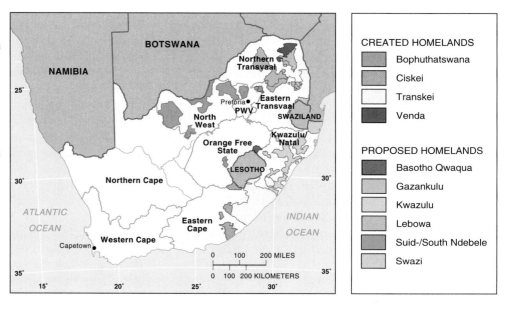

In sum, Detroit in 1950 contained about 1.7 million whites and 0.3 million blacks. The black population increased to 0.5 million in 1960, 0.7 million in 1970, and 0.8 million in 1980 and 1990, while the white population declined to 1.3 million in 1960, 0.9 million in 1970, 0.5 million in 1980, and 0.3 million in 1990.

White flight was encouraged by unscrupulous real estate practices, especially blockbusting. Under **blockbusting**, real estate agents convinced white homeowners living near a black area to sell their houses at low prices, preying on their fears that black families would soon move into the neighborhood and cause property values to decline. The agents then sold the houses at much higher prices to black families desperate to escape the overcrowded ghettos. Through blockbusting, a neighborhood could change from all-white to all-black in a matter of months, and real estate agents could start the process all over again in the next white area.

In the late 1960s, the National Advisory Commission on Civil Disorders, known as the Kerner Commission, concluded that U.S. cities were divided into two separate and unequal societies, one black and one white. Three decades later, despite serious efforts to integrate and equalize the two, segregation and inequality persist.

Division by Race in South Africa

Discrimination by race reached its peak in the late twentieth century in South Africa. While the United States was repealing laws that segregated people by race, South Africa was enacting them. The cornerstone of the South

South Africa's apartheid laws were designed to spatially segregate races as much as possible. Blacks and whites reached the platform at this train station in Johannesburg by walking up separate stairs. Whites waited at the front of the platform to get into cars at the head of the train, while blacks waited at the rear. (William Campbell/Sygma)

African policy was the creation of a legal system called apartheid. **Apartheid** was the physical separation of different races into different geographic areas. Although South Africa's apartheid laws were repealed during the 1990s, it will take many years for it to erase the impact of past policies.

Apartheid System. In South Africa, under apartheid, a newborn baby was classified as being of one of four races: black, white, colored (mixed white and black), or Asian. According to the most recent census, blacks constitute about 76 percent of South Africa's population, whites 13 percent, colored 9 percent, and Asians 3 percent.

Under apartheid, each of the four races had a different legal status in South Africa. The apartheid laws determined where different races could live, attend school, work, shop, and own land. Blacks were restricted to certain occupations and were paid far lower wages than were whites for similar work. Blacks could not vote or run for political office in national elections.

The apartheid system was created by descendants of whites who arrived in South Africa from Holland in 1652 and settled Cape Town at the southern tip of the territory. They were known either as *Boers*, from the Dutch word for farmer, or *Afrikaners*, from the word "Afrikaans," the name of their language, which is a dialect of Dutch.

In 1795, the British seized the Dutch colony at Cape Town for military reasons. To escape British administration and the freeing of slaves in 1833, about 12,000 Boers trekked northeast into the interior of South Africa and settled in the regions known as the Transvaal and the Orange Free State (Figure 7–10). After diamonds and gold were discovered in the Transvaal during the 1860s and 1870s, the British followed the Boers into South Africa's interior. A series of wars between the British and the Boers culminated in a British victory in 1902, and all of South Africa became part of the British Empire.

British descendants continued to control South Africa's government until 1948, when the Afrikaner-dominated Nationalist party won elections. The Afrikaners gained power at a time when colonial rule was being replaced in the rest of Africa by a collection of independent states run by the local black population. The Afrikaners vowed to resist pressures to turn over South Africa's government to blacks, and the Nationalist party created the apartheid laws in the next few years to perpetuate white dominance of the country.

Because they opposed apartheid, other countries cut off most relations with South Africa during the 1970s and 1980s. Foreign companies such as Ford and General Motors stopped operating factories in South Africa, and foreign athletes and teams refused to play in the country. However, neighboring countries felt compelled to maintain economic ties with South Africa, because they needed to ship their goods through South African ports. South Africa also played an important economic role in the entire southern Africa region, because it provided jobs for unemployed people from the much poorer neighboring countries, and it supplied the more developed countries with mineral resources critical for manufacturing and chemical processes, including chromium, platinum, and manganese.

To assure further geographic isolation of different races, the South African government designated ten so-called *homelands* for blacks. The white minority government expected every black to become a citizen of one of the homelands and to move there. More than 99 percent of the population in the ten homelands was black.

The first four homelands designated by the government were called Bophuthatswana, Ciskei, Transkei, and Venda (Figure 7–10). Bophuthatswana included six discontinuous areas, Transkei three discontinuous areas, and Venda two discontinuous areas. During the late 1970s, South Africa declared Bophuthatswana, Transkei, and Venda to be independent countries, but no other government in the world recognized the claim.

The first four homelands comprised about 9 percent of South Africa's land area and 19 percent of the population; if the government policy had been fully implemented, the ten black homelands together would have contained approximately 44 percent of South Africa's population on only 13 percent of the land.

Dismantling of Apartheid. In 1991, the white-dominated government of South Africa repealed the apartheid laws, including restrictions on property ownership and classification of people at birth by race. The principal anti-apartheid organization, the African National Congress, was legalized, and its leader, Nelson Mandela, was released from jail after 27-1/2 years. When all South Africans were permitted to vote in national elections for the first time, in April 1994, Mandela was overwhelmingly elected the country's first black president. Whites were guaranteed representation in the government during a 5-year transition period, until 1999. As of 1994, South Africa no longer considered the four homelands to be independent countries.

Now that South Africa's apartheid laws have been dismantled, and the country is governed by its black majority, other countries have re-established economic and cultural ties. However, the legacy of apartheid will linger for many years: South Africa's blacks have achieved political equality but they are much poorer than white South Africans. Average income among white South Africans is about ten times higher than for blacks.

Key Issue 2

Where Have Ethnicities Been Transformed into Nationalities?

- Rise of nationalities
- Nationalities in former colonies
- Revival of ethnic identity

With the end of apartheid in South Africa, blacks and whites voted together for the first time in 1994. (Haviv/Saba Press Photos, Inc.)

Ethnicity is distinct from race and nationality, two other terms commonly used to describe a group of people with shared traits. **Nationality** is identity with a group of people who share legal attachment and personal allegiance to a particular country. It comes from the Latin word *nasci*, which means *to have been born*.

A nation or nationality is a group of people tied together to a particular place through legal status and cultural tradition. Nationality and ethnicity are similar concepts in that membership in both is defined through shared cultural values. In principle, the cultural values shared with others of the same ethnicity derive from religion, language, and material culture, whereas those shared with others of the same nationality derive from voting, obtaining a passport, and performing civic duties.

In the United States, nationality is generally kept reasonably distinct from ethnicity and race in common usage. The American *nationality* identifies citizens of the United States of America, including those born in the country and those who immigrated and became citizens. *Ethnicity* identifies groups with distinct ancestry and cultural traditions, such as Hispanic-Americans, African-Americans, Chinese-Americans, or Polish-Americans. *Race* distinguishes blacks and other persons of color from whites.

Thus, every citizen living in the United States is a member of the American nationality, and every American is a member of a race, though only some Americans are identified with an ethnicity. A Moroccan living in the United States (perhaps attending a university or working at the United Nations) could be distinct from an American by all three features at the same time—dark skin color would distinguish by race; birth and citizenship in Morocco would distinguish by nationality; and follower of the Islamic religion and speaker of the Arabic language would distinguish by ethnicity.

In Canada, the Québécois are clearly distinct from other Canadians in language, religion, and other cultural traditions. But do the Québécois form a distinct ethnicity within the Canadian nationality or a second nationality separate altogether from Anglo-Canadian? The distinction is critical, because if Québécois is recognized as a separate nationality from Anglo-Canadian, the Québec government would have a much stronger justification for breaking away from Canada to form an independent country.

Outside North America, distinctions between ethnicity and nationality are even muddier. We have already seen in this chapter that confusion between ethnicity and race can lead to discrimination and segregation. Confusion between ethnicity and nationality can lead to violent conflicts.

Rise of Nationalities

Descendants of nineteenth-century immigrants to the United States from central and Eastern Europe identify themselves today by ethnicity rather than by nationality. When most Czechs, Germans, Poles, and Slovenes migrated to the United States, there were no countries called Czech Republic (or Czechoslovakia), Germany, Poland, or Slovenia. These ethnicities lived in Europe as subjects of the Austrian emperor, Russian czar, or Prussian kaiser.

U.S. immigration officials recorded the nationality of immigrants—that is, the place of birth and departure from Europe—and U.S. data concerning the origin of immigrants is organized by nationality. But immigrants considered ethnicity more important than nationality, and that is what they have preserved through distinctive social customs.

The United States forged a nation in the late eighteenth century out of a collection of ethnic groups gathered primarily from Europe and Africa, not through traditional means of issuing passports (African-Americans weren't considered citizens then) or voting (women and

African-Americans couldn't vote then), but through sharing the values expressed in the Declaration of Independence, Constitution, and Bill of Rights. To be an American meant believing in the "unalienable rights" of "life, liberty, and the pursuit of happiness."

Nation-States

Ethnic groups have been transformed into nationalities because desire for self-rule is a very important shared attitude for many ethnicities. To preserve and enhance distinctive cultural characteristics, ethnicities seek to govern themselves without interference. The concept that ethnicities have the right to govern themselves is known as **self-determination**.

During the nineteenth and twentieth centuries, political leaders have generally supported the right of self-determination for many ethnicities and have attempted to organize Earth's surface into a collection of nation-states. A **nation-state** is a state whose territory corresponds to that occupied by a particular ethnicity that has been transformed into a nationality. Yet despite continuing attempts to create nation-states, the territory of a state rarely corresponds precisely to the territory occupied by an ethnicity.

Nation-States in Europe. Throughout Europe, ethnicities were transformed into nationalities during the nineteenth century. The French nationality fused together French ethnic cultural traditions, including the French language and the Roman Catholic religion, with a belief in the values of the French Revolution of 1789, expressed in the phrase "liberté, égalité, fraternité" (liberty, equality, brotherhood). When France was ruled by kings, the French people went to war out of loyalty to the king. Under Napoleon Bonaparte, the French people went to war for the principles of the nation of France.

By around 1900, most of Western Europe was made up of nation-states. They disagreed over their boundaries, and competed to control territory in Africa and Asia. Eastern Europe included a mixture of empires and states that did not match the distribution of ethnicities. Following their defeat in World War I, the Austro-Hungarian and Ottoman empires were dismantled, and many European boundaries were redrawn according to the principle of nation-states.

During the 1930s, German National Socialists (Nazis) claimed that all German-speaking parts of Europe constituted one nationality and should be unified into one state. They pursued this goal forcefully, and other European powers did not attempt to stop the Germans from taking over Austria and the German-speaking portion of Czechoslovakia, known as the Sudetenland. But in 1939, when Germany invaded Poland (clearly not a German-speaking country), World War II began as England and France rose to defend themselves.

Denmark: There Are No Perfect Nation-States. Denmark is a fairly good example of a European nation-

state, because the territory occupied by the Danish ethnicity closely corresponds to the state of Denmark. The Danes have a strong sense of unity that derives from shared cultural characteristics and attitudes and a recorded history that extends back more than 1,000 years. Nearly all Danes speak the same language, Danish, and nearly all the world's speakers of Danish live in Denmark.

But even Denmark is not a perfect example of a nation-state. The country's 80-kilometer (50-mile) southern boundary with Germany does not divide Danish and German nationalities precisely. The border region, known as Schleswig-Holstein, historically was part of Denmark. Denmark lost the region to Germany during the nineteenth century, but after the German defeat in World War I, the people in North Schleswig voted to rejoin Denmark. As a result, some German speakers live in Denmark, while some Danish speakers live in Germany.

To dilute the concept of a nation-state further, Denmark controls two territories in the Atlantic Ocean that do not share Danish cultural characteristics. One is the Faeroe Islands, a group of 21 islands ruled by Denmark for more than 600 years. The nearly 50,000 inhabitants of the Faeroe Islands speak Faeroese (see red area in Figure 5–4).

Denmark also controls Greenland, the world's largest island, which is 50 times larger than Denmark proper. Only 14 percent of the residents of Greenland are considered Danish; the remainder are native-born Greenlanders, primarily Inuit. In 1979, the nearly 60,000 Greenlanders received more authority from Denmark to control their own domestic affairs. One decision was to change all place names in Greenland from Danish to the local Inuit language. Greenland is now officially known as Kalaallit Nunaat, and the capital city was changed from Godthaab to Nuuk.

Nationalism

A nationality, once established, must hold the loyalty of its citizens to survive. Politicians and governments try to instill loyalty through **nationalism**, which is loyalty and devotion to a nationality. Nationalism typically promotes a sense of national consciousness that exalts one nation above all others and emphasizes its culture and interests as opposed to those of other nations. People display nationalism by supporting a state that preserves and enhances the culture and attitudes of their nationality.

For many states, mass media are the most effective means of fostering nationalism. Americans regard independent news media as a strength and a watchdog over government. But most countries regard an independent source of news as more of a risk than a benefit to the stability of their government. Consequently, only a few states permit mass media to operate without government interference. Nearly all countries control, or at least regulate, most forms of communications, including mail, telephone, telegraph, television, radio, and satellite transmissions. The government either owns or controls newspapers in many countries.

States foster nationalism by promoting symbols of the nation-state, such as flags and songs. The symbol of the hammer and sickle on a field of red was long synonymous with the beliefs of communism. After the fall of communism, one of the first acts in a number of Eastern European countries was to redesign flags without the hammer and sickle. Legal holidays were changed from dates associated with Communist victories to those associated with historical events that preceded Communist takeovers. One of the strongest forms of political protest is to burn a state's flag, and there is wide support in the United States for laws to make burning the Stars and Stripes illegal.

Nationalism is also instilled through the creation of songs extolling the country's virtues. Nearly every state has a national anthem, which usually combines respect for the state with references to the nation's significant historic events or symbols of unity:

Oh, say does that star-spangled banner yet wave
O'er the land of the free and the home of the brave?

The unifying force of such a song is very powerful, especially to older people and to those who have served in a country's armed forces.

Nationalism can have a negative impact. The sense of unity within a nation-state is sometimes achieved through the creation of negative images of other nation-states. Travelers in southeastern Europe during the 1970s and 1980s found that jokes directed by one nationality against another recurred in the same form throughout the region, with only the name of the target changed. For example, "How many [fill in the name of a nationality] are needed to change a light bulb?" Such jokes seemed harmless, but in hindsight reflected the intense dislike for other nationalities that led to conflict in the 1990s.

Nationalism is an important example of a **centripetal force**, which is an attitude that tends to unify people and enhance support for a state. (The word centripetal means "directed toward the center"; it is the opposite of *centrifugal*, which means to spread out from the center.) Most nation-states find that the best way to achieve citizen support is to emphasize shared attitudes that unify the people.

Nationalities in Former Colonies

When most of the world consisted of colonies controlled by Britain, France, or other powers, ethnicity was the principal means of distinguishing among groups in the far-flung empires. When long-time colonies were carved up into independent countries, especially in the 1940s and 1950s, many new nationalities were suddenly created.

Boundaries of newly independent countries were often drawn to separate two ethnicities. However, boundary lines rarely can segregate two ethnicities completely. Members of an ethnicity caught on the "wrong" side of a boundary may be forced to migrate to the other side.

Creating Nationalities in South Asia

South Asia provides a vivid example of what happens when independence comes to a colony that contains two major ethnicities. When the British ended their colonial rule of the Indian subcontinent in 1947, they divided the colony into two irregularly shaped countries, India and Pakistan. Pakistan comprised two noncontiguous areas, West Pakistan and East Pakistan—1,600 kilometers (1,000 miles) apart, separated by India. East Pakistan became the independent state of Bangladesh in 1971. An eastern region of India was also practically cut off from the rest of the country, attached only by a narrow corridor north of Bangladesh that is less than 13 kilometers (8 miles) wide in some places.

The basis for this seemingly bizarre separation of West and East Pakistan from India was ethnic. The people living in the two areas of Pakistan were predominantly Muslim, while those in India were predominantly Hindu. Antagonism between the two religious groups was so great that the British decided to place the Hindus and Muslims in separate states.

Hinduism has become a great source of national unity in India. In modern India, with its hundreds of languages and ethnic groups, Hinduism has become the cultural trait shared by the largest percentage of the population.

Muslims have long fought with Hindus for control of territory, especially in South Asia. Around A.D. 1000, Mahmud, the Muslim king of Ghazni (modern-day Afghanistan), led raids on the Punjab area of northern India. His purpose originally was to acquire treasure from Hindu temples, but the raids turned into a Muslim-Hindu religious war. The Punjab became part of the Ghazni kingdom, with a governor at Lahore.

The fragmented Hindu kingdoms were unable to stop a second set of invasions by Muslims, who in the thirteenth century seized most of northern India as far east as Bengal. The population consisted primarily of Hindus and Buddhists, but the number of Muslims grew within a few generations as a result of intermarriage and further immigration from the west.

After the British took over India in the early 1800s, a three-way struggle began, with the Hindus and Muslims fighting each other as well as the British rulers. Muslims believed that the British discriminated more against them than against the Hindus. When the British granted independence to the region following World War II, Hindus and Muslims fought over the organization of the newly independent region. Mahatma Gandhi, the leading Hindu advocate of nonviolence and reconciliation with Muslims, was assassinated in 1948, ending the possibility of creating a single state in which Muslims and Hindus lived together peacefully.

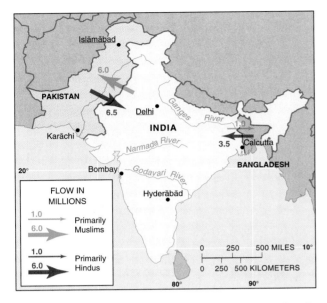

Figure 7–11 Ethnic division of South Asia. In 1947, British India was partitioned into two independent states, India and Pakistan, which resulted in the migration of an estimated 17 million people. The creation of Pakistan as two territories nearly 1,600 kilometers (1,000 miles) apart proved unstable, and in 1971 East Pakistan became the independent country of Bangladesh. The train station in Amristar, India, October 17, 1947, is crowded with Hindus who have been brought from Pakistan. (UPI/Bettmann)

Migration in South Asia. The partition of South Asia into two states resulted in massive migration, because the two boundaries did not correspond precisely to the territory inhabited by the two ethnicities. In the late 1940s, approximately 17 million people caught on the wrong side of a boundary felt compelled to migrate. Some 6 million Muslims moved from India to West Pakistan and about 1 million from India to East Pakistan. Hindus who migrated to India included approximately 6 million from West Pakistan and 3.5 million from East Pakistan (Figure 7–11).

Hindus in Pakistan and Muslims in India were killed attempting to reach the other side of the new border by people from the rival religion. Extremists attacked small groups of refugees traveling by road and halted trains to massacre the passengers.

Ethnic Disputes in Kashmir. Pakistan and India never agreed on the location of the boundary separating the two countries in the northern region of Kashmir. The original partition gave India two-thirds of Kashmir, even though a majority of its people were Muslims. In recent years, Muslims on the Indian side of Kashmir have begun a guerrilla war to secure independence. India blames Pakistan for the unrest and vows to retain its portion of Kashmir; Pakistan argues that Kashmiris on both sides of the border should choose their own future in a vote, confident that the majority Muslim population would break away from India (Figure 7–12).

India's religious unrest is further complicated by the presence of 17 million Sikhs, whose religion combines elements of Islam and Hinduism. Sikhs have long re-

sented that they were not given their own independent country when India was partitioned. Although they constitute only 2 percent of India's total population, Sikhs comprise a majority in the Indian state of Punjab, situated south of Kashmir along the border with Pakistan. Sikh extremists have fought for more control over the Punjab, or even complete independence from India.

Converting Ethnicities to Nationalities in Africa

Conflict is widespread in Africa largely because the present-day boundaries of states were drawn by European

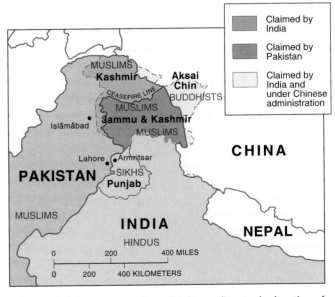

Figure 7–12 Kashmir. India and Pakistan dispute the location of their border. India claims Kashmir, in northernmost Pakistan, and India accuses Pakistan of encouraging unrest in India's state of Jammu and Kashmir, where the majority is Muslim.

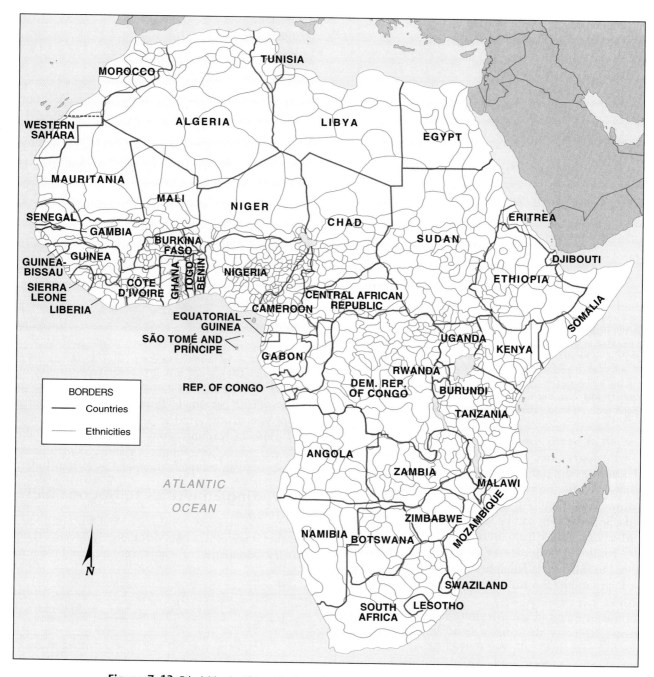

Figure 7–13 Ethnicities in Africa. The boundaries of modern African states do not match the territories long occupied by thousands of ethnic groups. State boundaries derive from the administrative units imposed by European colonial powers a century ago.

colonial powers about a hundred years ago without regard for the traditional distribution of ethnicities (Figure 7–13). Africa contains several thousand ethnicities (usually referred to as tribes) with a common sense of language, religion, social customs (refer to Figure 5–8 for a map of African languages). Some tribes are divided among more than one modern state, while others have been grouped with dissimilar tribes.

The precise number of tribes is impossible to determine, because boundaries separating them are not usually defined clearly. Further, it is hard to determine whether a particular group forms a distinct tribe or is part of a larger collection of very similar groups.

Pre-European States in Africa. The traditional unit of African society was the tribe rather than independent states with political and economic self-determination. Nonetheless, Africa has some tradition of state control, especially in West Africa. Important states in West Africa, based in present-day Mali and Mauritania, included Ghana (800 kilometers northwest of the present-day state of Ghana) between the eighth and twelfth centuries, Mali between the twelfth and fifteenth centuries, and Songhai during the fifteenth and sixteenth centuries.

Other kingdoms were located closer to the coast of West Africa. The Kongo kingdom, based near the mouth of the Congo (Zaire) River in present-day Angola and Zaire, flourished from the fourteenth to the seventeenth centuries. A group of Ewe-speaking people called the Aja established the Great Ardra kingdom in present-day Benin, which reached its height in the sixteenth and early seventeenth centuries. The Aja mixed with other local groups to form the Fon or Dahomey ethnic group. Four hundred kilometers (250 miles) west, the Ashanti ethnic group established a confederation in the seventeenth century in the central part of present-day Ghana, which survived until the late nineteenth century.

Impact of European Colonization. European exploration of the African coast began in the 1400s, but until the late nineteenth century, Africa was largely free of foreign control. Between the 1880s and the outbreak of World War I in 1914, European countries carved up the continent into a collection of colonies. The shapes of these colonies were dictated primarily by competition among the European colonial powers to control resources in the interior rather than the distribution of the thousands of tribes.

When the European colonies became independent states, especially during the 1950s and 1960s, the boundaries of the new states typically matched the colonial administrative units imposed by the Europeans. As a result, most African states contained large numbers of ethnicities. For example, the British colony of the Gold Coast became the independent state of Ghana in 1956. Ghana's territory includes the historic homelands of the Ashanti, Fanti, Mole-Dagbani, Ewe, and Ga-Adangme tribes.

Revival of Ethnic Identity

Ethnic identities never really disappeared in Africa, where loyalty to tribe often remained more important than loyalty to the nationality of a new country, perhaps controlled by another ethnicity. People in other regions, especially Europe, felt that they had "progressed" beyond ethnicity.

Europeans thought that ethnicity had been left behind as an insignificant relic, such as wearing quaint costumes to amuse tourists. Karl Marx wrote that nationalism was a means for the dominant social classes to maintain power over workers, and he believed that workers would identify with other working-class people instead of with an ethnicity. But Europeans were wrong, because in the late twentieth century ethnic identity once again became more important than nationality even in much of Europe.

Ethnicity and Communism

From the end of World War II in 1945 until the early 1990s, attitudes toward communism and economic cooperation were more important political factors in Europe than the nation-state principle. For example, the Communist government of Bulgaria repressed cultural differences by banning the Turkish language and the practice of some Islamic religious rites. The government took these steps to remove what it saw as obstacles to unifying national support for the ideology of communism. More than 1 million Bulgarian citizens of Turkish ancestry migrated to Turkey. The town of Bursa, about 100 kilometers (60 miles) south of Istanbul, became the largest settlement of Turkish refugees from Bulgaria.

Until they lost power in the late 1980s and early 1990s, Communist leaders in Eastern Europe and the former Soviet Union used centripetal forces to discourage ethnicities from expressing their cultural uniqueness. Writers and artists were pressured to conform to a style known as "socialist realism," which emphasized Communist economic and political values. Use of the Russian language was promoted as a centripetal device throughout the former Soviet Union. It was taught as the second language in other Eastern European countries. The role of organized religion was minimized, suppressing a cultural force that competed with the government.

Latvians celebrate independence from the Soviet Union in 1991 with a massive demonstration in the streets of the capital, Riga. (Ints Kalnins/Woodfin Camp & Associates)

The Communists did not completely suppress ethnicities in Eastern Europe: the administrative structures of the former Soviet Union and two other multi-ethnic Eastern European countries—Czechoslovakia and Yugoslavia—recognized the existence of ethnic groups. In the Soviet Union, 15 republics were created as principal units of local government. Six local units were created in Yugoslavia, and two in Czechoslovakia. All were designed to coincide as closely as possible with the territory occupied by the most numerous ethnicities. Ten of the Soviet Union's 15 republics and one in Yugoslavia were further divided into local government units to grant some autonomy to ethnicities that were too few to merit designation as republics.

Rebirth of Nationalism in Eastern Europe

Ethnic identity was effectively suppressed by Communists when they controlled the Soviet Union and other Eastern European countries. But during the 1990s, nationalism was resurgent and once again is important in forming peoples' cultural identities in the region.

In Eastern Europe, the breakup of the Soviet Union and Yugoslavia during the 1990s has given more numerous ethnicities the opportunity to organize nation-states. But the less numerous ethnicities still find themselves existing as minorities in multinational states, or divided among more than one of the new states. Especially severe problems have occurred in the Balkans, a rugged, mountainous region where nation-states could not be delineated peacefully.

With the fall of the Communist government in the early 1990s, Bulgaria's Turkish minority has pressed for more rights, including permission to teach the Turkish language as an optional subject in school. But many Bulgarians continue to oppose these efforts. Although communism has declined in importance in Bulgaria—as well as in other former Communist countries in Eastern Europe—it has been replaced by an ideology that encourages traditional cultural features, such as language and religion.

The Soviet Union, Yugoslavia, and Czechoslovakia were dismantled in the early 1990s largely because minority ethnicities opposed the long-standing dominance of the most numerous ones in each country—Russians in the Soviet Union, Serbs in Yugoslavia, and Czechs in Czechoslovakia. The dominance was pervasive, including economic, political, and cultural institutions.

No longer content to control a province or some other local government unit, ethnicities sought to be the majority in completely independent nation-states. Republics that once constituted local government units within the Soviet Union, Yugoslavia, and Czechoslovakia have generally made peaceful transitions into independent countries—as long as their boundaries have corresponded reasonably well with the territory occupied by a clearly defined ethnicity.

Slovenia is a good example of a nation-state that was carved from the former Yugoslavia in the 1990s. More than 90 percent of the residents of Slovenia are Slovenes, and nearly all the world's 2 million Slovenes live in Slovenia. The relatively close coincidence between the boundaries of the Slovene ethnic group and the country of Slovenia has promoted the country's relative peace and stability, compared to other former Yugoslavian republics.

For new nation-states in Eastern Europe such as Slovenia, sovereignty has brought difficulties in converting from Communist economic systems and fitting into the global economy (see Chapters 9 and 11). But their problems of economic reform are minor compared to the conflicts that have erupted in portions of Eastern Europe and the former Soviet Union where nation-states could not be created.

Key Issue 3

Why Do Ethnicities Clash?

- Ethnic competition to dominate nationality
- Overlapping of ethnicities and nationalities

A state that contains more than one ethnicity is a **multi-ethnic state**. In some multi-ethnic states, ethnicities all contribute cultural features to the formation of a single nationality. Belgium is a good example of a multi-ethnic state. As discussed in Chapter 5, Belgium is divided among the Dutch-speaking Flemish and the French-speaking Walloons. Both groups consider themselves belonging to the Belgian nationality.

Other multi-ethnic states, known as **multinational states**, contain two ethnic groups with traditions of self-determination that agree to coexist peacefully by recognizing each other as distinct nationalities. One example of a multinational state is the United Kingdom, which contains four main nationalities—England, Scotland, Wales, and Northern Ireland. The four display some ethnic differences, but the main reason for considering them as distinct nationalities is that each had very different historical experiences.

Wales was conquered by England in 1282 and formally united with England through the Act of Union of 1536. Welsh laws were abolished, and Wales became a local government unit. English became the official language of Wales, although Welsh is still spoken and is now being preserved (see Chapter 5).

Scotland was an independent country for nearly a thousand years, until 1603 when Scotland's King James VI also became King James I of England, thereby uniting the two countries. The Act of Union in 1707 formally merged the two governments, although Scotland was allowed to retain its own systems of education and local laws. England, Wales, and Scotland together comprise Great Britain, and the term British refers to the combined nationality of the three groups.

Northern Ireland, along with the rest of Ireland, was ruled by the British until the 1920s, as discussed in Chapter 6. The 1801 Act of Union created the United Kingdom of Great Britain and Ireland. During the 1920s, most of Ireland became a separate country, but the northern portion—with a majority of Protestants—remained under British control. The official name of the country was changed to the United Kingdom of Great Britain and Northern Ireland.

Today, the four nationalities hold little independent political power, although Scotland and Wales now have separately elected governments. The main element of distinct national identity comes from sports. England, Scotland, Wales, and Northern Ireland field their own national soccer teams and compete separately in major international tournaments, such as the World Cup. The most important annual rugby tournament, known as the Five Nations' Cup, includes teams from England, Scotland, and Wales, as well as Ireland, and France. Given the history of English conquest, the other nationalities typically root against England when it is playing teams from other countries.

Ethnicities do not always find ways to live together peacefully. In some cases, ethnicities compete in civil wars to dominate the national identity. In other cases, problems result from confusion between ethnic identity and national identity.

Ethnic Competition to Dominate Nationality

The Horn of Africa, in eastern Africa, has been a region especially plagued by conflicts among ethnic groups competing to become dominant within the various countries. Another example of this pattern of conflict is the island country of Sri Lanka.

Horn of Africa

The Horn of Africa encompasses the countries of Djibouti, Ethiopia, Eritrea, and Somalia. Especially severe problems have been found in Ethiopia, Eritrea, and Somalia, as well as the neighboring country of Sudan.

Ethiopia and Eritrea. Eritrea, located along the Red Sea, became an Italian colony in 1890. Ethiopia, an independent country for more than 2,000 years, was captured by Italy during the 1930s. After World War II, Ethiopia regained its independence, and the United Nations awarded Eritrea to Ethiopia (Figure 7–14).

The United Nations expected Ethiopia to permit Eritrea considerable authority to run its own affairs, but Ethiopia dissolved the Eritrean legislature and banned the use of Tigrinya, Eritrea's major local language. The Eritreans rebelled, beginning a 30-year fight for independence (1961–1991). During this civil war, an estimated 665,000 Eritrean refugees fled to neighboring Sudan, especially north to the city of Būr Sūdān (Port

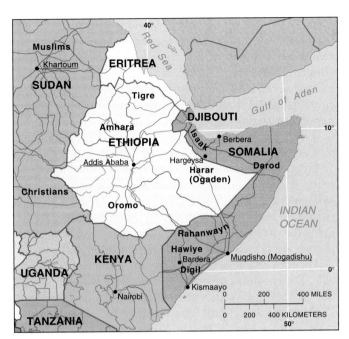

Figure 7–14 Ethnic diversity in the Horn of Africa. Conflicts have been widespread within the East African countries in the area known as the Horn of Africa, because each contains numerous ethnicities.

Sudan) along the Red Sea and west to Khartoum, the capital, as well as to Kassalā, a smaller border town.

In 1991, Eritrean rebels defeated the Ethiopian army, and in 1993 Eritrea became an independent state. A country of 4 million people split evenly between Christian and Muslim, Eritrea has nine major ethnic groups. At least in the first years of independence, a strong sense of national identity has united Eritrea's ethnicities, as a result of shared experiences during the 30-year war to break free of Ethiopia.

Even with the loss of Eritrea, Ethiopia remained a complex multi-ethnic state. From the late nineteenth century until the 1990s, Ethiopia was controlled by the Amharas, who are Christians. After the government defeat in the early 1990s, power passed to a combination of ethnic groups. The Oromo, who are Muslim fundamentalists from the south, are the largest ethnicity in Ethiopia, 40 to 50 percent of the population. Tigres live in the far north, the birthplace of the Ethiopian Orthodox Church. The Amhara had banned the languages and cultures of these groups since conquering Ethiopia in the late nineteenth century.

Ethiopia further suffered from warfare in its eastern region of Ogaden, a desert area also claimed by Somalia. The Ethiopian army uprooted several million native Somalis living in the Ogaden who preferred that the province be part of Somalia. According to international refugee organizations, approximately 365,000 Ethiopians fled to Somalia (Somali officials argue that more than 800,000 actually arrived). Half of these refugees lived in camps near the border, and the remainder wandered from village to village in Somalia.

Sudan. In Sudan, a civil war has raged since the 1980s between two ethnicities, the black Christian and animist rebels in the southern provinces and the Arab–Muslim-dominated government forces in the north. The black southerners have been resisting government attempts to convert the country from a multi-ethnic society to one nationality tied to Muslim traditions.

The government of Sudan has adopted laws designed to segregate the sexes in public. All schools are single sex, and men are prohibited from "lurking" outside all-female schools. Barriers must be erected to separate men and women at weddings, parties, and picnics. Women are not permitted to sit near the driver on buses. Sporting events involving women must be held in private, and female players are not allowed to wear tight-fitting clothes.

Where contact between men and women is unavoidable, laws prohibit provocative behavior. Women working in restaurants may not wear jewelry or perfume. Women shopping after dark must be accompanied by a male relative. Men as well as women must wear clothing that substantially covers the body, although women are allowed to wear their traditional colorful flowing gowns called *tobes* and do not have to wear veils. More street lights have been installed to prevent amorous couples from vanishing into the darkness.

More than 1 million Sudanese have been forced to migrate from the south to the north in the civil war, and another 350,000 have fled to Ethiopia. Sudan accuses Ethiopia and Eritrea, as well as Uganda, of helping the mostly Christian and animist rebels. For its part, Eritrea accuses Sudan of trying to undermine its government, and it has turned over Sudan's embassy to opposition groups.

Somalia. On the surface, Somalia should face fewer ethnic divisions than its neighbors in the Horn of Africa. Somalis are overwhelmingly Sunni Muslims and speak Somali. Most share a sense that Somalia is a nation-state, with a national history and culture.

Somalia contains six major ethnic groups known as clans, each of which is divided into a large number of sub-clans. Traditionally, the six major clans occupied different portions of Somalia—Isaak, Darod, and Dir to the north, Digil, Hawiye, and Rahanwayn to the south. Until 1991, a Darod sub-clan known as Mahareen ruled Somalia, but in 1991 rebels dominated by the Hawiye clan took control of southern Somalia, while the Isaak clan gained control of much of the north.

Through the 1990s, a complex series of splits within clans and alliances between clans has muddled the traditional geographic divisions. The Isaak clan has declared the north a separate state of Somaliland and has adopted its own flag and currency. During the colonial period, the territory that the Isaaks call Somaliland had been ruled by the British, whereas the rest of Somalia was an Italian colony.

With the collapse of a national government in Somalia, various clans and sub-clans have claimed control over portions of the country. As the armies of the individual clans and sub-clans seized food, property, and weapons, members of less-powerful clans and sub-clans migrated to refugee camps seek safety and food. In 1992, after an estimated 300,000 people, mostly women and children, died from famine and from warfare between clans, the United States sent several thousand troops to Somalia. The purpose of the mission was to protect delivery of food by international relief organizations to starving Somali refugees and to reduce the number of weapons in the hands of the clan and sub-clan armies. After peace talks among the clans collapsed in 1994, U.S. troops withdrew, but they returned briefly in 1995 to help evacuate United Nations forces.

Buddhists vs. Hindus in Sri Lanka

Sri Lanka (formerly Ceylon), an island country of 19 million inhabitants off the Indian coast, has been torn by fighting between the Sinhalese and the Tamils (Fig-

Somalis were forced to migrate during the early 1990s as a result of civil war among rival clans and sub-clans. U.S. military intervention in 1992 and 1993 helped international organizations deliver food to Somalis living in refugee camps, but violence increased again after the United States withdrew its forces in 1994. (Wendy Stone/Gamma-Liaison, Inc.)

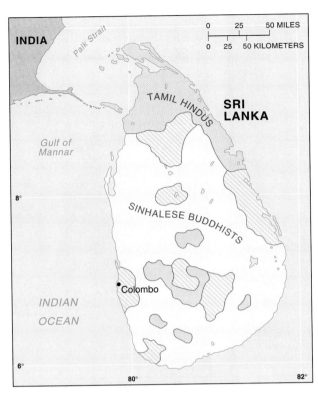

Figure 7–15 Sinhalese and Tamils in Sri Lanka. The Sinhalese are Buddhists who speak an Indo-European language, whereas the Tamils are Hindus who speak a Dravidian language. The striped areas show where the two groups intermingle.

ure 7–15). During the past two decades, 43,000 have died in ethnic conflict, and the Sinhalese president was assassinated by a Tamil in 1993.

Sinhalese, who comprise 74 percent of Sri Lanka's population, migrated from northern India about 543 B.C. Three hundred years later, the Sinhalese were converted to Buddhism, and Sri Lanka became one of that religion's world centers. Sinhalese is an Indo-European language, in the Indo-Iranian branch.

Tamils—18 percent of Sri Lanka's population—are Hindus, many of whom migrated from India. The Tamil language, in the Dravidian family, is also spoken by about 60 million people in India. The Tamils have repeatedly felt that they have been discriminated against by the Sinhalese majority, which controls the government, military, and most of the commerce. Many of Sri Lanka's Indian-born Tamils do not enjoy full protection of civil rights, although some have been permitted to become citizens of Sri Lanka. India, separated from Sri Lanka only by the 80-kilometer-wide (50-mile) Palk Strait, has sent soldiers to Sri Lanka to protect the Tamil minority.

Overlapping of Ethnicities and Nationalities

The Middle East is an especially complex region of conflict, because it contains overlapping ethnicities and nationalities. On one level, the region's problems stem from its importance to three religious groups—Jews,

Christians, and Muslims—as discussed in Chapter 6. At the same time, the current conflict in the Middle East is complicated by the recent emergence of strong nationalities.

Conflict over the Holy Land

Chapter 6 pointed out that the territory of Palestine and the city of Jerusalem, where shrines holy to Jews, Christians, and Muslims were clustered, gained independence from the United Kingdom in 1948. With the creation of independent countries in the region, long-standing ethnic conflicts among religious groups were redefined as conflicts among nationalities.

Israel became the world's only country with a Jewish majority, and the Israeli nationality was the only one that combined loyalty to a state with Jewish religious traditions. In all four of the states bordering Israel—Egypt, Jordan, Lebanon, and Syria—the people were predominantly Muslim. The Jordanian nationality was created in the twentieth century, but Egypt, Lebanon, and Syria had long traditions of nationality, extending back to ancient times, even before the people converted to Islam.

Current problems stem from the Six-Day War in 1967. From Jordan, Israel captured the West Bank (the territory west of the Jordan River taken by Jordan in the 1948–1949 war). From Syria, Israel acquired the Golan Heights. From Egypt came the Gaza Strip and Sinai Peninsula (Figure 7–16).

Egypt's President Anwar Sadat and Israel's Prime Minister Menachem Begin signed a peace treaty in 1979, following a series of meetings with U.S. President Jimmy Carter at Camp David, Maryland. In accordance with the treaty, Israel returned the Sinai Peninsula, and in return, Egypt recognized Israel's right to exist. Sadat was assassinated by Egyptian soldiers, who were extremist Muslims opposed to compromising with Israel. However, his successor, Hosni Mubarek, carried out the terms of the treaty.

More than a quarter-century after the Six-Day War, the status of the other territories occupied by Israel still has not been settled. In 1981, Israel formally annexed the Golan Heights, a sparsely inhabited, mountainous area from which Syria had launched attacks against Jewish settlements in the valley of the Sea of Galilee and the Jordan River. Israel is negotiating to return the Golan Heights to Syria in exchange for a peace treaty.

Palestinians: Conflict Between an Ethnicity and a Nationality. The series of wars between Israel and neighboring Arab states had been between different ethnic groups transformed into nationalities that disagreed over the location of the boundaries between their countries. Since the 1973 war, Egypt and Jordan have signed peace treaties with Israel, and Syria is no longer actively plotting an attack on Israel.

Despite the movement toward peace among the neighboring nationalities in the Middle East, unrest per-

Figure 7–16 (Upper left) Palestine under British control, 1922–1948.

(Upper right) The 1947 United Nations plan to partition Palestine. The plan was to create two countries, with the boundaries drawn to separate the predominantly Jewish areas from the predominantly Arab Muslim areas. Jerusalem was intended to be an international city, run by the United Nations.

(Lower left) Israel after the 1948–1949 War. The day after Israel declared its independence, several neighboring states began a war, which ended in an armistice. Israel's boundaries were extended beyond the U.N. partition to include the western suburbs of Jerusalem. Jordan gained control of the West Bank and East Jerusalem, including the Old City, where holy places are clustered.

(Lower right) The Middle East since the 1967 War. Israel captured the Golan Heights from Syria, the West Bank and East Jerusalem from Jordan, and the Sinai Peninsula and Gaza Strip from Egypt. Israel returned the Sinai to Egypt in 1979 and Gaza and a portion of the West Bank to the Palestinians in 1994. Israel still controls the Golan Heights, most of the West Bank, and East Jerusalem.

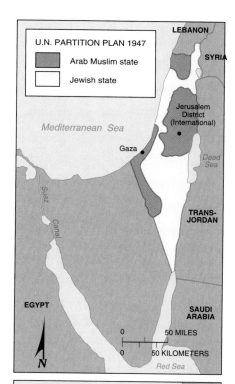

sists because of the emergence of a new nationality in the late 1960s, known as the Palestinians. To complicate the situation, five groups of people consider themselves Palestinians:

- People living in the West Bank, Gaza, and East Jerusalem.
- Citizens of Israel who are Muslims rather than Jews.
- People who fled from Israel to other countries after the 1948–1949 war.

- People who fled from the West Bank or Gaza to other countries after the 1967 war.
- Citizens of other countries, especially Jordan, Lebanon, Syria, Kuwait, and Saudi Arabia, who identify themselves as Palestinians.

During the 1990s, Israel turned over Gaza and most of the West Bank to the Palestinians. The Palestine Liberation Organization (PLO) became the governing organization for Gaza and most of the West Bank. Palestinians have achieved their ambition of controlling territory, essential to any national identity, but they do not agree on whether to settle for a country consisting of Gaza and most of the West Bank or continue fighting Israel for the entire territory between the Jordan River and the Mediterranean Sea.

For their part, Israeli Jews are divided between those who wish to retain some of the occupied territories and those who wish to make compromises with the Palestinians so that both nationalities can live peacefully in the region. However, peace between the two nationalities will be difficult to achieve as long as Israelis continue to hold East Jerusalem and Palestinians seek to obtain it (see Globalization and Local Diversity box).

Lebanon: Conflicts Among Several Nationalities and Ethnicities

Ethnic conflict in the Middle East is not confined to Israel. Israel's northern neighbor, Lebanon, was once known as a financial and entertainment center in the Middle East. Europeans flocked to Lebanon for its warm winter weather, sandy beaches, and cosmopolitan night life. Lebanon was a multi-ethnic state where different religious groups lived peacefully near each other, each enriching the Lebanese nationality. Lebanon managed to steer clear of the wars between Israel and its neighboring Arab states.

Smaller than Connecticut, Lebanon's 3.4 million people are members of 17 officially recognized ethnicities. Lebanon's ethnic groups coexisted fairly peacefully from the time of independence in 1943 until the mid-1970s. Since then, the country has been severely damaged by civil war among the factions.

Lebanon's Ethnic Diversity. The precise distribution of religions is unknown, because no census has been taken since 1932. Current estimate is about 55 percent Muslim, 38 percent Christian, and 7 percent Druze.

Of the Christian minority, the largest Lebanese sect is Maronite, which split from the Roman Catholic Church in the seventh century and was ruled by the patriarch of Antioch. Although some reconciliation has been made with the Roman Catholic Church, the liturgy is still performed in the ancient Syrian language. Approximately one-fifth of Lebanon's total population and 60 percent of its Christians are Maronites.

The second largest Christian sect is Greek Orthodox, one of the Eastern Orthodox churches that split from Roman Catholicism in the eleventh century. Lebanon's Greek Orthodox Christians use a Byzantine liturgy, because Eastern Orthodox Christians were led by the patriarch of Constantinople from the time of the split with Roman Catholicism in 1054 until the city of Constantinople was captured by Muslims in 1453. Greek Orthodox Christians comprise approximately 5 percent of Lebanon's total population.

Other relatively small and little-known Christian sects in Lebanon include Greek Catholic, Armenian, Syrian

Orthodox (Jacobites), and Chaldeans (Assyrian). Greek Catholics split from the Greek Orthodox Church and migrated to Lebanon because of persecution elsewhere. Armenians separated from the Roman Catholic Church in the fifth century and have a patriarch in Armenia.

Of the Muslim majority in Lebanon, about 34 percent belong to one of several Shiite Muslim sects. Largest is Mitwali, but in recent years more militant sects have gained power, especially Hezbollah, the Party of God, which has taken American and European hostages. The Sunni branch, accepted by about 90 percent of the world's Muslims, is practiced by 21 percent of Lebanon's population, including refugees from Palestine to the south.

Lebanon also has non-Christian and non-Muslim groups, most important of which is the Druze (about 7 percent of the population). The Druze religion combines elements of Islam and Christianity, but many of the rituals are kept secret from outsiders.

Lebanon's religious groups have tended to live in different regions of the country (Figure 7–17). Maronites are concentrated in the west central part, Sunnis in the northwest, and Shiites in the south and east. Beirut, the capital and largest city, has been divided between an eastern Christian zone and a Muslim western zone.

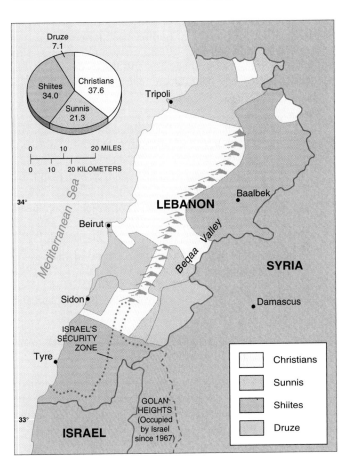

Figure 7–17 *Ethnicities in Lebanon. Christians dominate in the south and the northwest, Sunni Muslims in the far north, Shiite Muslims in the northeast and south, and Druze in the south-central and southeast.*

Globalization and Local Diversity

Physical Landscape of the Middle East

Human geographers refer to physical processes, such as landforms, to explain *where* people live. At a *global* scale, people prefer flatter land for agriculture, settlements, and most businesses. In understanding *why* the conflict in the Middle East is so intractable, geographers gain insight from *local diversity* in the area's physical landscape.

Israel's Perspective on the Physical Landscape.

Israel sees itself as a very small country—20,000 square kilometers (8,000 square miles)—with a Jewish majority, surrounded by a region of hostile Muslim Arabs encompassing more than 25 million square kilometers (10 million square miles). In dealing with its neighbors, Israel considers two elements of the local landscape especially meaningful.

First, the country's major population centers are quite close to international borders, making them vulnerable to surprise attack. The country's two largest cities, Tel Aviv and Haifa, are only 20 and 60 kilometers (12 and 37 miles), respectively, from Jordan, while the third largest city, Jerusalem, is adjacent to the Jordanian border.

The second geographical problem from Israel's perspective derives from local landforms. The northern half of Israel is a strip of land 80 kilometers (50 miles) wide between the Mediterranean Sea and the Jordan River. It is divided into three roughly parallel physical regions (Figure 1, left):

- A coastal plain along the Mediterranean, extending inland as much as 25 kilometers (15 miles) and as little as a few meters.
- A series of hills reaching elevations above 1,000 meters (3,300 feet).
- The Jordan River valley, much of which is below sea level.

The United Nations plan for the partition of Palestine in 1947, as modified by the armistice ending the 1948–1949 War, allocated most of the coastal plain to Israel, while Jordan took most of the hills between the coastal plain and the Jordan River valley, a region generally called the West Bank (of the Jordan River). Farther north, Israel's territory extended eastward to the Jordan River valley, but Syria controlled the highlands east of the valley, known as the Golan Heights.

Between 1948 and 1967, Jordan and Syria used the hills as staging areas to attack Israeli settlements on the adjacent coastal plain and in the Jordan River valley. During the 1967 War, Israel captured these highlands to stop attacks on the lowland population concentrations. A generation later, Israel still holds the Golan Heights and West Bank, and the security concerns of the past have faded. In part, technological changes have made obsolete the strategic benefits of controlling the highlands. During the 1991 Gulf War, Iraq was able to attack Israel with SCUD missiles from 400 kilometers (250 miles) away.

Palestinians' Perspective on the Physical Landscape.

To Palestinians, Middle Eastern geography looks very different. After capturing the West Bank from Jordan in 1967, Israel permitted Jewish settlers to construct more than 100 settlements in the territory (Figure 1, right). Some Israelis built settlements in the West Bank because they regarded the territory as an integral part of the biblical Jewish homeland, known as Judea and Samaria. Others migrated to the settlements because of a severe shortage of affordable housing inside Israel's pre-1967 borders. Although Jewish settlers comprise only about 7 percent of the West Bank population, Palestinians see their presence as a reflection of Israel's reluctance to grant independence to the occupied territory.

Given the long-standing mistrust among the ethnicities and nationalities trying to share space in the Middle East, future prospects for peace may rest on the ability of the antagonists to shape the physical landscape for mutual benefit. An area of limited agricultural land and water supply, cooperation may begin in such areas as sharing the water from the Jordan River, expanding programs to desalinate the water from the Dead and Mediterranean seas, and sensitively modifying the desert environment to expand agriculture.

Lebanon's Constitution.   In 1943, to preserve the identity of each religious group, Lebanon's constitution required that each religion be represented in the governing body according to its percentage in the 1932 census. Accordingly, the 99 members of the Chamber of Deputies comprised 54 Christians, including 30 Maronites, 11 Greek Orthodox, 6 Greek Catholics, 4 Armenian Christians, and 3 other Christians; 39 Muslims, including 20 Sunnis and 19 Shiites; and 6 Druzes. Thus, the 30 Maronite candidates with the largest vote totals were elected, as were the top vote-getters for the other religions.

By unwritten convention, the president of Lebanon was a Maronite Christian, the premier a Sunni Muslim, the speaker of the Chamber of Deputies a Shiite Muslim, and the foreign minister a Greek Orthodox Christian. Other cabinet members and civil servants were similarly apportioned among the various faiths.

When the governmental system was created, Christians constituted a majority and controlled the country's

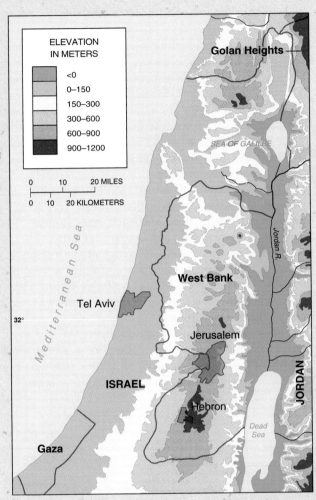

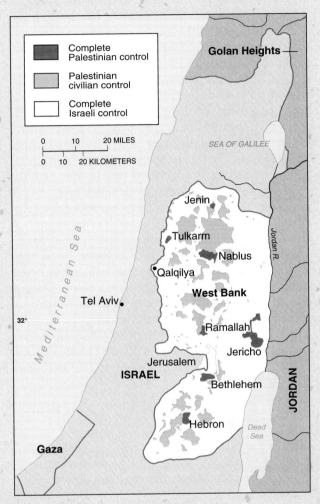

(Left) Physical geography of Israel and West Bank. The land between the Mediterranean Sea and the Jordan River is divided into three roughly parallel physical regions: a narrow coastal plain along the Mediterranean, a series of hills reaching elevations above 1,000 meters (3,300 feet), and the Jordan River valley. Between 1948 and 1967, Israel's boundaries encompassed primarily the coastal lowlands, while Jordan and Syria controlled the highlands. During the 1967 War, Israel captured these highlands and retained them to stop attacks on population concentrations in the Jordan River valley and the coastal plain.

(Right) Since Israel captured the West Bank in 1967, Jewish settlers have constructed more than 100 settlements in the territory. More than 50,000 Jews now live in the West Bank, about 7 percent of the territory's total population.

main businesses, but as the Muslims became the majority, they demanded political and economic equality. Lebanon's government was unable to deal with changing social and economic conditions.

Lebanon's Civil War. Lebanon's delicate balance was upset by the arrival of a large number of Palestinian refugees after the 1967 Arab-Israeli War. The Palestine Liberation Organization (PLO) controlled much of

southern Lebanon, where the refugee camps were clustered, and from which the PLO launched attacks against Israel. Lebanon's Muslims were generally sympathetic to the PLO's cause, while most Lebanese Christians supported Israel.

With the breakdown of the national government, each religious group formed a private army or militia to guard its territory. The territory controlled by each militia changed according to results of battles with other religious groups.

Most of Lebanon is now controlled by Syria, which has a historical claim over the territory. Syria sent peace-keeping forces into Lebanon in 1976, and most militias have supported a Syrian-imposed cease-fire. Israel invaded Lebanon in 1982 to clear out the PLO; after the PLO leaders were evacuated to Tunisia, Israel withdrew, but a PLO ally, the Southern Lebanese Army, continued to control southern Lebanon. The United States sent marines to Lebanon in 1982 to supervise the PLO's evac-uation, but they were removed in February 1984, four months after 241 of them died in their barracks from the explosion of a truck wired with a bomb.

Lebanon is relatively peaceful now. A 1990 consti-tution divided Parliament seats and cabinet positions equally between Christians and Muslims, although in reality the legislature is dominated by Muslims because most Christians boycotted the election. A Christian still is president, but most powers were transferred to the Muslim prime minister.

The real winner in the Lebanon civil war was Syria. The Syrian army controls much of Lebanon, once reli-gious groups disbanded their private militias, and the government now is closely allied to Syria. Beirut is being rebuilt, although it has a long way to go to reclaim its one-time reputation as a major banking and enter-tainment center in the Middle East. But after two decades of religious civil war, 150,000 Lebanese have died, and much of the country's economy and buildings lie in ruins.

Key Issue 4

Why Does Ethnic Cleansing Occur?

- Defining ethnic cleansing
- Ethnic cleansing in former Yugoslavia

Throughout history, ethnic groups have been forced to flee from other ethnic groups' more powerful armies. The largest ever level of forced migration came during World War II (1939–1945) because of events leading up to the war, the war itself, and postwar adjustments. Especially notorious was the deportation by the German Nazis of millions of Jews, gypsies, and other ethnic groups to the infamous concentration camps, where they exterminated most of them.

After World War II ended, millions of ethnic Ger-mans, Poles, Russians, and other groups were forced to migrate as a result of boundary changes (Figure 7–18). For example, when a portion of eastern Germany became part of Poland, the Germans living in the region were forced to move west to Germany and Poles were allowed to move into the area. Similarly, Poles were forced to move when the eastern portion of Poland was turned over to the Soviet Union.

The scale of forced migration during World War II has not been repeated, but in the 1990s a new term—"ethnic cleansing"—was invented to describe new prac-tices by ethnic groups against other ethnic groups.

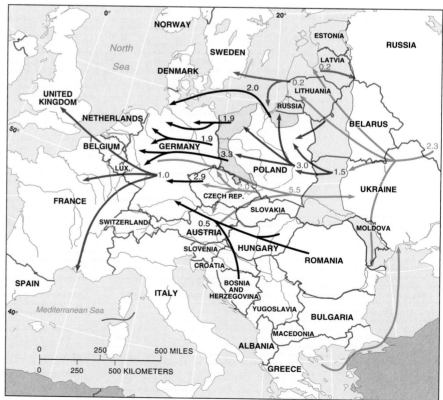

Figure 7–18 Forced migration of ethnicities as a result of territorial changes after World War II. The largest number were Poles forced to move from territory occupied by the Soviet Union, Germans forced to migrate from terri-tory taken over by Poland and the Soviet Union, and Russians forced to return to the Soviet Union from Western Europe.

Defining Ethnic Cleansing

Ethnic cleansing is a process in which a more powerful ethnic group forcibly removes a less powerful one in order to create an ethnically homogeneous region. The point of ethnic cleansing is not simply to defeat an enemy, or to subjugate them, as was the case in traditional wars.

Ethnic cleansing is undertaken to rid an area of an entire ethnicity, so that the surviving ethnic group can be the sole inhabitants. Rather than a clash between armies of male soldiers, ethnic cleansing involves the removal of every member of the less powerful ethnicity—women as well as men, children as well as adults, the frail elderly as well as the strong youth.

One recent example of ethnic cleansing illustrates how the process is undertaken. During the 1990s, ethnic cleansing was practiced primarily by Bosnian Serbs against Bosnian Muslims in Bosnia and Herzegovina.

The ethnic cleansing involved several steps. First, the powerful Bosnian Serb army took control of villages inhabited by Bosnian Muslims, and Bosnian Serbs were installed as local government officials. The Bosnian Serb local government officials passed laws to draft the Bosnian Muslim adult males into the Bosnian Serb army. When the Bosnian Muslims naturally refused to join an army dominated by Bosnian Serbs, the Bosnian Serb local government officials who had taken political control of the village stripped away the Bosnian Muslims' health insurance and other social benefits and removed both male and female Bosnian Muslims from responsible jobs.

According to independent journalists, intoxicated Bosnian Serb soldiers instigated individual acts of violence against Bosnian Muslims in the villages. When Bosnian Muslims retaliated, the Bosnian Serb army restored peace in the villages by killing the professional and able-bodied male Bosnian Muslims and deporting others to camps.

Deported Bosnian Muslims were divided into two groups: Potential trouble-makers were held in the detention camps, while Bosnian Muslims not considered a threat (essentially women, children, and elderly people) were relocated to open centers. Refugees in the open centers could receive visits from international relief agencies and were free to leave the camps, but most had no place to go, because their homes and property had been seized by the Bosnian Serbs. Other European countries only accepted a small number of the Bosnian Muslims. Bosnian Muslim women and children can be seen begging on the streets of major European cities.

Ethnic Cleansing in the Former Yugoslavia

Ethnic cleansing in Bosnia and Herzegovina stems from a complex pattern of ethnic diversity in the region of southeastern Europe known as the Balkan Peninsula. The region, about the size of Texas, is named for the Balkan Mountains (known in Slavic languages as Stara Planina), which extend east-west across the region. The Balkans includes Albania, Bulgaria, Greece, and Romania, as well as several countries that once comprised Yugoslavia (Figure 7–19).

Creation of Multi-Ethnic Yugoslavia

The Balkan Peninsula has long been a hotbed of unrest, a complex assemblage of ethnicities. Northern portions

Muslims living in Bosnia-Herzegovina have been victims of ethnic cleansing by Serbs. Muslim men have been rounded up and placed in detention centers, such as this one at Manjaca. (Marleen Daniels/Gamma-Liaison, Inc.)

Figure 7–19 The Balkans in 1914. At the outbreak of World War I, Austria-Hungary controlled the northern part of the region, including all or part of Croatia, Slovenia, and Romania. The Ottoman Empire controlled some of the south, although during the nineteenth century it had lost control of Albania, Bosnia-Herzegovina, Greece, Romania, and Serbia.

were incorporated into the Austria-Hungary Empire, whereas southern portions were ruled by the Ottomans. Austria-Hungary extended its rule farther south in 1878 to include Bosnia and Herzegovina, where the majority of the people had been converted to Islam by the Ottomans. In June 1914, the heir to the throne of Austria-Hungary was assassinated in Sarajevo by a Serb who sought independence for Bosnia. The incident sparked World War I.

After World War I, the allies created a new country, Yugoslavia, to unite several Balkan ethnicities that spoke similar South Slavic languages (Figure 7–20). The most numerous ethnicities brought into Yugoslavia were Serbs and Croats; others included Slovenes, Macedonians, and Montenegrens. The prefix "Yugo" in the country's name derives from the Slavic word for "south."

Ethnic Diversity in the Former Yugoslavia. Under the long leadership of Josip Broz Tito, who governed Yugoslavia from 1953 until his death in 1980, Yugoslavs liked to repeat a refrain that roughly translates as follows: "Yugoslavia has seven neighbors, six republics, five nationalities, four languages, three religions, two alphabets, and one dinar." Specifically:

- Yugoslavia's *seven* neighbors included three long-time democracies (Austria, Greece, and Italy) and four states then governed by Communists (Albania, Bulgaria, Hungary, and Romania). The diversity of neighbors reflected Yugoslavia's strategic location between the Western democracies and Communist Eastern Europe. Although a socialist country, Yugoslavia was militarily neutral after it had been expelled in 1948 from the Soviet-dominated military alliance for being too independent-minded. Yugoslavia's Communists permitted more communication and interaction with Western democracies than did other Eastern European countries.

- The *six* republics—Bosnia and Herzegovina, Croatia, Macedonia, Montenegro, Serbia, and Slovenia—had more autonomy from the national government to run their own affairs than was the case in other Eastern European countries.

- *Five* of the republics were named for the country's five recognized nationalities—Croats, Macedonians, Montenegrens, Serbs, and Slovenes. Bosnia and Herzegovina contained a mix of Serbs, Croats, and Muslims.

- Yugoslavia had *four* official languages—Croatian, Macedonian, Serbian, and Slovene (Montenegrens spoke Serbian).

- The *three* major religions included Roman Catholic in the north, Eastern Orthodox in the east, and Islam in the south. Croats and Slovenes were predominantly Roman Catholic, Serbs and Macedonians predominantly Eastern Orthodox, and the Bosnians and Montenegrens predominantly Muslim.

- *Two* of the four official languages—Croatian and Slovene—were written in the Roman alphabet, while Macedonian and Serbian were written in Cyrillic. Most linguists outside Yugoslavia considered Serbian and Croatian to be the same language except for different alphabets.

- The refrain concluded that Yugoslavia had *one* dinar, the national unit of currency. Despite cultural diversity, according to the refrain, common economic interests kept Yugoslavia's nationalities unified.

Creation of Yugoslavia brought stability that lasted for most of the twentieth century. Old animosities among ethnic groups were submerged, and younger people began to identify themselves as Yugoslavs, rather than as Serbs, Croats, or Montenegrens.

Destruction of Multi-Ethnic Yugoslavia

Rivalries among ethnicities resurfaced in Yugoslavia during the 1980s after Tito's death, leading to the breakup

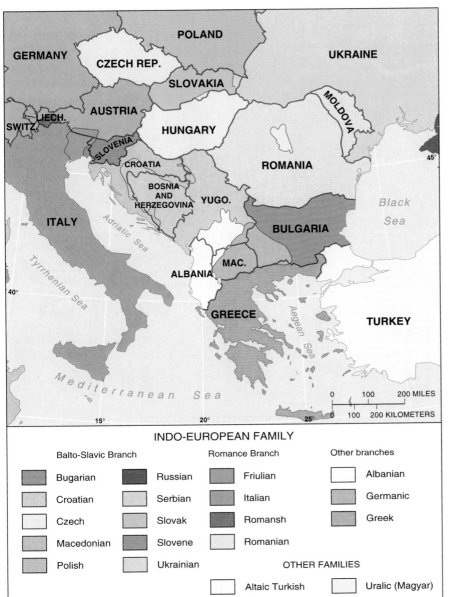

Figure 7–20 Languages in Southern and Eastern Europe. After World War I, world leaders created several new states and realigned the boundaries of existing ones, so that the boundaries of states matched language boundaries as closely as possible. These state boundaries proved to be relatively stable for much of the twentieth century. But in the 1990s, the region became a center of conflict among speakers of different languages.

INDO-EUROPEAN FAMILY

Balto-Slavic Branch

Bugarian
Croatian
Czech
Macedonian
Polish
Russian
Serbian
Slovak
Slovene
Ukrainian

Romance Branch

Friulian
Italian
Romansh
Romanian

Other branches

Albanian
Germanic
Greek

OTHER FAMILIES

Altaic Turkish
Uralic (Magyar)

of the country in the early 1990s. Breaking away to form independent countries were Bosnia and Herzegovina, Croatia, Macedonia, and Slovenia. Only Montenegro and Serbia remained in Yugoslavia.

As long as Yugoslavia comprised one country, ethnic groups were not especially troubled by the division of the country into six republics. But when Yugoslavia's republics were transformed from local government units into five separate countries, ethnicities fought to redefine the boundaries (Figure 7–21). Not only did the boundaries of Yugoslavia's six republics fail to match the territory occupied by the five major nationalities, but the country contained other important ethnic groups that had not received official recognition as nationalities.

Serbs attacked Croats to gain control of the Krajina region of eastern Croatia where Serbs outnumbered Croats. Albanians, who comprised 77 percent of the pop-

ulation in the Kosovo region of southern Serbia, fought to free themselves from cultural and political domination of Serbs, who retained control of Kosovo because it was their historic homeland.

Complexity in Bosnia-Herzegovina   The creation of a viable country proved especially difficult in the case of Bosnia and Herzegovina. Four of Yugoslavia's five officially recognized nationalities—Croats, Macedonians, Serbs, and Slovenes—were able to constitute majorities in four of the five independent countries carved out of Yugoslavia, and the fifth recognized nationality—Montenegrens—chose to remain in the smaller Yugoslavia with the Serbs.

In contrast, at the time of Yugoslavia's breakup, the largest group in Bosnia and Herzegovina—Bosnian Muslim—was considered an ethnicity rather than a

Figure 7–21 Yugoslavia, until its breakup in 1992. Yugoslavia comprised six republics (plus Kosovo and Vojvodina autonomous regions within the Republic of Serbia). According to the country's last census, taken in 1981, the territory occupied by the various nationalities did not match the boundaries of the republics or autonomous regions.

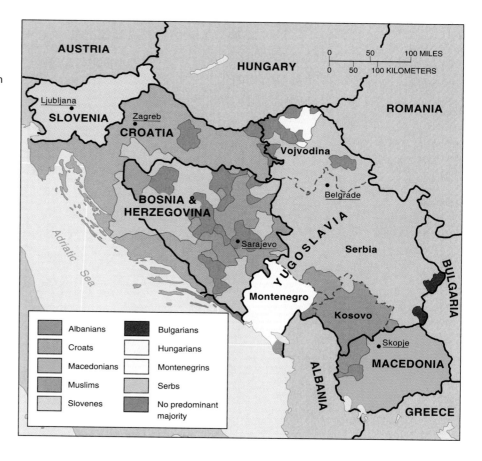

nationality. Bosnian Muslims comprised 40 percent of the population of Bosnia and Herzegovina. The remainder of Bosnia and Herzegovina consisted of 32 percent Serb and 18 percent Croat.

Rather than live in an independent multi-ethnic country with a Muslim plurality, Bosnia and Herzegovina's Serbs and Croats fought to unite the portions of the republic that they inhabited with Serbia and Croatia, respectively. To strengthen their cases for breaking away from Bosnia and Herzegovina, Serbs and Croats engaged in ethnic cleansing of Bosnian Muslims. Ethnic cleansing assured that areas did not merely have majorities of Bosnian Serbs and Bosnian Croats, but were ethnically homogeneous and therefore better candidates for union with Serbia and Croatia.

Ethnic cleansing by Bosnian Serbs against Bosnian Muslims was especially severe, because much of the territory inhabited by Bosnian Serbs was separated from Serbia by areas with Bosnian Muslim majorities. By ethnically cleansing Bosnian Muslims from intervening areas, Bosnian Serbs created one continuous area of Bosnian Serb domination rather than several discontinuous ones.

Accords reached in Dayton, Ohio, in 1996 by leaders of the various ethnicities divided Bosnia and Herzegovina into three regions, one each dominated by the Bosnian Croats, Muslims, and Serbs. The Bosnian Croat and Muslim regions were combined into a federation, with some cooperation between the two groups, but the Serb region has operated with almost complete independence in all but name from the others.

In recognition of the success of their ethnic cleansing, Bosnian Serbs received nearly half of the country, although they comprised less than one-third of the population, and Bosnian Croats got one-fourth of the land, although they comprised one-sixth of the population. Bosnian Muslims, 44 percent of the population before the ethnic cleansing, got 27 percent of the land.

Balkanization A century ago, the term **Balkanized** was widely used to describe a small geographic area that could not successfully be organized into one or more stable states because it was inhabited by many ethnicities with complex, long-standing antagonisms toward each other. World leaders at the time regarded **Balkanization**—the process by which a state breaks down through conflicts among its ethnicities—as a threat to peace throughout the world, not just in a small area. They were right: Balkanization directly led to World War I, because the various nationalities in the Balkans dragged into the war the larger powers with whom they had alliances.

At the end of the twentieth century—after two world wars and the rise and fall of communism—the Balkans have once again become Balkanized. Will the United States, Western Europe, and Russia once again be drawn reluctantly into conflict through entangled alliances in the Balkans?

If peace comes to the Balkans it will be because in a tragic way ethnic cleansing "worked." Millions of people were rounded up and killed or forced to migrate because they constituted ethnic minorities. Ethnic homogeneity may be the price of peace in areas that once were multi-ethnic.

Summary

Two major museums standing one block apart in Detroit illustrate the challenges of encouraging respect for different ethnic identities in the United States. One of the museums, the Detroit Institute of Arts (DIA), contains a major collection of paintings by medieval European artists, many of which were donated a century ago by rich Detroit industrialists. The DIA's most famous work is an enormous mural completed in 1932 by the Mexican muralist Diego Rivera glorifying workers in Detroit's auto factories. The 75-year-old building, the nation's fifth largest art museum, looks like a Greek temple.

The nearby Museum of African-American History houses the nation's largest exhibit devoted to the history and culture of African-Americans. Founded in 1965, the museum has moved twice to larger buildings, including the current one opened in 1997. The building is designed to reflect the cultural heritage of Africa, including an entry with large bronze doors topped by 14-carat gold plate decorative masks. The exhibits are primarily photographs, videos, and text.

The financially strapped city of Detroit has had difficulty adequately funding both museums, so it has had to make choices. Which museum should take priority, a crumbling temple of European masterpieces, or an emotionally powerful testimony to the rich cultural traditions of America's most numerous ethnic minority? Does it matter that Detroit's African-American population was 5 percent when the DIA was built and 75 percent when the Museum of African-American history was built?

Here again are the key issues for Chapter 7:

1. **Where are ethnicities distributed?** Major ethnicities in the United States include African-Americans, Hispanic-Americans and Asian-Americans. These ethnic groups are clustered in regions of the country and within urban neighborhoods. In the United States, race and ethnicity are often used interchangeably, because members of the African-American ethnic group are also distinguished as members of the black race (although not all blacks are African-Americans).

2. **Where have ethnicities been transformed into nationalities?** Nationalities are ethnic groups that possess among their cultural traditions attachment and loyalty to a particular country. A nationality combines an ethnic group's language, religion, and artistic expressions with a country's particular independence movement, history, and other patriotic events. During the past two centuries, many countries have been created that attempt to transform single ethnic groups into single nationalities.

3. **Why do ethnicities clash?** Conflicts can arise when countries contain more than one ethnicity. In the Horn of Africa, civil wars rage among various ethnicities competing to dominate the defining of a nationality. In the Middle East, long-standing ethnic conflicts have been complicated in modern times by emergence of nationalities.

4. **Why does ethnic cleansing occur?** Ethnic cleansing is an attempt by a more powerful ethnic group to create an ethnically homogeneous region by forcibly evicting all members of another ethnic group. The practice has been especially widespread in the countries that comprise the former country of Yugoslavia.

Case Study Revisited

Ethnic Cleansing in Central Africa

Line up five Hutus and five Tutsis, and the ethnic origin of perhaps six would be plain. The two ethnicities speak the same language, hold similar beliefs, and practice similar social customs, and intermarriage has lessened the physical differences between the two. Yet Hutus and Tutsis have engaged in ethnic cleansing on a scale greater than even in the former Yugoslavia. Forty percent of Rwanda's 8.5 million inhabitants have been killed or ethnically cleansed since 1994, and nearly that many from neighboring Burundi.

The Hutu were settled farmers, growing crops in the fertile hills and valleys of present-day Rwanda and Burundi, known as the Great Lakes region of central Africa. The Tutsi were cattle herders who migrated to present-day Rwanda and Burundi from the Rift Valley of western Kenya beginning 400 years ago.

Relations between settled farmers and cattle-herders are always uneasy, and the Tutsi took control of the kingdom of Rwanda and turned the Hutu into their serfs, although Tutsi comprised only about 15 percent of the population.

Rwanda, as well as Burundi, became a colony of Germany in the late nineteenth century, and after the Germans were defeated in World War I, a colony of Belgium. The Europeans reinforced differences between the two ethnicities. Colonial administrators permitted a few Tutsis to attend university and hold responsible government positions, while excluding the Hutu altogether.

In 1959, a few years before Rwanda gained its independence, Hutus killed or ethnically cleansed most of the Tutsis, out of fear that the Tutsis would seize control of the newly independent country. Those fears were realized in 1994, when the children of those ethnically cleansed Tutsis poured into Rwanda, defeated the Hutu army, and killed or ethnically cleansed a large percentage of the country's 7 million Hutus. But not before the Hutu-dominated military killed many Tutsi, as well as moderate Hutu government officials who favored reconciliation with the Tutsi after the president was killed in a plane crash.

■ Key Terms

Apartheid Laws (no longer in effect) in South Africa that physically separated different races into different geographic areas.

Balkanization Process by which a state breaks down through conflicts among its ethnicities.

Balkanized A small geographic area that could not successfully be organized into one or more stable states because it was inhabited by many ethnicities with complex, long-standing antagonisms toward each other.

Blockbusting A process by which real estate agents convince white property owners to sell their houses at low prices because of fear that black families will soon move into the neighborhood.

Centripetal force An attitude that tends to unify people and enhance support for a state.

Ethnicity Identity with a group of people that share distinct physical and mental traits as a product of common heredity and cultural traditions.

Ethnic cleansing Process in which more powerful ethnic group forcibly removes a less powerful one in order to create an ethnically homogeneous region.

Multi-ethnic state State that contains more than one ethnicity.

Multinational state State that contains two or more ethnic groups with traditions of self-determination that agree to coexist peacefully by recognizing each other as distinct nationalities.

Nationalism Loyalty and devotion to a particular nationality.

Nationality Identity with a group of people that share legal attachment and personal allegiance to a particular place as a result of being born there.

Nation-state A state whose territory corresponds to that occupied by a particular ethnicity that has been transformed into a nationality.

Race Identity with a group of people descended from a common ancestor.

Racism Belief that race is the primary determinant of human traits and capacities and that racial differences produce an inherent superiority of a particular race.

Racist A person who subscribes to the beliefs of racism.

Self-determination Concept that ethnicities have the right to govern themselves.

Sharecropper A person who works fields rented from a landowner and pays the rent and repays loans by turning over to the landowner a share of the crops.

Triangular slave trade A practice, primarily during the eighteenth century, in which European ships transported slaves from Africa to Caribbean islands, molasses from the Caribbean to Europe, and trade goods from Europe to Africa.

■ Thinking Geographically

1. The 2000 U.S. Census permits people to identify themselves as being of more than one race, in recognition that several million American children have parents of two races. Discuss the merits and difficulties of permitting people to choose more than one race.

2. Sarajevo, capital of Bosnia and Herzegovina, once contained concentrations of many ethnic groups. In retaliation for ethnic cleansing by the Serbs and Croats, the Bosnian Muslims now in control of Sarajevo have been forcing other ethnic groups to leave the city, and Sarajevo is now inhabited overwhelmingly by Bosnian Muslims. Discuss the merits and obstacles in restoring Sarajevo as a multi-ethnic city.

3. Despite the 1954 U.S. Supreme Court decision that racially segregated school systems are inherently unequal,

most schools remain segregated, with virtually none or virtually all African-American or Hispanic pupils. As long as most neighborhoods are segregated, how can racial integration in the schools be achieved?

4. A century ago, European immigrants to the United States had much stronger ethnic ties than today, including clustering in specific neighborhoods. Discuss the merits and disadvantages of retaining strong ethnic identity in the United States as opposed to full assimilation into the American nationality identity.

5. With the removal of the apartheid laws, South Africa now offers legal equality to all races in principle. Discuss obstacles that South Africa's blacks face in achieving cultural and economic equality.

■ Further Readings

Arreola, Daniel D. "Urban Ethnic Landscape Identity." *Geographical Review* 85 (1995): 518–34.

Bennett, D. Gordon, ed. *Tension Areas of the World: A Problem-Oriented World Regional Geography.* Champaign, IL: Park Press, 1982.

Boyd, Charles G. "Making Bosnia Work." *Foreign Affairs* 77 (1998): 42–55.

Cooke, Thomas J. "Geographic Access to Job Opportunities and Labor-Force Participation Among Women and African Americans in the Greater Boston Metropolitan Area." *Urban Geography* 18 (1997): 213–27.

Davis, Cary; Carl Haub, and JoAnne Willette. "U.S. Hispanics: Changing the Face of America." *Population Bulletin* 38 (3). Washington, DC: Population Reference Bureau, 1983.

Del Pinal, Jorge, and Audrey Singer. "Generations of Diversity: Latinos in the United States." *Population Bulletin* 52 (3). Washington, DC: Population Reference Bureau, 1997.

Dingemans, Dennis, and Robin Datel. "Urban Multiethnicity." *Geographical Review* 85 (1995): 458–77.

Falah, Ghazi. "Living Together Apart: Residential Segregation in Mixed Arab-Jewish Cities in Israel." *Urban Studies* 33 (1996): 823–57.

———. "The 1948 Israeli-Palestinian War and Its Aftermath: The Transformation and De-Signification of Palestine's Cultural Landscape." *Annals of the Association of American Geographers* 86 (1996): 256–85.

———. "Re-envisioning Current Discourse: Alternative Territorial Configurations of Palestinian Statehood." *Canadian Geographer* 41 (1997): 307–30.

Gabriel, Stuart A., and Stuart S. Rosenthal. "Commutes, Neighborhood Effects, and Earnings: An Analysis of Racial Discrimination and Compensating Differentials." *Journal of Urban Economics* 40 (1996): 61–83.

Haass, Richard N. "The Middle East: No More Treaties." *Foreign Affairs* 75 (1996): 53–63.

Haverluk, Terrence W. "The Changing Geography of U.S. Hispanics, 1850–1990." *Journal of Geography* 96 (1997): 134–45.

Keating, Michael. *Nations Against the State: The New Politics of Nationalism in Quebec, Catalonia and Scotland.* London: Macmillan, 1996.

Kellerman, Aharon. "Settlement Myth and Settlement Activity: Interrelationships in the Zionist Land of Israel." *Transactions of the Institute of British Geographers New Series* 21 (1996): 363–78.

Li, F.N.L.; A.J. Jowett, A.M. Findlay, and R. Skeldon. "Discourse on Migration and Ethnic Identity: Interviews with Professionals in Hong Kong." *Transactions of the Institute of British Geographers New Series* 20 (1995): 342–56.

Lian, Brad, and John R. O'Neal. "Cultural Diversity and Economic Development: A Cross-National Study of 98 Countries, 1920–1985." *Economic Development and Cultural Change* 42 (1997): 21–78.

Martinez-Vazquez, Jorge; Mark Rider, and Mary Beth Walker. "Race and the Structure of School Districts in the United States." *Journal of Urban Economics* 41 (1997): 281–300.

McLafferty, Sara, and Valerie Preston. "Gender, Race, and the Determinants of Commuting: New York in 1990." *Urban Geography* 18 (1997): 192–212.

Murphy, Alexander B. "Territorial Policies in Multiethnic States." *Geographical Review* 79 (1989): 410–21.

Newman, David. *Population, Settlement and Conflict: Israel and the West Bank.* Cambridge: Cambridge University Press, 1991.

———, and Ghazi Falah. "Bridging the Gap: Palestinian and Israeli Discourses on Autonomy and Statehood." *Transactions of the Institute of British Geographers New Series* 22 (1997): 111–29.

Pacione, Michael. "Ethnic Segregation in the European City: The Case of Vienna." *Geography* 81 (1996): 120–32.

Pincetl, Stephanie. "Immigrants and Redevelopment Plans in Paris, France: Urban Planning, Equity, and Environment Justice." *Urban Geography* 17 (1996): 440–55.

Pulido, Laura; Steve Sidawi, and Robert O. Vos. "An Archaeology of Environmental Racism in Los Angeles." *Urban Geography* 17 (1996): 419–39.

Richmond, Anthony H. "Ethnic Nationalism: Social Science Paradigm." *International Social Science Journal* 39 (1987): 3–18.

Robinson, Jennifer. "The Geopolitics of South African Cities: States, Citizens, Territory." *Political Geography* 16 (1997): 365–86.

Rogerson, C.M., and J.M. Rogerson. "The Changing Post-apartheid City: Emergent Black-owned Small Enterprises in Johannesburg." *Urban Studies* 34 (1997): 85–104.

Rose, Richard. "National Pride in Cross-National Perspective." *International Social Science Journal* 37 (1985): 85–96.

Slowe, Peter. *Geography and Political Power: The Geography of Nations and States.* London: Routledge, 1990.

Soffer, Arnon, and Julian V. Minghi. "Israel's Security Landscapes: The Impact of Military Considerations on Land Uses." *Professional Geographer* 38 (1986): 28–41.

United States National Advisory Commission on Civil Disorders, Otto Kerner, chairman. *Report.* New York: Dutton, 1968.

Wright, Richard, and Mark Ellis. "Immigrants and the Changing Racial/Ethnic Division of Labor in New York City, 1970–1990." *Urban Geography* 17 (1996): 317–53.

Yiftachel, Oren. "The Political Geography of Ethnic Protest: Nationalism, Deprivation and Regionalism Among Arabs in Israel." *Transactions of the Institute of British Geographers New Series* 22 (1997): 91–110.

CHAPTER 8

Political Geography

How many countries can you name? Old-style geography sometimes required memorization of countries and their capitals. Today, human geographers emphasize a thematic approach. We are concerned with the location of activities in the world, the reasons for particular spatial distributions, and the significance of the arrangements. Despite this change in emphasis, you still need to know the locations of countries. Without such knowledge, you lack a basic frame of reference: knowing where things are. It is like translating an article in a foreign language by looking up each word in a dictionary.

In recent years, we have repeatedly experienced military conflicts and revolutionary changes in once-obscure places. No one can predict where the next war will erupt, but political geography helps to explain the cultural and physical factors that underlie political unrest in the world. Political geographers study how people have organized Earth's land surface into countries and alliances, reasons underlying the observed arrangements, and the conflicts that result from the organization.

Key Issues

1. Where are states located?

2. Where are boundaries drawn between states?

3. Why do boundaries between states cause problems?

4. Why do states cooperate with each other?

Kurds who fled from Iraq to Turkey after the 1991 Gulf War dig snow from the mountain to use for drinking and cooking. (Anthony Suau/Gamma-Liaison)

265

Case Study

Changing Borders in Europe

Daniel Lenig lives in the village of Rittershoffen and works at a Mercedes-Benz truck factory in the town of Worth, about 50 kilometers (30 miles) away. Lenig's journey to work takes him across an international border, because Rittershoffen is in France, whereas Worth is in Germany.

As a citizen of France, Lenig has no legal difficulty crossing the German-French border twice a day; no guards ask him to show his passport or require him to pay customs duties on goods he purchases on the other side. If he is delayed, the cause is heavy traffic on the bridge that spans the Rhine River, which serves as the border between the two countries.

The boundary between France and Germany has not always been so easy to cross peacefully. The French long have argued that the Rhine River forms the logical physical boundary between France and Germany. But the Germans once claimed that they should control the Rhine, including the lowlands on the French side between the west bank of the river and the Vosges Mountains, an area known as Alsace. In fact, Alsace was initially inhabited by Germanic tribes, but was annexed by France in 1670.

Two centuries later, in 1870, Alsace and its neighboring province of Lorraine were captured by Prussia (which a year later formed the core of the newly proclaimed German Empire). France regained Alsace and Lorraine after Germany was defeated in World War I and has possessed them ever since, except between 1940 and 1945 when Germany controlled them during World War II.

With the end of the Cold War and the demise of communism in Eastern Europe, France and Germany now lie at the core of the world's wealthiest market area. Most French and German people consider the pursuit of higher standards of living to be more important than rehashing centuries-old boundary disputes.

Although old boundaries between France and Germany have been virtually eliminated, new ones have been erected elsewhere in Europe. Travelers between Ljubljana and Zagreb now must show their passports and convert their cash into a different currency. These two cities were once part of the same country—Yugoslavia—but now they are the capitals of two separate countries, Slovenia and Croatia. Similarly, travelers between Vilnius and Moscow—both once part of the Soviet Union—now must show their passports and change money when they cross the international boundary between Lithuania and Russia. Russians, who once made up a majority of the Soviet Union's population, now find themselves in the minority in such countries as Estonia, Turkmenistan, and Ukraine.

For several decades during the Cold War, many countries were polarized into two camps, one allied with the former Soviet Union and the other allied with the United States. But with the end of the Cold War in the 1990s, the global political landscape changed fundamentally. Geographic concepts help us to understand this changing political organization of Earth's surface. We can also use geographic methods to examine the causes of political change and instability and to anticipate potential trouble spots around the world.

When looking at satellite images of Earth, we easily distinguish landmasses and water bodies, mountains and rivers, deserts and fertile agricultural land, urban areas and forests. What we cannot see are *where* boundaries are located between countries. Boundary lines are not painted on Earth, but they might as well be, for these national divisions are very real. To many, national boundaries are more meaningful than natural features. One of Earth's most fundamental cultural characteristics—one that we take for granted—is the division of our planet's surface into a collection of countries.

In the post–Cold War era, the familiar division of the world into countries or states is crumbling. Geographers observe two principal reasons *why* this familiar division of the world is changing: globalization and cultural diversity. First, *globalization* is an important political trend: Individual countries have transferred military, economic, and political authority to regional and worldwide collections of states. This is quite a turnabout. Between the mid-1940s and the late 1980s, two superpowers—the United States and the Soviet Union—essentially "ruled" the world. As superpowers, they were involved in events around the globe.

But the United States is less dominant in the political landscape of the twenty-first century, and the Soviet Union no longer exists. Today, power is increasingly exercised by collections of states that have organized primarily for economic cooperation and to compete economically with countries in other regions.

Second, despite (or perhaps because of) greater global political cooperation, *local diversity* has increased in political affairs, as individual cultural groups demand more control over the territory they inhabit. States have transferred power to local governments, but this does not placate cultural groups who seek complete independence. Wars have broken out in recent years—both between small neighboring states and among cultural groups within countries—over political control of territory. Old countries have been broken up in a collection of smaller ones, some barely visible on world maps.

Key Issue 1

Where Are States Located?

- Defining states
- Development of the state concept

The question posed in this key issue may seem self-evident, because a map of the world shows that virtually all habitable land belongs to a country. But for most of history, until recently, this was not so. As recently as a half-century ago, the world contained only about 50 countries, compared to nearly 200 today (Figure 8–1). Included in the total are 185 countries that belong to the United Nations as of 1998, plus 7 other countries that are not U.N. members (Table 8–1, p. 270).

Defining States

A **state** is an area organized into a political unit and ruled by an established government that has control over its internal and foreign affairs. A state occupies a defined territory on Earth's surface and contains a permanent population. A state has **sovereignty**, which means independence from control of its internal affairs by other states. Because the entire area of a state is managed by its national government, laws, army, and leaders, it is a good example of a formal or uniform region. The term *country* is a synonym for *state*.

The term *state*, as used in political geography, does not refer to the 50 regional governments inside the United States. The 50 states of the United States are subdivisions within a single state: the United States of America.

Problems of Defining States

There is some disagreement about the actual number of sovereign states. Let us look at three interesting cases that test the definition of a state—Korea, China, and Antarctica.

Korea: One State or Two? A colony of Japan for many years, Korea was divided into two occupation zones by the United States and former Soviet Union after they defeated Japan in World War II. The country was divided into northern and southern sections along 38° north latitude.

The division of these zones became permanent in the late 1940s, when the two superpowers established separate governments and withdrew their armies. The new government of North Korea then invaded South Korea in 1950, touching off a 3-year war that ended with a cease-fire line near the 38th parallel. Both Korean governments are committed to reuniting the country into one sovereign state, but they do not agree on how this can be accomplished. Meanwhile, in 1992, North Korea and South Korea were admitted to the United Nations as separate countries.

China and Taiwan: One State or Two? Is the island of Taiwan a sovereign state? According to its government officials, Taiwan is not a separate sovereign state but is a part of China. The government of China agrees. Yet,

Figure 8–1 United Nations members. When it was organized in 1945, the U.N. had only 51 members, including 49 sovereign states plus Byelorussia (now Belarus) and Ukraine, then part of the Soviet Union. In 1998, the number had increased to 185. The greatest increase in sovereign states has occurred in Africa. Only four African states were original members of the U.N.—Egypt, Ethiopia, Liberia, and South Africa—and only six more joined during the 1950s. Beginning in 1960, however, a collection of independent states was carved from most of the remainder of the region. In 1960 alone, 16 newly independent African states became U.N. members. Creation of new sovereign states slowed during the 1980s. Among the states that joined the U.N. during the 1980s, only Zimbabwe had a population in excess of 200,000. The breakup of the Soviet Union and Yugoslavia stimulated the formation of more new states during the early 1990s.

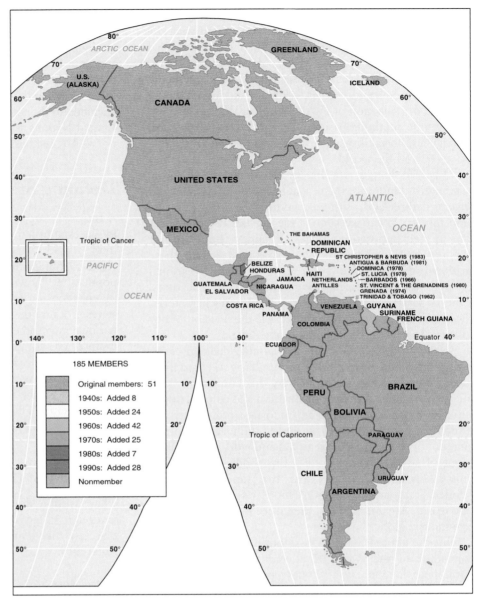

most other governments in the world consider China and Taiwan as two separate and sovereign states.

This confusing situation arose from a civil war between the Nationalists and the Communists in China during the late 1940s. After losing, Nationalist leaders in 1949 fled to the island of Taiwan, 200 kilometers (120 miles) off the Chinese coast. The Nationalists proclaimed that they were still the legitimate rulers of the entire country of China. Until some future occasion when they could defeat the Communists and recapture all of China, the Nationalists argued, at least they could continue to govern one island of the country.

The question of who constituted the legitimate government of China plagued U.S. officials during the 1950s and 1960s. The United States had supported the Nationalists during the civil war, so many Americans

opposed acknowledging that China was firmly under the control of the Communists. Consequently, the United States continued to regard the Nationalists as the official government of China until 1971, when U.S. policy finally changed, and the United Nations voted to transfer China's seat from the Nationalists to the Communists.

Antarctica. Antarctica is the only large landmass on Earth's surface that is not part of a sovereign state. Several states, including Argentina, Australia, Chile, France, New Zealand, Norway, and the United Kingdom, claim portions of Antarctica (Figure 8–2, p. 271). Argentina, Chile, and the United Kingdom make conflicting, overlapping claims. The United States, Russia, and a number of other states do not recognize the claims

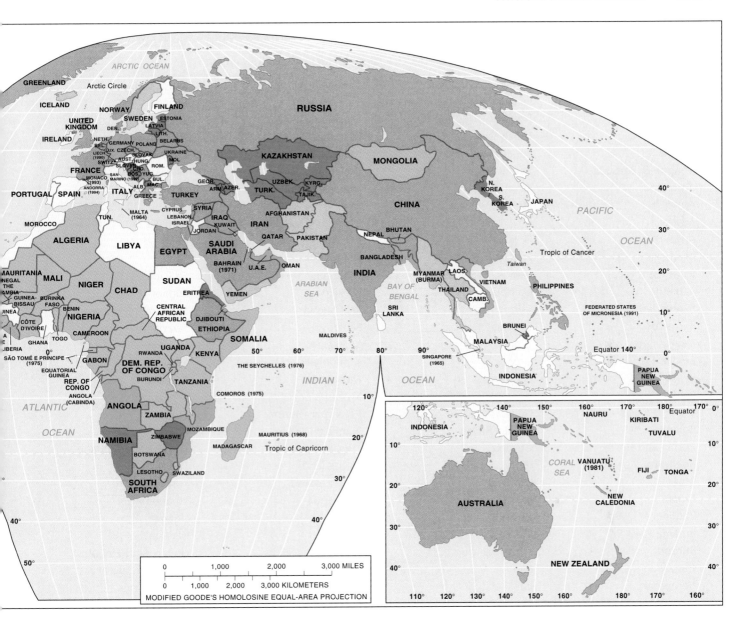

of any country to Antarctica. Several states have established research stations on Antarctica, but as yet no permanent settlement exists there.

Varying Size of States

The land area occupied by the states of the world varies considerably. The largest state is Russia, which encompasses 17.1 million square kilometers (6.6 million square miles), or 11 percent of the world's entire land area. The distance between the country's borders with Eastern European countries and the Pacific Ocean extends more than 7,000 kilometers (4,300 miles).

Other states with more than 5 million square kilometers (2 million square miles) include China (9.3 million square kilometers; 3.6 million square miles), Canada (9.2 million square kilometers; 3.6 million square miles), United States (9.2 million square kilometers; 3.5 million square miles), Brazil (8.5 million square kilometers; 3.3 million square miles), and Australia (7.6 million square kilometers; 2.9 million square miles).

At the other extreme are a number of **microstates**, which are states with very small land areas. The smallest microstate in the United Nations—Monaco—encompasses only 1.5 square kilometers (0.6 square miles), about the size of downtown Reno, Nevada. Other U.N. member states that are smaller than 1,000 square kilometers include Andorra, Antigua and Barbuda, Bahrain, Barbados, Dominica, Grenada, Liechtenstein, Maldives, Malta, Micronesia, Palau, St. Kitts and Nevis, St. Lucia, San Marino, St. Vincent and the Grenadines, São Tomé e Príncipe, the Seychelles, and Singapore. Many of these

Table 8–1 Sovereign States

Members of the United Nations (185)

Afghanistan	Dominican Republic	Lithuania	Saudi Arabia
Albania	Ecuador	Luxembourg	Senegal
Algeria	Egypt	Macedonia	Seychelles
Andorra	El Salvador	Madagascar	Sierra Leone
Angola	Equatorial Guinea	Malawi	Singapore
Antigua & Barbuda	Eritrea	Malaysia	Slovakia
Argentina	Estonia	Maldives	Slovenia
Armenia	Ethiopia	Mali	Solomon Islands
Australia	Fiji	Malta	Somalia
Austria	Finland	Marshall Islands	South Africa
Azerbaijan	France	Mauritania	Spain
Bahamas	Gabon	Mauritius	Sri Lanka
Bahrain	Gambia	Mexico	Sudan
Bangladesh	Germany	Micronesia	Suriname
Barbados	Georgia	Moldova	Swaziland
Belarus	Ghana	Monaco	Sweden
Belgium	Greece	Mongolia	Syria
Belize	Grenada	Morocco	Tajikistan
Benin	Guatemala	Mozambique	Tanzania
Bhutan	Guinea	Myanmar (Burma)	Thailand
Bolivia	Guinea-Bissau	Namibia	Togo
Bosnia and Herzegovina	Guyana	Nepal	Trinidad and Tobago
Botswana	Haiti	Netherlands	Tunisia
Brazil	Honduras	New Zealand	Turkey
Brunei	Hungary	Nicaragua	Turkmenistan
Bulgaria	Iceland	Niger	Uganda
Burkina Faso	India	Nigeria	Ukraine
Burundi	Indonesia	Norway	United Arab Emirates
Cambodia	Iran	Oman	United Kingdom
Cameroon	Iraq	Pakistan	United States
Canada	Ireland	Panama	Uruguay
Cape Verde	Israel	Papua New Guinea	Uzbekistan
Central African Republic	Italy	Paraguay	Vanuatu
Chad	Jamaica	Palau	Venezuela
Chile	Japan	Peru	Vietnam
China	Jordan	Philippines	Yemen
Colombia	Kazakhstan	Poland	Yugoslavia*
Comoros	Kenya	Portugal	Zambia
Congo Democratic Republic	Korea, North	Qatar	Zimbabwe
Congo Republic	Korea, South	Romania	**Not members of the**
Costa Rica	Kuwait	Russia	**United Nations (7)**
Côte d'Ivoire	Kyrgyzstan	Rwanda	Kiribiti
Croatia	Laos	Saint Kitts and Nevis	Nauru
Cuba	Latvia	Saint Lucia	Switzerland
Cyprus	Lebanon	Saint Vincent and the Grenadines	Taiwan
Czech Republic	Lesotho	Samoa (Western)	Tonga
Denmark	Liberia	San Marino	Tuvalu
Djibouti	Libya	São Tomé e Príncipe	Vatican
Dominica	Liechtenstein		

*The U.N. General Assembly voted to expel Yugoslavia from membership on September 22, 1992.

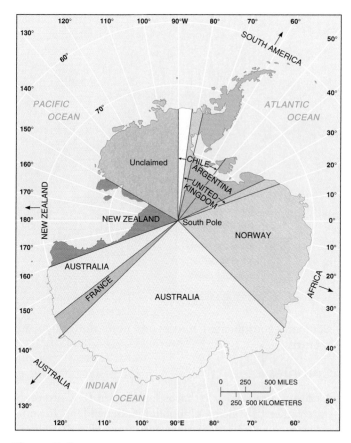

Figure 8–2 National claims to Antarctica. Antarctica is the only large landmass in the world that is not part of a sovereign state. It comprises 14 million square kilometers (5.4 million square miles), 50 percent larger than Canada. Portions are claimed by Argentina, Australia, Chile, France, New Zealand, Norway, and the United Kingdom; claims by Argentina, Chile, and the United Kingdom are conflicting. In 1959, these seven countries, plus Belgium, Japan, South Africa, the Soviet Union, and the United States, signed a treaty suspending any territorial claims for 30 years and establishing guidelines for scientific research. In 1991, 24 countries agreed to extend the treaty for another 50 years, established new pollution control standards, and banned mining and oil exploration for 50 years.

Black and white gentoo penguins cluster on Hovgaard Island, Antarctica. In the distance, a cruise ship brings humans to this uninhabited region of the world. (Richard Weiss/Silver Burdett Ginn)

are islands, which explains both their small size and sovereignty.

Development of the State Concept

The concept of dividing the world into a collection of independent states is recent. Prior to the 1800s, Earth's surface was organized in other ways, such as city-states, empires, and tribes. Much of Earth's surface consisted of unorganized territory.

Ancient and Medieval States

The modern movement to divide the world into states originated in Europe. However, the development of states can be traced to the ancient Middle East, in an area known as the Fertile Crescent.

Ancient States. The ancient Fertile Crescent formed an arc between the Persian Gulf and the Mediterranean Sea. The eastern end, Mesopotamia, was centered in the valley formed by the Tigris and Euphrates rivers, in present-day Iraq. The Fertile Crescent then curved westward over the desert, turning southward to encompass the Mediterranean coast through present-day Syria, Lebanon, and Israel. The Nile River valley of Egypt is sometimes regarded as an extension of the Fertile Crescent. Situated at the crossroads of Europe, Asia, and Africa, the Fertile Crescent was a center for land and sea communications in ancient times (Figure 8–3).

The first states to evolve in Mesopotamia were known as city-states. A **city-state** is a sovereign state that comprises a town and the surrounding countryside. Walls clearly delineated the boundaries of the city, and outside the walls the city controlled agricultural land to produce food for urban residents. The countryside also provided the city with an outer line of defense against attack by other city-states. Periodically, one city or tribe in Mes-

Figure 8–3 The Fertile Crescent is a crescent-shaped area of relatively fertile land situated between the Persian Gulf and the Mediterranean Sea. The Nile Valley of Egypt is sometimes included. Starting several thousand years ago, the territory has been organized into a succession of empires. As shown in Chapter 10, many important early developments in agriculture also originated in this region.

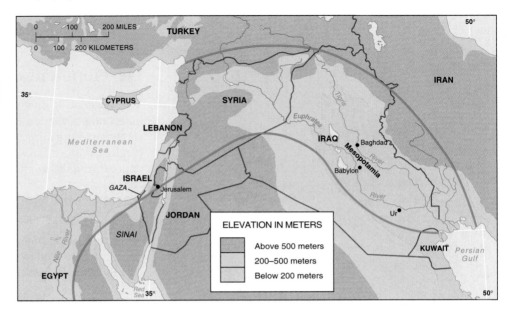

opotamia would gain military dominance over the others and form an empire. Mesopotamia was organized into a succession of empires by the Sumerians, Assyrians, Babylonians, and Persians.

Meanwhile, the state of Egypt emerged as a separate empire at the western end of the Fertile Crescent. Egypt controlled a long, narrow region along the banks of the Nile River, extending from the Nile Delta at the Mediterranean Sea southward for several hundred kilometers. Egypt's empire lasted from approximately 3000 B.C. until the fourth century B.C.

Early European States. Political unity in the ancient world reached its height with the establishment of the Roman Empire, which controlled most of Europe, North Africa, and Southwest Asia, from modern-day Spain to Iran and from Egypt to England. At its maximum extent, the empire comprised 38 provinces, each using the same set of laws that were created in Rome. Massive walls helped the Roman army defend many of the empire's frontiers. The Roman Empire collapsed in the fifth century A.D. after a series of attacks by people living on its frontiers, as well as internal disputes.

The European portion of the Roman Empire was fragmented into a large number of estates owned by competing kings, dukes, barons, and other nobles. Victorious nobles seized control of defeated rivals' estates, and after these nobles died others fought to take possession of their land. Meanwhile, most people were forced to live on an estate, working and fighting for the benefit of the noble.

Beginning about the year 1100, a handful of powerful kings emerged as rulers over large numbers of estates. The consolidation of neighboring estates under the unified control of a king formed the basis for the development of such modern Western European states as England, France, and Spain. However, much of central Europe—notably present-day Germany and Italy—remained fragmented into a large number of estates and

were not consolidated into states until the nineteenth century.

Colonies

A **colony** is a territory that is legally tied to a sovereign state rather than being completely independent. In some cases, a sovereign state runs only the colony's military and foreign policy. In others, it also controls the colony's internal affairs.

Colonialism. European states came to control much of the world through **colonialism**, which is the effort by one country to establish settlements and to impose its political, economic, and cultural principles on such territory. European states established colonies elsewhere in the world for three basic reasons:

1. European missionaries established colonies to promote Christianity.
2. Colonies provided resources that helped the economy of European states.
3. European states considered the number of colonies to be an indicator of relative power.

The three motives can be summarized as God, gold, and glory.

The colonial era began in the 1400s, when European explorers sailed westward for Asia but encountered and settled in the Western Hemisphere instead. The European states eventually lost most of their Western Hemisphere colonies: Independence was declared by the United States in 1776 and by most Latin American states between 1800 and 1824. European states then turned their attention to Africa and Asia (Figure 8–4). The European colonization of Africa and Asia is often termed **imperialism**, which is control of territory already occupied and organized by an indigenous society, whereas colonialism is control of previously uninhabited or sparsely inhabited land.

Seychelles, one of the smallest countries in the world, 453 square kilometers (175 square miles), one-sixth the size of Rhode Island. The country consists of about 100 islands situated in the Indian Ocean northeast of Madagascar. The second largest island, Praslin, shown here, is 38 square kilometers (15 square miles). This microstate was a British colony until 1976.

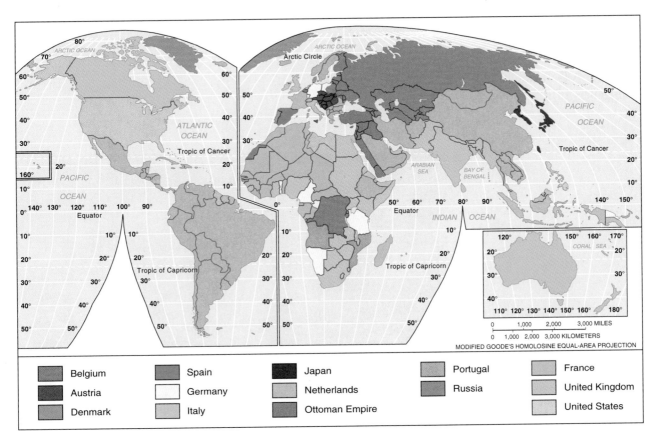

Figure 8–4 Colonial possessions, 1914. At the outbreak of World War I in 1914, European states held colonies in much of the world, especially in Africa and Asia. Most of the countries in the Western Hemisphere at one time had been colonized by Europeans but gained their independence in the eighteenth or nineteenth centuries.

The United Kingdom assembled by far the largest colonial empire. Britain planted colonies on every continent, including much of eastern and southern Africa, South Asia, the Middle East, Australia, and Canada. The British proclaimed that the "Sun never set" on their empire. France had the second largest overseas territory, although its colonies were concentrated in West Africa and Southeast Asia. Both the British and the French also took control of a large number of strategic islands in the Atlantic, Pacific, and Indian oceans.

Portugal, Spain, Germany, Italy, Denmark, the Netherlands, and Belgium all established colonies outside Europe, but they controlled less territory than the British and French. Germany tried to compete with Britain and France by obtaining African colonies that would interfere with communications in the rival European holdings.

Colonial Practices.   The colonial practices of European states varied. France attempted to assimilate its colonies into French culture and educate an elite group to provide local administrative leadership. After independence, most of these leaders retained close ties with France.

The British created different government structures and policies for various territories of their empire. This decentralized approach helped to protect the diverse cultures, local customs, and educational systems in their extensive empire. British colonies generally made peaceful transitions to independence, although exceptions can be found in the Middle East, Southern Africa, and Ireland.

Most African and Asian colonies became independent after World War II. Only 15 African and Asian states were members of the United Nations when it was es-

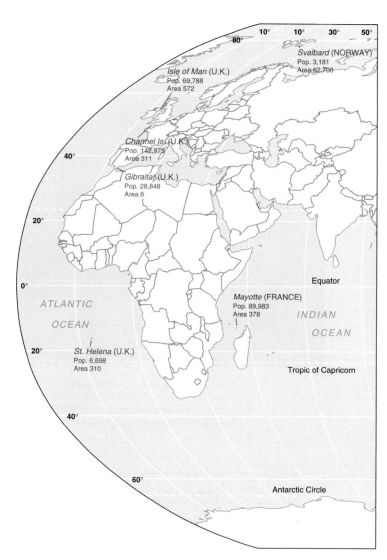

European countries carved up much of Africa into colonies during the late nineteenth century. The United Kingdom assembled the largest collection. This 1891 photograph shows the British commanders and governors asserting control in West Africa. (Mary Evans Picture Library/Photo Researchers, Inc.)

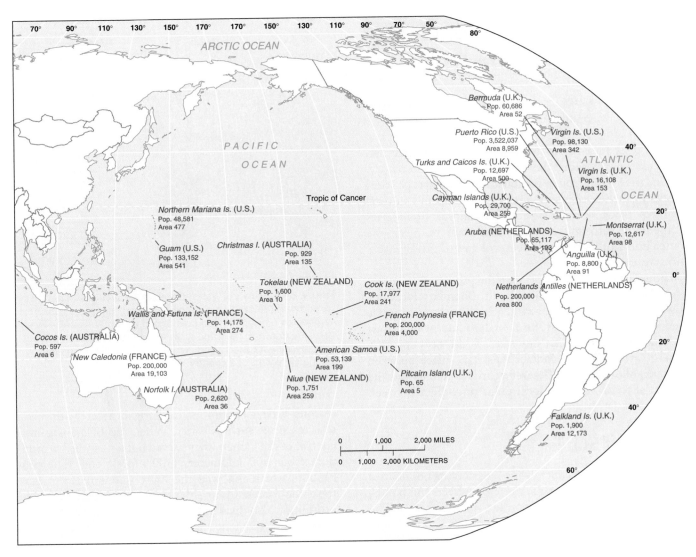

Figure 8–5 Colonial possessions, 1998. Most remaining colonies are tiny specks in the Pacific Ocean or the Caribbean Sea, too small to appear on the map. Svalbard, which belongs to Norway, is the only remaining colony with a land area greater than 10,000 square kilometers.

tablished in 1945, compared to 101 in 1998 (Table 8–1). The boundaries of the new states frequently coincide with former colonial provinces, although not always.

The Few Remaining Colonies. At one time, colonies were widespread over Earth's surface, but today only a handful remain. Nearly all are islands in the Pacific Ocean or Caribbean Sea (Figure 8–5).

With the return of Hong Kong to China by the British in 1997, the most populous remaining colony is Puerto Rico, which is a Commonwealth of the United States. Its 4 million residents are citizens of the United States, but they do not participate in U.S. elections, nor do they have a voting member of Congress. Puerto Ricans are split between those who want to maintain commonwealth status and those who want to see the island become a U.S. state.

Other than Puerto Rico, remaining colonies with populations between 100,000 and 300,000 include France's French Polynesia, Mayotte, and New Caledonia; the Netherlands' Netherlands Antilles; the United Kingdom's Channel Islands (Guernsey and Jersey); and the United States's Guam and U.S. Virgin Islands. Macao, a city of one-half million in southeastern China, controlled by Portugal since 1557, was turned over to China in 1999.

The world's least populated colony is Pitcairn Island, possessed by the United Kingdom. Pitcairn, in the South Pacific, has about 73 people on an island less than 5 square kilometers (2 square miles). The island was settled in 1790 by British mutineers from the ship *Bounty*, commanded by Captain William Bligh. Today, the islanders survive by selling fish and postage stamps to collectors.

Key Issue 2

Where Are Boundaries Drawn Between States?

- Shapes of states
- Types of boundaries

A state is separated from its neighbors by a **boundary**, an invisible line marking the extent of a state's territory. Boundaries result from a combination of natural physical features (such as rivers, deserts, mountains) and cultural features (such as language and religion). Boundaries completely surround an individual state to mark the outer limits of its territorial control and to give it a distinctive shape.

Boundaries interest us because the process of selecting their location is frequently difficult. Boundary locations also commonly generate conflict, both within a country and with its neighbors. The boundary line, which must be shared by more than one state, is the only location where direct physical contact must take place between two neighboring states. Therefore, the boundary has the potential to become the focal point of conflict between them.

Shapes of States

The shape of a state controls the length of its boundaries with other states. The shape therefore affects the potential for communications and conflict with neighbors. The shape of a state, such as the outline of the United States or Canada, is part of its unique identity. Beyond its value as a centripetal force, the shape of a state can influence the ease or difficulty of internal administration, and can affect social unity.

Five Basic Shapes

Countries have one of five basic shapes: compact, prorupted, elongated, fragmented, and perforated (Figure 8–6). Each shape displays distinctive characteristics and problems.

Compact States: Efficient.   In a **compact state**, the distance from the center to any boundary does not vary significantly. The ideal theoretical compact state would be shaped like a circle, with the capital at the center and the shortest possible boundaries to defend.

Compactness is a beneficial characteristic for most smaller states, because good communications can be more easily established to all regions, especially if the capital is located near the center. Examples of compact states include Bulgaria, Hungary, and Poland.

Prorupted States: Access or Disruption.   An otherwise compact state with a large projecting extension is a **prorupted state**. Proruptions are created for two principal

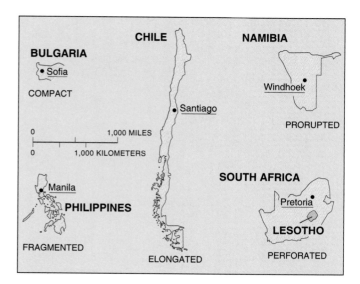

Figure 8–6 Shapes of states. Examples are shown of states that are compact (Bulgaria), prorupted (Namibia), elongated (Chile), fragmented (Philippines), and perforated (South Africa). The five states are drawn to the same scale. In general, compactness is an asset, because it fosters good communications and integration among all regions of a country.

reasons. First, a proruption can provide a state with access to a resource, such as water. When the Belgians gained control of the Congo, they carved out a westward proruption about 500 kilometers (300 miles) long. The proruption, which followed the Zaire (Congo) River, gave the colony access to the Atlantic Ocean (Figure 8–7). The proruption also divided the Portuguese colony of Angola (now an independent state) into two discontinuous fragments, 50 kilometers (30 miles) apart. The northern fragment, called Cabinda, constitutes less than 1 percent of Angola's total land area.

Proruptions can also separate two states that otherwise would share a boundary. When the British ruled the otherwise compact state of Afghanistan, they created a long, narrow proruption to the east, approximately 300 kilometers (200 miles) long and as narrow as 20 kilometers (12 miles) wide. The proruption prevented Russia from sharing a border with Pakistan (you can see this proruption on the map of Afghanistan in the Globalization and Local Diversity box).

In their former colony of South West Africa (now Namibia), the Germans in 1890 carved out a 500-kilometer (300-mile) proruption to the east. This proruption, known as the Caprivi Strip, provided the Germans with access to one of Africa's most important rivers, the Zambezi (Figure 8–7). The Caprivi Strip also disrupted communications among the British colonies of southern Africa. In recent years, South Africa, which controlled Namibia until its independence in 1990, stationed troops in the Caprivi Strip to fight enemies in Angola, Zambia, and Botswana.

Elongated States: Potential Isolation.   There are a handful of **elongated states**, or states with a long and

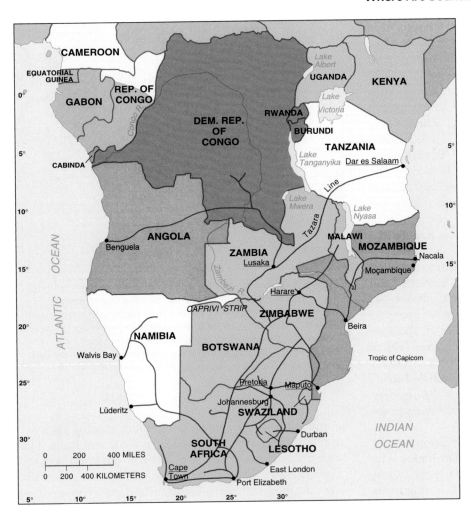

Figure 8-7 Southern Africa. Namibia and the Democratic Republic of Congo are prorupted states, South Africa is a perforated state. The map also shows landlocked African states, which must import and export goods by land-based transportation, primarily rail lines, to reach ocean ports in cooperating neighbor states. Colors show the European colonial rulers in 1914.

narrow shape. The best example is Chile, which stretches north-south for more than 4,000 kilometers (2,500 miles) but rarely exceeds an east-west distance of 150 kilometers (90 miles). Chile is wedged between the Pacific Coast of South America and the rugged Andes Mountains, which rise more than 6,700 meters (20,000 feet).

A less extreme example of an elongated state is Italy, which extends more than 1,100 kilometers (700 miles) from northwest to southeast, but is only approximately 200 kilometers (120 miles) wide in most places. In Africa, Malawi measures about 850 kilometers (530 miles) north-south but only 100 kilometers (60 miles) east-west.

In West Africa, Gambia is an elongated state extending along the banks of the Gambia River about 300 kilometers (200 miles) east-west but only between 20 and 50 kilometers north-south. Except for its short coast line along the Atlantic Ocean, Gambia is otherwise completely surrounded by Senegal. The shape of the two countries is a legacy of competition among European countries to establish colonies during the nineteenth century. Gambia became a British colony, whereas Senegal was French.

Elongated states may suffer from poor internal communications. A region located at an extreme end of the elongation might be isolated from the capital, which is usually placed near the center.

Fragmented States: Problematic. A **fragmented state** includes several discontinuous pieces of territory. Technically, all states that have offshore islands as part of their territory are fragmented. However, fragmentation is particularly significant for some states. There are two kinds of fragmented states: those with areas separated by water, and those separated by an intervening state.

The most extreme example is Indonesia, which comprises 13,677 islands that extend more than 5,000 kilometers (3,000 miles) across the Indian Ocean. Although more than 80 percent of the country's population live on two of the islands—Java and Sumatra—the fragmentation hinders communications and makes integration of people living on remote islands nearly impossible. To foster national integration, the Indonesian government has encouraged migration from the more densely populated islands to some of the sparsely inhabited ones. Other fragmented states that include more than one island are Japan and New Zealand.

A more difficult type of fragmentation occurs if the two pieces of territory are separated by another state. Picture the difficulty of communicating between Alaska and the lower 48 states if Canada were not a friendly neighbor. All land connections between Alaska and the rest of the United States must pass through a long expanse of Canada. The division of Angola into two pieces by Congo's proruption creates a fragmented state.

Panama—otherwise an example of an elongated state, 700 kilometers (450 miles) long and 80 kilometers (50 miles) wide—is fragmented by the Panama Canal, built in 1914 and owned by the United States. U.S. ownership of the canal and the surrounding Canal Zone was a source of tension for many years, but the United States and Panama signed a treaty in the late 1970s that transfers the canal to Panama on December 31, 1999. The treaty guarantees the neutrality of the canal and permits the United States to use force if necessary to keep the canal operating (Figure 8–8).

Even Russia, the world's largest state, is fragmented by other independent states. Kaliningrad (Konigsberg), an area measuring 16,000 square kilometers (6,000 square miles), is along the Baltic Sea. It is west of the remainder of Russia by 400 kilometers (250 miles), separated by the states of Lithuania and Belarus (see ahead to Figure 8–13). Until the end of World War II, the area was part of Germany, but the Soviet Union seized it after the German defeat. Virtually all of the area's 1 million residents are Russians; the German population fled westward after World War II. Russia wants Konigsberg because it has the country's largest naval base on the Baltic Sea.

Perhaps the most intractable fragmentation results from a tiny strip of land in India called Tin Bigha. The Tin Bigha corridor measures only 178 meters (about 600 feet) by 85 meters (about 300 feet). It fragments Dahagram and Angarpota from the rest of Bangladesh (Figure 8–9). The problem is a legacy of the late 1940s, when the British divided the region according to religion, allocating predominantly Hindu enclaves to India and predominantly Muslim ones to Bangladesh (formerly East Pakistan).

India agreed to lease the Tin Bigha corridor to Bangladesh in perpetuity, so that Dahagram and Angarpota could be connected to the rest of Bangladesh. But by

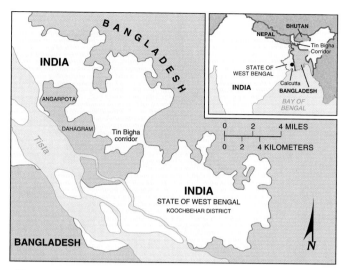

Figure 8–9 The Tin Bigha corridor. Smaller than three football fields, the Tin Bigha corridor is a part of India that fragments Dahagram and Angarpota from the rest of Bangladesh. India agreed to lease the corridor to Bangladesh in perpetuity, so that Dahagram and Angarpota could be connected to the rest of Bangladesh. But by eliminating one fragmentation, India created another one: Kuchlibari is now fragmented from the rest of India.

eliminating one fragmentation, India created its own: Kuchlibari is now fragmented from the rest of India. The agreement between the two countries gives Indians the right to move between Kuchlibari and the rest of India at certain times without submitting to passport inspection, customs declarations, and other international border controls. But given the long history of unrest between Hindus and Muslims, maintaining peace in the Tin Bigha corridor is difficult.

Perforated States: South Africa. A state that completely surrounds another one is a **perforated state**. The one good example of a perforated state is South Africa, which completely surrounds the state of Lesotho (Figure 8–7). Lesotho must depend almost entirely on South Africa for the import and export of goods. Dependency on South Africa was especially difficult for Lesotho when South Africa had a government controlled by whites who discriminated against the black majority population.

Landlocked States

Lesotho is unique in being completely surrounded by only one state, but it shares an important feature with several other states in southern Africa, as well as in other regions: It is landlocked. A **landlocked state** lacks a direct outlet to the sea because it is completely surrounded by several other countries (only one country in the case of Lesotho). Landlocked states are most common in Africa, where 14 of the continent's 54 states have no direct ocean access. The prevalence of landlocked states in Africa is a remnant of the colonial era, when Britain and France controlled extensive regions.

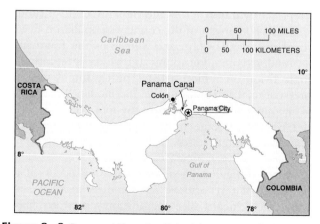

Figure 8–8 Panama. When the United States controlled the canal, Panama was an example of a fragmented state. With transfer of the canal to Panama in 1999, the country has instead become an example of an elongated state.

The European powers built railroads, mostly in the early twentieth century, to connect the interior of Africa with seaports. Railroads moved minerals from interior mines to seaports, and in the opposite direction, rail lines carried mining equipment and supplies from seaports to the interior. Now that the British and French empires are gone, and former colonies have become independent states, some important colonial railroad lines pass through several independent countries. This has created new landlocked states, which must cooperate with neighboring states that have seaports.

Direct access to an ocean is critical to states because it facilitates international trade. Bulky goods, such as petroleum, grain, ore, and vehicles, are normally transported long distances by ship. This means that a country needs a seaport where goods can be transferred between land and sea. To send and receive goods by sea, a landlocked state must arrange to use another country's seaport.

Landlocked States in Southern Africa. Cooperation between landlocked states in southern Africa has been complicated by racial patterns. Botswana, Lesotho, and Swaziland are landlocked states that ship 90 percent of their exports by rail through neighboring South Africa (Figure 8–7). Congo Democratic Republic, Zambia, and Zimbabwe must also transport most of their imports and exports through South Africa.

In the past, the states of southern Africa had to balance their economic dependency on South Africa with their dislike of the country's racial policies. Although they constitute more than 80 percent of South Africa's population, blacks suffered from discrimination (see Chapter 7). But if neighboring states had severed ties with South Africa because of its racial discrimination, they could have faced economic disaster.

Zimbabwe's particularly delicate problem can be understood by looking back about three decades, when it was a British colony called Southern Rhodesia. When the white minority in this landlocked colony unilaterally declared itself the independent country of Rhodesia in 1965, most other countries reduced or terminated trade with it. But the impact of trade sanctions on Rhodesia was limited because its major seaports were in South Africa, also ruled by a white minority government. As you can see in Figure 8–7, Rhodesia's main rail line ran through black-ruled Botswana to reach South Africa. Botswana was not cooperative, so the Rhodesian government completed a new rail line directly to South Africa in 1974, bypassing Botswana.

In 1979, the white-minority government of Rhodesia agreed to give blacks the right to vote, and blacks were elected to lead the government. The following year, Britain formally recognized the independence of the country, which was renamed Zimbabwe. The Zimbabwe government, now controlled by the black majority, faced a new set of relationships in southern Africa. Instead of working closely with South Africa, Zimbabwe tried to reduce its dependency on the neighboring white-minor-

ity government. The key element in Zimbabwe's strategy was to use railroads that connected to seaports outside South Africa. That turned into a very complex problem. Reference to Figure 8–7 will make this explanation easier to follow.

The closest seaport to Zimbabwe is Beira, in Mozambique. A railroad known as the Beira corridor runs west from the seaport to the Zimbabwean capital of Harare. Between 1976 and 1992, however, Mozambique was caught in a devastating civil war between its Marxist-oriented government and rebels backed by South Africa. Zimbabwe sent soldiers to Mozambique to keep the 500-kilometer (300-mile) Beira corridor repaired and protected from rebel attack, but the seaport of Beira itself was not well maintained.

More distant seaports were not reliable either. Mozambique's other two major deep-water ports—Nacala in the north and Maputo in the south—suffered even more than Beira from the civil war. The Benguela railway, which runs from the Atlantic Coast eastward across Angola to Zaire and Zambia, has also been disrupted since 1975 by a civil war in Angola.

The Tazara line, which runs from Zambia to Dar es Salaam in Tanzania, remains open, but service is unreliable. The equipment, much of it supplied by the Chinese in the 1970s, frequently breaks down, and landslides have periodically closed the line. As a result of these obstacles, Zimbabwe ships more than half of its freight through the South African seaport of Durban.

Types of Boundaries

Historically, frontiers rather than boundaries separated states. A **frontier** is a zone where no state exercises complete political control. A frontier is a tangible geographic area, whereas a boundary is an infinitely thin, invisible, imaginary line. A frontier provides an area of separation, often kilometers in width, but a boundary brings two neighboring states into direct contact, increasing the potential for violent face-to-face meetings. A frontier area is either uninhabited or sparsely settled by a few isolated pioneers seeking to live outside organized society.

Almost universally, frontiers between states have been replaced by boundaries. Modern communications systems permit countries to monitor and guard boundaries effectively, even in previously inaccessible locations. Once-remote frontier regions have become more attractive for agriculture and mining.

The only regions of the world that still have frontiers rather than boundaries are Antarctica and the Arabian Peninsula. Frontiers separate Saudi Arabia from Qatar, the United Arab Emirates, Oman, and Yemen. These frontier areas are inhabited by a handful of nomads who cross freely with their herds from one country to another. Until recently, part of Saudi Arabia's border with Iraq included an 8,000-square-kilometer (3,000-square-mile) frontier marked on maps as "Neutral Zone" (Figure 8–10). However, by stationing troops on either side

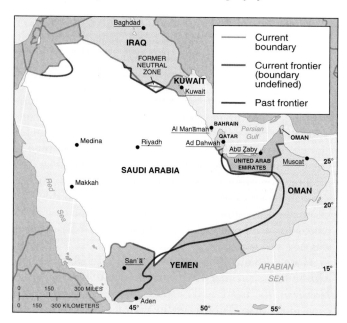

Figure 8–10 Frontiers in the Arabian peninsula. Several states in the Arabian peninsula are separated from each other by frontiers rather than by precisely drawn boundaries. The principal occupants of this desert area have been nomads, who have wandered freely through the frontier. A frontier known as the neutral zone existed between Saudi Arabia and Iraq until the two countries split it during the 1991 Gulf War.

of an east-west line across the Neutral Zone, Saudi Arabia and Iraq in 1990 transformed the frontier into a boundary, although not one officially ratified by the governments of the two countries.

Boundaries are of two types: physical and cultural. Physical boundaries coincide with significant features of the natural landscape (mountains, deserts, water), whereas cultural boundaries follow the distribution of cultural characteristics. Neither type of boundary is better or more "natural," and many boundaries are a combination of both types. The best boundaries are those to which all affected states agree, regardless of the rationale used to draw the line.

Physical Boundaries

Important physical features on Earth's surface can make good boundaries because they are easily seen, both on a map and on the ground. Three types of physical elements serve as boundaries between states: mountains, deserts, and water.

Mountain Boundaries. Mountains can be effective boundaries if they are difficult to cross. Contact between nationalities living on opposite sides may be limited, or completely impossible if passes are closed by winter storms. Mountains are also useful boundaries because they are rather permanent and usually are sparsely inhabited.

Mountains do not always provide for the amicable separation of neighbors. Argentina and Chile agreed to

be divided by the crest of the Andes Mountains but could not decide on the precise location of the crest. Was the crest a jagged line, connecting mountain peak to mountain peak? Or was it a curving line following the continental divide (the continuous ridge that divides rainfall and snowmelt between flow toward the Atlantic or Pacific)? The two countries almost fought a war over the boundary line. But with the help of U.S. mediators, they finally decided on the line connecting adjacent mountain peaks.

Desert Boundaries. A boundary drawn in a desert can also effectively divide two states. Like mountains, deserts are hard to cross and sparsely inhabited. Desert boundaries are common in Africa and Asia. In North Africa, the Sahara has generally proved to be a stable boundary separating Algeria, Libya, and Egypt on the north from Mauritania, Mali, Niger, Chad, and the Sudan on the south. (For an illustration, look back to Figure 1–12, the world climate map.) In the early 1980s, the Libyan army moved south across the desert to invade Chad, but retreated in 1987 following French intervention.

Water Boundaries. Rivers, lakes, and oceans are the physical features most commonly used as boundaries. Water boundaries are readily visible on a map and are relatively unchanging.

Water boundaries are especially common in East Africa (refer to Figure 8–7). For example:

- The boundary between Congo Democratic Republic and Uganda runs through Lake Albert.
- The boundary separating Kenya, Tanzania, and Uganda runs through Lake Victoria.
- The boundary separating Burundi, Congo Democratic Republic, Tanzania, and Zambia runs through Lake Tanganyika.
- The boundary between Congo Democratic Republic and Zambia runs through Lake Mwera.
- The boundary between Malawi and Mozambique runs through Lake Malawi (Lake Nyasa).

Boundaries are typically in the middle of the water, although the boundary between Malawi and Tanzania follows the north shore of Lake Malawi (Lake Nyasa). Again, the boundaries result from nineteenth-century colonial practices: Malawi was a British colony, whereas Tanzania was German.

Water boundaries can offer good protection against attack from another state, because an invading state must transport its troops by air or ship and secure a landing spot in the country being attacked. The state being invaded can concentrate its defense at the landing point.

The use of water as boundaries between states can cause difficulties, though. One problem is that the precise position of the water may change over time. Rivers, in particular, can slowly change their course. The Rio

Grande, the river separating the United States and Mexico, has frequently meandered from its previous course since it became part of the boundary in 1848. Land that had once been on the U.S. side of the boundary came to be on the Mexican side, and vice versa. The United States and Mexico have concluded treaties that restore land affected by the shifting course of the river to the country in control at the time of the original nineteenth-century delineation.

Ocean boundaries also cause problems because states generally claim that the boundary lies not at the coastline, but out at sea. The reasons are for defense and for control of valuable fishing industries. Beginning in the late eighteenth century, some states recognized a boundary, known as the territorial limit, that extended 3 nautical miles (about 5.5 kilometers or 3.5 land miles) from the shore into the ocean. Some states claimed more extensive territorial limits, and others identified a contiguous zone of influence that extended beyond the territorial limits.

The Law of the Sea, signed by 117 countries in 1983, standardized the territorial limits for most countries at 12 nautical miles (about 22 kilometers or 14 land miles). Under the Law of the Sea, states also have exclusive rights to the fish and other marine life within 200 miles (320 kilometers). Countries separated by less than 400 miles of sea must negotiate the location of the boundary between exclusive fishing rights. Disputes can be taken to a Tribunal for the Law of the Sea or to the International Court of Justice.

Cultural Boundaries

The boundaries between some states coincide with differences in ethnicity, especially language and religion. Other cultural boundaries are drawn according to geometry; they simply are straight lines drawn on a map, although good reasons always exist for where the lines are located.

Geometric Boundaries. Part of the northern U.S. boundary with Canada is a 2,100-kilometer (1,300-mile) straight line (more precisely, an arc) along 49° north latitude, running from Lake of the Woods between Minnesota and Manitoba to the Strait of Georgia between Washington State and British Columbia. This boundary was established in 1846 by treaty between the United States and Great Britain, which still controlled Canada.

At the time, some people in the United States wanted the boundary to be fixed 600 kilometers (400 miles) farther north, at 54°40′ north latitude. Before a compromise was reached, U.S. militants proclaimed "fifty-four forty or fight." The United States and Canada share an additional 1,100-kilometer (700-mile) geometric boundary between Alaska and the Yukon Territory along the north-south arc of 141° west longitude.

The 1,000-kilometer (600-mile) boundary between Chad and Libya is a straight line drawn across the desert in 1899 by the French and British to set the northern limit of French colonies in Africa (Figure 8–4). But subsequent actions by European countries created confusion over the boundary. In 1912, Italy seized Libya from the Turks and demanded that the boundary with French-controlled Chad be moved southward. In 1935, France agreed to move the boundary 100 kilometers (60 miles) to the south, but the Italian government was not satisfied and never ratified the treaty. The land that the French would have ceded is known as the Aozou Strip, named for the only settlement in this 100,000-square-kilometer (36,000-square-mile) area (Figure 8–11).

When Libya and Chad both became independent countries, the boundary was set at the original northern location. Claiming that it had been secretly sold the Aozou Strip by the president of Chad, Libya seized the territory in 1973, as well as a tiny bit of northeastern Niger that may contain uranium ore. In 1987, Chad expelled the Libyan army with the help of French forces and regained control of the strip.

Religious Boundaries. Religious differences often coincide with boundaries between states, but in only a few cases has religion been used to select the actual boundary line. The most notable example was in South Asia, when

The Great Wall of China historically served as one of the world's most visible boundaries. Originally built in the third century B.C. during the Qin (Ch'in) dynasty, the wall was extended the following century during the Han dynasty to keep out nomadic horsemen. The wall was partially reconstructed between the fourteenth and sixteenth centuries A.D. during the Ming dynasty. (R. Ian Lloyd/The Stock Market)

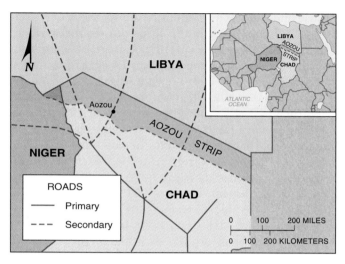

Figure 8–11 Aozou Strip. The boundary between Libya and Chad is a straight line, drawn by European countries early in the twentieth century when the area comprised a series of colonies. Libya, however, claims that the boundary should be located 100 kilometers to the south and that it should have sovereignty over the Aozou Strip.

the British partitioned India into two states on the basis of religion. The predominantly Muslim portions were allocated to Pakistan, while the predominantly Hindu portions became the independent state of India (see Figure 7–11).

Religion was also used to some extent to draw the boundary between two states on the island of Eire (Ireland). Most of the island became an independent country, but the northeast—now known as Northern Ireland—remained part of the United Kingdom. Roman Catholics comprise approximately 95 percent of the population in the 26 counties that joined the Republic of Ireland, whereas Protestants constitute the majority in the six counties of Northern Ireland (see Figure 6–15).

Language Boundaries. Language is an important cultural characteristic for drawing boundaries, especially in Europe. By global standards, European languages have substantial literary traditions and formal rules of grammar and spelling. Language has long been a significant means of distinguishing distinctive nationalities in Europe.

The French language was a major element in the development of France as a unified state in the seventeenth century. The states of England, Spain, and Portugal coalesced around distinctive languages. In the nineteenth century, Italy and Germany also emerged as states that unified the speakers of particular languages.

The movement to identify nationalities on the basis of language spread throughout Europe in the twentieth century. After World War I, leaders of the victorious countries met at the Versailles Peace Conference to redraw the map of Europe. One of the chief advisers to President Woodrow Wilson, the geographer Isaiah Bowman, played a major role in the decisions. Language was the most important criterion the allied leaders used to create new states in Europe and to adjust the boundaries of existing ones.

The conference was particularly concerned with Eastern and Southern Europe, regions long troubled by political instability and conflict. Boundaries were drawn around the states of Bulgaria, Hungary, Poland, and Romania to conform closely to the distribution of Bulgarian, Hungarian (Magyar), Polish, and Romanian speakers. Speakers of several similar South Slavic languages were placed together in the new country of Yugoslavia. Czechoslovakia was created by combining the speakers of Czech and Slovak, mutually intelligible West Slavic languages (refer to Figure 7–20).

Although the boundaries imposed by the Versailles conference on the basis of language were adjusted somewhat after World War II, they proved to be relatively stable, and peace ensued for several decades. However, during the 1990s, the map of Europe drawn at Versailles in 1919 collapsed. Despite speaking similar languages, Czechs and Slovaks found that they could no longer live together peacefully in the same state. Neither could Croats, Macedonians, Serbs, and Slovenes.

Key Issue 3

Why Do Boundaries Between States Cause Problems?

- One state with many nationalities
- One nationality in more than one state
- Internal organization of states

Boundaries between countries have been placed where possible to separate speakers of different languages or followers of different religions. As discussed in Chapter 7, a nation-state exists when the boundaries of a state match the boundaries of the territory inhabited by an ethnic group. Problems exist when the boundaries do not match.

One of two mismatches between the boundaries of states and ethnicities can occur: either one state contains more than one ethnic group, or one ethnic group is divided among more than one state. In either case, fashioning the population of a state into a single coherent nationality is difficult, if not impossible.

Chapter 7 discussed problems of forging a new nationality from competing ethnicities. In political geography, we are concerned with states that contain ethnicities with traditions of self-determination. These states may have difficulties in reconciling competing national traditions.

One State with Many Nationalities

A multinational state contains two or more nationalities with traditions of self-determination. Cyprus is a multinational state where the two nationalities were created in other locations. Russia (as well as the former Soviet Union) is a multinational state containing nationalities with traditions of self-determination before being taken over by the larger state.

Cyprus: Unfriendly Division of an Island

Cyprus, the third largest island in the Mediterranean Sea, contains two nationalities: Greek and Turkish (Figure 8–12). Although the island is physically closer to Turkey, Greeks comprise 78 percent of the country's population, whereas Turks account for 18 percent. When Cyprus gained independence from Britain in 1960, its constitution guaranteed the Turkish minority a substantial share of elected offices and control over its own education, religion, and culture.

Cyprus has never peacefully integrated the Greek and Turkish nationalities. In 1974, several Greek Cypriot military officers who favored unification of Cyprus with Greece seized control of the government. Shortly after the coup, Turkey invaded Cyprus to protect the Turkish Cypriot minority, occupying 37 percent of the island. The Greek coup leaders were removed within a few months, and an elected government was restored, but the Turkish army remained on Cyprus.

Traditionally, the Greek and Turkish Cypriots mingled, but after the coup and invasion, the two nationalities became geographically isolated. The northeastern part of the island is now overwhelmingly Turkish, while the southern part is overwhelmingly Greek. Approximately one-third of the island's Greeks were forced to move from the region controlled by the Turkish army, whereas nearly one-fourth of the Turks moved from the region now considered to be the Greek side. The percentage of one nationality living in the region dominated by the other nationality is now very low. The Turkish sector declared itself the independent Turkish Republic of Northern Cyprus in 1983, but only Turkey recognizes it as a separate state.

Cyprus. The photograph shows derelict conditions along the boundary between the two sectors, known as the Green Line, which runs through the heart of the capital, Nicosia. The Turkish sector, in the background, is demarcated by the flags of Turkey and the Turkish Republic of Northern Cyprus. (Raphael Gaillarde/Gamma-Liaison, Inc.)

A buffer zone patrolled by U.N. soldiers stretches across the entire island to prevent Greeks and Turks from crossing. The barrier even runs through the center of the capital, Nicosia. Only one official crossing point has been erected, and crossing is difficult except for top diplomats and U.N. personnel. Nevertheless, some cooperation continues between sectors: The Turks supply the Greek side with water and in return receive electricity.

Former Soviet Union: The Largest Multinational State

The Soviet Union was an especially prominent example of a multinational state until its collapse in the early 1990s. The 15 republics that once constituted the Soviet Union are now independent countries (Figure 8–13). These 15 newly independent states consist of five groups:

- 3 Baltic: Estonia, Latvia, and Lithuania.
- 3 European: Belarus, Moldova, and Ukraine.
- 5 Central Asian: Kazakhstan, Kyrgyzstan, Tajikistan, Turkmenistan, and Uzbekistan.
- 3 Caucasus: Azerbaijan, Armenia, and Georgia.
- Russia.

Reasonably good examples of nation-states have been carved out of the Baltic, European, and some Central Asian states. On the other hand, peaceful nation-states

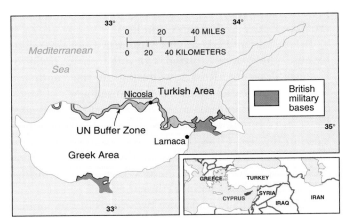

Figure 8–12 Cyprus. Since 1974, Cyprus has been divided into Greek and Turkish portions, with little mingling between the two groups. The Turkish sector has declared itself to be the Turkish Republic of Northern Cyprus, but only Turkey recognizes it as an independent country.

Figure 8–13 Union of Soviet Socialist Republics. The former Soviet Union included 15 republics, named for the country's largest ethnicities. With the breakup of the Soviet Union, the 15 republics became independent states.

have not been created in any of the small Caucasus states, and Russia is an especially prominent example of a state with major difficulties in keeping all of its ethnicities contented.

New Baltic Nation-States. Estonia, Latvia, and Lithuania are known as the Baltic states for their location on the Baltic Sea. They had been independent countries between the end of World War I in 1918 and 1940, when the former Soviet Union annexed them under an agreement with Nazi Germany.

Of the three Baltic states, Lithuania most closely fits the definition of a nation-state, because 80 percent of its population are ethnic Lithuanians. In Estonia, ethnic Estonians comprise only 62 percent of the population; in Latvia, only 53 percent are ethnic Latvians. In 1990, Russians comprised 9 percent of the population in Lithuania, 30 percent in Estonia, and 34 percent in Latvia.

These three small neighboring Baltic countries have clear cultural differences and distinct historical traditions. Most Estonians are Protestant (Lutherans), most Lithuanians are Roman Catholics, and Latvians are predominantly Lutheran with a substantial Roman Catholic minority. Estonians speak a Uralic language related to Finnish, whereas Latvians and Lithuanians speak languages of the Baltic group within the Balto-Slavic branch of the Indo-European language family.

New European Nation-States. To some extent, the former Soviet republics of Belarus, Moldova, and Ukraine now qualify as nation-states. Belarusians comprise 78 percent of the population of Belarus, Moldovans comprise 66 percent of the population of Moldova, and Ukrainians comprise 73 percent of the population of Ukraine.

The ethnic distinctions among Belarusians, Ukrainians, and Russians are somewhat blurred. The three groups speak similar East Slavic languages, and all are predominantly Eastern Orthodox Christians (some western Ukrainians are Roman Catholics).

Belarusians and Ukrainians became distinct ethnicities because they were isolated from the main body of Eastern Slavs—the Russians—during the thirteenth and fourteenth centuries. This was the consequence of Mongolian invasions and conquests by Poles and Lithuanians. Russians conquered the Belarusian and Ukrainian homelands in the late 1700s, but after five centuries of exposure to non-Slavic influences, the three Eastern Slavic groups displayed sufficient cultural diversity to consider themselves as three distinct ethnicities.

Russians actually constitute two-thirds of the population in the Crimea Peninsula of Ukraine. The Crimean Peninsula had been part of Russia until 1954, when the Soviet government turned over its administration to Ukraine, as a gift in honor of the 300th anniversary of Russian-Ukrainian friendship.

As long as both Russia and Ukraine were part of the Soviet Union, the Russians living in the Crimea were not concerned about the republic to which they were attached. After Russia and Ukraine became separate countries, a majority of the Crimeans voted to become independent of Ukraine. Control of the Crimean Peninsula was also important to both Russia and Ukraine because one of the Soviet Union's largest fleets was stationed there. The two countries agreed to divide the ships and to maintain jointly the naval base at Sevastopol.

Compounding the problem in the Crimea, 166,000 Tatars have migrated there from Central Asia in recent years. The Tatars once lived in the Crimea, but the Soviet leadership, suspecting them of sympathizing with the Germans during World War II, deported them to Central Asia. The Tatars prefer to be governed by Ukraine because of long-standing suspicion of the Russians, who dominated the government of the Soviet Union.

The situation is different in Moldova. Moldovans are ethnically indistinguishable from Romanians, and Moldova (then called Moldavia) was part of Romania until the Soviet Union seized it in 1940. When Moldova changed from a Soviet republic back to an independent country in 1992, many Moldovans pushed for reunification with Romania, both to reunify the ethnic group and to improve the region's prospects for economic development.

But it was not to be that simple. When Moldova became a Soviet republic in 1940, its eastern boundary was the Dniester River. The Soviet government increased the size of Moldova by about 10 percent, transferring from Ukraine a 3,000-square-kilometer (1,200-square-mile) sliver of land on the east bank of the Dniester. The majority of the inhabitants of this area, known as Trans-Dniestria, are Ukrainian and Russian. They, of course, oppose Moldova's reunification with Romania.

New Central Asian States.   The five states in Central Asia carved out of the former Soviet Union display varying degrees of conformance to the principles of nation-state. Together, the five provide an important reminder that multinational states can be more peaceful than nation-states.

In Turkmenistan and Uzbekistan, the leading ethnic group has an overwhelming majority—73 percent Turkmen and 71 percent Uzbek, respectively. Both ethnic groups are Muslims who speak an Altaic language. They were conquered by Russia in the nineteenth century, but Russians comprise only 9 percent of the population in Turkmenistan and 8 percent in Uzbekistan. Turkmen and Uzbeks are examples of ethnicities split into more than one country, the Turkmen between Turkmenistan and Russia, and Uzbeks among Kyrgyzstan, Tajikistan, and Uzbekistan.

Kyrgyzstan is 52 percent Kyrgyz, 22 percent Russian, and 13 percent Uzbek. The Kyrgyz—also Muslims who speak an Altaic language—resent the Russians for seizing the best farmland when they colonized this mountainous country early in the twentieth century.

In principle, Kazakhstan, twice as large as the other four Central Asian countries combined, is a recipe for ethnic conflict. The country is divided almost evenly between Kazakhs, who comprise 42 percent of the population, and Russians, at 37 percent. Kazakhs are Muslims who speak an Altaic language similar to Turkish, whereas the Russians are Eastern Orthodox Christians who speak an Indo-European language. Tensions do exist between the two groups, but Kazakhstan has been peaceful, in part because it has a somewhat less depressed economy than its neighbors.

In contrast, Tajikistan—65 percent Tajik, 25 percent Uzbek, and only 4 percent Russian—would appear to be a stable country, but it suffers from a civil war among the Tajik people, Muslims who speak a language in the Indic group of Indo-Iranian branch of Indo-European language. The civil war has been between Tajiks who are former Communists and an unusual alliance of Muslim fundamentalists and Western-oriented intellectuals. Fifteen percent of the population has been made homeless by the fighting.

Russia: Now the Largest Multinational State

Russia officially recognizes the existence of 39 nationalities, many of which are eager for independence. Russia's ethnicities are clustered in two principal locations (Figure 8–14). Some are located along borders with neighboring states, including Buryats and Tuvinian near Mongolia, and Chechen, Dagestan, Kabardin, and Ossetian near the two former Soviet republics of Azerbaijan and Georgia.

Other ethnicities are clustered in the center of Russia, especially between the Volga River basin and the Ural Mountains. Among the more numerous in this region are Bashkirs, Chuvash, and Tatars, who speak Altaic languages similar to Turkish, and Mordvins and Udmurts, who speak Uralic languages similar to Finnish. Most of these groups were conquered by the Russians in the sixteenth century under the leadership of Ivan IV ("Ivan the Terrible").

Independence movements are flourishing, because Russia is less willing to suppress these movements forcibly than the Soviet Union had once been. Particularly troublesome for the Russians are the Chechens, a group of Sunni Muslims who speak a Caucasian language and practice distinctive social customs.

Chechnya was brought under Russian control in the nineteenth century only after a 50-year fight. When the Soviet Union broke up into 15 independent states in 1991, the Chechens declared their independence and refused to join the newly created country of Russia. Russian leaders ignored the declaration of independence for 3 years, but in late 1994 they sent in the Russian army in an attempt to regain control of the territory.

Russia fought hard to prevent Chechnya from gaining independence because it feared that other ethnicities

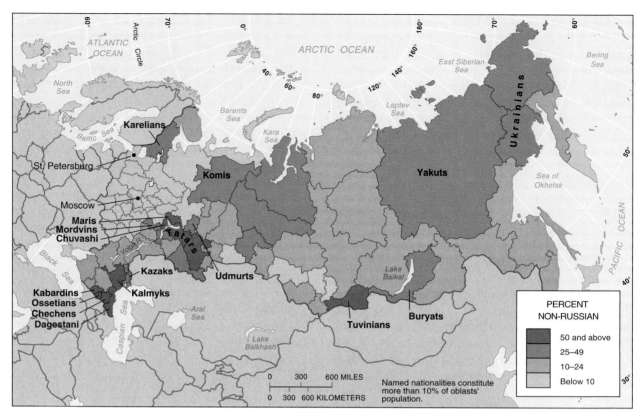

Figure 8-14 Ethnicities in Russia. Russians are clustered in the western portion of Russia, and the percentage declines to the south and east. The largest numbers of non-Russians are found in the center of the country between the Volga River and the Ural Mountains and near the southern boundaries. (Below) The Russian army attacked Chechnya to supress the independence movement. When Chechens offered still resistance, the Russians destroyed much of Chechnya, including this town of Argun, a strategic spot on the road to the capital Grozny 10 kilometers to the west. (Blanche/Gamma)

would follow suit. Chechnya was also important to Russia because the region contained deposits of petroleum. Russia viewed political stability in the area as essential for promoting economic development and investment by foreign petroleum companies.

Russians in Other States. Decades of Russian domination has left a deep reservoir of bitterness among other ethnicities once part of the Soviet Union. Because Russians were the dominant ethnicity in the Soviet Union, they were blamed for confiscating property and prohibiting the use of local languages in schools, hospitals, and factories.

Years after the demise of the Soviet Union, Russian soldiers have remained stationed in other countries, in part because Russia cannot afford to rehouse them. Other ethnicities fear that the slow withdrawal of Russian troops indicates that the Russians are trying to reassert the dominance over the economies and governments of other countries that they once exercised as the most numerous ethnicity in the Soviet Union.

For their part, Russians claim that they are now subject to discrimination as minorities in countries that were once part of the Soviet Union. Some of the countries once part of the Soviet Union have passed laws making it difficult for Russians to vote or to qualify as citizens with full civil rights. Russians are being passed over for hiring and promotion unless they learn the local languages. Yet, despite local hostility, Russians living in other countries of the former Soviet Union feel that they cannot migrate to Russia, because they have no jobs, homes, or land awaiting them there.

Turmoil in the Caucasus

The Caucasus region, an area about the size of Colorado, situated between the Black and Caspian seas, gets its name from the mountains that separate Russia from Azerbaijan and Georgia (Figure 8–15). The region is home to several ethnicities, with Azeris, Armenians, and Georgians the most numerous. Other important ethnicities include Abkhazians, Chechens, Ingush, and Ossetians (Table 8–2). Kurds and Russians—two ethnicities that are more numerous in other regions—are also represented in the Caucasus.

When the entire Caucasus region was part of the Soviet Union, the Soviet government promoted allegiance to communism and the Soviet state and quelled disputes among ethnicities, by force if necessary. But with the breakup of the region into several independent countries, long-simmering conflicts among ethnicities have erupted into armed conflicts.

Each ethnicity has a long-standing and complex set of grievances against others in the region. But from a political geography perspective, every ethnicity in the Caucasus has the same aspiration: to carve out a sovereign nation-state. The region's ethnicities have had varying success in achieving this objective, but none have fully achieved it.

Azeris. Azeris (or Azerbaijanis) trace their roots to Turkish invaders who migrated from Central Asia in the eighth and ninth centuries and merged with the existing Persian population. An 1828 treaty allocated northern Azeri territory to Russia and southern Azeri territory to Persia (now Iran). In 1923, the Russian portion became the Azerbaijan Soviet Socialist Republic within the Soviet Union. With the Soviet Union's breakup in 1991, Azerbaijan became an independent country again.

Approximately 6 million Azeris now live in Azerbaijan, nearly 80 percent of the country's total population. Another 6 million Azeris are clustered in northwestern Iran, where they constitute 10 percent of that country's population. Azeris hold positions of responsibility in Iran's government and economy, but Iran restricts teaching of the Azeri language.

Azerbaijan is a good example of a fragmented state: the western part of the country, Nakhichevan (named for the area's largest city), is separated from the rest of Azerbaijan by a 40-kilometer (25-mile) corridor belonging to Armenia.

Armenians. More than 3,000 years ago, Armenians controlled an independent kingdom in the Caucasus.

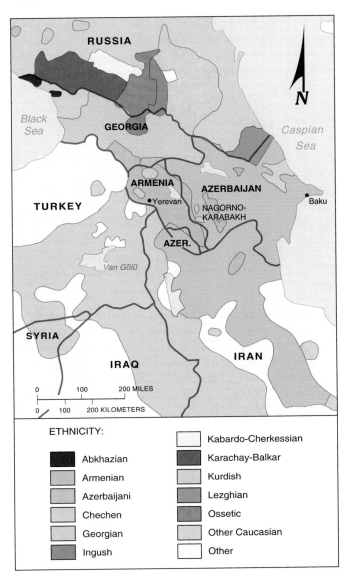

Figure 8–15 Ethnicities in the Caucasus. Armenians, Azeris, and Georgians are examples of ethnicities that were able to dominate new states during the 1990s, following the breakup of the Soviet Union. But the boundaries of the states of Armenia, Azerbaijan, and Georgia do not match the territories occupied by the Armenian, Azeri, and Georgian ethnicities. The Abkhazians, Chechens, Kurds, and Ossetians are examples of ethnicities in this region that have not been able to organize nation-states.

Converted to Christianity in A.D. 303, they lived for many centuries as an isolated Christian enclave under the rule of Turkish Muslims. During the late nineteenth and early twentieth centuries, hundreds of thousands of Armenians were killed in a series of massacres organized by the Turks. Others were forced to migrate to Russia, which had gained possession of eastern Armenia in 1828.

After World War I, the allies created an independent state of Armenia, but it was soon swallowed by its neighbors. In 1921, Turkey and the Soviet Union agreed to divide Armenia between them. The Soviet portion became the Armenian Soviet Socialist Republic and then an independent country in 1991. More than 90 percent of the population in Armenia are Armenians, making it the most ethnically homogeneous country in the region.

Table 8–2	Principal Ethnicities in the Caucasus		
Ethnicity	**Population (millions)**	**Religion**	**Language Family**
Abkhazians	0.2	Sunni Muslim	Caucasian
Ajars	0.1	Sunni Muslim	Caucasian
Armenians	5	Eastern Orthodox Christian	Indo-European
Azeris	15	Shiite Muslims	Altaic
Georgians	4	Eastern Orthodox Christian	Caucasian
Kurds	0.2[a]	Sunni Muslim	Indo-European
Ossetians	0.4	Eastern Orthodox Christian	Indo-European
Russians	0.6[b]	Eastern Orthodox Christian	Indo-European

[a]Includes only Kurds in the former Soviet Union.
[b]Includes only Russians in Armenia, Azerbaijan, and Georgia.

Armenians and Azeris both have achieved long-held aspirations of forming nation-states, but the two have been at war with each other since 1988 over the boundaries between the two nationalities. The conflict concerns possession of Nagorno-Karabakh, a 5,000-square-kilometer (2,000-square mile) enclave within Azerbaijan that is inhabited primarily by Armenians but placed under Azerbaijan's control by the Soviet Union during the 1920s.

Georgians.   The population of Georgia is more diverse than that in Armenia and Azerbaijan. Only 69 percent of the people living in Georgia are ethnic Georgians. The country includes about 9 percent each Armenian and Russian and 5 percent Azeri, 3 percent Ossetian, 2 percent Abkhazian, and 1 percent Ajars.

Georgia's cultural diversity has been a source of unrest, especially among the Ossetians and Abkhazians. The Abkhazians have fought for control of the northwestern portion of Georgia and would like to form an independent nation-state. Rather than a sovereign nation-state, the Ossetians want South Ossetia transferred from Georgia to Russia and united with North Ossetia, already part of Russia.

One Nationality in More Than One State

When the Soviet Union existed, its 15 republics were based on the 15 largest ethnicities. Less numerous ethnicities were not given the same level of recognition. With the breakup of the Soviet Union into 15 independent countries, a number of these less numerous ethnicities are now divided among more than one state.

The Kurds

An example of an ethnicity divided among several states is the Kurds, who live in the Caucasus south of the Armenians and Azeris. The Kurds are Sunni Muslims who speak a language in the Iranian group of the Indo-Iranian branch of Indo-European and have distinctive literature, dress, and other cultural traditions.

During the 1920s, the Kurds lived in an independent nation-state called Kurdistan, but today Kurds are split among six countries. Ten million live in eastern Turkey, 5 million in western Iran, 4 million in northern Iraq, and smaller numbers in Armenia, Azerbaijan, and northeastern Syria (Figure 8–15). Kurds comprise a fifth of the population in Iraq, a sixth in Turkey, and nearly a tenth in Iran.

When the victorious European allies carved up the Ottoman Empire after World War I, they created an independent state of Kurdistan to the south and west of Van Gölü (Lake Van) under the 1920 Treaty of Sèvres. Before the treaty was ratified, however, the Turks, under the leadership of Mustafa Kemal (later known as Kemal Ataturk), fought successfully to expand the territory under their control beyond the small area the allies had allocated to them. The Treaty of Lausanne in 1923 established the modern state of Turkey, with boundaries nearly identical to the current ones. Kurdistan became part of Turkey and disappeared as an independent state.

To foster the development of Turkish nationalism, the Turks have tried repeatedly to suppress Kurdish culture. Use of the Kurdish language was illegal in Turkey until 1991, and laws banning its use in broadcasts and classrooms remain in force. Kurdish nationalists, for their part, have waged a guerrilla war since 1984 against the Turkish army.

Kurds in other countries have fared just as poorly as those in Turkey. Iran's Kurds secured an independent republic in 1946, but it lasted less than a year. Iraq's Kurds have made several unsuccessful attempts to gain independence, including in the 1930s, 1940s, and 1970s. A few days after Iraq was defeated in the 1991 Gulf War, the country's Kurds launched another unsuccessful rebellion. The United States and its allies decided not to resume their recently concluded fight against Iraq on behalf of the Kurdish rebels, but after the revolt was crushed, they did send troops to protect the Kurds from further attacks by the Iraqi army.

Thus, despite their numbers, the Kurds are an ethnicity with no corresponding Kurdish state today. Instead, they are forced to live under the control of the region's more powerful nationalities.

Pan-Arab Nation-State

The world maps of language (Figure 5–1) and religion (Figure 6–1) portray a large, homogeneous region from northern Africa to southwestern Asia, where Arabic is the predominant language and Islam the predominant religion. The fragmentation of the Middle East into two dozen countries is a legacy of centuries of domination by the Ottomans (based in present-day Turkey), British, French, and other colonial powers. The British, in particular, made conflicting promises of land to Middle Eastern groups, in exchange for their support in fighting the Ottomans during World War I.

For some Arabs, this vast region of apparent cultural homogeneity comprises one unified nationality. Supporters of pan-Arab unity believe that fragmentation of the Middle East into two dozen states has resulted in an unfair distribution of wealth, because some states possess petroleum whereas others do not. Unifying the region into one nation-state would encourage sharing the wealth generated from the sale of petroleum.

The most prominent exception to ethnic unity in the Middle East is Israel, where the majority of the population is Jewish. For those who support the creation of one unified Arab nationality, the presence of Israel and its Jewish population represents a major obstacle.

Attempts have been made to unify the Arab nationality. In 1958, Egypt and Syria formed the United Arab Republic, but 3 years later after a revolution Syria broke away and became independent again. During the 1970s, the Organization of Petroleum Exporting Countries (OPEC), dominated by Arab countries, met frequently to set common policies for pricing and production of petroleum (see Chapter 14).

Middle East unity was shattered in 1980 when Iraq attacked Iran to gain control of the Shatt-al-Arab waterway near the border between the two countries. The United States and Western Europe had little sympathy for either side. A 1979 revolution had brought to power in Iran strongly anti-American Shiite Muslim fundamentalists who held 50 Americans hostage in the U.S. Embassy for 77 days during 1979–80. For its part, Iraq employed chemical weapons during the war and built a nuclear reactor that Israeli planes destroyed in 1981 before it was operational. The Iran-Iraq war ended in a stalemate in 1988, after the death of 100,000 Iranians and 50,000 Iraqis.

In 1990, Iraq invaded Kuwait, claiming that the small oil-rich country belonged to it historically until taken away by the colonial powers earlier in the twentieth century. Iraq also launched missile attacks against Israel. Under the leadership of the Ba'ath Party, headed by Saddam Hussein, Iraq was committed to the creation of a pan-Arab nationality by force if necessary, as well as to the destruction of Israel.

The United Nations, led by the United States, launched Desert Storm in 1991 to expel Iraq from Kuwait. Arab states supported Desert Storm because Iraq's takeover of Kuwait was a threat to their own independence. Desert Storm was also supported as a way to protect Israel from Iraqi attacks. For many other countries, protection of petroleum reserves in Kuwait and Saudi Arabia was the most important consideration in opposing Iraq. Iraq was defeated, but its leader Saddam Hussein remained in power.

Internal Organization of States

In the face of increasing demands by ethnicities for more self-determination, states have restructured their governments to transfer some authority from the national government to local government units. An ethnicity that is not sufficiently numerous to gain control of the national government may be content with control of a regional or local unit of government.

Unitary and Federal States

The governments of states are organized according to one of two approaches: the unitary system or the federal system. The **unitary state** places most power in the hands of central government officials, whereas the **federal state** allocates strong power to units of local government within the country. A country's cultural and physical characteristics influence the evolution of its governmental system.

In principle, the unitary government system works best in nation-states characterized by few internal cultural differences and a strong sense of national unity. Because the unitary system requires effective communications with all regions of the country, smaller states are more likely to adopt it. Unitary states are especially common in Europe.

In reality, multinational states often have adopted unitary systems, so that the values of one nationality can be imposed on others. In a number of African countries, such as Ghana, Kenya, and Rwanda, for instance, the mechanisms of a unitary state have enabled one ethnic group to extend dominance over weaker groups. When Communist parties controlled the governments, most Eastern European countries had unitary systems, to promote diffusion of Communist values.

In a federal state, such as the United States, local governments possess more authority to adopt their own laws. Multinational states may adopt a federal system of government to empower different nationalities, especially if they live in separate regions of the country. Under a federal system, local government boundaries can be drawn to correspond with regions inhabited by different ethnicities.

The federal system is also more suitable for very large states because the national capital may be too remote to provide effective control over isolated regions. Most of the world's largest states are federal, including Russia (as well as the former Soviet Union), Canada, the United States, Brazil, and India. However, the size of the state is not always an accurate predictor of the form of government: tiny Belgium is a federal state (to accommodate the two main cultural groups, the Flemish and Waloons, as discussed in Chapter 5), whereas China is a unitary state (to promote Communist values).

Trend Toward Federal Government

In recent years, there has been a strong global trend toward federal government. Unitary systems have been sharply curtailed in a number of countries and scrapped altogether in others.

France: Curbing a Unitary Government. A good example of a nation-state, France has a long tradition of unitary government in which a very strong national government dominates local government decisions. Their basic local government unit is the *département* (department). Each of the 100 departments has an elected general council, but its administrative head is a powerful *préfet* appointed by the national government rather than directly elected by the people. Engineers, architects, planners, and other technical experts working in the department are actually employed by national government ministries.

A second tier of local government in France is the *commune*. Each of the 36,000 communes has a locally elected mayor and council, but the mayor can be a member of the national parliament at the same time. Further, the average commune has only 1,500 inhabitants, too small to govern effectively, with the possible exception of the largest ones, such as in Paris, Lyon, Lille, and Marseille.

During the 1980s, the French government granted additional legal powers to the departments and communes. Local governments could borrow money freely to finance new projects without explicit national government approval, formerly required. The national government agreed to give a block of funds to localities with no strings attached. In addition, 22 regional councils that previously held minimal authority were converted into full-fledged local government units, with elected councils and the power to levy taxes.

Poland: A New Federal Government. Poland switched from a unitary to a federal system after control of the national government was wrested from the Communists. The federal system was adopted to dismantle legal structures by which Communists had maintained unchallenged power for more than 40 years.

Under the Communists' unitary system, local governments held no legal authority. The national government appointed local officials and owned public property. This system led to deteriorated buildings, roads, and water sys-

tems, because the national government did not allocate sufficient funds to maintain property, and no one had clear responsibility for keeping property in good condition.

Poland's 1989 constitution called for a peaceful revolution: creation of 2,400 municipalities, to be headed by directly elected officials. To these new municipalities, the national government turned over ownership of housing, water supplies, transportation systems, and other publicly owned structures. For existing schools, each local authority decided case-by-case whether to operate the school, let the national government continue to run it, or hand it over to a private group, such as a church. Similarly, businesses owned by the national government, such as travel agencies, were either turned over to the municipalities or converted into private enterprises. Local authorities were allowed for the first time to levy income and property taxes; as in France, the national government also allocated funds for localities to use as they see fit.

The transition to a federal system of government proved difficult in Poland and other Eastern European countries. In May 1990, Poles elected 52,000 municipal councilors; given the absence of local government for a half-century, not one of these officials had experience in governing a community. The first task for many newly elected councilors was to attend a training course in how to govern.

To compound the problem of adopting a federal system, Poland's locally elected officials had to find thousands of qualified people to fill appointed positions, such as directors of education, public works, and planning. Municipalities had the option of hiring some of the 95,000 national government administrators who previously looked after local affairs under the unitary system. However, many of these former officials were rejected by the new local governments because of their close ties to the discredited Communist party. The national government was not allowed to intervene in local decisions on whether to retain or replace the former administrators.

Key Issue 4

Why Do States Cooperate With Each Other?

- Political and military cooperation
- Economic cooperation

The previous section illustrated examples of threats to the survival of states from the trend toward local diversity. The principal threat has been the desire of ethnicities for the right of self-determination as an expression of unique cultural identity. In a number of cases, the inability to accommodate the diverse aspirations of ethnicities has led to the breakup of states into smaller ones.

The future of the world's current collection of sovereign states is also threatened by the trend toward globalization. All but a handful of states have joined the United Nations, although it has limited authority. But states are willingly transferring authority to regional organizations, established primarily for economic cooperation.

Political and Military Cooperation

During the Cold War era (late 1940s until early 1990s) most states joined the United Nations, as well as regional organizations. These international and regional organizations were established primarily to prevent a third world war in the twentieth century and to protect countries from a foreign attack.

The United Nations

The most important international organization is the United Nations, created at the end of World War II by the victorious Allies. When established in 1945, the United Nations comprised 49 states, but by the early 1990s, membership had grown to 185, making it a truly global institution.

Switzerland and Taiwan are the most populous territories on Earth that are not in the United Nations. Fearing loss of sovereignty, Switzerland has traditionally avoided membership in most international organizations. Taiwan resigned when the United Nations voted to admit the People's Republic of China in 1971, because the government of Taiwan still considered itself the proper ruler of the Chinese mainland.

The number of countries in the United Nations has increased rapidly on three occasions: 1955, 1960, and the early 1990s. Sixteen countries joined in 1955, mostly European countries that had been liberated from Nazi Germany during World War II. Seventeen new members were added in 1960, all but one a former African colony of Britain or France. Twenty-six countries were added between 1990 and 1993, primarily from the breakup of the Soviet Union and Yugoslavia. U.N. membership also increased in the 1990s because of the admission of several microstates.

The United Nations was not the world's first attempt at international peacemaking. The U.N. replaced an earlier organization known as the League of Nations, which was established after World War I. The League was never an effective peace-keeping organization. The United States did not join, despite the fact that President Woodrow Wilson initiated the idea, because the U.S. Senate refused to ratify the membership treaty. By the 1930s, Germany, Italy, Japan, and the Soviet Union had all withdrawn, and the League could not stop aggression by these states against neighboring countries.

U.N. members can vote to establish a peace-keeping force and request states to contribute military forces. During the Cold War era, U.N. peace-keeping efforts were often stymied because any one of the five permanent members of the Security Council—China, France, Russia (formerly the Soviet Union), the United Kingdom, and the United States—could veto the operation. In the past, the United States and Soviet Union often used the veto to prevent undesired U.N. intervention. The major exception came in 1950, when the U.N. voted to send troops to support South Korea, after the Soviet Union's delegate walked out of a Security Council meeting.

During the 1990s, the United Nations has played an increasingly important role in trying to separate warring groups in a number of regions, especially in Eastern Europe, the Middle East, and sub-Saharan Africa. Because it must rely on individual countries to supply troops, the U.N. often lacks enough of them to keep peace effectively. The U.N. tries to maintain strict neutrality in separating warring factions, but this has proved difficult in places such as Bosnia-Herzegovina, where most of the world sees one ethnicity (Bosnian Serbs) as a stronger aggressor and another (Bosnian Muslims) as a weaker victim. Despite its shortcomings, though, the United Nations represents a forum where, for the first time in history, virtually all states of the world can meet and vote on issues without resorting to war.

United Nations soldiers overlook Sarajevo, the capital of Bosnia-Herzegovina. When the photo was taken in September 1994, Sarajevo had been without running water for 14 days, and residents had to carry water from a collection point. (Rikard Larma/AP/Wide World Photos)

Globalization and Local Diversity

Afghanistan: A Cold War Conflict Becomes a Local Civil War

In 1979, the Soviet Union invaded Afghanistan. At the time, the United States and its NATO allies viewed the invasion as one more event in the Cold War. Two decades later, war continues in Afghanistan, even though Soviet troops are no longer there, and the Soviet Union itself no longer exists. Afghanistan's war has been transformed from an outpost of the Cold War to a local struggle over cultural diversity.

Afghanistan has long been contested by foreign powers because of *where* the country is situated, between South Asia and Europe. The Khyber Pass, in eastern Afghanistan, is the major route through the rugged Hindu Kush mountain range, an extension of the Himalayas, which runs east-west across Afghanistan. The United Kingdom fought two wars in Afghanistan during the nineteenth century to keep the Russians from invading its South Asian colonies, and the British created the proruption in northeastern Afghanistan to prevent Russia from having a boundary with India.

The reason *why* the Soviet Union sent 115,000 troops to Afghanistan was in response to a request for help from its pro-Soviet government. Fundamentalist Muslims, known as *mujahedeen*, or *holy warriors*, had started a rebellion against the pro-Soviet government, which had come to power in Afghanistan a year earlier in a bloody coup.

During the Cold War era, when local conflicts were placed in the context of the *globalization* of the struggle between the two superpowers, the philosophy of the two sides was often "the enemy of my enemy is my friend." Consistent with that philosophy, the United States supplied arms and military equipment to the mujahedeen, not because the United States supported the principles of the mujahedeen, but simply because they were fighting against the Soviet Union.

After the Soviet invasion, more than 5 million Afghans fled to refugee camps set up in neighboring countries, including 3 million in Pakistan and 2 million in Iran. Because of a very high natural increase rate, the population in the refugee camps swelled to more than 6 million. Half returned home during the 1990s, but half remain in the camps. About 1.5 million Afghans died during the war.

Although heavily outnumbered by Soviet troops and possessing much less sophisticated equipment, the mujahedeen offset the Soviet advantage by waging a guerrilla war in the country's rugged mountains, where they were more comfortable than the Soviet troops and where Soviet air superiority was ineffective. Unable to subdue the mujahedeen, the Soviet Union withdrew its troops in 1989, and the Soviet-installed government in Afghanistan collapsed in 1992.

The end of the Cold War and superpower intervention caused fighting in Afghanistan to intensify rather than subside. Afghanistan's *local diversity* was no longer suppressed by Cold War rivalries. The most numerous ethnic group in Afghanistan were Pashtun (or Pathan), followed by Tajik, Hazara, Uzbek, Turkmen, and Baluchi.

Pashtun comprise one-third to one-half of Afghanistan's population, Tajik about one-fourth, and Hazara one-fifth. The precise percentages are unknown because the total population of Afghanistan, as well as the number in refugee camps, is unknown. Pashtun live primarily in the south and east, Tajik in the northeast and northwest, Hazara in the center, Uzbek in the north, Turkmen in the northwest, and Baluchi in the south (Figure 1).

Pashtun speak a language in the Iranian group of the Indo-Iranian branch of Indo-European. Hazara and some Tajik speak a dialect of Persian, also an Iranian language, as is Baluchi. Tajik, Turkmen, and Uzbek are Altaic languages.

Regional Military Alliances

In addition to joining the United Nations, many states joined regional military alliances after World War II. The division of the world into military alliances resulted from the emergence of two states as superpowers—the United States and the Soviet Union.

Era of Two Superpowers. During the Cold War era, the United States and the Soviet Union were the world's two superpowers. Before then, the world typically contained more than two superpowers. For example, during the Napoleonic Wars in the early 1800s, Europe boasted eight major powers: Austria, France, Great Britain, Poland, Prussia, Russia, Spain, and Sweden.

By the outbreak of World War I, eight great powers again existed. Germany, Italy, Japan, and the United States replaced Poland, Prussia, Spain, and Sweden on the list. By the late 1940s, most of the former great powers were beaten or battered by the two world wars, and only the United States and the Soviet Union remained as superpowers.

When a large number of states ranked as great powers of approximately equal strength, no single state could dominate. Instead, major powers joined together to form temporary alliances. A condition of roughly equal strength between opposing alliances is known as a **balance of power**.

Historically, the addition of one or two states to an alliance could tip the balance of power. The British in particular entered alliances to restore the balance of power and prevent any other state from becoming too strong. In contrast, the post–World War II balance of power was bipolar between the United States and the Soviet Union. Because the power of these two states was so much greater than all others, the world comprised

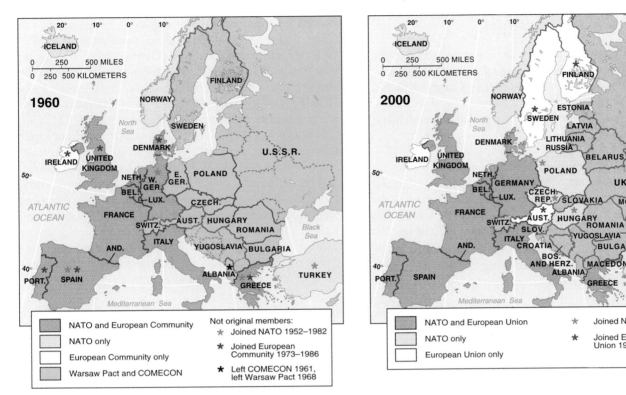

Figure 8–16 (left) Economic and military alliances in Europe during the Cold War. Western European countries joined the European Union and the North Atlantic Treaty Organization (NATO), whereas Eastern European countries joined COMECON and the Warsaw Pact. (right) Economic and military alliances in Europe, 1990s. COMECON and the Warsaw Pact have been disbanded, whereas the European Union and NATO have accepted or plan to accept new members.

expanded its membership in the late 1990s to include several former Warsaw Pact countries, including Czech Republic, Hungary, and Poland. Membership in NATO offers Eastern European countries an important sense of security against any future Russian threat, no matter how remote that appears at the moment (Figure 8–16, right).

The Conference on Security and Cooperation in Europe (CSCE), which had been founded in 1975 by Western European countries, was expanded to more than 50 countries, including the United States, Canada, and former republics of the Soviet Union. The CSCE had played a limited role during the Cold War era, but during the 1990s became a forum for all countries concerned with ending conflicts in Europe, especially in the Balkans and Caucasus. Although the CSCE does not directly command armed forces, it can call upon member states to supply troops if necessary.

Other Regional Organizations. The Organization of American States (OAS) includes 32 of the 33 states in the Western Hemisphere (except Canada). Cuba is a member but was suspended from most OAS activities in 1962. The organization's headquarters, including the permanent council and general assembly, are located in Washington, DC. The OAS promotes social, cultural, political, and economic links among member states.

A similar organization in Africa is the Organization for African Unity (OAU). Founded in 1963, OAU includes every African state except South Africa. The organization's major effort has been to eliminate minority white-ruled governments in southern Africa.

The Commonwealth of Nations includes the United Kingdom and 48 other states that were once British colonies, including Australia, Bangladesh, Canada, India, Nigeria, and Pakistan. Most other members are African states or island countries in the Caribbean or Pacific. Commonwealth members seek economic and cultural cooperation.

The Nonaligned Movement, founded in 1961, has approximately a hundred members, including nearly every country in Africa and the Middle East. The organization was established as a haven for countries that did not wish to be forced into an alliance with one of the superpowers. With the end of the Cold War, the initial purpose of the organization has disappeared. The principal focus now has become to represent the interests of less developed countries. Such a change reflects the growing importance of economic competition among blocs of countries rather than military conflict. The Nonaligned Movement may merge with the Group of 77, another organization of less developed countries established in the 1960s.

Economic Cooperation

The era of a bipolar balance of power formally ended when the Soviet Union was disbanded in 1992. Instead, the world has returned to the pattern of more than two superpowers that predominated before World War II.

But the contemporary pattern of global power displays two key differences:

1. The most important elements of state power are increasingly economic rather than military; Japan and Germany have joined the ranks of superpowers on their economic success, whereas Russia has slipped in strength because of economic problems.

2. The leading superpower in the 1990s is not a single state, such as the United States or Russia, but an economic union of European states led by Germany.

European Union

With the decline in the military-oriented alliances, European states increasingly have turned to economic cooperation. Western Europe's most important economic organization is the European Union (formerly known as the European Economic Community, the Common Market, and the European Community).

When it was established in 1958, the European Union included six countries: Belgium, France, West Germany, Italy, Luxembourg, and the Netherlands. Membership was widened to include Denmark, Ireland, and the United Kingdom in 1973, Greece in 1981, Portugal and Spain in 1986, and Austria, Finland, and Sweden in 1995 (Figure 8–16). A European Parliament is elected by the people in each of the member states simultaneously.

The main task of the European Union is to promote development within the member states through economic cooperation. At first, the European Union played a limited role, such as providing subsidies to farmers and to depressed regions like southern Italy. Most of the European Union's budget still goes to these purposes.

However, the European Union has taken on more importance in recent years, as member states seek greater economic and political cooperation. It has removed most barriers to free trade: with a few exceptions, goods, services, capital, and people can move freely through Europe. Trucks can carry goods across borders without stopping, and a bank can open branches in any member country with supervision only by the bank's home country. The introduction of the Euro in 1999 as the common currency in most European Union members will eliminate most of the remaining differences within Europe in prices, interest rates, and other economic policies. The effect of these actions has been to turn Western Europe into the world's wealthiest market.

Former Communist Countries and the European Union. In 1949, during the Cold War, the seven Eastern European Communist states in the Warsaw Pact formed an organization for economic cooperation, the Council for Mutual Economic Assistance (COMECON). Cuba, Mongolia, and Vietnam were also members of the alliance, which was designed to promote trade and sharing of natural resources. Like the Warsaw Pact, COMECON disbanded in the early 1990s after the fall of communism in Eastern Europe.

The former Communist Eastern European countries that have made the most progress in converting to market economies—Poland, Hungary, the Czech Republic, and the Baltic states—want to join the European Union. However, current European Union members are wary of admitting a large number of relatively poor Southern and Eastern European countries. Admission would create administrative nightmares—such as expanding the number of official languages—and dilute the economic benefits that current members enjoy.

German Domination in Western Europe. Although economic and political unity may have reduced the importance of nation-states in Western Europe, many Europeans, especially those old enough to remember World War II, fear that Germany has become more powerful than the region's other nation-states.

Germany is a newer nation-state than the others of Western Europe, for a state known as Germany was not created until 1871. Prior to that time, the map of the central European area now called Germany was just a patchwork of small states—more than 300 during the seventeenth century, for example.

Under Frederick the Great (1740–1786), the previously obscure state of Prussia gained control of a continuous stretch of territory abutting the Baltic Sea from Memel on the east to beyond the Elbe River on the west. Other consolidations reduced the number of states in the area to approximately two dozen by 1815. In 1871, Prussia's prime minister, Otto von Bismarck, was instrumental in forcing most of the remaining states in the area to join a Prussian-dominated German Empire, which extended westward beyond the Rhine River (Figure 8–17, upper left). Bismarck failed to consolidate all German speakers into the empire, as Austria, Switzerland, and Bohemia were excluded. The German Empire lasted less than 50 years.

Germany lost much of its territory after World War I (Figure 8–17, upper right). While the boundaries of states in Southern and Eastern Europe were fixed to conform when possible to those of ethnicities, Germany's new boundaries were arbitrary. Germany became a fragmented state, with East Prussia separated from the rest of the country by the Danzig Corridor, created to give Poland a port on the Baltic Sea. Nazi takeovers of Austria, Poland, and portions of Czechoslovakia during the 1930s were justified by the Germans as attempts to reconstruct a true German nation-state.

After Germany was defeated in World War II, the victorious Allies carved the country, and its capital city of Berlin, into four zones. Each zone was controlled by one

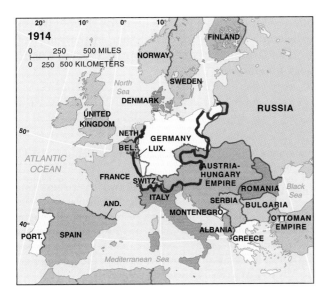

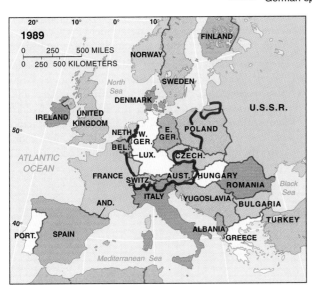

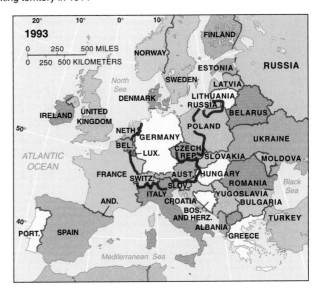

——— German-speaking territory in 1914

Figure 8–17 Europe's twentieth-century boundary changes. (Upper left) In 1914, at the outbreak of World War I, Germany extended 1,300 kilometers (800 miles) from east to west. Germany's boundaries at the time coincided fairly closely to the German-speaking area of Europe, although German was also spoken in portions of Switzerland and the Austria-Hungary Empire. (Upper right) After losing World War I, Germany was divided into two discontinuous areas, separated by the Danzig Corridor, part of the newly created state of Poland. (Lower left) Germany's boundaries changed again after World War II, as eastern portions of the country were taken by Poland and the Soviet Union. (Lower right) With the collapse of communism in Eastern Europe, East Germany and West Germany were united. Because of forced migration of Germans (as well as other peoples) after World War II, the territory occupied by German-speakers today is much farther west than the location a century ago.

of the victors: the United States, France, the United Kingdom, and the former Soviet Union. When sharp political differences at the start of the Cold War between the Soviet Union and the other three made reestablishment of a single Germany impossible, two new countries were created: East Germany (the German Democratic Republic) in the Soviet zone, and West Germany (the Federal Republic of Germany) in the other three zones (Figure 8–17, lower left).

Two Germanys existed from 1949 until 1990. East Germany was a Soviet ally and member of the Warsaw Pact and COMECON, and West Germany was a U.S. ally and member of NATO and the European Union. In 1990, at the fall of Communist governments in Eastern Europe, the German Democratic Republic ceased to exist, and its territory became part of the German Federal Republic (Figure 8–17, lower right).

As the most populous and economically strongest member of the European Union, Germany has taken the lead in setting the political agenda for a united Europe. When the European Union was founded, Germany was a quiet member, content to subsidize inefficient French farmers and impoverished southern Italians in exchange for acceptance as a respectable ally and reliable trading partner. A half–century later, Germany has succeeded through economic competition in achieving what previ-

ous generations failed to obtain through military means: to become the most powerful state in the midst of the world's largest market. This prospect worries its European neighbors.

■ Summary

Two political trends dominate the start of the twenty-first century. First, after a half-century dominated by the Cold War between two superpowers—the United States and the former Soviet Union—the world has entered a period characterized by an unprecedented increase in the number of new states created to satisfy the desire of nationalities for self-determination as an expression of cultural distinctiveness. Turmoil has resulted because in many cases the boundaries of the new states do not precisely match the territories occupied by distinct nationalities.

At the same time, with the end of the Cold War, military alliances have become less important than patterns of global and regional economic cooperation and competition among states. Economic cooperation has increased among neighboring states in Western Europe and North America, whereas competition among these two blocs, as well as Japan, has increased.

Here is a review of issues raised at the beginning of the chapter.

1. **Where are states located?** A state is a political unit, with an organized government and sovereignty, whereas a nation is a group of people with a strong sense of cultural unity. Most of Earth's surface is allocated to states, and only a handful of colonies and tracts of unorganized territory remain.

2. **Where are boundaries drawn between states?** Boundaries between states, where possible, are drawn to coincide either with physical features, such as mountains, deserts, and bodies of water, or with such cultural characteristics as geometry, religion, and language. Boundaries affect the shape of countries and affect the ability of a country to live peacefully with its neighbors.

3. **Why do boundaries between states cause problems?** Problems arise when the boundaries of states do not coincide with the boundaries of ethnicities. In some cases, one state contains more than one ethnicity, whereas in other cases one ethnicity is split among more than one state.

4. **Why do states cooperate with each other?** Following World War II, the United States and the Soviet Union, as the world's two superpowers, formed military alliances with other countries. With the end of the Cold War, nationalities now are cooperating with each other, especially in Western Europe, primarily to promote economic growth rather than to provide military protection.

Case Study Revisited

The Future of the Nation-State in Europe

The importance of the nation-state has diminished in Western Europe, the very world region most closely associated with development of the concept during the past 200 years. European nation-states have put aside their centuries-old rivalries to forge the world's most powerful economic union.

European economic integration has been pushed further in recent years with the introduction of a single currency, the Euro. Within a few years, France's franc, Germany's mark, and Italy's lira—powerful symbols of sovereign nation-states—will disappear. European leaders have bet that every country in the region will have a stronger economy if national currencies are replaced with the Euro.

Opposition to the Euro persists in Europe, especially in the United Kingdom. Efforts to block the Euro are especially strong in communities where inefficient companies have lost business to more efficient competitors based in other countries. National identity may still matter to Europeans who suffer eco-

nomic hardship after introduction of the Euro, but boundaries where hundreds of thousands of soldiers once stood guard now have little more economic significance than boundaries between states inside the United States.

Western Europeans may one day carry European Union rather than national passports, even though they won't need to show them when traveling within Western Europe. But they will still observe cultural differences when they cross borders. For example, highways in the Netherlands are more likely than those in neighboring Belgium to be flanked by well-manicured vegetation and paths reserved for bicycles.

The most fundamental obstacle to Western European integration is the multiplicity of languages. The European Union must spend a very high percentage of its annual budget translating documents and speeches into all nine of the community's official languages: Danish, Dutch, English, French, German, Greek, Italian, Portuguese, and Spanish. Businesses must figure out how to effectively advertise their products in several languages. English, understood by an increasing number of Europeans, may in the future become the principal language of business—the lingua franca—within the European Union, despite French opposition.

At the same time that residents of Western European countries are displaying increased tolerance for cultural values of neighboring nationalities, opposition has increased to the immigration of people from the south and east, especially those who have darker skins and adhere to Islam. Immigrants from poorer regions of Europe, Africa, and Asia fill low-paying jobs, such as cleaning streets and operating buses, that Western Europeans are not willing to perform. Nonetheless, many Western Europeans fear that large-scale immigration will transform their nation-states into multiethnic societies.

Underlying this fear of immigration is recognition that natural increase rates are higher in most African and Asian countries than in Western Europe as a result of higher crude birth rates. Many Western Europeans believe that Africans and Asians who immigrate to their countries will continue to maintain relatively high crude birth rates and consequently will constitute even higher percentages of the population in Western Europe in the future.

Even more troubling to Western Europeans is the prolonged war in the Balkans. Bosnia-Herzegovina lies only 250 kilometers (150 miles) from the borders of the European Union states of Austria and Italy. The barbaric practices of the combatants, such as ethnic cleansing, and the primitive conditions under which survivors must live, stand in stark contrast to the prosperity of European Union members.

For Europeans, Bosnia represents an uncomfortably nearby reminder that nationalities are still willing to throw away the prospect of economic prosperity through regional and global economic cooperation in order to preserve their individual cultural identities.

■ Key Terms

Balance of power Condition of roughly equal strength between opposing countries or alliances of countries.

Boundary Invisible line that marks the extent of a state's territory.

City-state A sovereign state that comprises a town and surrounding countryside.

Colonialism Attempt by one country to establish settlements and to impose its political, economic, and cultural principles in another territory.

Colony A territory that is legally tied to a sovereign state rather than completely independent.

Compact state A state in which the distance from the center to any boundary does not vary significantly.

Elongated state A state with a long, narrow shape.

Federal state An internal organization of a state that allocates most powers to units of local government.

Fragmented state A state that includes several discontinuous pieces of territory.

Frontier A zone separating two states in which neither state exercises political control.

Imperialism Control of territory already occupied and organized by an indigenous society.

Landlocked state A state that does not have a direct outlet to the sea.

Microstate A state that encompasses a very small land area.

Perforated state A state that completely surrounds another one.

Prorupted state An otherwise compact state with a large projecting extension.

Sovereignty Ability of a state to govern its territory free from control of its internal affairs by other states.

State An area organized into a political unit and ruled by an established government with control over its internal and foreign affairs.

Unitary state An internal organization of a state that places most power in the hands of central government officials.

■ Thinking Geographically

1. In his book *1984*, George Orwell envisioned the division of the world into three large unified states, held together through technological controls. To what extent has Orwell's vision of a global political arrangement been realized?

2. In the Winter 1992/93 issue of *Foreign Policy*, Gerald Helman and Steven Ratner identified countries that they called failed nation-states, including Cambodia, Liberia, Somalia, and Sudan, and others that they predicted would fail. Helman and Ratner argue that the governments of these countries were maintained in power during the Cold War era through massive military and economic aid from the United States or the Soviet Union. With the end of the Cold War, these failed nation-states have sunk into civil wars, fought among groups who share language, religion, and other cultural characteristics. What obligations do other countries have to restore order in failed nation-states?

3. Given the movement toward increased local government autonomy on the one hand and increased authority for international organizations on the other hand, what is the future of the nation-state? Have political and economic trends in the 1990s strengthened the concept of nation-state, or weakened it?

4. The world has been divided into a collection of countries on the basis of the principle that ethnicities have the right of self-determination. National identity, however, derives from economic interests as well as from such cultural characteristics as language and religion. To what extent should a country's ability to provide its citizens with food, jobs, economic security, and material wealth, rather than the principle of self-determination, become the basis for dividing the world into independent countries?

5. A century ago, the British geographer Halford J. Mackinder identified a heartland in the interior of Eurasia (Europe and Asia) that was isolated by mountain ranges and the Arctic Ocean. Surrounding the heartland was a series of fringe areas, which the geographer Nicholas Spykman later called the *rimland*, oriented toward the oceans. Mackinder argued that whoever controlled the heartland would control Eurasia and hence the entire world. To what extent has Mackinder's theory been validated during the twentieth century by the creation and then the dismantling of the Soviet Union?

■ Further Readings

Argenbright, Robert. "The Soviet Agitational Vehicle: State Power on the Social Frontier." *Political Geography* 17 (1998): 253–72.

Arlinghaus, Sandra L., and John D. Nystuen. "Geometry of Boundary Exchanges." *Geographical Review* 80 (1990): 21–31.

Brown, Curtis M.; Walter G. Robillard, and Donald A. Wilson. *Boundary Control and Legal Principles*. New York: John Wiley, 1986.

Burghart, A. F. "The Bases of Territorial Claims." *Geographical Review* 63 (1973): 225–45.

Burnett, Alan D., and Peter J. Taylor, eds. *Political Studies from Spatial Perspectives*. Chichester: John Wiley, 1981.

Busteed, M. A., ed. *Developments in Political Geography*. London: Academic Press, 1983.

Ching, Frank "Misreading Hong Kong." *Foreign Affairs*. 76 (1997): 53–66.

Christopher, A.J. *The British Empire at Its Zenith*. London: Croom Helm, 1988.

Cohen, Saul B. "Global Political Change in the Post–Cold War Era." *Annals of the Association of American Geographers* 81 (1991): 551–80.

———. "The World Geopolitical System in Retrospect and Prospect." *Journal of Geography* 89 (1990): 2–12.

Cope, Meghan. "Participation, Power, and Policy: Developing a Gender-Sensitive Political Geography." *Journal of Geography* 96 (1997): 91–97.

Cox, Kevin R. *Location and Public Problems: A Political Geography of the Contemporary World*. Chicago: Maaroufa Press, 1979.

Dale, E. H. "Some Geographical Aspects of African Land-Locked States." *Annals of the Association of American Geographers* 58 (1968): 485–505.

Davidson, Fiona M. "Integration and Disintegration: A Political Geography of the European Union." *Journal of Geography* 96 (1997): 69–75.

Dicken, Peter. "Transnational Corporations and Nation-states." *International Social Science Journal* 151 (1997): 77–90.

Dikshit, R. D. "Geography and Federalism." *Annals of the Association of American Geographers* 61 (1971): 97–130.

Drucker, Peter F. "The Global Economy and the Nation-State." *Foreign Affairs* 72 (1997): 159–71.

Gottmann, Jean, ed. *Centre and Periphery: Spatial Variation in Politics*. Beverly Hills, CA: Sage, 1980.

Grundy-Warr, Carl, and Richard N. Schofield. "Man-Made Lines That Divide the World." *Geographical Magazine* 62 (1990): 10–15.

Helman, Gerald B., and Steven R. Ratzner. "Saving Failed States." *Foreign Policy* 89 (1992): 3–20.

Johnston, R. J. *Geography and the State*. New York: St. Martin's Press, 1982.

———; Peter J. Taylor, and Michael Watts. *Geographies of Global Change*. Cambridge, MA: Blackwell, 1996.

Kidron, Michael, Ronald Segal, and Angela Wilson. *The State of the World Atlas: A Unique Visual Survey of Global Political, Economic and Social Trends*, 5th ed. New York: Penguin, 1995.

Kliot, Nurit, and Stanley Waterman, eds. *Pluralism and Political Geography—People, Territory and State*. New York: St. Martin's Press, 1983.

Kliot, N., and Y. Mansfield. "The Political Landscape of Partition: The Case of Cyprus." *Political Geography* 16 (1997): 495–521.

Mathieson, R. S. "Nuclear Power in the Soviet Bloc." *Annals of the Association of American Geographers* 70 (1980): 271–79.

Mathews, Jessica T. "Power Shift." *Foreign Affairs* 76 (1977): 50–66.

Mellor, Roy E.H. *Nation, State and Territory: A Political Geography*. London and New York: Routledge, 1989.

Merrett, Christopher D. "Research and Teaching in Political Geography: National Standards and the Resurgence of Geography's 'Wayward Child'." *Journal of Geography* 96 (1997): 50–54.

Morgenthau, Hans J. *Politics Among Nations*, 4th ed. New York: Knopf, 1967.

Murphy, Alexander B. "Historical Justifications for Territorial Claims." *Annals of the Association of American Geographers* 80 (1990): 531–48.

Newhouse, John. "Europe's Rising Regionalism." *Foreign Affairs* 76 (1977): 67–84.

Nijman, Jan. "The Limits of Superpower: The United States and the Soviet Union since World War II." *Annals of the Association of American Geographers* 82 (1992): 681–95.

———, et al. "The Political Geography of the Post–Cold War World." *Professional Geographer* 44 (1992): 1–29.

O'Loughlin, John, and Herman van der Wusten. "Political Geography of Panregions." *Geographical Review* 80 (1990): 1–20.

O'Sullivan, Patrick. *Geopolitics*. New York: St. Martin's Press, 1986.

_____, and Jesse W. Miller. *The Geography of Warfare*. London: Croom Helm, 1983.

Ó Tuathail, Gearóid. "Political Geography II: (Counter) Revolutionary Times." *Progress in Human Geography* 20 (1996): 404–12.

Pacione, Michael, ed. *Progress in Political Geography*. London: Croom Helm, 1985.

Parker, W. H. *Mackinder: Geography as an Aid to Statecraft*. Oxford: The Clarendon Press, 1982.

Pickles, John, and Jeff Woods. "South Africa's Homelands in the Age of Reform: The Case of QwaQwa." *Annals of the Association of American Geographers* 82 (1992): 629–52.

Prescott, J. R. V. *Boundaries and Frontiers*. London: Croom Helm, 1978.

_____. *The Geography of State Politics*. Chicago: Aldine Publishing, 1968.

_____. *Political Geography*. London: Methuen, 1972.

_____. *The Political Geography of the Oceans*. Newton Abbot, UK: David and Charles, 1975.

Taylor, Peter J. *Political Geography of the Twentieth Century: A Global Analysis*. London: Belhaven, 1993.

_____, and John W. House, eds. Political Geography: *Recent Advances and Future Directions*. London: Croom Helm, 1984.

Williams, Allan M. *The European Community: The Contradictions of Integration*. Cambridge, MA: Blackwell, 1991.

Zelinsky, Wilbur. *Nation into State*. Chapel Hill: University of North Carolina Press, 1988.

Also consult the following journals: *American Journal of Political Science*; *American Political Science Review*; *Foreign Affairs*; *Foreign Policy*; *International Affairs*; *International Journal*; *International Journal of Middle East Studies*; *Political Geography*; *Post-Soviet Geography*.

CHAPTER 9

Development

Have you ever traveled to a Caribbean island? Even if you haven't, you have probably seen advertisements for resorts featuring a bronzed couple sipping exotic drinks, lying on a deserted beach surrounded by palm trees.

Beyond this paradise is another world, fleetingly glimpsed by tourists traveling between the resort and the airport. The permanent residents of the islands may live in poverty, earning less money in a year than a night's hotel bill. They are ill-fed, ill-clothed, and underemployed.

This depressing view of conditions on the islands is shielded from tourists, of course. They do not travel hundreds of kilometers to encounter misery on their vacation or honeymoon. Tourists bring money to the islands and in the process help pay for whatever improvements can be made to the squalid living conditions.

But can you imagine the feelings of the local residents? What would you think if a very expensive and exclusive resort were built in your neighborhood, and you and your family, who were economically disadvantaged, were expected to work there (for good wages, perhaps) to serve the needs of the vacationers? You might welcome the money, but would you resent the wealthy tourists?

The world is divided between relatively rich and relatively poor countries. Geographers try to understand the reasons for this division and learn what can be done about it.

Key Issues

1. Where are more and less developed countries distributed?

2. Why does development vary among countries?

3. Why do less developed countries face obstacles to development?

Unpaved road near Kumasi, Ghana, an obstacle to development. (James Strachan/Robert Harding Picture Library)

303

Case Study

Bangladesh's Development Problems

Rabea Rahman lives in the village of Bathoimuri, Bangladesh, with her three children—a son, 18, and two daughters, ages 10 and 7. Rahman's two other children died in infancy. Her husband died of tuberculosis.

Rahman's husband was a tenant farmer, or sharecropper. Under this arrangement, he shared a portion of his crops with the landowner, instead of paying rent. After he died, Rahman went to work as a domestic servant and water carrier, working from 7 A.M. to 4 P.M. and from 6 P.M. to 11 P.M., seven days a week. Her son sells bread and prepares a mid-day meal for his two sisters. Total household income is $16 per month (compared to a monthly household average of more than $3,000 in the United States).

Their house has a dirt floor and leaky roof, but the rent is only $2 per month, plus $3 per month for fuel. The remaining $11 a month goes for food. The sum is sufficient to provide each member of the household with 100 grams (about a quarter pound) of rice per day, but little else. The diet is supplemented by leftover food Rahman receives from her employer. After paying for rent, fuel, and food, the family has no money left for other necessities. Because they cannot afford shoes, the family members often go barefoot. Rahman suffers from a gastric ulcer but cannot afford treatment.

Underlying the impoverished condition of the Rahman household is the role of women in a predominantly Muslim country like Bangladesh. In rural villages, fewer than 10 percent of the women can read and write. Typically, a woman is married as a teenager and bears six babies in her lifetime, although typically one of the six does not survive infancy. A woman like Rahman, who is forced to find a job, is limited to working as a servant or farm laborer. The condition of women—poor, illiterate, overburdened with children—is one of the most important factors holding back economic development in South Asian countries like Bangladesh.

In previous chapters, we examined global demographic and cultural patterns. We saw that birth, death, and natural increase rates vary among regions of the world. People in different regions also have different social customs, languages, religions, and ethnic identities. We saw how political problems arise when the distribution of cultural characteristics does not match the boundaries between states. The political geography chapter concluded that in the contemporary world, global military confrontation and alliances have been replaced by global economic competition and cooperation.

The second half of the book concentrates on economic rather than cultural elements of human geography. This chapter examines the most fundamental global economic pattern—the division of the world into relatively wealthy regions and relatively poor ones. Subsequent chapters look at the three basic ways that humans earn their livings—growing food, manufacturing objects, and providing services.

Earth's nearly 200 countries can be classified according to their level of **development**, which is the process of improving the material conditions of people through diffusion of knowledge and technology. The development process is continuous, involving never-ending actions to constantly improve the health and prosperity of the people. Every country lies at some point along a continuum of development.

Because many countries cluster at the high or low end of the continuum of development, they can be divided into two groups. A **more developed country** (abbreviated **MDC**, also known as a **relatively developed country** or simply a **developed country**) has progressed further along the development continuum. A country in an earlier stage of development is frequently called a **less developed country** (**LDC**), although many analysts prefer the term **developing country**. *Developing* implies that the country has already made some progress and expects to continue.

The first geographic task is to identify *where* more developed and less developed countries are located. Geographers observe that more developed countries cluster in a handful of regions of the world, and less developed countries cluster in other regions. Next, geographers are concerned with *why* some regions are more developed than others. A number of economic, social, and demographic indicators distinguish more and less developed regions.

For more developed regions, the economic challenge is to maintain a high level of development in a *global economy*. For less developed countries, the challenge is to improve the level of development by finding a role in a global economy that takes advantage of *local diversity* in skills and resources.

Key Issue 1

Where Are More and Less Developed Countries Distributed?

- More developed regions
- Less developed regions

The distribution of more and less developed countries reflects a clear global pattern. If we draw a circle around the world at about 30° north latitude, we find that nearly all of the more developed countries are situated to the north, whereas nearly all of the less developed countries lie south of the circle. This division of the world between more and less developed and developing countries is known as the *north-south split*.

The north-south split between more and less developed countries shows up clearly in maps of measures of development, such as the Human Development Index (HDI) created by the United Nations (Figure 9–1). More developed countries in the north have relatively high HDIs; southern countries have lower indexes.

The HDI recognizes that a country's level of development is a function of three factors: economic, social, and demographic. The United Nations selects one economic factor, two social factors, and one demographic factor that in the opinion of an international team of analysts best reveal a country's level of development. The economic factor is gross domestic product per capita; the social factors are the literacy rate and amount of education; and the demographic factor is life expectancy. The four factors are combined to produce a country's HDI. The U.N. has computed HDIs for countries every year since 1990, although it has tinkered a few times with the method of computation. The highest HDI possible is 1.0, or 100 percent.

In 1997, the highest ranking country was Canada, with an HDI of 0.960, or 96.0 percent. Some years, Japan has a higher HDI than Canada. The United States has never ranked first, although it invariably falls within the top half dozen. The other highest ranking countries are typically in Western Europe. The lowest ranked country in 1997 was Sierra Leone, with an HDI of 0.176, or 17.6 percent. The 19 lowest-ranking countries in 1997 were all in sub-Saharan Africa.

The countries of the world can be categorized into nine major regions according to their level of development. These regions also have distinctive demographic and cultural characteristics that have been discussed in earlier chapters (Figure 9–2). Subsequent chapters will show that the nine major regions also differ from each other in how people earn their living, how the societies use their wealth, and other economic characteristics. As we move toward a global economy, geographers increasingly study the similarities and differences in the economic patterns of the various regions.

In the Western Hemisphere, two regions—Anglo-America (Canada and the United States) and Latin America—can be distinguished on the basis of dominant languages, religions, and natural increase rates. Despite the considerable diversity within these regions, at a global scale the individual countries within these regions display cultural similarities.

Europe can be divided into two regions, Western and Eastern. Although they share many cultural traditions, distinctive political experiences have produced different levels of economic development.

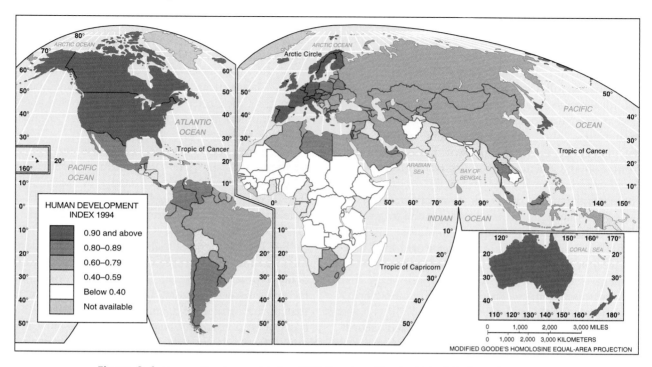

Figure 9–1 Human Development Index (HDI). Developed by the United Nations, the HDI combines several measures of development: life expectancy at birth, adjusted GDP per capita, and knowledge (schooling and literacy). Each country received an index figure for the various measures, which range between minimum and desirable levels. The minimum for each index was set at the lowest level actually observed. The desirable levels were 100 percent for literacy and the maximum observed for life expectancy and mean years of schooling.

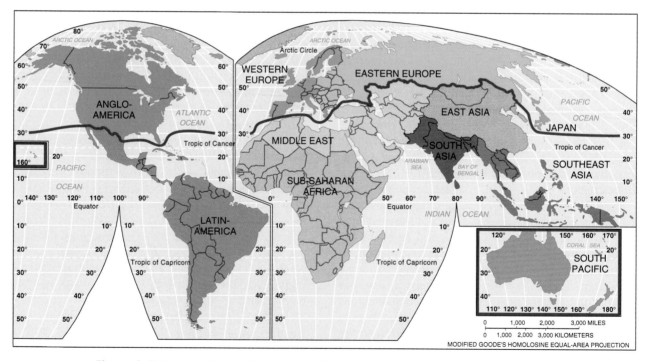

Figure 9–2 More and less developed regions. Earth's six less developed regions are Latin America, Southeast Asia, Middle East, East Asia, South Asia, and sub-Saharan Africa. The world's more developed regions are Anglo-America, Western Europe, and Eastern Europe, plus Japan and the South Pacific.

Asia comprises four major cultural regions: East, South, Southeast, and Southwest. Demographic, religious, linguistic, ethnic, and political characteristics distinguish these four regions. Because of similarities in language, religion, and population growth, Southwest Asia can be combined with North Africa to form the Middle East region. Africa south of the Sahara comprises the ninth major region.

In addition to those nine major regions, two other important areas can be identified: Japan and the South Pacific. Japan is a populous country with cultural and demographic characteristics that contrast sharply with neighboring states in East Asia. The South Pacific, primarily Australia and New Zealand, covers an extensive area of Earth's surface, but is much less populous than the nine major regions.

More Developed Regions

Three of the nine major cultural regions—Anglo-America, Western Europe, and Eastern Europe, plus Japan and the South Pacific—are considered more developed. The other six regions are considered less developed. This section examines the more developed regions.

Anglo-America

HDI 0.95

Language and religious patterns are less diverse in Anglo-America than in other world regions. About 95 percent of the region's people use English as their first language, and about 95 percent are Christian (excluding those with no religion). Cultural diversity generates some tensions in the region, including discrimination against ethnic minorities, intolerance of other Christian sects as well as non-Christian faiths, and uncertain status of French-speaking Québécois. However, Anglo-America's relative homogeneity reduces the possibility that a large minority will be excluded from participating in the region's economy on the basis of cultural characteristics.

Well endowed with minerals and natural resources important for manufacturing, Anglo-America was once the world's major producer of steel, automobiles, and other goods, but in the past quarter-century Japan, Western Europe, and less developed countries have eroded the region's dominance. Americans remain the leading consumers and world's largest market for many of these products.

Despite the loss of manufacturing jobs, the region has adapted relatively successfully to the global economy, in part because it is the leading provider of many computing, information, and other high-tech services, as well as entertainment, mass media, sports, recreation equipment, and other services that promote use of leisure time. In addition, Anglo-America is the world's most important food exporter and the only region that could significantly expand the amount of land devoted to agriculture.

The United States is the world's major producer of television programs and movies. For the movie *The Peacemaker*, George Clooney and Nicole Kidman prepare to shoot a scene on the streets of New York. (Gamma/Liaison, Inc.)

Few Americans are farmers, but a large percentage of the region's workforce is engaged in some aspect of producing or serving food.

Western Europe

HDI 0.93

On a global scale, Western Europe displays cultural unity, because nearly all Western Europeans speak an Indo-European language and practice Christianity. However, the diversity of individual languages and religious practices has been a long-time source of conflict in Western Europe, especially when strong national identities were forged out of distinctive ethnic traditions and historical experiences.

Competition among Western European nationalities caused many wars, most notably the two world wars fought in the twentieth century. Since the end of World War II in 1945, Western Europe has become much more unified politically, militarily, economically, and culturally. Offsetting the increased cultural unity within Western Europe is greater diversity through migration of Muslims and Hindus from less developed countries in search of jobs. With natural increase rates at or below zero in most Western European countries, immigrants are responsible for much of the region's population growth, and they have become scapegoats for the region's economic problems according to many Europeans.

Within Western Europe, the level of development is the world's highest in a core area that includes western Germany, northeastern France, northern Italy, Switzerland, southern Scandinavia, southeastern United Kingdom, Belgium, the Netherlands, and Luxembourg. Because the region's peripheral areas—southern Italy, Portugal, Spain, and Greece—lag somewhat in development, Western Europe as a whole has a slightly lower development level than Anglo-America.

To maintain its high level of development, Western Europe must import food, energy, and minerals. In past centuries, Western Europeans explored and mapped the rest of the world and established colonies on every continent. These colonies supplied many resources needed to foster European economic development. Colonization also diffused Western European languages, religions, and social customs worldwide.

Now that most colonies have been granted independence, Western Europeans must buy raw materials from other countries. To pay for their imports, Western Europeans provide high-value goods and services, such as insurance, banking, and luxury motor vehicles, such as the Mercedes-Benz and Rolls-Royce.

The elimination of most economic barriers within the European Union makes Western Europe potentially the world's largest and richest market. Restructuring of the region's economy has lagged behind Anglo-America, in part because most governments have been willing to sacrifice some economic growth in exchange for protection of existing jobs and social services.

Western European manufacturers provide luxury goods, such as Rolls-Royce cars, and shops selling luxury goods cluster along streets such as Bond Street in London. (William Strode/Woodfin Camp & Associates)

Eastern Europe

HDI 0.78 Eastern Europe has the dubious distinction of being the only region where the HDI has declined significantly since the United Nations created the index in 1990. In 1990, Eastern Europe clearly ranked among the world's more developed regions, and it had an HDI only slightly behind those of Western Europe and Anglo-America. A decade later, Eastern Europe has an HDI even lower than Latin America, which is classified as a less developed region.

Eastern Europe's rapidly declining HDI is a legacy of the region's history of Communist rule. Winston Churchill declared in a 1946 speech that an "Iron Curtain" had descended across Europe, from the Baltic Sea (near Germany) in the north to the Adriatic Sea (east of Italy) in the south. This became the dividing line between Western and Eastern Europe. Eastward of 15° east longitude, the Soviet Union during the late 1940s imposed or inspired Communist governments in Albania, Bulgaria, Czechoslovakia, East Germany (the German Democratic Republic), Hungary, Poland, Romania, and Yugoslavia.

Under the Communists, the national governments exercised strong control over development (see Globalization and Local Diversity box). The system turned out large supplies of weapons and tractors but neglected production of clothing, televisions, and other products that individual consumers desire.

During the 1990s, Eastern European countries dismantled the economic structure inherited from the Communists. The Czech Republic, Hungary, and Slovenia have converted more rapidly to market economies, taking advantage of their proximity to the relatively developed core region of Western Europe. Because workers in these countries are comparably skilled yet much lower-paid compared to their counterparts in Western Europe, some manufactured goods are being exported to wealthier countries in the West. As memories of the Communist era fade, these countries will display social and economic characteristics similar to such Western European countries as Greece, Ireland, and Portugal.

On the other hand, restructuring to market economies has proved painful in Russia and a number of other Eastern European countries. Closing inefficient businesses has increased unemployment, and prices for many goods skyrocketed with the elimination of government subsidies. Most Russians have suffered declining standards of living since the end of communism, while a handful—including gangsters—have become very rich.

The dismantling of the Communist system led to the breakup of Czechoslovakia, the Soviet Union, and Yugoslavia. In Czechoslovakia, Czechs were willing to bear a short-term decline in their standard of living, because they believed that rapid conversion to a market economy would bring long-term benefits. Slovaks

Eastern Europe: An "Un-Developing" Region

Eastern Europe is the only example of a region **where** the development process is moving backwards. A decade ago, Eastern Europe clearly ranked as one of the world's more developed regions. By most social and demographic indicators, the region is still highly developed. It has high levels of literacy and low rates of birth and natural increase. But economic conditions have deteriorated so rapidly in Eastern Europe during the past decade that it is hard to justify including it with the other more developed regions.

The reason **why** the development process is running backwards in Eastern Europe is a legacy of the region's Communist rule. When Communist parties gained control of Russia in 1917 (the Bolshevik Revolution) and other Eastern European countries after World War II, they achieved rapid development, especially during the 1950s and 1960s. Annual per capita GDPs increased from a few hundred dollars to several thousand, and most social and demographic indicators became comparable to Western European countries.

Early Communist theorists, such as Karl Marx and Friedrich Engels, believed that communism would triumph in more developed countries because exploited factory workers would lead a revolution and overthrow their governments. The social and economic programs of these theorists were based on conditions in advanced industrial societies. Because few of these states had modern industries (Czechoslovakia, East Germany, and Poland were exceptions), the Communists had to figure out how to apply their theories to those of poor, agricultural societies.

The Communists promoted development during the 1950s and 1960s through **globalization** on a scale far greater than in the contemporary world. As centrally planned societies, Eastern European countries typically had economies directed by government officials rather than private entrepreneurs. In the Soviet Union, for example, a national planning commission called *Gosplan* developed five-year plans to guide economic development.

The plans prescribed production goals for the entire country by economic sector and region. They specified the type and quantity of minerals, manufactured goods, and agricultural commodities to be produced, and the factories, railways, roads, canals, and houses to be built in each part of the country.

The five-year plans featured three main development policies. First, Soviet planners emphasized heavy industry—iron and steel, machine tools, petrochemicals, mining equipment, locomotives, and armaments. To allow industrial growth, the country also promoted development of mining, electric power, and transportation.

Second, the plans dispersed production facilities from the European to the Asian portion of the Soviet Union. Soviet decision-makers considered the concentration of industry in the west to be a liability, and with cause: the country had been invaded from the west by the French under Napoleon Bonaparte in the nineteenth century and the Germans under Adolf Hitler in the twentieth century,

and they wanted to reduce the vulnerability of their vital industries to attack. Planners also wished to promote equal development throughout the country and believed that dispersal of industries would accomplish this goal.

Third, Soviet planners preferred to locate manufacturing facilities near sources of raw materials, rather than near markets. This policy reflected both the needs of industries emphasized in Soviet plans and the lack of effective consumer demand. By locating heavy industry near the raw materials, Soviet planners gave lower priority to producing consumer goods, such as telephones, washing machines, shoes, and dishes.

During the late 1980s and early 1990s, Communist parties lost power throughout Eastern Europe, and the national governments now exercise less control over the economies. Aside from the desire for freedom, the principal reason that Eastern Europeans rejected communism was that central planning proved to be disastrous at running national economies:

- Scarce funds were used to meet annual production targets rather than to invest in long-term improvements in productivity, such as modernizing equipment and redeploying workers to other tasks.

- Despite an abundant supply of productive farmland, Eastern Europe had to import food from the West because of inefficient agricultural practices.

- Orders sent from national government offices hundreds of kilometers away were often not implemented in the factories.

- Some targets were impossible to achieve. Others were simply ignored: Why work hard when your job is guaranteed and your supervisor cannot fire you?

- Factories polluted the air and water, and citizens were unable to pressure their governments into investing in pollution-control devices.

But for many Eastern Europeans, the most fundamental problem was that, by concentrating on basic industry, the Communists neglected consumer products like automobiles, refrigerators, and clothing. Severe housing shortages forced entire families to live in dwellings the size of a college dormitory room. Although restricted from visiting Western countries, many Eastern Europeans could see on television the much higher level of comfort on the other side of the Iron Curtain.

To convert from communism to Western-style market-oriented economies, Eastern Europeans have concluded that the essential priority is to replace central planning with **local diversity** in decision making. Consequently, individuals and corporations have been given the government-owned houses, shops, and factories, so that they can make their own investment decisions.

wanted to slow the pace of change; they feared high levels of unemployment in the large, inefficient factories that the Communists had clustered there to promote economic development during the 1950s.

Similarly, the Soviet Union and Yugoslavia broke up in part because republics such as Russia and Slovenia preferred more rapid economic change than did Belarus and Serbia. However, the end of communism in these two countries also unleashed long-suppressed friction among ethnicities. As a multiethnic state, Russia is especially vulnerable to further unrest among ethnic minorities suffering from the conversion to a market economy.

The region's HDI may have declined because of production cutbacks, higher death rates, and other stresses associated with the end of communism. Alternatively, higher mortality and lower wealth may be because the Communists inflated statistics when they were in power.

Because of Eastern Europe's tradition of economic development, the region is classified here as more developed. But the low HDI shows the region's distinctive history, as well as difficulties in comparing levels of development among regions.

Japan

HDI 0.94

Anglo-America and Western Europe share many cultural characteristics. Anglo-America was colonized by European immigrants, so the regions share language, religion, and other political, economic, and cultural traditions. From the perspective of LDCs, the economic influence wielded by these two regions is

With by far the highest per capita income in Asia, Japan is one of the world's three major centers for retail sales, along with Western Europe and North America. Shops selling luxury goods cluster in central Tokyo. (Alan G. Dejecacion/Gamma-Liaison)

closely intertwined with the global influence of European and American culture. Japan, the third major center of development, has a different cultural tradition.

Japan's development is especially remarkable because it has an extremely unfavorable ratio of population to resources. The country has some of the world's most intensively farmed land and one of the highest physiological densities (refer to Table 2–1). The Japanese consume relatively little meat and grain other than rice but still must import these products. Japan also lacks many key raw materials for basic industry. For example, although Japan is one of the world's leading steel producers, it must import virtually all the coal and iron ore needed for steel production.

How has Japan become such a great industrial power? At first, the Japanese economy developed by taking advantage of the country's one asset, an abundant supply of people willing to work hard for low wages. The Japanese government encouraged manufacturers to sell their products in other countries at prices lower than domestic competitors. Having gained a foothold in the global economy by selling low-cost products, Japan then began to specialize in high-quality, high-value products, such as electronics, motor vehicles, and cameras.

Japan's dominance was achieved in part by concentrating resources in rigorous educational systems and training programs to create a skilled labor force. Japanese companies spend twice as much as U.S. firms on research and development, and the government provides further assistance to develop new products and manufacturing processes.

Under communism, factories such as this Russian auto parts manufacturer were inefficient and unresponsive to market demands. Faced with the need to make a profit as well as compete against more modern plants in Western Europe, Japan, and North America, many Eastern European plants closed. A large percentage of factory workers in Russia are women, in part because of the Communist tradition of encouraging female participation in the labor force. (Bill Swersey/Gamma-Liaison, Inc.)

South Pacific

HDI 0.93

The South Pacific has a relatively high HDI, but is much less central to the global economy because of its small number of inhabitants and peripheral location. The HDIs of Australia and New Zealand are comparable to those of other MDCs, whereas Papua New Guinea is less developed. The area's remaining people are scattered among sparsely inhabited islands that generally are less developed.

As former British colonies, Australia and New Zealand share many cultural characteristics with the United Kingdom. Over 90 percent of the residents are descendants of nineteenth-century British settlers, although indigenous populations remain. Australia and New Zealand are net exporters of food and other resources, especially to the United Kingdom. Increasingly, their economies are tied to Japan and other Asian countries.

Less Developed Regions

Six regions are classified as less developed. The following section briefly describes these six regions in descending order of development level.

The level of development varies widely among the six regions. Latin America has the highest HDI among the regions, exceeding Eastern Europe, where the HDI has been falling in recent years. Southeast Asia, the Middle East, and East Asia have virtually identical HDIs, well behind those of Latin America and Eastern Europe. South Asia has a much lower HDI than do the other Asian regions, and sub-Saharan Africa lags behind all other regions.

Latin America

HDI 0.80

Most Latin Americans speak one of two Romance languages, Spanish or Portuguese, and adhere to Roman Catholicism. These cultural characteristics resulted from the fact that Brazil was a colony of Portugal and most of the remaining states once belonged to Spain. In reality, the region is culturally diverse. A large percentage of the population are descendants of inhabitants living in the region prior to the European conquest, while others trace their ancestors to African slaves.

Latin Americans are more likely to live in urban areas than people in other developing regions. Mexico City, São Paulo, and Buenos Aires rank among the world's ten largest, according to the United Nations. The region's population is highly concentrated along the Atlantic Coast, while population density remains low in most of the region, especially the tropical interior of South America. Large areas of interior rain forest are being destroyed to sell the timber or to clear the land for settled agriculture.

The level of development is relatively high along the South Atlantic Coast from Curitiba, Brazil, to Buenos Aires, Argentina. This area enjoys high agricultural productivity and ranks among the world's leaders in production and export of wheat and corn (maize). Mexico's development has been aided by proximity to the United States. Development is lower in Central America, several Caribbean islands, and the interior of South America.

Overall development in Latin America is hindered by inequitable income distribution. In many countries, a handful of wealthy families control much of the land and rent parcels to individual farmers. Many tenant farmers grow coffee, tea, and fruits for export to relatively developed countries rather than food for domestic consumption. Latin American governments encourage redistribution of land to peasants but do not wish to alienate the large property owners, who generate much of the national wealth.

Southeast Asia

HDI 0.67

Southeast Asia's most populous country, Indonesia, includes 13,667 islands. Nearly two-thirds of the population live on the island of Java, which has one of the world's highest arithmetic densities. People have concentrated on Java partly because the island's soil, derived from volcanic ash, is more fertile than elsewhere in the region and partly because the Dutch established their colonial headquarters there.

Other than Indonesia, Southeast Asia's most populous countries are Vietnam and Thailand on the Asian mainland and the Philippines, like Indonesia situated on a series of islands. The region has suffered from a half-century of nearly continuous warfare. Japan, the Netherlands, France, and the United Kingdom were all forced to withdraw from colonies they had established in the region. In addition, France and the United States both fought unsuccessfully to prevent Communists from controlling Vietnam during the Vietnam War, which ran from the 1950s to 1975. Wars have also devastated neighboring Laos and Cambodia.

The region's tropical climate limits intensive cultivation of most grains. The heat is nearly continuous, the rainfall abundant, and the vegetation dense. Soils are generally poor, because the heat and humidity rapidly destroy nutrients when land is cleared for cultivation. Economic development is also limited in Southeast Asia by several mountain ranges, active volcanoes, and frequent typhoons.

This inhospitable environment traditionally kept population growth low in Southeast Asia. But the injection of Western medicine and technology resulted in one of the most rapid rates of increase in the world during the second half of the twentieth century.

Rice, the region's most important food, is exported in large quantities from some countries, such as Thailand and Vietnam, but must be imported to other countries in the region, such as Malaysia and the Philippines. Because of distinctive vegetation and climate, farmers in Southeast Asia concentrate on harvesting products that are used in manufacturing. The region produces a large percentage of the world's supply of palm oil and copra (coconut oil), natural rubber, kapok (fibers from the ceiba tree used for insulation and filling), and abaca (fibers from banana leafstalks used in fabrics and ropes). Southeast Asia also contains a large percentage of the world's tin as well as some petroleum reserves.

Development has been rapid in some Southeast Asian countries, notably Thailand, Singapore, Malaysia, and the Philippines. The region has become a major manufacturer of textiles and clothing, taking advantage of cheap labor. Thailand has become the region's center for manufacturing of automobiles and other consumer goods.

Southeast Asia's HDI is likely to decline because of economic difficulties in the late 1990s. Earlier economic growth had been achieved through very close cooperation among manufacturers, financial institutions, and government agencies. In the absence of independent watchdogs and regulators, funds for development were sometimes invested unwisely or stolen by corrupt officials. To restore economic confidence among international investors, Southeast Asian countries have been forced to undertake painful reforms that reduce the people's standard of living.

Middle East

HDI 0.66

Much of the Middle East is desert that can sustain only sparse concentrations of plant and animal life. Most products must be imported. However, the region possesses one major economic asset: a large percentage of the world's petroleum reserves.

Because of petroleum exports, the Middle East is the only one of the nine major world regions that enjoys a trade surplus. In every other major region, the value of imports exceeds exports. To a considerable extent, this is because countries in these other regions must purchase large quantities of petroleum from Middle Eastern states.

Government officials in Middle Eastern states, such as Saudi Arabia and the United Arab Emirates, have used the billions of dollars generated from petroleum sales to finance economic development. The Middle East is the only region in which development is not hindered by lack of capital for new construction. To the contrary, many governments in the region have access to more money than they can use to finance development.

However, not every country in the region has abundant petroleum reserves. Most are concentrated in states that border the Persian Gulf. Development possibilities are limited in countries that lack significant petroleum reserves: Egypt, Jordan, Syria, and others (refer to Figure 14–3 for a map of petroleum production and reserves).

The large gap in per capita income between the petroleum-rich countries and those that lack resources causes

Muslim women at a keyboard. Exposure to modern technology does not necessarily destroy traditional culture. Women in predominantly Muslim countries have been urged to wear the chador, a combination head covering and veil, as a sign of adherence to traditional Islamic religious principles. (Esaias Baitel/Gamma-Liaison, Inc.)

great tension in the Middle East. People in poorer states held little sympathy for wealthy Kuwait when Iraq invaded it in 1990, triggering the Gulf War. Kuwait was accused of not sharing its petroleum-generated wealth and failing to provide good living conditions for guest workers from poorer Arab countries.

The challenge for many Middle Eastern states is to promote development without abandoning the traditional cultural values of Islam, the religion of more than 95 percent of the region's population. Many Middle Eastern countries sharply restrict the role of women in business. They also prevent diffusion of financial practices that are considered incompatible with Islamic principles. The low level of literacy among women is the main reason the United Nations considers the development among these petroleum-rich states to be lower than the region's wealth would indicate.

The region also suffers from serious internal cultural disputes, as discussed in Chapters 6 and 7. Iraq's long war with Iran and attempted annexation of Kuwait split the Arab world. Countries dominated by Shiite Muslims, especially Iran, have promoted revolutions elsewhere in the region to sweep away elements of development and social customs they perceive to be influenced by Europe or Anglo-America.

Most Middle Eastern states have refused to recognize the existence of Israel, the region's only state controlled by Jews. Israel has successfully repelled several attacks by neighboring states and, since 1967, has occupied territory captured from its adversaries. Money that could be used to promote development is diverted to military funding and rebuilding war-damaged structures.

East Asia

HDI 0.63 China, the largest country in East Asia, ranks among the world's poorest. However, this does not accurately portray the region's potential for development. Within a few years, China is projected to exceed the United States as the world's largest economy, although the U.S. economy would still be much larger on a per capita basis

Traditionally, most Chinese farmers were forced to pay high rents and turn over a percentage of their crops to a property owner. Farmers in a typical year produced enough food to survive but frequently suffered from famines, epidemics, floods, and other disasters. Exploitation of the country's resources by Europe and Japan further retarded China's development.

China's watershed year was 1949, when the Communist party won a civil war and created the People's Republic of China. The old Nationalist government fled to the island of Taiwan, setting up a government in exile. Since then, dramatic changes have been made in China's economy.

To ensure the production and distribution of enough food, the Communist government took control of most agricultural land. In some villages, officials assigned specific tasks to each farmer, distributed food to each family according to individual needs, and sold any remaining food to urban residents. In other cases, farmers rented land from the local government, received orders to grow specific amounts of particular crops, and sold for their own profit any crops above the minimum production targets.

In recent years, such strict control has been loosened. Individuals again are able to own land and control their own production. Farmers have an incentive to work hard, because the sale of surplus crops is their main source of revenue to buy household goods. However, agricultural land must be worked intensively to produce enough food for China's large population, and farmers in the country's less fertile areas may not be able to produce a large surplus.

The Chinese government controls the daily lives of the citizenry more than in other countries, and the people have difficulty obtaining some goods. Nonetheless, most Chinese recognize that they are better off now than before the 1949 revolution, because they have less fear of famine. Because of government controls, China has a much lower natural increase rate than other LDCs, so more of the country's economic growth can contribute to improving the standard of living of the existing population rather than meeting the needs of a rapidly expanding population.

South Asia

HDI 0.44 South Asia includes India, Pakistan, Bangladesh, Sri Lanka, and the small Himalayan states of Nepal and Bhutan. The region has the world's second-highest population and second-lowest per capita income. Population density is very high throughout the region, and the natural increase rate is among the world's highest.

India, South Asia's largest country, is the world's leading producer of jute (used to make burlap and twine), peanuts, sugar cane, and tea. India has mineral reserves including uranium, bauxite (aluminum ore), coal, manganese, iron ore, and chromite (chromium ore). However, the overall ratio of population to resources is unfavorable because of the region's huge population.

India is one of the world's leading rice and wheat producers. The region was a principal beneficiary of the Green Revolution, a series of inventions beginning in the 1960s that dramatically increased agricultural productivity. As a result of the Green Revolution, "miracle" rice and wheat seeds were widely diffused through South Asia (see Chapter 14).

Agricultural productivity in South Asia also depends on climate. The region receives nearly all its precipitation from rain that falls during the monsoon season between May and August. Agricultural output declines sharply if the monsoon rains fail to arrive. In a typical year, farmers in South Asia produce a grain surplus that is stored for distribution during dry years. However, several consecutive years without monsoon rains produce widespread hardship in South Asia.

Noodle-making in Kashgar's Sunday Market, in China's Xinjiang province. Economic activities in China require a lot of labor, possible in a country with such a large workforce. (James Strachan/Tony Stone Images)

Sub-Saharan Africa

HDI 0.35

Africa has been allocated to two regions. Countries north of the Sahara Desert share economic and cultural characteristics with the Middle East. South of the desert is called sub-Saharan Africa.

Sub-Saharan Africa has a number of assets. Population density is lower than in any other less developed region. The region contains many resources important for economic development, including bauxite in Guinea, cobalt and copper in Congo Democratic Republic and Zambia, iron ore in Liberia, manganese in Gabon, petroleum in Nigeria, and uranium in Niger (Figure 9–3). Wealth is comparable to levels found in other LDCs.

Despite these assets, sub-Saharan Africa has the least favorable prospect for development. The region has the world's highest percentage of people living in poverty and suffering from poor health and low education levels.

Some of the region's economic problems are a legacy of the colonial era. Mining companies and other businesses were established to supply European industries with needed raw materials rather than to promote overall economic development in sub-Saharan Africa. Africa's many landlocked states have difficulties shipping out raw materials through neighboring countries (see Figure 8-7). In recent years, African countries have suffered because world prices for their resources have fallen.

Political problems have also plagued sub-Saharan Africa. European colonies were converted to states without regard for the distribution of ethnicities (see Figure 7-13). After independence, leaders of many countries in the region pursued personal economic gain and local wars rather than policies to promote development of the national economy. Frequent wars within and between countries in sub-Saharan Africa have retarded development.

The fundamental problem in many countries of sub-Saharan Africa is a dramatic imbalance between the number of inhabitants and the capacity of the land to feed the population. Nearly all of the region consists of either tropical or dry climate. Both climate regions can support some people, but not large concentrations. Yet, because sub-Saharan Africa has by far the world's highest rate of natural increase, the region's land is more and more overworked, and agricultural output has declined.

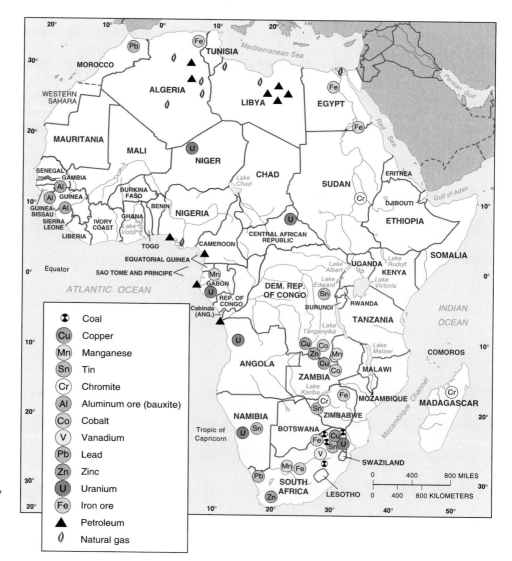

Figure 9–3 Minerals in Africa. Several African countries contain minerals important for industrial development. World prices for many of these minerals have declined or failed to rise at the same rate as the prices for industrial products, transportation, and energy.

Why Does Development Vary Among Countries?

- Economic indicators of development
- Social indicators of development
- Demographic indicators of development

The characteristics that distinguish a country's level of development can be sorted into three types: *economic*, *social*, and *demographic*. In this key issue, we will examine the distribution of these development indicators.

Economic Indicators of Development

The United Nation's HDI includes one economic indicator of development, gross domestic product per capita. Four other economic indicators distinguish more developed from less developed countries—economic structure, worker productivity, access to raw materials, and availability of consumer goods.

Gross Domestic Product Per Capita

The average individual earns a much higher income in a more developed country than in a less developed one. The typical worker receives $10 to $15 per hour in more developed countries, compared to less than $0.50 per hour in less developed ones. MDCs generally mandate a minimum wage of at least several dollars per hour.

Per capita income is a difficult figure to obtain in many countries, so to get a sense of average incomes in various countries, geographers substitute per capita gross domestic product, a more readily available indicator. The **gross domestic product (GDP)** is the value of the total output of goods and services produced in a country, normally during a year. Dividing the GDP by total population measures the contribution the average individual makes to generating a country's wealth in a year. For example, GDP in the United States is currently about $8 trillion, and its population is about 270 million, so the GDP per capita is about $30,000. The gross national product (GNP) is similar to the GDP, except that it includes income people earn abroad, such as a Canadian working in the United States.

Annual per capita GDP averages about $20,000 in more developed countries, compared to about $1,000 in less developed countries (Figure 9–4). GDP per capita exceeds $40,000 in Luxembourg, Switzerland, and Japan. The level is about $30,000 in several Western European countries, as well as the United States. The lowest per capita GDPs are found in sub-Saharan Africa, South Asia, and Southeast Asia. Nearly every country in these regions has a per capita GDP of less than $1,000 per year, and the level in many countries is below $500.

The gap in per capita GDP between more and less developed countries has been widening. Since the early 1980s, per capita GDP has increased by about $10,000 in MDCs, compared with about $200 in LDCs. Per capita GDP has actually declined over the two decades in many African and Latin American countries.

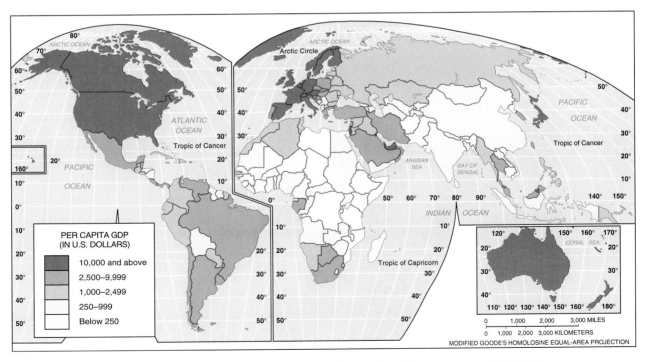

Figure 9–4 Annual gross domestic product (GDP) per capita. This measure exceeds $20,000 in most MDCs, compared to under $1,000 in most LDCs.

Per capita GDP—or, for that matter, any other single indicator—cannot measure perfectly the level of a country's development. Few people are starving in less developed countries with per capita GDPs of a few hundred dollars. And not everyone is wealthy in a developed country like the United States, with its per capita GDP of $30,000. In fact, about one-seventh of the U.S. population is officially classified as poor, including about one-third of African-Americans and Hispanics.

Per capita GDP measures average (mean) wealth, not its distribution. If only a few people receive much of the GDP, then the standard of living for the majority may be lower than the average figure implies. On the other hand, the higher the per capita GDP, the greater is the potential for ensuring that all citizens enjoy a comfortable life.

Types of Jobs

Average per capita income is higher in MDCs because people typically earn their living by different means than in LDCs. Jobs fall into three categories: primary (including agriculture), secondary (including manufacturing), and tertiary (including services). To compare the types of economic activities found in more and less developed countries, we can compute the percentage of people working in each of these three sectors.

Workers in the **primary sector** directly extract materials from Earth through agriculture, and sometimes by mining, fishing, and forestry. The **secondary sector** includes manufacturers that process, transform, and assemble raw materials into useful products. Other sec-

ondary-sector industries take manufactured goods and fabricate them into finished consumer goods. The **tertiary sector** involves the provision of goods and services to people in exchange for payment. Tertiary sector activities include retailing, banking, law, education, and government.

At one time, the practice was to identify quaternary and quinary sectors, as well. Quaternary-sector jobs were in business services, such as trade, insurance, banking, advertising, and wholesaling, whereas quinary-sector jobs were in health, education, research, government, retailing, tourism, and recreation. Current practice is to consider all of these jobs as groups within the tertiary sector (see Chapter 12).

The distribution of workers among primary, secondary, and tertiary sectors varies sharply between more and less developed countries. The percentage of people working in agriculture exceeds 75 percent in many LDCs of Africa and Asia, compared to less than 5 percent in Anglo-America and many Western European countries (Figure 9–5).

The first priority for all people is to secure food for survival. A high percentage of agricultural workers in a country indicates that most of its people are spending their days producing food for their own survival. In contrast, a low percentage of primary-sector workers indicates that a handful of farmers can produce enough food for the rest of society. Freed from the task of growing their own food, most people in a more developed country can contribute to an increase in the national wealth by working in the secondary and tertiary sectors.

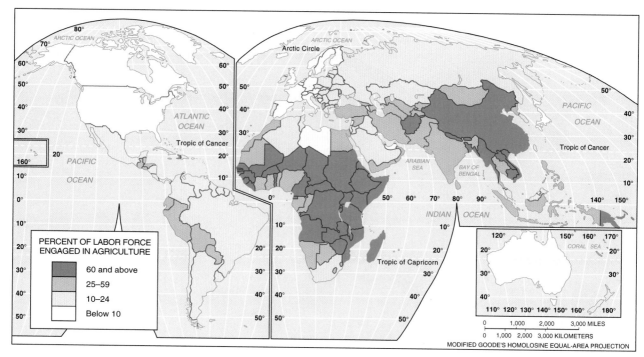

Figure 9–5 Percent primary-sector workers. A priority for all people is to secure the food they need to survive. In LDCs, most people work in agriculture to produce the food they and their families require. In MDCs, few people are farmers, and most people buy food with money earned by working in factories, offices, or performing other services.

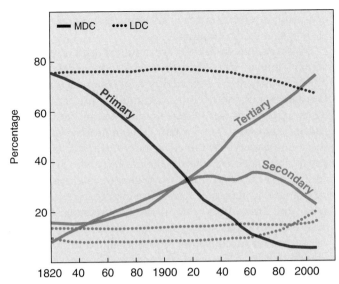

Figure 9–6 Changes in percent employed in primary, secondary, and tertiary sectors. In more developed countries, the percentage of agricultural workers has sharply declined, while the percentage of service workers has sharply increased. The percentage engaged in industry increased during the nineteenth and early twentieth centuries and decreased during the late twentieth century.

Within MDCs, the number of jobs has decreased in the primary and secondary sectors and increased in the tertiary sector (Figure 9–6). The decline in manufacturing jobs reflects not only greater efficiency inside the factories but also increased global competition in many industries. At the same time, employment in the service sector continues to expand as a result of increased consumer demand for many goods and services.

Productivity

Workers in MDCs are more productive than in LDCs. **Productivity** is the value of a particular product compared to the amount of labor needed to make it.

Workers in more developed countries produce more with less effort because they have access to more machines, tools, and equipment to perform much of the work. On the other hand, production in less developed countries must rely more on human and animal power. The larger per capita GDP in developed countries in part pays for the manufacture and purchase of machinery, which in turn makes workers more productive and generates more wealth.

Productivity can be measured by the value added per worker. The **value added** in manufacturing is the gross value of the product minus the costs of raw materials and energy. The value added per worker is 30 times greater in more developed countries than in less developed ones. The average worker adds a value of $40,000 in the United States compared to $2,500 in LDCs.

Raw Materials

Development requires access to raw materials, such as minerals and trees, that can be fashioned into useful products. It also requires energy to operate the factories, whether in the form of water power, coal, oil, natural gas, or uranium for nuclear power. In the twentieth century, both the United States and Russia (formerly the Soviet Union) became powerful industrial states, partly because both possessed a wide variety of raw materials and energy resources essential to development.

The United Kingdom, the first country to be transformed into a developed society late in the eighteenth century, had abundant supplies of coal and iron ore, the most important industrial raw materials at the time because they were used to make steel for tools. During the 1800s, other European countries took advantage of domestic coal and iron ore to promote industrial development.

European countries in the nineteenth century ran short of many raw materials essential for development and began to import them from other regions of the world. To ensure an adequate supply of these materials, European countries established colonies, especially in Africa and Asia (see Chapter 8). The international flow of raw materials sustained development in Europe, but retarded it in Africa and Asia. Although most former colonies have become independent states, they still export raw materials to more developed countries and import finished goods and services.

As certain raw materials become more important, a country's level of development can advance. The LDCs that possess energy resources, especially petroleum, have been able to use revenues from the sale of these resources to finance development. Prices for other raw materials, such as cotton and copper, have fallen because of excessive global supply and declining industrial demand. Less developed countries depending on the sale of these resources have had difficulty achieving development.

In a global economy, availability of raw materials and energy resources measures a country's development potential, rather than its actual development. A country with abundant resources has a better chance of developing. Yet some countries that lack resources—such as Japan, Singapore, South Korea, and Switzerland—have developed through world trade.

Consumer Goods

Part of the wealth generated in more developed countries goes for essential goods and services (food, clothing, and shelter). But the rest is available for consumer goods and services (cars, telephones, entertainment). The wealth used to buy "nonessentials" promotes expansion of manufacturing, which in turn generates additional wealth in the society.

The quantity and type of goods and services purchased in a society is a good measure of the level of development. Among the thousands of things that consumers buy, three are particularly good indicators of a society's development: motor vehicles, telephones, and televisions.

These products are accessible to virtually all residents in more developed countries and are vital to the health of the economy. In MDCs, the ratio of people to motor vehicles, telephones, and televisions is approaching 1:1. In other words, there is nearly one motor vehicle, telephone, and television set for each person in more developed countries.

The motor vehicle, telephone, and television all play important economic roles. Motor vehicles provide individuals with access to jobs and services and permit businesses to distribute their products. Telephones enhance communications with suppliers and customers of goods and services. Televisions provide exposure to activities in different locations.

In contrast, in less developed countries, these products do not play a central role in daily life for many people. Motor vehicles are not essential to people who live in a small village and work all day growing food in nearby fields. Telephones are not essential for those who live in the same village as their friends and relatives. Televisions are not essential to people who have little leisure time.

The number of individuals per telephone and motor vehicle exceeds 100 in most LDCs. This indicates that people are much less likely to have access to these products (Figure 9–7). The number of persons per television set varies widely among less developed countries, from less than ten in China and some Latin American countries to several hundred in Bangladesh and many African countries (see Figure 4–15). The variation reflects the rapid diffusion of television in recent years in LDCs:

acquiring a television set is an important priority in the early stages of development.

Most people in LDCs are familiar with these consumer goods, even though they cannot afford them. These objects may be desired as symbols of development. Because possession of consumer goods is not universal in developing countries, a gap might emerge between the "haves" and the "have-nots." The minority who have these goods may include government officials, landowners, and other elites, whereas the majority who are denied access to these goods may provoke political unrest.

In many LDCs, the "haves" are concentrated in urban areas; the "have-nots" live in the countryside. Technological innovations tend to diffuse from urban to rural areas. Access to consumer goods is more important in urban areas because of the dispersion of homes, factories, offices, and shops.

Motor vehicles, telephones, and televisions also contribute to social and cultural elements of development. These consumer goods provide people with access to leisure activities and exposure to new ideas. A person can explore new places in a motor vehicle, talk to people in distant locations by telephone, and see what life is like elsewhere by television. As a result of greater exposure to cultural diversity, people in developed countries display different social characteristics from people in LDCs.

Social Indicators of Development

More developed countries use part of their greater wealth to provide schools, hospitals, and welfare services.

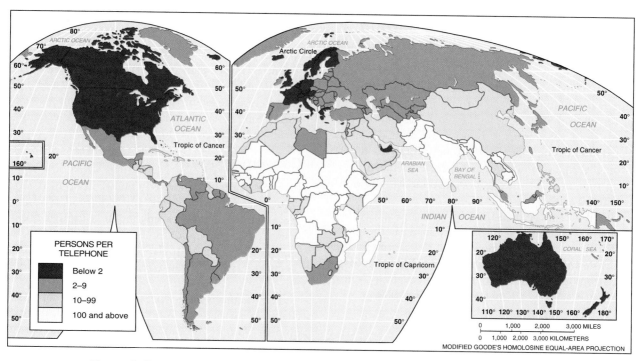

Figure 9–7 Persons per telephone. MDCs have fewer than two people for every telephone, compared to several hundred in LDCs.

As a result, their people are better educated, healthier, and better protected from hardships. Infants are more likely to survive, and adults are more likely to live longer. In turn, this well-educated, healthy, and secure population can be more economically productive.

Education and Literacy

In general, the higher the level of development, the greater are both the quantity and the quality of a country's education. A measure of the quantity of education is the average number of school years attended. The assumption is that, no matter how poor the school, the longer pupils attend, the more likely they are to learn something. The quality of education is measured in two ways—student/teacher ratio and literacy rate. The fewer pupils a teacher has, the more likely that each student will receive instruction.

The average pupil attends school for about 10 years in more developed countries, compared to only a couple of years in LDCs. The student/teacher ratio is twice as high in less developed countries as in more developed ones (Figure 9–8).

Women are less likely to attend school in LDCs. Globally, 73 women attend secondary schools (high schools) for every 100 men. The ratio of women to men in high school is 99/100 in more developed countries, but only 60/100 in less developed countries. Stated another way, in LDCs, females are roughly half of the total population of high school age, but constitute less than 40 percent of the students. But the ratio in LDCs has improved over the past quarter-century: in 1970 it was 45/100.

The MDCs publish more books, newspapers, and magazines per person because more of their citizens read and write. More developed countries dominate scientific and nonfiction publishing worldwide; this textbook is an example. Students in less developed countries must learn technical information from books that usually are not in their native language, but in English, German, Russian, or French.

The **literacy rate** is the percentage of a country's people who can read and write. It exceeds 95 percent in developed countries, compared to less than one-third in many LDCs (Figure 9–9, top).

If we compare literacy rates for women rather than for both sexes, the gap between more and less developed countries is greater. In the Middle East and South Asia, literacy rates generally fall between 25 and 75 percent for both sexes combined, but are less than 25 percent for women (Figure 9–9, bottom). Elsewhere in Asia and Latin America, gender differences are less, but in nearly every LDC, the literacy rate is higher for men. In contrast, literacy rates for men and women are virtually the same in MDCs. Low female literacy rates are an obstacle to development in some countries.

For many in LDCs, education is the ticket to better jobs and higher social status. Improved education is a major goal of many developing countries, but funds are scarce. Education may receive a higher percentage of the GDP in less developed countries, but their GDP is far

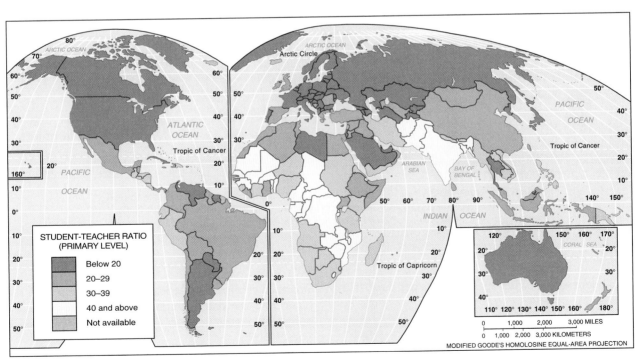

Figure 9–8 Students per teacher, primary school. Primary-school teachers must deal with much larger average class sizes in LDCs than in MDCs.

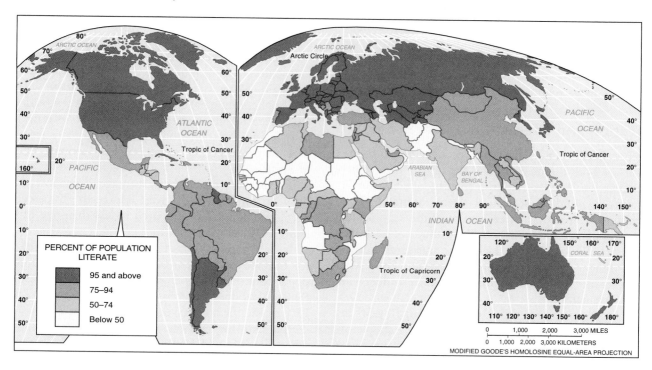

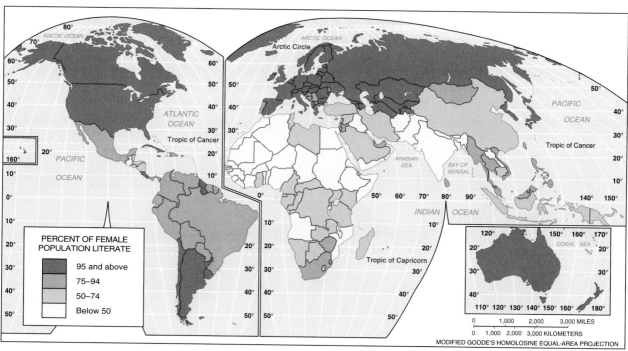

Figure 9–9 (top) Literacy rate. In most MDCs, at least 95 percent of adults are able to read and write. The percentage ranges from about 50 percent to only 10 percent in most LDCs, especially in Africa and Asia. (bottom) Female literacy rate. In some LDCs, the percentage of women who can read and write is much lower than that for men. Compared with total literacy, the gender gap is relatively high in South Asia and the Middle East.

lower to begin with, so they spend far less per pupil than do MDCs.

Health and Welfare

People are healthier in more developed countries than in less developed ones. The health of a population is influenced by diet. On average, people in MDCs receive more calories and proteins daily than they need. But in less developed countries of Africa and Asia, most people receive less than the daily minimum allowance of calories and proteins recommended by the United Nations (Figure 9–10).

When people get sick, more developed countries possess the resources to care for them. These states have better ratios of people to hospitals, doctors, and nurses (Figure 9–11). In many wealthier countries, health care is

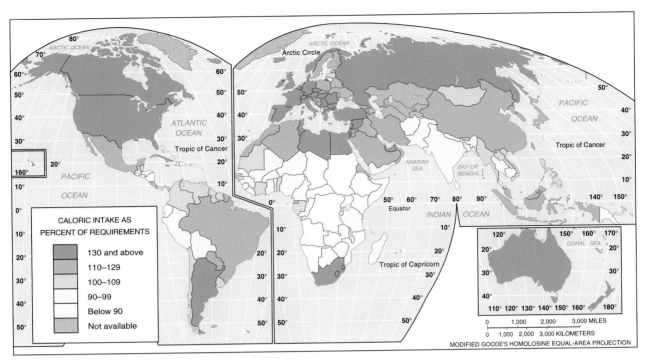

Figure 9–10 Daily available calories per capita as a percentage of requirements. Daily available calories per capita (food supply) is the domestic agricultural production plus imports, minus exports and nonfood uses. To maintain a moderate level of physical activity, an average individual requires at least 2,360 calories a day, according to the United Nations Food and Agricultural Organization. The figure must be adjusted for age, sex, and region of the world. In more developed countries, the average citizen consumes about one-third more calories than the minimum needed. The typical resident of a less developed country receives almost precisely the minimum number of calories needed to maintain moderate physical activity—on average. At first glance, this does not reveal a serious problem. However, because these figures are means, a substantial proportion of the population must be receiving less than the necessary daily minimum. The problem is especially severe in Africa, where most people consume less than the needed minimum.

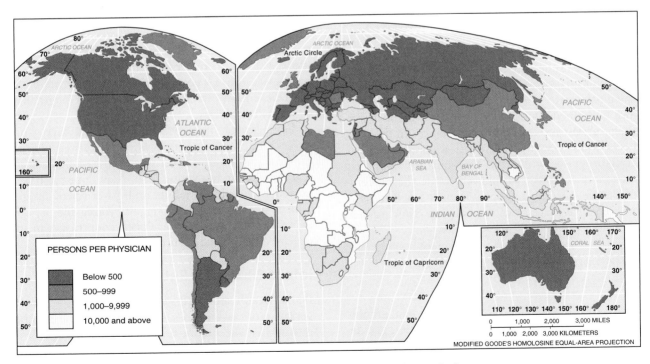

Figure 9–11 Persons per physician. People in more developed countries have more access to health care, as shown in ratios between people and hospital beds, nurses, doctors, and other medical indicators. In MDCs, for example, each doctor is available to an average of 500 people, but in LDCs each doctor is shared by thousands.

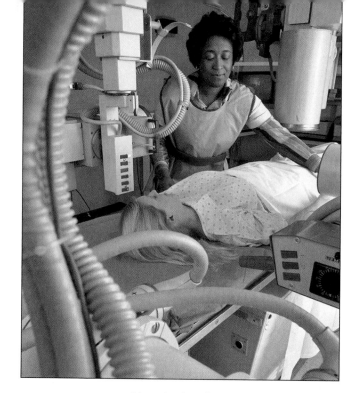
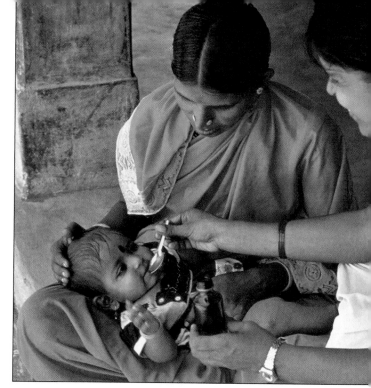

More developed countries possess better equipped hospitals and more extensive medical technology to diagnose and treat people's illnesses, such as the CAT-scan, than is the case in less developed countries, such as this clinic in India, where the nurse is administering the polio vaccine to the child. (Photo left: Gabe Palmer/The Stock Market; Photo right: S. Nagendra/Photo Researchers, Inc.)

a public service that is available at little or no cost. The United States, being an exception, still considers health care to be an activity best performed by private, profit-making enterprises.

The MDCs use part of their wealth to protect people who, for various reasons, are unable to work. In these states, some public assistance is offered to those who are sick, elderly, poor, disabled, orphaned, veterans of wars, widows, unemployed, or single parents. Countries in northwestern Europe, such as Denmark, Norway, and Sweden, typically provide the highest level of public assistance payments.

More developed countries are hard-pressed to maintain their current levels of public assistance. In the past, rapid economic growth permitted these states to finance generous programs with little hardship. But in recent years, economic growth has slowed, while the percentage of people needing public assistance has increased. Governments have faced a choice between reducing benefits or increasing taxes to pay for them.

Demographic Indicators of Development

More developed countries display many demographic differences compared to less developed countries. We described several demographic characteristics in Chapter 2. The U.N. HDI utilizes life expectancy as a measure of development (see Figure 2–11). Others that distinguish more and less developed countries include infant mortality, natural increase, and crude birth rates.

Life Expectancy

Better health and welfare in developed countries permit people to live longer. Life expectancy at birth was defined in Chapter 2 as the average number of years a newborn infant can expect to live at current mortality levels. Babies born today can expect to live into their early forties in less developed countries compared to their mid-seventies in more developed countries (see Figure 2–11). The gap in life expectancy is greater for females than for males. Males can expect to live 9 years longer in MDCs than in LDCs, whereas females can expect to live 13 years longer in more developed countries.

With longer life expectancies, MDCs have a higher percentage of elderly people who have retired and receive public support and a lower percentage of children under age 15, who are too young to work and must also be supported by employed adults and government programs. The combination of percentage of young and old dependents is lower in more developed countries than in less developed ones (see Figure 2–15).

Infant Mortality Rate

Better health and welfare in developed countries also permit more babies to survive infancy. The number of infants that die before reaching 1 year of age is fewer than 10/1,000 per year in many developed countries, compared to more than 100/1,000 in many LDCs (see Figure 2–10).

The infant mortality rate is greater in the LDCs for several reasons. Babies may die from malnutrition or lack of medicine needed to survive illness, such as dehydration from diarrhea. They may also die from poor medical practices that arise from lack of education. For example, the use of a dirty knife to cut the umbilical cord is a major cause of fatal tetanus in India.

Natural Increase Rate

The natural increase rate averages more than 2 percent annually in less developed countries and less than 1 percent in more developed ones. Greater natural increase strains a country's ability to provide hospitals, schools, jobs, and other services that can make its people healthier and more productive. Many LDCs must allocate increasing percentages of their GDPs just to care for the rapidly expanding population, rather than to improve care for the current population (see Figure 2–6).

Crude Birth Rate

Less developed countries have higher natural increase rates because they have higher crude birth rates. The annual crude birth rate exceeds 40 per 1,000 in many LDCs, compared to less than 15 per 1,000 in MDCs. Women in more developed countries choose to have fewer babies for various economic and social reasons, and they have access to varied birth control devices to achieve this goal (see Figure 2–8).

Crude death rate does not indicate a society's level of development. More developed and less developed countries both have annual crude death rates of about 10 per 1,000. Two reasons account for the lack of difference. First, diffusion of medical technology from more developed countries has eliminated or sharply reduced the incidence of several diseases in less developed countries. Second, MDCs have higher percentages of older people, who have high mortality rates, as well as lower percentages of children, who have low mortality rates once they survive infancy.

The mortality rate for women in childbirth is significantly higher in LDCs. For every 100,000 babies born, fewer than 10 mothers die giving birth in most MDCs, compared with several hundred in LDCs.

Key Issue 3

Why Do Less Developed Countries Face Obstacles to Development?

- Development through self-sufficiency
- Development through international trade
- Financing development

The indicators presented in the previous key issue reflect sharp differences in the levels of development of more developed and less developed countries. To promote development, LDCs seek improvements in these indicators. LDCs have in fact made progress, but for many of

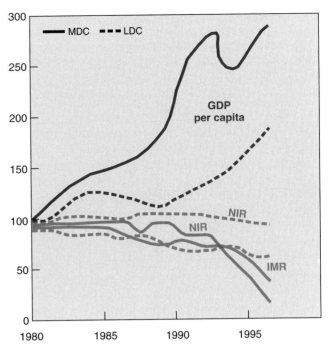

Figure 9–12 Progress toward development. All indicators are set to equal 100 in 1980. During the past two decades, GDP per capita has increased more rapidly, and natural increase and infant mortality rates have decreased more rapidly in MDCs than in LDCs.

the indicators the gap between the two is widening rather than narrowing.

Since 1980, GDP per capita has doubled in less developed countries, but tripled in more developed ones. Natural increase has dropped by 5 percent in LDCs, but by 83 percent in more developed ones. Infant mortality has dropped by one-third in LDCs, but by one-half in MDCs (Figure 9–12).

To reduce disparities between rich and poor countries, LDCs must develop more rapidly. This means increasing per capita GDP more rapidly and using the additional funds to make more rapid improvements in people's social and economic conditions. LDCs face two fundamental obstacles in trying to encourage more rapid development:

- Adopting policies that successfully promote development
- Finding funds to pay for development

Less developed countries chose one of two models to promote development. One approach emphasizes international trade; the other advocates self-sufficiency. Each has important advantages and serious problems. We will examine examples of countries that have tried each alternative, successfully and unsuccessfully.

Development Through Self-Sufficiency

For most of the twentieth century, self-sufficiency, or balanced growth, was the more popular of the development alternatives. The world's two most populous countries, China and India, adopted this strategy, as did most African and Eastern European countries.

Elements of Self-Sufficiency Approach

According to the balanced growth approach, a country should spread investment as equally as possible across all sectors of its economy, and in all regions. The pace of development may be modest, but the system is fair because residents and enterprises throughout the country share the benefits of development. Under self-sufficiency, incomes in the countryside keep pace with those in the city, and reducing poverty takes precedence over encouraging a few people to become wealthy consumers.

The approach nurses fledgling businesses in a less developed country by isolating them from competition of large international corporations. A country's fragile businesses can be independent and insulated from potentially adverse impacts of decisions made by businesses and governments in the more developed countries.

Countries promote self-sufficiency by setting barriers that limit the import of goods from other places. Three widely used barriers include setting high taxes on imported goods to make them more expensive than domestic goods, fixing quotas to limit the quantity of imported goods, and requiring licenses to restrict the number of legal importers. The approach also restricts local businesses from exporting to other countries.

India: Example of the Self-Sufficiency Approach. For many years, India made major use of all of the trade barriers. To import goods into India, most foreign companies had to secure a license. The process was long and cumbersome, because several dozen government agencies had to approve the request. Once a company received an import license, the government severely restricted the quantity it could sell in India. The government also imposed heavy taxes on imported goods, which doubled or even tripled the price to consumers.

At the same time, Indian businesses were discouraged from producing goods for export to more developed or other less developed countries. Indian money could not be converted to other currencies.

Businesses were supposed to produce goods for consumption inside India. Effectively cut off from the world economy, businesses required government permission to sell a new product, modernize a factory, expand production, set prices, hire or fire workers, and change the job classification of existing workers.

If private companies were unable to make a profit selling goods only inside India, the government provided subsidies, such as cheap electricity, or wiped out debts. The government owned not just communications, transportation, and power companies, a common feature around the world, but also businesses such as insurance companies and automakers, left to the private sector in most countries.

Problems with the Self-Sufficiency Alternative

The experience of India and other LDCs has revealed two major problems with self-sufficiency. They are described below.

Inefficiency. Self-sufficiency protects inefficient industries. Because businesses can sell all they make, at high government-controlled prices, to customers culled from long waiting lists, they have little incentive to improve quality, lower production costs, reduce prices, or increase production. Companies protected from international competition do not feel pressure to keep abreast of rapid technological changes.

India's auto industry, for example, was dominated by Hindustan Motors, which sold only one model, called the Ambassador, painted white, still based in the 1990s on a 1950s-era British design. Yet the waiting list for these outdated Ambassadors was long, and Hindustan Motors sold all it made, because foreign cars were prohibitively expensive and difficult to import.

Large Bureaucracy. The second problem with the self-sufficiency approach was the large bureaucracy needed to administer the controls. A complex administrative system encouraged abuse and corruption. Potential entrepreneurs found that struggling to produce goods or offer services was less rewarding financially than advising others how to get around the complex government regulations. Other potential entrepreneurs earned more money by illegally importing goods and selling them at inflated prices on the black market.

Development Through International Trade

The international trade model of development calls for a country to identify its distinctive or unique economic assets. What animal, vegetable, or mineral resources does the country have in abundance that other countries are willing to buy? What product can the country manufacture and distribute at a higher quality and a lower cost than other countries?

According to the international trade approach, a country can develop economically by concentrating scarce resources on expansion of its distinctive local industries. The sale of these products in the world market brings funds into the country that can be used to finance other development.

Rostow's Development Model

A leading advocate of this approach was W. W. Rostow, who in the 1950s proposed a five-stage model of development. Several countries adopted the approach during the 1960s, although most continued to follow the self-sufficiency approach.

1. **The traditional society.** This term defines a country that has not yet started a process of development. A traditional society contains a very high percentage of people engaged in agriculture and a high percentage of national wealth allocated to what Rostow called "nonproductive" activities, such as the military and religion.

2. **The preconditions for take-off.** Under the international trade model, the process of development begins when an elite group initiates innovative economic activities. Under the influence of these well-educated leaders, the country starts to invest in new technology and infrastructure, such as water supplies and transportation systems. These projects will ultimately stimulate an increase in productivity.

3. **The take-off.** Rapid growth is generated in a limited number of economic activities, such as textiles or food products. These few take-off industries achieve technical advances and become productive, while other sectors of the economy remain dominated by traditional practices.

4. **The drive to maturity.** Modern technology, previously confined to a few take-off industries, diffuses to a wide variety of industries, which then experience rapid growth comparable to the take-off industries. Workers become more skilled and specialized.

5. **The age of mass consumption.** The economy shifts from production of heavy industry, such as steel and energy, to consumer goods, like motor vehicles and refrigerators.

According to the international trade model, each country is in one of these five stages of development. More developed countries are in stage 4 or 5, whereas less developed ones are in one of the three earlier stages. The model also asserts that today's more developed countries have already passed through the early stages. The United States, for example, was in stage 1 prior to independence, stage 2 during the first half of the nineteenth century, stage 3 during the middle of the nineteenth century, and stage 4 during the late nineteenth century, before entering stage 5 during the early twentieth century. The model assumes that less developed countries will achieve development by moving along from an earlier to a later stage.

A country that concentrates on international trade benefits from exposure to consumers in other countries. To remain competitive, the take-off industries must constantly evaluate changes in international consumer preferences, marketing strategies, production engineering, and design technologies. This concern for international competitiveness in the exporting take-off industries will filter through less advanced economic sectors.

Rostow's optimistic development model was based on two factors. First, the developed countries of Western Europe and Anglo-America had been joined by others, notably Japan. If Japan could become more developed by following this model, why not other countries?

Second, many LDCs contain an abundant supply of raw materials sought by manufacturers and producers in more developed countries. In the past, European colonial powers extracted many of these resources without paying compensation to the colonies. In a global economy, the sale of these raw materials could generate funds for LDCs to promote development.

Examples of International Trade Approach

During the 1960s and 1970s, when most LDCs were following the self-sufficiency approach, two groups of countries chose the international trade approach. One such group was along the Arabian Peninsula near the Persian Gulf; the others were in East and Southeast Asia.

Petroleum-Rich Persian Gulf States. Saudi Arabia is the most prominent country in the Persian Gulf area; others include Kuwait, Bahrain, Oman, and the United Arab Emirates. Until the 1970s, this region was one of the world's least developed, but escalation of petroleum prices during the 1970s transformed these countries overnight into some of the wealthiest per capita.

These countries have used petroleum revenues to finance large-scale projects, such as housing, highways, airports, universities, and telecommunications networks. Recently built steel, aluminum, and petrochemical factories compete on world markets with the help of government subsidies.

The landscape has been further changed by the diffusion of consumer goods. Large motor vehicles, color TVs, audio equipment, and motorcycles are readily available and affordable. Supermarkets are stocked with food imported from Europe and Anglo-America.

With stagnating petroleum prices during the 1980s and 1990s, petroleum-rich Middle Eastern countries no longer have per capita GDPs at the same level as in more developed countries. Rapid natural increase rates and several wars have also eroded the region's wealth.

Some Islamic religious principles, which dominate the culture of the Middle East, conflict with business practices in more developed countries. Women are excluded from holding most jobs and visiting public places, such as restaurants and swimming pools. In some places they are expected to wear traditional black clothes, a shroud, and a veil. All business halts several times a day when Muslims are called to prayers. Shops close their checkout lines and permit people to unwrap their prayer rugs and prostrate themselves on the floor.

The Four Asian Dragons. Also adopting the international trade alternative during the 1960s were South Korea, Singapore, Taiwan, and the former British colony of Hong Kong. These four areas were given several nicknames, including the "four dragons," the "four little tigers," and "the gang of four."

Singapore and Hong Kong, British colonies until 1965 and 1997, respectively, have virtually no natural resources. Both comprise large cities surrounded by very small amounts of rural land. South Korea and Taiwan have traditionally taken their lead from Japan, which occupied both countries until after World War II. Their adoption of the international trade approach was strongly influenced by Japan's success.

Lacking natural resources, the four dragons promoted development by concentrating on producing a handful of manufactured goods, especially clothing and electronics. Low labor costs enabled these countries to sell products inexpensively in more developed countries.

Problems with the International Trade Alternative

Three problems have hindered countries outside the Persian Gulf and the four Asian dragons from developing through the international trade approach:

1. **Uneven Resource Distribution.** Middle Eastern countries successfully developed because petroleum prices skyrocketed during the 1970s. Other countries found that the prices of their commodities did not increase, and in some cases actually decreased. LDCs that depended on the sale of one product suffered because the price of their leading commodity did not rise as rapidly as the cost of the products they needed to buy. For example, Zambia has a large percentage of the world's copper reserves, but it has been unable to use this asset to promote development because of declining world prices for copper.

2. **Market Stagnation.** Countries such as the four dragons that depend on selling low-cost manufactured goods find that the world market for many products is expanding slower than in the past. More developed countries have limited growth in population, consumer purchasing power, and market size. To increase sales, less developed countries may need to capture sales from established competitors rather than share in an expanding market.

3. **Increased Dependence on MDCs.** Building up a handful of take-off industries that sell to people in more developed countries may force LDCs to cut back on production of food, clothing, and other necessities for their own people. Rather than finance new development, funds generated from the sale of products to other countries may have to be used to buy these necessities from MDCs for the employees of the take-off industries.

Recent Triumph of the International Trade Approach

Despite problems with the international trade approach, it has been embraced by most countries as the preferred alternative for stimulating development. Long-time advocates of the self-sufficiency approach are scrambling to convert to international trade as rapidly as possible. During the past quarter-century, world wealth (as measured by GDP) has doubled, whereas world trade has tripled.

India, for example, during the 1990s, dismantled its formidable collection of barriers to international trade. Foreign companies were allowed to set up factories and sell in India; tariffs and restrictions on the import and export of goods were reduced or eliminated. Monopolies in communications, insurance, and other industries were eliminated. With increased competition, Indian companies improved the quality of their products.

Countries converted from self-sufficiency to international trade during the 1990s for one simple reason: overwhelming evidence that international trade better promoted development. The World Bank found that between 1963 and 1985, per capita GDP increased 7 percent annually in countries strongly oriented toward international trade, compared to 1 percent for countries strongly oriented toward self-sufficiency.

In the case of India, under self-sufficiency between 1960 and 1990, GDP grew by 4 percent per year, much lower than in Asian countries that had embraced international trade. In comparison, during the same period, GDP increased 7 percent per year in Thailand, 8 percent in Taiwan, and 9 percent in South Korea. After adopting the international trade alternative in the early 1990s, India's GDP grew during the 1990s by 7 percent annually.

Financing Development

Regardless of whether self-sufficiency or international trade is preferred, less developed countries lack the money needed to finance development. The LDCs generally must obtain funds from more developed countries. These funds come from two primary sources: loans from banks and international organizations, and direct investment by transnational corporations.

Loans. The LDCs borrow money to build new infrastructure, such as hydroelectric dams, electric transmission lines, flood protection systems, water supplies, roads, and hotels. The two major lenders are international lending organizations controlled by the MDC governments—the World Bank and the International Monetary Fund—which together lend about $50 billion a year to LDCs for development. Money is also lent by commercial banks in more developed countries. The total value of all outstanding loans to LDCs was about

Japan built economic development on international trade. Nissan cars are awaiting export from Japan to other Asian countries, as well as to North America. (J. Nordell/The Image Works)

$2.1 trillion in 1996, an increase of about $1 trillion in a decade.

The theory behind borrowing money to build infrastructure is that new roads and dams will make conditions more favorable for domestic and foreign businesses to open or expand. After all, no business wants to locate in a place that lacks paved roads, running water, and electricity. In principle, new or expanded businesses attracted to an area because of improved infrastructure will contribute additional taxes that the LDC uses in part to repay the loans and in part to improve its citizens' living conditions.

The problem is that many of the new infrastructure projects are expensive failures. In Mali, for example, a French-sponsored project to pump water from the Niger River using solar energy functioned for only a month. Even when it worked, the project, which cost over $1 million, produced no more water than could two diesel pumps that together cost $6,000. The World Bank has judged half of the projects it has funded in Africa to be failures. In other cases, roads are opened, and equipment operates correctly, but new businesses are still not attracted to the area.

In recent years, many LDCs have been unable to repay the interest on their loans, let alone the principal. Brazil, Mexico, Argentina, and several other Latin American countries have accumulated the largest debts, although several African countries have very high ratios of debt to GDP (Figure 9–13). When these countries cannot repay their debts, financial institutions in more developed countries refuse to make further loans, so construction of needed infrastructure stops. The inability of many LDCs to repay loans also damages the financial stability of banks in the more developed countries.

The MDCs have become more cautious in granting loans. Less developed countries judged likely to repay debts receive more of the loans and at lower rates of interest. In exchange for canceling or refinancing debts, the international lending agencies require LDCs to impose economic policies consistent with those in MDCs. To create conditions that encourage trade, LDCs are forced to raise taxes, reduce government spending, control inflation, sell publicly owned utilities to private corporations, and charge more for services. These programs can be unpopular with the voters and can encourage political unrest. For their part, LDCs demand an increased role in loan-making decisions made by international agencies.

Transnational Corporations. A transnational corporation operates in countries other than the one in which its headquarters are located. Initially, transnational corporations were primarily U.S.-owned, but in recent years active transnational corporations have been based in other MDCs, especially Japan, Germany, France, and the United Kingdom.

International investment has grown from very little in the 1980s to $200 billion annually by the mid-1990s. About one-third of the investment involves transfers within transnational corporations, such as automotive engines and transmissions manufactured in one country and shipped to a final assembly plant in another country. Refer back to Figure 1–22, which shows the scope of international operations of a major Japanese auto parts maker, Denso.

Foreign investment does not flow equally around the world. In 1996, only 35 percent of foreign investment went from a more developed country to a less developed

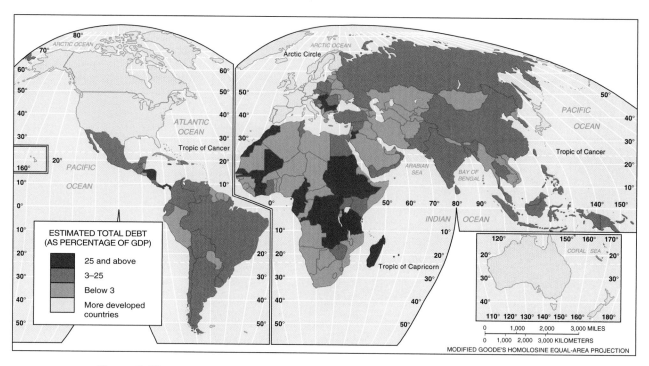

Figure 9–13 Debt as percentage of GDP. To finance development, many developing countries have accumulated large foreign debts relative to their annual GDPs. As a result, a large percentage of their national budgets must be used to repay loans. When LDCs cannot repay their debts, financial institutions in relatively developed countries suffer because they were a major source of the loans.

country, whereas the other 65 percent went from one MDC to another MDC. This level represents rapid progress for LDCs, because as recently as 1990 only 10 percent of foreign investment went from a more developed to a less developed country.

Nearly all of the international investment went to just eight less developed countries. China received about one-third of all investment in LDCs; South Korea, Malaysia, Indonesia, and Thailand combined received about one-third; and Argentina, Mexico, and Brazil received most of the remaining one-third. (Figure 1–18 shows the major flows of foreign investment by transnational corporations.)

■ Summary

The relationship between the more and less developed regions—described at the beginning of the chapter as a north-south split (refer to Figure 9–2)—appears somewhat different on a north polar projection. More developed countries form a triangular-shaped inner core area, whereas less developed countries occupy peripheral locations (Figure 9–14). This unorthodox world map projection emphasizes the central role played by MDCs in the world economy, and the secondary role of LDCs.

In an increasingly unified world economy, the MDCs clustered in the core play dominant roles in forming the economies of the LDCs on the periphery. Anglo-America, Western Europe, and Japan account for a high percentage of the world's economic activity and wealth. The LDCs in the periphery have less access to the world centers of consumption, communications, wealth, and power, which are clustered in the core. Development prospects of Latin America are tied to governments and businesses in Anglo-America, those of Africa, the Middle

East, and Eastern Europe to Western Europe, and those of Asia to Japan and to a lesser extent Western Europe and Anglo-America.

To some geographers, the economies of more developed core regions appear to be exploiting the people and resources of less developed peripheral regions. But from the perspective of people in less developed regions, integration into a world economy through trade with MDCs may be a small price to pay to receive material benefits of development, such as a steady job and a television.

By the year 2020, the World Bank projects that China will have the world's largest economy, ahead of the United States and Japan. India and Indonesia are expected to follow, then Germany, South Korea, France, Taiwan, Brazil, Italy, Russia, the United Kingdom, and Mexico. Thus, seven of the top 14 economies are expected to be in countries currently considered less developed. For that to happen, the world will have to make considerable progress in the next few years toward achieving

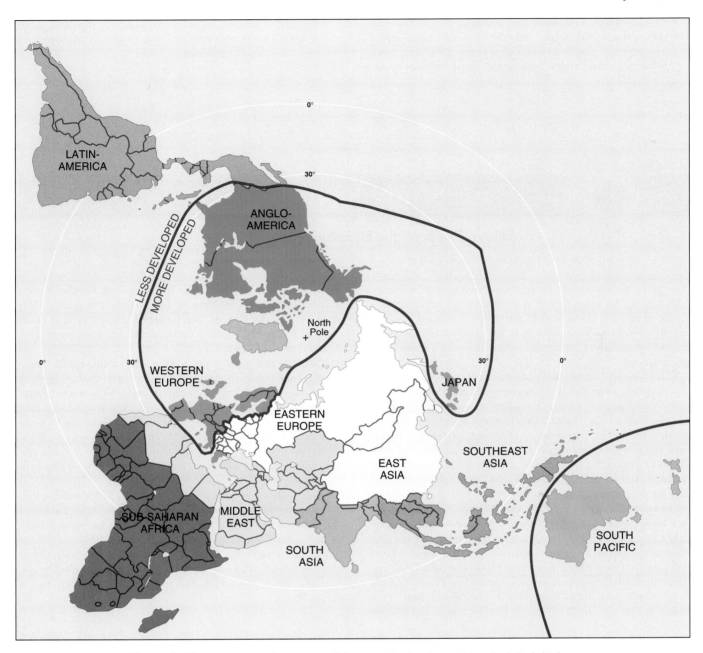

Figure 9–14 Core and periphery. Most of the countries that have achieved relatively high levels of development are located above 30º north latitude. Viewed from this north polar projection, more developed countries appear clustered in an inner core, while less developed countries are generally relegated to a peripheral or outer-ring location.

what the United Nations called for in 1974—"a new international economic order," based on greater equality and economic interdependence between more and less developed countries.

Here again are the key issues concerning development:

1. **Where are more and less developed countries distributed?** We can identify three more developed regions—Anglo-America, Western Europe, and Eastern Europe—plus two other developed areas—Japan and South Pacific. Six less developed regions include Latin America, Southeast Asia, Middle East, East Asia, South Asia, and sub-Saharan Africa. These less developed regions face different prospects for promoting development.

2. **Why does development vary among countries?** Development is the process by which the material conditions of a country's people are improved. A more developed country has a higher level of per capita GDP, achieved through a transformation in the structure of the economy from a predominantly agricultural to an industrial and service-providing society. More developed countries use their wealth in part to provide better health, education, and welfare services. Conversely, LDCs must use their additional wealth primarily to meet the needs of a rapidly growing population.

3. **Why do less developed countries face obstacles to**

development? Less developed countries choose between the international trade and the self-sufficiency paths toward development. In either alternative, LDCs may need to borrow considerable sums of money to promote devel-

opment. The inability of many LDCs to pay back these loans is a source of considerable tension between them and more developed countries.

Case Study Revisited

Future Prospects for Development

The most fundamental obstacle to development in many LDCs is gender inequality. A precondition for effective nurturing of take-off industries and effective use of loans is assuring an effective role for women in the development process. Excluding women is not merely unfair; it wastes a major economic asset.

The United Nations has constructed the Gender-Related Development Index (GDI) to measure the condition of women and the extent of gender inequality in various countries. Components of the Human Development Index that vary by gender, such as life expectancy and literacy, are computed for each country for women only. A low GDI means that the condition of women is relatively unfavorable. A GDI much lower than the HDI means that women have a much lower status than men in the country.

As was the case with the HDI, sub-Saharan Africa and South Asia have the lowest GDIs. In these regions, women have relatively low literacy rates, low levels of education, and low life expectancy, but so do men. Southeast Asia, the Middle East, and East Asia have virtually identical HDIs, but their GDIs are different. Among the three regions, the GDI is relatively high in Southeast Asia and relatively low in the Middle East. This indicates that the status of women is quite low in the Middle

East, whereas in Southeast Asia the position of the two sexes is relatively close. Latin America has a somewhat higher HDI than has Eastern Europe, but Eastern Europe's GDI is much higher, indicating equality between the sexes in the former Communist countries.

One country with a legacy of gender inequality is trying to do something about it. The Grameen Bank, based in Bangladesh, specializes in making loans to women, three-fourths of the borrowers since the bank was established in 1977. The bank has made several hundred thousand loans to women in Bangladesh and neighboring South Asian countries, and only 1 percent of the borrowers—an extraordinarily low percentage for a bank—have failed to make their weekly loan repayments.

Rabea Rahman borrowed $90 from the Grameen Bank to buy a cow. Earnings from selling the cow's milk enabled her to buy her son an $85 rickshaw bicycle so that he could make a living. The smallest loan the bank has made was $1, to a woman who wanted to sell plastic bangles door-to-door. Other women have borrowed money to make perfume, bind books, and sell matches, mirrors, and bananas. The average loan is about $60.

■ Key Terms

Development A process of improvement in the material conditions of people through diffusion of knowledge and technology.

Gross domestic product (GDP) The value of the total output of goods and services produced in a country in a given time period (normally one year).

Less developed country (LDC) Also known as a **developing country**, a country that is at a relatively early stage in the process of economic development.

Literacy rate The percentage of a country's people who can read and write.

More developed country Also known as a **relatively developed country** or a **developed country (MDC)**, a country that has progressed relatively far along a continuum of development.

Primary sector The portion of the economy concerned with the direct extraction of materials from Earth's surface, generally through agriculture, although sometimes by mining, fishing, and forestry.

Productivity The value of a particular product compared to the amount of labor needed to make it.

Secondary sector The portion of the

economy concerned with manufacturing useful products through processing, transforming, and assembling raw materials.

Tertiary sector The portion of the economy concerned with transportation, communications, and utilities, sometimes extended to the provision of all goods and services to people in exchange for payment.

Value added The gross value of the product minus the costs of raw materials and energy.

Thinking Geographically

1. Review the major economic, social, and demographic characteristics that contribute to a country's level of development. Which indicators can vary significantly by gender within countries and between countries at various levels of development? Why?

2. Some geographers have been attracted to the concepts of Immanuel Wallerstein, who argued that the modern world consists of a single entity, the capitalist world economy, that is divided into three regions: the core, semi-periphery, and periphery. How have the boundaries among these three regions changed?

3. China has relied on self-sufficiency to promote development, whereas Hong Kong has been a prominent practi-

tioner of international trade. Explain how these two approaches have been reconciled since Hong Kong became part of China in 1997.

4. Some LDCs claim that the requirements placed on them by lending organizations such as the World Bank impede rather than promote development. Should LDCs be given a greater role in deciding how much the international organizations should spend and how such funds should be spent? Why or why not?

5. What obstacles do Eastern European countries face as they dismantle 40 years of communism and convert to market economies?

Further Readings

Afxentiou, Panos C., and Apostolos Serletis. "Growth and Foreign Indebtedness in Developing Countries: An Empirical Study Using Long-Term Cross-Country Data." *Journal of Developing Areas* 31 (1996): 25–40.

Austin, Patricia. "Women and the Rural Idyll." *Journal of Rural Studies* 12 (1996): 101–12

Ayres, Ed. "The Shadow Economy." *World Watch* 9 (1996): 10–23.

Azad, Nandini. "Gender and Equity: Experience of the Working Women's Forum, India. *International Social Science Journal* 48 (1996): 219–30.

Ballance, R.; J. Ansari, and H. Singer. *The International Economy and Industrial Development: Trade and Investment in the Third World.* Totowa, NJ: Allanheld, Osmun, 1982.

Barker, Randolph, and Robert W. Herdt. *The Rice Economy of Asia.* Washington, DC: Resources for the Future, 1985.

Barros, Ricardo; Louise Fox, and Rosane Mendonca. "Female-Headed Households, Poverty, and the Welfare of Children in Urban Brazil." *Economic Development and Cultural Change* 45 (1997): 231–58.

Bater, James H. *The Soviet Scene: A Geographical Perspective.* New York: Routledge, 1989.

Bebbington, Anthony J.; Hernan Carrasco, Lourdes Peralbo, Galo Ramon, Jorge Trujillo, and Victor Torres. "Fragile Lands, Fragile Organizations: Indian Organizations and the Politics of Sustainability in Ecuador." *Transactions of the Institute of British Geographers, New Series* 18 (1993): 179–96.

Berry, Brian J. L.; Edgar C. Conkling, and D. Michael Ray. *Economic Geography.* Englewood Cliffs, NJ: Prentice-Hall, 1987.

Blakemore, Harold, and Clifford T. Smith, eds. *Latin America: Geographical Perspectives*, 2nd ed. London: Methuen, 1983.

Bobek, Hans. "The Main Stages in Socioeconomic Evolution from a Geographic Point of View." In *Readings in Cultural Geography,* ed. by Philip L. Wagner and Marvin W. Mikesell. Chicago: University of Chicago Press, 1962.

Bradbury, I.; R. Kirkby, and S. Gaunbao. "Development and Environment: The Case of Rural Industrialization and Small-Town Growth in China." *Ambio* 25 (1996): 204–209.

Buvinic, Mayra, and Geeta Rao Gupta. "Female-Headed Households and Female-Maintained Families: Are They Worth Targeting to Reduce Poverty in Developing Countries?" *Economic Development and Cultural Change* 45 (1997): 259–80.

Chang, Sen-dou. "Modernization and China's Urban Development." *Annals of the Association of American Geographers* 71 (1981): 572–79.

Chisholm, Michael. *Modern World Development: A Geographical Perspective.* Totowa, NJ: Barnes & Noble, 1982.

———. "The Wealth of Nations." *Transactions of the Institute of British Geographers, New Series* 5 (1980): 255–76.

Cole, John P. *The Development Gap: A Spatial Analysis of World Poverty and Inequality.* New York: John Wiley, 1980.

Crow, Ben, and Alan Thomas. *Third World Atlas.* Philadelphia: Open University Press, 1985.

Cruz, Wilfrido; Mohan Munasinghe, and Jeremy Warford. "Greening Development: Environmental Implications of Economic Policies." *Environment* 38 (1996): 6–11.

Dunford, Michael, and Diane Perrons. "Regional Inequality, Regimes of Accumulation and Economic Development in Contemporary Europe." *Transactions of the Institute of British Geographers New Series* 19 (1994): 163–82.

Demko, George, ed. *Regional Development: Problems and Policies in Eastern and Western Europe.* New York: St. Martin's Press, 1984.

DeSouza, Anthony R., and Phillip Porter. *The Underdevelopment and Modernization of the Third World.* Washington, DC: Association of American Geographers, 1974.

DeSouza, Anthony R., and Frederick P. Stutz. *The World Economy*: *Resources, Location, Trade, and Development*. 2nd ed. New York: Macmillan, 1994.

Dickenson, J. P.; C. G. Clarke, W. T. S. Gould, R. M. Prothero, D. J. Siddle, C. T. Smith, E. M. Thomas-Hope, and A. G. Hodgkiss. *A Geography of the Third World*. New York: Methuen, 1983.

Doe, Lubin. "The Economic Development of Francophone Africa: A Comparison with the Republic of Korea." *International Social Science Journal* 49 (1997): 105–22.

Dott, Ashok K., ed. *Southeast Asia*: *Realm of Contrasts*. 3rd ed. Boulder, CO: Westview Press, 1985.

Flavin, Christopher. "Electricity for a Developing World: New Directions." *Worldwatch Paper* 70. Washington, DC: Worldwatch Institute, June 1986.

Forbes, D. K. *The Geography of Underdevelopment*: *A Critical Survey*. Baltimore: The Johns Hopkins University Press, 1984.

Fryer, Donald D. "The Political Geography of International Lending by Private Banks." *Transactions of the Institute of British Geographers, New Series* 12 (1987): 413–32.

Ginsburg, Norton S. *Atlas of Economic Development*. Chicago: University of Chicago Press, 1961.

_____, ed. *Essays on Geography and Economic Development*. Chicago: University of Chicago Press, 1960.

Grossman, Larry. "The Cultural Ecology of Economic Development." *Annals of the Association of American Geographers* 71 (1981): 220–36.

Hoffman, George W., ed. *A Geography of Europe: Problems and Prospects*. 5th ed. New York: John Wiley, 1983.

Holloway, Steve R., and Kavita Pandit. "The Disparity Between the Level of Economic Development and Human Welfare." *Professional Geographer* 44 (1992): 57–71.

Illuwishewa, Rohana. "Rural Development and its Consequences for Women: A Case Study of a Development Project in Sri Lanka." *Geography* 82 (1997): 110–47.

James, Preston E. *Latin America*. 4th ed. New York: Odyssey House, 1969.

Jones, D. B., ed. *Oxford Economic Atlas of the World*. 4th ed. London and New York: Oxford University Press, 1972.

Jumper, Sidney R.; Thomas L. Bell, and Bruce A. Ralston. *Economic Growth and Disparities*: *A World View*. Englewood Cliffs, NJ: Prentice-Hall, 1980.

Kammen, Daniel M., and Michael R. Dove. "The Virtues of Mundane Science." *Environment* 39, no. 6 (1997): 10–15.

Mabogunje, Akinlawon L. *The Development Process*: *A Spatial Perspective*. London: Hutchinson University Library, 1981.

Momsen, Janet Henshall. *Women and Development in the Third World*. London: Routledge, 1991.

_____; Janet Henshall, and Janet Townsend. *Geography of Gender in the Third World*. Albany: State University of New York Press, 1987.

Murphy, Alexander B. "Western Investment in East-Central Europe: Emerging Patterns and Implications for State Stability." *Professional Geographer* 44 (1992): 249–59.

Myrdal, Gunnar. *Rich Lands and Poor*. New York: Harper and Bros., 1957.

O'Connor, A. M. *The Geography of Tropical African Development*. 2nd ed. Oxford: Pergamon Press, 1978.

_____. *Poverty in Africa*: *A Geographical Approach*. London: Belhaven Press, 1991.

Ó'hUállachain, Breandan, and Neil Reid. "Source Country Differences in the Spatial Distribution of Foreign Direct Investment in the United States." *Professional Geographer* 44 (1992): 272–85.

O'Sullivan, Patrick, and Alasdair McNelly. "World Trade in the 1990s." *Geography* 81 (1996): 275–78.

Ó Tuathail, Gearóid, and Timothy W. Lake. "Present at the (Dis)integration: Deterritorialization and Reterritorialization in the New Wor(l)d Order." *Annals of the Association of American Geographers* 84 (1994): 381–98.

Patchell, Jerry. "Kaleidoscope Economies: The Processes of Cooperation, Competition, and Control in Regional Economic Development." *Annals of the Association of American Geographers* 86 (1996): 481–506.

Porter, Doug; Bryant Allen, and Gaye Thompson. *Development in Practice*: *Paved with Good Intentions*. London and New York: Routledge, 1991.

Rostow, Walter W. *The Stages of Economic Growth*. Cambridge: Cambridge University Press, 1960.

Seager, Joni, and Ann Olson. *Women in the World: An International Atlas*. New York: Simon & Schuster, 1986.

Smith, David M. *Where the Grass Is Greener*: *Living in an Unequal World*. London: Croom Helm, 1979.

Straussfogel, Debra. "World-Systems Theory: Toward a Heuristic and Pedagogic Conceptual Tool." *Economic Geography* 73 (1997): 118–30.

Szentes, Tamas. *The Political Economy of Underdevelopment*. 4th ed. Budapest, Hungary: Akademiai Kiado, 1983.

Taylor, Peter J. "World-Systems Analysis and Regional Geography." *Professional Geographer* 40 (1988): 259–65.

Timberg, Thomas A. "Big and Free Is Beautiful: China and India, the Past 40 Years and the Next." *Economic Development and Cultural Change* 45 (1997): 435–42.

Wallenstein, Immanuel. *The Capitalist World-Economy*. Cambridge: Cambridge University Press, 1979.

_____. *Geopolitics and Geoculture*: *Essays on the Changing World-System*. Cambridge: Cambridge University Press, 1991.

_____ *The Politics of the World-Economy*. Cambridge: Cambridge University Press, 1984.

Wheeler, James O., and Peter O. Muller. *Economic Geography*. New York: John Wiley, 1981.

Whyte, Martin King. "The Chinese Family and Economic Development: Obstacle or Enigma." *Economic Development and Cultural Change* 45 (1996): 1–30.

Wilbanks, Thomas J. *Location and Well-Being: An Introduction To Economic Geography*. San Francisco: Harper & Row, 1980.

World Commission on Environment and Development. *Our Common Future*. London: Oxford University Press, 1987.

Yapa, Lakshman. "What Causes Poverty?: A Postmodern View." *Annals of the Association of American Geographers* 86 (1996): 707–28.

Also consult these journals: *Economic Development and Cultural Change*; *Economic Geography*; *International Development Review*; *International Economic Review*; *International Journal of Political Economy*; *Journal of Developing Areas*; *Netherlands Journal of Economic and Social Geography*; *Regional Studies*.

CHAPTER 10

Agriculture

When you buy food in the supermarket, are you reminded of a farm? Not likely. The meat is carved into pieces that no longer resemble an animal and is wrapped in paper or plastic film. Often the vegetables are canned or frozen. The milk and eggs are in cartons.

Providing food in the United States and Canada is a vast industry. Only a few people are full-time farmers, and they may be more familiar with the operation of computers and advanced machinery than the typical factory or office worker.

The mechanized, highly productive American or Canadian farm contrasts with the subsistence farm found in much of the world. The most "typical" human—if there is such a person—is an Asian farmer who grows enough food to survive, with little surplus. This sharp contrast in agricultural practices constitutes one of the most fundamental differences between the more developed and less developed countries of the world.

Key Issues

1. Where did agriculture originate?

2. Where are agricultural regions in less developed countries?

3. Where are agricultural regions in more developed countries?

4. Why does agriculture vary among regions?

Idaho farmer selling wheat to Asian buyers. (David R. Frazier/Photo Researchers, Inc.)

Case Study

Wheat Farmers in Kansas and Pakistan

The Iqbel family grows wheat on their 1–hectare (2.5–acre) plot of land in the Punjab province of Pakistan in a manner similar to that of their ancestors. They perform most tasks by hand or with the help of animals. To irrigate the land, for example, they lift water from a 20-meter (65-foot) well by pushing a water wheel. More prosperous farmers in Pakistan use bullocks to turn the wheel.

The farm produces about 1,500 kilograms (3,300 pounds) of wheat per year, enough to feed the Iqbel family. Some years, they produce a small surplus, which they can sell. They can then use that money to buy other types of food or household items. In drought years, however, the crop yield is lower, and the Iqbel family must receive food from government and international relief organizations.

A world away in Kansas, the McKinley's farm the prairie sod. Like the Iqbels, they grow wheat in a climate that receives little rain. Otherwise, the two farm families lead very different lives. The McKinley family's farm is 200 times as large—200 hectares (500 acres). The McKinleys derive several hundred times more income from the sale of wheat than do the Iqbels.

The wheat grown on the McKinley's farm is not consumed directly by them. Instead, it is sold to a processing company and ultimately turned into bread wrapped in plastic and sold in a supermarket hundreds of kilometers away. Most of the wheat from the Iqbel's farm is consumed in the village where it is grown.

Approximately two-thirds of the people in the world are farmers. The overwhelming majority of them are like the Iqbels, growing enough food to feed themselves, but little more. In most African and Asian countries, more than 60 percent of the people are farmers. In contrast, fewer than 2 percent of the people in the United States and Canada are farmers. Yet the advanced technology used by these farmers allows them to produce enough food for people in the United States and Canada at a very high standard, plus food for many people elsewhere in the world.

The previous chapter divided economic activities into primary, secondary, and tertiary sectors. This chapter is concerned with the principal form of primary-sector economic activity—agriculture. The next two chapters look at the secondary and tertiary sectors.

Geographers study *where* agriculture is distributed across Earth. The most important distinction is what happens to farm products. In less developed countries the farm products are most often consumed on or near the farm where they are produced, whereas in more developed countries farmers sell what they produce.

Geographers observe a wide variety of agricultural practices. The reason *why* farming varies around the world relates to culture and the environment. Elements of the physical environment, such as climate, soil, and topography, set broad limits on agricultural practices, and farmers make choices to modify the environment in a variety of ways.

Farming is an economic activity that still depends very much on *local diversity* of environmental and cultural conditions. Despite increased knowledge of alternatives, farmers practice distinctive agriculture in different regions, and in fact on neighboring farms. Broad climate patterns influence the crops planted in a region, and local soil conditions influence the crops planted on an individual farm.

In each society, farmers possess very specific knowledge of their environmental conditions and certain technology for modifying the landscape. Within the limits of their technology, farmers choose from a variety of agricultural practices, based on their perception of the value of each alternative. These values are partly economic and partly cultural.

How farmers deal with their physical environment varies according to dietary preferences, availability of technology, and other cultural traditions. Farmers select agricultural practices based on cultural perceptions, because a society may hold some foods in high esteem while avoiding others

In a *global* economy, farmers increasingly pursue the most profitable agriculture. Agriculture is big business in more developed countries, and a major component of international trade in less developed countries.

After examining the origins and diffusion of agriculture, we will consider the agricultural practices used in less developed and more developed regions. We also will examine the problems farmers face in each type of region. Although each farm has a unique set of physical conditions and choice of crops, geographers group farms into several types by their distinctive environmental and cultural characteristics.

Key Issue 1

Where Did Agriculture Originate?

- Origins of agriculture
- Location of agricultural hearths
- Classifying agricultural regions

We cannot document the origins of agriculture with certainty, because it began before recorded history. Scholars try to reconstruct a logical sequence of events based on fragments of information about ancient agricultural practices and historical environmental conditions. Improvements in cultivating plants and domesticating animals evolved over thousands of years. This section offers an explanation for the process of origin and diffusion of agriculture.

Origins of Agriculture

Determining the origin of agriculture first requires a definition of what it is—and agriculture is not easily defined. We will use this definition: **Agriculture** is deliberate modification of Earth's surface through cultivation of plants and rearing of animals to obtain sustenance or economic gain. Agriculture thus originated when humans domesticated plants and animals for their use. The word *cultivate* means "to care for," and a **crop** is any plant cultivated by people.

Hunters and Gatherers

Before the invention of agriculture, all humans probably obtained the food they needed for survival through hunting for animals, fishing, or gathering plants (including berries, nuts, fruits, and roots). We call this *hunting and gathering*. Hunters and gatherers lived in small groups, usually fewer than 50, because a larger number would quickly exhaust the available resources within walking distance. They survived by collecting food often, perhaps daily. The food search might take only a short time or much of the day, depending on local conditions. The men hunted game or fished, and the women collected berries, nuts, and roots. This division of labor sounds like a stereotype, but is based on evidence from archaeology and anthropology.

The group traveled frequently, establishing new home bases or camps. The direction and frequency of migration depended on the movement of game and the seasonal growth of plants at various locations. We can assume that groups communicated with each other concerning hunting rights, intermarriage, and other specific subjects. For the most part, they kept the peace by steering clear of each other's territory.

Contemporary Hunting and Gathering. Today, only about 250,000 people, or less than 0.005 percent (one two-hundredth of 1 percent) of the world's population, still survive by hunting and gathering, rather than by agriculture. These people live in isolated locations, including the Arctic and the interior of Africa, Australia, and South America. Examples include African Bushmen of Namibia and Botswana and Aborigines in Australia.

Contemporary hunting and gathering societies are isolated groups living on the periphery of world settlement. But they provide insight into human customs that prevailed in prehistoric times, before the invention of agriculture.

Invention of Agriculture

Why did nomadic groups convert from hunting, gathering, and fishing to agriculture? In gathering wild vegetation, people inevitably cut plants and dropped berries, fruits, and seeds. These hunters probably observed that, over time, damaged or discarded food produced new plants. They may have deliberately cut plants or dropped berries on the ground to see if they would produce new plants. Subsequent generations learned to pour water over the site and to introduce manure and other soil improvements. Over thousands of years, plant cultivation apparently evolved from a combination of accident and deliberate experiment.

Prehistoric people may have originally domesticated animals for noneconomic reasons, such as sacrifices and other religious ceremonies. Other animals probably were domesticated as household pets, surviving on the group's food scraps.

Two Types of Cultivation. The earliest form of plant cultivation, according to prominent cultural geographer Carl Sauer, was **vegetative planting**, which is the reproduction of plants by direct cloning from existing plants, such as cutting stems and dividing roots. Plants found growing wild were deliberately divided and transplanted.

Coming later, according to Sauer, was **seed agriculture**, which is the reproduction of plants through annual planting of seeds that result from sexual fertilization. Seed agriculture is practiced by most farmers today.

Location of Agricultural Hearths

Agriculture probably did not originate in one location, but began in multiple, independent hearths, or points of origin. From these hearths, agricultural practices diffused across Earth's surface.

Location of First Vegetative Planting

Sauer believes that vegetative planting probably originated in Southeast Asia (Figure 10–1). The region's diversity of climate and topography probably encouraged growth of a wide variety of plants suitable for dividing and transplanting. Also, the people obtained food primarily by fishing, rather than by hunting and gathering, so they may have been more sedentary and therefore able to devote more attention to growing plants.

The first plants domesticated in Southeast Asia through vegetative planting probably included roots such as the taro and yam, and tree crops such as the banana and palm. Vegetative planting diffused from the Southeast Asian hearth northward and eastward to China and Japan, and westward through India to Southwest Asia, tropical Africa, and the Mediterranean lands. As for livestock, the dog, pig, and chicken probably were domesticated first in Southeast Asia.

Other early hearths of vegetative planting also may have emerged independently in West Africa and northwestern South America. It may have begun with the oil-palm tree and yam in West Africa and the manioc, sweet potato, and arrowroot in South America. The practice diffused from northwestern South America to Central America and eastern portions of South America.

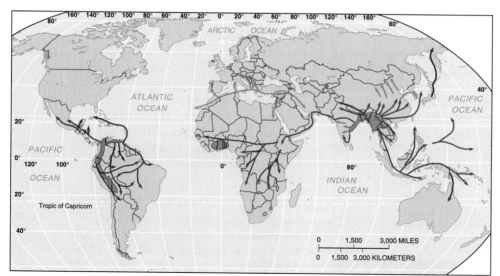
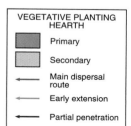

Figure 10–1 Origin and diffusion of vegetative planting. Vegetative planting is the reproduction of plants by direct cloning from existing plants. The practice originated primarily in Southeast Asia, according to Carl Sauer. Two other early centers of vegetative planting were in West Africa and northwestern South America. From these hearths, the practice diffused to other regions. (Adapted from Carl O. Sauer, *Agricultural Origins and Dispersals*, with the permission of the American Geographical Society.)

Location of First Seed Agriculture

Seed agriculture also originated in more than one hearth. Sauer identified three hearths in the Eastern Hemisphere: western India, northern China, and Ethiopia (Figure 10–2). Seed agriculture diffused quickly from western India to Southwest Asia, where important early advances were made, including the domestication of wheat and barley, two grains that became particularly important thousands of years later in European and American civilizations.

Apparently, inhabitants of Southwest Asia also were first to integrate seed agriculture with domestication of herd animals such as cattle, sheep, and goats. These animals were used to plow the land before planting seeds and, in turn, were fed part of the harvested crop. Other animal products, such as milk, meat, and skins, were first exploited at a later date, according to Sauer. This integration of plants and animals is a fundamental element of modern agriculture.

Diffusion of Seed Agriculture. Seed agriculture diffused from Southwest Asia across Europe and through North Africa. Greece, Crete, and Cyprus display the earliest evidence of seed agriculture in Europe. From these countries, agriculture may have diffused northwestward through the Danube River basin, eventually to the Baltic Sea and North Sea, and northeastward to Ukraine. Most of the plants and animals domesticated in Southwest Asia spread into Europe, although barley and cattle became more important farther north, perhaps because of cooler and moister climatic conditions.

Seed agriculture also diffused eastward from Southwest Asia to northwestern India and the Indus River plain. Again, various domesticated plants and animals were brought from Southwest Asia, although other plants, such as cotton and rice, arrived in India from different hearths.

From the northern China hearth, millet diffused to South Asia and Southeast Asia. Rice, which ultimately became the most important crop in much of Asia, has an unknown hearth, although some geographers consider Southeast Asia to be its most likely location. Sauer identified a third independent hearth in Ethiopia, where millet and sorghum were domesticated early. However, he argued that agricultural advances in Ethiopia did not diffuse widely to other locations. That Ethiopia is an ancient hearth for seed agriculture is ironic, because rapid population growth, devastating civil wars, and adverse environmental conditions have combined to make Ethiopia the site of widespread starvation.

Two independent seed agriculture hearths originated in the Western Hemisphere: southern Mexico and northern Peru. The hearth in southern Mexico, which extended into Guatemala and Honduras, was the point of origin for squash and maize (corn). Squash, beans, and cotton may have been domesticated in northern Peru. From these two hearths, agricultural practices diffused to other parts of the Western Hemisphere, although agriculture was not widely practiced until European colonists began to arrive some 500 years ago. The only domesticated animals were the llama, alpaca, and turkey; herd animals were unknown until European explorers brought them in the sixteenth century.

That agriculture had multiple origins means that, from earliest times, people have produced food in distinctive ways in different regions. This diversity derives from a unique legacy of wild plants, climatic condi-

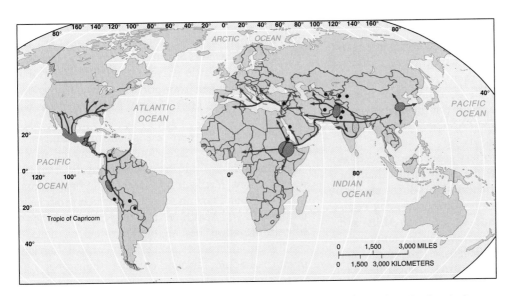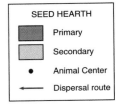

Figure 10–2 Origin and diffusion of seed agriculture and livestock herding. Seed agriculture may have originated in several hearths, including western India, northern China, and Ethiopia. Southern Mexico and northwestern South America may have been other early hearths. Early advances were made in Southwest Asia. (Adapted from Carl O. Sauer, *Agricultural Origins and Dispersals*, with the permission of the American Geographical Society.)

tions, and cultural preferences in each region. Improved communications in recent centuries have encouraged the diffusion of some plants to varied locations around the world. Many plants and animals thrive across a wide portion of Earth's surface, not just in their place of original domestication. Only after A.D. 1500, for example, were wheat, oats, and barley introduced to the Western Hemisphere, and maize to the Eastern Hemisphere.

Classifying Agricultural Regions

The most fundamental differences in agricultural practices are between those in less developed countries and those in more developed countries. Farmers in LDCs generally practice subsistence agriculture, whereas farmers in MDCs, including the United States, Canada, Western Europe, Australia, and New Zealand, practice commercial agriculture.

Differences Between Subsistence and Commercial Agriculture

Subsistence agriculture, found in less developed countries, is the production of food primarily for consumption by the farmer's family. **Commercial agriculture,** found in more developed countries, is the production of food primarily for sale off the farm. Five principal features distinguish commercial agriculture from subsistence agriculture:

- Purpose of farming
- Percentage of farmers in the labor force
- Use of machinery
- Farm size
- Relationship of farming to other businesses

Purpose of Farming. Subsistence and commercial agriculture are undertaken for different purposes. In LDCs, most people produce food for their own consumption. Some surplus may be sold to the government or to private firms, but the surplus product is not the farmer's primary purpose and may not even exist some years because of growing conditions.

In commercial farming, farmers grow crops and raise animals primarily for sale off the farm rather than for their own consumption. Agricultural products are not sold directly to consumers but to food-processing companies. Large processors, such as General Mills and Ralston Purina, typically sign contracts with commercial farmers to buy their grain, chickens, cattle, and other output. Farmers may have contracts to sell sugar beets to sugar refineries, potatoes to distilleries, and oranges to manufacturers of concentrated juices.

Percentage of Farmers in the Labor Force. In more developed countries, less than 5 percent of the workers are engaged directly in farming, compared to 60 percent in less developed countries. (Refer to Figure 9–5, which shows the world distribution of percentage of workers engaged in farming.)

The percentage of farmers is even lower in the United States and Canada, only 2 percent. Yet the small percentage of farmers in the United States and Canada produces enough food, not only for themselves and the rest of the region, but also a surplus to feed people elsewhere.

The number of farmers has declined dramatically in more developed societies during the twentieth century. Both push and pull migration factors have been responsible: people have been pushed away from farms by lack of opportunity to earn a decent income, and at the same time they have been pulled to higher-paying jobs in urban areas. The United States had about 6 million farms in 1945, 4 million in 1960, and 2 million in 1995.

Use of Machinery. A small number of farmers in more developed societies can feed many people because they rely on machinery to perform work, rather than people or animals (Figure 10–3). In less developed countries, farmers do much of the work with hand tools and animal power.

Traditionally, the farmer or local craftspeople made equipment from wood, but beginning in the late eighteenth century, factories produced farm machinery. The first all-iron plow was made in the 1770s and was followed in the nineteenth and twentieth centuries by inventions that made farming less dependent on human or animal power. Tractors, combines, corn pickers, planters, and other factory-made farm machines have replaced or supplemented manual labor.

Transportation improvements also aid commercial farmers. The building of railroads in the nineteenth century and highways and trucks in the twentieth century have enabled farmers to transport crops and livestock farther and faster. Cattle arrive at market heavier and in better condition when transported by truck or train than when driven on hoof. Crops reach markets without spoiling.

Commercial farmers use scientific advances to increase productivity. Experiments conducted in university laboratories, industry, and research organizations generate new fertilizers, herbicides, hybrid plants, animal breeds, and farming practices, which produce higher crop yields and healthier animals. Access to other scientific information has enabled farmers to make more intelligent decisions concerning proper agricultural practices. Some farmers conduct their own on-farm research.

Farm Size. The average farm size is relatively large in commercial agriculture, especially in the United States and Canada. U.S. farms average about 181 hectares

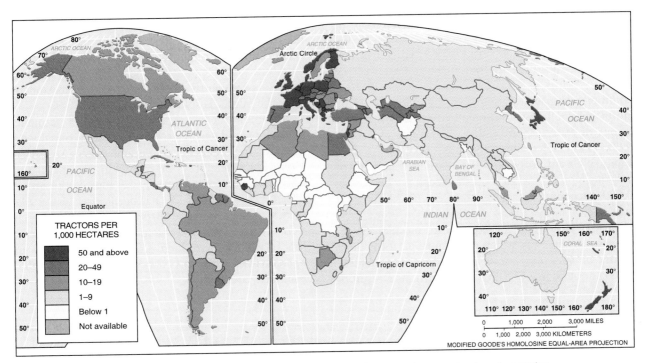

Figure 10–3 Tractors per 1,000 hectares of farmland. Farmers in more developed countries possess more machinery, such as tractors, than do farmers in less developed ones. The machinery makes it possible for commercial farmers to farm extensive areas, a necessary practice to pay for the expensive machinery.

Commercial agriculture depends heavily on expensive machinery to manage large farms efficiently. These combine machines are reaping, threshing, and cleaning wheat in Sherman County, Kansas. (Cotton Coulson/Woodfin Camp & Associates)

(469 acres). Despite their size, most commercial farms in developed countries are family owned and operated—98 percent in the United States. Commercial farmers frequently expand their holdings by renting nearby fields.

Commercial agriculture is increasingly dominated by a handful of large farms. In the United States, the largest 5 percent of farms (those with cash receipts exceeding $250,000 per year) account for more than one-half of the country's total output (and the largest 2 percent of farms—those with annual cash receipts exceeding $500,000—account for more than one-third of total output).

Large size is partly a consequence of mechanization. Combines, pickers, and other machinery perform most efficiently at very large scales, and their considerable expense cannot be justified on a small farm. As a result of the large size and the high level of mechanization, commercial agriculture is an expensive business. Farmers spend hundreds of thousands of dollars to buy or rent land and machinery before beginning operations. This money is frequently borrowed from a bank and repaid after the output is sold.

Although there are fewer farms and farmers, the amount of land devoted to agriculture has remained fairly constant in Western Europe and Anglo-America since 1900. The annual loss of farmland in the United States is only 0.01 percent—primarily because of the growth of urban areas—but this loss has been offset by the creation of new agricultural land through irrigation and reclamation. A more serious problem in the United States is the loss of the most productive farmland, known as **prime agricultural land**, as urban areas sprawl into the surrounding countryside.

Figure 10–4 Agricultural regions. The major agricultural practices of the world can be divided into subsistence and commercial regions.

Subsistence regions include:
- *Shifting cultivation*—primarily the tropical regions of South America, Africa, and Southeast Asia.
- *Pastoral nomadism*—primarily the drylands of North Africa and Asia.
- *Intensive subsistence, wet rice dominant*—primarily the large population concentrations of East and South Asia.
- *Intensive subsistence, crops other than rice dominant*—primarily the large population concentrations of East and South Asia where growing rice is difficult.

Commercial regions include:
- *Mixed crop and livestock*—primarily U.S. Midwest and central Europe.
- *Dairying*—primarily near population clusters in northeastern United States, southeastern Canada, and northwestern Europe.
- *Grain*—primarily north-central United States and Eastern Europe.
- *Ranching*—primarily the drylands of western United States, southeastern South America, Central Asia, southern Africa, and Australia.
- *Mediterranean*—primarily lands surrounding the Mediterranean Sea, western United States, and Chile.
- *Commercial gardening*—primarily southeastern United States and southeastern Australia.
- *Plantation*—primarily the tropical and subtropical regions of Latin America, Africa, and Asia.

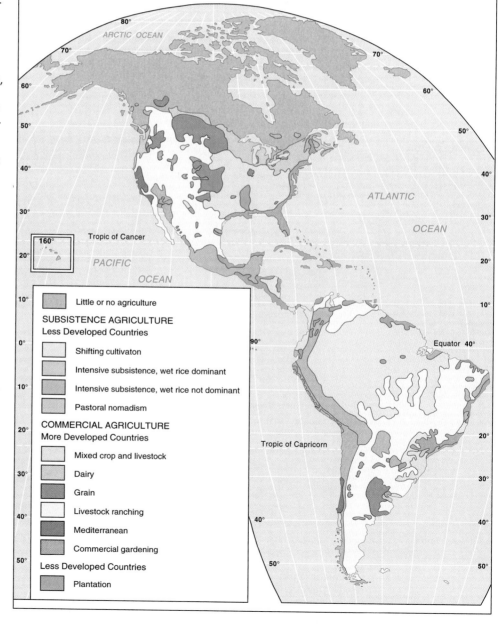

Relationship of Farming to Other Businesses. Commercial farming is closely tied to other businesses. The system of commercial farming found in the United States and other relatively developed countries has been called **agribusiness**, because the family farm is not an isolated activity but is integrated into a large food production industry. Commercial farmers make heavy use of modern communications and information technology to stay in touch and keep track of prices, yields, and expenditures.

Although farmers are less than 2 percent of the U.S. labor force, more than 20 percent of U.S. labor works in food production related to agribusiness: food processing, packaging, storing, distributing, and retailing. Agribusiness encompasses such diverse enterprises as tractor manufacturing, fertilizer production, and seed distribution. Although most farms are owned by individual fam-

ilies, many other aspects of agribusiness are controlled by large corporations.

Mapping Agricultural Regions

Several attempts have been made to outline the major types of subsistence and commercial agriculture currently practiced in the world, but few of these classifications include maps that show regional distributions. The most widely used map of world agricultural regions was prepared by geographer Derwent Whittlesey in 1936.

Whittlesey identified 11 main agricultural regions, plus an area where agriculture was nonexistent. Whittlesey's 11 regions are divided between five that are important in less developed countries and six that are important in more developed countries (Figure 10–4).

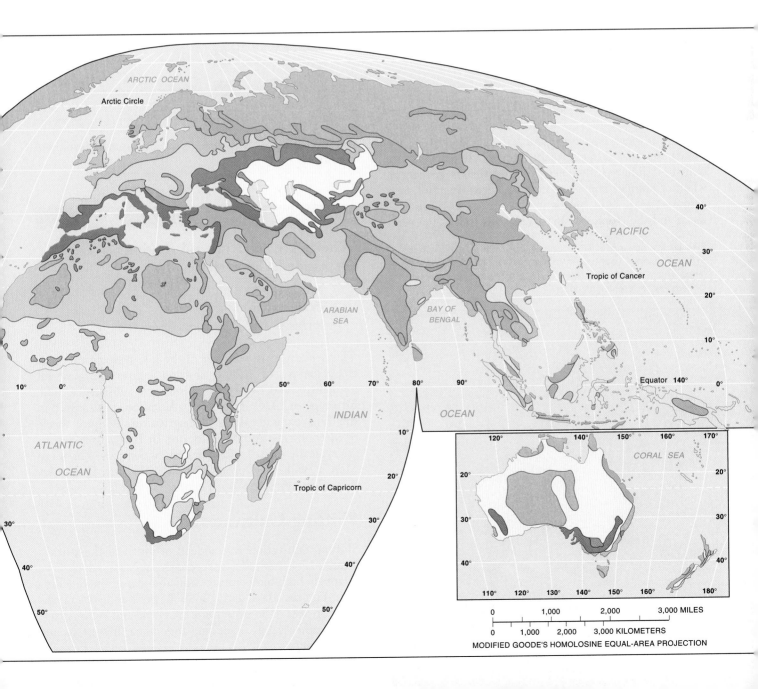

Figure 10–5 Simplified climate regions. Figure 1–12 shows more detail. Compare the broad distribution of the major climate regions with the distinctive types of agriculture in more developed and less developed countries shown in Figure 10-4. Shifting cultivation is found in much of the world's Humid Low-Latitude climate region. Pastoral nomadism is important in the Dry climates of the less developed countries, whereas ranching is important in the Dry climates of the more developed countries. In China, rice is the most important crop in the Warm Mid-Latitude climate region, whereas other crops are most important in the Cold Mid-Latitude climate region.

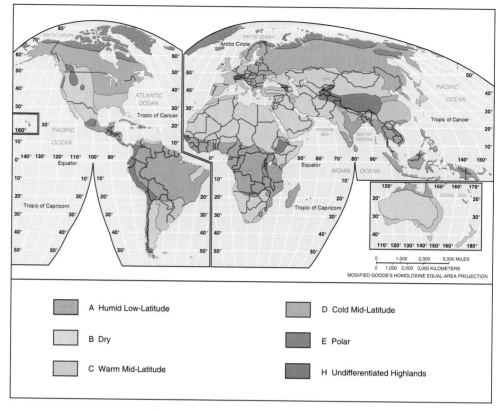

A Humid Low-Latitude

B Dry

C Warm Mid-Latitude

D Cold Mid-Latitude

E Polar

H Undifferentiated Highlands

Within more developed and less developed regions, Whittlesey sorted out agricultural practices primarily by climate. Climate influences the crop that is grown, or whether animals are raised instead of growing any crop (Figure 10–5). Thus, agriculture varies between the drylands and the tropics within LDCs—as well as between the drylands of less developed and more developed countries.

Farmers live on large isolated farms in the United States, whereas in less developed countries most farmers live near others in villages. This is a dairy farm in southwestern Wisconsin. (Francois Gohier/Photo Researchers, Inc.)

Key Issue 2

Where Are Agricultural Regions in Less Developed Countries?

- Shifting cultivation
- Pastoral nomadism
- Intensive subsistence agriculture

This section considers four of the five agricultural types characteristic of LDCs: shifting cultivation, pastoral nomadism, and two types of intensive subsistence. The fifth type of LDC agriculture, plantation, is discussed in the third key issue along with agriculture in more developed countries. Although located in LDCs, plantations are typically owned and operated by companies based in MDCs, and they produce principally for consumers located in more developed countries.

Shifting Cultivation

Shifting cultivation is practiced in much of the world's Humid Low-Latitude, or A, climate regions, which have relatively high temperatures and abundant rainfall (compare Figures 10–4 and 10–5). It predominates in the Amazon area of South America, Central and West Africa, and Southeast Asia, including Indochina, Indonesia, and New Guinea.

Why is it called shifting cultivation, rather than shifting agriculture? It is essentially a matter of scale. We use the term "cultivation" (as in "cultivate a garden") because "agriculture" implies greater use of tools and animals and more sophisticated modification of the landscape. Shifting cultivation bears little relation to the agriculture practiced in the more developed regions of Western Europe and North America, or even in other LDCs such as China.

Characteristics of Shifting Cultivation

Shifting cultivation has two distinguishing hallmarks:

- Farmers clear land for planting by slashing vegetation and burning the debris (**slash-and-burn agriculture**).
- Farmers grow crops on a cleared field for only a few years until soil nutrients are depleted, and then leave it fallow (nothing planted) for many years, so the soil can recover.

People who practice shifting cultivation generally live in small villages and grow food on the surrounding land, which the village controls. Well-recognized boundaries usually separate neighboring villages.

The Process of Shifting Cultivation. Each year, villagers designate for planting an area surrounding the settlement. Before planting, they must remove the dense vegetation that typically covers tropical land. Using axes, they cut most of the trees, sparing only those that are economically useful. An efficient strategy they employ is to cut selected large trees, which bring down smaller trees that may have been weakened by notching.

The undergrowth is cleared away with a machete or other long knife. On a windless day, the debris is burned under carefully controlled conditions. The rains wash the fresh ashes into the soil, providing needed nutrients. The cleared area is known by a variety of names in different regions, including **swidden**, *ladang*, *milpa*, *chena*, and *kaingin*.

Before planting, fields are prepared by hand, perhaps with the help of a simple implement such as a hoe; plows and animals are rarely used. The only fertilizer generally available is potash (potassium) from burning the debris when the site is cleared. Little weeding is done the first year that a cleared patch of land is farmed; weeds may be cleared with a hoe in subsequent years.

The cleared land can support crops only briefly, usually 3 years or less. In many regions, the most productive harvest comes in the second year after burning. Thereafter, soil nutrients are rapidly depleted, and the land becomes too infertile to nourish crops. Rapid weed growth also contributes to the abandonment of a swidden after a few years.

When the swidden is no longer fertile, villagers identify a new site and begin clearing it. They leave the old site uncropped for many years, allowing it to become overrun again by natural vegetation. The field is not actually abandoned: the villagers will return to the site someday, perhaps as few as 6 years or as many as 20 years later, to begin the process of clearing the land again. In the meantime, they may still care for fruit-bearing trees on the site.

If a cleared area outside a village is too small to provide food for the population, then some of the people may establish a new village and practice shifting cultivation there. Some farmers may move temporarily to another settlement if the field they are clearing that year is distant.

Crops of Shifting Cultivation. The precise crops grown by each village vary by local custom and taste. The predominant crops include upland rice in Southeast Asia, maize (corn) and manioc (cassava) in South America, and millet and sorghum in Africa. Yams, sugar cane, plantain, and vegetables also are grown in some regions. These crops may have begun in one region of shifting cultivation, but then diffused to other areas in recent years.

These shifting cultivation farmers in Peru are preparing fields for planting by slashing and burning the vegetation. The dense vegetation is chopped down, and the debris is burned in order to provide the soil with needed nutrients. (Asa C. Thoresen/Photo Researchers, Inc.)

The Kayapo people of Brazil's Amazon tropical rainforest do not arrange crops in the rectangular fields and rows that are familiar to us. They plant in concentric rings. At first, they plant sweet potatoes and yams in the inner area. In successive rings go corn and rice, manioc, and more yams. The outermost ring contains papaya, banana, pineapple, mango, cotton, and beans. Plants that require more nutrients are located in the outer ring. It is here that the leafy crowns of cut trees fall when the field is cleared, and their rotting releases more nutrients into the soil. In subsequent years, the inner area of potatoes and yams expands to replace corn and rice.

Most families grow only for their own needs, so one swidden may contain a large variety of intermingled crops, which are harvested individually at the best time. In shifting cultivation, a "farm field" appears much more chaotic than do fields in more developed countries, where a single crop like corn or wheat may grow over an extensive area. In some cases, families may specialize in a few crops and trade with villagers who have a surplus of others.

Ownership and Use of Land in Shifting Cultivation. Traditionally, land is owned by the village as a whole rather than separately by each resident. The chief or ruling council allocates a patch of land to each family and allows it to retain the output. Individuals may also have the right to own or protect specific trees surrounding the village. Private individuals now own the land in some communities, especially in Latin America.

Shifting cultivation occupies approximately one-fourth of the world's land area, a higher percentage than any other type of agriculture. However, only 5 percent of the world's population engage in shifting cultivation. The gap between the percentage of people and land area is not surprising, because the practice of moving from one field to another every couple of years requires more land per person than do other types of agriculture.

The percentage of land devoted to shifting cultivation is declining in the tropics, and its future role in world agriculture is not clear. Shifting cultivation is being replaced by logging, cattle ranching, and cultivation of cash crops. The reason is development (see Globalization and Local Diversity box). Alternatives to shifting cultivation require cutting of vast expanses of forest.

Pastoral Nomadism

Pastoral nomadism is a form of subsistence agriculture based on the herding of domesticated animals. The word *pastoral* refers to sheep herding. It is adapted to dry climates where planting crops is impossible. Pastoral nomads live primarily in the large belt of arid and semi-arid land that includes North Africa, the Middle East, and parts of Central Asia (refer to Figures 10–4 and 10–5). The Bedouins of Saudi Arabia and North Africa and the Masai of East Africa are examples of nomadic groups. Only approximately 15 million people are pastoral nomads, but they sparsely occupy approximately 20 percent of Earth's land area.

Characteristics of Pastoral Nomadism

In contrast to other subsistence farmers, pastoral nomads depend primarily on animals rather than crops for survival. The animals provide milk, and their skins and hair are used for clothing and tents. Like other subsistence farmers, though, pastoral nomads consume mostly grain rather than meat. Their animals are commonly not slaughtered, although dead ones may be consumed. To nomads, the size of their herd is both an important measure of power and prestige and their main security during adverse environmental conditions.

Some pastoral nomads obtain grain from sedentary subsistence farmers in exchange for animal products. More often, part of a nomadic group—perhaps the

Pastoral nomadism is practiced in the drylands of less developed countries. A sheperd herds sheep in Taza Province, Morocco. (Lauren Goodsmith/The Image Works)

Globalization and Local Diversity

Shifting Cultivation and Deforestation

Geographers are concerned with **where** humans are transforming the landscape, because these changes can have global impacts. One area of particular concern is the tropical rain forests, because in recent years they have been disappearing at the rate of 10 million to 20 million hectares (25 million to 50 million acres or 40,000 to 80,000 square miles) per year.

An area somewhere between the size of Virginia and Kansas is cleared annually, a very significant environmental problem. The amount of Earth's surface allocated to tropical rain forests has already been reduced to less than half of its original area, and unless drastic measures are taken, the area will be reduced by another 20 percent within a decade.

The reason **why** governments in less developed countries have supported the destruction of rain forests is promotion of economic development. Selling timber to builders or raising beef cattle for fast-food restaurants is a more effective development strategy than maintaining shifting cultivation. Until recent years, the World Bank supported deforestation with loans to finance development schemes that required clearing forests. Less developed countries also see shifting cultivation as an inefficient way to grow food in a hungry world. Indeed, compared to other forms of agriculture, shifting cultivation can support only a small population in an area without causing environmental damage.

To its critics, shifting cultivation is at best a preliminary step in economic development. Pioneers use shifting cultivation to clear forests in the tropics and to open land for development where permanent agriculture never existed. People unable to find agricultural land elsewhere can migrate to the tropical forests and initially practice shifting cultivation. Critics say it then should be replaced by more sophisticated agriculture that yields more per land area.

But defenders of shifting cultivation consider it the most environmentally sound approach for the tropics. Practices used in other forms of agriculture, like using fertilizers and pesticides and permanently clearing fields, may damage the soil, cause severe erosion, and upset balanced ecosystems.

Large-scale destruction of the rain forests also may contribute to **global** warming. When large numbers of trees are cut, their burning and decay release large volumes of carbon dioxide. This gas can build up in the atmosphere, acting like the window glass in a greenhouse to trap solar energy in the atmosphere—the "greenhouse effect," discussed in Chapter 14.

Elimination of shifting cultivation could also upset the traditional **local diversity** of cultures in the tropics. The activities of shifting cultivation are intertwined with other social, religious, political, and various folk customs. A drastic change in the agricultural economy could disrupt other activities of daily life.

As the importance of tropical rain forests to the global environment has become recognized, LDCs have been pressured to restrict further destruction of them. In one innovative strategy, Bolivia agreed to set aside 1.5 million hectares (3.7 million acres) in a forest reserve in exchange for cancellation of $650,000,000 of its debt to developed countries.

In Brazil's Amazon rain forest, deforestation is increasing. From 21,000 square kilometers (6,000 square miles) per year during the 1980s, deforestation declined nearly half to 11,000 square kilometers (4,000 square miles) in 1991, but increased to 29,000 square kilometers (11,000 square miles) in 1995. A 1997 U.S. government study placed deforestation even higher, at 58,000 square kilometers (22,000 square miles) per year.

women and children—may plant crops at a fixed location while the rest of the group wanders with the herd. Nomads might hire workers to practice sedentary agriculture in return for grain and protection. Other nomads might sow grain in recently flooded areas and return later in the year to harvest the crop. Yet another strategy is to remain in one place and cultivate the land when rainfall is abundant; then, during periods that are too dry to grow crops, the group can increase the size of the herd and migrate in search of food and water.

Choice of Animals. Nomads select the type and number of animals for the herd according to local cultural and physical characteristics. The choice depends on the relative prestige of animals and the ability of species to adapt to a particular climate and vegetation. The camel is most frequently desired in North Africa and the Middle East, followed by sheep and goats. In Central Asia, the horse is particularly important.

The camel is well suited to arid climates, because it can go long periods without water, carry heavy baggage, and move rapidly. However, the camel is particularly bothered by flies and sleeping sickness and has a relatively long period—12 months—from conception to birth. Goats need more water than do camels but are tough and agile and can survive on virtually any vegetation, no matter how poor. Sheep are relatively slow-moving and are more affected by climatic changes. They require more water and are more selective in which plants they will eat. The minimum number of animals necessary to support each family adequately varies according to the particular group and animal. The typical nomadic family needs 25 to 60 goats or sheep or 10 to 25 camels.

Movements of Pastoral Nomads. Pastoral nomads do not wander randomly across the landscape but have a strong sense of territoriality. Every group controls a piece of territory and will invade another group's territory only in an emergency or if war is declared. The goal of each group is to control a territory large enough to contain the forage and water needed for survival. The actual amount of land a group controls depends on its wealth and power.

The precise migration patterns evolve from intimate knowledge of the area's physical and cultural characteristics. Groups frequently divide into herding units of five or six families and choose routes depending on experience concerning the most likely water sources during the various seasons of the year. The selection of routes varies in unusually wet or dry years, and is influenced by the condition of their animals and the area's political stability.

Some pastoral nomads practice **transhumance**, which is seasonal migration of livestock between mountains and lowland pasture areas. **Pasture** is grass or other plants grown for feeding grazing animals, as well as land used for grazing. Sheep or other animals may pasture in alpine meadows in the summer and be herded back down into valleys for winter pasture.

The Future of Pastoral Nomadism

Agricultural experts once regarded pastoral nomadism as a stage in the evolution of agriculture, between the hunters and gatherers who migrated across Earth's surface in search of food and sedentary farmers who cultivate grain in one place. Because they had domesticated animals but not plants, pastoral nomads were considered more advanced than hunters and gatherers but less advanced than settled farmers.

Pastoral nomadism is now generally recognized as an offshoot of sedentary agriculture, not a primitive precursor of it. It is simply a practical way of surviving on land that receives too little rain for cultivation of crops. The domestication of animals—the basis for pastoral nomadism—probably was achieved originally by sedentary farmers, not by nomadic hunters. Pastoral nomads therefore had to be familiar with sedentary farming, and in many cases practiced it.

Today, pastoral nomadism is a declining form of agriculture, a victim in part of modern technology. Before recent transportation and communications inventions, pastoral nomads played an important role as carriers of goods and information across the sparsely inhabited drylands. Nomads used to be the most powerful inhabitants of the drylands, but now, with modern weapons, national governments can control the nomadic population more effectively.

Government efforts to resettle nomads have been particularly vigorous in China, Kazakhstan, and several Middle Eastern countries, including Egypt, Israel, Saudi Arabia, and Syria. Nomads are reluctant to cooperate, so these countries have experienced difficulty in trying to force settlement in collectives and cooperatives. Governments force groups to give up pastoral nomadism because they want the land for other uses. Land that can be irrigated is converted from nomadic to sedentary agriculture. In some instances, the mining and petroleum industries now operate in drylands formerly occupied by pastoral nomads.

Some nomads are encouraged to try sedentary agriculture or to work for mining or petroleum companies. Others are still allowed to move about, but only within ranches of fixed boundaries. In the future, pastoral nomadism will be increasingly confined to areas that cannot be irrigated or that lack valuable raw materials.

Intensive Subsistence Agriculture

Shifting cultivation and pastoral nomadism are forms of subsistence agriculture found in regions of low density. But three-fourths of the world's people live in LDCs, and another form of subsistence agriculture is needed to feed most of them: **intensive subsistence agriculture**. The term *intensive* implies that farmers must work more intensively to subsist on a parcel of land.

In densely populated East, South, and Southeast Asia, most farmers practice intensive subsistence agriculture. The typical farm in Asia's intensive subsistence agriculture regions is much smaller than elsewhere in the world. Many Asian farmers own several fragmented plots, frequently a result of dividing individual holdings among several children over several centuries.

Because the agricultural density—the ratio of farmers to arable land—is so high in parts of East and South Asia, families must produce enough food for their survival from a very small area of land. They do this through careful agricultural practices, refined over thousands of years in response to local environmental and cultural patterns. Most of the work is done by hand or with animals rather than with machines, in part due to abundant labor, but largely from lack of funds to buy equipment.

To maximize food production, intensive subsistence farmers waste virtually no land. Corners of fields and irregularly shaped pieces of land are planted rather than left idle. Paths and roads are kept as narrow as possible to minimize the loss of arable land. Livestock are rarely permitted to graze on land that could be used to plant crops, and little grain is grown to feed the animals.

Intensive Subsistence with Wet Rice Dominant

The intensive agriculture region of Asia can be divided between areas where wet rice dominates and areas where it does not (refer to Figure 10–4). The term **wet rice** refers to the practice of planting rice on dry land in a nursery, and then moving the seedlings to a flooded field to promote growth. Wet rice occupies a relatively small percentage of Asia's agricultural land but is the region's most important source of food. Intensive wet rice farming is the dominant type of agriculture in Southeast China, East India, and much of Southeast Asia (Figure 10–6).

Successful production of large yields of rice is an elaborate process, time-consuming and done mostly by hand.

The consumers of the rice also perform the work, and all family members, including children, contribute to the effort.

Growing rice involves several steps: First, a farmer prepares the field for planting, using a plow drawn by water buffalo or oxen. The use of a plow and animal power is one characteristic that distinguishes subsistence agriculture from shifting cultivation.

Then, the plowed land is flooded with water. The water is collected from rainfall, river overflow, or irrigation. Too much or too little can damage the crop—a par-

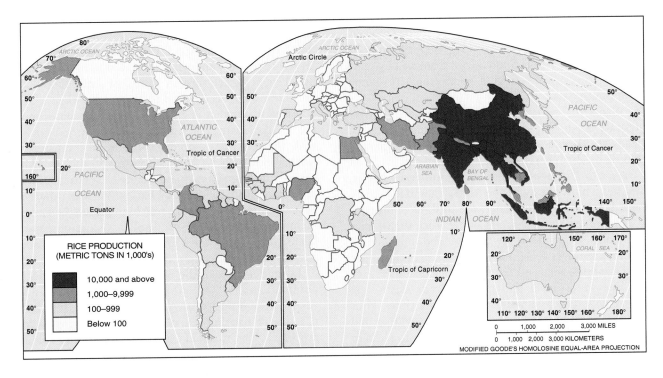

RICE PRODUCTION
(METRIC TONS IN 1,000's)

- 10,000 and above
- 1,000–9,999
- 100–999
- Below 100

MODIFIED GOODE'S HOMOLOSINE EQUAL-AREA PROJECTION

Figure 10–6 Rice production. Rice is the most important crop in the large population concentrations of East, South, and Southeast Asia. Asian farmers grow more than 90 percent of the world's rice, and two countries—China and India—account for more than half of world production. Growing rice is a labor-intensive operation, done mostly by hand. In Yunnan Province, China, rice seedlings grown in a nursery are transplanted to the field. (Ken Straiton/The Stock Market.)

ticular problem for farmers in South Asia who depend on monsoon rains, which do not always arrive at the same time each summer. Before planting, dikes and canals are repaired to assure the right quantity of water in the field. The flooded field is called a **sawah** in the Austronesian language widely spoken in Indonesia, including Java. Europeans and North Americans frequently, but incorrectly, call it a **paddy**, the Malay word for wet rice.

The customary way to plant rice is by growing seedlings on dryland in a nursery, and then transplanting the seedlings into the flooded field. Typically, one-tenth of a sawah is devoted to cultivation of seedlings. After about a month, they are transferred to the rest of the field. Rice plants grow submerged in water for approximately three-fourths of the growing period. Another method of planting rice is to broadcast dry seeds by scattering them through the field, a method used to some extent in South Asia.

Rice plants are harvested by hand, usually with knives. To separate the husks, known as **chaff**, from the seeds, the heads are **threshed** by beating them on the ground or treading on them barefoot. The threshed rice is placed in a tray, and the lighter chaff is **winnowed**, that is, allowed to be blown away by the wind. If the rice is to be consumed directly by the farmer, the **hull**, or outer covering, is removed by mortar and pestle. Rice that is sold commercially is frequently whitened and polished, a process that removes some nutrients but leaves rice more pleasing in appearance and taste to many consumers.

Wet rice is most easily grown on flat land, because the plants are submerged in water much of the time. Thus most wet-rice cultivation is located in river valleys and deltas. But the pressure of population growth in parts of East Asia has forced expansion of areas under rice cultivation. One method of developing additional land suitable for growing rice is to terrace the hillsides of river valleys.

Land is used even more intensively in parts of Asia by obtaining two harvests per year from one field, a process known as **double cropping**. Double cropping is common in places having warm winters, such as South China and Taiwan, but is relatively rare in India, where most areas have dry winters. Normally, double cropping involves alternating between wet rice, grown in the summer when precipitation is higher, and wheat, barley, or another dry crop, grown in the drier winter season. Crops other than rice may be grown in the wet-rice region in the summer on nonirrigated land.

Intensive Subsistence with Wet Rice Not Dominant

Climate prevents farmers from growing wet rice in portions of Asia, especially where summer precipitation levels are too low and winters are too harsh (refer to Figure 10–5). Agriculture in much of interior India and northeast China is devoted to crops other than wet rice.

Aside from what is grown, this region shares most of the characteristics of intensive subsistence agriculture with the wet-rice region. Land is used intensively and worked primarily by human power with the assistance of some hand implements and animals. Wheat is the most important crop, followed by barley. Various other grains and legumes are grown for household consumption, including millet, oats, corn, kaoliang, sorghum, and soybeans. Also grown are some crops sold for cash, such as cotton, flax, hemp, and tobacco.

In milder parts of the region where wet rice does not dominate, more than one harvest can be obtained some years through skilled use of **crop rotation**, which is the practice of rotating use of different fields from crop to crop each year, to avoid exhausting the soil. In colder climates, wheat or another crop is planted in the spring and harvested in the fall, but no crops can be sown through the winter.

Since the Communist Revolution in 1949, private individuals have owned little agricultural land in China. Instead, the Communist government organized agricul-

Wet rice terraces. Seedlings are transplanted into flooded fields. Because wet rice should be grown on flat land, hillsides may be terraced in places such as Sikkim, India, in the foothills of the Himalayas, to increase the area of rice production. (U.N. Food and Agriculture Organization)

tural producer communes, which typically consisted of several villages of several hundred people. By combining several small fields into a single large unit, the government hoped to promote agricultural efficiency, because scarce equipment and animals could be shared, and larger improvement projects, such as flood control, water storage, and terracing, could be completed. In reality, productivity did not increase as much as the government had expected, because people worked less efficiently for the commune than when working for themselves.

China has dismantled the agricultural communes. The communes still hold legal title to agricultural land, but villagers sign contracts entitling them to farm portions of the land as private individuals. Chinese farmers may sell to others the right to use the land and to pass on the right to their children. Reorganization has been difficult because irrigation systems, equipment, and other infrastructure were developed to serve large communal farms rather than small individually managed ones, which cannot afford to operate and maintain the machinery. But production has increased greatly.

Key Issue 3

Where Are Agricultural Regions in More Developed Countries?

- Mixed crop and livestock farming
- Dairy farming
- Grain farming
- Livestock ranching
- Mediterranean agriculture
- Commercial gardening and fruit farming
- Plantation agriculture

Commercial agriculture in more developed countries can be divided into six main types: mixed crops and livestock; dairying; grain farming; livestock ranching; Mediterranean agriculture; and gardening and fruit culture. Each type is predominant in distinctive regions within MDCs, depending largely on climate. The end of this section examines plantation farming, a form of commercial agriculture in LDCs.

Mixed Crop and Livestock Farming

Mixed crop and livestock farming is the most common form of commercial agriculture in the United States west of the Appalachians and east of 98° west longitude and in much of Europe from France to Russia (Figure 10–4).

Characteristics of Mixed Crop and Livestock Farming

The most distinctive characteristic of mixed crop and livestock farming is its integration of crops and livestock. Most of the crops are fed to animals rather than consumed directly by humans. In turn, the livestock supply manure to improve soil fertility to grow more crops. A typical mixed commercial farm devotes nearly all land area to growing crops but derives more than three-fourths of its income from the sale of animal products, such as beef, milk, and eggs. In the United States, pigs are often bred directly on the farms, while cattle may be brought in to be fattened on corn.

Mixed crop and livestock farming permits farmers to distribute the workload more evenly through the year. Fields require less attention in the winter than in the spring, when crops are planted, and in the fall, when they are harvested. Livestock, on the other hand, require year-long attention. A mix of crops and livestock also reduces seasonal variations in income; most income from crops comes during the harvest season, but livestock products can be sold throughout the year.

Crop Rotation Systems. Mixed crop and livestock farming typically involves crop rotation. The farm is divided into a number of fields, and each field is planted on a planned cycle, often of several years. The crop planted changes from one year to the next, typically going through a cycle of two or more crops and a year of fallow before the cycle is repeated. Crop rotation helps maintain the fertility of a field, because various crops deplete the soil of certain nutrients but restore others.

Crop rotation contrasts with shifting cultivation, in which nutrients depleted from a field are restored only by leaving the field fallow (uncropped) for many years. In any given year, crops cannot be planted in most of an area's fields, so overall production in shifting cultivation is much lower than in mixed commercial farming.

A two-field crop rotation system was developed in Northern Europe as early as the fifth century A.D. A **cereal grain**, such as oats, wheat, rye, or barley, was planted in Field A one year, while Field B was left fallow. The following year, Field B was planted, but A left fallow, and so forth. Beginning in the eighth century, a three-field system was introduced. The first field was planted with a winter cereal, the second with a spring cereal, and the third was left fallow. As a result, each field yielded four harvests every 6 years, compared to three every 6 years under the two-field system.

By the eighteenth century, a four-field system was used in Northwest Europe. The first year, the farmer could plant a root crop (such as turnips) in Field A, a cereal in Field B, a "rest" crop (such as clover, which helps restore the field) in Field C, and a cereal in Field D. The second year, the farmer might select a cereal for Field A, a rest crop for Field B, a cereal for Field C, and a root for Field D. The rotation would continue for 2 more years before the cycle would start again. Each field thus passed through a cycle of four crops: root, cereal, rest crop, and another cereal.

Cereals such as wheat and barley were sold for flour and beer production, and straw, which is the stalks that remain after the heads of wheat are threshed, was retained for animal bedding. Root crops such as turnips were fed to the animals during the winter. Clover and other "rest" crops were used for cattle grazing and restoration of nitrogen to the soil.

Choice of Crops

In the United States, mixed crop and livestock farmers select corn most frequently because of higher yields per area than other crops. Some of the corn is consumed by people either directly or as oil, margarine, and other food products, but most is fed to pigs and cattle (Figure 10–7). The most important mixed crop and livestock farming region in the United States—extending from Ohio to the Dakotas, with its center in Iowa—is frequently called the Corn Belt, because approximately half of the crop land is planted in corn (maize).

Soybeans have become the second most important crop in the U.S. mixed commercial farming region. Like corn, soybeans are sometimes used to make products consumed directly by people, but mostly to make animal feed. Tofu (made from soybean milk) is a major food source, especially for people in China and Japan. Soybean oil is widely used in U.S. foods, but as a hidden ingredient.

Dairy Farming

Dairy farming is the most important type of commercial agriculture practiced on farms near the large urban areas of the Northeast United States, Southeast Canada, and Northwest Europe. It accounts for approximately 20 per-

cent of the total value of agricultural output throughout Western Europe and North America. Russia, Australia, and New Zealand also have extensive areas devoted to dairy farming. Nearly 90 percent of the world's supply of milk is produced and consumed in these developed regions (Figure 10–8).

Traditionally, fresh milk was rarely consumed except directly on the farm or in nearby villages. With the rapid growth of cities in relatively developed countries during the nineteenth century, demand for the sale of milk to urban residents increased. Rising incomes permitted urban residents to buy milk products, which were once considered luxuries. Average weekly consumption of milk per person in England, for example, rose from 0.8 liters (0.2 U.S. gallons) in the 1870s to 2.8 liters (0.7 U.S. gallons) by the 1950s.

Why Dairy Farms Locate Near Urban Areas

Dairying has become the most important type of commercial agriculture in the first ring outside large cities because of transportation factors. Dairy farms must be closer to their market than other products because milk is highly perishable. The ring surrounding a city from which milk can be supplied without spoiling is known as the **milkshed**.

Improvements in transportation have permitted dairying to be undertaken farther from the market. Until the 1840s, when railroads were first used for transporting dairy products, milksheds rarely had a radius beyond 50

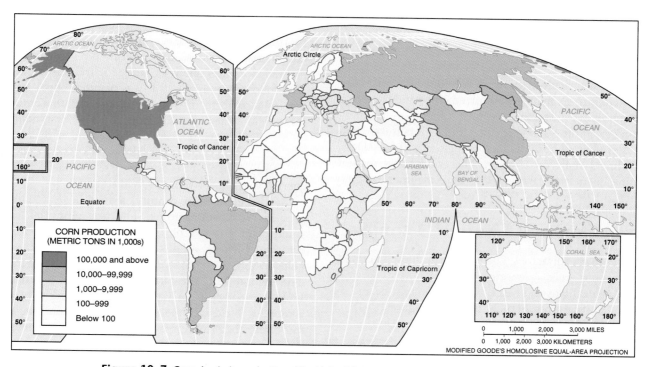

Figure 10–7 Corn (maize) production. The United States accounts for about 40 percent of the world's corn production. China is the second leading producer. Outside North America, corn is called maize.

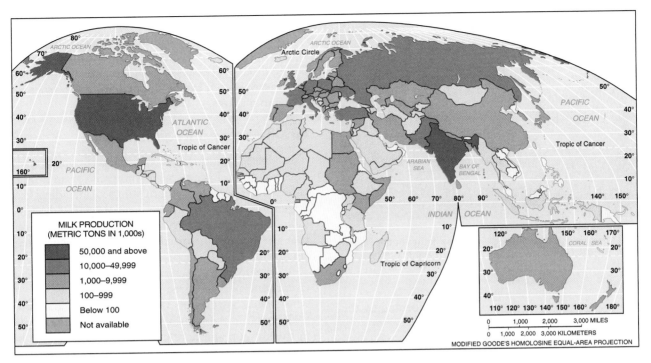

Figure 10–8 Milk production. The distribution of milk production closely matches the division of the world into more and less developed regions. Consumers in MDCs have the income to pay for milk products, and farmers in these countries can afford the high cost of establishing dairy farms. Two very populous countries—Brazil and India—rank among the world leaders in total milk production, but not in production per capita.

kilometers (30 miles). Today, refrigerated rail cars and trucks enable farmers to ship milk more than 500 kilometers (300 miles). As a result, nearly every farm in the U.S. Northeast and Northwest Europe is within the milkshed of at least one urban area.

Some dairy farms specialize in products other than milk. Originally, butter and cheese were made directly on the farm, primarily from the excess milk produced in the summer, before modern agricultural methods evened the flow of milk through the year. In the twentieth century, dairy farmers have generally chosen to specialize either in fresh milk production or other products such as butter and cheese.

Regional Differences in Dairy Products

The choice of product varies within the U.S. dairy region, depending on whether the farms are within the milkshed of a large urban area. In general, the farther the farm is from large urban concentrations, the smaller is the percentage of output devoted to fresh milk. Farms located farther from consumers are more likely to sell their output to processors who make butter, cheese, or dried, evaporated, and condensed milk. The reason is that these products keep longer than does milk and therefore can be safely shipped from remote farms.

In the East, virtually all milk is sold to consumers living in New York, Philadelphia, Boston, and the other large urban areas. Farther west, most milk is processed into cheese and butter. Virtually all of the milk in

Wisconsin is processed, for example, compared to only 5 percent in Pennsylvania. The proximity of northeastern farmers to several large markets accounts for these regional differences.

Countries likewise tend to specialize in certain products. New Zealand, the world's largest producer of dairy products, devotes only 8 percent to liquid milk, compared to 68 percent in the United Kingdom. New Zealand farmers do not sell much liquid milk because the country is too far from North America and Northwest Europe, the two largest relatively wealthy population concentrations.

Dairy farmers, like other commercial farmers, usually do not sell their products directly to consumers. Instead, they generally sell milk to wholesalers, who distribute it in turn to retailers. Retailers then sell milk to consumers in shops or at home. Farmers also sell milk to butter and cheese manufacturers.

Distribution of milk to consumers differs between the United States and the United Kingdom. Home delivery of milk has become rare in the United States but is still common in the United Kingdom. Many British families have a small card that looks like a clock with one hand. Before they go to sleep, they set the hand to the number of pints of milk they want delivered the next morning and place the card outside the front door. Early the next morning the milk is delivered, in bottles rather than cartons. The cream usually rises to the top of the bottle and can either be poured off or mixed in. Empty bottles are

Milk and other dairy products are delivered to homes in the United Kingdom in small, silent, electric trucks. (Robert Harding Picture Library)

then set outside and taken away the next morning by the delivery person.

Problems for Dairy Farmers

Like other commercial farmers, dairy farmers face economic problems because of declining revenues and rising costs. Distinctive features of dairy farming have exacerbated the economic difficulties. First, dairy farming is labor-intensive, because the cows must be milked twice a day, every day. Although the actual milking can be done by machines, dairy farming nonetheless requires constant attention throughout the year.

Dairy farmers also face the expense of feeding the cows in the winter, when they may be unable to graze on grass. In Northwest Europe and New England, farmers generally purchase hay or grain for winter feed. In the western part of the U.S. dairy region, crops are more likely to be grown in the summer and stored for winter feed on the same farm.

A recent survey by the Minnesota Department of Agriculture found that the state is losing 800 dairy farms a year, a decline of about 6 percent annually. Departing dairy farmers most often cite lack of profitability and excessive workload as reasons for leaving.

Grain Farming

Grain is the seed from various grasses, like wheat, corn, oats, barley, rice, millet, and others. Some form of grain is the major crop on most farms. Commercial grain agriculture is distinguished from mixed crop and livestock farming because crops on a grain farm are grown primarily for consumption by humans rather than by livestock. Farms in less developed countries also grow crops for human consumption, but the output is directly consumed by the farmers. Commercial grain farms sell their output to manufacturers of food products, like breakfast cereals and snack-food makers.

The most important crop grown is wheat, used to make bread flour. Wheat generally can be sold for a higher price than other grains, such as rye, oats, and barley, and it has more uses as human food. It can be stored relatively easily without spoiling and can be transported a long distance. Because wheat has a relatively high value per unit weight, it can be shipped profitably from remote farms to markets.

Grain-Farming Regions

The United States is by far the largest commercial producer of grain. Large-scale commercial grain production is found in only a few other countries, including Canada, Argentina, Australia, France, and the United Kingdom. Commercial grain farms are generally located in regions that are too dry for mixed crop and livestock agriculture (Figure 10–9).

Within North America, large-scale grain production is concentrated in three areas. The first is the winter wheat belt that extends through Kansas, Colorado, and Oklahoma. In the **winter wheat** area, the crop is planted in the autumn and develops a strong root system before growth stops for the winter. The wheat survives the winter, especially if it is insulated beneath a snow blanket, and is ripe by the beginning of summer.

The second important grain-producing region in North America is the **spring wheat** belt of the Dakotas, Montana, and southern Saskatchewan in Canada. Because winters are usually too severe for winter wheat in this region, spring wheat is planted in the spring and harvested in the late summer. Approximately two-thirds of the wheat grown in the United States comes either from the winter or the spring wheat belt. A third important grain-growing region is the Palouse region of Washington State.

Large-scale grain production, like other commercial farming ventures in more developed countries, is heavily mechanized, conducted on large farms, and oriented to consumer preferences. The McCormick **reaper** (a machine that cuts grain standing in the field), invented in the 1830s, first permitted large-scale wheat production. Today, the **combine** machine performs in one operation the three tasks of reaping, threshing, and cleaning.

Unlike work on a mixed crop and livestock farm, the effort required to grow wheat is not uniform throughout the year. Some individuals or firms may therefore have two sets of fields, one in the spring-wheat belt and one in the winter-wheat belt. Because the planting and harvesting in the two regions occur at different times of the year, the workload can be distributed throughout the year. In addition, the same machinery can be used in the two regions, thus spreading the cost of the expensive equipment. Combine companies start working in Oklahoma in early summer and work their way northward.

Importance of Wheat

Wheat's significance extends beyond the amount of land or number of people involved in growing it. Unlike other agricultural products, wheat is grown to a considerable

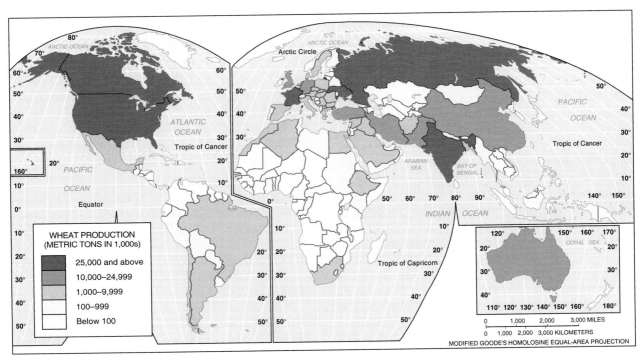

Figure 10–9 Wheat production. The United States is by far the world's leading producer of wheat for sale off the farm. China is the world's leading wheat producer, and India is a major producer, but wheat grown in the Asian countries is used principally to feed the local population. A large percentage of the wheat grown commercially in North America is exported to other countries.

extent for international trade and is the world's leading export crop. As the United States and Canada account for about half of the world's wheat exports, the North American prairies are accurately labeled the world's "breadbasket." The ability to provide food for many people elsewhere in the world is a major source of economic and political strength for the United States and Canada.

Livestock Ranching

Ranching is the commercial grazing of livestock over an extensive area. This form of agriculture is adapted to semiarid or arid land. It is practiced in more developed countries where the vegetation is too sparse and the soil too poor to support crops.

Cattle Ranching in U.S. Popular Culture

The importance of ranching in the United States extends beyond the number of people who choose this form of commercial farming because of its prominence in popular culture, especially in Hollywood films and television. Cattle ranching in Texas, though, as glamorized in popular culture, actually dominated commercial agriculture for a short period—from 1867 to 1885.

Beginning of U.S. Cattle Ranching. Cattle were first brought to the Americas by Columbus on his second voyage, because they were sufficiently hardy to survive the ocean crossing. Living in the wild, the cattle multiplied

and thrived on abundant grazing lands on the frontiers of North and South America. Immigrants from Spain and Portugal—the only European countries with a tradition of cattle ranching—began ranching in the Americas. They taught the practice to settlers from Northern Europe and the Eastern United States who moved to Texas and other frontier territories in the nineteenth century.

Cattle ranching in the United States expanded because of demand for beef in the East Coast cities during the 1860s. The challenge for ranchers was to transport the cattle from Texas to eastern markets. Ranchers who could get their cattle to Chicago were paid $30 to $40 per head, compared to only $3 or $4 per head in Texas. Once in Chicago, the cattle could be slaughtered and processed by meat-packing companies and sold to consumers in the East.

Transporting Cattle to Market. To reach Chicago, cattle were driven on hoof by cowboys over trails from Texas to the nearest railhead. Distances were several hundred kilometers. There they were driven into cattle cars for the rest of their journey. In 1867, the western terminus of the rail line reached Abilene, Kansas. That year, a man named Joseph G. McCoy (on whom the expression "the real McCoy" was based) launched a massive construction effort to provide Abilene with homes, shops, and stockyards. As a result, the number of cattle brought into Abilene increased from 1,000 in 1867 to 35,000 in 1868 and 150,000 in 1869. McCoy became the first mayor of the city of Abilene.

Like other frontier towns, Abilene became a haven where cowboys let off steam. Gunfights, prostitution, gambling, and alcoholism were rampant until McCoy hired James B. ("Wild Bill") Hickock as sheriff to clean up the town. After a few years, the terminus of the railroad moved farther west. Wichita, Caldwell, Dodge City, and other towns in Kansas took their turn as the main destination for cattle driven north on trails from Texas. Abilene became a ghost town for a while. Eventually, though, use of the surrounding land changed from cattle grazing to crop growing, and Abilene became a prosperous market center.

The most famous route from Texas northward to the rail line was the Chisholm Trail, which began near Brownsville at the Mexican border and extended northward through Texas, Indian Territory (now the state of Oklahoma), and Kansas. The trail had many branches, but the main line extended through Austin, Waco, Fort Worth, and Caldwell (Figure 10–10). The Western Trail became more important in the 1870s when the railroad terminus moved farther west. Today, U.S. Route 81 roughly follows the course of the Chisholm Trail.

Fixed Location Ranching

Cattle ranching declined in importance during the 1880s, after it came in conflict with sedentary agriculture. Most early U.S. ranchers adhered to "The Code of the West," although the system had no official legal status. Under the code, ranchers had range rights—that is, their cattle could graze on any open land and had access to scarce water sources and grasslands. The early cattle ranchers in the West owned little land, only cattle.

Range Wars. The U.S. government, which owned most of the land used for open grazing, began to sell it to farmers to grow crops, leaving cattle ranchers with no legal claim to it. For a few years, the ranchers tried to drive out the farmers by cutting fences and then illegally erecting their own fences on public land, and "range wars" flared.

The farmers' most potent weapon proved to be barbed wire, first commercially produced in 1873. The farmers eventually won the battle, and ranchers were compelled to buy or lease land to accommodate their cattle. Large

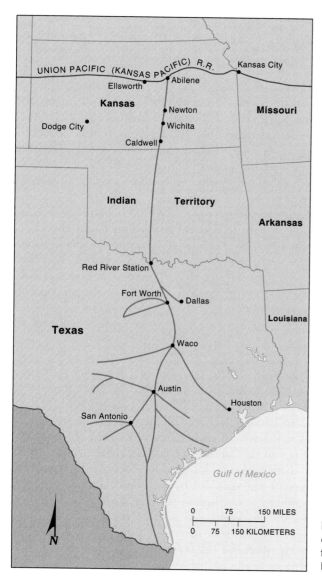

ON THE TRAIL.

(North Wind Picture Archives)

Figure 10–10 The Chisholm Trail. Although actively used for only a few years, the Chisholm Trail became famous in American folklore as the main route for cattle drives, from Texas ranches to Kansas railheads.

cattle ranches were established, primarily on land that was too dry to support crops. Ironically, 60 percent of cattle grazing today is on land leased from the U.S. government.

Changes in Cattle Breeding. Ranchers were also induced to switch from cattle drives to fixed-location ranching by a change in the predominant breed of cattle. Longhorns, the first cattle used by ranchers, were hardy animals, able to survive the long-distance drive along the trails with little weight loss. But longhorns were susceptible to cattle ticks, parasitic insects that carried a fever and were difficult to remove, and the meat of longhorns was of poor quality.

New cattle breeds introduced from Europe, such as the Hereford, offered superior meat but were not adapted to the old ranching system. The new breeds could not survive the winter by open grazing, as could the longhorns. Instead, crops had to be grown or feed purchased for them. The cattle could not be driven long distances, and they required more water. However, these breeds thrived once open grazing was replaced by fixed ranching, and long-distance trail drives and rail journeys to Chicago gave way to short rail or truck trips to nearby meat packers.

With the spread of irrigation techniques and hardier crops, land in the United States has been converted from ranching to crop growing. Ranching generates lower income per area of land, although it has lower operating costs. Cattle are still raised on ranches but are frequently sent for fattening to farms or to local feed lots along major railroad and highway routes rather than directly to meat processors. The average size of a ranch is large, because the capacity of the land to support cattle is low in much of the semiarid West. Large ranches may be owned by meat-processing companies rather than individuals.

Cattle Ranching Outside the United States

Commercial ranching is conducted in other more developed regions of the world (Figure 10–11). Ranching is rare in Europe, except in Spain and Portugal. In South America, a large portion of the pampas of Argentina, southern Brazil, and Uruguay are devoted to grazing cattle and sheep. The cattle industry grew rapidly in Argentina in part because the land devoted to ranching was relatively accessible to the ocean, and meat could be transported to overseas markets.

The relatively humid climate on the pampas provides more shoots and shrubs on a given area of land than in the U.S. West. This growth of ranching in South America was stimulated, because more cattle could graze on a given area of land in the pampas than in the U.S. West. Land was divided into large holdings in the nineteenth century, in contrast to the U.S. practice of permitting common grazing on public land. Ranching has declined in Argentina, as in the United States, because growing crops is more profitable except on very dry lands.

The interior of Australia was opened for grazing in the nineteenth century, although sheep are more common than cattle. Ranches in the Middle East, New Zealand,

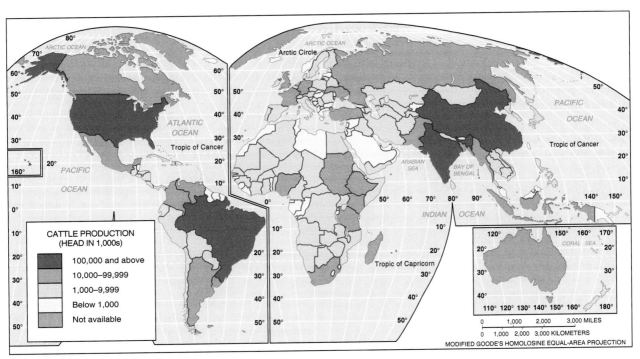

Figure 10–11 Cattle production. Cattle outnumber people in Argentina, Australia, and New Zealand, where commercial ranching is an important type of agriculture. Commercial ranching is not widely practiced in the more developed countries of Western Europe, a region that lacks extensive drylands.

and South Africa are also more likely to have sheep. Like the U.S. West, Australia's drylands went through several land-use changes. Until the 1860s, shepherding was practiced on the open range. Then large ranches with fixed boundaries were established, stock was improved, and water facilities were expanded. Eventually, ranching was confined to drier lands, and wheat—which yielded greater profits per hectare than ranching—was planted where precipitation levels permitted.

Thus, ranching has followed similar stages around the world. First was the herding of animals over open ranges, in a semi-nomadic style. Then, ranching was transformed into fixed farming by dividing the open land into ranches. Many of the farms converted to growing crops, and ranching was confined to the drier lands. To survive, the remaining ranches experimented with new methods of breeding and sources of water and feed. Ranching became part of the meat-processing industry rather than an economic activity carried out on isolated farms. In this way, commercial ranching differs from pastoral nomadism, the form of animal herding practiced in less developed regions.

Mediterranean Agriculture

Mediterranean agriculture exists primarily in the lands that border the Mediterranean Sea in Southern Europe, North Africa, and western Asia (refer to Figure 10–4). Farmers in California, central Chile, the southwestern part of South Africa, and southwestern Australia practice Mediterranean agriculture as well.

These Mediterranean areas share a similar physical environment (refer to Figure 10–5). Every Mediterranean area borders a sea. Mediterranean areas are on west coasts of continents (except for some lands surrounding the Mediterranean Sea). Prevailing sea winds provide moisture and moderate the winter temperatures. Summers are hot and dry, but sea breezes provide some relief. The land is very hilly, and mountains frequently plunge directly to the sea, leaving very narrow strips of flat land along the coast.

Farmers derive a smaller percentage of income from animal products in the Mediterranean region than in the mixed crop and livestock region. Livestock production is hindered during the summer by the lack of water and good grazing land. Some farmers living along the Mediterranean Sea traditionally used transhumance to raise animals, although the practice is now less common. Under transhumance, animals—primarily sheep and goats—are kept on the coastal plains in the winter and transferred to the hills in the summer.

Mediterranean Crops

Most crops in Mediterranean lands are grown for human consumption rather than for animal feed. **Horticulture**—which is the growing of fruits, vegetables, and

Mediterranean agriculture accounts for most of the world's grapes. This woman is harvesting grapes on Thira (Santorini), a Greek island. (Granitsas/The Image Works)

flowers—and tree crops form the commercial base of the Mediterranean farming. Most of the world's olives, grapes, fruits, and vegetables are grown in Mediterranean agriculture areas. A combination of local physical and cultural characteristics determines which crops are grown in each area. The hilly landscape encourages farmers to plant a variety of crops within one farming area.

In the lands bordering the Mediterranean Sea, the two most important cash crops are olives and grapes. Two-thirds of the world's wine is produced in countries that border the Mediterranean Sea, especially Italy, France, and Spain. Mediterranean agricultural regions elsewhere in the world produce most of the remaining one-third. The lands near the Mediterranean Sea are also responsible for a large percentage of the world's supply of olives, an important source of cooking oil.

Despite the importance of olives and grapes to commercial farms bordering the Mediterranean Sea, approximately half of the land is devoted to growing cereals, especially wheat for pasta and bread. As in the U.S. winter-wheat belt, the seeds are sown in the fall and harvested in early summer. After cultivation, cash crops are planted on approximately 20 percent of the land, while the remainder is left fallow for a year or two to conserve moisture in the soil.

Cereals occupy a much lower percentage of the cultivated land in California than in other Mediterranean climates. Instead, 30 percent of California farmland is devoted to fruit and vegetable horticulture. California supplies much of the citrus fruits, tree nuts, and deciduous fruits consumed in the United States. Horticulture is

358

practiced in other Mediterranean climates, but not to the extent found in California.

The rapid growth of urban areas in California, especially Los Angeles, has converted high-quality agricultural land into housing developments. Thus far, the loss of farmland has been offset by expansion of agriculture into arid lands. However, farming in drylands requires massive irrigation to provide water. In the future, agriculture may face stiffer competition to divert the Southwest's increasingly scarce water supply.

Commercial Gardening and Fruit Farming

Commercial gardening and fruit farming is the predominant type of agriculture in the U.S. Southeast. The region has a long growing season and humid climate and is accessible to the large markets of New York, Philadelphia, Washington, and the other eastern U.S. urban areas. The type of agriculture practiced in this region is frequently called **truck farming**, because "truck" was a Middle English word meaning bartering or the exchange of commodities.

Truck farms grow many of the fruits and vegetables that consumers demand in more developed societies, such as apples, asparagus, cherries, lettuce, mushrooms, and tomatoes. Some of these fruits and vegetables are sold fresh to consumers, but most are sold to large processors for canning or freezing.

Truck farms are highly efficient, large-scale operations that take full advantage of machines at every stage of the growing process. Truck farmers are willing to experiment with new varieties, seeds, fertilizers, and other inputs to maximize efficiency. Labor costs are kept down by hiring migrant farm workers, some of whom are undocumented immigrants from Mexico who work for very low wages. Farms tend to specialize in a few crops, and a handful of farms may dominate national output of some fruits and vegetables.

A form of truck farming called *specialty farming* has spread to New England. Farmers are profitably growing crops that have limited but increasing demand among affluent consumers, such as asparagus, peppers, mushrooms, strawberries, and nursery plants. Specialty farming represents a profitable alternative for New England farmers, at a time when dairy farming is declining because of relatively high operating costs and low milk prices.

Plantation Farming

The plantation is a form of commercial agriculture found in the tropics and subtropics, especially in Latin America, Africa, and Asia. Although generally situated in less developed countries, plantations are often owned or operated by Europeans or North Americans and grow crops for sale primarily in more developed countries.

A **plantation** is a large farm that specializes in one or two crops. Among the most important crops grown on plantations are cotton, sugar cane, coffee, rubber, and tobacco. Also produced in large quantities are cocoa, jute, bananas, tea, coconuts, and palm oil. Latin American plantations are more likely to grow coffee, sugar cane, and bananas, whereas Asian plantations may provide rubber and palm oil.

Because plantations are usually situated in sparsely settled locations, they must import workers and provide them with food, housing, and social services. Plantation managers try to spread the work as evenly as possible throughout the year to make full use of the large labor force. Where the climate permits, more than one crop is planted and harvested during the year. Rubber-tree plantations try to spread the task of tapping the trees through the year.

Crops such as tobacco, cotton, and sugar cane, which can be planted only once a year, are less likely to be grown on large plantations today than in the past. Crops are normally processed at the plantation before shipping, because processed goods are less bulky and therefore cheaper to ship long distances to the North American and European markets.

Until the Civil War, plantations were important in the U.S. South, where the principal crop was cotton, followed by tobacco and sugar cane. Demand for cotton increased dramatically after the establishment of textile factories in England at the start of the industrial revolution in the late eighteenth century. Cotton production was stimulated by the improvement of the cotton gin by Eli Whitney in 1793 and the development of new varieties of cotton that were hardier and easier to pick. Slaves brought from Africa performed most of the labor until the abolition of slavery and the defeat of the South in the Civil War. Thereafter, plantations declined in the United States; they were subdivided and either sold to individual farmers or worked by tenant farmers.

Key Issue 4

Why Does Agriculture Vary Among Regions?

- Environmental and cultural factors
- Economic issues for subsistence farmers
- Economic issues for commercial farmers

Three types of reasons help to explain differences among agricultural regions: environmental, cultural, and economic. Environmental and cultural issues are mentioned here only briefly, because the major environmental factor—climate—was discussed at the beginning of this chapter, and cultural factors were addressed in Chapter 4. This section looks primarily at the distinctive economic issues faced by subsistence farmers, followed by those faced by commercial farmers.

Environmental and Cultural Factors

Regions of distinctive agricultural practices exist in part because of differences in climate. Look again at Figures 10–4 and 10–5. Figure 10–4 shows Whittlesey's 11 agricultural regions, divided between subsistence and commercial agriculture. Figure 10–5 is a simplified version of the world climate map presented in Chapter 1 (Figure 1–12). The similarities between the two maps are striking.

For example, in the Middle East, which has a dry climate (Figure 10–5), pastoral nomadism is the predominant type of agriculture (Figure 10–4). In central Africa, which has a tropical climate, shifting cultivation is the predominant type of agriculture.

Note the division between southeastern China (warm midlatitude climate, intensive subsistence agriculture with wet rice dominant) and northeastern China (cold midlatitude climate, intensive subsistence agriculture with wet rice not dominant). In the United States, much of the west is distinguished from the rest of the country according to climate (dry) and agriculture (livestock ranching).

The correlation between agriculture and climate is by no means perfect, but clearly some relationship exists between climate and agriculture. Because of the problems with environmental determinism discussed in Chapter 1, geographers are wary of placing too much emphasis on the role of climate.

Cultural preferences, discussed in Chapter 4, explain some agricultural differences in areas of similar climate. Hog production is virtually nonexistent in predominantly Muslim regions, because of that religion's taboo against consuming pork products (refer to Figure 4–5). Wine production is relatively low in Africa and Asia, even where climate is favorable for growing grapes, because of alcohol avoidance in predominantly non-Christian countries (refer to Figure 4–14).

Economic Issues for Subsistence Farmers

Two economic issues discussed in earlier chapters influence the choice of crops planted by subsistence farmers. First, because of rapid population growth in less developed countries (discussed in Chapter 2), subsistence farmers must feed an increasing number of people. Second, because of adopting the international trade approach to development (discussed in Chapter 9), subsistence farmers must grow food for export instead of for direct consumption.

Subsistence Farming and Population Growth

Ester Boserup, an economist, has offered an explanation for why population growth influences the distribution of types of subsistence farming. According to the Boserup thesis, population growth compels subsistence farmers to consider new farming approaches that produce enough food to take care of the additional people.

For hundreds if not thousands of years, subsistence farming in less developed countries yielded enough food for people living in rural villages to survive, assuming no drought, flood, or other natural disaster. Suddenly in the late twentieth century, the LDCs needed to provide enough food for a rapidly increasing population, as well as for the growing number of urban residents who cannot grow their own food.

According to the Boserup thesis, subsistence farmers increase the supply of food through intensification of production, achieved in two ways. First, land is left fallow for shorter periods, resulting in an expansion in the amount of land area devoted to growing crops at any given time. Boserup identified five basic stages in the intensification of farmland:

- **Forest Fallow.** Fields are cleared and utilized for up to 2 years and left fallow for more than 20 years, long enough for the forest to grow back.
- **Bush Fallow.** Fields are cleared and utilized for up to 8 years and left fallow for up to 10 years, long enough for small trees and bushes to grow back.
- **Short Fallow.** Fields are cleared and utilized for perhaps 2 years (Boserup was uncertain) and left fallow for up to 2 years, long enough for wild grasses to grow back.
- **Annual Cropping.** Fields are used every year and left fallow for a few months by planting legumes and roots.
- **Multicropping.** Fields are used several times a year and never left fallow.

Contrast shifting cultivation, practiced in regions of low population density, such as central Africa, with intensive subsistence agriculture, practiced in regions of high population density, such as East Asia. Under shifting cultivation, cleared fields are utilized for a couple of years, then left fallow for 20 years or more. This type of agriculture supports a small population living at low density. As the number of people living in an area increases (that is, the population density increases), and more food must be grown, fields will be left fallow for shorter periods of time. Eventually, farmers achieve the very intensive use of farmland characteristic of areas of high population density.

The second way that subsistence farmers intensify production, according to the Boserup thesis, is through adopting new farming methods. Ploughs replace axes and sticks. More weeding is done, more manure applied, more terraces carved out of hillsides, more irrigation ditches dug. The additional labor needed to perform these operations comes from the population growth. The farmland yields more food per area of land, but with the growing population output per person remains about the same.

Subsistence Farming and International Trade

To expand production, subsistence farmers need higher-yield seeds, fertilizer, pesticides, and machinery. Some needed supplies can be secured through trading food with urban dwellers. For many African and Asian countries, though, the main source of agricultural supplies is importing from other countries. These countries lack the money to buy agricultural equipment and materials from more developed countries.

To generate the funds they need to buy agricultural supplies, less developed countries must produce something they can sell in more developed countries. The LDCs sell some manufactured goods (see Chapter 11), but most raise funds through the sale of crops in MDCs. Consumers in more developed countries are willing to pay high prices for fruits and vegetables that would otherwise be out of season, or for crops such as coffee and tea that cannot be grown there because of the climate.

In a less developed country such as Kenya, families may divide by gender between traditional subsistence agriculture and contributing to international trade. Women practice most of the subsistence agriculture—that is, growing food for their families to consume—in addition to the tasks of cooking, cleaning, and carrying water from wells. Men may work for wages, either growing crops for export or in jobs in distant cities. Because men in Kenya frequently do not share the wages with their families, many women try to generate income for

the household by making clothes, jewelry, baked goods, and other objects for sale in local markets.

The sale of export crops brings a less developed country foreign currency, a portion of which can be used to buy agricultural supplies. But governments in LDCs face a dilemma: The more land that is devoted to growing export crops, the less that is available to grow crops for domestic consumption. Rather than helping to increase productivity, the funds generated through the sale of export crops may be needed to feed the people who switched from subsistence farming to growing export crops.

Drug Crops. The export crops chosen in some LDCs, especially in Latin America and Asia, are those that can be converted to drugs (Figure 10–12). Various drugs, such as coca leaf, marijuana, opium, and hashish, have distinctive geographic distributions.

Coca leaf is grown principally in four contiguous countries in northwestern South America. One-half of the supply comes from Peru, more than one-fourth from Bolivia, and most of the remainder from Colombia and Ecuador. Eighty percent of the processing of cocaine, as well as its distribution to the United States and other more developed countries, is based in Colombia.

Mexico grows the overwhelming majority of the marijuana that reaches the United States, followed by Colombia, Jamaica, and Belize. Mexico is also responsible for some of the opium, but most opium originates in Asia. Southeast Asia is the center of opium production, with more than half produced in Myanmar, followed by Laos. Thailand produces some opium, as well, but its main role is to serve as the transportation hub for

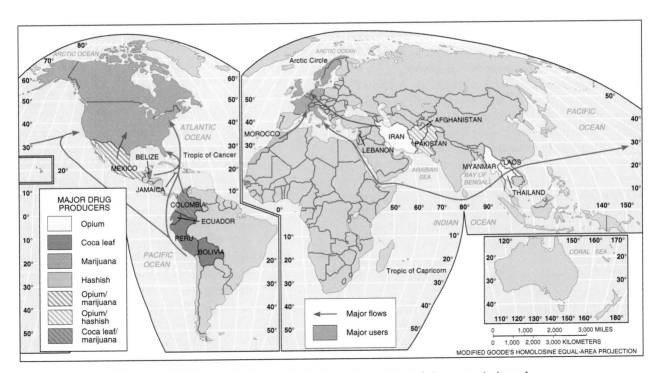

Figure 10–12 Sources of drugs. Instead of concentrating on subsistence agriculture, farmers in some LDCs make far more money growing crops that are converted to drugs sold in MDCs.

Police raid a poppy field in Colombia. Opium is made from the dried juice of the opium poppy. (Timothy Ross/The Image Works)

distribution to more developed countries. Farther west in Asia, Afghanistan, Iran, and Pakistan are also major opium producers. Afghanistan and Pakistan are also major producers of hashish, as are Lebanon and Morocco.

Economic Issues for Commercial Farmers

Two economic factors influence the choice of crops (or livestock) by commercial farmers: access to markets and overproduction.

Access to Markets

Because the purpose of commercial farming is to sell produce off the farm, the distance from the farm to the market influences the farmer's choice of crop to plant. The clearest example of the importance of proximity to the market is dairy farming, because milk spoils quickly. Crops that can be shipped long distances without spoiling are grown farther from the market. Geographers use the von Thünen model to help explain the importance of proximity to market in the choice of crops on commercial farms.

Von Thünen's Model. The von Thünen model was first proposed in 1826 by Johann Heinrich von Thünen, a farmer in northern Germany, in a book titled *The Isolated State*. According to the model, which was later modified by geographers, a commercial farmer initially considers which crops to cultivate and animals to raise based on market location.

In choosing an enterprise, a commercial farmer compares two costs: the cost of the land versus the cost of transporting products to market. First, a farmer identifies a crop that can be sold for more than the land cost. Assume that a farmer's land costs $100 per hectare per year. The farmer would consider planting wheat if the output from 1 hectare could be marketed for more than $100 that year. Another crop, such as corn, will also be considered if the yield from 1 hectare can sell for more than $100.

A farmer will not necessarily plant the crop that sells for the highest price per hectare. The choice further depends on the distance of the farmer's land from the central market city. Distance to market is critical because the cost of transporting each product is different.

Example of Von Thünen's Model. The following example illustrates the influence of transportation cost on the profitability of growing wheat:

1. Gross profit from sale of wheat grown on 1 hectare of land *not* including transportation costs:
   a. Wheat can be grown for $0.25 per kilogram
   b. Yield per hectare of wheat is 1,000 kilograms
   c. Gross profit is $250 per hectare ($0.25 per kilogram × 1,000 kilograms per hectare)
2. Net profit from sale of wheat grown on 1 hectare of land *including* transportation costs:
   a. Cost of transporting 1,000 kilograms of wheat to the market is $62.50 per kilometer
   b. Net profit from sale of 1,000 kilograms of wheat grown on a farm located 1 kilometer from the market is $187.50 ($250 gross profit − $62.50 per kilometer transport costs)
   c. Net profit from sale of 1,000 kilograms of wheat grown on a farm located 4 kilometers from the market is $0 ($250 gross profit − ($62.50 per kilometer × 4 kilometers)

The example shows that a farmer would make a profit growing wheat on land located less than 4 kilometers from the market. Beyond 4 kilometers, wheat is not profitable, because the cost of transporting it exceeds the gross profit.

The von Thünen model shows that a commercial farmer must combine two sets of monetary values to determine the most profitable crop:

- the value of the yield per hectare
- the cost of transporting the yield per hectare

These calculations demonstrate that farms located closer to the market tend to select crops with higher transportation costs per hectare of output, whereas more distant farms are more likely to select crops that can be transported less expensively.

Application of Von Thünen's Model. Von Thünen based his general model of the spatial arrangement of different crops on his experiences as owner of a large estate in northern Germany during the early nineteenth century (Figure 10–13). He found that specific crops were grown in different rings around the cities in the area. Market-oriented gardens and milk producers were located in the first ring out from the cities. These products are expensive to deliver and must reach the market quickly because they are perishable.

The next ring out from the cities contained wood lots, where timber was cut for construction and fuel; closeness to market is important for this commodity because of its weight. The next rings were used for var-

ious crops and for pasture; the specific commodity was rotated from one year to the next. The outermost ring was devoted exclusively to animal grazing, which requires lots of space.

Von Thünen did not consider site or human factors in his model. The model assumed that all land in a study area had similar site characteristics and was of uniform quality, although he recognized that the model could vary according to topography and other distinctive physical conditions. For example, a river might modify the shape of the rings because transportation costs change when products are shipped by water routes rather than over roads. The model also failed to understand that social customs and government policies influence the attractiveness of plants and animals for a commercial farmer.

Although von Thünen developed the model for a small region with a single market center, it also applies to a national or global scale. Farmers in relatively remote locations who wish to sell their output in the major markets of Western Europe and North America, for example, are less likely to grow highly perishable and bulky products.

Overproduction in Commercial Farming

Commercial farmers suffer from low incomes because they produce too much food, rather than too little. A surplus of food has been produced in part because of widespread adoption of efficient agricultural practices. New seeds, fertilizers, pesticides, mechanical equipment,

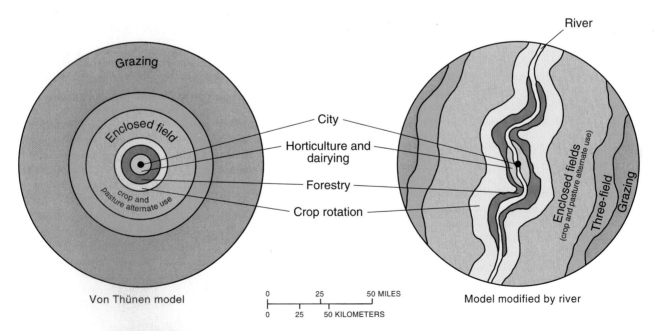

Figure 10–13 (left) Von Thünen model of role of situation factors in choice of crop. According to the von Thünen model, in the absence of topographic factors, different types of farming are conducted at different distances from a city, depending on the cost of transportation and the value of the product. (right) Von Thünen recognized that his model would be modified by site factors, such as a river in this sketch, which changes the accessibility of different land parcels to the market center. Agricultural uses that seek highly accessible locations need to locate nearer the river.

and management practices have enabled farmers to obtain greatly increased yields per area of land.

Commercial farmers have dramatically increased the capacity of the land to produce food. For example, during the 1960s about 20 million dairy cows produced 57 million metric tons (63 million tons) of milk a year in the United States. The number of dairy cows in the United States has declined to 10 million during the 1990s, but they produce 68 million metric tons (75 million tons).

While the food supply has increased in more developed countries, demand has remained constant, because the market for most products is already saturated. In MDCs, consumption of a particular commodity may not change significantly if the price changes. Americans, for example, do not switch from wheat to corn products if the price of corn falls more rapidly than wheat. Demand is also stagnant for most agricultural products in more developed countries because of low population growth.

U.S. Government Policies. Agricultural production is also increasing in the United States because of government programs. The U.S. government has three policies to attack the problem of excess productive capacity. First, farmers are encouraged to avoid producing crops that are in excess supply. Because soil erosion is a constant threat, the government encourages planting fallow crops, such as clover, to restore nutrients to the soil and to help hold the soil in place. These crops can be used for hay, forage for pigs, or producing seeds for sale.

Second, the government pays farmers when certain commodity prices are low. The government sets a target price for the commodity and pays farmers the difference between the price they receive in the market and a target price set by the government as a fair level for the commodity. The target prices are calculated to give the farmers the same price for the commodity today as in the past, when compared to other consumer goods and services.

Third, the government buys surplus production and sells or donates it to foreign governments. In addition, low-income Americans receive food stamps in part to stimulate their purchase of additional food.

The United States spends about $10 billion a year on farm subsidies, including an average of $6 billion for feed grains, such as corn and soybeans, $2 billion for wheat, and $2 billion for dairy products. Annual spending varies considerably from one year to the next: subsidy payments are lower in years when market prices rise and production is down, typically as a result of poor weather conditions in the United States or political problems in other countries.

Government policies point out a fundamental irony in worldwide agricultural patterns. In a more developed country such as the United States, farmers are encouraged to grow less food, while less developed countries struggle to increase food production to match the rate of the growth in population.

■ Summary

A country's agricultural system remains one of the best measures of its level of development and standard of material comfort. Despite major changes, agriculture in less developed countries still employs the majority of the population, and producing food for local survival is still paramount.

Farming in more developed countries directly employs few people, but when manufacturers of food products, supermarkets, restaurants, and other businesses that handle food are considered, then the food industry is actually the largest employer. The production and distribution of food are not primary-sector or agricultural activities, though; they are part of the industrial and service sectors of the economies in more developed countries.

Even farming itself may one day no longer be considered a distinct primary-sector activity in MDCs. True, farmers still deliberately modify the land by planting seeds or grazing animals, but they spend more time sitting at computers, operating sophisticated machinery, and reviewing finances, and devising marketing strategies.

Here again are the key questions concerning agricultural geography:

1. **Where did agriculture originate?** Prior to the development of agriculture, people survived by hunting animals, gathering wild vegetation, or fishing. Agriculture was not simply invented but was the product of thousands of years of experiments and accidents. Current agricultural practices vary between more developed and less developed countries.

2. **Where are agricultural regions in less developed countries?** Most people in the world, especially those in LDCs, are subsistence farmers, growing crops primarily to feed themselves. Important types of subsistence agriculture include shifting cultivation, pastoral nomadism, and intensive farming. Regions where subsistence agriculture is practiced are characterized by a large percentage of the labor force engaged in agriculture, with few mechanical aids.

3. **Where are agricultural regions in more developed countries?** The most common type of farm found in MDCs is mixed crop and livestock. Where mixed crop and livestock farming is not suitable, commercial farmers practice a variety of other types of agriculture, including dairying, commercial grain, and ranching.

4. **Why does agriculture vary among regions?** Environmental, cultural, and economic factors help to explain the distinctive form of agriculture practiced in various regions. Agriculture in LDCs faces distinctive economic problems resulting from rapid population growth and pressure to adopt international trade strategies to promote development. Agriculture in MDCs faces problems resulting from access to markets and overproduction.

Case Study Revisited

Uncertain Future for Farming

The future is uncertain for both subsistence farmers in less developed countries such as Pakistan and commercial farmers in more developed countries such as the United States. In one respect, the uncertainty stems from a similar problem: Farming in neither location produces sufficient income to support the standard of living farm families desire. However, the underlying cause of low incomes differs significantly between more developed and less developed countries.

In LDCs, people migrate from the farms to the cities in search of higher-paying jobs and a better life. Given the high natural increase rate and pressure to produce more for international trade, the migrants are not missed on the farms. The need for more food will not be met by adding more workers on the farms, but from more intensive use of existing farms and purchase of food from abroad.

In MDCs, people also migrate from the farms to the cities in search of higher-paying jobs. However, these migrants are missed. Small farming communities in the United States are dying, and their death causes a loss of rural-based culture and values.

In many ways, the current migration from the farms in MDCs is simply the continuation of long-term trends. With farms becoming ever larger and more mechanized, the number of farmworkers continually declines. Farm communities are suffering because most of the emigrants are young. But this has always been the case, as well: Nearly a century ago, World War I veterans returning to the United States from Europe sang "How ya gonna keep 'em down on the farm after they've seen Paree (Paris)."

The current decline in the farm population in MDCs has an especially strong impact on the rural landscape because with so few farmers left, each further loss registers a large decline in percentage terms. Farming is the backbone of many small-town economies. Without farmers, banks and shops lose their main sources of income. For every five people that give up farming, one business closes in a small town. People still live on farms but work in factories, offices, or businesses in the nearest big city. And they shop at the big-city Wal-Mart instead of the small-town Main Street.

■ Key Terms

Agribusiness Commercial agriculture characterized by integration of different steps in the food processing industry, usually through ownership by large corporations.

Agriculture The deliberate effort to modify a portion of Earth's surface through the cultivation of crops and the raising of livestock for sustenance or economic gain.

Cereal grain A grass yielding grain for food.

Chaff Husks of grain separated from the seed by threshing.

Combine A machine that reaps, threshes, and cleans grain while moving over a field.

Commercial agriculture Agriculture undertaken primarily to generate products for sale off the farm.

Crop Grain or fruit gathered from a field as a harvest during a particular season.

Crop rotation The practice of rotating use of different fields from crop to crop each year, to avoid exhausting the soil.

Double cropping Harvesting twice a year from the same field.

Grain Seed of a cereal grass.

Horticulture The growing of fruits, vegetables, and flowers.

Hull The outer covering of a seed.

Intensive subsistence agriculture A form of subsistence agriculture in which farmers must expend a relatively large amount of effort to produce the maximum feasible yield from a parcel of land.

Milkshed The area surrounding a city from which milk is supplied.

Paddy Malay word for wet rice, commonly but incorrectly used to describe a sawah.

Pastoral nomadism A form of subsistence agriculture based on herding domesticated animals.

Pasture Grass or other plants grown for feeding grazing animals, as well as land used for grazing.

Plantation A large farm in tropical and subtropical climates that specializes in the production of one or two crops for sale, usually to a more developed country.

Prime agricultural land The most productive farmland.

Ranching A form of commercial agriculture in which livestock graze over an extensive area.

Reaper A machine that cuts grain standing in the field.

Sawah A flooded field for growing rice.

Seed agriculture Reproduction of plants through annual introduction of seeds, which result from sexual fertilization.

Slash-and-burn agriculture Another name for shifting cultivation, so named because fields are cleared by slashing the vegetation and burning the debris.

Shifting cultivation A form of subsistence agriculture in which people shift activity from one field to another; each field is used for crops for a relatively few years and left fallow for a relatively long period.

Spring wheat Wheat planted in the spring and harvested in the late summer.

Subsistence agriculture Agriculture designed primarily to provide food for direct consumption by the farmer and the farmer's family.

Swidden A patch of land cleared for planting through slashing and burning.

Thresh To beat out grain from stalks by trampling it.

Transhumance The seasonal migration of livestock between mountains and lowland pastures.

Truck farming Commercial gardening and fruit farming, so named because *truck* was a Middle English word meaning *bartering* or the exchange of commodities.

Vegetative planting Reproduction of plants by direct cloning from existing plants.

Wet rice Rice planted on dryland in a nursery, then moved to a deliberately flooded field to promote growth.

Winnow To remove chaff by allowing it to be blown away by the wind.

Winter wheat Wheat planted in the fall and harvested in the early summer.

■ Thinking Geographically

1. Assume that the United States constitutes one agricultural market, centered around New York City, the largest metropolitan area. To what extent can the major agricultural regions of the United States be viewed as irregularly shaped rings around the market center, as von Thünen applied to southern Germany?

2. New Zealand once sold nearly all its dairy products to the British, but since the United Kingdom joined the European Union in 1973, New Zealand has been forced to find other markets. What are some other examples of countries that have restructured their agricultural production in the face of increased global interdependence and regional cooperation?

3. Review the concept of overpopulation (the number of people in an area exceeds the capacity of the environment to support life at a decent standard of living). What agricultural regions have relatively limited capacities to support intensive food production? Which of these regions face rapid population growth?

4. Compare world distributions of corn, wheat, and rice production. To what extent do differences derive from environmental conditions and to what extent from food preferences and other social customs?

5. How might the loss of farmland on the edge of rapidly growing cities alter the choice of crops that other farmers make in a commercial agricultural society?

■ Further Readings

Archer, J. Clark, and Richard E. Lonsdale, "Geographical Aspects of US Farmland Values and Changes During the 1978–1992 Period." *Journal of Rural Studies* 13 (1997): 399–414.

Babbington, Anthony, and Judith Carney. "Geography in the International Agricultural Research Centers: Theoretical and Practical Concerns." *Annals of the Association of American Geographers* 80 (1990): 34–48.

Bascom, Jonathan B. "Border Pastoralism in Eastern Sudan." *Geographical Review* 80 (1990): 416–30.

Bayliss-Smith, T.P. *The Ecology of Agricultural Systems.* Cambridge: Cambridge University Press, 1982.

Blaikie, P. "The Theory of the Spatial Diffusion of Agricultural Innovations; a Spacious Cul-de-sac." *Progress in Human Geography* 2 (1978): 268–95.

Boserup, Ester. *The Conditions of Agricultural Change: The Economics of Agrarian Change Under Population Pressure.* London: Allen and Unwin, 1965.

Cochran, Willard W., and Mary E. Ryan. *American Farm Policy 1948–73.* Minneapolis: University of Minnesota Press, 1976.

Cromley, Robert G. "The Von Thünen Model and Environmental Uncertainty." *Annals of the Association of American Geographers* 72 (1982): 404–10.

Dahlberg, Kenneth A., ed. *New Directions for Agriculture and Agricultural Research: Neglected Dimensions and Emerging Alternatives.* Totowa, NJ: Rowman and Allanheld, 1986.

Dove, Michael R. *Swidden Agriculture in Indonesia: The Subsistence Strategies of the Kalimantan Kantu.* Amsterdam: Mouton, 1985.

Duckham, A. N., and G. B. Masefield. *Farming Systems of the World.* New York: Praeger, 1970.

Durand, Loyal, Jr. "The Major Milksheds of the Northeastern Quarter of the United States." *Economic Geography* 40 (1964): 9–33.

Ebeling, Walter. *The Fruited Plain: The Story of American Agriculture.* Berkeley: University of California Press, 1979.

Furuseth, Owen J., and John T. Pierce. *Agricultural Land in an Urban Society.* Washington, DC: Association of American Geographers, 1982.

Galaty, John G., and Douglas L. Johnson, eds. *The World of Pastoralism: Herding Systems in Comparative Perspective.* New York: Guilford Press, 1990.

Gardner, Gary. "Shrinking Cropland." *World Watch* 9 (1996): 18–27.

Grigg, David B. *The Agricultural Systems of the World: An Evolutionary Approach.* London: Cambridge University Press, 1974.

_____. *An Introduction to Agricultural Geography.* London: Hutchinson Education, 1984.

_____. *English Agriculture: An Historical Perspective.* Oxford: Basil Blackwell, 1989.

_____. "Ester Boserup's Theory of Agrarian Change: A Critical Review." *Progress in Human Geography* 3 (1979): 64–84.

_____. "The Starchy Staples in World Food Consumption." *Annals of the Association of American Geographers* 86 (1996): 412–31.

_____. *The Transformation of Agriculture in the West.* Cambridge, MA: Blackwell, 1992.

Hart, John Fraser. "Change in the Corn Belt." *Geographical Review* 76 (1986): 51–73.

_____. *The Land That Feeds Us.* New York: W. W. Norton & Co., 1991.

Hayami, Y. and Ruttan, V.W. *Agricultural Development: An International Perspective.* 2d ed. Baltimore: The Johns Hopkins University Press, 1985

Heathcote, R. L. *The Arid Lands: Their Use and Abuse.* London: Longman, 1983.

Hewes, Leslie, and Christian I. Jung. "Early Fencing on the Middle Western Prairie." *Annals of the Association of American Geographers* 71 (1981): 177–201.

Holloway, L.E., and B.W. Ilbery. "Farmers' Attitudes Towards Environmental Change, Particularly Global Warming, and the Adjustment of Crop Mix and Farm Management." *Applied Geography* 16 (1996): 159–71.

Hsu, Song-ken. "The Agroindustry: A Neglected Aspect of the Location Theory of Manufacturing." *Journal of Regional Science* 37 (1997): 259–74.

Ilbery, Brian W. *Agricultural Geography: A Social and Economic Analysis.* New York: Oxford University Press, 1985

Kline, Jeffrey, and Dennis Wichelns, "Public Preferences and Farmland Preservation Programs." *Land Economics* 72 (1996): 538–49.

Kostrowicki, J. "A Hierarchy of World Types of Agriculture." *Geographica Polonica* 43 (1980): 125–48.

_____. "The Types of Agriculture Map of Europe." *Geographica Polonica* 48 (1982): 79–91.

Kramer, Randall A., and D. Evan Mercer. "Valuing a Global Environmental Good: U.S. Residents' Willingness to Pay to Protect Tropical Rain Forests." *Land Economics* 73 (1997): 196–210.

Lewthwaite, G. R. "Wisconsin and the Waikato: A Comparative Study of Dairy Farming in the United States and New Zealand." *Annals of the Association of American Geographers* 54 (1974): 59–87.

Nelson, Toni. "Urban Agriculture." *World Watch* 9 (1996): 10–17.

Pannell, Clifton. "Recent Chinese Agriculture." *Geographical Review* 75 (1985): 170–85.

Peters, William J., and Leon F. Neuenschwander. *Slash and Burn: Farming in the Third World Forest.* Moscow: University of Idaho Press, 1988.

Potter, Clive, Paul Burnham, Angela Edwards, Ruth Gasson, and Bryn Green. *The Diversion of Land: Conservation in a Period of Farming Contraction.* New York: Routledge, 1991.

Sauer, Carl O. *Agricultural Origins and Dispersals*, 2d ed. Cambridge, MA: M.I.T. Press, 1969.

Singh, Ram D. "Female Agricultural Workers' Wages, Male-Female Wage Differentials, and Agricultural Growth in a Developing Country, India." *Economic Development and Cultural Change* 45 (1996): 89–123.

Smith, Everett G., Jr. "America's Richest Farms and Ranches." *Annals of the Association of American Geographers* 70 (1980): 528–41.

Syers, J. K.; J. Lingard, and G. Faure. "Sustainable Land Management for the Semiarid and sub-humid Tropics." *Ambio* 25 (1996): 484–536.

Symons, Leslie. *Agricultural Geography*, rev. ed. London: G. Bell, 1979.

Tarrant, John R. *Agricultural Geography.* New York: John Wiley, 1974.

Tivy, Joy. *Agricultural Ecology.* Harlow, UK: Longman, 1990.

Turner, B. L., II, and Stephen B. Brush, eds. *Comparative Farming Systems.* New York: Guilford Press, 1987.

von Thünen, Johann Heinrich. *Von Thünen's Isolated State: An English Edition of "Der Isolierte Staat."* Translated by Carla M. Wartenberg. Elmsford, NY: Pergamon Press, 1966.

Warren, Andrew. "Changing Understandings of African Pastoralism and the Nature of Environmental Paradigms." *Transactions of the Institute of British Geographers New Series* 20 (1995): 193–203.

Watts, Michael. "Development III: The Global Agrofood System and Late Twentieth-Century Development (or Kautsky Redux)." *Progress in Human Geography* 20 (1996): 230–45.

Whittlesey, Derwent. "Major Agricultural Regions of the Earth." *Annals of the Association of American Geographers* 26 (1936): 199–240.

Also consult this journal: *Journal of Rural Studies*.

CHAPTER 11

Industry

Japanese products, including televisions, cars, and cameras, have deluged the United States and Canada. Although manufactured thousands of kilometers away, these products often sell below domestic competitors and, according to many people, are of better quality. This accomplishment of Japanese industry is even more remarkable when you recall that, only 50 years ago, Japan was a defeated and battered enemy from World War II.

The recent success of Japan, South Korea, Taiwan, and other Asian countries is a dramatic change from the historic dominance of world industry by Western countries. The industrial revolution originated in the United Kingdom during the 1700s and diffused to Europe and North America in the 1800s. The high standard of living enjoyed by most Western Europeans and North Americans is based on industrial power.

While the success of Asian countries is admirable, their profit may be North America's loss. For every new Toyota sold in the United States and Canada, one less Chevrolet may be manufactured. As Asian manufacturers expand their factories, North American companies lay off workers. Thus, the preference of North American consumers for Japanese motor vehicles hurts the U.S. and Canadian economies.

On the other hand, if we truly believe in international cooperation, then people should be encouraged to buy products regardless of national origin. If the Japanese or Koreans can build better cars at lower prices, that presents a challenge to North American companies.

Key Issues

1. Where did industry originate?

2. Where is industry distributed?

3. Why do industries have different distributions?

4. Why do industries face problems?

Port of Yokohama, Japan (Bavaria/Viesti Associates, Inc.)

Case Study

Maquiladoras *in* Mexico

Edi Bencomo is a factory worker in Ciudad Juárez, Mexico. Her job is to clip together several color-coded wires for Alambrados y Circuitos Eléctricos, a factory that is owned by the Packard Electric Division of General Motors.

Bencomo migrated to Ciudad Juárez 4 years ago, at age 16, from Madera, a village in the Sierra Madre Occidental, a mountain range 300 kilometers (200 miles) to the southwest. One of seven children, Bencomo saw no future for herself in remaining on her parents' corn farm. Had she remained in Madera, Bencomo probably would have been unemployed, along with 25 percent of the villagers.

In Ciudad Juárez, Bencomo lives with her husband in a two-room shack more than an hour from the plant. They can afford to rent a somewhat better dwelling, but none are available in this rapidly growing city. She leaves home each weekday at 4 A.M. to battle hordes of workers who crowd onto buses that serve the factory area.

Bencomo earns Mexico's minimum wage, approximately 50 cents an hour. She also receives two important benefits by working for Alambrados: a bus pass so that she can reach the plant at no cost, and two meals in the cafeteria paid for almost entirely by the company. She considers her job to be superior to that of her husband, who makes piñatas; both are paid minimum wage, but he receives no benefits.

Packard's Ciudad Juárez plant is known as a **maquiladora**, from the Spanish verb *maquilar*, which means to receive payment for grinding or processing corn. The term originally applied to a tax when Mexico was a Spanish colony. Under U.S. and Mexican laws, companies receive tax breaks if they ship materials from the United States, assemble components at a *maquiladora* plant in Mexico, and export the finished product back to the United States. More than 1,000 U.S. companies have *maquiladoras* in Mexico. General Motors alone has two dozen *maquiladoras* employing more than 25,000 people and is one of Mexico's largest employers.

In January 1985, General Motors revealed that it was designing an entirely new car called Saturn, and it would need a factory somewhere in the United States to build it. The announcement touched off a fierce competition among states and localities to become the home for the Saturn plant. All 1,700 schoolchildren in New Hampton, Iowa, wrote letters to GM executives urging that their town be selected. Thousands of Cleveland residents sent GM "We Want Saturn" coupons clipped from their local newspapers. Seven governors appeared on Phil Donahue's popular daytime television show to explain why their state should be chosen.

Swamped with material from competing communities, GM took 7 months to select a factory site. The choice was Spring Hill, Tennessee, then a village of 1,000 inhabitants, 50 kilometers (30 miles) south of Nashville.

GM's process of selecting a location for its Saturn factory raises several issues that geographers address:

1. *What factors did GM consider in evaluating locations?* Geographers recognize two critical elements in deciding *where* to locate factories: where the markets are (where the automobiles will be sold), and where the resources are that are needed to make the product.

2. *Why did communities nationwide compete for the Saturn factory?* Government officials everywhere recognize the powerful role of such an industry in the economic health of a community. In the global competition to attract new industries—or, in many places, to retain existing ones—communities possess *locally diverse* locational characteristics. Geographers identify a community's assets that enable it to compete successfully for industries, as well as locational handicaps that must be overcome to retain older companies.

3. *Why did GM feel compelled to build a new plant at all?* As GM was building the new factory in Tennessee, the corporation closed more than a dozen others elsewhere in the country. Why didn't the corporation modernize one of its closed factories, or perhaps construct a new factory on land cleared by demolishing an older one? The answer is very geographic: To succeed in an intensely competitive *global* market for products like automobiles, corporations must find optimal factory locations.

Key Issue 1

Where Did Industry Originate?

- The industrial revolution
- Diffusion of the industrial revolution

The modern concept of industry—meaning the manufacturing of goods in a factory—began in the United Kingdom in the late 1700s. This process of change is called the *industrial revolution*, discussed in Chapter 2 as a cause of population growth between 1750 and 1950. The industrial revolution transformed how goods are produced for society and the way people obtain food, clothing, and shelter.

Today, the industrial revolution has penetrated virtually all economic, social, and political elements of society. This section examines the sequence of changes that occurred with the industrial revolution, and its diffusion from the United Kingdom to the rest of the world.

The Industrial Revolution

The root of the industrial revolution was technology, involving several inventions that transformed the way in which goods were manufactured. The revolution in industrial technology created an unprecedented expansion in productivity, resulting in substantially higher standards of living. From its beginnings in the north of the United Kingdom around 1750, the industrial revolution diffused to Europe and North America in the nineteenth century and to the rest of the world in the twentieth century.

The term *industrial revolution* is somewhat misleading, because it was far more than industrial, and it didn't happen overnight. The industrial revolution resulted in new social, economic, and political inventions, not just industrial ones. The changes involved a gradual diffusion of new ideas and techniques over decades, rather than an overnight revolution. Nonetheless, the term is commonly used to define the events of the late 1700s to early 1800s in Western Europe and North America.

Prior to the industrial revolution, industry was geographically dispersed across the landscape. People made household tools and agricultural equipment in their own homes, or obtained them in the local village. Home-based manufacturing was known as the **cottage industry** system. One important cottage industry was textile manufacturing. People known as putters-out were hired by merchants to drop off wool at homes, where women and children sorted, cleaned, and spun it. The putters-out then picked up the finished work and paid according to the number of pieces that were completed ("piece-rate").

The industrial revolution was the collective invention of hundreds of mechanical devices. But the one invention most important to the development of factories was the steam engine, patented in 1769 by James Watt, a maker of mathematical instruments in Glasgow, Scotland.

When water is boiled into steam, its volume increases about 1,600 times, producing a force that can be used to move a piston back and forth inside a cylinder. If it is attached to a crankshaft, a piston's back-and-forth motion can be converted into rotary motion suitable for driving machinery.

Inventors as far back as the ancient Greek Hero of Alexandria had built engines operated by steam, but steam engines built by Watt's predecessors were not practical, because virtually all of the energy they generated was used in their own operation. Watt built the first useful steam engine that could pump water far more efficiently than the watermills then in common use, let alone human or animal power. Watt's steam engine provided a separate chamber for condensing the steam and used the steam pressure to move the piston in both directions.

Diffusion of the Industrial Revolution

The iron industry was first to increase production through extensive use of Watt's steam engine, plus other inventions. The textile industry followed. From these two pioneering industries, new industrial techniques diffused during the nineteenth century.

Diffusion from the Iron Industry

The first industry to benefit from Watt's steam engine was iron. To produce iron, iron ore is mined from the ground. The ore is not in a useful form for making tools, so it has to be smelted (melted) in a blast furnace (blasted with air to make its fires burn hotly). The molten iron metal is poured into crude molds, where it hardens into pigs, fancifully named for their shape. This pig iron then can be transported and remelted to form useful tools and objects of cast iron, wrought iron, or steel.

The usefulness of iron and steel had been known for centuries, but the scale of production was small. The process demanded constant heating and cooling of the iron, a time-consuming and skilled operation, because energy could not be generated to keep the ovens hot for a sufficiently long period of time. The Watt steam engine provided a practical way to keep the ovens constantly heated.

Henry Cort, a navy agent, established an iron forge near Fareham, England, where iron was shaped into useful objects. He patented two processes, known as puddling and rolling, in 1783. Puddling involved reheating pig iron until it was pasty and then stirring it with iron rods until carbon and other impurities burned off. The rolling process involved passing pig iron between iron rollers to remove remaining dross (a scum of impurities that forms on the surface of melted metal). The combination of Watt's engine and Cort's iron purification process increased iron-manufacturing capability.

The needs of the iron industry in turn generated innovations in coal mining, engineering, transportation, and other industries. These inventions, in turn, permitted the modernization of other industrial activities.

Coal. Iron and steel manufacturing required energy to operate the blast furnaces and the steam engines. Wood, the main energy source prior to the industrial revolution, became increasingly scarce because it was needed for construction of ships, buildings, and furniture, as well as for heat. An obvious solution was to use high-energy coal, which was plentiful. Then, Abraham Darby of Coalbrookdale in Shropshire, England, produced high-quality iron smelted not with ordinary coal, but with purified carbon made from coal, known as coke. This invention provided an abundant source of energy for the iron industry.

Because of the need for large quantities of bulky, heavy coal, the iron industry's geographic pattern changed from dispersed to clustered. Blast furnaces, forges, and mills, which had been scattered in separate small plants, were combined into large, integrated factories. These factories clustered at four locations: Staf-

The industrial revolution transformed the production of textiles beginning in Great Britain in the late eighteenth century. Many looms to weave yarn into fabric could operate at the same time by being connected to a steam engine. (North Wind)

POWER-LOOMS IN AN ENGLISH MILL, 1820

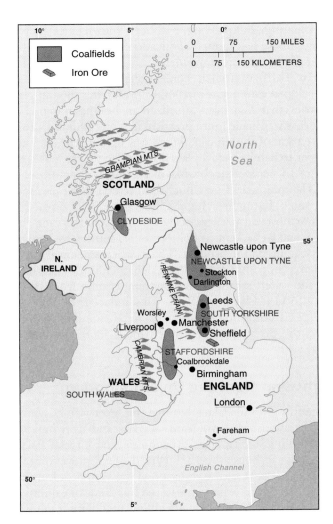

Figure 11–1 The hearth of the industrial revolution. The industrial revolution originated in northern England and southern Scotland in the late eighteenth century. Factories clustered near productive coalfields.

fordshire, South Yorkshire, Clydeside, and South Wales. Each site was near a productive coalfield (Figure 11–1).

Engineering. In 1795, James Watt decided to go into business for himself rather than serve as a consultant to industrialists. He and Matthew Boulton established the Soho Foundry at Birmingham, England, and produced hundreds of new machines to improve industrial processes still further. From this operation came our modern engineering and manufacture of machine parts. Their technical expertise was required to invent new machines, apply existing ones to new situations, and repair worn and broken equipment.

Transportation. The new engineering profession made its biggest impact on transportation, especially canals and railways. Transportation inventions played a critical role in diffusing the industrial revolution. New transportation systems enabled factories to attract large numbers of workers, bring in bulky raw materials like iron ore and coal, and ship finished goods to consumers.

In 1759, Francis Egerton, the second Duke of Bridgewater, decided to build a canal between Worsley and Manchester (center of map in Figure 11–1). He hired James Brindley to direct the project, which took 2 years to complete. This feat launched a generation of British canal construction that enabled industrial goods and workers to be moved longer distances quickly and inexpensively. An extension of the duke's canal in 1767 permitted ships to travel between the sea and Manchester, which is 80 kilometers (50 miles) inland.

The canals soon were superseded by the invention of another transportation system, the railway, or "iron horse." More than any other invention, the railway symbolized the impact of the new engineering profession on the industrial revolution. The railway was not invented by one individual, but through teamwork. Two separate but coordinated engineering improvements were required: the locomotive, and iron rails for it to run on.

A locomotive using Watt's steam engine was invented by William Symington and William Murdoch in 1784. However, it was impractical to operate on bumpy, congested city streets made of dirt, brick, or stone. A few years earlier, Richard Reynolds had constructed an iron track for horse-drawn wagons to cross an uneven surface from the Coalbrookdale coal mines to the Severn River.

Many thought that running a steam locomotive on iron rails was impossible because the wheels would slip off the rails. But William Hedley demonstrated in 1812 that the steam locomotive could run on rails if the wheels had rims. The first public railway was opened between Stockton and Darlington in the north of England in 1825, using a locomotive named the Rocket. Designed by George Stephenson, the Rocket firmly established the benefits of steam locomotives when, in 1829, it won a race against a horse on the Liverpool and Manchester Railway, averaging 38 kilometers (24 miles) per hour.

Diffusion from the Textile Industry

As the engineering industries were developing, a revolution also was underway in the manufacturing of **textiles**, which are woven fabric. A series of inventions between 1760 and 1800 transformed textile production from a dispersed cottage industry to a concentrated factory system.

Richard Arkwright, a barber and wigmaker from the city of Preston, improved the process of spinning yarn. Spinning turns the short threads from cotton plants into the continuous yarn needed to weave cloth. First, he produced a spinning frame in 1768. It used rollers to untangle the twisted cotton fibers before it was spun around a spindle. Arkwright then patented a process for carding (untwisting the fibers prior to spinning). Because these two operations required more power than human beings could provide, the textile industry joined the iron industry early in adopting Watt's steam engine as a power source.

Like iron, the textile industry was transformed from a large collection of dispersed home-based enterprises, each performing a separate task, to a small number of large, integrated firms clustered in a few locations. The large supply of steam power available from Watt's engines induced firms to concentrate all steps in one building attached to the same power source.

The changes in the textile industry just discussed relate to early stages in the process: spinning rough cotton fibers into usable thread, followed by weaving, or lacing together strands of yarn to form cloth. But cotton cloth also had to be bleached and dyed before it was cut into patterns to make into clothing. From the clothing industry's need for new bleaching techniques emerged another industry that is characteristic of the industrial revolution: chemicals.

Chemicals. The traditional method of bleaching cotton involved either exposing the fabric to the Sun or boiling it. In the boiling technique, the cloth first was treated in a solution of ashes and then in sour milk. In 1746, John Roebuck and Samuel Garbett established a factory in which sulfuric acid, obtained from burning coal, was used instead of sour milk. Bleaching was further modernized in Glasgow by Charles Tennant, who in 1798 produced a bleaching powder made from chlorine gas and lime, a safer product than sulfuric acid.

Meanwhile, sulfuric acid was also used to dye clothing. When combined with various metals, sulfuric acid produced another acid, called vitriol, the color of which varied with the metal. Sulfuric acid produced a blue vitriol when combined with copper, green with iron, and white with zinc.

The chemical industry has greatly expanded its role in textile manufacturing. Natural-fiber cloth, such as cotton and wool, is now combined with chemically produced synthetic fibers. They are made from petroleum or coal derivatives, and include nylon, Dacron, and Orlon. Today, the largest textile factories are owned by chemical companies.

Food Processing. Another industry derived from the chemical industry: food processing. An increasing number of urban factory workers, who could not grow their own food or obtain fresh produce, required preserved food. While some preserving techniques, such as drying, fermenting, and pickling, had been known since ancient times, these had limited application to the needs of nineteenth-century urban residents.

In 1810, a French confectioner, Nicholas Appert, developed canning, a method of preserving food in glass bottles that had been sterilized in boiling water. The process was made more practical by Peter Durand's 1839 invention of the tin can, which was lighter, cheaper, and easier to handle than a glass bottle. The tin can was actually 98.5 percent steel, with only a thin coating of tin.

Canning works by killing the bacteria that cause food to spoil. It requires high temperature over time. The major obstacle to large-scale canning was the time that cans had to be kept in boiling water, some 4 to 5 hours, depending on the product. This is where chemical experiments contributed. In 1861, calcium chloride was added to the water, raising its boiling temperature from 100°C to 116°C (212°F to 240°F). This reduced the time for proper sterilization to only 25 to 40 minutes. Consequently, production of canned foods increased tenfold that year.

Diffusion from the United Kingdom

Britain's Crystal Palace became the most visible symbol of the industrial revolution. A glass and iron building resembling a very large greenhouse, the Crystal Palace was built to house the 1851 World's Fair, more formally known as the "Great Exhibition of the Works of Industry of All Nations." The fair featured hundreds of exhibits of modern machinery, virtually all invented within the preceding 100 years.

When Queen Victoria opened the Crystal Palace, the United Kingdom was the world's dominant industrial power. The country produced more than half of the world's cotton fabric and iron, and mined two-thirds of its coal. As the first state to be transformed by the industrial revolution, the U.K.'s production systems far outpaced the rest of the world.

From the United Kingdom, the industrial revolution diffused eastward through Europe and westward across the Atlantic Ocean to North America. From these places, industrial development continued diffusing to other parts of the world.

The Crystal Palace, designed by Sir Joseph Paxton, was erected in London's Hyde Park to house the Great Exhibition of 1851. The glass and iron building was longer than six football fields and enclosed two giant elm trees. After the fair closed, the structure was rebuilt in a South London park. It burned in 1936. (AP/Wide World Photos)

Diffusion to Europe.   Europeans developed many early inventions of the industrial revolution in the late 1700s. The Belgians led the way in new coal-mining techniques, the French had the first coal-fired blast furnace for making iron, and the Germans made the first industrial cotton mill. However, the industrial revolution did not make a significant impact elsewhere in Europe until the late 1800s.

Political instability delayed the diffusion of the industrial revolution in Europe. The French Revolution (1789–1799) and Napoleonic Wars (1796–1815) disrupted Europe, and Germany did not become a unified country until the 1870s. Other revolutions and wars plagued Europe throughout the nineteenth century.

Europe's political problems retarded development of modern transportation systems, especially the railway. Cooperation among small neighboring states was essential to build an efficient rail network, and to raise money for constructing and operating the system. Because such cooperation could not be attained, railways in some parts of Europe were delayed 50 years after their debut in Britain (Figure 11–2).

The industrial revolution reached Italy, the Netherlands, Russia, and Sweden in the late 1800s. However, their industrial development did not match the level in Belgium, France, and Germany until the twentieth century. Other Southern and Eastern European countries joined the industrial revolution during the twentieth century.

Diffusion to the United States.   Industry arrived a bit later in the United States than in Western European countries like France and Belgium, but it grew much faster. At the time of independence in 1776, the United States was a predominantly agricultural society, dependent on the import of manufactured goods from Britain. Manufacturing was more expensive in the United States than in Britain because labor and capital were scarce, and shipping to European markets was expensive.

The first U.S. textile mill was built in Pawtucket, Rhode Island, in 1791, by Samuel Slater, a former worker at Arkwright's factory in England. The textile industry grew rapidly after 1808, when the U.S. government imposed an embargo on European trade to avoid entanglement in the Napoleonic Wars. The textile industry grew rapidly from 8,000 spindles in 1808 to 31,000 in 1809 and 80,000 in 1811.

By 1860, the United States had become a major industrial nation, second only to the United Kingdom. However, except for textiles, leading U.S. industries did not widely use the new industrial processes. Instead, many engaged in processing North America's abundant food and lumber resources. Industries such as iron and steel did not apply new manufacturing techniques on a large scale in the United States until the final third of the nineteenth century.

In the twentieth century, industry has diffused to other parts of the world, including Japan, Eastern Europe, and many former British colonies, such as Canada, Australia, New Zealand, South Africa, and India. Although industrial development has diffused across Earth's surface, much of the world's industry is concentrated in four regions. We will examine this concentration next.

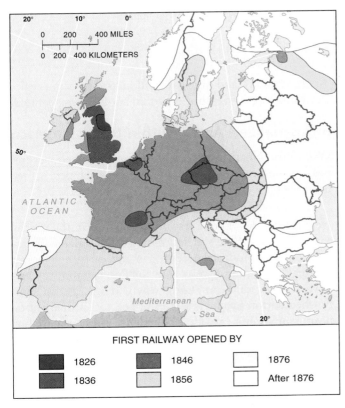

Figure 11–2 Year by which first railway was constructed in area. The diffusion of the railway from the United Kingdom to the European continent reflects the diffusion of the industrial revolution. More than 50 years passed between the construction of the first railways in Britain and the first ones in some Eastern European countries.

Key Issue 2

Where Is Industry Distributed?

- North America
- Western Europe
- Eastern Europe
- East Asia

Approximately three-fourths of the world's industrial production is concentrated in four regions: eastern North America, northwestern Europe, Eastern Europe, and East Asia (Figure 11–3).

Industrial distribution differs from that of agriculture. Agriculture occupies one-fourth of Earth's land area and covers extensive areas throughout the inhabited world. In

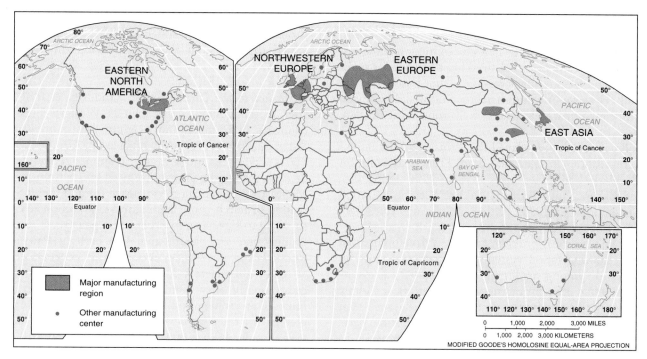

Figure 11–3 Manufacturing is clustered in four main regions: *Northeastern United States-Southeastern Canada*—primarily along the U.S. East Coast from Boston to Baltimore and along the Great Lakes from Toronto and Buffalo west to Chicago and Milwaukee. *Western Europe*—primarily near the Rhine and Ruhr rivers in Germany, France, Belgium, the Netherlands, and Luxembourg; also including southern United Kingdom and northern Italy. *Eastern Europe*—primarily in western Russia and eastern Ukraine, although more recent industrial activities have been in Asian Russia, east of the Urals. *East Asia*—primarily in southern Japan, eastern China, South Korea, and Taiwan.

contrast, less than 1 percent of Earth's land is devoted to industry. This section describes each of the four industrial regions, as well as important industrial subareas within each region.

North America

Manufacturing in North America is concentrated in the northeastern quadrant of the United States and in southeastern Canada. The region comprises only 5 percent of the land area of these countries but contains one-third of the population and nearly two-thirds of the manufacturing output.

This manufacturing belt has achieved its dominance through a combination of historical and environmental factors. As the first area of European settlement in the Western Hemisphere, the U.S. East Coast was tied to European markets and industries during the first half of the nineteenth century. The early date of settlement gave eastern cities an advantage in creating the infrastructure needed to become the country's dominant industrial center.

The Northeast also had essential raw materials, including iron and coal. Good transportation moved raw materials to factories and manufactured goods to markets. The Great Lakes and major rivers (Mississippi, Ohio, St. Lawrence) were supplemented in the 1800s by canals, railways, and highways. All helped to connect the westward-migrating frontier with manufacturing centers.

Industrialized Areas Within North America

Within the North American manufacturing belt, several heavily industrialized areas have developed (Figure 11–4).

New England. The oldest industrial area in the northeastern United States is southern New England. It developed as an industrial center in the early nineteenth century, beginning with cotton textiles. Cotton was imported from southern states, where it was grown, and finished cotton products were shipped to Europe. European immigrants provided abundant, inexpensive labor throughout the 1800s. Today, New England is known for relatively skilled but expensive labor.

Middle Atlantic. The Middle Atlantic area, between New York City and Washington, is the largest U.S. market. It has long attracted industries that need proximity to a large number of consumers. Many industries that depend on foreign markets or imported raw materials have located near one of this region's main ports: New York City (the country's largest port), Baltimore, Philadelphia, and Wilmington, Delaware. Other firms

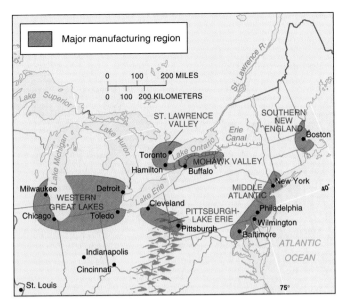

Figure 11–4 Major industrial regions of North America. Manufacturing in North America is highly clustered in several regions within the northeastern United States and southeastern Canada, although important manufacturing centers exist elsewhere in the two countries.

seek locations near the financial, communications, and entertainment industries, which are highly concentrated in New York.

Mohawk Valley. A linear industrial belt developed in upper New York State along the Hudson River and Erie Canal, which connects New York City and the Great Lakes. Buffalo, near the confluence of the Erie Canal and Lake Erie, is the region's most important industrial center, especially for steel and food processing. Inexpensive, abundant electricity, generated at nearby Niagara Falls, has attracted aluminum, paper, and electrochemical industries to the region.

Pittsburgh–Lake Erie. The area between Pittsburgh and Cleveland is the nation's most important steel-producing area. Steel manufacturing originally concentrated in the region because of its proximity to Appalachian coal and iron ore. When northern Minnesota became the main source of iron ore, the Pittsburgh-Lake Erie region could bring in ore via the Great Lakes.

Western Great Lakes. The western Great Lakes area extends from Detroit and Toledo, Ohio, on the east to Chicago and Milwaukee, Wisconsin, on the west. Chicago, the third-largest U.S. urban area, is the dominant market center between the Atlantic and Pacific coasts and the hub of the nation's transportation network. Because road, rail, air, and sea routes converge on Chicago, the city has become a transfer point among transportation systems (water, rail, truck, air) or between routes within the same type of transportation system.

Automobile manufacturers and other industries that have a national market locate in the western Great Lakes

region to take advantage of this convergence of transportation routes. The region's industries are also the main suppliers of machine tools, transportation equipment, clothing, furniture, agricultural machinery, and food products to people living in the interior of the country.

St. Lawrence Valley–Ontario Peninsula. Canada's most important industrial area is the St. Lawrence Valley–Ontario Peninsula area, which stretches across southern Canada along the U.S. border. The region has several assets: centrality to the Canadian market, proximity to the Great Lakes, and access to inexpensive hydroelectric power from Niagara Falls. Most of Canada's steel production is concentrated in Hamilton, Ontario, while most automobiles are assembled in the Toronto area. Inexpensive electricity has attracted aluminum manufacturing, paper making, flour mills, textile manufacturing, and sugar refining.

Changing Distribution of U.S. Manufacturing

Industry has grown in areas outside the main U.S. manufacturing belt (Figure 11–5). Steel, textiles, tobacco products, and furniture industries have become dispersed through smaller communities in the South. The Gulf Coast is becoming an important industrial area because of access to oil and natural gas. Along the Gulf Coast are oil refining, petrochemical manufacturing, food processing, and aerospace product manufacturing.

The Southeast attracts manufacturers that seek a location where few workers have joined labor unions. Southeastern states are known as **right-to-work states**, because they have passed laws preventing a union and company from negotiating a contract that requires workers to join a union as a condition of employment.

On the West Coast, Los Angeles first built its manufacturing base on aircraft and military equipment, but it is now a center for low-tech manufacturing. Los Angeles has become the country's largest area of clothing and textile production, the second-largest furniture producer, and a major food-processing center. San Diego has attracted many industries that support naval operations. High-tech manufacturing is located farther north along the coast, including the famous Silicon Valley in the San Francisco Bay area (see page 394), the nation's center for production of semiconductors and computers, and Seattle, home of the leading aircraft company Boeing and computer firm Microsoft.

Western Europe

Like the North American manufacturing belt, the Western European industrial region appears as one region on a world map. In reality, four distinct districts have emerged, primarily because European countries competed with each other to develop their own industrial

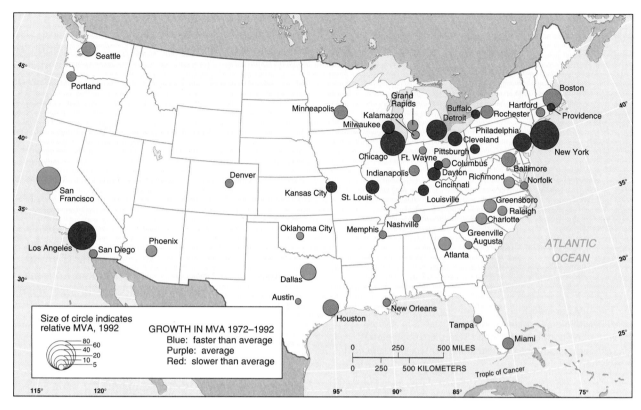

Figure 11–5 Amount of manufacturing and changes in largest U.S. metropolitan areas, 1972–1992. Manufacturing is measured by the value added during the process. Manufacturing value added (MVA) is the gross value of manufactured products minus the costs of raw materials and energy. MVAs are in billions of dollars.

areas: the Rhine–Ruhr Valley, the mid-Rhine, the United Kingdom, and northern Italy (Figure 11–6). Each of these areas is divided into subareas. These four became important for industry because of their proximity to raw materials (coal and iron ore) and markets (large concentrations of wealthy European consumers).

The Rhine–Ruhr Valley

Western Europe's most important industrial area is the Rhine–Ruhr Valley. The region lies mostly in northwestern Germany, but extends into nearby Belgium, France, and the Netherlands. Because of each country's political uniqueness, each country established its own industrial complex.

At the heart of the region lie two rivers: the Rhine, which flows northward through Germany and westward through the Netherlands, and the Ruhr, which flows westward across Germany into the Rhine. Within the region, industry is dispersed rather than concentrated in one or two cities. Although more than 20 million people live in the region, no individual city has more than 1 million inhabitants. Larger cities in the German portion include Dortmund, Düsseldorf, and Essen. The city of Duisburg is located near where the Ruhr flows into the Rhine.

The Rhine divides into multiple branches as it passes through the Netherlands. The city of Rotterdam is near to where several major branches flow into the North Sea. This location at the mouth of Europe's most important river has made Rotterdam the world's largest port.

Iron and steel manufacturing has concentrated in the Rhine–Ruhr Valley because of proximity to large coalfields. Access to iron and steel production stimulated the location of other heavy-metal industries, such as locomotives, machinery, and armaments.

The Mid-Rhine

The second most important industrial area in Western Europe includes southwestern Germany, northeastern France, and the small country of Luxembourg. In contrast to the Rhine–Ruhr Valley, the German portion of the Mid-Rhine region lacks abundant raw materials, but it is at the center of Europe's most important consumer market. The Mid-Rhine region became a major industrial center when Germany was split into two countries after World War II. Then, it was close to the population center of West Germany (Federal Republic of Germany). Although the Mid-Rhine region is once again on the periphery of a reunified Germany, it remains the most central industrial area within the European Union.

The three largest cities in the German portion are Frankfurt, Stuttgart, and Mannheim. Frankfurt became West Germany's most important financial and commercial center and the hub of its road, rail, and air networks.

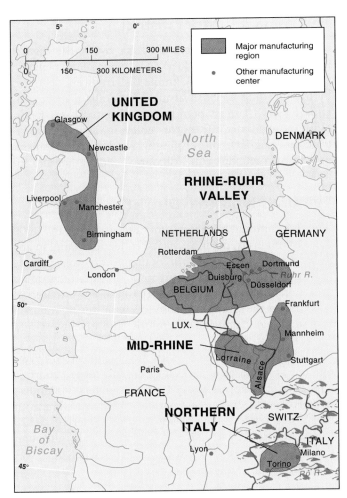

Figure 11–6 Manufacturing centers in Western Europe. A large percentage of manufacturing in Western Europe extends in a north-south belt, from the United Kingdom on the north to Italy on the south. At the core of the European manufacturing region lie Germany, France, and the Benelux countries (Belgium-Netherlands-Luxembourg).

Consequently, Frankfurt attracted industries that produce goods for consumers countrywide, and the city is well situated to play a comparable role in the European Community. Stuttgart's industries specialize in high-value goods and require skilled labor; Mercedes-Benz and Audi automobiles are among the city's best-known products. Mannheim, an inland port along the Rhine, has a large chemical industry that manufactures synthetic fibers, dyes, and pharmaceuticals.

The French portion of the Mid-Rhine region—Alsace and Lorraine—contains Europe's largest iron-ore field, and is the production center for two-thirds of France's steel. Tiny Luxembourg is also one of the world's leading steel producers, because the Lorraine iron-ore field extends into the southern part of the country.

United Kingdom

The industrial revolution originated in the Midlands and northern England and southern Scotland, as described earlier. The industrial revolution began in the northern regions of England and southern Scotland in part because those areas contained a remarkable concentration of innovative engineers and mechanics during the late eighteenth century. Through the nineteenth century, this region dominated world production of iron and steel, textiles, and coal mining.

The United Kingdom lost its international industrial leadership in the twentieth century. As the first country to enter the industrial revolution, Britain was saddled with what became outmoded and deteriorating factories and support services. The British have referred ironically to their "misfortune" of winning World War II. The losers, Germany and Japan, became industrial powers in part because they received American financial assistance to build modern factories, replacing those destroyed during the war.

Although no longer a world leader in steel, textiles, and other early industrial revolution industries, the United Kingdom expanded industrial production in the late twentieth century by attracting new high-tech industries that serve the European market. Japanese companies have built more factories in the United Kingdom than has any other European country. The British have done more than the other major European countries to lower taxes on businesses, reduce government regulations, convert government monopolies to private ownership, and utilize computers.

Today, British industries are more likely to locate in southeastern England, near London. Southeastern England has the country's largest concentrations of population and wealth, and is closest to the Channel Tunnel, which has substantially reduced the time needed to ship goods to Europe.

Northern Italy

A fourth European industrial region of some importance lies in the Po River Basin of northern Italy. It contains about one-fifth of Italy's land area, but approximately half of the country's population and two-thirds of its industries.

Modern industrial development in the Po Basin began with establishment of textile manufacturing during the nineteenth century. The Po Basin has attracted textiles and other industries because of two key assets, compared to Europe's other industrial regions: numerous workers willing to accept lower wages, and inexpensive hydroelectricity from the nearby Alps. Industries concentrated in this region include raw-material processors and mechanical-parts assemblers.

Eastern Europe

Eastern Europe has six major industrial regions. Four are entirely in Russia, one is in Ukraine, and one is southern Poland and northern Czech Republic (Figure 11–7). Central industrial district, St. Petersburg, Eastern Ukraine, and Silesia became manufacturing centers in the 1800s.

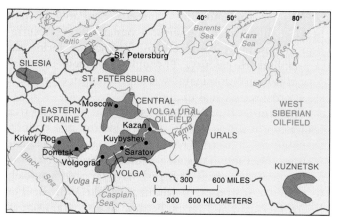

Figure 11–7 Manufacturing centers in Eastern Europe and Russia. In Eastern Europe, manufacturing is clustered in western (European) Russia and Ukraine. The former Soviet government encouraged development of manufacturing regions in the center of the country east of the Ural Mountains.

The Volga and Urals regions were established by the Communists during the twentieth century. In addition to these six industrial regions, Russia also contains another major industrial region, Kuznetsk, in the eastern or Asian portion of the country.

Central Industrial District

Russia's oldest industrial region is centered around Moscow, the country's capital and largest city. Although not well endowed with natural resources, the Central industrial district produces one-fourth of Russian industrial output, primarily because it is situated near the country's largest market. Products of this region tend to be of high value relative to their bulk and require a large pool of skilled labor. Of Moscow's industrial workforce, 30 percent is employed making linen, cotton, wool, and silk fabrics. Moscow factories also specialize in chemicals and light industrial goods.

St. Petersburg Industrial District

St. Petersburg, Eastern Europe's second largest city, was one of Russia's early nodes of industrial innovation. Railways were built in the St. Petersburg area several decades earlier than in the rest of Russia (Figure 11–2). Given its proximity to the Baltic Sea, the St. Petersburg area specializes in shipbuilding and other industries serving Russia's navy and ports. The area also produces goods that meet the needs of the local market, such as food-processing, textiles, and chemicals.

Eastern Ukraine Industrial District

The Donetsk coalfield, in the far eastern portion of Ukraine, contains one of the world's largest coal reserves. Eastern Ukraine also possesses large deposits of iron ore, manganese, and natural gas. These assets make the region Eastern Europe's largest producer of pig iron and steel. Major plants are located at Krivoy Rog, near iron ore fields, and Donetsk, near coalfields.

The Volga Industrial District

Situated along the Volga and Kama rivers, the district grew rapidly during World War II, when many plants in the Central and Eastern Ukraine districts were occupied by the invading German army. The Volga district contains Russia's largest petroleum and natural gas fields. Within the district, the motor vehicle industry is concentrated in Togliatti, oil refining in Kuybyshev, chemicals in Saratov, metallurgy in Volgograd, and leather and fur in Kazan.

The Urals Industrial District

The Ural mountain range contains more than 1,000 types of minerals, the most varied collection found in any mining region in the world. Valuable deposits include iron, copper, potassium, manganese, bauxite (aluminum ore), salt, and tungsten. Proximity to these raw materials encouraged the Communists to locate in this region iron and steel manufacturing, chemicals, and machinery and metal fabricating.

Although well endowed with metals, industrial development is hindered by a lack of nearby energy sources. Coal must be shipped nearly 1,500 kilometers (900 miles) from Kuznetsk, and oil and natural gas are piped in from the Volga-Ural, Bukhara, and central Siberian fields. Russia controls nearly all the Urals minerals, although the southern portion of the region extends into Kazakhstan.

Kuznetsk Industrial District

Kuznetsk is Russia's most important manufacturing district east of the Ural Mountains. The region contains the country's largest reserves of coal and an abundant supply of iron ore. Soviet planners took advantage of these natural assets to invest considerable capital in constructing iron, steel, and other factories in the region.

Silesia

Outside the former Soviet Union, Eastern Europe's leading manufacturing area is in Silesia, which includes southern Poland and the northern Czech Republic. It is an important steel production center because it is near coalfields, although iron ore must be imported.

East Asia

East Asia is the most heterogeneous industrial region from the perspective of level of development. Japan is one of the world's wealthiest countries thanks to industrial development. China—the world's most populous country—has the world's second largest economy, behind the United States, although GDP per capita is low. South Korea and Taiwan, prominent examples of countries that have employed international trade to become important industrial powers, have per capita GDPs substantially above that of China but substantially below that of Japan.

China has abundant reserves of coal, iron ore, and other important minerals, but Japan and the other East Asian countries have few natural resources. East Asia is far from wealthy consumers in North America and Western Europe, yet the region has become a major exporter of consumer goods.

Faced with isolation from world markets and a shortage of nearly all essential resources, East Asia has taken advantage of its most abundant resource: a large labor force. Although its industries were devastated during World War II, Japan became an industrial power in the 1950s and 1960s, initially by producing goods that could be sold in large quantity at cut-rate prices to consumers in other countries. (During those days, the label "Made in Japan" was a joke in America, signifying cheap, shoddy goods.) Prices were kept low, despite high shipping costs, because workers received much lower wages in Japan than in North America or Western Europe.

Aware that South Korea, Taiwan, and other Asian countries were building industries based on even lower-cost labor, Japan started training workers for highly skilled jobs. Because wages remained lower than in other more developed countries, Japan could manufacture high-quality products at a lower cost than those in North America or Western Europe. As a result, during the

1970s and 1980s, Japan earned a reputation for high-quality electronics, precision instruments, and other products that required well-trained workers. The country became the world's leading manufacturer of automobiles, ships, cameras, stereos, and televisions.

In recent years, China has become a major manufacturer of steel, farm machinery, and construction materials, as well as the leader exporter of clothing to the United States. The combination of a potentially enormous domestic market, as well as the availability of low-cost labor, has attracted foreign investment in China's industries.

As in other regions, industry is not distributed uniformly within East Asia. Japan's manufacturing is concentrated in the central region between Tokyo and Nagasaki, especially the two large urban areas of Tokyo-Yokohama and Osaka-Kobe-Kyoto (Figure 11–8). Industry in China in clustered in three areas along the east coast: near Hong Kong, the Yangtze River valley between Shanghai and Wuhan, and along the Gulf of Bo Hai from Tianjin and Beijing to Shenyang.

Although industry is located elsewhere in the world, the four industrial regions of North America, Western Europe, Eastern Europe, and Japan account for most of the world's industrial production. Having looked at the

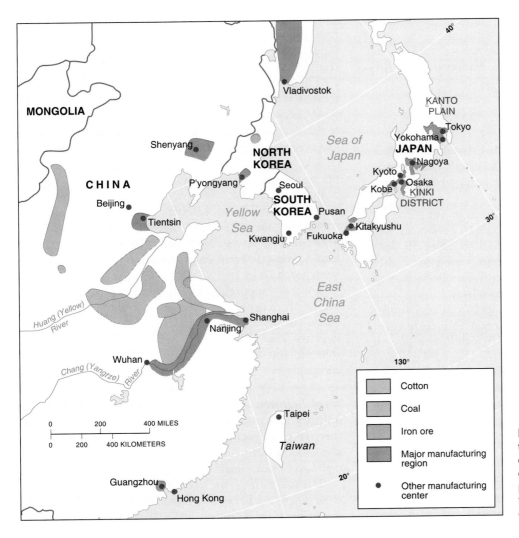

Figure 11–8 Manufacturing centers in East Asia. Within Japan, production is clustered along the southeast coast. Within China, a large percentage of industries are clustered in three centers along the east coast.

"where" question for industrial location, we can next consider the "why" question: Why are industries located where they are?

Why Do Industries Have Different Distributions?

- Situation factors
- Site factors
- Obstacles to optimum location

Industry seeks to maximize profits by minimizing production costs. Geographers try to explain why one location may prove more profitable for a factory than others. A company ordinarily faces two geographical costs: situation and site.

- **Situation factors** involve transporting materials to and from a factory. A firm seeks a location that minimizes the cost of transporting inputs to the factory and finished goods to the consumers.
- **Site factors** result from the unique characteristics of a location. Land, labor, and capital are the three traditional production factors that may vary among locations.

While a variety of situation and site costs explain the location of factories, the particular combination of critical factors varies among firms.

Situation Factors

Every manufacturer buys and sells. Manufacturers buy from companies and individuals who supply manufacturing inputs (materials, energy, machinery, services). They sell to companies and individuals who purchase the product. One objective of every company is to minimize the aggregate cost of transporting inputs to its factory and transporting finished products from its plant to consumers. The farther something is transported, the higher the cost, so a manufacturer tries to locate its factory as close as possible to both buyers and sellers.

A company that obtains all inputs from one source and sells all products to one customer can easily compute the optimal location for its factory. If the cost of transporting the product exceeds the cost of transporting inputs, then optimal plant location is as close as possible to the customer. Conversely, if inputs are more expensive to transport, a factory should locate near the source of inputs.

Location Near Inputs

Every industry uses some inputs. These may be resources from the physical environment (minerals, wood, or animals), or they may be parts or materials made by other companies. If the weight and bulk of any one input is particularly great, the firm may locate near the source of that input to minimize transportation cost.

Copper Industry. The North American copper industry is a good example of locating near the source of heavy, bulky inputs to minimize transportation cost. In copper production, the first step is mining the copper ore. Much copper ore mined in North America is low-grade, less than 0.6 percent copper, whereas the rest is waste, known as gangue. Obviously, the weight and bulk of this low-grade ore are considerable.

The next step is to concentrate the copper. Concentration mills crush and grind the ore into fine particles, mix them with water and chemicals, and filter and dry them. Copper concentrate is about 25 percent copper. Concentration mills are always near the mines, because concentration transforms the heavy, bulky copper ore into a product of much higher value per weight. Copper concentration is a **bulk-reducing industry**, an economic activity in which the final product weighs less than its inputs.

The concentrated copper then becomes the input for smelters, which remove more impurities from the copper. Smelters produce copper matte (about 60 percent copper), blister copper (about 97 percent copper), and anode copper (about 99 percent copper). As another bulk-reducing industry, smelters are built near their main inputs—the concentration mills—again to minimize transportation cost.

The purified copper produced by smelters is further treated at refineries to produce copper cathodes, about 99.99 percent pure copper. Little further weight loss occurs, so proximity to the mines, mills, and smelters is a less critical factor in the locating refineries.

A U.S. map demonstrates the locational needs for copper processing (Figure 11–9). Two-thirds of U.S. copper is mined in Arizona, so the state also has most of the concentration mills and smelters. Arizona also contains refineries, but others are located in Maryland, New Mexico, New Jersey, New York, Texas, and Utah. The East Coast refineries import much of its copper anode from abroad.

Another important locational consideration is the source of energy to power these energy-demanding operations. In general, metal processors such as the copper industry also try to locate near economical electrical sources, and to negotiate favorable rates from power companies.

Steel Industry. Steelmaking is another bulk-reducing industry that traditionally has been located to minimize the cost of transporting inputs. The U.S. steel industry also demonstrates how locations change when the source and cost of raw materials change.

The main inputs for steel production are iron ore and coal. These are heavy and bulky, contain a high percentage of impurities, and must be used in large quantities.

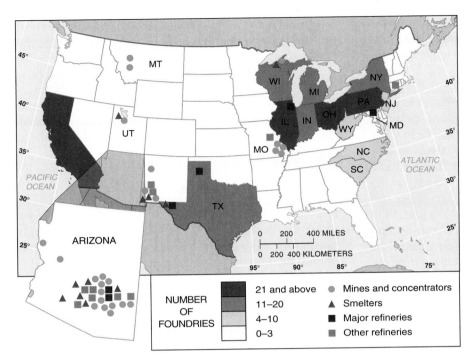

Figure 11–9 Copper mining, concentrating, smelting, and refining, examples of bulk-reducing industries. In the United States, most plants that concentrate, smelt, and refine copper are in or near Arizona, where most copper mines are located. Copper refining plants in coastal locations use imported materials.

These characteristics influence steel processors to minimize transportation cost through location.

In the mid-1800s, the U.S. steel industry concentrated in southwestern Pennsylvania around Pittsburgh, where iron ore and coal were both mined. Later in the 1800s, more steel mills were built around Lake Erie, in the Ohio cities of Cleveland, Youngstown, and Toledo, and around Detroit, and in other communities (Figure 11–10). The locational shift was largely influenced by the discovery of rich iron ore in the Mesabi Range, a series of low mountains in northern Minnesota. This area soon became the source for virtually all iron ore used in the U.S. steel industry. The ore was transported by way of Lake Superior, Lake Huron, and Lake Erie. Coal was shipped from Appalachia by train.

Around 1900, new steel mills began to be located farther west, near the southern end of Lake Michigan—Gary in Indiana, Chicago, and other communities. The main raw materials continued to be iron ore and coal, but changes in steelmaking required more iron ore in proportion to coal. Thus, new steel mills were built closer to the Mesabi Range to minimize transportation cost. Coal was available from nearby southern Illinois, as well as from Appalachia.

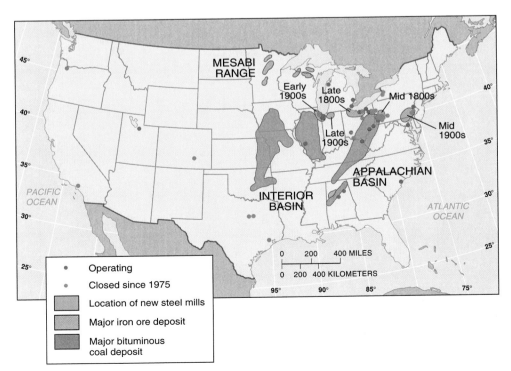

Figure 11–10 Integrated steel mills in the United States. Integrated steel mills are highly clustered near the southern Great Lakes, especially Lake Erie and Lake Michigan. Historically, the most critical factor in siting a steel mill was to minimize transportation cost for raw materials, especially heavy, bulky iron ore and coal. In recent years, many integrated steel mills have closed. Most surviving mills are in the Midwest to maximize access to consumers.

Most large U.S. steel mills built during the first half of the twentieth century were located in communities near the East and West coasts, such as Trenton in New Jersey, Baltimore, and Los Angeles. These coastal locations partly reflected further changes in transportation cost. Iron ore increasingly came from other countries, especially Canada and Venezuela, and locations near the Atlantic and Pacific oceans were more accessible to those foreign sources. Further, scrap iron and steel—widely available in the large metropolitan areas of the East and West coasts—had become an important input in the steel-production process.

Recently, more steel plants have closed than opened in the United States. Among the survivors, plants around southern Lake Michigan and along the East Coast have significantly increased their share of national production. This success derives primarily from market access, rather than input access. In contrast with the main historical locational factor—transportation cost of raw materials—successful steel mills today are located increasingly near major markets. Coastal plants provide steel to large East Coast population centers, and southern Lake Michigan plants are centrally located to distribute their products countrywide.

The growth of steel minimills also demonstrates the increasing importance of access to markets rather than to inputs. Traditionally, most steel was produced at large,

integrated mills. They processed iron ore, converted coal into coke, converted the iron into steel, and formed the steel into sheets, beams, rods, or other shapes. Minimills, generally limited to one step in the process—steel production—have captured one-fourth of the U.S. steel market. Less expensive than integrated mills to build and operate, minimills can locate near their markets because their main input—scrap metal—is widely available (Figure 11–11).

Location Near Markets

For many firms, the optimal location is close to markets, where the product is sold. The cost of transporting goods to consumers is a critical locational factor for three types of industries: bulk-gaining, perishable, and single-market.

Bulk-Gaining Industries. A **bulk-gaining industry** makes something that gains volume or weight during production. Soft-drink bottling is a good example of an industry that gains weight. Empty cans or bottles are brought to the bottler, filled with the soft drink, and shipped to consumers. Two main inputs are placed in the container: syrup (relatively concentrated and easy to transport) and water (relatively bulky, heavy, and expensive to transport). Major soft-drink companies like Coca-

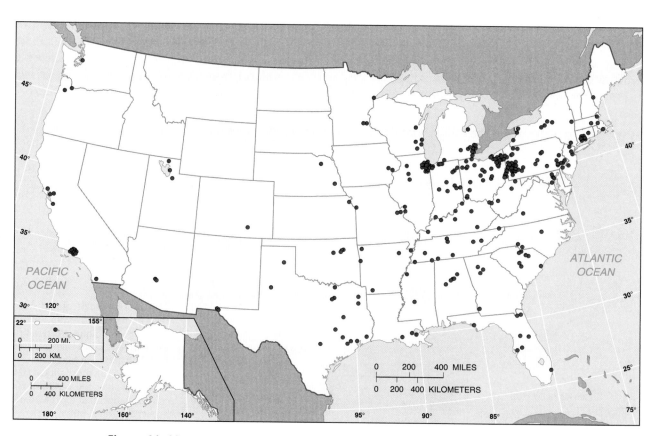

Figure 11–11 Minimills for steel production in the United States. Minimills, which produce steel from scrap metal, are more numerous than integrated steel mills, and they are distributed around the country near local markets.

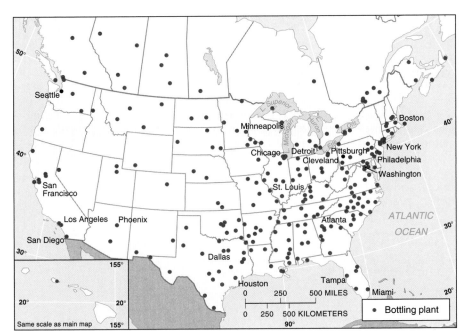

Figure 11–12 Coca-Cola bottling plants in the United States and Canada. A soft-drink bottling plant is a good example of a bulk-gaining industry, which needs to be located near consumers. Consequently, there are more than 200 soft drink bottlers in the United States and Canada, situated near all major population concentrations.

Cola and Pepsico manufacture syrups according to proprietary recipes and ship them to bottlers in hundreds of communities (Figure 11–12).

If water were only available in a few locations around the country, then bottlers might cluster near the source of such a scarce, bulky input. But because water is available where people live, bottlers can minimize costs by producing soft drinks near their consumers instead of shipping water (their heaviest input) long distances.

A filled container has the same volume as an empty one, but it is much heavier—the container itself accounts for less than 5 percent of the weight of a filled 355 ml (12 fl oz) can or 1 liter (33.8 fl oz) bottle. Because they are heavier, the filled containers are more expensive to ship than the empty ones, and bottlers locate near their customers rather than the manufacturers of the containers.

Most major bottlers of beer, such as Anheuser-Busch and Miller, follow a similar pattern of locating a number of facilities around the country, near major population centers, to minimize the cost of shipping to consumers. Production of one of the major brewers, Coors, remains in one location—Golden, Colorado—because the company advertises that the community's water imparts a distinctive flavor to the beer.

Scotch whiskey is another weight-gaining product, but its spatial distribution differs from that of soft drinks. Although the product is mostly water, it does not have sufficient consumers to justify a bottling plant in each city. One Scotch distiller must serve more than one market and charge higher prices to cover the delivery cost to dispersed consumers.

More commonly, bulk-gaining industries manufacture products that gain volume rather than weight. A prominent example is the fabricated-metals industry. A fabricated-metals factory brings together a number of previously manufactured parts as the main inputs and assembles them into a more complex product. Many common products are so fabricated, including TV sets, refrigerators, and automobiles.

If the fabricated product occupies a much larger volume than its individual parts, like a car or freezer, then the cost of shipping the final product to consumers is likely to be a critical factor. A fabricated products indus-

Manufacture of potato chips is a bulk-gaining industry: potatoes are much bulkier after they have been sliced, fried until they curl, and placed in large bags. Further, potato chips are best consumed when fresh. As a result, potato chips are produced closer to the consumers than are other snack foods. (David Joel/Tony Stone Images)

Globalization and Local Diversity

Shifting Geography of Motor Vehicle Production

Cars and trucks are manufactured at two types of factories. Several thousand components plants make one or more parts that go into vehicles. These parts are then combined into finished vehicles at about 70 assembly plants in the United States and Canada.

The locations in North America *where* cars and trucks are assembled have changed. Historically, producers divided North America into regions, and an assembly plant was located in or near a large metropolitan area within each region (Figure 1). During the past two decades, this long-standing distribution changed. Assembly plants near East Coast and West Coast population centers have closed. New ones have opened in the interior, especially along interstate highways 65 and 75 (Figure 2).

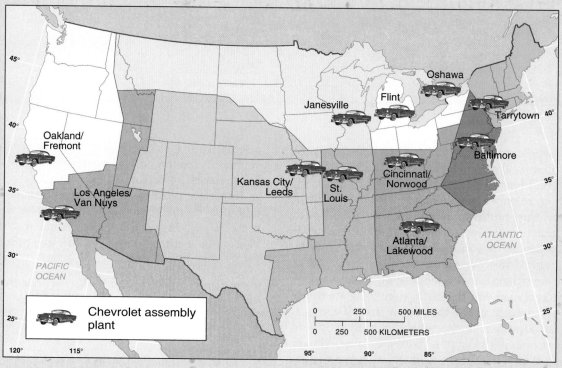

Chevrolet assembly plants, 1955. In 1955, General Motors assembled identical Chevrolets at ten U.S. final assembly plants located near major population centers. This distribution enabled GM to minimize the cost of distributing its relatively bulky products to consumers.

try locates to minimize the cost of shipping its bulky product to the market. The Globalization and Local Diversity box discusses a major fabricated product, the automobile.

Perishable Products. To deliver their products to consumers as rapidly as possible, perishable-product industries must be located near their markets. Food producers such as bakers and milk bottlers must locate near their customers to assure rapid delivery, because few people want stale bread or sour milk.

Processors of fresh food into frozen, canned, and preserved products can locate far from their customers. Cheese and butter, for example, are manufactured in Wisconsin because rapid delivery to the urban markets is not critical for products with a long shelf life, and the area is well-suited agriculturally for raising dairy cows.

The daily newspaper is an example of a product other than food that is highly perishable because it contains dated information. People demand their newspaper as soon after its printing as possible. Therefore, newspaper publishers must locate near markets to minimize transportation cost.

Difficulty with timely delivery is one of the main factors in the demise of afternoon newspapers. Morning newspapers are printed between 9 P.M. and 6 A.M. and delivered during the night, when traffic is light. After-

The reason **why** assembly plant locations have changed reflects the importance of situation factors. For a fabricated product like a motor vehicle, the critical location factor is minimizing transportation to the market, in this case the 15 million North Americans who buy new vehicles each year.

But the market has changed through greater **diversity** of products. The number of distinct vehicle models made in North America has increased from about 100 in the 1950s to 700 in the 1990s, with the addition of sport utility vehicles, minivans, and a wide variety of large and small passenger cars. So instead of building the identical Ford or Chevrolet model at several assembly plants for regional distribution, producers now operate specialized assembly plants that build single models for distribution throughout North America.

The diversity of motor vehicles produced in North America has also increased through **globalization** of the industry—the entry of Japanese and German companies, which have also located assembly plants in the interior. In geographic terms, if a company has a product that is made at only one plant, and the critical locational factor is to minimize the cost of distributing it to U.S. and Canadian consumers, then the optimal factory location is in the U.S. interior, rather than on the East or West Coast.

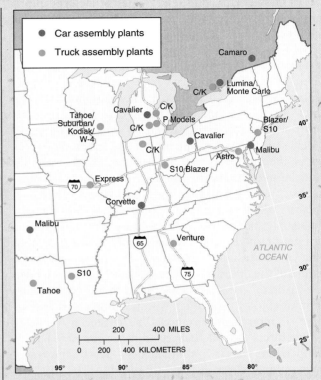

Chevrolet assembly plants 1998. During the past two decades, motor vehicle producers have located new assembly plants in the U.S. interior rather than near coastal population concentrations, which had been preferred in the past. Most coastal plants were closed during the 1980s and 1990s.

noon newspapers, published between 9 A.M. and 5 P.M., must be delivered in heavy daytime traffic, which slows delivery and thereby raises total production cost.

In European countries, national newspapers are printed in the largest city during the evening and delivered by train throughout the country overnight. This has been possible because of the comparatively compact size of most European states.

Until recently, publishers considered the United States to be too large to make a national newspaper feasible. With satellite technology, however, the *New York Times*, *Wall Street Journal*, and *USA Today* have moved in the direction of national delivery. These newspapers are composed in New York or Washington. Digitized page images are transmitted by satellite to other locations, such as Atlanta and Chicago, where the papers are printed. The papers are then delivered by air and surface transport to consumers nearest each city where printing is done.

Single-Market Manufacturers. Single-market manufacturers make products sold primarily in one location, so they also cluster near their markets. For example, several times a year, buyers from individual clothing stores and department-store chains come to New York from all over the United States to select high-style apparel they will sell in the coming season. Manufacturers of fashion clothing then receive large orders for certain garments to be delivered in a short time. Consequently, high-style clothing manufacturers concentrate around New York.

New York-based high-style clothing manufacturers in turn demand rapid delivery of specialized components, such as clasps, clips, pins, and zippers. The specialized component manufacturers, therefore, also concentrate in New York.

The manufacturers of parts for motor vehicles are also specialized manufacturers with only one or two customers—the major motor vehicle producers such as General Motors and Toyota. In the past, most motor vehicle parts were made in Michigan and shipped to nearby warehouses and distribution centers maintained in that state by the major producers. From the warehouses, the producers sent the parts to branch assembly plants located around the country near major metropolitan areas (see Globalization and Local Diversity box).

Parts makers now ship most of their products directly to assembly plants, most of which are now clustered in the interior of the country. Proximity to the assembly plant is increasingly important for parts producers because of the diffusion of "just-in-time" delivery. Under "just-in-time," parts are delivered to the assembly plant just in time to be used, often within minutes, rather than weeks or months in advance. The clustering of parts manufacturers around their customers—the new Japanese-operated U.S. assembly plants—clearly illustrates the adoption of "just-in-time" (Figure 11–13).

Ship, Rail, Truck, or Air?

Inputs and products are transported in one of four ways: ship, rail, truck, or air. Firms seek the lowest-cost mode of transport, but the cheapest of the four alternatives changes with the distance that goods are being sent.

The farther something is transported, the lower is the cost per kilometer (or mile). Longer-distance transportation is cheaper per kilometer in part because firms must pay workers to on-load and off-load goods onto vehicles, whether the material travels 10 kilometers or 10,000. The cost per kilometer decreases at different rates for each of the four modes, because the loading and unloading expenses differ for each mode.

Trucks are most often used for short-distance delivery and trains for longer distances, because trucks can be loaded and unloaded more quickly and cheaply than trains. If a water route is available, ship transport is attractive for very long distances, because the cost per kilometer is even less.

Air is normally the most expensive alternative for all distances, but an increasing number of firms transport by air to ensure speedy delivery of small-bulk, high-value packages. Air transport companies like FedEx, Airborne, and UPS promise overnight delivery for most packages. They pick up packages in the afternoon and transport them by truck to the nearest airport. Late at night, planes filled with packages are flown to a central hub airport in the interior of the country, such as

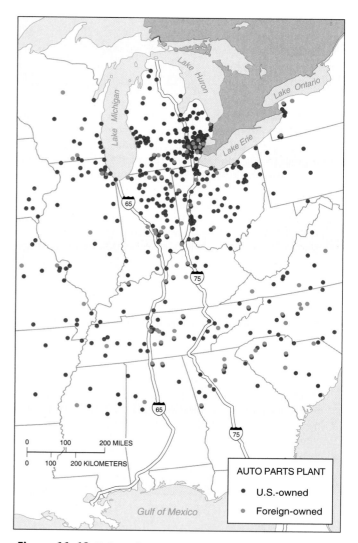

Figure 11–13 U.S.- and Japanese-owned motor vehicle parts plants. Plants are clustered in the interior of the country, near the major customers, the final assembly plants. Japanese-owned plants are more likely to be farther south, where workers are less likely to join a union.

Memphis, Tennessee, and Dayton, Ohio. The packages are then transferred to other planes, flown to airports nearest their destination, transferred to trucks, and delivered the next morning.

Break-of-Bulk Points. Regardless of transportation mode, cost rises each time that inputs or products are transferred from one mode to another. For example, workers must unload goods from a truck and then reload them onto a plane. The company may need to build or rent a warehouse to store goods temporarily after unloading from one mode and before loading to another mode.

Some companies may calculate that the cost of one mode is lower for some inputs and products, while another mode may be cheaper for other goods. Many companies that use multiple transport modes locate at a break-of-bulk point. A **break-of-bulk point** is a location where transfer among transportation modes is possible.

Containers are widely used to transfer goods efficiently between ships and trucks or trains. (Claudia Dhimitri/Viesti Associates, Inc.)

Important break-of-bulk points include seaports and airports. For example, a steel mill near the port of Baltimore receives iron ore by ship from South America and coal by train from Appalachia.

Site Factors

Situation factors are important for many firms, but their relative importance has changed. Locations near markets or break-of-bulk points have become more important than locations near raw materials for firms in developed countries. Consumers concentrated in large urban areas have greater wealth with which to buy products. Communications improvements have increased demand for rapid access to products.

However, situation factors do not explain the growing importance of Japanese and other Asian manufacturers. Japan not only lacks key raw materials needed by industries, but is far from the most important North American and European markets as well. Manufacturing has grown in Japan and other Asian countries primarily because site factors have become increasingly important in industrial location decisions. The cost of conducting business varies among locations, depending on the cost of three production factors: land, labor, and capital.

Land

Modern factories are more likely to be suburban or rural than near the center city. Contemporary factories generally require large tracts of land, because they usually operate more efficiently when laid out in one-story buildings. The land needed to build one-story factories is more likely to be available in suburban or rural locations.

Also, land is much cheaper in suburban or rural locations than near the center city. A hectare (or an acre) of land in the United States may cost only a few thousand dollars in a rural area, tens of thousands in a suburban location, and hundreds of thousands near a center city.

Industries may be attracted to specific parcels of land that are accessible to energy sources. Prior to the industrial revolution, many economic activities were located near rivers and close to forests, because running water and burning of wood were the two most important sources of energy. When coal became the dominant form of industrial energy in the late eighteenth century, location near coalfields became more important. Because coalfields were less ubiquitous than streams or forests, industry began to concentrate in fewer locations.

In the twentieth century, electricity became an important source of energy for industry. Electricity is generated in several ways, using coal, oil, natural gas, running water (hydroelectricity), nuclear fuel, and solar and wind to a very limited degree. In the United States, electricity usually is purchased from utility companies, which are either publicly owned or privately owned but regulated by the state government.

Like home consumers, industries are charged a certain rate per kilowatt hour of electricity consumed, although large industrial users usually pay a lower rate than do home consumers. Each utility company sets its own rate schedule, subject to approval by its state's regulatory agency. Industries with a particularly high demand for energy may select a location with lower electrical rates.

The aluminum industry, for example, requires a large amount of electricity to separate pure aluminum from bauxite ore. The first aluminum plant was located near Niagara Falls to take advantage of the large amount of cheap hydroelectric power generated there. Aluminum plants have been built near other sources of inexpensive hydroelectric power, including the Tennessee Valley and the Pacific Northwest.

Industry may also be attracted to a particular location because of amenities at the site. Not every location has the same climate, topography, recreational opportunities, cultural facilities, and cost of living. Some executives select locations in the U.S. South and West because they are attracted to the relatively mild climates and opportunities for year-round outdoor recreation activities. Others prefer locations that are accessible to cultural facilities or major-league sports franchises.

Labor

The cost of labor varies considerably, not only among countries but also within regions of one country. A **labor-intensive** industry is one in which labor cost is a high percentage of expense. Some labor-intensive industries require highly skilled labor to maximize profit, whereas others need less skilled, inexpensive labor.

Textile and Clothing Industries. Textile and clothing production are prominent examples of labor-intensive industries that generally require less skilled, low-cost workers. Textile production involves three principal steps:

- Spinning of fibers to make yarn
- Weaving or knitting yarn into fabric (as well as finishing of fabric by bleaching or dyeing)
- Cutting and sewing of fabric into clothing or other products (such as carpets and towels)

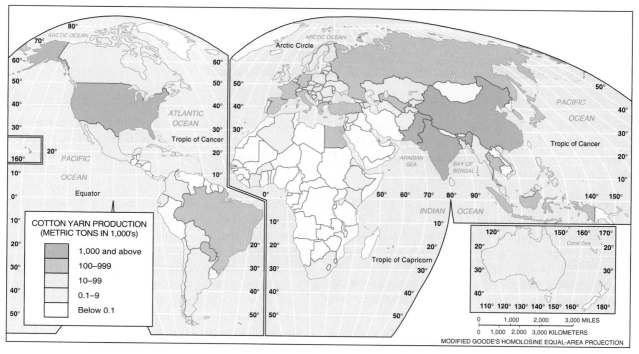

Figure 11–14 Cotton yarn production. Spinning of cotton fiber into yarn is clustered in a handful of countries where cotton is grown, including the United States, Russia, China, India, and Pakistan. The spindles of spun cotton become inputs into the weaving of cotton fabric, which is done primarily in Asia. Three-fourths of the yarn is produced in less-developed countries, shown in green on the pie chart. MDCs are shown in purple. (Henri Bureau/The Image Works)

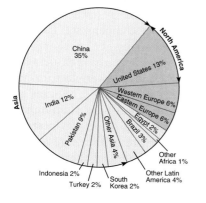

The global distributions of spinning, weaving, and sewing plants are not the same, because the three steps are not equally labor-intensive.

Fibers can be spun from natural or synthetic elements. The distribution of plants that spin natural fiber responds primarily to situation factors: manufacturing is concentrated in countries where the principal input—cotton—is grown. China, India, Pakistan, the United States, and Uzbekistan grow more than half of the world's cotton and produce more than half of the world's cotton fiber (Figure 11–14). The other major natural fiber—wool—is not typically produced in proximity to sheep farms.

Synthetic fibers—produced from petroleum and other chemical processes—account for an increasing share of textile production. Production was once dominated by a handful of more developed countries, where the chemical industry was concentrated, but less developed countries now account for about half of global production of synthetic fibers. Fiber production has expanded especially rapidly in China and Indonesia.

The LDCs are responsible for 86 percent of the world's woven cotton fabric and 76 percent of its spun yarn (Figure 11–15). Weaving is more likely to locate in LDCs, because labor is a higher percentage of total production cost. Despite their remoteness from European

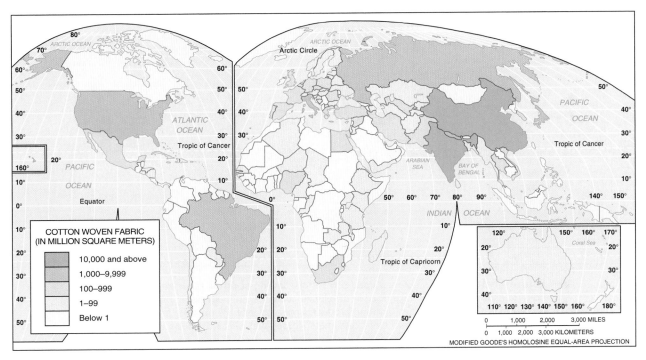

Figure 11–15 Woven cotton fabric production. Woven cotton fabric is likely to be produced in LDCs, because the process is more labor-intensive than the other major processes in textile and clothing manufacturing. (Luiz C. Marigo/Peter Arnold, Inc.)

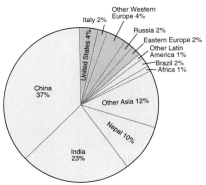

and North American markets, Asian countries have become major fabric producers because lower labor cost offsets the expense of shipping inputs and products long distances.

Most of the world's cotton clothing, such as shirts, trousers, and underwear, still is produced in the MDCs of Europe and North America (Figure 11–16). During the 1980s, shirt production declined more than one-third in the United States and Europe, while remaining about the same in the LDCs. As a result, the percentage of shirts produced in LDCs increased during the 1980s and 1990s from about 45 percent to 62 percent.

U.S. Textile and Clothing Industries. U.S. textile weavers and clothing manufacturers have changed locations to be near sources of low-cost employees. During

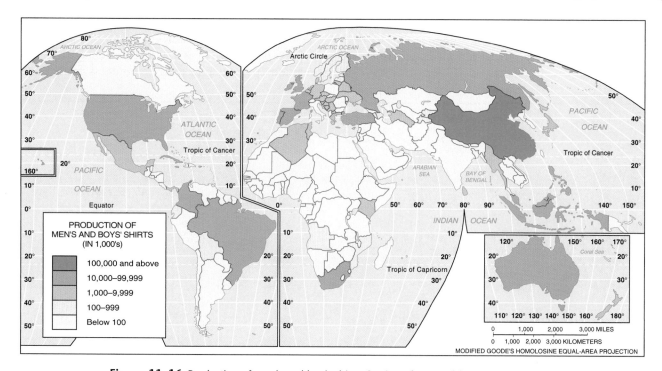

Figure 11–16 Production of men's and boy's shirts. Sewing of cotton fabric into men's and boys' shirts is more likely to take place in developed countries, although some production has moved to LDCs in recent years. Clothing producers must balance the need for low-wage workers with the need for proximity to customers. (Joseph Nettis/Photo Researchers, Inc.)

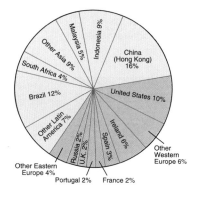

most of the 1800s, U.S. textile and clothing firms were concentrated in the Northeast. The region's major attraction was a large supply of European immigrants willing to sew long hours in sweatshops for low pay. In the late 1800s and early 1900s, textile and clothing workers began to demand better working conditions and higher wages. They formed unions to represent their interests.

Their claim was bolstered by tragic events, such as the 1911 Triangle Shirtwaist Company fire in New York City. In the fire, 146 workers, mostly women, died because the owners locked the doors to the eighth-floor workroom, where the fire originated. This was done to prevent them from taking breaks and stealing company property.

Employers argued that they could not afford to pay high wages and still make a profit. Because so many workers were needed in the industry, the wages of each individual worker had to be kept low. Faced with union demands for higher salaries in the Northeast, cotton textile and clothing manufacturers moved to the Southeast, where people were willing to work longer hours for lower wages.

Although they earned less than their northeastern counterparts, southeastern workers cooperated because wages were higher than those paid for other types of work in the region. With better working conditions and higher salaries than previously found in the region, workers were less likely to vote in the unions, thus keeping costs to industry low.

Cotton textile and clothing manufacturing in the United States is now located in the Appalachian Mountains and Piedmont of the Southeast, especially western North and South Carolina and northern Georgia and Alabama (Figure 11–17). Firms are dispersed among many communities rather than being concentrated in a few cities. They are in the same general region to take advantage of lower labor costs and consequently do not need to be located in the same city.

But the clothing industry has not completely abandoned the Northeast. The wool industry has remained because its labor demands are different from those of the cotton textile industry. Wool clothing, such as knit outerwear, requires more skill to cut and assemble the material, and skilled textile workers are more plentiful in the Northeast (Figure 11–18).

Skilled Labor Industries. More firms are requiring workers to perform highly skilled tasks, using complex equipment or performing precise cutting and drilling. Companies may become more successful by paying higher wages for skilled labor than by producing an inferior product made by lower-paid, less skilled workers.

One industry that demands highly skilled workers is electronics. Computer manufacturers have concentrated in the highest-wage regions in the United States, especially New York, Massachusetts, and California. These

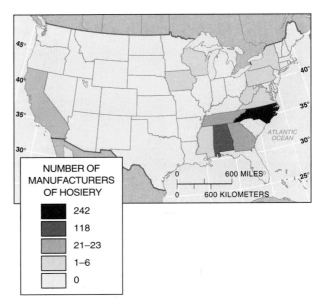

Figure 11–17 Hosiery manufacturers. To support their labor-intensive industry, hosiery manufacturers locate where a low-cost workforce exists. In the United States, lowest-cost labor is concentrated in the Southeast. The U.S. Bureau of the Census classifies these manufacturers as Standard Industrial Classification (SIC) 2252.

regions have a large concentration of skilled workers because of proximity to major university centers (Figure 11–19).

Many industries are attracted to locations with relatively skilled labor to introduce new work rules. Traditionally in large factories, each worker was assigned one specific task to perform repeatedly. Some geographers call this approach **Fordist**, because the Ford Motor Company was one of the first to organize its production this way early in the twentieth century. In recent years, companies have adopted more flexible rules, such as the allocation of workers to teams that must perform a vari-

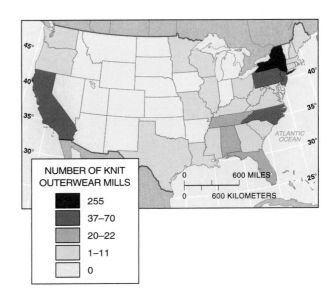

Figure 11–18 Knit outerwear manufacturers (SIC 2253). Products that require more skilled workers, such as knit outerwear, are still produced primarily in or near New York City.

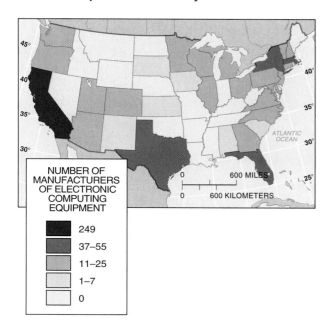

NUMBER OF
MANUFACTURERS
OF ELECTRONIC
COMPUTING
EQUIPMENT

249

37–55

11–25

1–7

0

Figure 11–19 Electronic computing equipment manufacturers (SIC 3571). Manufacturers of computing equipment need access to highly skilled workers to perform precision tasks. They are willing to pay relatively high wages to attract the workers. The largest clusters of skilled workers are in the Northeast and on the West Coast.

ety of tasks. Relatively skilled workers are needed to master the wider variety of assignments given them under more flexible **post-Fordist** work rules.

Capital

Manufacturers typically borrow funds to establish new factories or expand existing ones. The U.S. motor vehicle industry concentrated in Michigan early in the twentieth century largely because this region's financial institutions were more willing than eastern banks to lend money to the industry's pioneers.

The most important factor in the clustering of high-tech industries in California's Silicon Valley—even more

important than proximity to skilled labor—was the availability of capital (Figure 11–20). Banks in Silicon Valley have long been willing to provide money for new software and communications firms even though lenders elsewhere have hesitated. High-tech industries have been risky propositions—roughly two-thirds of them fail—but Silicon Valley financial institutions have continued to lend money to engineers with good ideas so that they can buy the software, communications, and networks they need to get started. One-fourth of all capital in the United States is spent on new industries in the Silicon Valley.

The ability to borrow money has become a critical factor in the distribution of industry in LDCs. Financial institutions in many LDCs are short of funds, so new industries must seek loans from banks in MDCs. But enterprises may not get loans if they are located in a country that is perceived to have an unstable political system, a high debt level, or ill-advised economic policies.

Local and national governments increasingly attempt to influence the location of industry by providing financial incentives. These include grants, low-cost loans, and tax breaks. Communities compete to offer new factories the most attractive financial package. Generally, the cost of the financial package is less than the additional revenues the new firm will generate overall in taxes and employment.

Obstacles to Optimum Location

The location that a firm chooses cannot always be explained by situation and site factors. Many industries have become "footloose," meaning they can locate in a wide variety of places without a significant change in their cost of transportation, land, labor, and capital. An industry may be especially footloose if it uses facsimile machines, electronic mail, the Internet, and other communications systems to move inputs and products, and if

The Ford Motor Company innovated mass-production techniques, including the moving assembly line, which it installed in 1913 at its Highland Park factory, near Detroit. Each worker performed a specific task, such as attaching the running boards to a Ford Model T. (Courtesy of the Ford Archives, Henry Ford Museum, Dearborn, Michigan)

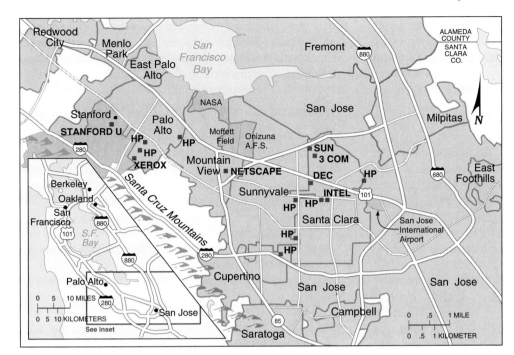

Figure 11–20 Silicon Valley. The major center of computer software, semiconductors, and other computer products is situated south of the San Francisco Bay area. The region benefits from a pool of skilled labor and the willingness of local financial institutions to invest in risky businesses.

its workers do not need to work at desktop computers clustered in one building.

The process by which corporate executives make decisions can explain the location of a firm's factories. An executive operating with high levels of knowledge and power may be able to identify precisely the location that maximizes the company's profits, whereas a less skilled official might select an inferior location.

An individual may choose a location on the basis of a corporate goal other than to maximize profits—for example, to promote growth or to assure survival of the firm. Personal preferences of the owner are especially important in influencing the location of a smaller firm. The location may be dictated by where the owner was born, went to school, participates in leisure activities, or visits friends.

The search for an optimal location can be time-consuming and costly. Consequently, the selected plant location might be the first acceptable alternative encountered, rather than the best possible one. The firm might select its location on the basis of inertia and history. Once a firm is located in a particular community, expansion in the same place is likely to be cheaper than moving operations to a new location. A large corporation may operate plants in inferior locations that were inherited through mergers and acquisitions.

Key Issue 4

Why Do Industries Face Problems?

- Industrial problems from a global perspective
- Industrial problems in more developed countries
- Industrial problems in less developed countries

Leaders worldwide consider industrial growth to be a most fundamental priority in a global economy. Each government defines problems of industrial development from its own local perspective. But geographers point out that diverse local constraints on industrial growth faced by individual countries are related to conditions in the global economy.

Industrial Problems from a Global Perspective

From a global perspective, the most basic industrial problem is a gap between the world demand for products and the world capacity to supply them. Global capacity to produce manufactured goods has increased more rapidly than demand for many products.

Stagnant Demand

From the industrial revolution's beginnings in the late 1700s until the 1970s, industrial growth in more developed countries was fueled by long-term increases in population and wealth. The growth formula was simple: more people with more wealth demanded more industrial goods. Demand was met by building more factories, which hired more people, who became wealthier and therefore demanded more goods. Times of major world conflict or economic depression were temporary exceptions to long-term growth in wealth, demand, and production.

However, since the 1970s, demand for many manufactured goods has slowed in MDCs. More developed countries now have little, if any, population growth. Because wages have not risen as fast as prices during the past two decades, individuals typically have not increased their level of spending, when adjusted for inflation.

Demand has also been flat for many consumer goods in MDCs because of market saturation. Nearly every household already has a color television, refrigerator, and motor vehicle. Most contemporary purchasers of these products are replacing older models rather than buying for the first time.

Industrial output is also stagnant because of the increasing quality of products. Consumers in MDCs increasingly select specific goods for high-quality and reliability rather than low price, and then replace them less frequently. For example, during the 1980s Japanese companies expanded their share of the North American motor vehicle market to more than one-fourth by selling products that were comparably priced with American models but widely acknowledged to be better built.

By the late 1990s, the quality of all vehicles had markedly improved since the 1980s. The average number of defects per vehicle declined by more than one-half during the 1990s. The gap in quality between American and Japanese products has narrowed—if not disappeared altogether—although some American consumers still perceive that Japanese models are superior.

Improvements in quality were expensive to achieve, so the price of motor vehicles rose faster than salaries during the 1990s. Japanese vehicles went from being less expensive to being more expensive than American-made models. With the improved quality and higher prices, North American consumers are holding on to their vehicles longer than at any time since the 1940s, when production was halted for nearly four years because of World War II. To retain their share of the North American market during the 1990s, Japanese producers have encouraged their satisfied customers to trade in their smaller, less expensive models for larger, more luxurious ones.

Changing technology has resulted in declining demand for some industrial products. For example, the global steel demand is less than in the mid-1970s. Today's typical motor vehicle uses one-fourth less steel than those built a quarter-century ago. Producers now build smaller, lighter vehicles and have replaced steel with plastic and ceramic products in the body, chassis, passenger compartment, and trim.

Increased Capacity Worldwide

While demand for products such as steel has stagnated since the 1970s, global capacity to produce them has increased. Higher industrial capacity is primarily a result of two trends: the global diffusion of the industrial revolution and the desire by individual countries to maintain their production despite a global overcapacity.

Historically, manufacturing was concentrated in a few locations. From the beginning of the industrial revolution until recently, demand for products manufactured in MDCs increased in part through sales to new markets—countries that lacked competing industries. Such industrial growth through increased international sales was feasible when most of the world was organized into colonies and territories controlled by MDCs.

For much of the nineteenth century, the United Kingdom's output in some industrial sectors exceeded the rest of the world combined. From the late 1800s until recently, the British were joined in dominating global industrial production by the United States, Russia (and the former Soviet Union), Germany, and several other countries in Europe. Then, Asian countries like Japan and South Korea blossomed into major industrial producers. Few colonies remain in the world today, and nearly every independent country wants to establish its own industrial base.

The steel industry illustrates the changing distribution of the global economy. Twenty years ago, the MDCs of North America, Western Europe, and Japan accounted for nearly two-thirds of the world's steel production, compared to approximately one-fourth in Eastern Europe and the Soviet Union, and less than 10 percent in LDCs (Figure 11–21, top).

The United States Steel (now USX) plant on the South Side of Chicago, along Lake Michigan, closed in the early 1970s. The company's nearby Gary Works is still open, but employment at the plant has dropped from about 30,000 during the 1970s to 10,000 today. (Jeffrey D. Smith/Woodfin Camp & Associates)

The overall level of world steel production in the early 1990s remained virtually the same as in the mid-1970s, but the proportion changed significantly among various regions. Production declined by nearly one-fourth in MDCs and more than doubled in LDCs. In two decades, the share of the world's steel production concentrated in MDCs has declined from 90 percent to 65 percent, while the LDCs have increased from nearly 10 percent to 40 percent of the world's output. LDCs such as Brazil, South Korea, Taiwan, India, and the People's Republic of China have substantially increased steel production, while MDCs—even Japan—have reduced production (Figure 11–21, bottom).

This global diffusion of steel mills has allowed capacity to exceed demand by a wide margin. Many companies have been unable to sell enough steel to make a profit and have closed. However, because the governments of many MDCs have been reluctant to let their steel mills close, the problem of excess capacity and unprofitable operations persists.

Steel mills in many countries receive substantial government financial support to remain open. Many European governments heavily subsidize the continued operation of their steel mills. The reason is economic: if the mills closed, governments would have to pay unemployment compensation to laid-off workers and deal with the social problems of increased unemployment. Maintaining a steel industry also ensures a domestic steel source in times of crisis.

Industrial Problems in More Developed Countries

Countries at all levels of development face a similar challenge: to make their industries competitive in an increasingly integrated global economy. Although they share this same overall goal, each state faces distinctive geographical issues in ensuring that their industries compete effectively. Industries in more developed regions must protect their markets from new competitors, whereas less developed countries of Africa, Asia, and Latin America must identify new markets and sources of capital to generate industrial growth.

Impact of Trading Blocs

Industrial competition in the more developed world increasingly occurs not among individual countries, but within regional **trading blocs**. The three most important trading blocs are the Western Hemisphere, Western Europe, and East Asia. Within each bloc, countries cooperate in trade. Each bloc then competes against the other two.

Cooperation Within Trading Blocs. In the Western Hemisphere, most trade barriers between the United States and Canada had been eliminated over the past several decades. The North Atlantic Free Trade Agreement (NAFTA), implemented in 1994, brought Mexico into the free trade zone with the United States and Canada. Since then, the three NAFTA partners have been negotiating with other Latin American countries to extend the free trade provisions further.

The European Union has eliminated most barriers to trade through Western Europe, as discussed in Chapter 8. European countries not members of the European Union, such as Switzerland and several former Communist Eastern European countries, depend heavily on trade with the European Union.

Japanese companies play leading roles in the economies of other East Asian countries. But cooperation among countries is less formal in East Asia, in part because Japan's neighbors have much lower levels of economic development and unpleasant memories of Japanese military aggression during the 1930s and 1940s.

The free movement of most products across the borders has led to closer integration of industries within North America and within Western Europe. For example, most cars sold in Canada used to be manufactured in Canada, but today, most are assembled in the United States. Every Ford Taurus sold in Canada is actually assembled in the United States, but every Ford Crown Victoria sold in the United States is actually assembled in Canada. Canada exports twice as many automobiles to its southern neighbor as it imports. While recognizing that they assemble a disproportionately large share of North American motor vehicles, Canadians complain that the United States has virtually all of the high-skilled engineering, design, and executive jobs.

A similar integration of motor vehicle production exists within Europe. For example, Volkswagen manufactures cars in Germany, Spain, and the Czech Republic and sells them throughout the Continent. Some Germans buy Italian-made Fiats, whereas some Italians buy German-made VWs.

Competition among Trading Blocs. The three trading blocs have promoted internal cooperation, yet they have erected trade barriers to restrict other regions from competing effectively. European Union members slap a tax on goods produced in other countries. Japan has lengthy permit procedures that effectively hinder foreign companies from selling there. The Japanese government maintains quotas on the number of automobiles that Japanese companies can export to the United States to counter charges of unfair competition.

Faced with a decline in domestic steel production of about one-third during the late 1970s, the U.S. government negotiated a series of voluntary export-restraint agreements with other major steel-producing countries. These quotas limited the sales of foreign-made steel to about 20 percent of the U.S. market.

When these quotas were in effect—from 1982 until 1992—U.S. steel companies spent $24 billion modernizing their plants and buying more efficient equipment. This restructuring stabilized U.S. steel production levels,

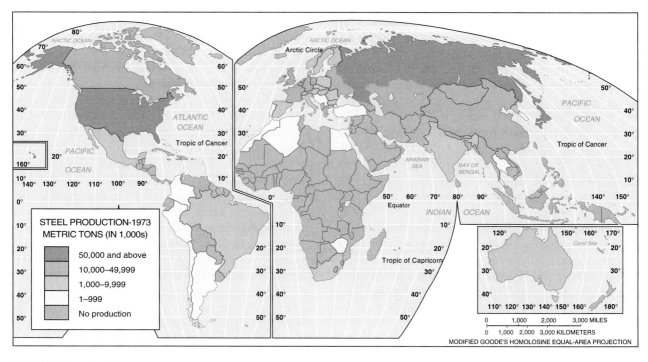

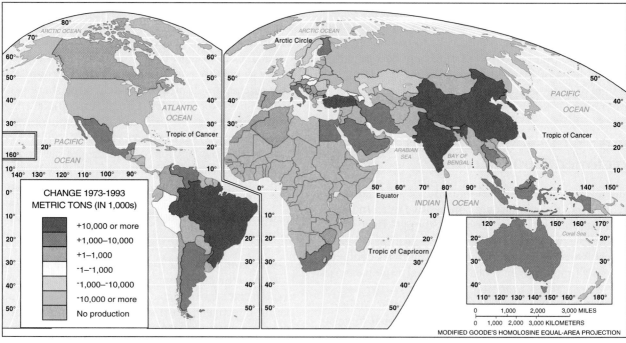

Figure 11–21 World steel production, 1973 and 1993. In 1973, MDCs in Europe and North America accounted for 90 percent of global production. Global steel production is about the same in the 1990s as in the early 1970s, but the distribution has changed. Production has declined in MDCs and increased in LDCs, especially China and Brazil.

but the number of steelworkers fell by two-thirds. Because of declining employment, the number of hours of labor needed to produce a ton of steel—a widely used measure of industrial efficiency—is now lower in the United States than in Japan or Europe.

Steel towns have suffered severely from this decline. The USX (formerly U.S. Steel) mill at Gary, Indiana—the country's largest—employed nearly 30,000 workers

during the 1970s, but had fewer than 8,000 by the early 1990s. Youngstown, Ohio, had more than 26,000 steel-industry jobs in the mid-1970s, but lost 80 percent of them. Some unemployed steelworkers have taken lower-paying jobs in other businesses, some have migrated else-where in search of jobs, and some have retired or remained unemployed. The steel industry's problems have affected the economy and morale of communities

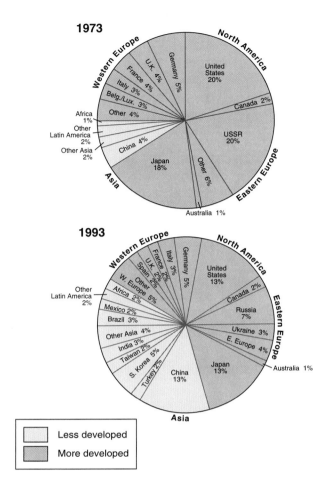

1973

1993

Less developed

More developed

like Gary and Youngstown. Declines in other manufacturing sectors in MDCs have had similar impacts in their communities.

Transnational Corporations.   Cooperation and competition within and among trading blocs take place primarily through the actions of large transnational corporations, sometimes called multinational corporations. A transnational corporation operates factories in countries other than the one in which its headquarters are located. Initially, transnational corporations were primarily American-owned, but in recent years corporations with headquarters in other more developed countries (especially Japan, Germany, France, and the United Kingdom) have been active as well (refer to Figures 1–18 and 1–22).

Some transnational corporations locate factories in other countries to expand their markets. Manufacturing the product where it is to be sold is done to overcome restrictions that many countries place on import. Further, given the lack of economic growth in many MDCs, a corporation may find that the only way it can increase sales is to move into another country. Transnational corporations also open factories in countries with lower site factors, to reduce their production cost. The site factor that varies most dramatically among countries is labor.

Japanese transnational corporations have been especially active in the United States in recent years. Several

hundred Japanese-owned corporations have built U.S. factories, primarily to develop new markets for electronics, automotive components, and metal products. Most plants have been located in a handful of interior states, including Ohio, Indiana, Kentucky, Michigan, Tennessee, and Illinois. German transnationals have clustered in the Carolinas.

Disparities Within Trading Blocs

Within the major trading blocs, industries are concentrated in some regions and sparse elsewhere. One country or region within a country may have lower levels of income and amenities because it has less industry than other countries or regions within the trading bloc. The lack of a uniform internal distribution of industry has caused trouble for the trading blocs, as well as for individual countries.

Disparities Within Western Europe.   Within Western Europe, the relationship between wealth and industrialization can be seen by comparing the map of major industrial regions (Figure 11–6) with the maps of per capita gross domestic product (Figure 3–12). Europe's most important industrial areas, such as western Germany and northern Italy, are relatively wealthy.

Disparities exist at the scale of the individual country, as well. For example, French industry and wealth are concentrated in the Paris region, while the south and west suffer. Per capita income is three times higher in the north of Italy than in the south. Within England, unemployment is 50 percent higher in the north and west, while average incomes are 25 percent higher in the south and east. Sweden's south is much more developed than its north. In each case, industry is concentrated in the regions most accessible to Western Europe's core of population, wealth, and industry, whereas more problems are found in peripheral regions.

Germany has had a particularly difficult problem with regional disparities. The eastern portion of the country has required massive financial assistance to modernize its industries. This is a legacy of the 40 years during which the region was Communist-run East Germany (German Democratic Republic).

The European Union, through its European Regional Development Fund, assists its three least industrialized member countries—Greece, Ireland, and Portugal—as well as regions in three other countries that lack industrial investment—Northern Ireland (part of the United Kingdom), southern Italy, and most of Spain (Figure 11–22). Funds also aid a number of declining industrial areas, including the northern areas of Denmark, England, France, Italy, and Spain. Regions eligible for support must submit five-year development plans explaining how the funds would be used.

A number of Western European countries use incentives to lure industry into poorer regions and discourage growth in the richer regions. In the United Kingdom, to aid development in the less prosperous north and west,

Figure 11–22 Assisted areas in the European Union. The European Union assists regions that have either relatively few industries or high concentrations of declining industries. Less industrialized regions are primarily in Southern European countries; regions suffering from industrial decline are primarily in the north. Compare to distribution of wealth (Figure 3–2).

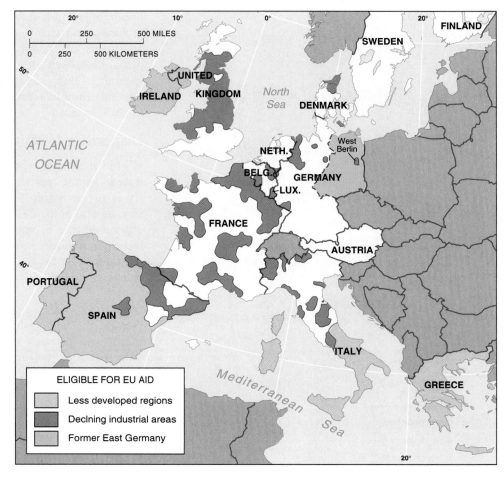

ELIGIBLE FOR EU AID
- Less developed regions
- Declning industrial areas
- Former East Germany

the government has designated several Development Areas and Intermediate Areas. Industries that locate in one of these areas may be entitled to receive loans, grants, tax reductions, and other government aid. On the other hand, to discourage industries from locating in an Unassisted Area, they may be required to obtain government permission. Other European countries also use regional incentives and regulations to encourage industrial location in peripheral regions and to discourage it in the congested core.

Disparities Within the United States. The problem of regional disparity is somewhat different in the United States. The South, historically the poorest U.S. region, has had the most rapid growth since the 1930s, stimulated partly by government policy and partly by changing site factors. The Northeast, traditionally the wealthiest and most industrialized region, claims that development in the South has been at the expense of old industrialized communities in New England and the Great Lakes states.

Regional development policies scored some successes as long as national economies were expanding overall, because the lagging regions shared in the national growth. But in the 1990s, an era of limited economic growth for MDCs, governments increasingly questioned policies that strongly encourage industrial location in poorer regions. Excessive control of industrial location could harm the overall national economy. MDCs have

not completely abandoned policies that aid poorer regions, but the level of financial commitment has been severely reduced.

Industrial Problems in Less Developed Countries

Less developed countries of Africa, Asia, and Latin America seek to reduce the disparity in wealth between themselves and Western Europe and North America. Knowing that their agriculture-based economies offer limited economic growth, the leaders of virtually every LDC encourage new industry. Industrial development not only can raise the value of exports but can generate money these countries need to buy other products. Development also can supply goods that are currently imported. If Western countries have built their wealth on industrial modernization, why can't the LDCs?

Old Problems for LDCs

In some respects, LDCs face obstacles similar to those once experienced by today's MDCs.

Distance from Markets. As in the past, today's newly industrializing countries are far from wealthy consumers in MDCs. In the early 1800s, U.S. and Central European factories were far from England, then the world's most

With one-fifth of the world's population and rising incomes, China has become attractive for market-oriented producers. If only one percent of the Chinese people traded in their bicycles for cars, for example, China would become the world's second-largest car market, behind only the United States. (Alain le Garsmeur/Tony Stone Worldwide)

important concentration of wealthy consumers. In the twentieth century, wealthy consumers in North America and Western Europe are distant from the LDCs of Africa, Asia, and Latin America. To minimize geographic isolation, industrializing countries invest scarce resources in constructing and subsidizing transportation facilities.

Inadequate Infrastructure. As in the past, today's LDCs lack support services critical to industrial development. These include transportation, communications, and domestic sources of equipment, tools, and machines needed to build and operate new factories. The LDCs also lack universities capable of training factory managers, accountants, and other technical people needed for industrial development. The LDCs obtain support services by importing advisers and materials from other countries, or by borrowing money to develop domestic sources.

New Problems for LDCs

In addition, industrializing countries now face a new obstacle. New factories once could count on selling in countries that lacked competing industries. But few untapped foreign markets remain to be exploited. New industries must sell primarily to consumers inside their own country—often a market too small to support them—or compete with existing manufacturers in other countries.

Considering the obstacles to launching new industries for which market access is critical, what kind of factories can LDCs attract? According to principles of economic geography, there are two other critical locational factors: access to raw materials and site factors. In fact, new African factories generally are those for which these factors are critical:

1. *Raw material access.* Bauxite in Guinea, uranium in Niger, iron ore in Mauritania and Liberia, and copper in Zambia are processed for industrial uses elsewhere in the world. African countries also process food and agricultural products, such as palm and peanut oil, flour, and beer. Fertilizer is produced from phosphate or nitrate deposits in Côte d'Ivoire, Mozambique, Senegal, Uganda, Zambia, and Zimbabwe.

2. *Site factors.* Most critical usually is cheap, abundant labor. The textile and clothing industries still consider low-cost labor to be their most critical site-selection factor. Consequently, manufacturers that migrated from New England to the Southeast earlier this century have migrated again to Asia, Latin America, and Africa. Workers in LDCs receive a fraction of U.S. wages. For example, Nike's shoe factories in China, Indonesia, South Korea, Taiwan, Thailand, and Vietnam pay workers about $2 a day.

Transnational corporations have been especially aggressive in using low-cost labor in LDCs. To remain competitive in the global economy, they carefully review their production processes to identify steps that can be performed by low-paid, low-skilled workers in LDCs. Given the substantial difference in wages between MDCs and LDCs, transnational corporations can profitably transfer some work to LDCs, despite greater transportation cost. At the same time, operations that require highly skilled workers remain in factories in MDCs. This selective transfer of some jobs to LDCs is known as the **new international division of labor**.

Many African countries possess iron ore. But steel, perhaps the most important industry for a less developed country, has had difficulty getting a foothold in Africa. The only large, integrated steel mill in Africa south of the Sahara and north of South Africa is in Zimbabwe. Small plants have been established in Congo, Ethiopia, Ghana, Nigeria, and Uganda, using scrap metal as the input. Without cooperation among several small states, steel manufacturing is not likely to develop further in Africa.

■ Summary

Three recent changes in the structure of manufacturing have geographic consequences:

1. Factories have become more productive through introduction of new machinery and processes. A factory may continue to operate at the same location but require fewer workers to produce the same output. Faced with meager prospects of getting another job in the same community, workers laid off at these factories migrate to other regions.

2. Companies are locating production in communities where workers are willing to adopt more flexible work rules.

Firms are especially attracted to smaller towns where low levels of union membership and high visibility reduce vulnerability to work stoppages, even if wages are kept low and lay-offs become necessary.

3. By spreading production among many countries, or among many communities within one country, large corporations have increased their bargaining power with local governments and labor forces. Production can be allocated to locations where the local government is especially helpful and generous in subsidizing the costs of expansion, and the local residents are especially eager to work in the plant.

These, again, are the key issues in the geography of industry:

1. **Where did industry originate?** The industrial revolution dates from the late 1700s in the United Kingdom, when a series of inventions transformed industrial production. By 1900, only four other countries could be classified as industrial: Belgium, France, Germany, and the United States. During the twentieth century, industrialization diffused to several dozen other countries in Europe, Asia, and the Western Hemisphere.

2. **Where is industry distributed?** In contrast to agriculture, which covers a large percentage of Earth's land area, industry is highly concentrated. Approximately three-fourths of the world's industrial output is concentrated in four regions: the North American manufacturing belt, Western Europe, Eastern Europe, and Japan.

3. **Why do industries have different distributions?** Factories try to identify a location where production cost is minimized. Critical industrial location costs include situation factors for some firms and site factors for others. Situation factors involve the cost of transporting both inputs into the factory and products from the factory to consumers. Site factors—land, labor, and capital—control the cost of doing business at a location.

4. **Why do industries face problems?** The entire world faces a problem with industry because global capacity to produce many goods now exceeds demand. The MDCs in North America and Western Europe have a distinctive problem that results from an uneven internal distribution of industry and wealth. The LDCs, located farther from markets, must attract industries for which access to inputs or low-cost labor are critical.

Case Study Revisited

Free Trade in North America

Competition to attract new industries and to retain existing ones extends across international borders. The governments of Canada, Mexico, and the United States have agreed to eliminate barriers to free trade among the three countries. As competition increases among regional blocs of countries, U.S. and Canadian business and government leaders see substantial benefits to including Mexico in a free trade zone. With the addition of Mexico and other Latin American countries, the North American free trade area can rival the European Union as the world's most populous and wealthy market.

But creating an integrated North American economy is a formidable task, given the substantially lower standard of living in Mexico and Latin America than in the United States and Canada. U.S. and Canadian labor union leaders are concerned that with the removal of barriers, more manufacturers will relocate production to Mexico to take advantage of lower wage rates. Such labor-intensive industries as food processing and textile manufacturing may be especially attracted to a region where prevailing wage rates are lower.

Environmentalists fear that under a free trade agreement, firms will move production to Mexico, where laws governing air and water quality standards are less stringent than in the United States and Canada. Mexico has adopted regulations to reduce air pollution in Mexico City; catalytic converters were required on Mexican automobiles beginning in 1991. But enforcement of environmental protection laws is still lax in Mexico.

According to industrial location theory, firms select locations for various situation and site factors. Wage rates and environmental controls are two important site factors, but such factors as access to markets and to skilled workers are also critical. Geography's global perspective in analyzing industrial location reinforces the fact that the problems of an unemployed steelworker in Gary or Youngstown are not just local, but are related to worldwide characteristics of the steel industry. The future health of industry in the United States depends on a national commitment to a combination of competition and cooperation in a global economy.

To recapture competitiveness with other countries' industries, North American business leaders must learn more about the culture, politics, and economy of other nations. The success enjoyed by Japanese and Korean businesses in North America derives to a considerable extent from the fact that executives in those countries know more about U.S. society than Americans know about Asia. Asian officials are likely to speak English and are familiar with the tastes and preferences of American consumers, whereas few American officials speak Japanese or Korean, and they have relatively little knowledge of the buying habits of Asians.

At the same time, global industrial development depends on increased cooperation among different countries. As a result of lower transportation cost, more people worldwide have access to more goods at lower prices than in the past. Given this trend, consumers in more developed countries are increasingly challenged to choose between the purchase of the highest-quality, lowest-cost goods regardless of where they were made, and support for local industries against foreign competitors at any price.

Key Terms

Break-of-bulk point A location where transfer is possible from one mode of transportation to another.

Bulk-gaining industry An industry in which the final product weighs more or comprises a greater volume than the inputs.

Bulk-reducing industry An industry in which the final product weighs less or comprises a lower volume than the inputs.

Cottage industry Manufacturing based in homes rather than in a factory, commonly found before the industrial revolution.

Fordist Form of mass production in which each worker is assigned one specific task to perform repeatedly.

Labor-intensive industry An industry for which labor costs comprise a high percentage of total expenses.

Maquiladora Factories built by U.S. companies in Mexico near the U.S. border, to take advantage of much lower labor costs in Mexico.

New international division of labor Transfer of some types of jobs, especially those requiring low-paid less skilled workers, from more developed to less developed countries.

Post-Fordist Adoption by companies of flexible work rules, such as the allocation of workers to teams that perform a variety of tasks.

Right-to-work state A U.S. state that has passed a law preventing a union and company from negotiating a contract that requires workers to join a union as a condition of employment.

Site factors Location factors related to the costs of factors of production inside the plant, such as land, labor, and capital.

Situation factors Location factors related to the transportation of materials into and from a factory.

Textile A fabric made by weaving, used in making clothing.

Trading bloc A group of neighboring countries that promote trade with each other and erect barriers to limit trade with other blocs.

Thinking Geographically

1. What have been the benefits and costs to Canada of its free trade agreement with the United States? How are the benefits and costs to Canada likely to change with the implementation of NAFTA?

2. To induce Toyota to build its U.S. production facilities in Kentucky, the state spent $49 million to buy the 1,500-acre site ($32,667/acre), $40 million to construct roads and sewers, and $68 million to train new workers. Kentucky also agreed to spend up to $168 million to pay the interest on loans should Toyota decide to borrow money to finance the project. Did Kentucky overpay to win Toyota's business? Explain.

3. Foreign cars account for one-fourth of the sales in the midwestern United States, compared to half in California and other West Coast states. What factors might account for this regional difference?

4. Draw a large triangle on a map of Russia, with one point near Moscow, one point in the Ural Mountains, and one point in Central Asia. What are the principal economic assets of the three regions at each side of the triangle? How do the distributions of markets, resources, and surplus labor vary within Russia?

5. What are the principal manufacturers in your community or area? How have they been affected by increasing global competition?

Further Readings

Amin, Ash, and John Goddard, eds. *Technological Change, Industrial Restructuring, and Regional Development.* London: Allen & Unwin, 1986.

Ashton, Thomas S. *The Industrial Revolution.* New York: Oxford University Press, 1964.

Bagchi-Sen, Sharmistha, and Bruce W. Pigozzi. "Occupational and Industrial Diversification in the Metropolitan Space Economy in the United States, 1985–1990." *Professional Geographer* 45 (1993): 44–54.

Behrman, Jack N. *Industrial Policies: International Restructuring and Transnationals.* Lexington, MA: Lexington Books, 1984.

Bell, Michael E., and Paul S. Lande, eds. *Regional Dimensions of Industrial Policy.* Lexington, MA: Lexington Books, 1982.

Bialeschki, M. Deborah, and Kathryn Lynn Walbert. "'You Have To Have Some Fun To Go Along With Your Work': The Interplay of Race, Class, Gender, and Leisure in the Industrial New South." *Journal of Leisure Research* 30 (1998): 79–100.

Birdsall, Stephen S., and John W. Florin. *Regional Landscape of the United States and Canada.* 2nd ed. New York: John Wiley, 1981.

Blackbourn, Anthony, and Robert G. Putnam. *The Industrial Geography of Canada.* New York: St. Martin's Press, 1984.

Bluestone, Barry, and Bennett Harrison. *The Deindustrialization of America: Plant Closings, Community Abandonment, and the Dismantling of Basic Industry.* New York: Basic Books, 1982.

Brotchie, John F., Peter Hall, and Peter W. Newton, eds. *The Spatial Impact of Technological Change.* London: Croom Helm, 1987.

Casetti, Emilio, and John Paul Jones III. "Spatial Aspects of the Productivity Slowdown: An Analysis of U.S. Manufacturing Data." *Annals of the Association of American Geographers* 77 (1987): 76–88.

Caves, Richard E. *Multinational Enterprise and Economic Analysis.* Cambridge: Cambridge University Press, 1982.

Conway, Dennis, and Jeffrey H. Cohen. "Consequences of Migration and Remittances for Mexican Transnational Communities." *Economic Geography* 74 (1998): 26–44.

de Bartolome, Charles A. M., and Mark M. Spiegel. "Does State Economic Development Spending Increase Manufacturing Employment?" *Journal of Urban Economics* 41 (1997): 153–75.

Dicken, Peter. *Global Shift: Industrial Change in a Turbulent World.* 2nd ed. London: Harper & Row, 1991.

_____. "Transnational Corporations and Nation-States." *International Social Science Journal* 49 (1997): 77–90.

_____, and Nigel Thrift. "The Organization of Production and the Production of Organization: Why Business Enterprises Matter in the Study of Geographical Industrialization." *Transactions of British Geographers New Series* 17 (1992): 279–91.

Duncan, Simon. "The Geography of Gender Divisions of Labour in Britain." *Transactions of British Geographers New Series* 16 (1991): 420–29.

Earney, F. C. F. "The Geopolitics of Minerals." *Focus* 31 (1981): 1–16.

Erickson, Rodney A., and David J. Hayward. "The International Flows of Industrial Exports from U.S. Regions." *Annals of the Association of American Geographers* 81 (1991): 371–90.

Ethier, Wilfred J. and James R. Markusen. "Multinational Firms, Technology Diffusion and Trade." *Journal of International Economics* 41 (1996): 1–28.

Ettlinger, Nancy. "The Roots of Competitive Advantage in California and Japan." *Annals of the Association of American Geographers* 81 (1991): 391–407.

Feldman, Maryann P., and Richard Florida. "The Geographic Sources of Innovation: Technological Infrastructure and Product Innovation in the United States." *Annals of the Association of American Geographers* 84 (1994): 210–29.

Gibbs, David, and Michael Healey. "Industrial Geography and the Environment." *Applied Geography* 17 (1997): 193–202.

Gillespie, A. E., ed. *Technological Change and Regional Development.* London: Pion Ltd., 1983.

Glasmeier, Amy K. *The High-Tech Potential: Economic Development in Rural America.* New Brunswick, NJ: Center for Urban Policy Research, 1991.

_____; Jeffery W. Thompson, and Amy J. Kays. "The Geography of Trade Policy: Trade Regimes and Location Decisions in the Textile and Apparel Complex." *Transactions of the Institute of British Geographer, New Series* 18 (1993): 19–35.

Gould, Peter. *Spatial Diffusion.* Washington, DC: Association of American Geographers, 1969.

Habakkuk, H. J., and M. M. Postan, eds. *The Cambridge Economic History of Europe,* Vol. 6. Cambridge: Cambridge University Press, 1965.

Hamilton, F.E. Ian, ed. *Contemporary Industrialization.* London and New York: Longman, 1978.

_____, and G.J.R. Linge, eds. *Spatial Analysis, Industry and the Industrial Environment: Progress in Research and Applications. Volume I: Industrial Systems* (1979); *Volume II: International Industrial Systems* (1981); *Volume 3: Regional Economies and Industrial Systems* (1983). Chichester, UK: John Wiley.

Hanson, Gordon H. "Agglomeration, Dispersion, and the Pioneer Firm." *Journal of Urban Economics* 39 (1996): 255–81.

Harris, C. D. "The Market as a Factor in the Localization of Industry in the United States." *Annals of the Association of American Geographers* 44 (1954): 315–48.

Hirsh, Michael, and E. Keity Henry. "The Unraveling of Japan Inc." *Foreign Affairs* 76 (1997): 11–32.

Hoare, Anthony G. "What Do They Make, Where, and Does It Matter Any More? Regional Industrial Structures in Britain Since the Great War." *Geography* 7 (1986): 289–304.

_____. *The Location of Industry in Britain.* New York: Cambridge University Press, 1983.

Hoffman, George W., ed. *Eastern Europe: Essays in Geographical Problems.* London: Methuen, 1971.

Hogan, William T. *Minimills and Integrated Mills: A Comparison of Steelmaking in the United States.* Lexington, MA: Lexington Books, 1987.

_____. *Global Steel in the 1990s: Growth or Decline.* Lexington, MA: Lexington Books, 1991.

HOST Network, Pascal Byé, and Alain Mounier. "Growth Patterns and the History of Industrialization." *International Social Science Journal* 48 (1996): 537–49.

Langton, John. "The Industrial Revolution and the Regional Geography of England." *Transactions of the Institute of British Geographers New Series* 9 (1984): 145–67.

Langton, John, and R. J. Morris, eds. *Atlas of Industrializing Britain, 1780–1914.* London: Methuen, 1986.

Law, Christopher M, ed. *Restructuring the Global Automobile Industry.* London: Routledge, 1991.

Leichenko, Robin M., and Rodney A. Erickson. "Foreign Direct Investment and State Export Performance." *Journal of Regional Science* 37 (1997): 307–29.

Malmberg, Anders. "Industrial Geography: Agglomeration and Local Milieu." *Progress in Human Geography* 20 (1996): 392–403.

Massey, Doreen, and Richard Meegan, eds. *Politics and Method: Contrasting Studies in Industrial Geography.* New York: Methuen, 1986.

Meyer, Stephen P., and Milford B. Green. "Foreign Direct Investment from Canada: An Overview." *The Canadian Geographer* 40 (1996): 219–37.

Oakey, Raymond P. *High Technology Small Firms: Innovation and Regional Development in Britain and the United States.* New York: St. Martin's Press, 1984.

Ó hUallacháin, Breandan. "Restructuring the American Semiconductor Industry: Vertical Integration of Design Houses and Wafer Fabricators." *Annals of the Association of American Geographers* 87 (1997): 217–37.

_____, and Richard A. Matthews. "Economic Restructuring in Primary Industries: Transaction Costs and Corporate Vertical Integration in the Arizona Copper Industry, 1980–1991." *Annals of the Association of American Geographers* 84 (1994): 399–417.

Oxford University Cartographic Department. *Oxford Economic Atlas: The United States and Canada.* London: Oxford University Press, 1975.

Pattie, Charles J., and R. J. Johnston. "One Nation or Two? The Changing Geography of Unemployment in Great Britain, 1983–1988." *Professional Geographer* 42 (1990): 288–98.

Peet, Richard, ed. *International Capitalism and Industrial Restructuring.* Boston: Allen & Unwin, 1987.

Rich, D.C., and G.J.R. Linge, eds. *The State and the Spatial Management of Industrial Change*. New York: Routledge, Chapman, and Hall, 1991.

Rubenstein, James M. *The Changing US Auto Industry*. London: Routledge, 1992.

Sagers, Matthew J. "The Iron and Steel Industry in Russia and the CIS in the Mid-1990s." *Post-Soviet Geography and Economics* 37 (1996): 195–263.

Sanna-Randaccio, Francesca. "New Protectionism and Multinational Companies." *Journal of International Economics* 41 (1996): 29–52.

Sayer, Andrew, and Richard Walker. *The New Social Economy*. Cambridge, MA: Blackwell, 1991.

Schmenner, Roger W. *Making Business Location Decisions*. Englewood Cliffs, NJ: Prentice-Hall, 1982.

Scott, Allen J., and Michael Storper, eds. *Production, Work, Territory*. Boston: Allen & Unwin, 1986.

Simmons, C., and C. Kalantaridis. "Making Garments in Southern Europe: Entrepreneurship and Labour in Rural Greece." *Journal of Rural Studies* 12 (1996): 169–86.

Smith, David M. *Industrial Location: An Economic Geographical Analysis*. 2nd ed. New York: John Wiley, 1981.

———. "A Theoretical Framework for Geographical Studies of Industrial Location." *Economic Geography* 42 (1966): 95–113.

South, Robert B. "Transnational 'Maquiladora' Location." *Annals of the Association of American Geographers* 80 (1990): 529–70.

Stobart, Jon. "Geography and Industrialization: The Space Economy of Northwest England, 1701–1760." *Transactions of the Institute of British Geographers New Series* 21 (1996): 681–96.

Storper, Michael, and Richard Walker. *The Capitalist Imperative: Territory, Technology, and Industrial Growth*. New York: Basil Blackwell, 1989.

Sweeney, Stuart H., and Edward J. Feser. "Plant Size and Clustering of Manufacturing Activity." *Geographical Analysis* 30 (1998): 45–64.

Thrift, Nigel, and Kris Olds. "Refiguring the Economic in Economic Geography." *Progress in Human Geography* 20 (1996): 311–37.

Toyne, Brian, Jeffrey S. Arpan, David A. Ricks, Terence A. Shimp, and Andy Barnett. *The Global Textile Industry*. London: Allen & Unwin, 1984.

Warren, Kenneth. "World Steel: Change and Crisis." *Geography* 70 (1985): 106–17.

Webber, Michael J. *Industrial Location*. Beverly Hills, CA: Sage Publications, 1984.

Wright, Melissa W. "Crossing the Factory Frontier: Gender, Place and Power in the Mexican Maquiladora." *Antipode* 29 (1997): 278–302.

ZumBrunnen, Craig, and Jeffrey Osleeb. *The Soviet Iron and Steel Industry*. Totowa, NJ: Rowmand and Allanheld, 1986.

Also consult these journals: *Journal of Industrial Economics*; *Journal of International Economics*; *Journal of Marketing*; *Journal of Transport Economics and Policy*; *Journal of Transport History*; *Journal of Urban Economics*.

CHAPTER 12

Services

Flying across the United States on a clear night, you look down on the lights of settlements, large and small. You see small clusters of lights from villages and towns, and large, brightly lit metropolitan areas. It may appear that the light clusters are random, but geographers discern a regular pattern in them. These regularities have been documented, and concepts from economic geography can be applied to understand why this pattern exists.

However, the regular distribution observed over North America, and over other more developed countries, is not seen in less developed countries. Geographers explain this difference, and why the absence of a regular pattern is significant.

The regular pattern of settlement in more developed countries reflects where services are provided. In more developed countries, the majority of the workers are employed in the tertiary sector of the economy, defined in Chapter 9 as the provision of goods and services to people in exchange for payment. In contrast, less than 10 percent of the labor force in less developed countries provide services.

Everyone needs food for survival. In less developed countries, most people work in the primary sector, growing food. In more developed countries, people purchase food at supermarkets or restaurants. The people employed at the supermarkets and restaurants are examples of service sector workers, and the customers pay for the food with money earned in other service-sector jobs, such as retailing, banking, law, education, and government.

Key Issues

1. Where did services originate?

2. Why are consumer services distributed in a regular pattern?

3. Why do business services locate in large settlements?

4. Why do services cluster downtown?

Babar, Garfield, Mickey Mouse, and other stuffed animals await customers at a Toys 'Я' Us.

407

Case Study

Obtaining Goods in Romania

The Preda family lives in Comena, a Romanian village of 800 inhabitants. The Predas are farmers, working the fields outside the village, earning just enough to survive without hardship. But some goods are hard to obtain. Because the village has only a few shops, Elena Preda must travel for an hour by bus to a larger town to buy everything she needs.

Romania lacks cities of certain sizes. The largest city, Bucharest, has 2 million inhabitants, and the second-largest, Braşov, has 350,000. In a country with a high level of economic development, geographers would expect to find four cities between 350,000 and 2 million inhabitants, according to the rank-size rule, discussed later in this chapter. But Romania has no city of this size. Geographers also expect to find more towns with population between 1,000 and 10,000 than Romania has.

The absence of cities between 350,000 and 2 million inhabitants, and between 1,000 and 10,000, constitutes a hardship for people who must travel long distances to reach an urban settlement with shops and such services as hospitals. Because most Romanians do not have cars, the government must provide bus service for citizens to reach larger towns. A trip to a shop or a doctor that takes a few minutes in the United States could take several hours in Romania.

The state of Colorado is comparable in land area to Romania. Colorado's largest metropolitan area, Denver (including Boulder), has about the same number of inhabitants as does Bucharest, and Colorado Springs, the state's second-largest metropolitan area, has about the same number of inhabitants as Braşov. But the absence of cities between 350,000 and 2 million inhabitants is not a hardship in Colorado; access to urban services is much greater in Colorado than in Romania, because virtually all Coloradans—including those living in rural areas—either own or have access to a car or truck. Life in rural Colorado would be much harder if most residents depended on public buses to transport them to shops and jobs.

In more developed countries, most people work in such places as shops, offices, restaurants, universities, and hospitals. These are examples of the tertiary, or service, sector of the economy. A **service** is any activity that fulfills a human want or need and returns money to those who provide it. Relatively few people work in factories or farms, the primary and secondary sectors.

In sorting out *where* services are distributed, geographers see a close link between services and settlements, because services are located in settlements. A **settlement** is a permanent collection of buildings, where people reside, work, and obtain services. Settlements range in size from tiny rural villages with barely a hundred inhabitants to teeming metropolitan areas with 20 million people. They occupy a very small percentage of Earth's surface, well under 1 percent, but settlements are home to nearly all humans, because few people live in isolation.

Explaining *why* services are clustered in settlements is at one level straightforward for geographers. In geographic terms, only one locational factor is critical for a service: proximity to the market. The optimal location of industry, described in the last chapter, requires balancing a number of site and situation factors, but the optimal location for a service is simply near its customers.

On the other hand, locating a service calls for far more precise geographic skills than locating a factory. The optimal location for a factory may be an area of several hundred square kilometers, whereas the optimal location for a service may be a specific street corner. Service providers often say that the three critical factors in selecting a suitable site are "location, location, and location." Although geographically imprecise, the expression is a way for nongeographers to appreciate that a successful service must carefully select its precise location.

Industries can locate in remote areas, confident that workers, water, and highways will be brought to the location if necessary. The distribution of services must follow to a large extent the distribution of where people live, within a city, country, or world region. However, if services were located merely where people lived, then China and India would have the most, rather than the United States and other more developed countries. Services cluster in more developed countries because more people able to buy services live there. Within more developed countries, larger cities offer more services than do small towns, because more customers reside there.

As in other economic and cultural features, geographers observe trends toward both globalization and local diversity in the distribution of services. In terms of *globalization*, the provision of services is increasingly uniform from one urban settlement to another, especially within MDCs. Every urban settlement in the United States above a certain size has a branch of a large retail chain, such as a McDonald's restaurant, and the larger cities have several. In England, every city above a certain size has a Sainsbury's supermarket, and the larger cities have several. In a more developed country, the demand for many types of services produces a regular distribution of settlements.

Despite the strong globalization trend so clearly visual on the landscape, *local diversity* is alive and well in the provision of services. Within more developed countries, fast-food restaurants may be located in every settlement, but other services cluster in particular locations. A settlement may offer a service such as a medical clinic, advertising agency, or a film studio not found in other settlements of comparable size. And every place—more developed or less developed countries alike—offers distinctive services that attract tourists and visitors.

Key Issue 1

Where Did Services Originate?

- Types of services
- Origin of services
- Services in rural settlements

Services are provided in all societies, but in more developed countries a majority of workers are engaged in the provision of services (Figure 12–1). In North America, three-fourths of workers are in services. The percentage of service workers varies widely in less developed countries, but is typically less than one-fourth. One reason for the wide variation is that in a number of LDCs, workers engaged in agriculture or manufacturing are counted in the service sector because they are employed by the government. Logically, the distribution of service workers is opposite that of percent primary workers (see Figure 9–5).

Types of Services

The service sector of the economy is subdivided into three types: consumer services, business services, and public services. The first two groups are divided into two subgroups. This division of the service sector has largely replaced earlier approaches that identified tertiary, quaternary, and quinary sectors in various ways.

Consumer Services

The principal purpose of **consumer services** is to provide services to individual consumers who desire them and can afford to pay for them. Retail services and personal services are the two main types of consumer services.

Retail Services. About 25 percent of all jobs in the United States are in **retail services**, which provide goods for sale to consumers. Within the group, about 25 percent of the jobs are in wholesale, 25 percent in restaurants, 25 percent in shops selling food, motor vehicles, or clothing, and 25 percent in shops selling other goods.

Figure 12–1 Percentage of workers in services. The percentage is approximate in less developed countries, because some workers engaged in primary- and secondary-sector activities are employed by the government and may be counted in the service sector.

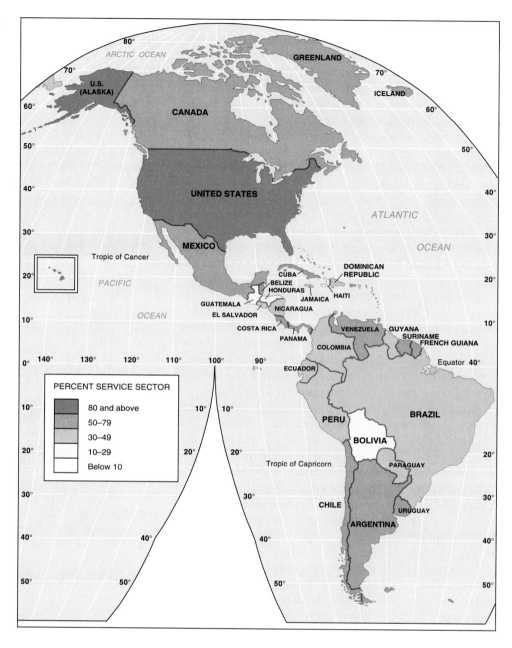

Personal Services.

Personal services provide services for the well-being and personal improvement of individual consumers. In the United States, about 20 percent of all jobs are in personal and social services. Health care comprises 40 percent of these types of services, including hospitals, nursing homes, clinics, and doctors' offices. About 10 percent each are in education, social services, recreation, hotels, membership organizations (such as churches), and other personal services (primarily automotive repair, dry cleaners, and beauty salons).

Business Services

The principal purpose of **business services** is to facilitate other businesses. Producer services and transportation services are the two main types of business services.

Producer Services.

Producer services provide services primarily to help people conduct other business—either agriculture, manufacturing, or other services. About 15 percent of U.S. jobs are in producer services. About 45 percent of the producer service jobs are in financial services, including 15 percent each in banks and insurance companies, 10 percent in real estate, and 5 percent in other financial institutions. About 20 percent are in professional services, primarily law, engineering, and management. The remaining 35 percent are in other business services, including 10 percent in employment agencies and temporary help, 5 percent in computer and data processing services, and 20 percent in a variety of supporting services, such as advertising.

Transportation and Similar Services.

Businesses that diffuse and distribute services are grouped as **transportation, communications, and utilities services**. In the United States, about 5 percent of all jobs are in this group of services. About 60 percent of these services are

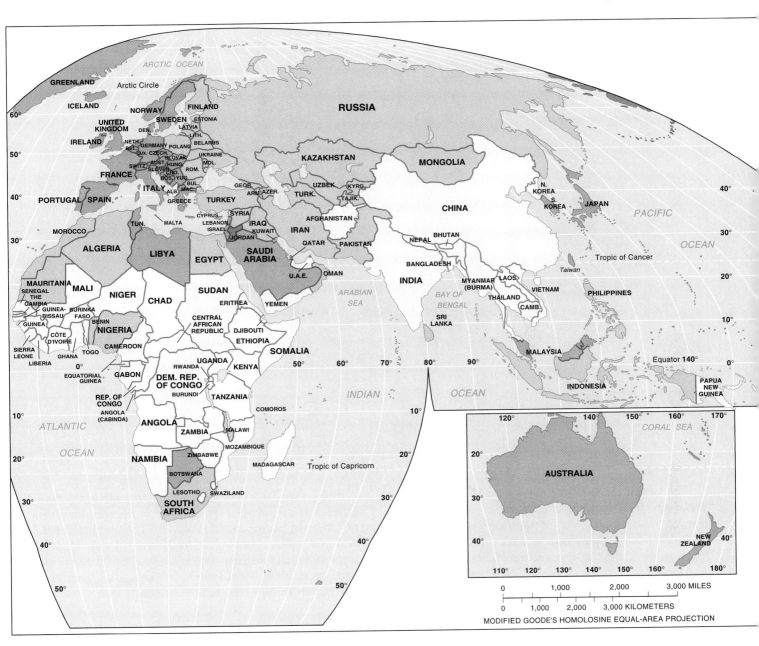

in transportation, including 30 percent in trucking, 10 percent in aviation, and 20 percent in railroads, public transit, and other transportation. About 25 percent are in communications, including telephone and broadcasting, and the remaining 15 percent are in utilities, including gas, electric, and sanitary services.

Public Services

The purpose of **public services** is to provide security and protection for citizens and businesses. In the United States, about 15 percent of all workers are in the public sector. About 15 percent of public service workers are employed by the federal government, 25 percent by one of the 50 state governments, and the remaining 60 percent by one of the tens of thousands of local governments.

The distinction among services is not absolute. For example, individual consumers use business services, such as consulting lawyers and keeping money in banks, and

businesses use consumer services, such as purchasing stationery and hiring janitorial services. A public service worker at a National Park may provide the same service as a consumer service worker at Disneyland. Geographers find the classification useful, because the various types of services have different distributions, and different factors influence locational decisions.

Changes in Number of Employees

Figure 12–2 shows changes in employment in the United States between 1960 and 1995. Employment declined in agriculture and other primary-sector activities by 46 percent, and increased in manufacturing and other secondary-sector activities by a modest 13 percent. In comparison, employment in services increased 143 percent during the period. Within the service sector, employment grew relatively rapidly in producer and personal services and relatively slowly in transportation services.

Figure 12–2 Employment change in the United States by sector. Since 1960, the greatest increase has occurred in the number of employees in professional services and personal and social services. The greatest decrease has occurred in the number of employees in the primary sector.

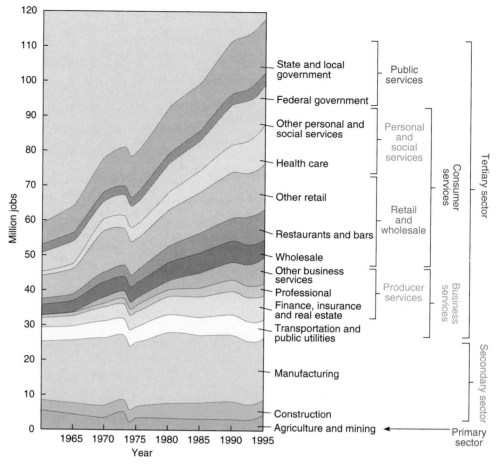

The largest increase in employment has been in producer services. Especially large increases have been experienced in computer and data processing, advertising, and temporary employment agencies. Professional services, such as engineering, management, and law, have also increased. Finance, insurance, and real estate—often grouped into one category under the acronym FIRE—have had relatively modest increases.

The personal service sector has increased rapidly primarily because of a very large increase in provision of health care services, including hospital staff, clinics, nursing homes, and home health care programs. Other large increases have been recorded in recreation, repair, and social services. Services offering personal care have grown more modestly, because increases in such services as dry cleaners and beauty salons have been offset by declines in home help and cleaning services.

Retail services have increased moderately overall. Much of the increase in retailing is attributable to the growth of restaurants. General merchandise and department stores have increased more slowly than have specialty shops, such as clothing, furniture, automotive, and electronics.

Transportation services have increased moderately, because large increases in new technology have been offset by declines in older technology. Employment has increased in trucking and air services and declined in railroads. Television broadcasting has expanded, whereas fewer employees are needed in other communications systems, such as telephone and telegraph.

Public services have expanded more modestly than other services, despite widespread misperception that government has expanded massively. The number of federal government employees has increased only slightly since 1960, and actually declined in the 1990s. State and local government employment has expanded more rapidly in part to offset the federal decline.

Origin of Services

Services are clustered in settlements. To understand why, picture conditions before the establishment of permanent settlements as places that provided services. People lived as nomads, migrating in small groups across the landscape in search of food and water (see Chapter 10). They gathered wild berries and roots, or killed wild animals for food.

At some point, groups decided to build permanent settlements. Several families clustered together in a rural settlement and obtained food in the surrounding area. What services would these nomads require? Why would they establish permanent settlements to provide these services?

No one knows the precise sequence of events through which settlements were established to provide services. Based on archeological research, settlements probably originated to provide personal services, especially religion and education, as well as public services such as government and police protection. Transportation, producer, and retail services came later.

Early Personal Services

The early permanent settlements may have been established to offer personal services, specifically places to bury the dead. Perhaps nomadic groups had rituals honoring the deceased, including ceremonies commemorating the anniversary of a death. Having established a permanent resting place for the dead, the group might then install priests at the site to perform the service of saying prayers for the deceased.

This would have encouraged the building of structures—places for ceremonies and dwellings. By the time recorded history began about 5,000 years ago, many settlements existed, and some featured a temple. In fact, until the invention of skyscrapers in the late nineteenth century, religious buildings were often the tallest structures in a community.

Settlements also may have been places to house families, permitting unburdened males to travel farther and faster in their search for food. Women kept "home and hearth," making household objects, such as pots, tools, and clothing—the origin of industry. The education of children became an important service provided in settlements. Over thousands of years, making pots and educating children evolved into a wide variety of services, including schools, libraries, theaters, and museums, which create and store a group's values and heritage and transmit them from one generation to the next.

People also needed tools, clothing, shelter, containers, fuel, and other material goods. Settlements therefore became manufacturing centers. Men gathered the materials needed to make a variety of objects, including stones for tools and weapons, grass for containers and matting, animal hair for clothing, and wood for shelter and heat. Women used these materials to manufacture household objects and maintain their dwellings.

The variety of personal services expanded as people began to specialize in particular ones. One person could be skilled at repairing tools, another at training horses. People could then trade these services with each other.

Early Public Services

Public services probably followed the religious activities into early permanent settlements. The group's political leaders also chose to live permanently in the settlement, which may have been located for strategic reasons, to protect the group's land claims. Everyone in a settlement was vulnerable to attack from other groups, so for protection, some members became soldiers, stationed in the settlement. The settlement likely was a good base from which the group could defend nearby food sources against competitors.

For defense, the group might surround the settlement with a wall. Defenders were stationed at small openings or atop the wall, giving them a great advantage over attackers. Thus, settlements became citadels—centers of military power. Walls proved an extremely effective defense for thousands of years, until warfare was revolutionized by the introduction of gunpowder in Europe in the fourteenth century.

Although modern settlements no longer have walls, their military and political services continue to be important. The largest structure in our nation's capital—the Pentagon—houses the U.S. Department of Defense. Similarly, Russian military leaders work in the Kremlin, which is the medieval walled area of central Moscow.

Origin of Other Services

Everyone in settlements needed food, which was supplied by the group through hunting or gathering. At some point, someone probably wondered: why not bring in extra food for hard times, such as drought or conflict? This perhaps was the origin of transportation services.

Medieval European cities, such as Carcassonne, in southwestern France, were often surrounded by walls for protection. The walls have been demolished in most places, but they still stand around the old center of Carcassonne. A small portion of the modern industrial city of 40,000 can be seen at far right. (Jonathan Blair/Woodfin Camp & Associates)

Settlements took on a retail service function. Not every group had access to the same resources, because of the varied distribution of vegetation, animals, fuelwood, and mineral resources across the landscape. People brought objects and materials they collected or produced into the settlement and exchanged them for items brought by others. Settlements became warehousing centers to store the extra food.

The settlement served as neutral ground where several groups could safely come together to trade goods and services. To facilitate this trade, officials in the settlement provided producer services, such as regulating the terms of transactions, setting fair prices, keeping records, and creating a currency system.

Services in Rural Settlements

Through centuries of experiments and accidents, residents of early settlements realized that some of the wild vegetation they had gathered could generate food if deliberately placed in the ground and nursed to maturity—in other words, agriculture, as described in Chapter 10. Settlements were surrounded by fields, where people produced most of their food by planting seeds and raising animals rather than by hunting and gathering.

Most people in the world still live in rural settlements that have changed little in purpose since ancient times. They are known as **clustered rural settlements**, where a number of families live in close proximity to each other, with fields surrounding the collection of houses and farm buildings. **Dispersed rural settlements**, characteristic of the contemporary North American rural landscape, are characterized by farmers living on individual farms isolated from neighbors, rather than alongside other farmers in settlements.

Clustered Rural Settlements

A clustered rural settlement typically includes homes, barns, tool sheds, other farm structures, plus personal services, such as religious structures and schools. A handful of public, retail, and producer services may also be present in the clustered rural settlement. In common language, such a settlement is called a hamlet or village.

Each person living in a clustered rural settlement is allocated strips of land in the surrounding fields. The fields must be accessible to the farmers and are thus generally limited to a radius of 1 or 2 kilometers (half a mile to a mile) from the buildings. The strips of land are allocated in different ways. In some places, individual farmers own or rent the land; in other places, the land is owned collectively by the settlement or by a lord, and farmers do not control the choice of crops or use of the output.

Parcels of land surrounding the settlement may be allocated to specific agricultural activities, either because of land characteristics or because of decisions by the inhabitants. Consequently, farmers typically own, or have responsibility for, a collection of scattered parcels in

several fields. This pattern of controlling several fragmented parcels of land has encouraged living in a clustered rural settlement to minimize travel time to the various fields.

Traditionally, when the population of a settlement grew too large for the capacity of the surrounding fields, new settlements were established nearby. This was possible because not all land was under cultivation.

The establishment of satellite settlements often is reflected in place-names. For example, the parish of Offley, in Hertfordshire, England, contains these rural settlements: Great Offley (the largest), Little Offley, Offley Grange (barn), Offley Cross, Offley Bottom, Offley Place, Offleyhoo (house), and Offley Hole (Figure 12–3). All are within a few kilometers of each other. The name "Offley"" means the wooded clearing of Offa, who was a ruler of Mercia (see Figure 5–14) during the eighth century and is said to have died at the site of the settlement.

Homes, public buildings, and fields in a clustered rural settlement are arranged according to local cultural and physical characteristics. Clustered rural settlements are often arranged in one of two types of patterns: circular and linear.

Circular Rural Settlements. The circular form consists of a central open space surrounded by structures (Figure 12–4). The kraal villages in southern Africa have enclosures for livestock in the center, surrounded by a ring of houses (compare our English word, corral). In East Africa, the Masai people, who are pastoral nomads, built kraal settlements as camps; women had principal responsibility for constructing them.

The German *Gewandorf* settlement consisted of a core of houses, barns, and churches, encircled by different

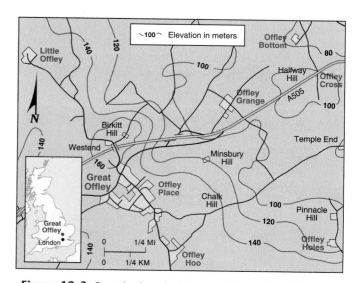

Figure 12–3 Growth of rural settlements. The rural landscape reflects the historical pattern of growth through establishment of satellite settlements. On the map, note the numerous places with "Offley" in their name: Great Offley (the largest and the original settlement), Little Offley, Offley Grange (barn), Offley Cross, Offley Bottom, Offley Place, Offley Hoo (house), and Offley Hole. These are satellite rural settlements in the parish of Offley, in Hertfordshire (Hertford County), England.

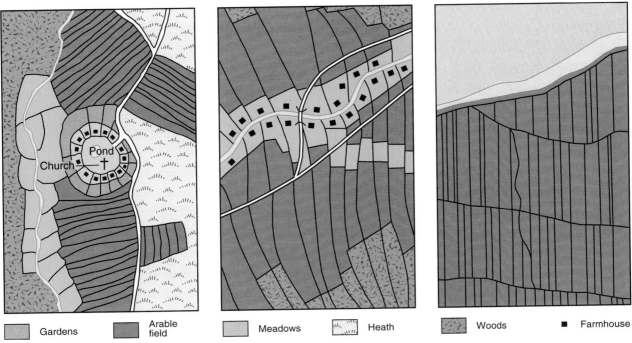

Figure 12–4 Rural settlement patterns. At left is the circular arrangement common in Germany. At center is a linear arrangement called "long-lot," used in France, which gives everyone access to the river. When French settlers came to America, the long-lot system came with them, as shown at right, in Québec.

Legend: Gardens | Arable field | Meadows | Heath | Woods | ■ Farmhouse

types of agricultural activities. Small garden plots were located in the first ring surrounding the village, with cultivated land, pastures, and woodlands in successive rings. Von Thünen observed this circular rural pattern in his landmark agricultural studies in the early nineteenth century (see Chapter 10).

Most rural settlements in Africa, like this one in Côte d'Ivoire, are clustered. Houses and farm structures are built close to each other, and the fields and grazing land surround the settlement. (Marc & Evelyne Bernheim/Woodfin Camp & Associates)

Linear Rural Settlements. Linear rural settlements feature buildings clustered along a road, river, or dike, to facilitate communications. The fields extend behind the buildings in long, narrow strips. Today in North America, linear rural settlements exist in areas settled by the French. The French settlement pattern, called *long-lot* or *seigneurial*, was commonly used along the St. Lawrence River in Québec and the lower Mississippi River (Figure 12–4).

In the French long-lot system, houses were erected along a river, which was the principal water source and means of communication. Narrow lots from 5 to 100 kilometers deep (3 to 60 miles) were established perpendicular to the river, so that each original settler had river access. This created a linear settlement along the river.

Eventually, these long, narrow lots were subdivided. French law required that each son inherit an equal portion of an estate, so the heirs established separate farms in each division. Roads were constructed parallel to the river for access to inland farms. In this way, a new linear settlement emerged along each road, parallel to the original riverfront settlement.

Dispersed Rural Settlements

In the past 200 years, dispersed rural settlements have become more common, especially in Anglo-America and the United Kingdom, because in more developed societies they are generally considered more efficient than clustered settlements (see Globalization and Local Diversity box).

To improve agricultural production, a number of European countries converted their rural landscapes from

From Clustered to Dispersed Rural Settlements in the American Colonies

The first European colonists settled along the East Coast in three regions: New England, the Southeast, and the Middle Atlantic. The rural settlements *where* they lived in the three regions differed. Colonial rural settlements were clustered in New England and the Southeast and dispersed in the Middle Atlantic. The reason *why* the three sets of rural settlements differed related to patterns of migration to the American colonies.

New England colonists built clustered settlements centered around an open area called a common. Settlers grouped their homes and public buildings, such as the church and school, around the common. In addition to their houses, each settler had a home lot of 1 to 5 acres (1/2 to 2 hectares), which contained a barn, garden, and enclosures for feeding livestock.

Clustered settlements were favored by New England colonists for a number of reasons. Typically, they traveled to the American colonies in a group. The English government granted the group an area of land in New England perhaps 4 to 10 square miles (10 to 25 square kilometers). Members then traveled to America to settle the land. The settlement was usually built near the center of the land grant.

New England settlements were also clustered to reinforce common cultural and religious values. Most of the group came from the same English village and belonged to the same church. Many of the early New England colonists left England in the 1600s to gain religious freedom. The settlement's leader often was an official of the Puritan church, and the church played a central role in daily activities. Colonists also favored clustered settlements for defense against Indian attacks.

Outsiders could obtain land in the settlement only through permission of the town's residents. Land was not sold, but rather awarded to an individual after the town's residents felt confident that the recipient would work hard.

The southeastern colonies were first settled in the 1600s with small, dispersed farms. Then a different style emerged, called a plantation, a large farm that used many workers to produce tobacco and cotton for sale in Europe and the northern colonies. Plantations grew more profitable in the 1700s when the tobacco and cotton markets expanded and two large sources of labor were identified. These included indentured whites, who were legally bound to work for the plantation for a period of time, and black slaves forcibly transported from Africa and sold to the plantation owner.

The plantation's wealthy owner lived in a large mansion, frequently fronting on a body of water. Surrounding the mansion were service buildings, including a laundry, kitchen, dairy, and bakery. Other buildings on the estate included a flour mill, carpenter shop, stables, coach house, and living quarters for the slaves.

The Middle Atlantic colonies were settled by a more heterogeneous group of people. In addition to English, they included immigrants from Germany, Holland, Ireland, Scotland, and Sweden. Further, most Middle Atlantic colonists came as individuals rather than as members of a cohesive religious or cultural group. Some bought tracts of land from speculators. Others acquired land directly from individuals who had been given large land grants by the British government, including William Penn (Pennsylvania), Lord Baltimore (Maryland), and Sir George Carteret (the Carolinas).

The *local diversity* in settlement types within the American colonies eventually gave way to a more uniform rural landscape. The New England rural settlement pattern underwent the most radical change: in the 1700s, a dispersed distribution began to replace the clustered settlements. In part, the cultural bonds that had created clustered rural settlements had weakened. Descendants of the original settlers grew less interested in the religious and cultural values that had unified the original immigrants. They permitted people to buy land regardless of their religious affiliation.

More importantly, economic factors induced the change from clustered to dispersed rural settlements in New England. The clustered rural settlement pattern worked when the population was low, but settlements had no spare land to meet the needs of a population that was growing through natural increase and net in-migration.

At first, settlements accommodated the growing population by establishing new settlements nearby. As in the older settlements, the newer ones contained central commons surrounded by houses and public buildings, home lots, and outer fields. However, the shortage of land eventually forced immigrants and children to strike out alone and claim farmland on the frontier.

At the same time, demand for more efficient agricultural practices led to a redistribution of farmland. At first, each villager owned several discontinuous parcels on the periphery of the settlement, to provide the variety of land types needed for different crops. Beyond the fields, the town held pastures and woodland for the common use of all residents.

Owning several discontinuous fields had several disadvantages: farmers lost time moving between fields, villagers had to build more roads to connect the small lots, and farmers had been restricted in what they could plant. Eventually, people bought, sold, and exchanged land to create large, continuous holdings instead of several isolated pieces.

The *globalization* of settlement patterns, at least at the scale of North America, was sealed when the region's principal agricultural production moved from the East Coast to the interior. Dispersed settlement patterns dominated in the American Midwest in part because the early settlers came primarily from the Middle Atlantic colonies. The pioneers crossed the Appalachian Mountains and established dispersed farms on the frontier. Land was plentiful and cheap, and people bought as much as they could manage. However, dispersed settlement patterns dominated in the Midwest primarily because new agricultural practices favored larger farms, as discussed in Chapter 10.

Meanwhile, the contemporary New England landscape contains remnants of the old clustered rural settlement pattern. Many New England towns still have a central common surrounded by the church, school, and various houses. However, quaint New England towns are little more than picturesque shells of clustered rural settlements, because today's residents work in factories, shops, and offices, rather than on farms.

clustered settlements to dispersed patterns. A prominent example was the **enclosure movement** in Great Britain, between 1750 and 1850. The British government transformed the rural landscape by consolidating individually owned strips of land surrounding a village into a single large farm, owned by an individual. When necessary, the government forced people to give up their former holdings.

The benefit of enclosure was greater agricultural efficiency, because a farmer did not have to waste time scurrying among discontinuous fields. With the introduction of farm machinery, farms operated more efficiently at a larger scale. Because the enclosure movement coincided with the industrial revolution, villagers who were displaced from farming moved to urban settlements and became workers in factories and services.

The enclosure movement brought greater agricultural efficiency, but it destroyed the self-contained world of village life. Village populations declined drastically as displaced farmers moved to urban settlements. Some villages became the centers of the new, larger farms, but villages that were not centrally located to a new farm's extensive land holdings were abandoned and replaced with entirely new farmsteads at more strategic locations. As a result, the isolated, dispersed farmstead, unknown in medieval England, is now a common feature of that country's rural landscape.

Key Issue 2

Why Are Consumer Services Distributed in a Regular Pattern?

- Central place theory
- Market area analysis
- Hierarchy of services and settlements

Within more developed countries, consumer services and business services do not have the same distributions. This key issue shows how consumer services generally follow a regular pattern based on size of settlements, with larger settlements offering more consumer services than smaller ones. In the next section of the chapter, business services will be seen to cluster in specific settlements, creating a specialized pattern.

Central Place Theory

A **central place** is a market center for the exchange of goods and services by people attracted from the surrounding area. The central place is so-called because it is centrally located to maximize accessibility from the surrounding region. Central places compete against each other to serve as markets for goods and services. This competition creates a regular pattern of settlements, according to central place theory.

The geographic concept of **central place theory** explains not only the distribution of services, but why a regular pattern of settlements exists—at least in more developed countries like the United States. Central place theory was first proposed in the 1930s by German geographer Walter Christaller, based on his studies of southern Germany. August Lösch in Germany and Brian Berry and others in the United States further developed the concept during the 1950s. The theory applies most clearly in regions such as the Great Plains, which are neither heavily industrialized nor interrupted by major physical features like rivers or mountain ranges.

Market Area of a Service

The area surrounding a service from which customers are attracted is the **market area** or **hinterland**. A market area is a good example of a nodal region—a region with a core where the characteristic is most intense. To establish the market area, a circle is drawn around the node of service on a map. The territory inside the circle is its market area.

Because most people prefer to get services from the nearest location, consumers near the center of the circle obtain services from local establishments. The closer to the periphery of the circle, the greater is the percentage of consumers who will choose to obtain services from other nodes. People on the circumference of the market area circle are equally likely to use the service, or go elsewhere.

Circles can be drawn to designate market areas of entire urban settlements, not just individual services. But circles cause a geometric problem. When drawn to represent adjacent market areas, they either overlap or have gaps between them (Figure 12–5, left). Neither pattern is consistent with the theory that people usually go to the nearest sources.

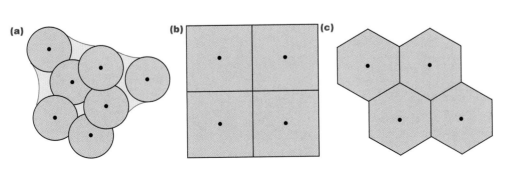

Figure 12–5 Why hexagons are used in theory to delineate market areas: (a) Circles are equidistant from center to edge, but they overlap or leave gaps. (b) Squares nest together without gaps, but their sides are not equidistant from the center. (c) Geographers use hexagons to depict the market area of a good or service, because hexagons offer a compromise between the geometric properties of circles and squares.

An arrangement of circles that leaves gaps indicates that people living in the gaps are outside the market area of any service, obviously not true. On the other hand, overlapping circles are unsatisfactory, for one service or another will be closer, and people will tend to patronize it. Therefore, market areas must be separated by a line that does not overlap territories.

Central place theory requires a geometric shape without gaps or overlaps, so circles are out. Squares fit without gaps, but they have a different problem. If the market area is a circle, the radius—the distance from the center to the edge—can be measured, because every point around a circle is the same distance from the center. But in a square, the distance from the center varies among points along a square (Figure 12–5, center).

Therefore, to represent a market area, the hexagon is the best compromise between circles and squares (Figure 12–5, right). Like squares, hexagons nest without gaps. Although all points along the hexagon are not the same distance from the center, the variation is less than with a square. Consequently, geographers draw hexagons around settlements to indicate market areas.

Size of Market Area

The market area of every service varies. To determine the extent of a market area, geographers need two pieces of information about a service: its range and its threshold.

Range of a Service.   How far are you willing to drive for a pizza? How far would you travel to see a doctor for a serious problem? How far would you travel for a weekend vacation? The **range** is the maximum distance people are willing to travel to use a service. People are willing to go only a short distance for everyday services, like groceries, laundromats, or video rentals. But they will travel a long distance for other services, such as a major league baseball game or concert. Thus, a convenience store has a small range, whereas a stadium has a large range. The range is the radius of the circle drawn to delineate a service's market area.

If firms at other locations compete by providing the service, the range must be modified. As a rule, people tend to go to the nearest available service: someone in the mood for a McDonald's hamburger is likely to go to the nearest McDonald's. Therefore, the range of a service must be determined from the radius of a circle that is irregularly shaped rather than perfectly round. The irregularly shaped circle takes in the territory for which the proposed site is closer than competitors.

For example, on a map of Dayton, Ohio, we can indicate the location of all Kroger supermarkets and draw irregularly shaped circles around each of them (Figure 12–6). The radius of each circle shows the range for each store. The median radius for Kroger supermarkets in Dayton is approximately 2 kilometers (1.2 miles).

Figure 12–6 Market area, range, and threshold for Kroger supermarkets in the Dayton, Ohio, metropolitan area. Fewer stores are in the southwest and northeast, which are predominantly industrial areas, and in the west, which contains lower-income residents.

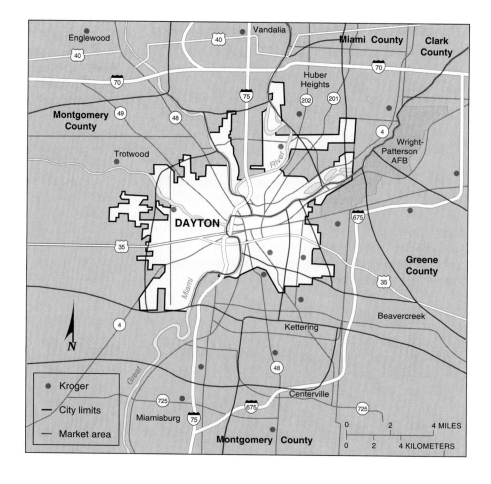

The range must be modified further because most people think of distance in terms of time, rather than a linear measure like kilometers or miles. If you ask people how far they are willing to travel to a restaurant or a baseball game, they are more likely to answer in minutes or hours than in distance. If the range of a good or service is expressed in travel time, then the irregularly shaped circle must be drawn to acknowledge that travel time varies with road conditions. "One hour" may translate into 90 kilometers (55 miles) on an expressway but only 50 kilometers (30 miles) driving congested city streets.

To determine the range of a service, geographers observe consumer behavior. We can ask people in a laundromat, supermarket, or stadium where they came from. We can ask people at home where they normally go to buy food, have their clothes cleaned, or attend a sporting event. The result shows how far the typical customer is willing to travel for various services.

Threshold of a Service. The second piece of geographic information needed to compute a market area is the **threshold**, which is the minimum number of people needed to support the service. Every enterprise has a minimum number of customers required to generate enough sales to make a profit. Once the range has been determined, a service provider must determine whether a location is suitable by counting the potential customers inside the irregularly shaped circle. For example, the median threshold needed to support a Kroger supermarket in Dayton is about 30,000 people. Census data help us determine the population within the circle.

How potential consumers inside the range are counted depends on the product. Convenience stores and fast-food restaurants appeal to nearly everyone, whereas other goods and services appeal primarily to certain consumer groups. Movie theaters attract younger people; chiropractors attract older folks. Poorer people are drawn to thrift stores; wealthier ones might frequent upscale department stores. Amusement parks attract families with children, but nightclubs appeal to singles. If a good or service appeals to certain customers, then only their type should be counted inside the range.

Developers of shopping malls, department stores, and large supermarkets typically count only higher-income people, perhaps those whose annual incomes exceed $50,000. Even though the stores may attract individuals of all incomes, higher-income people are likely to spend more and purchase items that carry higher profit margins for the retailer. Hence, in the Dayton area, Kroger operates more supermarkets in the southern part, where higher-income people are clustered, and fewer in the western part, a lower-income area.

Market Area Analysis

Retailers and other service providers make use of market area studies to determine whether locating in the market would be profitable, and if so the best location within the market area.

Profitability of a Location

The range and threshold together determine whether a good or service can be profitable in a particular location. To illustrate, consider this: Would a convenience store be profitable in your community? First, compute the range, the maximum distance people are willing to travel. You might survey local residents and determine that people are generally willing to travel up to 15 minutes to reach a convenience store.

Then, compute the threshold. Suppose a convenience store must sell at least $10,000 worth of goods per week to make a profit, and the average customer spends $2 a week. The store needs at least 5,000 customers each week, spending $2 each, to achieve the break-even sales level of $10,000. If the average customer goes to a convenience store once a week, the threshold in this example would be 5,000.

Finally, on a map, draw a circle around your community with a 15-minute travel radius, adjusting the boundaries to account for any competitors. Count the number of people within the irregularly shaped circle. If more than 5,000 people are within the radius, then the threshold may be high enough to justify locating the new convenience store in your community.

Optimal Location Within a Market

Having determined that the threshold and range justify providing a particular service, the next geographic question is: Where should the service be located within the market area to maximize profit? According to geographers, the best location is the one that minimizes the distance to the service for the largest number of people.

Best Location in a Linear Settlement. Suppose that you want to establish your hot business idea, *Geographers' Pizza*, in your community. Where is the best place to build it? Assume for a moment that you are seeking the optimal location for your business in an elongated community like Miami Beach, Florida, or Atlantic City, New Jersey, or Ocean City, Maryland. The community has only one major north-south street and a number of short east-west streets that are numbered consecutively.

The best location will be the one that minimizes the distance that your van must travel to deliver to all potential customers. That location can be determined precisely rather than through trial-and-error: It corresponds to the median, which mathematically is the middle point in any series of observations. In a linear community like an Atlantic Ocean resort, the service should be located where half of the customers are to the north and half to the south.

Suppose *Geo Pizza* has seven potential customers, families A through G, distributed in the community as

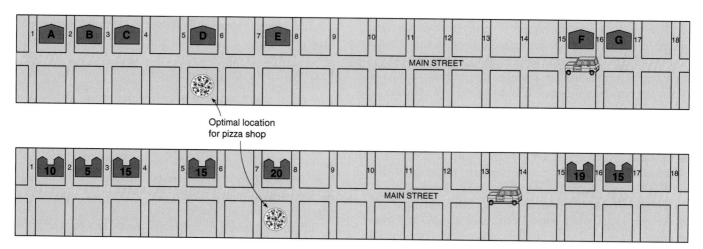

Figure 12–7 (top) Best location for a pizza delivery service in a linear settlement with seven families. (bottom) Best location for a pizza delivery service in a linear settlement with 99 families. Numbers in each apartment building represent families at each location.

shown in Figure 12–7, top. If the shop were between 5th and 6th streets, the delivery van would travel 4 blocks to deliver a pizza to Family A (between 1st and 2nd streets), 3 blocks to Family B (between 2nd and 3rd), 2 blocks to Family C (between 3rd and 4th), 0 blocks to Family D (between 5th and 6th), 2 to Family E (between 7th and 8th), 10 to Family F (between 15th and 16), and 11 to Family G (between 16th and 17th). The van would have to travel a total of $4 + 3 + 2 + 0 + 2 + 10 + 11$ blocks, or 32 blocks, to deliver a pizza to each of the seven customers.

Compare this location to any other possibility. For example, if *Geo Pizza* were between 7th and 8th streets, then the van would travel 6 blocks to reach Family A, 5 blocks to Family B, 4 blocks to Family C, 2 blocks to Family D, 0 blocks to Family E, 8 to Family F, and 9 to Family G. The van would have to travel $6 + 5 + 4 + 2 + 0 + 8 + 9$ blocks, or 34 blocks, to deliver pizza to all seven customers. This is a greater total distance than locating at 5th Street.

In fact, no other location results in a lower aggregate distance than between 5th and 6th streets, which in this example is the median observation. Between 5th and 6th streets, an equal number of potential customers are located on either side—three to the west (Families A, B, C), and three to the east (Families E, F, G).

What if a different number of customers live at each block of the city? What if the buildings are apartments, each housing a different number of families? To compute the optimal location in these cases, geographers have adapted the gravity model from physics. The **gravity model** predicts that the optimal location of a service is directly related to the number of people in the area, and inversely related to the distance people must travel to access it.

According to the gravity model, consumer behavior reflects two patterns. First, the greater the number of people living in a particular place, the greater is the number of potential customers for a service. A city block or apartment building that contains 100 families will generate more customers than a house containing only one family. Second, the farther people are from a particular service, the less likely they are to use it. People who live 1 kilometer from a store are more likely to patronize it than people who live 10 kilometers away.

In the case of *Geo Pizza*, the optimal location along Main Street will again be the median location (Figure 12–7, bottom). Note the settlement has 99 families. The median location is the middle observation among these 99 families, the place where 49 families live to the west and 49 families live to the east. *Geo Pizza* should locate between 7th and 8th streets.

Best Location in a Nonlinear Settlement. Most settlements are more complex than a single main street. Geographers still apply the gravity model to find the best location, following these steps:

1. Identify a possible site for a new service.
2. Within the range of the service, identify where every potential user lives.
3. Measure the distance from the possible site of the new service to every potential user.
4. Divide each potential user by the distance to the potential site for the service.
5. Sum all of the results of potential users divided by distances.
6. Select a second possible location for the new service, and repeat steps 2, 3, 4, and 5.
7. Compare the results of step 5 for all possible sites. The site with the highest score has the highest potential number of users, and is therefore the optimal location for the service.

Hierarchy of Services and Settlements

Small settlements are limited to services that have small thresholds, short ranges, and small market areas, because too few people live in small settlements to support many services. A large department store or specialty store cannot survive in a small settlement because the minimum number of people needed exceeds the population within range of the settlement.

Larger settlements provide services having larger thresholds, ranges, and market areas. However, neighborhoods within large settlements also provide services having small thresholds and ranges. Services patronized by a small number of locals can coexist in a neighborhood ("mom & pop stores") along with services that attract many from throughout the settlement. This difference is vividly demonstrated by comparing the *Yellow Pages* for a small settlement with one for a major city. The major city's *Yellow Pages* are plump with more services, and diverse headings show widely varied services unavailable in small settlements.

We spend as little time and effort as possible in obtaining services, and thus go to the nearest place that fulfills our needs. There is no point in traveling to a dis-

tant department store if the same merchandise is available at a nearby one. We travel greater distances only if the price is much lower or if the item is unavailable locally.

Nesting of Services and Settlements

According to central place theory, market areas across a more developed country would be a series of hexagons of various sizes, unless interrupted by physical features such as mountains and bodies of water. More developed countries have numerous small settlements with small thresholds and ranges, and far fewer large settlements with large thresholds and ranges.

The nesting pattern can be illustrated with overlapping hexagons of different sizes (Figure 12–8). The figure shows four different levels of market area, for hamlet, village, town, and city. Hamlets with very small market areas are represented by the smallest contiguous hexagons. Larger hexagons represent the market areas of larger settlements, and are overlaid on the smaller hexagons, because consumers from smaller settlements shop for some goods and services in larger settlements.

In his original study, Walter Christaller showed that the distances between settlements in southern Germany followed a regular pattern. He identified seven sizes of

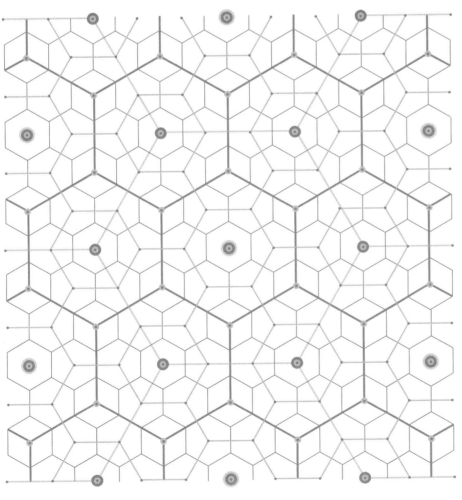

◎ City   ◉ Town   • Village   · Hamlet

Figure 12–8 Central place theory. According to central place theory, market areas are arranged in a regular pattern. Larger market areas, based in larger settlements, are fewer in number and farther apart from each other than smaller market areas and settlements. However, larger settlements also provide goods and services with smaller market areas; consequently, larger settlements have both larger and smaller market areas drawn around them.

settlements (market hamlet, township center, county seat, district city, small state capital, provincial head capital, and regional capital city). For example, the smallest (market hamlet) had an average population of 800 and a market area of 45 square kilometers (17 square miles). The average distance between market hamlets was 7 kilometers (4.4 miles). The figures were higher for the average settlement at each increasing level in the hierarchy. Brian Berry has documented a similar hierarchy of settlements in parts of the U.S. Midwest.

The principle of nesting market areas also works at the scale of services within cities. For example, compare the market areas within Dayton of Kroger (Figure 12–6) with those of United Dairy Farmers (UDF) and Elder-Beerman (Figure 12–9). The UDF convenience stores are more numerous than Krogers and have smaller thresholds, ranges, and market areas. The Elder-Beerman department stores are less numerous than the Kroger stores and have larger thresholds, ranges, and market areas.

Rank-Size Distribution of Settlements

In many MDCs, geographers observe that ranking settlements from largest to smallest (population) produces a regular pattern or hierarchy. This is the **rank-size rule**, in which the country's *n*th largest settlement is 1/n the population of the largest settlement. In other words, the second-largest city is one-half the size of the largest,

the fourth-largest city is one-fourth the size of the largest, and so on. When plotted on logarithmic paper, the rank-size distribution forms a fairly straight line. The distribution of settlements closely follows the rank-size rule in the United States and a handful of other countries (Figure 12–10, upper line).

If the settlement hierarchy does not graph as a straight line, then the society does not have a rank-size distribution of settlements. Several MDCs in Europe follow the rank-size distribution among smaller settlements, but not among the largest ones (Figure 12–10, lower line). Instead, the largest settlement in these countries follows the primate city rule. According to the **primate city rule**, the largest settlement has more than twice as many people as the second-ranking settlement. In this distribution, the country's largest city is called the **primate city**.

In France, for example, Paris is a primate city, because it has about 9 million inhabitants, whereas the second-largest settlement, Marseille, has only about 2 million, instead of the 4.5 million that the rank-size rule predicts. The primate city in the United Kingdom—London—has some 9 million, while Birmingham—the second-largest—has only about 2 million inhabitants.

Many LDCs also follow the primate city rule. However, in these countries, the rank-size rule tends to fail at other levels in the hierarchy as well. As previously noted, Romania has no settlement between 350,000 and 2 million inhabitants and too few settlements of less than 10,000 inhabitants.

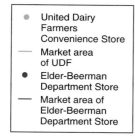

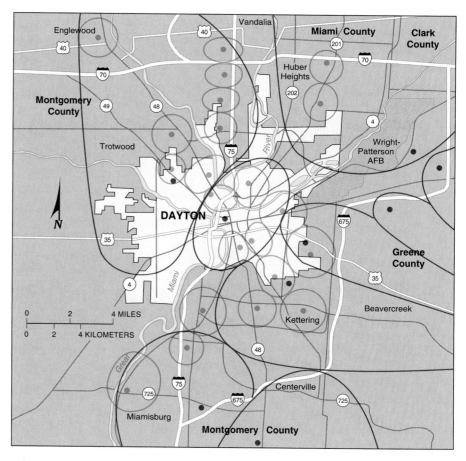

Figure 12–9 Market areas, ranges, and thresholds for United Dairy Farmers (UDF) convenience stores and Elder-Beerman department stores in the Dayton, Ohio, metropolitan area. Compared to the Kroger supermarkets (see Figure 12–6), UDF stores are more numerous and have smaller market areas, ranges, and thresholds, whereas Elder-Beerman stores are less numerous and have larger market areas, ranges, and thresholds.

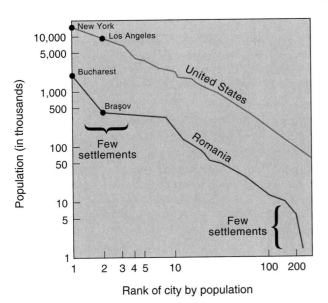

Figure 12–10 Rank-size distribution of settlements in the United States and Romania. (top line) U.S. settlements generally follow the rank-size distribution, as reflected by their nearly straight line on this logarithmic display. (bottom line) Romania has a shortage of settlements in two size groups: between 350,000 and 2 million population, and fewer than 10,000 inhabitants. These gaps are reflected in the irregular shape of the line.

The existence of a rank-size distribution of settlements is not merely a mathematical curiosity. It has a real impact on the quality of life for a country's inhabitants. A regular hierarchy—as in the United States—indicates that the society is sufficiently wealthy to justify provision of goods and services to consumers throughout the country. The absence of the rank-size distribution in a less developed country indicates that there is not enough wealth in the society to pay for a full variety of services.

Key Issue 3

Why Do Business Services Locate in Large Settlements?

- World cities
- Hierarchy of business services
- Economic base of settlements

Every settlement in a more developed country like the United States provides consumer services to people in a surrounding hinterland of varying area, but not every settlement of a given size has the same number and types of business services. Business services disproportionately cluster in a handful of settlements, and individual settlements specialize in particular business services.

World Cities

Prior to modern times, virtually all settlements were rural, because the economy was based on the agriculture

of the surrounding fields. Providers of personal services and a handful of other types of services met most of the needs of farmers living in the village. Even in ancient times, a handful of urban settlements provided producer and public services, as well as retail and personal services with large market areas.

Ancient World Cities

Urban settlements date from the beginning of documented history in the Middle East and Asia. They may have originated in Mesopotomia, part of the Fertile Crescent of the Middle East (see Figure 8–3), and diffused at an early date to Egypt, China, and South Asia's Indus Valley. Or they may have originated independently in each of the four hearths. In any case, from these four hearths, the concept of urban settlements diffused to the rest of the world.

Earliest Urban Settlements. Among the oldest well-documented urban settlements is Ur in Mesopotamia (present-day Iraq). Ur, which means "fire," was where Abraham lived prior to his journey to Canaan in approximately 1900 B.C., according to the Bible. Archaeologists have unearthed ruins in Ur that date from approximately 3000 B.C. (Figure 12–11).

Ancient Ur was compact, perhaps covering 100 hectares (250 acres), and was surrounded by a wall. The most prominent structure was a temple, known as a *ziggurat*, the command center for the ancient settlement and surrounding hinterland. The ziggurat was originally a three-story structure with a base of 64-by-46 meters (210-by-150 feet) and the upper stories stepped back. Four more stories were added in the sixth century B.C. Surrounding the ziggurat were residential areas containing a dense network of narrow winding streets and courtyards.

Recent evidence unearthed at Titris Hoyuk, in present-day Turkey, from about 2500 B.C. suggests that early urban settlements were well-planned communities. Houses were arranged in a regular pattern, because walls and streets were apparently laid out first. Palaces, temples, and other public buildings were placed at the center, cemeteries beyond the walls.

Houses varied in size but were of similar design, built around a central courtyard that contained a crypt where some of the family members were buried. Houses were apparently occupied by an extended family, because they contained several cooking areas. Evidence of wine production and weaving has been found in the houses.

Titris Hoyuk occupied a 50-hectare (125-acre) site and apparently had a population of about 10,000. The site is especially well-preserved today because after 300 years the settlement was abandoned and never covered by newer buildings.

Ancient Athens. Settlements were first established in the eastern Mediterranean about 2500 B.C. The oldest include Knossos on the island of Crete, Troy in Asia Minor (Turkey), and Mycenae in Greece. These settle-

Figure 12–11 Ancient Ur. The remains of Ur, in present-day Iraq, provide evidence of early urban civilization. The most prominent building was the stepped temple at left, called a *ziggurat*. Surrounding the ziggurat was a dense network of small residences built around courtyards and opening onto narrow passageways. The excavation site was damaged during the 1991 war in the Persian Gulf.

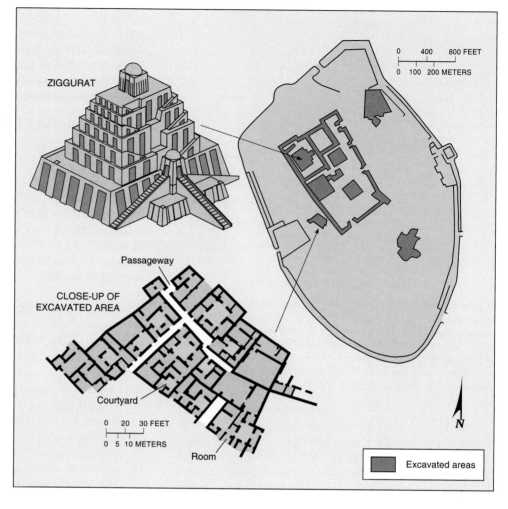

ments were trading centers for the thousands of islands dotting the Aegean Sea and the eastern Mediterranean. They were organized into **city-states**—independent self-governing communities that included the settlement and nearby countryside. The settlement provided the government, military protection, and other public services for the surrounding hinterland.

The number of urban settlements grew rapidly during the eighth and seventh centuries B.C. Hundreds of new towns were founded throughout the Mediterranean lands. The residents of one settlement would establish a new settlement to fill a gap in trading routes and to open new markets for goods. The diffusion of urban settlements from the eastern Mediterranean westward is well documented. For example, the city-state of Syracuse (in the southeastern part of the island of Sicily) established new settlements in Italy and Sicily between 750 and 700 B.C. Farther west at Marseille, France (then known as Massilia), about 600 B.C., Massilians founded settlements along the coast of present-day Spain during the sixth century B.C.

Athens, the largest city-state in ancient Greece, was probably the first city to attain a population of 100,000. Athens made substantial contributions to the development of culture, philosophy, and other elements of Western civilization. This demonstrates that urban set-

tlements have been traditionally distinguished from rural ones not only by public services but by a concentration of personal services, notably cultural activities, not found in smaller settlements (Figure 12–12).

Ancient Rome. The rise of the Roman Empire encouraged urban settlement. With much of Europe, North Africa, and Southwest Asia under Roman rule, settlements were established as centers of administrative, military, and other public services, as well as trading and other retail services. Trade was encouraged through transportation and utility services, notably construction of many roads and aqueducts, and the security the Roman legions provided. The city of Rome—the empire's center for administration, commerce, culture, and all other services—grew to at least a quarter-million inhabitants, although some claim that the population may have reached a million. The city's centrality in the empire's communications network was reflected in the old saying, "All roads lead to Rome" (see Figure 6–5).

With the Fall of the Roman Empire in the fifth century A.D., urban settlements declined. Their prosperity had rested on trading in the secure environment of imperial Rome. With the empire fragmented under hundreds of rulers, trade diminished. Large urban settlements shrank or were abandoned. For several hundred years,

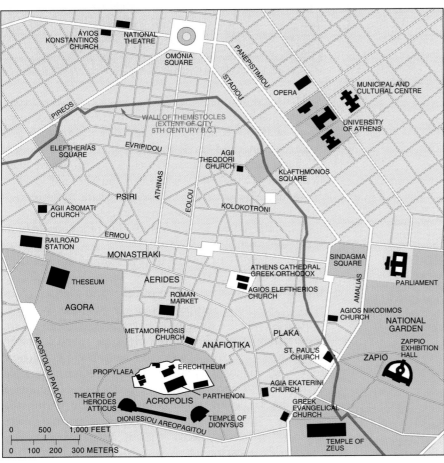

Figure 12–12 Athens, Greece. Dominating the skyline of modern Athens is the original hilltop site of the city, the Acropolis. Ancient Greeks selected this high place because it was defensible, and as a place to erect shrines to their gods. The most prominent structure on the Acropolis is the Parthenon, built in the fifth century B.C. to honor the goddess Athena. The structure to the left of the Parthenon, dating from the same time, is the Propylaea, the only opening in the wall surrounding the Acropolis. At the bottom of the hill is the Theater of Herodes Atticus, named for a wealthy Roman who built it as a memorial to his wife in A.D. 161. (David Pollack/The Stock Market)

Europe's cultural heritage was preserved largely in monasteries and isolated rural areas.

Medieval World Cities

Urban life began to revive in Europe in the eleventh century as feudal lords established new urban settlements. They gave residents charters of rights to establish independent cities, in exchange for fighting for the lord. Both the lord and urban residents benefited from this arrangement. The lord obtained people to defend his territory at less cost than maintaining a standing army. For their part, urban residents preferred periodic military service to the burden faced by rural serfs, who farmed the lord's land and could keep only a small portion of their own agricultural output.

With their newly won freedom from the relentless burden of rural serfdom, the urban dwellers set about expanding trade. Surplus from the countryside was brought into the city for sale or exchange, and markets were expanded through trade with other free cities. Trade among different urban settlements was enhanced by new roads and more use of rivers. By the fourteenth century,

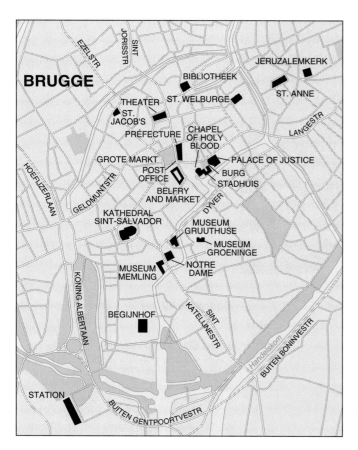

Figure 12–13 Brugge, Belgium. Modern Brugge (Bruges in French) is a town of more than 100,000 in the western part of Belgium, near the North Sea coast. Beginning in the twelfth century, Brugge was the most important port in northwestern Europe and a major center for manufacturing wool. However, three events forced the city's decline during the fifteenth century: foreign competitors captured much of the wool industry; the Belgian city of Antwerpen developed a better port; and the River Zwin silted, stranding the town 13 kilometers (8 miles) inland from the North Sea. Typical of medieval towns, the center of Brugge is dominated by squares surrounded by public buildings, churches, and markets. The tower on the right contains the market. The building on the left is the post office and prefecture (government office). (Larry Lee/West Light)

Europe was covered by a dense network of small market towns serving the needs of particular lords.

The largest medieval European urban settlements served as power centers for the lords and church leaders, as well as major market centers. The major public services occupied palaces, churches, and other prominent buildings arranged around a central market square (Figure 12–13). The tallest and most elaborate structures were usually the church, many of which still dominate the landscape of smaller European towns.

European urban settlements were usually surrounded by walls in medieval times, even though by then cannonballs could destroy them. Paris, for example, surrounded itself with new fortifications as recently as the 1840s and did not completely remove them until 1932 (Figure 12–14).

Dense and compact within the walls, medieval urban settlements lacked space for construction, so ordinary shops and houses nestled into the side of the walls and large buildings. In modern times, most of the modest medieval shops and homes, as well as the walls, have been demolished, with only the massive churches and palaces surviving. Modern tourists can appreciate the architectural beauty of these medieval churches and palaces, but they do not receive an accurate image of a densely built medieval town.

From the collapse of the Roman Empire until the diffusion of the industrial revolution across Europe during the nineteenth century, most of the world's largest cities were in Asia, not Europe. About A.D. 900, the five most populous cities are thought to have included Baghdad (in present-day Iraq), Constantinople (now called Istanbul, in Turkey), Kyoto (in Japan), and Changan and Hangchow (in China). Beijing (China) competed with Constantinople as the world's most populous city for several hundred years, until London claimed the distinction during the early 1800s. Agra (India), Cairo (Egypt), Canton (China), Isfahan (Iran), and Osaka (Japan) also ranked among the world's most populous cities prior to the industrial revolution.

Modern World Cities

In modern times, several world cities have emerged where a high percentage of the world's business is transacted and political power is concentrated. These world cities are centers of business services, but they stand at the top of the central place hierarchy in the provision of consumer services, and many also serve as public service centers.

New forms of transportation and communications were expected to reduce the need for clustering of eco-

426

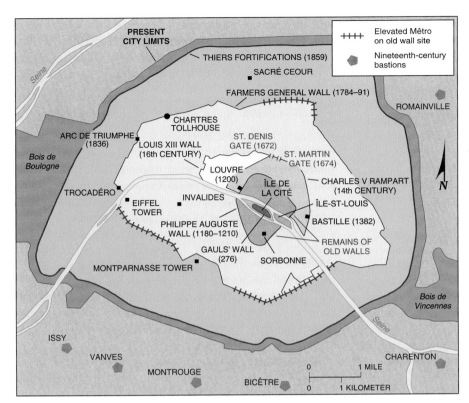

Figure 12–14 Walls around Paris. Periodically, a new wall (*barrière*) would be constructed to encompass new neighborhoods that had grown on the periphery. Highways and parks have been built on the sites of the nineteenth-century walls. The old gates of St. Denis and St. Martin still stand, although the walls have been demolished.

nomic activities in large cities. The telegraph and telephone in the nineteenth century and the computer in the twentieth century have both made it possible to communicate immediately with co-workers, clients, and customers around the world. The railroad in the nineteenth century and the automobile and airplane in the twentieth century made it possible to deliver people, inputs, and products quickly. To some extent, economic activities have decentralized, especially manufacturing, but modern inventions reinforce rather than diminish the primacy of world cities in the global economy.

Business Services in World Cities. The clustering of business services in the modern world city is a product of

the industrial revolution. Factories are operated by large corporations formed to minimize the liability to any individual owner. A board of directors located far from the factory building makes key decisions concerning what to make, how much to produce, and what prices to charge. Support staff also far from the factory account for the flow of money and materials to and from the factories. This work is done in offices in world cities.

World cities offer many financial services to these businesses. As centers for finance, world cities attract the headquarters of the major banks, insurance companies, and specialized financial institutions where corporations obtain and store funds for expansion of production. Shares of major corporations are bought and sold on the

The New York Stock Exchange is one of the financial activities that makes New York one of the world's three most important centers for business services, along with London and Tokyo. The photographer used a slow shutter speed to capture the swirl of traders on the Exchange floor. (L.M. Otero/AP/Wide World Photos)

stock exchanges, which are located in world cities. Obtaining information in a timely manner is essential to buy and sell shares at attractive prices.

Lawyers, accountants, and other professionals cluster in world cities to provide advice to major corporations and financial institutions. Advertising agencies, marketing firms, and other services concerned with style and fashion locate in world cities to help corporations anticipate changes in taste and to help shape those changes.

Transportation services converge on world cities. They tend to have busy harbors and airports, and lie at the junction of rail and highway networks.

Consumer Services in World Cities. Because of their large size, world cities have retail services with extensive market areas, but they may even have more retailers than large size alone would predict. A disproportionately large number of wealthy people live in world cities, so luxury and highly specialized products are especially likely to be sold there.

Personal services of national significance are especially likely to cluster in world cities, in part because they require large thresholds and large ranges, and in part because of the presence of wealthy patrons. World cities typically offer the most plays, concerts, operas, night clubs, restaurants, bars, and professional sports events. They contain the largest libraries, museums, and theaters. London presents more plays than the rest of the United Kingdom combined, and New York nearly has more theaters than the rest of the United States combined.

Public Services in World Cities. World cities may be centers of national or international political power. Most are national capitals, so they contain mansions or palaces for the head of state, imposing structures for the national legislature and courts, and offices for the government agencies. Also clustered in the world cities are offices for groups having business with the government, such as representatives of foreign countries, trade associations, labor unions, and professional organizations.

Unlike other world cities, New York is not a national capital. But as the home of the world's major international organization, the United Nations, it attracts thousands of U.N. diplomats and bureaucrats, as well as employees of organizations with business at the United Nations. Brussels is a world city because it is the most important center for European Union activities.

Hierarchy of Business Services

Geographers distinguish four levels of cities that play a major role in the provision of producer and other business services in the global economy. Atop the hierarchy are a handful of world cities, which can be subdivided into three groups. Below the world cities, but still playing major roles in the global business service economy, are regional command and control centers, specialized producer service centers, and dependent centers. Differences also exist among cities within each level. Other cities—some with large populations—play less important roles in the provision of business services.

World Cities

As described in the previous section, world cities are most closely integrated into the global economic system, because they are at the center of the flow of information and capital. Business services concentrate in disproportionately large numbers in world cities, including law, banking, insurance, accounting, and advertising.

Three world cities stand out in a class of their own: London, New York, and Tokyo (Figure 12–15). Each is the largest city in one of the three main regions of the more developed world (Western Europe, North America, and East Asia), as discussed in Chapter 9. The world's most important stock exchanges operate in these three cities, and they contain large concentrations of financial and related business services.

A second tier of major world cities includes Chicago, Los Angeles, and Washington in North America, and Brussels, Frankfurt, Paris, and Zurich in Western Europe. Only two of the nine second-tier world cities—São Paulo and Singapore—are in less developed regions. Major corporations and banks may have their headquarters in these world cities, rather than in London, New York, or Tokyo.

A third tier of secondary world cities includes four in North America (Houston, Miami, San Francisco, and Toronto), seven in Asia (Bangkok, Bombay, Hong Kong, Manila, Osaka, Seoul, and Taipei), five in Western Europe (Berlin, Madrid, Milan, Rotterdam, and Vienna), four in Latin America (Buenos Aires, Caracas, Mexico City, and Rio de Janeiro), one each in Africa (Johannesburg) and the South Pacific (Sydney).

Command and Control Centers

The second level of cities—command and control centers—contains the headquarters of many large corporations, well-developed banking facilities, and concentrations of other business services, including insurance, accounting, advertising, law, and public relations. Important educational, medical, and public institutions can be found in these command and control centers. Two levels of command and control centers can be identified: regional centers and subregional centers.

In the United States, regional command centers include Atlanta, Baltimore, Boston, Cincinnati, Cleveland, Columbus, Denver, Indianapolis, Kansas City, Minneapolis, New Orleans, Philadelphia, Phoenix, Portland (Oregon), St. Louis, and Seattle (Figure 12–16). Subregional centers include Biloxi, Birmingham, Charlotte, Des Moines, Jackson, Jacksonville, Little Rock, Memphis, Nashville, Oklahoma City, Omaha, Richmond, Salt Lake City, Shreveport, Spokane, and Syracuse.

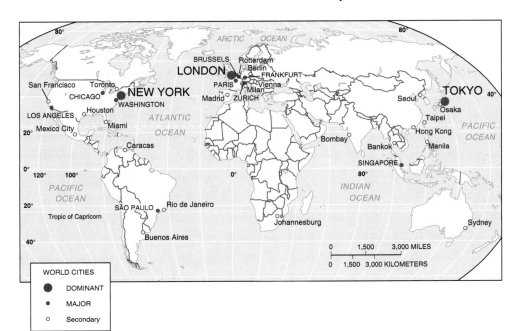

Figure 12–15 Hierarchy of world cities. World cities are centers for provision of services in the global economy. London, New York, and Tokyo are the three dominant world cities. Major and secondary world cities play somewhat less central roles in the provision of services than the three dominant world cities.

Specialized Producer-Service Centers

The third level of cities, specialized producer service centers, offers a more narrow and highly specialized variety of services. One group of these cities specializes in the management and R&D (research and development) activities related to specific industries, such as motor vehicles in Detroit, steel in Pittsburgh, office equipment in Rochester, New York, and semiconductors in San Jose, California. A second group of these cities specializes as centers of government and education, notably state capitals that also have a major university, such as Albany, Lansing, Madison, and Raleigh-Durham.

Dependent Centers

The fourth-level cities, dependent centers, provide relatively unskilled jobs and depend for their economic health on decisions made in the world cities, regional command and control centers, and specialized producer-service centers. Four subtypes of dependent centers can be identified in the United States:

- Resort, retirement, and residential centers, such as Albuquerque, Fort Lauderdale, Las Vegas, and Orlando, clustered in the South and West.

London has long been a center for financial institutions, including Lloyd's of London, which began in a coffee house more than three hundred years ago to insure ships. In 1986, Lloyd's moved into an ultra-modern building designed by Richard Rogers in London's financial district, known as the City. In the lobby is the Lutine bell, traditionally rung to signal information about an overdue ship—once for bad news and twice for good news. The bell was salvaged in 1858 from the wreck of HMS Lutine, which had sunk in 1799 carrying a cargo of gold and silver bullion insured by Lloyd's. (Gillian Allen/AP)

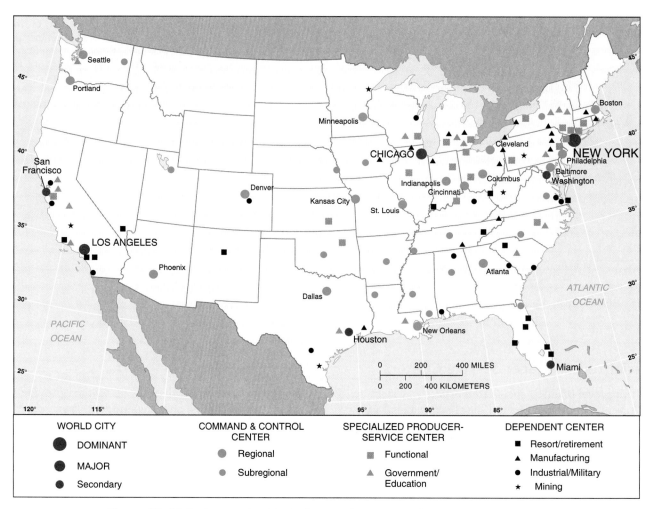

Figure 12–16 Business service cities in the United States. Atop the hierarchy are New York and the major and secondary world cities. Below the world cities in the hierarchy are regional command and control centers, specialized producer service centers, and dependent centers.

- Manufacturing centers, such as Buffalo, Chattanooga, Erie, and Rockford, clustered mostly in the old northeastern manufacturing belt.
- Industrial and military centers, such as Huntsville, Newport News, and San Diego, clustered mostly in the South and West.
- Mining and industrial centers, such as Charleston (West Virginia) and Duluth, located in mining areas.

Economic Base of Settlements

A settlement's distinctive economic structure derives from its **basic industries**, which export primarily to consumers outside the settlement. **Nonbasic industries** are enterprises whose customers live in the same community, essentially consumer services. A community's unique collection of basic industries defines its **economic base**.

A settlement's economic base is important, because exporting by the basic industries brings money into the local economy, thus stimulating the provision of more nonbasic consumer services for the settlement. New basic industries attract new workers to a settlement, and they bring their families with them. The settlement attracts additional consumer services to meet the needs of the new workers and their families. Thus, a new basic industry stimulates establishment of new supermarkets, laundromats, restaurants, and other consumer services. But a new nonbasic service, such as a supermarket, will not induce construction of new basic industries.

A community's basic industries can be identified by computing the percentage of the community's workers employed in different types of businesses. The percentage of workers employed in a particular industry in a settlement is then compared to the percentage of all workers in the country employed in that industry. If the percentage is much higher in the local community, then that type of business is a basic economic activity (Figure 12–17).

Settlements in the United States can be classified by their type of basic activity. Each type of basic activity has a different spatial distribution. Compared to the national average, some settlements have a very high percentage of workers employed in the primary sector, notably mining. Mining settlements are located near reserves of coal, petroleum, and other resources.

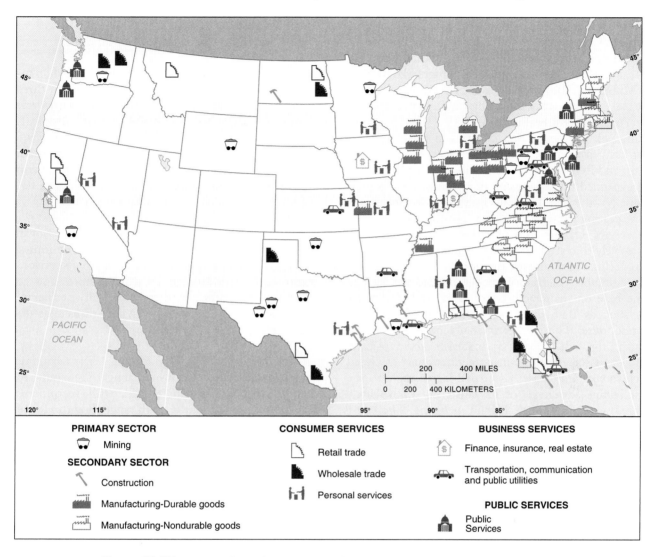

Figure 12–17 Economic base of U.S. cities. Symbols on this map represent cities that have a significantly higher percentage of their labor force engaged in the type of economic activity shown. Other cities also engage in such activities, but are not shown because they specialize in multiple activities, or are near the national average for all sectors. (Mathematically, a city was included if the percentage of its labor force in one sector was more than two standard deviations above the mean for all U.S. cities.)

The economic base of some settlements is in the secondary sector. Some of these specialize in durable manufactured goods, such as steel and automobiles, others in nondurable manufactured goods, such as textiles, apparel, food, chemicals, and paper. Most communities that have an economic base of manufacturing durable goods are clustered between northern Ohio and southeastern Wisconsin, near the southern Great Lakes. Nondurable manufacturing industries, such as textiles, are clustered in the Southeast, especially in the Carolinas.

Specialization of Cities in Different Services

The concept of basic industries originally referred to manufacturing. Detroit specializes in manufacturing motor vehicles; Gary, Indiana, in steel. But in a postindustrial society like the United States, increasingly the basic economic activities are in business, consumer, or public services. Steel was once the most important basic industry of Cleveland and Pittsburgh, but now medical services such as hospitals and clinics and medical high-technology research are more important.

Ó hUallacháin and Reid have documented examples of settlements that specialize in particular types of business services. Boston and San Jose specialize in the provision of services related to computing and data-processing. Austin, Orlando, and Raleigh-Durham specialize in the provision of services related to high-technology industries. Albuquerque, Colorado Springs, Huntsville, Knoxville, and Norfolk specialize in the provision of services related in support of military activity. Washington, DC, specializes in the provision of management consulting services.

Settlements specializing in public services are dispersed around the country, because these communities typically include a state capital, large university, or military base. Consumer service settlements include entertainment and recreation centers like Atlantic City, Las Vegas, and Reno, as well as medical centers like Rochester, Minnesota. Business services are concentrated in large metropolitan areas, especially Chicago, Los Angeles, New York, and San Francisco.

Although the population of cities in the South and West has grown more rapidly in recent years, Ó hUallacháin and Reid found that cities in the North and East have expanded their provision of business services more rapidly. Northern and eastern cities that were once major manufacturing centers have been transformed into business service centers. These cities have moved more aggressively to restructure their economic bases to offset sharp declines in manufacturing jobs.

For example, Baltimore once depended for its economic base on manufacturers of fabricated steel products, such as Bethlehem Steel, General Motors, and Westinghouse. The city's principal economic asset was its port, through which raw materials and fabricated products passed. As these manufacturers have declined, the city's economic base has turned increasingly to services, taking advantage of its clustering of research-oriented universities, especially in medicine. The city is trying to become a center for the provision of services in biotechnology.

Key Issue 4

Why Do Services Cluster Downtown?

- Central Business District
- Suburbanization of services

Historically, services of all types clustered in the center of the city, commonly called downtown, and known to geographers by the more precise term **central business district (CBD)**. Recently, services, especially retail, have moved from the CBD to suburban locations.

Central Business District

The center is the best-known and most visually distinctive area of most cities. It is usually one of the oldest districts in a city, often the original site of the settlement. The CBD is compact—less than 1 percent of the urban land area—but contains a large percentage of the shops, offices, and public institutions (Figure 12–18).

Consumer and business services are attracted to the CBD because of its accessibility. The center is the easiest part of the city to reach from the rest of the region and is the focal point of the region's transportation network.

Retail Services in the CBD

Three types of retail services concentrate in the center, because they require accessibility to everyone in the region—shops with a high threshold, shops with a long range, and shops that serve people who work in the center.

Retail Services with a High Threshold.   A shop may be in the center if it has a high threshold. High threshold shops, such as department stores, traditionally preferred a central location to be accessible to many people. Large department stores in the CBD often clustered across the street from one another. Retailers referred to the intersection nearest such a cluster as the "100 percent corner." Rents were highest there because this location had the highest accessibility for the most customers.

In recent years, many high threshold shops such as large department stores have closed their downtown branches. CBDs that once boasted three or four stores now have none, or perhaps one struggling survivor. The customers for downtown department stores now consist of downtown office workers, inner-city residents, and tourists. Department stores with high thresholds are now more likely to be in suburban malls.

Retail Services with a High Range.   The second type of shop in the center has a high range. Generally, a high range shop is very specialized, with customers who patronize it infrequently. High range shops prefer central locations because their customers are scattered over a wide area. For example, an expensive jewelry or fur shop attracts shoppers from all over the urban area, but each customer visits infrequently.

Many high range shops have moved with department stores to suburban shopping malls. These shops can still thrive in some CBDs if they combine retailing with recreational activities. People are willing to make a special trip to a specific destination downtown for unusual shops in a dramatic setting, perhaps a central atrium with a fountain or a view of a harbor.

Entirely new large shopping malls have been built in several downtown areas in North America in recent years. In Boston, the eighteenth-century market, known as Faneuil Hall, was transformed from a derelict area into a modern shopping center of more than 150 stores, covering 34,000 square meters (362,000 square feet). Philadelphia's Gallery at Market East, a downtown four-level shopping center of 125,000 square meters (1,348,000 square feet), has more than 200 stores. It is anchored by three large department stores and provides direct access to a subway station and multistory parking garage.

Harbor Place in Baltimore (13,000 square meters or 135,000 square feet) includes two shopping pavilions with about 75 stores integrated into a collection of waterfront museums, tourist attractions, hotels, and cultural facilities. These downtown malls attract suburban shoppers as well as out-of-town tourists because they offer unique recreation and entertainment experiences, besides shops.

A number of cities have preserved their old downtown markets. These markets feature a large number of stalls, each operated by individual merchants. They may have a high range, because they attract customers who willingly

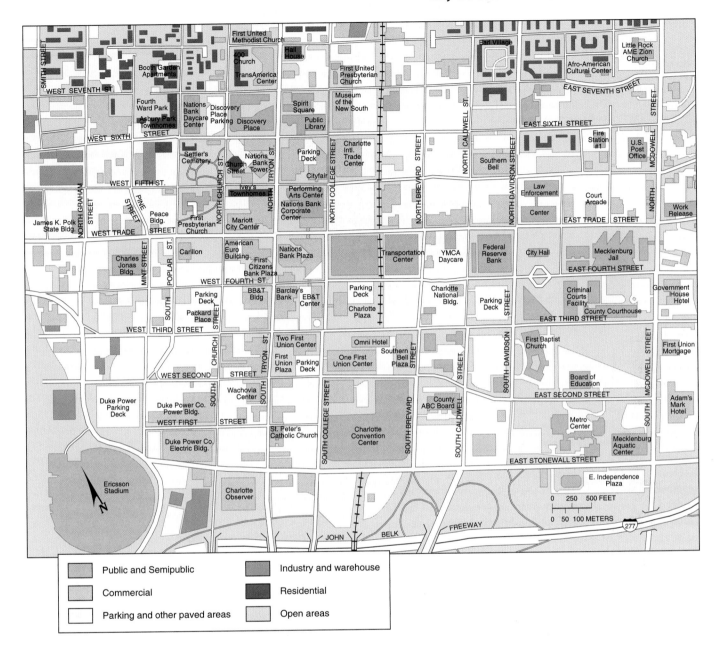

Public and Semipublic

Commercial

Parking and other paved areas

Industry and warehouse

Residential

Open areas

Figure 12–18 The CBD of Charlotte, North Carolina. Charlotte's CBD is dominated by retail and office buildings. Also clustered in the downtown area are public and semipublic buildings, like City Hall, government office buildings, and the central post office (Map adapted from Department of Geography and Earth Sciences, University of North Carolina at Charlotte; photo Jeff Greenberg/Peter Arnold, Inc.)

Rotting warehouses and derelict docks in Baltimore's Inner Harbor have been replaced with shops, restaurants, museums, and office buildings. Many of the shops and restaurants are in Harbor Place, the two two-story green-roofed buildings along the waterfront. The tall building to the right is the World Trade Center. (Chromosohm/Joe Sohm/Photo Researchers, Inc.)

travel far to find more exotic or higher-quality products. At the same time, inner-city residents may use these markets for their weekly grocery shopping.

Retail Services Serving Downtown Workers. A third type of retail activity in the center serves the many people who work in the center and shop during lunch or working hours. These businesses sell office supplies, computers, and clothing, or offer shoe repair, rapid photocopying, dry cleaning, and so on.

The CBDs in cities outside North America are more likely to contain supermarkets, bakeries, butchers, and other food stores. However, these stores may be open for limited hours in the evenings or on weekends. The 24-hour supermarket is rare outside North America, because of preferences of shopkeepers, government regulations, and long-time shopping habits.

In contrast to the other two types of retailers, shops that appeal to nearby office workers are expanding in the CBD, in part because the number of downtown office workers has increased and in part because downtown offices require more services. Patrons of downtown shops tend increasingly to be downtown employees who shop during the lunch hour. Thus, while the total volume of sales in downtown areas has been stable, the pattern of demand has changed. Large department stores have difficulty attracting their old customers, whereas smaller shops that cater to the special needs of the downtown labor force are expanding.

Many cities have attempted to revitalize retailing in the CBD and older neighborhoods. One popular method is to ban motor vehicles from busy shopping streets. By converting streets to pedestrian-only walkways, cities emulate one of the most attractive attributes of large shopping malls. Shopping streets reserved for pedestrians are widespread in Northern Europe, including the Netherlands, Germany, and Scandinavian countries.

Producer Services. Offices cluster in the center for accessibility. Despite the diffusion of modern telecommunications, many professionals still exchange information with colleagues primarily through face-to-face contact. Financial analysts discuss attractive stocks or impending corporate takeovers. Lawyers meet to settle disputes out-of-court. Offices are centrally located to facilitate rapid communication of fast-breaking news through spatial proximity. Face-to-face contact also helps to establish a relationship of trust based on shared professional values.

People in such business services as advertising, banking, finance, journalism, and law particularly depend on proximity to professional colleagues. Lawyers, for example, locate downtown to be near government offices and courts. Services such as temporary secretarial agencies and instant printers locate downtown to be near lawyers, forming a chain of interdependency that continues to draw offices to the center city.

A central location also helps businesses that employ workers from a variety of neighborhoods. Top executives may live in one neighborhood, junior executives in another, secretaries in another, and custodians in still another. Only a central location is readily accessible to all groups. Firms that need highly specialized employees are more likely to find them in the central area, perhaps currently working for another company downtown.

High Land Costs in the CBD

The center's accessibility produces extreme competition for the limited sites available. As a result, land value in the center is very high. In a rural area, a hectare of land might cost several thousand dollars. In a suburb, it might run tens of thousands of dollars. But in a large CBD like New York or London, if a hectare of land were even available, it would cost several hundred million dollars.

Tokyo's CBD probably contains Earth's most expensive land. Transactions have exceeded $250,000 per square meter ($1,000,000,000 per acre). If this page were a parcel of land in Tokyo, it would sell for $15,000.

Tokyo's high prices result from a severe shortage of buildable land. Buildings in most areas are legally restricted to less than 10 meters height (normally three stories) for fear of earthquakes, even though recent earthquakes have demonstrated that modern, well-built skyscrapers are safer than older three-story structures. Further, Japanese tax laws favor retention of agricultural land. Although it is the world's most populous urban area, Tokyo contains 36,000 hectares (90,000 acres) of farmland.

Two distinctive characteristics of the central city follow from the high land cost. First, land is used more intensively in the center than elsewhere in the city. Second, some activities are excluded from the center because of the high cost of space.

Intensive Land Use. The intensive demand for space has given the central city a three-dimensional character, pushing it vertically. Compared to other parts of the city, the central area uses more space below and above ground level.

A vast underground network exists beneath most central cities. The typical "underground city" includes multistory parking garages, loading docks for deliveries to offices and shops, and utility lines (water, sewer, phone, electric, and some heating). Subways run beneath the streets of larger central cities. Cities such as Minneapolis, Montreal, and Toronto have built extensive pedestrian passages and shops beneath the center. These underground areas segregate pedestrians from motor vehicles and shield them from harsh winter weather.

Typically, telephone, electric, and cable television wires run beneath the surface in central areas. Not enough space is available in the center for the large number of telephone poles that would be needed for such a dense network, and the wires are unsightly and hazardous.

Skyscrapers. Demand for space in the central city has also made high-rise structures economically feasible. Downtown skyscrapers give a city one of its most distinctive images and unifying symbols. Suburban houses, shopping malls, and factories look much the same from one city to another, but each city has a unique downtown skyline, resulting from the particular arrangement and architectural styles of its high-rise buildings.

The first skyscrapers were built in Chicago in the 1880s, made possible by two inventions: the elevator and iron-frame building construction. The first high-rises caused great inconvenience to neighboring structures, because they blocked light and air movement. Artificial lighting, ventilation, central heating, and air conditioning have helped solve these problems. Most North American and European cities enacted zoning ordinances early in the twentieth century in part to control the location and height of skyscrapers.

A recent building boom in CBDs of many North American cities is generating problems again. Too many skyscrapers built near each other causes traffic congestion in the narrow streets. Skyscrapers may prevent sunlight from penetrating to the sidewalks and small parcels of open space, and high winds can be channeled through the deep artificial canyons created between buildings.

Construction of high-rises may also be affected in the future by the need to conserve energy. As the Sun and natural air movement are increasingly relied upon again for light and ventilation, the old complaints about high-rises may return.

Many recently built skyscrapers are only partially occupied, because private developers overbuilt during the 1980s. In other CBDs, tenants have moved to the new skyscrapers, leaving a high percentage of vacancies in the older ones.

Skyscrapers are an interesting example of "vertical geography." The nature of an activity influences which floor it occupies in a typical high-rise. Some shop owners demand street-level space to entice the most customers and are willing to pay higher rents. Professional offices are less dependent on walk-in trade and can occupy the higher levels at lower rents. Hotel rooms and apartments may be included in the upper floors of a skyscraper to take advantage of lower noise levels and more panoramic views. Residents of the world's highest apartments, on the upper floors of the 97-story John Hancock Center in Chicago, sometimes are above the clouds. They may telephone the doorman to find out about the weather at street level.

The one large U.S. CBD without skyscrapers is Washington, D.C., where no building is allowed to be higher than the U.S. Capitol dome. Consequently, offices in downtown Washington rise no more than 13 stories. As a result, the typical Washington office building uses more horizontal space—land area—than in other cities. The city's CBD spreads over a much wider area than those in comparable cities.

Activities Excluded from the CBD

High rents and land shortage discourage two principal activities in the central area: manufacturing and residence.

Declining Manufacturing in the CBD. The typical modern industry requires a large parcel of land to spread operations among one-story buildings. Suitable land is generally available in suburbs. In central Paris, manufacturing jobs declined from more than 500,000 in the late 1960s to 400,000 around 1980 and 200,000 around 1990. The Citroën automobile factory, situated along the River Seine, barely 1 kilometer from the Eiffel Tower, has been replaced by high-rise offices and apartments. Warehouses on the southeast edge of the central area have also been replaced by office and apartment towers. Slaughterhouses in the northeastern part of central Paris have been replaced by a park and museum complex.

Port cities in North America and Europe have transformed their waterfronts from industry to commercial and recreational activities. Once, ships docked at piers that jutted out into the water, and warehouses lined the waterfront to facilitate loading and unloading of goods. But today's large oceangoing vessels are unable to maneuver in the tight, shallow waters of the old inner-city harbors. Consequently, port activities have moved to more modern facilities downstream. Cities have demolished derelict warehouses and rotting piers along their waterfronts and replaced them with new offices, shops, parks, and museums.

Once-rotting downtown waterfronts have become major tourist attractions in a number of North American cities, including Boston, Toronto, Baltimore, and San Francisco, as well as in European cities like Barcelona and London. The cities took the lead in clearing the sites and constructing new parks, docks, walkways, museums, and parking lots. They also have built large convention centers to house professional meetings and trade shows. Private developers have added hotels, restaurants, boutiques, and entertainment centers to accommodate tourists and conventioneers.

Lack of Residents in CBDs.   Few people live in U.S. CBDs, because offices and shops can afford to pay higher rents for the scarce space. Monthly rents in downtown high-rise office buildings are several dollars per square foot, depending on the city and the quality of the building. Typical apartments in most U.S. cities rent for less than $1 per square foot per month.

The shortage of affordable space is especially critical in Europe, because Europeans prefer living near the center city more than Americans do. Prohibitions on constructing new high-rises induce developers to convert older houses into offices.

Abandoned warehouses have been converted into residences in a number of CBDs. City officials welcome the additional downtown residents but are reluctant to see the permanent loss of industrial space, although realistically, manufacturers are highly unlikely to occupy multistory downtown lofts. The warehouses may require expensive alterations to meet local code standards for emergency exits and ventilation.

Many people used to live downtown. For example, the City of London—the region's one-square-mile financial center and the site of its earliest settlement—contained 128,000 residents in 1851. The residential population declined to 72,000 in 1871, to 13,000 in 1921, to about 5,000 today. The CBDs of North American cities have witnessed comparable population losses.

People have migrated from central areas for a combination of pull and push factors. First, people have been lured to suburbs, which offer larger homes with private yards and modern schools. Second, people have sought to escape from the dirt, crime, congestion, and poverty of the central city.

European CBDs

The central area is less dominated by commercial considerations in Europe than in the United States. In addition to retail and office functions, many European cities display a legacy of low-rise structures and narrow streets, built as long ago as medieval times. Today these buildings are protected from the intrusion of contemporary development. The most prominent structures may be churches and former royal palaces, situated on the most important public squares, at road junctions, or on hilltops. Parks in the center of European cities often were first laid out as private gardens for aristocratic families, and later were opened to the public.

Some European cities have tried to preserve their historic core by limiting high-rise buildings and the number of cars. During the early 1970s, several high-rise offices were built in Paris, including Europe's tallest office building (the 210-meter or 688-foot Tour Montparnasse). The public outcry over this disfigurement of the city's historic skyline was so great that officials reestablished lower height limits. In Rome, officials periodically try to ban private automobiles from the center city, because they cause pollution and congestion and damage ancient monuments.

The central area of Warsaw, Poland, represents an extreme example of preservation. The Nazis completely destroyed Warsaw's medieval core during World War II, but Poland rebuilt the area exactly as it had appeared, working from old photographs and drawings. The reconstruction of central Warsaw served as a powerful symbol of cultural tradition for the Polish people after the upheavals of World War II and the postwar Communist takeover.

Although constructing large new buildings is difficult, many shops and offices still wish to be in the center of European cities. The alternative to new construction is renovation of older buildings. However, renovation is more expensive and does not always produce enough space to meet the demand. As a result, rents are much higher in the center of European cities than in U.S. cities of comparable size.

Suburbanization of Businesses

Businesses have moved to suburbs. Manufacturers have selected peripheral locations because land costs are lower. Service providers have moved to the suburbs because most of their customers are there.

Suburbanization of Retailing

Suburban residential growth has fostered change in traditional retailing patterns. Historically, urban residents bought food and other daily necessities at small neighborhood shops in the midst of housing areas and shopped in the CBD for other products. But, since the end of World War II, downtown sales have not increased, whereas suburban sales have risen at an annual rate of 5 percent.

Downtown sales have stagnated because suburban residents who live far from the CBD won't make the long journey there. At the same time, small corner shops do not exist in the midst of newer residential suburbs. The low density of residential construction discourages people from walking to stores, and restrictive zoning practices often exclude shops from residential areas.

Shopping Malls. Instead, retailing has been increasingly concentrated in planned suburban shopping malls of varying sizes. Corner shops have been replaced by supermarkets in small shopping centers. Larger malls contain department stores and specialty shops traditionally reserved for the CBD. Generous parking lots surround the stores. Shopping malls require as many as 40 hectares (100 acres) of land and are frequently near key road junctions, like the interchange of two interstate highways (Figure 12–19).

Some shopping malls are elaborate multilevel structures exceeding 100,000 square meters (1 million square feet), with more than 100 stores arranged along covered walkways. Malls have become centers for activities in suburban areas that lack other types of community facilities. Retired people go to malls for safe, vigorous walking exercises, or they sit on the benches to watch the passing scene. Teenagers arrive after school to meet their friends. Concerts and exhibitions are frequently set up in the malls.

Large shopping malls, such as Newport Center Fashion Island, in Newport Beach, California, are surrounded by a sea of parking. (Spencer Grant/Photo Researchers, Inc.)

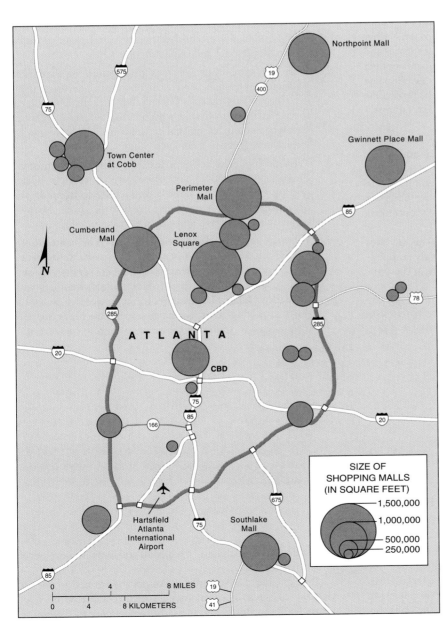

Figure 12–19 Major retail centers in Atlanta. Most shopping malls in the Atlanta metropolitan area, as elsewhere in North America, are in the suburbs, not the inner city. The optimal location for a large shopping mall is near an interchange on an interstate highway "beltway." These encircle many American cities, like I-285 around Atlanta.

A shopping mall is built by a developer, who buys the land, builds the structures, and leases space to individual merchants. Typically, a merchant's rent is a percentage of sales revenue. The key to a successful large shopping mall is the inclusion of one or more *anchors*, usually large department stores. Typically, consumers go to a mall to shop at an anchor and, while there, patronize the smaller shops. In smaller shopping centers, the anchor is frequently a supermarket or discount store.

Suburbanization of Factories and Offices

Factories and warehouses have migrated to suburbia for more space, cheaper land, and better truck access. Modern factories and warehouses demand more land for efficient operation because conveyor belts, forklift trucks,

loading docks, and machinery are spread over a single level. Suburban locations facilitate truck shipments with good access to main highways and no central city traffic congestion. Industries increasingly receive inputs and distribute products by truck.

Offices that do not require face-to-face contact increasingly are moving to suburbs where rents are much lower than in the CBD. Executives can drive on uncongested roads to their offices from their homes in nearby suburbs and park their cars without charge.

For other employees, however, suburban office locations can pose a hardship. Secretaries, custodians, and other lower-status office workers may not have cars, and public transportation may not serve the site. Other office workers might miss the stimulation and animation of a central location, particularly at lunch time.

■ Summary

Geographers do not merely observe the distribution of services; they play a major role in creating it. Shopping center developers, large department store and supermarket chains, and other retailers employ geographers to identify new sites for stores and assess the performance of existing stores. Geographers conduct statistical analyses based on the gravity model to delineate underserved market areas where new stores could be profitable, as well as to identify overserved market areas where poorly performing stores are candidates for closure.

Developers of new retail services obtain loans from banks and financial institutions to construct new stores and malls. Lending institutions want assurance that the proposed retailing has a market area with potential to generate sufficient profits to repay the loan. They employ geographers to make objective market area analyses independent of the excessively optimistic forecasts submitted by the retailer.

Many service providers make location decisions on the basis of instinct, intuition, and tradition. In an increasingly competitive market, retailers and other services that locate in the optimal location secure a critical advantage.

1. **Where did services originate?** Services are divided into five types, including two consumer services (retail and personal), two business services (producer and transportation), and public services. Services originated in settlements. Early settlements, established to serve rural areas, provided primarily personal and public services.

2. **Why are consumer services distributed in a regular pattern?** Consumer services attract customers from market areas of varying size. Geographers calculate whether a service can be profitable within a market area. In more developed countries, market areas form a regular hierarchy by size and distance from each other.

3. **Why do business services locate in large settlements?** Financial, professional, and other business services cluster disproportionately in large world cities to support the operations of major corporations. World cities also play major retail, personal, and public service functions.

4. **Why do services cluster downtown?** The central business district contains a large percentage of a settlement's retail and producer services. Producer services cluster downtown to facilitate face-to-face contact. Retailers with large thresholds or large ranges locate downtown, as well as those attracting producer service workers. Many services have moved to suburban locations in recent years.

Case Study Revisited

Romanian Policy

Virtually all residents of more developed countries are functionally tied to the services offered by an urban settlement. Not all residents of MDCs choose to work and shop in a large urban settlement, read a big-city newspaper, and watch television programs from the nearest metropolis. However, the opportu-

nity to do so is available to virtually all residents in more developed societies.

Recognizing that virtually all Americans are tied to the services of a large urban settlement, studies conducted by C. A. Doxiadis, Brian Berry, and the U.S. Department of Commerce

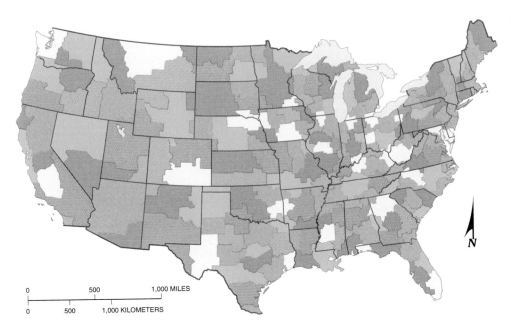

Figure 12–20 "Daily urban systems." The U.S. Department of Commerce divided the 48 contiguous states into "daily urban systems," delineated by functional ties, especially commuting, to the nearest metropolitan area. Dividing the country into "daily urban systems" demonstrates that everyone in the United States has access to services in at least one large settlement.

divided the 48 contiguous states into 171 functional regions or "daily urban systems," which are centered around commuting hubs (Figure 12–20).

In LDCs, most people are not part of a "daily urban system." They live in rural settlements that lack access to many services. Recognizing the absence of a regular hierarchy of settlements by size, as found in MDCs, the government of Romania for a number of years had a policy of improving the rank-size distribution of their settlements. Restrictions were placed on the growth of Bucharest, and people needed a permit to move there. The government built new apartments and shops in cities such as Braşov.

At the other end of the spectrum, rural Romanian settlements were designated for upgrading to small urban settlements. Government policy called for increasing the population of these small settlements from a few hundred to several thousand inhabitants. New apartments, schools, hospitals, and shops were planned, as well as electricity, paved roads, and sanitation. In this way, families living in rural areas would have greater access to the services essential for achieving a higher standard of living.

Romanian government planners formulated the policy for what they saw as logical economic geography reasons. But

Nicolae Ceaucescu, Romania's long-time leader, turned the policy into a nightmare for many people. Population was controlled in the center of Bucharest by razing entire historic neighborhoods. Eastern Orthodox churches and individual homes were demolished and replaced with apartment buildings, as well as massive squares and monuments honoring Ceaucescu and his family.

Believing that rural residents did not wholeheartedly support his programs for modernizing Romania, Ceaucescu ordered small villages to be destroyed, not expanded. Ironically, because of the isolation of many rural villages, Ceaucescu's policies could not be fully implemented. Typically, rural village leaders demolished only a small number of buildings, to show visiting government officials that they had done something to implement Ceaucescu's policy.

After Ceaucescu was overthrown in late 1989, the new Romanian government terminated the policy of indiscriminate demolition. But the challenge remained to bring electricity, paved roads, and other improvements to rural settlements, and to integrate them into a market-oriented economy.

■ Key Terms

Basic industries Industries that sell their products or services primarily to consumers outside the settlement.

Business services Services that primarily meet the needs of other businesses.

Central business district (CBD) The area of the city where retail and office activities are clustered.

Central place A market center for the exchange of services by people attracted from the surrounding area.

Central place theory A theory that explains the distribution of services, based on the fact that settlements serve as centers of market areas for services; larger settlements are fewer and farther apart than smaller settlements and provide services for a larger number of people who are willing to travel farther.

City-state An independent state comprising a city and its immediate hinterland.

Clustered rural settlement A rural settlement in which the houses and farm

buildings of each family are situated close to each other and fields surround the settlement.

Consumer services Businesses that provide services primarily to individual consumers, including retail services and personal services.

Dispersed rural settlement A rural settlement pattern characterized by isolated farms rather than clustered villages.

Economic base A community's collection of basic industries.

Enclosure movement The process of consolidating small landholdings into a smaller number of larger farms in England during the eighteenth century.

Gravity model A model that holds that the potential use of a service at a particular location is directly related to the number of people in a location and inversely related to the distance people must travel to reach the service.

Market area (or **hinterland**) The area surrounding a central place, from which people are attracted to use the place's goods and services.

Nonbasic industries Industries that sell their products primarily to consumers in the community.

Personal services Services that provide for the well-being and personal improvement of individual consumers.

Primate city The largest settlement in a country, if it has more than twice as many people as the second-ranking settlement.

Primate city rule A pattern of settlements in a country, such that the largest settlement has more than twice as many people as the second-ranking settlement.

Producer services Services that primarily help people conduct business.

Public services Services offered by the government to provide security and protection for citizens and businesses.

Range The maximum distance people are willing to travel to use a service.

Rank-size rule A pattern of settlements in a country, such that the *n*th largest settlement is 1/n the population of the largest settlement.

Retail services Services that provide goods for sale to consumers.

Service Any activity that fulfills a human want or need and returns money to those who provide it.

Settlement A permanent collection of buildings and inhabitants.

Transportation, communications, and utilities services Services that diffuse and distribute services.

Threshold The minimum number of people needed to support the service.

■ Thinking Geographically

1. Determine the economic base of your community. Consult the *U.S. Census of Manufacturing* or *County Business Patterns*. To make a rough approximation of your community's basic industries, compute the decimal fraction of the nation's population that lives in your community. It will be a small number, such as 0.0005. Then, find the total number of U.S. firms (or employees) in each industrial sector that is present in your community. Multiply these national figures by your local population fraction. Subtract the result from your community's actual number of firms (or employees) for that type of industry. If the difference is positive, you have identified one of your community's basic industries.

2. Your community's economy is expanding or contracting as a result of the performance of its basic industries. Two factors can explain the performance of your community's basic industries. One is that the sector is expanding or con-

tracting nationally. The second is that the industry is performing much better or worse in the community than in the nation as a whole. Which of the two factors better explains the performance of your community's basic industries?

3. Rural settlement patterns along the U.S. East Coast were influenced by migration during the colonial era. To what extent do distinctive rural settlement patterns elsewhere in the United States result from international or internal migration?

4. Compare the CBDs of Toronto and Detroit. What might account for differences?

5. What evidence can you find in your community of economic ties to world cities located elsewhere in North America, Western Europe, or Japan?

■ Further Readings

Archer, Clark J., and Ellen R. White. "A Service Classification of American Metropolitan Areas." *Urban Geography* 6 (1985): 122–51.

Bairoch, Paul. *Cities and Economic Development: From the Dawn of History to the Present.* Chicago: University of Chicago Press, 1988.

Bagchi-Sen, Sharmistha. "Service Employment in Large, Medium, and Small Metropolitan Areas in the United States." *Urban Geography* 18 (1997): 264–81.

Benevolo, Leonardo. *The History of the City*, 2d ed. Cambridge, MA: M.I.T. Press, 1991.

Beaverstock, Jonathan V., and Joanne Smith. "Lending Jobs to Global Cities: Skilled International Labour Migration, Investment Banking and the City of London." *Urban Studies* 33 (1996): 1377–94.

Berry, Brian J. L. *The Geography of Market Centers and Retail Distribution.* Englewood Cliffs, NJ: Prentice-Hall, 1967.

Bourne, L. S., and J. W. Simmons. *Systems of Cities.* New York: Oxford University Press, 1978.

_____; R. Sinclair, and K. Dziewonski, eds. *Urbanization and Settlement Systems: International Perspectives.* New York: Oxford University Press, 1984.

Brunn, Stanley D., and Jack L. Williams, eds. *Cities of the World: World Regional Urban Development.* New York: Harper and Row, 1983.

Chisholm, Michael. *Rural Settlement and Land Use*, 3d ed. London: Hutchinson, 1979.

Christaller, Walter. *The Central Places of Southern Germany.* Englewood Cliffs, NJ: Prentice-Hall, 1966.

Crane, Randall. "The Influence of Uncertain Job Location on Urban Form and the Journey to Work." *Journal of Urban Economics* 39 (1996): 342–56.

Daniels, P. W. *Service Industries: A Geographical Appraisal.* London: Methuen, 1986.

Davis, Kingsley. *Cities: Their Origin, Growth, and Human Impact*. San Francisco: W. H. Freeman, 1973.

Demangeon, Albert. "The Origins and Causes of Settlement Types." In *Readings in Cultural Geography*, ed. by P. L. Wagner and M. W. Mikesell. Chicago: University of Chicago Press, 1962.

Dickinson, Robert E. "Rural Settlements in the German Lands." *Annals of the Association of American Geographers* 39 (1949): 239–63.

Ford, Larry R. "Reading the Skylines of American Cities." *Geographical Review* 82 (1992): 180–200.

Frieden, Bernard J., and Lynne B. Sagalyn. *Downtown Inc.: How America Rebuilds Cities*. Cambridge, MA: M.I.T. Press, 1989.

Fuguitt, Glenn V.; David L. Brown, and Calvin L. Beale. *Rural and Small Town America*. New York: Russel Sage Foundation, 1989.

Gos, John. "The 'Magic of the Mall': An Analysis of Form, Function, and Meaning in the Contemporary Retail Built Environment." *Annals of the Association of American Geographers* 83 (1993): 18–47.

Green, Milford B. "A Geography of Institutional Stock Ownership in the United States." *Annals of the Association of American Geographers* 83 (1993): 66–89.

Hamnett, Chris. "Social Polarisation, Economic Restructuring and Welfare State Regimes." *Urban Studies* 33 (1996): 1407–30.

Harris, Chauncey D. "A Functional Classification of Cities in the United States." *Geographical Review* 33 (1943): 86–99.

Harris, Nigel. "Cities in a Global Economy: Structural Change and Policy Reactions." *Urban Studies* 34 (1997): 1693–1704.

Hauser, Philip M., and Leo F. Schnore, eds. *The Study of Urbanization*. New York: John Wiley, 1965.

Jacobs, Jane. *The Economy of Cities*. New York: Vintage Books, 1970.

Jones, Kenneth G., and James W. Simmons. *The Retail Environment*. London and New York: Routledge, 1990.

King, Leslie J. *Central Place Theory*. Beverly Hills, CA: Sage Publications, 1984.

Kirn, Thomas J. "Growth and Change in the Service Sector of the U.S.: A Spatial Perspective." *Annals of the Association of American Geographers* 77 (1987): 353–72.

Krider, Robert E., and Charles B. Weinberg. "Spatial Competition and Bounded Rationality: Retailing at the Edge of Chaos." *Geographical Analysis* 29 (1997): 16–34.

Leyshorn, Andrew, and Nigel Thrift. "Spatial Financial Flows and the Growth of the Modern City." *International Social Science Journal* 151 (1997): 41–54.

Longcore, Travis R. and Peter W. Rees. "Information Technology and Downtown Restructuring: The Case of New York City's Financial District." *Urban Geography* 17 (1996): 354–72.

Longley, Paul A.; Michael Batty, and John Shepherd. "The Size, Shape, and Dimensions of Urban Settlements." *Transactions of the Institute of British Geographers New Series* 16 (1991): 75–94.

Lösch, August. *The Economics of Location*. New Haven, CT: Yale University Press, 1954.

Lowe, John M. and Ashish Sen. "Gravity Model Applications in Health Planning: Analysis of an Urban Hospital Market." *Journal of Regional Science* 36 (1996): 437–61.

Marshall, J. N. "Services in a Postindustrial Economy." *Environment and Planning A* 17 (1985): 1155–67.

Marshall, John U. "Beyond the Rank-Size Rule: A New Descriptive Model of City Sizes." *Urban Geography* 18 (1997): 36–55

Michie, R.C. "Friend or Foe? Information Technology and the London Stock Exchange Since 1700." *Journal of Historical Geography* 23 (1997): 304–26.

Mitchelson, Ronald L., and James O. Wheeler. "The Flow of Information in a Global Economy: The Role of the American Urban System in 1990." *Annals of the Association of American Geographers* 84 (1994): 87–107.

Morrill, Richard. "The Structure of Shopping in a Metropolis." *Urban Geography* 8 (1987): 97–128.

Mosher, Anne E.; Barry D. Keim, and Susan A. Franques. "Downtown Dynamics." *Geographical Review* 85 (1995): 497–517.

Mun, Se-Il. "Transport Network and System of Cities." *Journal of Urban Economics* 42 (1997): 205–21.

Noyelle, T.J., and T.M. Stanback, Jr. *The Economic Transformation of American Cities*. Totowa, NJ: Rowman & Allanheld, 1984.

Ó hUallacháin, Breandan, and Neil Reid. "The Location and Growth of Business and Professional Services in American Metropolitan Areas, 1976–1986." *Annals of the Association of American Geographers* 81 (1991): 254–70.

O'Kelly, M. E. "Multipurpose Shopping Trips and the Size of Retail Facilities." *Annals of the Association of American Geographers* 73 (1983): 231–39.

Riley, Ray. "Retail Change in Post-Communist Poland." *Geography* 82 (1997): 1–37.

Scofield, Edna. "The Origin of Settlement Patterns in Rural New England." *Geographical Review* 28 (1938): 652–63.

Scott, Peter. *Geography and Retailing*. London: Hutchinson University Press, 1970.

Trewartha, Glen T. "Types of Rural Settlements in North America." *Geographical Review* 36 (1946): 568–96.

Van Marrewijk, Charles; Joachim Stibora, and Jean-Marie Viaene. "Producer Services, Comparative Advantage, and International Trade Patterns." *Journal of International Economics* 42 (1997): 195–220.

Wheeler, James O., and Ronald L. Mitchelson. "Information Flows among Major Metropolitan Areas in the United States." *Annals of the Association of American Geographers* 79 (1989): 523–43.

Wood, Peter. "Business Services, the Management of Change and Regional Development in the U.K.: A Corporate Client Perspective." *Transactions of the Institute of British Geographers New Series* 21 (1996): 649–65.

Also consult these journals: *Journal of Historical Geography*; *Journal of Regional Science*; *Journal of Rural Studies*; *Journal of Urban Economics*; *Urban Geography*.

CHAPTER 13

Urban Patterns

Suppose as a geography class assignment you were dropped off on a street corner in a very large city and told to meet your instructor and classmates in one hour at City Hall. How would you find it? In a small town you could simply ask for directions, but in an unfamiliar neighborhood of a large city would you hesitate to ask strangers?

Your destination is probably downtown, because that's where public services such as City Hall cluster. Which direction is downtown? The skyscrapers far in the distance are probably a clue, and house numbers on major streets get lower as you head toward downtown.

In a small town, everything is within easy walking distance, but in a large city your destination is too far to walk. How would you get there without a car? Hitch-hiking is too dangerous, and you don't have enough money to hire a taxi. What about the bus? Where does the bus stop? What route does it follow? How much is the fare? Do you have the exact change, as required on most big-city buses?

Once on the bus, you sit down next to another passenger. Is your neighbor of the same ethnicity as you? In fact, are you the only person on the bus of your ethnicity? Have you been in other large groups where you were the only person of your ethnicity? Do the other passengers smile at you and chat, or do they mind their own business?

A large city is stimulating and agitating, entertaining and frightening, welcoming and cold. A city has something for everyone, but a lot of those things are for people who are different from you. Urban geography helps to sort out the complexities of familiar and unfamiliar patterns in urban areas.

Children from Puerto Rico sitting in stairwell, Brooklyn, New York.
(Katherine McGlynn/The Image Works)

1. Where have urban areas grown?

2. Where are people distributed within urban areas?

3. Why do inner cities have distinctive problems?

4. Why do suburbs have distinctive problems?

Case Study

Two Families in New Jersey

Ruth Merritt lives in the city of Camden, New Jersey. She is 24, a single parent with three children (ages 7, 2, and 1). Her income, derived from the community's program of aid to families with dependent children, is $235 per month. That works out to $2,820 a year.

The Merritt family lives in a four-room apartment in a row house that was divided some years ago into six dwelling units. The apartment has generally adequate plumbing and kitchen facilities, but the residents sometimes see rats in the building. The rent is $75 per month, plus an average of $50 per month for electricity and other utilities. Ruth Merritt receives food stamps, but her monthly expenses for food, clothing, and shelter exceed her income. In cold weather, she must sometimes reduce the food budget to pay for heat.

Just 10 kilometers away, east of Camden, the Johnson family lives in Cherry Hill, New Jersey. William Johnson is a lawyer. He commutes to downtown Philadelphia, across the Delaware River from Camden. Diane Johnson works for a nonprofit organization with offices in the suburban community where they live. Their two children attend a recently built school in the community.

The Johnson family's dwelling is a detached house with three bedrooms, a living room, dining room, family room, and kitchen. The attached garage contains two cars, one for each parent to get to work. The half-acre lawn surrounding the house provides ample space for the children to play. The Johnsons bought their house 5 years ago for $150,000. The monthly payments for mortgage and utilities are $2,000, but the family's combined annual income of $100,000 is more than adequate to pay the housing costs.

The Merritt and Johnson households illustrate the contrasts that exist today in U.S. urban areas. As you have seen throughout this book, dramatic differences in material standards exist around the world. However, the picture drawn here is based on families living in the same urban area, only a few kilometers apart.

Were these examples taken from an urban area elsewhere in the world, the spatial patterns might be reversed. In most of the world, the higher-status Johnsons would live near the center of the city, while the lower-status Merritts would live in the suburbs.

When you stand at the corner of Fifth Avenue and 34th Street in New York City, staring up at the Empire State Building, you know that you are in a city. When you are standing in an Iowa cornfield, you have no doubt that you are in the country. Geographers help explain what makes city and countryside different.

Urban geographers are interested in the *where* question at two scales. First, geographers examine the global distribution of urban settlements. Having a high percentage of people living in urban areas is a distinctive feature of life in more developed countries, a consequence of the shift from agricultural to manufacturing and later services economy.

Geographers are also interested in where people and activities are distributed within urban areas. Models have been developed to explain *why* differences occur within urban areas. The major physical, social, and economic contrasts are between inner-city and suburban areas.

We all experience the interplay between *globalization* and *local diversity* of urban settlements. If you were transported to the downtown of another city, you might be able to recognize the city from its skyline. Many downtowns have a collection of high-rise buildings, towers, and landmarks that are identifiable even to people who have never visited them.

On the other hand, if you were transported to a suburban residential neighborhood, you would have difficulty identifying the urban area. Suburban houses, streets, schools, and shopping centers look very much alike from one American city to another.

In more developed countries, people are increasingly likely to live in suburbs. This changing structure of our cities is a response to conflicting desires. We wish to spread across the landscape to avoid urban problems, but at the same time, we want convenient access to the city's jobs, shops, culture, and recreation.

This chapter examines the causes and consequences of today's evolving urban patterns. Although different internal structures characterize urban areas in the United States and elsewhere, the problems arising from current spatial trends are quite similar. Geographers describe where different types of people live, and they try to explain the reasons for the observed patterns.

Key Issue 1

Where Have Urban Areas Grown?

- Urbanization
- Defining urban settlements

As recently as 1800, only 3 percent of Earth's population lived in cities, and only a handful of cities had more than 100,000 inhabitants. Two centuries later, nearly half of the world's people live in cities, and more than 100 cities have at least 2 million inhabitants. This rapid growth has made it difficult to define the extent of cities.

Urbanization

The process by which the population of cities grows, known as **urbanization**, has two dimensions: an increase in the *number* of people living in cities and an increase in the *percentage* of people living in cities. The distinction between the two factors is important, because they occur for different reasons and have different global distributions.

Increasing Percentage of People in Cities

In 1800, only 3 percent of the world's population was urban. The percentage increased to 6 percent in 1850, 14 percent in 1900, 30 percent in 1950, and 43 percent in 1997. The United Nations forecasts that within a decade the population of urban settlements will exceed that of rural settlements for the first time in human history.

A large percentage of people living in urban areas is a measure of a country's level of development. In more developed countries, about three-fourths of the people live in urban areas, compared to about one-third in less developed countries (Figure 13–1). The major exception to the global pattern is Latin America, where the urban percentage is closer to the level of more developed countries.

The higher percentage of urban residents in MDCs is a consequence of changes in economic structure during the past two centuries, first the industrial revolution in the nineteenth century and then the growth of services in the twentieth. The world map of percent urban looks very much like the world map of percent of workers in services (refer to Figure 12–1).

The percentage of urban dwellers is high in more developed countries because over the past 200 years, rural residents have migrated from the countryside to work in the factories and services that are concentrated in cities. The need for fewer farm workers has pushed people out of rural areas, and rising employment opportunities in manufacturing and services have lured them into urban areas. Because everyone resides either in an urban settlement or a rural settlement, an increase in the percentage living in urban areas has produced a corresponding decrease in the percentage living in rural areas.

In more developed countries, the process of urbanization that began around 1800 has largely ended, because the percentage living in urban areas simply cannot increase much more. Nearly everyone interested in migrating from farms to cities has already done so, leaving those who choose to live in rural areas. We can now speak of more developed countries as being fully urbanized, because the percentage of urban residents is so high.

The percentage living in cities has begun to rise in recent years in less developed countries because of migration of rural residents to the cities in search of jobs in manufacturing or services. As in more developed countries, people in less developed countries are pushed off the farms by declining opportunities. However, urban jobs are by no means assured in LDCs experiencing rapid overall population growth.

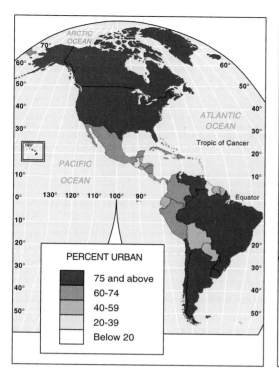

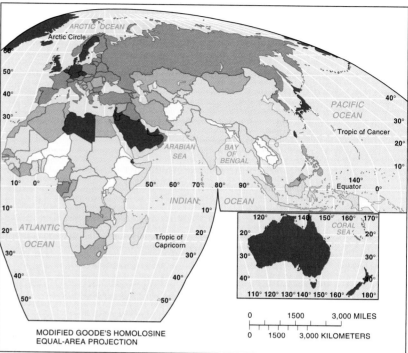

Figure 13–1 Urban percentages worldwide. More developed countries have a higher percentage of people living in urban areas.

Increasing Number of People in Cities

More developed countries have a higher percentage of urban residents, but less developed countries have more of the large urban settlements (Figure 13–2). According to the U.S. Bureau of the Census, seven of the ten most populous cities are currently in LDCs: Mexico City, São Paulo, Seoul, Bombay, Calcutta, Rio de Janeiro, and Buenos Aires. New York and Japan's two large urban areas—Tokyo-Yokohama and Osaka-Kobe-Kyoto—are the only ones in more developed countries among the ten most populous.

The United Nations also estimates that seven of the ten largest urban areas are in less developed countries, but disagrees with the U.S. Bureau of the Census on the specific list. Among the largest urban areas in LDCs, the United Nations includes Shanghai and Beijing instead of Rio de Janeiro and Buenos Aires, and in MDCs substitutes Los Angeles for Osaka.

The United Nations and U.S. Bureau of the Census also disagree sharply on the population of some urban areas, especially in Latin America and China. The Census Bureau estimates the population of Mexico City at 28 million, the United Nations "only" 16 million. The Census Bureau estimates 25 million in São Paulo, the United Nations 16 million. On the other hand, the United Nations estimates 15 million in Shanghai and 12 million in Beijing, much higher than the Census Bureau estimates of 8 million and 6 million, respectively.

How can two respected organizations disagree by 12 million on the population of Mexico City, despite using similar estimation techniques? That the Census Bureau and United Nations arrive at very different populations demonstrates the difficulty in counting people in rapidly growing less developed countries. Where does the urban area end and the rural area begin? How can the number of inhabitants be counted in a densely packed apartment building?

That urban areas in LDCs dominate both lists of largest urban areas is remarkable, because urban growth historically has resulted from diffusion of the industrial revolution. In 1800, as the industrial revolution began to diffuse from Great Britain to Western Europe, only three of the world's ten most populous cities were in Europe—London, Paris, and Naples—and the remainder were in Asia. But by 1900, nine of the world's ten most populous cities were in countries that had rapidly industrialized during the nineteenth century.

In 1900, London, capital of the world's first industrial state, was by far the world's largest city. The world's ten largest cities in 1900 included five others in Europe (Paris, Berlin, Vienna, St. Petersburg, Manchester) and three in the United States (New York, Chicago, Philadelphia). Tokyo was the only top-ten city then in a preindustrial country. As recently as 1950, seven of the ten largest cities in the world remained clustered in MDCs that had industrialized.

The rapid growth of cities in the LDCs is a reversal of the historical trend in Western Europe and North America created by the industrial revolution, and not a measure of an improved level of development. Migration from the countryside is fueling some of the increase in population in urban areas of LDCs, even though job opportunities may not be available.

São Paulo, Brazil, is one of the world's ten largest cities. High-rise offices and apartment buildings extend for several kilometers around the central business district. (Will & Deni McIntyre/Photo Researchers, Inc.)

Most of the growth in the population of urban areas in less developed countries, however, results from high natural increase rates. In India, for example, where total population increases by about 20 million annually, natural increase contributes about 5 million to the annual growth of urban areas. Urban areas account for 4 million of sub-Saharan Africa's annual natural increase of 16 million and 4 million of China's annual natural increase of 12 million.

Defining Urban Settlements

Defining where urban areas end and rural areas begin is difficult. Geographers and other social scientists have formulated definitions that distinguish between urban and rural areas according to social and physical factors.

Social Differences Between Urban and Rural Settlements

A century ago, social scientists observed striking differences between urban and rural residents. During the 1930s, Louis Wirth argued that an urban dweller follows a different way of life from a rural dweller. Thus, Wirth defined a city as a permanent settlement that has three characteristics: large size, high population density, and socially heterogeneous people. These characteristics produced differences in the social behavior of urban and rural residents.

Large Size. If you live in a rural settlement, you know most of the other inhabitants, and may even be related to many of them. The people with whom you relax are probably the same ones you see in local shops and at church.

In contrast, if you live in an urban settlement, you can know only a small percentage of the other residents. You meet most of them in specific roles: your supervisor, your lawyer, your supermarket cashier, your electrician. Most of these relationships are contractual: you are paid wages according to a contract, and you pay others for goods and services. Consequently, the large size of an urban settlement influences different social relationships from those found in rural settlements.

The large size and high density of very large cities influences social relationships. People compete to occupy very little amounts of space in such places as the Tokyo subway. At the same time, to prevent chaos the people of Tokyo cooperate by standing in line for their turn to be pushed into a crowded subway car. (Kim Newton/Woodfin Camp & Associates)

Figure 13–2 Cities having a population of 2 million or more. The proportion of urban dwellers is greater in more developed countries. However, the largest urban areas now are mostly in less developed countries. Rapid city growth in LDCs reflects increasing overall population, plus migration from rural areas.

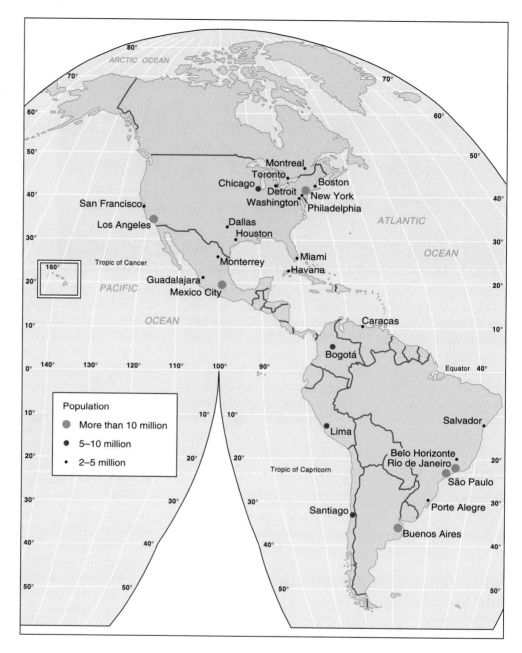

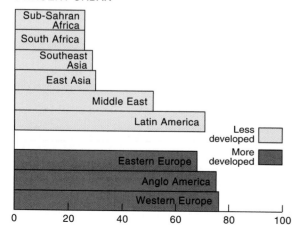

PERCENT URBAN

High Density. According to Louis Wirth, high density also produces social consequences for urban residents. The only way that a large number of people can be supported in a small area is through specialization. Each person in an urban settlement plays a special role or performs a specific task to allow the complex urban system to function smoothly.

At the same time, high density also encourages people to compete for survival in limited space. Social groups compete to occupy the same territory, and the stronger group dominates. This behavior distinguishes an urban settlement from a rural one.

Social Heterogeneity. The larger the settlement, the greater the variety of people. A person has greater freedom in an urban settlement than in a rural settlement to pursue an unusual profession, sexual orientation, or cul-

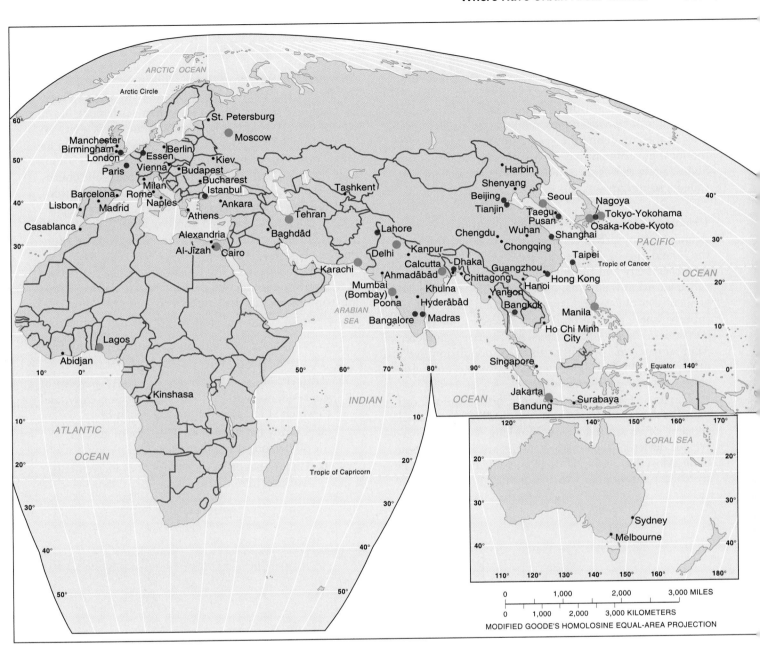

tural interest. In a rural settlement, unusual actions might be noticed and scorned, but urban residents are more tolerant of diverse social behavior. Regardless of values and preferences, in a large urban settlement individuals can find people with similar interests.

Yet despite the freedom and independence of an urban settlement, people may also feel lonely and isolated. Residents of a crowded urban settlement often feel that they are surrounded by people who are indifferent and reserved.

Wirth's three-part distinction between urban and rural settlements may still apply in LDCs. But in more developed societies, social distinctions between urban and rural residents have blurred. According to Wirth's definition, nearly everyone in a developed society now is urban. More than 95 percent of workers in developed societies hold "urban" types of jobs. Nearly universal own-

ership of automobiles, telephones, televisions, and other modern communications and transportation also has reduced the differences between urban and rural lifestyles in more developed societies. Almost regardless of where you live in the United States, you have access to urban jobs, services, culture, and recreation. So geographers look for physical definitions to distinguish between urban and rural areas.

Physical Definitions of Urban Settlements

Historically, physical differences between urban and rural settlements were easy to define, because cities were surrounded by walls. The removal of walls and the rapid territorial expansion of cities have blurred the traditional physical differences. Urban settlements today can be physically defined in three ways: by legal boundary, as continuously built-up area, and as a functional area.

Legal Definition of a City. The term *city* defines an urban settlement that has been legally incorporated into an independent, self-governing unit. Virtually all countries have a local government system that recognizes cities as legal entities with fixed boundaries. A city has locally elected officials, the ability to raise taxes, and responsibility for providing essential services. The boundaries of the city define the geographic area within which the local government has legal authority. In the United States, a city that is surrounded by suburbs is sometimes called a *central city*.

Urbanized Area. With the rapid growth of urban settlements, many urban residents live in suburbs, beyond the boundaries of the central city. In the United States, the central city and the surrounding built-up suburbs is called an **urbanized area**. More precisely, an urbanized area consists of a central city plus its contiguous built-up suburbs where population density exceeds 1,000 persons per square mile (400 persons per square kilometer). Approximately 60 percent of the U.S. population live in urban areas, divided about equally between central cities and surrounding jurisdictions.

The U.S. Census Bureau and United Nations estimates at the beginning of this section were based on the concept of urbanized areas. Working with urbanized areas is difficult because few statistics are available about them. Most data in the United States and other countries are collected for cities, counties, and other local government units, but urbanized areas do not correspond to government boundaries.

Metropolitan Statistical Area. The urbanized area also has limited applicability because it does not accurately reflect the full influence that an urban settlement has in contemporary society. The area of influence of a city extends beyond legal boundaries and adjacent built-up jurisdictions. For example, commuters may travel a long distance to work and shop in the city or built-up suburbs. People in a wide area watch the city's television stations, read the city's newspapers, and support the city's sports teams. Therefore, we need another definition of urban settlement to account for its more extensive zone of influence (Figure 13.3).

The U.S. Bureau of the Census has created a method of measuring the functional area of a city, known as the **metropolitan statistical area (MSA)**. An MSA includes the following:

1. A central city with a population of at least 50,000.
2. The county within which the city is located.
3. Adjacent counties in which at least 15 percent of the residents work in the central city's county, and to which at least two of these tests apply:
   a. County has a residential density of at least 60 persons per square mile.
   b. County has at least 65 percent of its residents working in nonfarm jobs.

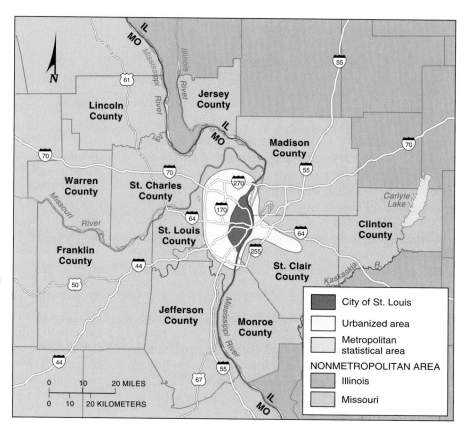

Figure 13–3 St. Louis city, urbanized area, and metropolitan statistical area. Surrounding the city of St. Louis is an urbanized area that spreads westward into St. Louis County and eastward across the Mississippi River into Illinois. The St. Louis metropolitan statistical area includes six Missouri counties and five in Illinois, as well as the city of St. Louis. The situation of St. Louis makes it a diversified trade center, for it is at the confluence of the Missouri and Mississippi rivers and several federal highways.

c. County had a population growth rate during the 1970s of at least 20 percent.

d. County has at least 10 percent of its population, or at least 5,000 persons, living in an urbanized area.

Studies of metropolitan areas in the United States are usually based on information about MSAs. The MSAs are widely used because many statistics are published for counties, the basic MSA building block (Figure 13–3). Older studies may refer to SMSAs, or standard metropolitan statistical areas, which the census used before 1980 to designate metropolitan areas in a manner fairly similar to MSAs.

An MSA is not the perfect tool for measuring the functional area of a city. One problem is that some MSAs include extensive land area that is not urban. For example, Great Smokies National Park is partly in the Knoxville, Tennessee, MSA, and Sequoia National Park is in the Visalia-Tulare-Porterville, California, MSA. The MSAs comprise some 20 percent of total U.S. land area, compared to only 2 percent for urbanized areas. The urbanized area typically occupies only 10 percent of an MSA land area, but contains over 75 percent of its population.

Consolidated Metropolitan Statistical Areas. Some adjacent MSAs overlap. A county between two central cities may send a large number of commuters to jobs in each.

In the northeastern United States, large metropolitan areas are so close together that they now form one continuous urban complex, extending from north of Boston to south of Washington, D.C. In 1961, the geographer Jean Gottmann named this region "Megalopolis," a Greek word meaning great city; others have called it the "Boswash" corridor (Figure 13–4).

Other continuous urban complexes exist in the United States: the southern Great Lakes between Chicago and Milwaukee on the west and Pittsburgh on the east, and southern California from Los Angeles to Tijuana. Among important examples in other MDCs are the German Ruhr (including the cities of Dortmund, Düsseldorf, and Essen), Randstad in the Netherlands (including the cities of Amsterdam, the Hague, and Rotterdam), and Japan's Tokaido (including the cities of Tokyo and Yokohama).

Within Megalopolis, the downtown areas of individual cities like Baltimore, New York, and Philadelphia retain distinctive identities, and the urban areas are visibly separated from each other by open space used as parks, military bases, and dairy or truck farms. But at the periphery of the urban areas, the boundaries overlap. Washingtonians attend major league baseball games in downtown Baltimore, and Baltimoreans attend major league hockey and basketball games in downtown Washington. Once considered two separate areas, Washington and Baltimore were combined into a single metropolitan statistical area after the 1990 census.

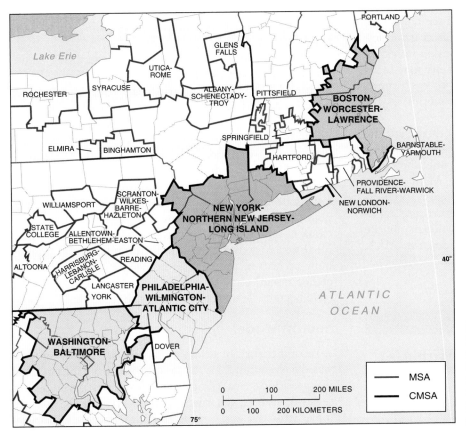

Figure 13–4 Megalopolis. Also known as the *Boswash* corridor, Megalopolis extends more than 700 kilometers (440 miles) from Boston on the northeast to Washington, D.C., on the southwest. Megalopolis contains more than one-fourth of the U.S. population on less than 2 percent of the country's total land area.

Two adjacent MSAs with overlapping commuting patterns are combined into a **consolidated metropolitan statistical area (CMSA)**. Examples of CMSAs are New York–Northern New Jersey–Long Island, Los Angeles–Anaheim–Riverside, and Chicago–Gary–Lake County. Within a CMSA, an MSA that exceeds 1 million population may be classified as a **primary metropolitan statistical area (PMSA)**. Again, the building block is the county.

A PMSA consists of at least one county that has all of the following characteristics:

1. A population of more than 100,000.
2. At least 60 percent of the residents work in non-farm jobs.
3. Less than 50 percent of the county's workers commute to jobs outside the county.

Importance of MSAs. Difficulties involved in designating MSAs, CMSAs, and PMSAs can be seen in the southeastern corner of Wisconsin. Kenosha and Racine counties each have a city with more than 50,000 people and therefore qualify as separate MSAs. However, these two metropolitan counties are sandwiched between the much larger Chicago and Milwaukee metropolitan areas. Northern Racine County is within one-half hour of downtown Milwaukee, and southern Kenosha County is within an hour of downtown Chicago. The U.S. Bureau of the Census decided to call Kenosha County a separate MSA and to designate Racine County as a PMSA within the Milwaukee-Racine CMSA.

Why do officials in Kenosha and Racine counties care whether they are a separate MSA or merely part of Milwaukee or Chicago's CMSA? The reason is money: several types of federal assistance are allocated to MSAs. The separate designation of Kenosha County may bring more funds to be used at the discretion of the county, rather than to be shared with other counties in the Milwaukee or Chicago region.

Local officials also wish to preserve their separate MSA designation because it increases the county's visibility. The U.S. Bureau of the Census publishes considerable information at the MSA level. Many private companies also compile information and make initial investment decisions at the MSA level. Advertising agencies select MSAs as test markets, and developers choose MSAs for new shopping center sites. As a separate MSA, a county like Kenosha increases its likelihood of being selected by an investor.

Key Issue 2

Where Are People Distributed Within Urban Areas?

- Three models of urban structure
- Use of the models outside North America

People are not distributed randomly within an urban area. They concentrate in particular neighborhoods, depending on their social characteristics. Geographers describe where people with particular characteristics are likely to live within an urban area, and they offer explanations for why these patterns occur.

Three Models of Urban Structure

Sociologists, economists, and geographers have developed three models to help explain where different types of people tend to live in an urban area: the concentric zone, sector, and multiple nuclei models. The three models describing the internal social structure of cities were all developed in Chicago, a city on a prairie.

Except for Lake Michigan to the east, few physical features have interrupted the region's growth. Chicago includes a CBD, known as the Loop because elevated railway lines loop around it. Surrounding the Loop are residential suburbs to the south, west, and north. The three models were later applied to cities elsewhere in the United States and in other countries.

Concentric Zone Model

The concentric zone model was the first to explain the distribution of different social groups within urban areas. It was created in 1923 by sociologist E. W. Burgess. According to the **concentric zone model**, a city grows outward from a central area in a series of concentric rings, like the growth rings of a tree. The precise size and width of the rings vary from one city to another, but the same basic types of rings appear in all cities in the same order (Figure 13–5).

The innermost of the five zones is the CBD, where nonresidential activities are concentrated. The CBD is surrounded by the second ring, the zone in transition, which contains industry and poorer-quality housing. Immigrants to the city first live in this zone in small dwelling units, frequently created by subdividing larger houses into apartments. The zone also contains rooming houses for single individuals.

The third ring, the zone of working-class homes, contains modest older houses occupied by stable, working-class families. The fourth zone has newer and more spacious houses for middle-class families. Finally, Burgess identified a commuters' zone, beyond the continuous built-up area of the city. Some people who work in the center nonetheless choose to live in small villages that have become dormitory towns for commuters.

Sector Model

A second theory of urban structure, the **sector model**, was developed in 1939 by land economist Homer Hoyt (Figure 13–6). According to Hoyt, the city develops in a series of sectors, not rings. Certain areas of the city are more attractive for various activities, originally because

1 Central business district
2 Zone of transition
3 Zone of independent workers' homes
4 Zone of better residences
5 Commuter's zone

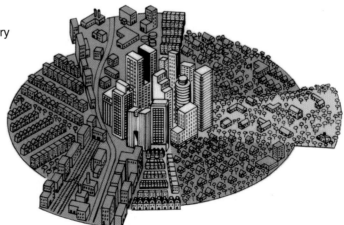

Figure 13–5 Concentric zone model. According to the model, a city grows in a series of rings that surround the central business district (CBD).

of an environmental factor or even by mere chance. As a city grows, activities expand outward in a wedge, or sector, from the center. Once a district with high-class housing is established, the most expensive new housing is built on the outer edge of that district, farther out from the center. The best housing is therefore found in a corridor extending from downtown to the outer edge of the city. Industrial and retailing activities develop in other sectors, usually along good transportation lines.

To some extent, the sector model is a refinement of the concentric zone model rather than a radical restatement. Hoyt mapped the highest-rent areas for a number of U.S. cities at different times and showed that the highest social class district usually remained in the same sector, although it moved farther out along that sector over time.

Hoyt and Burgess both claimed that social patterns in Chicago supported their model. According to Burgess, Chicago's CBD was surrounded by a series of rings, broken only by Lake Michigan on the east. Hoyt argued that the best housing in Chicago developed north from the CBD along Lake Michigan, while industry located along major rail lines and roads to the south, southwest, and northwest.

Multiple Nuclei Model

Geographers C. D. Harris and E. L. Ullman developed the multiple nuclei model in 1945. According to the **multiple nuclei model**, a city is a complex structure that includes more than one center around which activities revolve. Examples of these nodes include a port, neighbor-

1. Central business district
2. Transportation and industry
3. Low-class residential
4. Middle-class residential
5. High-class residential

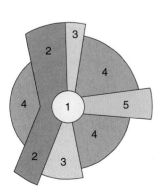

Figure 13–6 Sector model. According to the model, a city grows in a series of wedges or corridors, which extend out from the CBD.

Figure 13–7 Multiple nuclei model. According to the model, a city consists of a collection of individual nodes, or centers, around which different types of people and activities cluster.

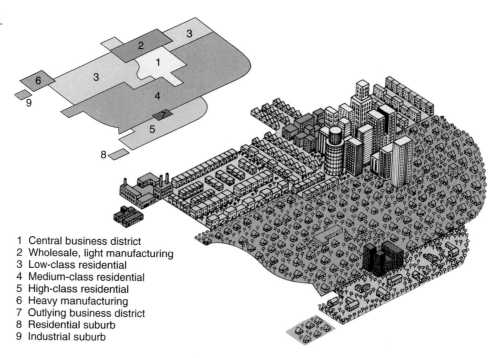

1 Central business district
2 Wholesale, light manufacturing
3 Low-class residential
4 Medium-class residential
5 High-class residential
6 Heavy manufacturing
7 Outlying business district
8 Residential suburb
9 Industrial suburb

hood business center, university, airport, and park (Figure 13–7).

The multiple nuclei theory states that some activities are attracted to particular nodes while others try to avoid them. For example, a university node may attract well-educated residents, pizzerias, and bookstores, whereas an airport may attract hotels and warehouses. On the other hand, incompatible land use activities will avoid clustering in the same locations. Heavy industry and high-class housing, for example, rarely exist in the same neighborhood.

Geographic Applications of the Models

The three models help us understand where people with different social characteristics tend to live within an urban area. They can also help to explain why certain types of people tend to live in particular places.

Effective use of the models depends on the availability of data at the scale of individual neighborhoods. In the United States and many other countries, that information comes from a national census. Urban areas in the United States are divided into **census tracts**, which contain approximately 5,000 residents and correspond where possible to neighborhood boundaries. Every decade, the U.S. Bureau of the Census publishes data summarizing the characteristics of the residents living in each tract. Examples of information the bureau publishes include the number of nonwhites, the median income of all families, and the percentage of adults who finished high school.

Social Area Analysis. The spatial distribution of any of these social characteristics can be plotted on a map of the community's census tracts. Computers have become in-

valuable in this task, because they permit rapid creation of maps and storage of voluminous data about each census tract. Social scientists can compare the distributions of characteristics and create an overall picture of where various types of people tend to live. This kind of study is known as *social area analysis*.

None of the three models by itself completely explains why different types of people live in distinctive parts of the city. Critics point out that the models are too simple and fail to consider the variety of reasons that lead people to select particular residential locations. Because the three models are all based on conditions that existed in U.S. cities between the two world wars, critics also question their relevance to contemporary urban patterns in the United States or in other countries.

But if the models are combined, rather than considered independently, they do help geographers explain where different types of people live in a city. People tend to reside in certain locations depending on their particular personal characteristics. This does not mean that everyone with the same characteristics must live in the same neighborhood, but the models say that most people prefer to live near others having similar characteristics:

- Consider two families with the same income and ethnic background. One family includes married parents with young children and the other is an unmarried couple with no children. The concentric zone model suggests that the married household is much more likely to live in an outer ring and the unmarried one in an inner ring (Figure 13–8).

- The sector theory suggests that, given two families of the same age with the same number of children, the family with the higher income will not live in

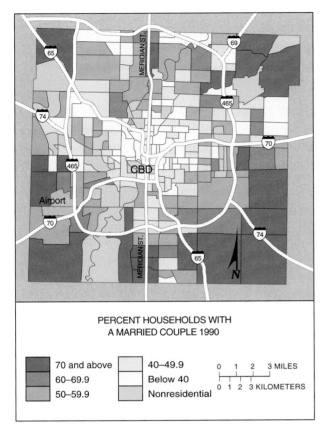

Figure 13–8 Example of concentric zone model in Indianapolis, the distribution of married couples. The percentage of households with a married couple is less near the CBD and greater in the outer rings of the city.

PERCENT HOUSEHOLDS WITH
A MARRIED COUPLE 1990

70 and above   40–49.9
60–69.9   Below 40
50–59.9   Nonresidential

0 1 2 3 MILES
0 1 2 3 KILOMETERS

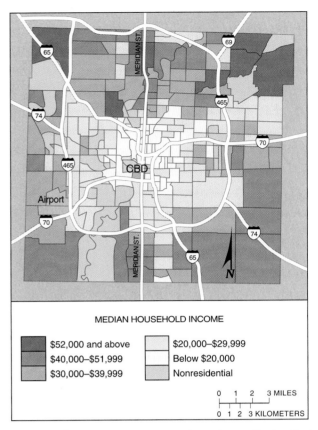

Figure 13–9 Example of sector model in Indianapolis, the distribution of high-income households. The median household income is the highest in a sector to the north, which extends beyond the city limits to the adjacent county.

MEDIAN HOUSEHOLD INCOME

$52,000 and above $20,000–$29,999
$40,000–$51,999   Below $20,000
$30,000–$39,999   Nonresidential

0 1 2 3 MILES
0 1 2 3 KILOMETERS

the same sector of the city as the poorer one (Figure 13–9).

- The multiple nuclei theory suggests that people with the same ethnic or racial background are likely to live near each other (Figure 13–10).

Putting the three models together, we can identify, for example, the neighborhood in which a childless, high-income, Asian-American family is most likely to live.

Use of the Models Outside North America

The three models may describe the spatial distribution of social classes in the United States, but American urban areas differ from those elsewhere in the world. These differences do not invalidate the models. But they do point out that social groups in other countries may not have the same reasons for selecting particular neighborhoods within their cities.

European Cities

As in the United States, wealthier people in European cities cluster along a sector extending out from the CBD. In Paris, for example, the rich moved to the southwestern hills to be near the royal palace (the Louvre beginning in

the twelfth century, and the Palace of Versailles from the sixteenth century until the French Revolution in 1789). The preference of the wealthy to cluster in the southwest was reinforced in the nineteenth century during the industrial revolution. Factories were built to the south, east, and north along the Seine and Marne River valleys, but relatively few were built on the southwestern hills (Figure 13–11, p. 457). Similar high-class sectors developed in other European cities, typically on higher elevation and near royal palaces.

However, in contrast to most U.S. cities, wealthy Europeans still live in the inner rings of the high-class sector, not just in the suburbs. A central location provides proximity to the region's best shops, restaurants, cafes, and cultural facilities. Wealthy people are also attracted by the opportunity to occupy elegant residences in carefully restored, beautiful old buildings.

By living in high-density, centrally located townhouses and apartments, wealthy people in Europe do not have large private yards and must go to public parks for open space. To meet the desire for large tracts of privately owned land, some wealthy Europeans purchase abandoned farm buildings in clustered rural settlements for use as second homes on weekends and holidays. Some of the worst traffic jams in Paris occur on summer Sunday nights, when families return from their weekend homes. A trip from the weekend home to the city that

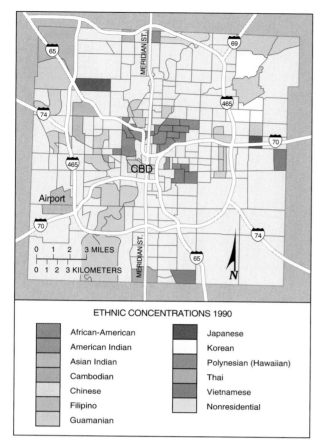

ETHNIC CONCENTRATIONS 1990

African-American	Japanese
American Indian	Korean
Asian Indian	Polynesian (Hawaiian)
Cambodian	Thai
Chinese	Vietnamese
Filipino	Nonresidential
Guamanian	

Figure 13–10 Example of multiple nuclei model in Indianapolis, the distribution of minorities. The black concentration consists of census tracts that are 90 percent or more African-American. The other groups are clustered in tracts that contain at least 5 percent of the total Indianapolis population of that ethnic group.

normally takes an hour can consume 4 hours on Sunday night.

In the past, poorer people also lived in the center of European cities. Before the invention of electricity in the nineteenth century, social segregation was vertical: richer people lived on the first or second floors, while poorer people occupied the dark, dank basements, or they climbed many flights of stairs to reach the attics. As the city expanded during the industrial revolution, housing for poorer people was constructed in sectors near the factories and away from the rich.

Today, poorer people are less likely to live in European inner-city neighborhoods. Poor-quality housing has been renovated for wealthy people, or demolished and replaced by offices or luxury apartment buildings. Building and zoning codes prohibit anyone from living in basements, and upper floors have become attractive for wealthy individuals once elevators are installed.

Poorer people have been relegated to the outskirts of European cities (Figure 13–12). Vast suburbs containing dozens of high-rise apartment buildings house the poorer people displaced from the inner city. European suburban residents face the prospect of long commutes by public transportation to reach jobs and other downtown amenities. Shops, schools, and other services are

worse than in inner neighborhoods, and the suburbs are centers for crime, violence, and drug-dealing. Because the housing is mostly in high-rise buildings, people lack large private yards. Many residents of these dreary suburbs are persons of color or recent immigrants from Africa or Asia who face discrimination and prejudice by "native" Europeans.

European officials encouraged the construction of high-density suburbs to help preserve the countryside from development and avoid the inefficient sprawl that characterizes American suburbs, as discussed in the last section of this chapter. And tourists are attracted to the historic, lively centers of European cities. But these policies have resulted in the clustering of people with social and economic problems in remote suburbs rarely seen by wealthier individuals.

Less Developed Countries

In LDCs as in Europe, the poor are accommodated in the suburbs, whereas the rich live near the center of cities, as well as in a sector extending from the center. The similarity between European and LDC cities is not a coincidence: past European colonial policies have left a heavy mark on the development of cities in many less developed countries. In fact, most cities in less developed countries have passed through three stages of development—before European colonization, during the European colonial period, and since independence.

Pre-Colonial Cities. Before the Europeans established colonies, few cities existed in Africa, Asia, and Latin America, and most people lived in rural settlements. The principal cities in Latin America were located in Mexico and the Andean highlands of northwestern South America (see Globalization and Local Diversity box). In Africa, cities could be found along the western coast, Egypt's Nile River valley, and Islamic empires in the north and east (as well as in Southwest Asia). Cities were also built in South and East Asia, especially India, China, and Japan.

Cities were often laid out surrounding a religious core, such as a mosque in Muslim regions. The center of Islamic cities also had a bazaar or marketplace, which served as the commercial core. Government buildings and the homes of wealthy families surrounded the mosque and bazaar. Narrow, winding streets led from the core to other quarters. Families with less wealth and lower status located farther from the core, and recent migrants to the city lived on the edge.

Commercial activities were arranged in a concentric and hierarchical pattern: higher-status businesses directly related to religious practices (such as selling religious books, incense, and candles) were located closest to the mosque. In the next ring were secular businesses, such as leather works, tailors, rug shops, and jewelers. Food products were sold in the next ring; then came blacksmiths, basket makers, and potters. A quarter would be reserved for Jews, a second for Christians, and a third for foreigners.

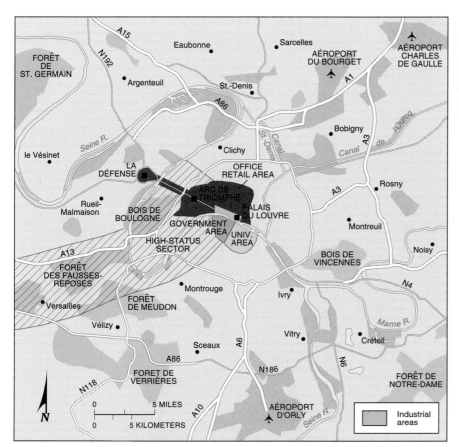

Figure 13–11 Social areas in Paris. Wealthy people moved to a southwestern sector, near the king's residences, first in the Louvre and later at Versailles. The eastern neighborhoods of Paris once housed poorer families, but in recent years these have been gentrified, and poorer people have moved to suburban high-rise apartments.

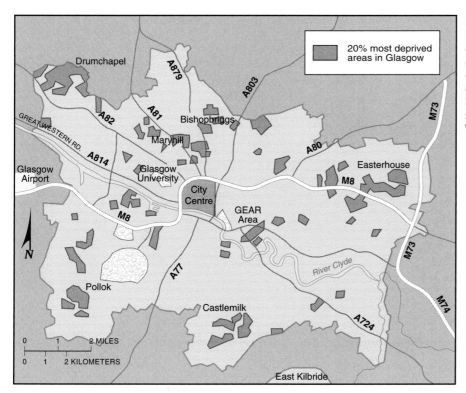

Figure 13–12 Deprived neighborhoods in a European city, Glasgow, Scotland. Glasgow's deprived neighborhoods are primarily in the outer areas, just opposite of the case in U.S. cities. Areas of social deprivation contain high concentrations of unemployed people receiving public assistance. Most of these areas consist of massive housing projects built after World War II.

Colonial Cities. When Europeans gained control of Africa, Asia, and Latin America, they expanded existing cities to provide colonial services, such as administration, military command, and international trade, as well as housing for Europeans who settled in the colony.

Existing native towns were either left to one side or demolished because they were totally at variance with European ideas.

Fez, Morocco, consists of two separate and distinct towns, one that existed before the French gained control

Globalization and Local Diversity

Three Eras in Mexico City

The social geography of any city is a product of a combination of global patterns and local diversity. In Mexico City, **where** different groups of people live is a product of the three eras typical of cities in less developed countries—precolonial, colonial, and post-independence.

Why do the rich people live southwest of downtown Mexico City? Why do poor people live to the east? These patterns reflect **local diversity** in a city's specific history of growth. But these local patterns fit the **global** model of urban structure typical of Latin American cities

Precolonial Aztec City (A.D. 1325–1521)

The Aztecs migrated from an unknown location in southwestern Mexico and settled west of present-day downtown Mexico City on a hill known as Chapultepec ("the hill of the grasshopper"). Forced by other people to leave the hill, the Aztecs first migrated a few kilometers south, near the present-day site of the University of Mexico, and then in 1325 to a marshy 10-square-kilometer (4-square-mile) island in Lake Texcoco (Figure 1). They named the city Tenochtitlán.

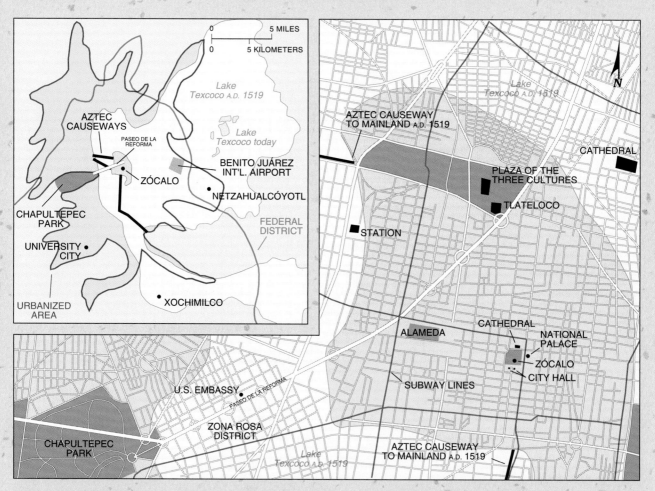

Mexico City. The Aztec city of Tenochtitlán was built on an island in Lake Texcoco. The elite live in a sector to the west, while poorer people live on landfill in the former lake bed.

and one built by the French colonialists (Figure 13–13). Similarly, the British built New Delhi near the existing city of Delhi, India. On the other hand, the French colonial city of Saigon, Vietnam (now Ho Minh City), was built by completing demolishing the existing city without leaving a trace (Figure 13–14).

Compared to the existing cities, the European districts typically contain wider streets and public squares, larger houses surrounded by gardens, and much lower density. In contrast, the old quarters have narrow, winding streets, little open space, and cramped residences. Colonial cities followed standardized plans. All Spanish cities in Latin

At first, Tenochtitlán consisted of a small temple and a few huts of thatch and mud. Over the next two centuries, the Aztecs conquered the neighboring peoples and extended their control through much of present-day Mexico. As their wealth and power grew, the Aztecs built elaborate stone houses and temples in Tenochtitlán.

The node of Aztec religious life in Tenochtitlán was the Great Temple, a massive multicolor structure containing two shrines, one for the rain god (painted blue) and one for the god of war (painted blood red). The main market center, Tlatelolco, was located at the north end of the island.

Most food, merchandise, and building materials crossed from the mainland to the island by canoe, barge, or other boat. The island itself was laced with canals to facilitate pick-up and delivery of people and goods. Three causeways with drawbridges linked Tenochtitlán to the mainland and also helped to control flooding. An aqueduct brought fresh water from Chapultepec.

Colonial Spanish City (1521–1821)

When 400 Spanish soldiers, under the leadership of Hernando Cortés, reached Tenochtitlán in 1519, they were shocked to find a teeming city of more than 500,000, probably the largest in the world. After a 2-year siege, the Spanish conquered Tenochtitlán with the aid of 100,000 natives who believed Cortés to be the reincarnation of a former ruler-priest.

The Spanish destroyed Tenochtitlán and dispersed or killed most of the inhabitants. Cortés ordered the surviving Aztecs to move out of the center and divided up the land among his soldiers. The population declined to 30,000. The city, renamed Mexico City, was rebuilt in accordance with Spanish preferences.

A main square, called the Zócalo, was placed in the center of the island, on the site of the Aztecs' sacred precinct. The Spanish reconstructed the streets in a grid pattern, with the main ones extending from the Zócalo. A Roman Catholic cathedral was built on the north side of the square, near the site of the demolished Great Temple, and the National Palace was erected on the east side, on the site of the Aztec emperor Moctezuma's destroyed palace. The Spanish placed a church and monastery on the site of the Tlatelolco market.

Independent City (1821–present)

Mexico City grew slowly during the three centuries of colonial rule, to about 100,000 at the time of independence in 1821. Although again the most populous city in the Western Hemisphere, Mexico City had been five times larger when Cortés conquered it.

Although no longer a colony of Spain, Mexico remained under strong European influence during the nineteenth century. The Mexican emperor Maximilian (1864–1867), who was the brother of Francis Joseph I of Austria, designed a 14-lane tree-lined boulevard for Mexico City patterned after the Champs-Elysées in Paris. The boulevard (now known as the Paseo de la Reforma) extended 3 kilometers southwest from the center to Chapultepec.

The Reforma between downtown and Chapultepec became the spine of an elite sector, as shown in the generalized model of Latin American cities (Figure 13–15). During the regime of Porfirio Díaz (president of Mexico 1877–1880 and 1884–1911), the wealthy built pretentious *palacios* (palaces) along the Reforma. Physical factors influenced the movement of wealthy people toward the west along the Reforma. Because elevation was higher than elsewhere in the city, sewage flowed eastward and northward away from Chapultepec.

In 1903, most of Lake Texcoco was drained by a gigantic canal and tunnel project, allowing the city to expand to the north and east. However, the lake bed was a less desirable residential location than the west side, because prevailing winds from the northeast stirred up dust storms from the dried-up lakebed.

As Mexico City's population has grown rapidly during the twentieth century, the social patterns inherited from the nineteenth century have been reinforced. The wealthy push out farther to the west along the spine of the Reforma, replacing the *palacios* with high-rise apartment buildings, offices, and American franchise stores. The poorest people occupy squatter settlements on the dried lake bed and in the inaccessible mountains surrounding the city.

America, for example, were built according to the Laws of the Indies, drafted in 1573. The laws explicitly outlined how colonial cities were to be constructed: a gridiron street plan centered on a church and central plaza, walls around individual houses, and neighborhoods centered around smaller plazas with parish churches or monasteries.

Cities Since Independence. Following independence, cities have become the focal points of change in less developed countries. Millions of people have migrated to the cities in search of work.

Geographers Ernest Griffin and Larry Ford show that in Latin American cities, wealthy people push out from

Figure 13–13 Layout of Fez, Morocco. The French laid out an entirely new district in the west (background in the photograph), separate and distinct from the existing city to the east, characterized by narrow, winding streets, and high density (foreground). (Gilles Mingasson/Gamma-Liaison)

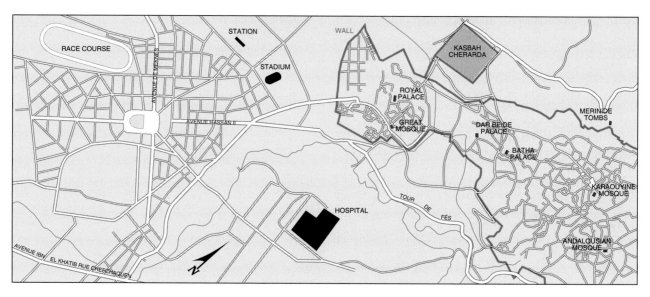

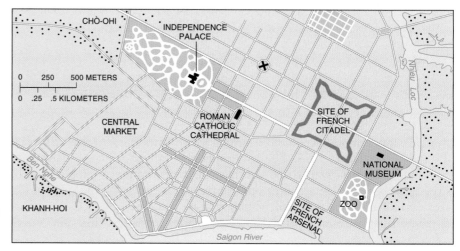

Figure 13–14 Layout of Ho Chi Minh City (formerly Saigon), Vietnam. The French demolished the existing city and replaced it with one built according to colonial principles, with wide boulevards and public squares.

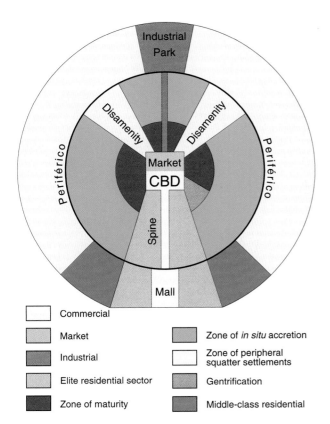

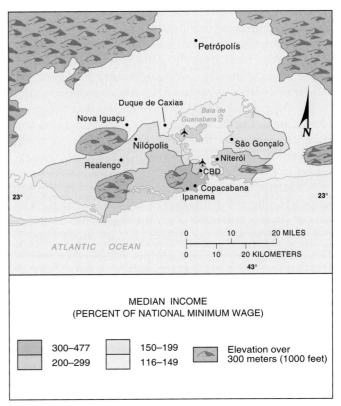

Figure 13–15 Model of a Latin American city. Wealthy people live in the inner city and a sector extending along a commercial spine. (Adapted from Larry R. Ford, "A New and Improved Model of Latin American City Structure," *Geographical Review* 86 (1996): 438. Used by permission of the publisher.)

Figure 13–16 High- and low-income households in Rio de Janeiro, Brazil. The highest income areas are near the CBD and in a spine along the ocean, whereas low-income people are more likely to live in peripheral areas.

the center in a well-defined elite residential sector. The elite sector forms on either side of a narrow spine that contains offices, shops, and amenities attractive to wealthy people, such as restaurants, theaters, parks, and zoos (Figure 13–15). The rich are also attracted to the center and spine, because services like water and electricity are more readily available and reliable.

For example, in Brazil, Rio de Janeiro's high-income people are clustered in the center of the city and to the south, whereas low-income residents are in the northern suburbs (Figure 13–16). The distribution of income groups coincides with other social characteristics, such as the percent of households with a telephone, automobile, or television. High-income groups are clustered near the center in part because of greater access to services, like electricity and city sewers (Figure 13–17).

Physical geography also influences the distribution of social classes within Rio. The original site of the city was along the west shore of Guanabara Bay, a protected harbor. Residents were attracted to the neighborhoods immediately south of the central area, like Copacabana and Ipanema, to enjoy spectacular views of the Atlantic Ocean and access to beaches. On the other hand, low-income households have clustered along the northern edge of the city, where steep mountains have restricted construction of other types of buildings. Development on the eastern side of Guanabara Bay was restricted until a bridge was constructed in the 1970s.

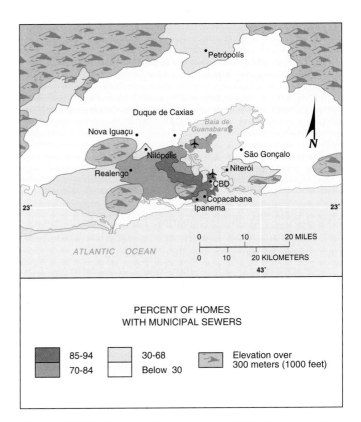

Figure 13–17 Sewers in Rio de Janeiro. In LDC cities such as Rio de Janeiro, high-income individuals are attracted to central areas because services, like municipal sewers, are more widely available than in peripheral areas.

Squatter Settlements. The LDCs are unable to house the rapidly growing number of poor. Their cities are growing because of overall population increase and migration from rural areas for job opportunities. Because of the housing shortage, a large percentage of poor immigrants to urban areas in LDCs live in squatter settlements. **Squatter settlements** are known by a variety of names, including *barrios, barriadas,* and *favelas* in Latin America, *bidonvilles* in North Africa, *bustees* in India, *gecekondu* in Turkey, *kampongs* in Malaysia, and *barung-barong* in the Philippines.

A squatter settlement is typically initiated by a group of people who move together onto land outside the city that is owned either by a private individual or (more frequently) by the government. People move literally overnight with all their possessions, which usually are so few that they can be easily carried. The leaders of the invasion allocate small parcels of the seized land to each participating family.

At first, squatters do little more than camp on the land or sleep in the street. In severe weather, they may take shelter in markets and warehouses. Families then erect primitive shelters with scavenged cardboard, wood boxes, sackcloth, and crushed beverage cans. As they find new bits of material, they add them to their shacks. After a few years, they may build a tin roof and partition the space into rooms, and the structure acquires a more permanent appearance.

Squatter settlements have few services because neither the city nor the residents can afford them. Latrines are usually designated by the settlement's leaders, and water is carried from a central well or dispensed from a truck. The settlements generally lack schools, paved roads, telephones, or sewers. Electricity service may be stolen by running a wire from the nearest power line. In the absence of bus service or available private cars, a resident may have to walk 2 hours to reach a place of employment.

To improve their housing conditions, squatters have two basic choices. One alternative is to move illegally into better-quality, vacant housing close to the center of the city. The second alternative is to rent slum housing legally from a landlord. Squatters rarely have the financial means to move directly from a squatter settlement into decent housing on legally owned land.

The percentage of people living in squatter settlements, slums, and other illegal housing ranges from 33 percent in São Paulo, Brazil, to 85 percent in Addis Ababa, Ethiopia, according to a United Nations study. The U.N. estimates that more than half of the residents live in some form of illegal housing in Lusaka, Zambia; Ankara, Turkey; Bogotá, Colombia; Dar es Salaam, Tanzania; and Luanda, Angola.

Governments in LDCs face a difficult choice regarding squatter settlements. If the government sends in the police or army to raze the settlement, it risks sparking a violent confrontation with the displaced people. On the

In Bombay, India, squatter settlements are erected in sight of modern high-rises occupied by middle-class families. (Viviane Moos/The Stock Market)

other hand, if the government decides that improving and legalizing squatter settlements is cheaper than building the necessary new apartment buildings, it may encourage other poor rural people to migrate to the city to live as squatters.

Immigrants to New York and London have lived in squalid conditions, but an expanding economy has at least provided them with jobs, even if rather menial and poorly paid. An adequate supply of jobs is simply not available in Mexico City, São Paulo, and the other large urban settlements of today's LDCs.

Key Issue 3

Why Do Inner Cities Have Distinctive Problems?

- Inner-city physical problems
- Inner-city social problems
- Inner-city economic problems

Most of the land in urban areas is devoted to residences, where people live. Within U.S. urban areas, the most fundamental spatial distinction is between inner-city residential neighborhoods that surround the central business district (CBD) and suburban residential neighborhoods on the periphery. Inner cities in the United States contain concentrations of low-income people with a variety of physical, social, and economic problems very different from those faced by suburban residents.

Inner-City Physical Problems

The major physical problem faced by inner-city neighborhoods is the poor condition of the housing, most of which was built before 1940. Deteriorated housing can either be demolished and replaced with new housing, or it can be rehabilitated.

Process of Deterioration

As the number of low-income residents increase in the city, the territory they occupy expands. Neighborhoods can shift from predominantly middle-class to low-income occupants within a few years. Middle-class families move out of a neighborhood to newer housing farther from the center and sell or rent their houses to lower-income families.

Filtering. Large houses built by wealthy families in the nineteenth century are subdivided by absentee landlords into smaller dwellings for low-income families. This process of subdivision of houses and occupancy by successive waves of lower-income people is known as **filtering**. The ultimate result of filtering may be abandonment of the dwelling.

Like a car, tape player, or any other object, the better a house is maintained, the longer it will last. Landlords stop maintaining houses when the rent they collect becomes less than the maintenance cost. In such a case, the building soon deteriorates and grows unfit for occupancy. Not even the poorest families will rent the dwelling. At this point in the filtering process the owner may abandon the property, because the rents that can be collected are less than the cost of taxes and upkeep.

Cities have codes that require owners to maintain houses in good condition. But governments that aggressively go after landlords to repair deteriorated properties may in fact hasten abandonment, because landlords will not spend money on repairs that they are unable to recoup in rents. Thousands of vacant houses stand in the inner areas of American cities because the landlords have abandoned them.

One hundred years ago, low-income inner-city neighborhoods in the United States teemed with throngs of recent immigrants from Europe. These inner-city neighborhoods that housed perhaps 100,000 a century ago contain less than 10,000 inhabitants today. Schools and shops close because they are no longer needed in inner-city neighborhoods with rapidly declining populations. Through the filtering process, many poor families have moved to less-deteriorated houses farther from the center.

Redlining. Some banks engage in **redlining**, drawing lines on a map to identify areas in which they will refuse to loan money. As a result of redlining, families who try to fix up houses in the area have difficulty borrowing money. Although redlining is illegal, enforcement of laws against it is frequently difficult.

The Community Reinvestment Act requires banks to document by census tract where they make loans. A bank must demonstrate that inner-city neighborhoods within its service area receive a fair share of its loans.

Urban Renewal

North American and European cities have demolished much of their substandard inner-city housing through urban renewal programs. Under **urban renewal**, cities identify blighted inner-city neighborhoods, acquire the properties from private owners, relocate the residents and businesses, clear the site, and build new roads and utilities. The land is then turned over to private developers or to public agencies, such as the board of education or the parks department, to construct new buildings or services. National government grants help cities pay for urban renewal.

Public Housing. Many substandard inner-city houses have been demolished and replaced with public housing. In the United States, **public housing** is reserved for low-

The zone in transition in many U.S. cities includes a large number of vacant and abandoned houses, such as these in Philadelphia. (Chromosohm/Joe Sohm/Photo Researchers, Inc.)

income households, who must pay 30 percent of their income for rent. A housing authority, established by the local government, manages the buildings, while the federal government pays the cost of construction and the maintenance, repair, and management that is not covered by rent.

In the United States, public housing accounts for only 2 percent of all dwellings, although it may account for a high percentage of housing in inner-city neighborhoods. In the United Kingdom, more than one-third of all housing is publicly owned, and the percentage is even higher in northern cities such as Liverpool, Manchester, and Glasgow. Private landlords control only a small percentage of housing in the United Kingdom, for the most part confined to central London and resort communities.

Elsewhere in Western Europe, governments typically do not own the housing. Instead, they subsidize construction cost and rent for a large percentage of the privately built housing. Developers of low-cost housing may be either nonprofit organizations, such as church groups and labor unions, or profit-making corporations that agree to build some low-cost housing in exchange for permission to build higher-cost housing elsewhere. The U.S. government has also provided subsidies to private developers, but on a much smaller scale than in Europe.

A number of high-rise public housing projects built in the United States and Europe during the 1950s and early 1960s are now considered unsatisfactory environments for families with children. The elevators are frequently broken, juveniles terrorize other people in the hallways, and drug use and crime rates are high. Some observers claim that the high-rise buildings caused the problem, because too many low-income families are concentrated into a high-density environment. Because of poor conditions, public housing authorities have demolished high-rise public housing projects in recent years in Dallas, Newark, St. Louis, Liverpool in England, Glasgow in Scotland, and other U.S. and European cities.

More recent public housing projects have consisted primarily of two- or three-story apartment buildings and row houses, with high-rise apartments reserved for elderly people. Cities have also experimented with "scattered-site" public housing, in which dwellings are dispersed throughout the city rather than clustered in a large project.

In recent years, the U.S. government has stopped funding new public housing altogether. Some federal support is available to renovate older buildings and to help low-income households pay their rent, but the overall level of funding is much lower today than in the late 1970s. As a result, the supply of public housing and other government-subsidized housing diminished by approximately 1 million units between the early 1980s and the early 1990s. But during the same period, the number of households needing low-rent dwellings increased by more than 2 million.

In Britain, the supply of public housing, known as council estates, has also declined, because the govern-

Beacon Hill is an inner-city neighborhood in Boston that has been renovated by well-to-do people. (Gene Peach/Liaison International)

ment has forced local authorities to sell some of the dwellings to the residents. But at the same time, the British have expanded subsidies to nonprofit housing associations, which build housing for groups with special needs, including single mothers, immigrants, the disabled, and the elderly, as well as the poor.

Urban renewal has been criticized for destroying the social cohesion of older neighborhoods and reducing the supply of low-cost housing. Because African-Americans comprised a large percentage of the displaced population in U.S. cities, urban renewal was often called "Negro Removal" during the 1960s. Most North American and European cities have turned away from urban renewal since the 1970s, and national governments, including the United States, have stopped funding it.

Renovated Housing. An alternative to demolishing deteriorated inner-city houses is to renovate them. In some cases, nonprofit organizations renovate housing and sell or rent them to low-income people. But more often, the renovated housing attracts middle-class people.

Most cities have at least one substantially renovated inner-city neighborhood where middle-class people live. In a few cases, inner-city neighborhoods never deteriorated, because the community's social elite maintained them as enclaves of expensive property. In most cases, inner-city neighborhoods have only recently been renovated by the city and by private investors.

The process by which middle-class people move into deteriorated inner-city neighborhoods and renovate the housing is known as **gentrification**. Middle-class families are attracted to deteriorated inner-city housing for a number of reasons. First, houses may be larger, more substantially constructed, yet cheaper in the inner-city than in the suburbs. Inner-city houses may also possess attractive architectural details such as ornate fireplaces, cornices, high ceilings, and wood trim.

Gentrified inner-city neighborhoods also attract middle-class individuals who work downtown. Inner-city living eliminates the strain of commuting on crowded freeways or public transit. Others seek proximity to theaters, bars, restaurants, and other cultural and recreational facilities located downtown. Renovated inner-city housing appeals to single people and couples without children, who are not concerned with the quality of inner-city schools.

Because renovating an old inner-city house can be nearly as expensive as buying a new one in the suburbs, cities encourage the process by providing low-cost loans and tax breaks. Public expenditures for renovation have been criticized as subsidies for the middle class at the expense of poor people, who are forced to move out of the gentrified neighborhoods because the rents in the area are suddenly too high for them.

Cities try to reduce the hardship on poor families forced to move. First, U.S. law requires that they be reimbursed both for moving expenses and for rent increases over a 4-year period. Western European countries have similar laws. Second, cities renovate old houses specifically for lower-income families through public housing or other programs. By renting renovated houses, the city also helps to disperse low-income families throughout the city, instead of concentrating them in large inner-city public housing projects.

Inner-City Social Problems

Beyond the pockets of gentrified neighborhoods, inner-cities contain primarily low-income people who face a variety of social problems. Inner-city residents constitute a permanent underclass who live in a culture of poverty.

Underclass

Inner-city residents frequently are referred to as a permanent **underclass** because they are trapped in an unending cycle of economic and social problems. The underclass suffers from relatively high rates of unemployment, alcoholism, drug addiction, illiteracy, juvenile delinquency, and crime. Their schools are deteriorated, and affordable housing is increasingly difficult to find. Their neighborhoods lack adequate police and fire protection, shops, hospitals, clinics, or other health-care facilities.

Lack of Job Skills.   The future is especially bleak for the underclass because they are increasingly unable to compete for jobs. Inner-city residents lack the technical skills needed for most jobs because fewer than half complete high school. Despite the importance of education in obtaining employment, many in the underclass live in an atmosphere that ignores good learning habits, such as regular school attendance and completion of homework.

The gap between skills demanded by employers and the training possessed by inner-city residents is widening. In the past, people with limited education could become factory workers or filing clerks, but today these jobs require skills in computing and handling of electronics. Meanwhile, inner-city residents do not even have access to the remaining low-skilled jobs, such as custodians and fast-food servers, because they are increasingly in the distant suburbs.

Homeless.   An increasing number of the underclass are homeless. Several million Americans sleep in doorways, on heated street grates, and in bus and subway stations. Los Angeles County alone has an estimated 35,000 homeless individuals, attracted by the area's relatively mild climate. Homelessness is an even more serious problem in less developed countries. An estimated 300,000 people in Calcutta sleep, bathe, and eat on sidewalks and traffic islands.

A homeless woman pushes a cart filled with her belongings through the traffic in Boston. (Robert Harbison)

Most people are homeless because they cannot afford housing and have no regular income. Homelessness may have been sparked by family problems or job loss. However, roughly one-third of U.S. homeless are individuals who are unable to cope in society after being released from hospitals or other institutions.

Culture of Poverty

Inner-city residents are trapped as permanent underclass because they live in a culture of poverty. Unwed mothers give birth to two-thirds of the babies in U.S. inner-city neighborhoods, and 90 percent of children in the inner-city live with only one parent. Because of inadequate child-care services, single mothers may be forced to choose between working to generate income and staying at home to take care of the children.

In principle, government officials would like to see more fathers living with their wives and children, but they provide little incentive for them to do so. Only a small percentage of "deadbeat dads" are tracked down for failing to provide required child-care support. If the husband moves back home, his wife may lose welfare benefits, leaving the couple financially worse off together than apart.

Crime. Trapped in a hopeless environment, some inner-city residents turn to drugs. Although drug use is a problem in both the suburbs and rural areas, rates of use in recent years have increased most rapidly in the inner cities. Some drug-users obtain money through criminal activities. Gangs form in inner-city neighborhoods to control lucrative drug distribution. Violence erupts when two gangs fight over the boundaries between their drug distribution areas.

For example, the locations in the Dayton urban area with the highest number of arrests on felony charges for drug violations are clustered just west of the CBD, where the percentage of low-income African-American households is very high (Figure 13–18). The higher incidence of arrests in low-income African-American areas does not necessarily mean that drug usage is higher, or that African-Americans are more involved in drug trafficking than whites. Some studies have shown that among male high school students, rates of drug use may actually be higher among whites.

In high-density inner-city areas, people are more likely to sell drugs while standing on street corners, under the clear view of neighborhood residents, who may call police. In contrast, drug sales in low-density automobile-oriented suburbs may occur discreetly behind closed doors, and arrests may require elaborate undercover operations.

Ethnic and Racial Segregation. Many neighborhoods in the United States are segregated by ethnicity, as discussed in Chapter 7. African-Americans and Hispanics concentrate in one or two large continuous areas of the inner city, while whites live in the suburbs.

Even small cities display strong social distinctions among neighborhoods. A frequently noticed division is

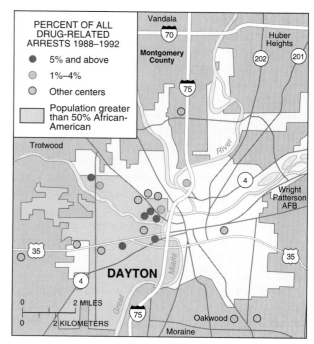

Figure 13–18 The locations in the Dayton urban area that have the highest numbers of drug-related felony arrests. Arrests are clustered in the predominantly low-income, African-American, inner-west side of the city.

between the east and west sides of a city, or between the north and south sides, with one side attracting the higher-income residents and the other left to lower-status and minority families.

A family seeking a new residence usually considers only a handful of districts, where the residents' social and financial characteristics match their own. Residential areas designed for wealthy families are developed in scenic, attractive areas, possibly a hillside or near a water body, while flat, dull land closer to industry becomes built up with cheaper housing.

Segregation by ethnicity explains voting patterns in many American urban areas. The winning candidate for mayor of Dayton in 1989 gained a majority of the votes in every ward on the predominantly African-American west side and lost every ward on the predominantly white east side. He was black, and his opponent was white (Figure 13–19).

Inner-City Economic Problems

The concentration of low-income residents in inner-city neighborhoods of central cities has produced financial problems. These people require public services, but they can pay very little of the taxes to support the services. Central cities face a growing gap between the cost of needed services in inner-city neighborhoods and the availability of funds to pay for them.

A city has two choices to close the gap between the cost of services and the funding available from taxes. One alternative is to reduce services by closing libraries, eliminating some public transit routes, collecting trash less frequently, and delaying replacement of outdated school

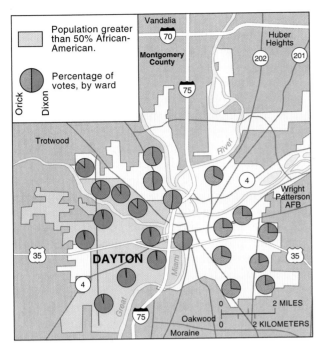

Figure 13–19 Race and voting in Dayton. In the 1989 mayoral election, the winner was an African-American who carried every ward on the west side, in some cases with more than 95 percent of the vote. These wards are predominantly African-American. The losing candidate, who was white, carried every ward on the east side. Compare to figures 12–6, 12–9, and 13–18, also of Dayton. Many inner-city areas suffer from high incidence of social problems and lack of services.

equipment. Aside from the hardship imposed on individuals laid off from work, cutbacks in public services also encourage middle-class residents and industries to move from the city.

The other alternative is to raise tax revenues. Because higher tax rates can drive out industries and wealthier people, cities prefer instead to expand their tax base, especially through construction of new CBD projects. Even with generous subsidies and tax breaks, a new downtown high-rise pays far more taxes than the buildings demolished to make way for it. Luxury hotels, restaurants, shops, and offices in the new downtown buildings provide minimum-wage personal service jobs for low-income inner-city residents. Still, spending public money to increase the downtown tax base can take scarce funds away from projects in inner-city neighborhoods, such as subsidized housing and playgrounds.

From the 1950s through the 1970s, inner-city fiscal problems were alleviated by increasing contributions from the federal government. The percentage of the budgets of the 50 largest U.S. cities supplied by the federal government increased from 1 percent in 1950 to 18 percent in 1980. But during the 1980s the percentage shrank substantially, to 6 percent in 1990. Federal aid to U.S. cities declined from about $44 billion in 1981 to about $23 billion in 1993, a 66 percent decline when adjusted for inflation. To offset a portion of these lost federal funds, some state governments increased financial assistance to cities.

Annexation

For many cities, economic problems are exacerbated by their inability to annex peripheral land. **Annexation** is the process of legally adding land area to a city. Until recently in the United States, as cities grew, they expanded by annexing peripheral land. Rules concerning annexation vary among states. Normally, land can be annexed into a city only if a majority of residents in the affected area vote in favor of doing so.

In the nineteenth century, peripheral residents generally desired annexation because the city offered better services, such as water supply, sewage disposal, trash pickup, paved streets, public transportation, and police and fire protection. Thus, while U.S. cities grew rapidly in the nineteenth century, the problem of defining a city seldom arose, because the legal boundaries frequently changed to accommodate newly developed areas. For example, the city of Chicago expanded from 26 square kilometers (10 square miles) in 1837 to 492 square kilometers (190 square miles) in 1900 (Figure 13–20).

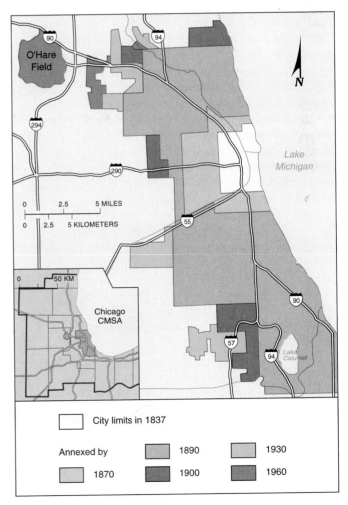

Figure 13–20 Growth of Chicago. During the nineteenth century, the city of Chicago grew rapidly through annexation of peripheral land. Relatively little land was annexed during the twentieth century; the major annexation was on the northwest side for O'Hare Airport. The inset shows that the city of Chicago covers only a small portion of the Chicago CMSA.

Today, however, cities are less likely to annex peripheral land because the residents prefer to organize their own services rather than pay city taxes for them. As a result, today's cities are surrounded by a collection of suburban jurisdictions, whose residents prefer to remain legally independent of the large city. Originally, some of these peripheral jurisdictions were small, isolated towns that had a tradition of independent local government before being swallowed up by urban growth. Others are newly created communities whose residents wish to live close to the large city but not be legally a part of it.

Key Issue 4

Why Do Suburbs Have Distinctive Problems?

- The peripheral model
- Contribution of transportation to suburbanization

Since 1950, overall population has declined more than 40 percent in the central cities of Buffalo, Cleveland, Detroit, and St. Louis, and by more than one-fourth in other cities like Boston, Cincinnati, Dayton, Jersey City, Louisville, Minneapolis, and Newark. The number of tax-paying middle-class families and industries has invariably declined by much higher percentages in these cities.

The suburban population has grown much faster than the overall population, especially in the United States. According to the U.S. Bureau of the Census, it has increased from 20 percent in 1950 to 60 percent today. To put it another way, since 1950, the U.S. population has increased by about 125 million people. During that time, population increased by some 165 million in the suburbs and declined by 40 million in central cities and rural areas.

Public opinion polls in the United States and Western Europe show people's strong desire for suburban living. In most polls, more than 90 percent of respondents prefer the suburbs to the inner city. About 60 percent of Americans now live in suburbs, compared to about 15 percent in inner cities and 25 percent in rural areas.

Suburbs offer varied attractions: a detached single-family dwelling rather than a row house or apartment, private land surrounding the house, space to park cars, and a greater opportunity for home ownership. The suburban house provides space and privacy, a daily retreat from the stress of urban living.

Families with children are especially attracted to suburbs, which offer more space for play and protection from the high crime rates and heavy traffic that characterize inner-city life. As incomes have risen in the twentieth century, first in the United States and more recently in Western Europe, more families can afford to buy suburban homes.

The Peripheral Model

North American urban areas follow what Chauncey Harris (creator of the multiple nuclei model) calls the peripheral model. According to the **peripheral model**, an urban area consists of an inner city surrounded by large suburban residential and business areas tied together by a beltway or ring road (Figure 13–21). Peripheral areas lack the severe physical, social, and economic problems of inner-city neighborhoods. But the peripheral model

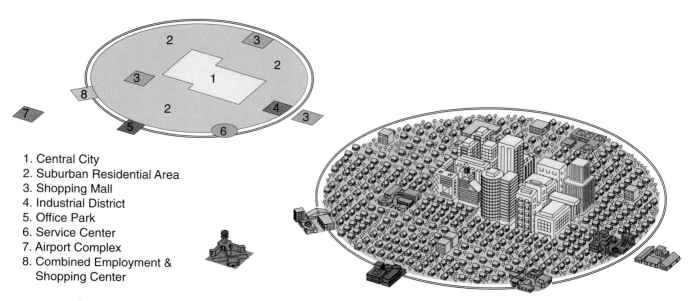

1. Central City
2. Suburban Residential Area
3. Shopping Mall
4. Industrial District
5. Office Park
6. Service Center
7. Airport Complex
8. Combined Employment & Shopping Center

Figure 13–21 Peripheral model of urban areas. The central city is surrounded by a beltway or ring road. Around the beltway are suburban residential areas and nodes, or edge cities, where consumer and business services and manufacturing cluster. (Adapted from Chauncy D. Harris, "The Nature of Cities' and Urban Geography in the Last Half Century." Reprinted with permission from *Urban Geography*, vol. 18, No. 1, p. 17. © V. H. Winston & Son, Inc., 360 South Ocean Blvd., Palm Beach, FL 33480. All rights reserved.)

points to problems of sprawl and segregation that characterize many suburbs.

Around the beltway are nodes of consumer and business services, called **edge cities**. Edge cities originated as suburban residences for people who worked in the central city, and then shopping malls were built to be near the residents. Now edge cities contain manufacturing centers spread out over a single story for more efficient operations and office parks where producer services cluster. Specialized nodes emerge in the edge cities: a collection of hotels and warehouses around an airport, a large theme park, a distribution center near the junction of the beltway and a major long-distance interstate highway.

Density Gradient

As you travel outward from the center of a city, you can watch the decline in the density at which people live. Inner-city apartments or row houses may pack as many as 250 dwellings on a hectare of land (100 dwellings per acre). Older suburbs have larger row houses, semidetached houses, and individual houses on small lots, at a density of about ten houses per hectare (four houses per acre). A detached house typically sits on a lot of one-quarter to one-half hectare (0.6 to 1.2 acres) in new suburbs, and a lot of one hectare or greater (2.5 acres) on the fringe of the built-up area.

This density change in an urban area is called the **density gradient**. According to the density gradient, the number of houses per unit of land diminishes as distance from the center city increases.

Changes in Density Gradient. Two changes have affected the density gradient in recent years. First, the number of people living in the center has decreased. The density gradient thus has a gap in the center where few live.

Second is the trend toward less density difference within urban areas. The number of people living on a hectare of land has decreased in the central residential areas through population decline and abandonment of old housing. At the same time, density has increased on the periphery through construction of apartment and row-house projects and diffusion of suburbs across a larger area (Figure 13–22).

In European cities, density gradient has also been affected by low-income high-rise apartments in the suburbs and by stricter control over construction of detached houses on large lots. The result of the two changes is to flatten the density gradient and reduce the extremes of inner and outer areas traditionally found within cities.

Cost of Suburban Sprawl

U.S. suburbs are characterized by **sprawl**, which is the progressive spread of development over the landscape. When private developers select new housing sites, they seek cheap land that can be easily prepared for construction—land often not contiguous to the existing built-up area. Sprawl is also fostered by the desire of many families to own large tracts of land.

Suburban Development Process. As long as demand for single-family detached houses remains high, land on the fringe of urbanized areas must be converted from open space to residential land use. The current system for developing land on urban fringes is inefficient, especially in the United States.

Land is not transformed immediately from farms to housing developments. Instead, developers buy farms for future construction of houses by individual builders. Developers frequently reject land adjacent to built-up areas in favor of detached isolated sites, depending on the price and physical attributes of the alternatives. The periphery of U.S. cities therefore looks like Swiss cheese, with pockets of development and gaps of open space.

Urban sprawl has some undesirable traits. Roads and utilities must be extended to connect isolated new developments to nearby built-up areas. The cost of these new roads and utilities is either funded by taxes, or the services are installed by the developer, who passes the cost on to new residents through higher home prices.

Sprawl also wastes land. Some prime agricultural land may be lost through construction of isolated housing developments; in the interim, other sites lie fallow, while speculators await the most profitable time to build homes on them. In reality, sprawl has little impact on the total farmland in the United States, but it does reduce the ability of city dwellers to get to the country for recreation, and it can affect the supply of local dairy products and vegetables. The low-density suburb also wastes more energy, especially because the automobile is required for most trips.

The supply of land for construction of new housing is more severely restricted in European urban areas. Officials attack sprawl by designating areas of mandatory open space. London, Birmingham, and several other British cities are surrounded by **greenbelts**, or rings of open space. New housing is built either in older suburbs inside the greenbelts or in planned extensions to small towns and new towns beyond the greenbelts (Figure 13–23). However, restriction of the supply of land on the urban periphery has driven up house prices in Europe.

Suburban Segregation

The modern residential suburb is segregated in two ways. First, residents are separated from commercial and manufacturing activities, which are confined to compact, distinct areas. Second, housing in a given suburban community is usually built for people of a single social class, with others excluded by virtue of the cost, size, or location of the housing.

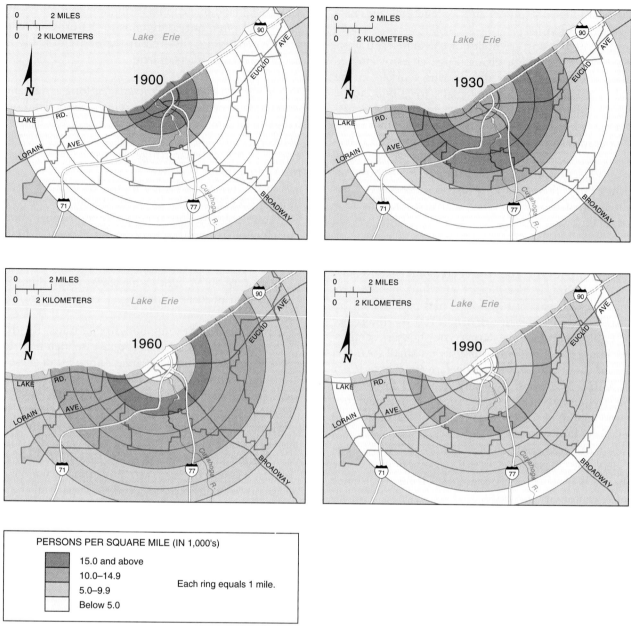

Figure 13–22 Density gradient in Cleveland. In 1900, the population was highly clustered in and near the CBD. By 1930 and 1960, the population was spreading, leaving the original core less dense. By 1990, population was distributed over a much larger area, the variation in the density among different rings was much less, and the area's lowest densities existed in the rings near the CBD. The current boundary of the city of Cleveland is shown. (First three maps adapted from Avery M. Guest. "Population Suburbanization in American Metropolitan Areas, 1940–1970." *Geographical Analysis 7* (1975): 267–83, table 4. Used by permission of the publisher.)

The homogeneous suburb is a twentieth-century phenomenon. In older cities, activities and classes were more likely to be separated vertically rather than horizontally. In a typical urban building, shops were on the street level, with the shop owner or another well-to-do family living on one or two floors above the shop.

Poorer people lived on the higher levels or in the basement, the least attractive parts of the building. The basement was dark and damp, and before the elevator was invented, the higher levels could be reached only by climbing many flights of stairs. Rich families lived in houses with space available in the basement or attic to accommodate servants.

Once cities spread out over much larger areas, the old pattern of vertical separation was replaced by territorial segregation. Large sections of the city were developed with houses of similar interior dimension, lot size, and cost, appealing to people with similar incomes and lifestyles.

Zoning ordinances, developed in Europe and North America in the early decades of the twentieth century, encouraged spatial separation. They prevented mixing of

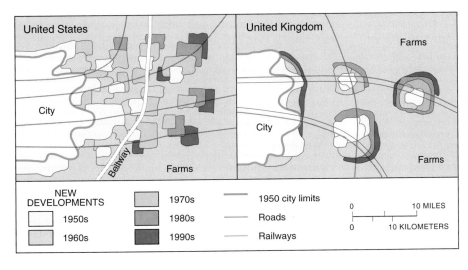

NEW DEVELOPMENTS			
1950s		1970s	1950 city limits
1960s		1980s	Roads
		1990s	Railways

0 | 10 MILES
0 | 10 KILOMETERS

Figure 13–23 Suburban development patterns in the United Kingdom and the United States. The United States has much more sprawl than the United Kingdom. In the United Kingdom, new housing is more likely to be concentrated in new towns or planned extensions of existing small towns, whereas in the United States growth occurs in discontinuous developments. In the United Kingdom, the boundary is sharp between a new town, such as Letchworth (top) and the surrounding rural areas. Compare to suburban sprawl outside Las Vegas (bottom). (top: Anthony Howarth, Woodfin Camp & Associates; bottom: J. Marshall/The Image Works)

land uses within the same district. In particular, single-family houses, apartments, industry, and commerce were kept apart, because the location of one activity near another was considered unhealthy and inefficient.

The strongest criticism of U.S. residential suburbs is that low-income and minority people are unable to live in them because of the high cost of the housing and the unfriendliness of established residents. Suburban communities discourage the entry of lower-income and minority individuals because of fear that property values will decline if the high-status composition of the neighborhood is altered. Legal devices, such as requiring each house to sit on a large lot and the prohibition of apartments, prevent low-income families from living in many suburbs.

School Busing.   The segregation of residential communities by race has led to a difficult situation in many school systems. Since 1954, when the U.S. Supreme Court ruled that racially segregated school systems are inherently unequal, cities have been faced with the need to improve racial balance in their schools. However, the goal of integration has conflicted with the strong desire of many parents and educators that children should live within walking distance of school. As long as neighborhoods have a homogeneous population, it is impossible to achieve both goals.

Some school districts have tried to promote integration by busing students from their homes to schools elsewhere in the city. Busing has been unpopular, but in some communities no other system has been found to integrate schools. In many communities, white parents have chosen to send their children to private schools rather than have them attend integrated public schools. Ironically, because the private school is frequently farther away than the public school, children are bused anyway.

Contribution of Transportation to Suburbanization

Urban sprawl makes people more dependent on transportation for access to work, shopping, and leisure activities. People do not travel aimlessly: their trips have a precise point of origin, destination, and purpose. More than half of all trips are work-related—commuting between work and home, business travel, or deliveries. Shopping or other personal business and social journeys each account for approximately one-fourth of all trips.

Historically, the growth of suburbs was constrained by transportation problems. People lived in crowded cities because they had to be within walking distance of shops and places of employment. The invention of the railroad in the nineteenth century enabled people to live in suburbs and work in the central city. Cities then built street railways—frequently known as trolleys, streetcars, or trams—and underground railways (subways) to accommodate commuters.

Many so-called streetcar suburbs built in the nineteenth century still exist and retain unique visual identities. They consist of houses and shops clustered near a station or former streetcar stop at a much higher density than is found in newer suburbs.

Motor Vehicles

The suburban explosion in the twentieth century has relied on motor vehicles rather than railroads, especially in the United States. In the nineteenth century, rail and trolley lines restricted suburban development to narrow ribbons within walking distance of the stations. Cars and trucks have permitted large-scale development of suburbs at greater distances from the center, in the gaps between the rail lines. Motor vehicle drivers have much greater flexibility in the choice of residence than was ever before possible.

Motor vehicle ownership is nearly universal among American households, with the exception of some poor families, older individuals, and people living in the center of large cities like New York. More than 95 percent of all trips within U.S. cities are made by car, compared to fewer than 5 percent by bus or rail. Outside the big cities, public transportation service is extremely rare or nonexistent.

The U.S. government has encouraged the use of cars and trucks by paying 90 percent of the cost of limited-access high-speed interstate highways, which stretch for 70,000 kilometers (44,000 miles) across the country. The use of motor vehicles is also supported by policies that limit the price of fuel to less than one-half the level found in Western Europe.

A few years ago, Senator Daniel Patrick Moynihan, a former Harvard professor, quipped that the United States has a national urban growth policy—the interstate highway system. He was referring to the dominant role of new expressway construction in fostering decentralization of U.S. cities.

The motor vehicle is an important user of land in the city. An average city allocates about one-fourth of its land to roads and parking lots. Valuable land in the central city is devoted to parking cars and trucks, although expensive underground and multistory parking structures can reduce the amount of ground-level space needed. Modern six-lane freeways cut a 23-meter (75-foot) path through the heart of cities, and elaborate interchanges consume even more space. European and Japanese cities have been especially disrupted by attempts to insert new roads and parking areas in or near to the medieval central areas.

Technological improvements may help traffic flow. Computers mounted on the dashboards alert drivers about traffic jams and suggest alternate routes. On freeways, vehicle speed and separation from other vehicles can be controlled automatically rather than by the driver. Motorists can be charged for using congested roads or pay high tolls to drive on uncongested roads. The inevitable diffusion of such technology in the twenty-first century will reflect the continuing preference of most people in MDCs to use private motor vehicles rather than switch to public transportation.

Public Transportation

Because few people in the United States live within walking distance of their place of employment, urban areas are characterized by extensive commuting. The heaviest flow of commuters is into the CBD in the morning and out of it in the evening.

Rush Hour Commuting. The intense concentration of people in the center during working hours strains transportation systems, because a large number of people must reach a small area of land at the same time in the morning and disperse at the same time in the afternoon. As much as 40 percent of all trips made into or out of a CBD occur during four hours of the day, two in the morning and two in the afternoon. **Rush hour**, or peak hour, is the four consecutive 15-minute periods that have the heaviest traffic.

In larger cities, public transportation is better suited than motor vehicles to moving large numbers of people, because each traveler takes up far less space. But most Americans still prefer to commute by car. One-third of the high-priced central land is devoted to streets and parking lots, although multistory and underground garages also are constructed.

Public transportation is cheaper, less polluting, and more energy-efficient than the automobile. It also is particularly suited to rapidly bringing a large number of people into a small area. Consequently, its use is increasingly confined in the United States to rush-hour commuting by workers in the CBD. A bus can accommodate 30 people in the amount of space occupied by one automobile, while a double-track rapid transit line can transport the same number of people as 16 lanes of urban freeway.

Automobiles have costs beyond their purchase and operation: delays imposed on others, increased need for highway maintenance, construction of new highways, and pollution. Most people overlook these important costs because they place higher value on the car's privacy and flexibility of schedule. Yet despite the obvious advantages of public transportation for commuting, ridership in the United States has declined from 23 billion per year in the 1940s to 7 billion in the 1990s.

At the end of World War I, U.S. cities had 50,000 kilometers (30,000 miles) of street railways and trolleys that carried 14 billion passengers a year, but only a few hundred kilometers of track remain. The number of U.S. and Canadian cities with trolley service declined from approximately 50 in 1950 to 8 in the 1960s. General Motors acquired many of the privately owned streetcar companies and replaced the trolleys with buses that the company made. You can still ride the trolley in Boston, Cleveland, Newark, New Orleans, Philadelphia, Pittsburgh, San Francisco, and Toronto.

Buses offer a more flexible service than do trolleys because they are not restricted to fixed tracks. However, bus ridership has declined from a peak of 11 billion riders annually in the late 1940s to 6 billion in the 1990s. Commuter railroad service, like trolleys and buses, has also been drastically reduced in most U.S. cities.

New Rapid Transit Lines. The one exception to the downward trend in public transportation is rapid transit. They are now known to transportation planners as either fixed heavy rail (such as subways) or fixed light rail (such as streetcars).

Cities such as Boston and Chicago have attracted new passengers through construction of new subway lines and modernization of existing service. Chicago has been a pioneer in the construction of heavy rail rapid transit lines in the median strips of expressways. Entirely new subway systems have been built in recent years in U.S. cities, including Atlanta, Baltimore, Miami, San Francisco, and Washington, D.C.

The federal government has permitted Boston, New York, and other cities to use funds originally allocated for interstate highways to modernize rapid transit service instead. New York's subway cars, once covered with graffiti spray-painted by gang members, have been cleaned, so that passengers can ride in a more hospitable environment. As a result of these improvements, subway ridership in the United States has increased 2 percent each year since 1980.

The trolley—now known by the more elegant term of fixed light rail transit—is making a modest comeback in North America. Once relegated almost exclusively as a tourist attraction in New Orleans and San Francisco, new trolley lines have been built or are under construction in Baltimore, Buffalo, Calgary, Edmonton, Los Angeles, Portland (Oregon), Sacramento, St. Louis, San Diego, and San Jose. However, new construction in all ten cities amounted to about 200 kilometers (130 miles) during the 1980s and 1990s.

California, the state that most symbolizes the automobile-oriented American culture, leads in construction of new fixed light rail transit lines. San Diego has added more kilometers than any other city. One line that runs from the center south to the Mexican border has been irreverently dubbed the "Tijuana trolley" because it is heavily used by residents of nearby Tijuana, Mexico.

Los Angeles—the city perhaps most associated with the motor vehicle—has planned the most extensive new light-rail system. The city had a rail network exceeding 1,600 kilometers (1,000 miles) as recently as the late 1940s, but the lines were abandoned when freeways were built to accommodate rising automobile usage. Now Los Angeles wants to entice motorists out of their cars and trucks with new light rail lines, but construction is very expensive, and the lines serve only a tiny percentage of the region.

The Metromover loops around the central business district of Miami, Florida. It enables people to get from one end of the CBD to the other without walking outside in Miami's oppressive summer heat and humidity, and it provides an easy connection to the city's subway, called Metrorail. (Joe Sohm/The Image Works)

Service Versus Cost. People who are too poor to own an automobile may still not be able to reach places of employment by public transportation. Low-income people tend to live in inner-city neighborhoods, but the job opportunities, especially those requiring minimal training and skill in personal services, are in suburban areas not well served by public transportation. Inner-city neighborhoods have high unemployment rates at the same time that suburban firms have difficulty attracting workers. In some cities, governments and employers subsidize vans to carry low-income inner-city residents to suburban jobs.

Despite modest recent successes, most public transportation systems are caught in a vicious circle, because fares do not cover operating costs. As patronage declines and expenses rise, the fares are increased, which drives away passengers and leads to service reduction and still higher fares. Public expenditures to subsidize construction and operating costs have increased, but the United States does not fully recognize that public transportation is a vital utility deserving of subsidy to the degree long assumed by European governments.

Public Transit in Other Countries. In contrast, even in more developed Western European countries and Japan,

where automobile ownership rates are high, extensive networks of bus, tram, and subway lines have been maintained, and funds for new construction have been provided in recent years (Figure 13–24). Since the late 1960s, London has opened 27 kilometers (17 miles) of subways, including two new lines, plus 18 kilometers (11 miles) in light rail transit lines to serve the docklands area, which has been transformed from industrial to residential and office uses. During the same period, Paris has built 65 kilometers (40 miles) of new subway lines, including a new system, known as the Réseau Express Régional (R.E.R.) to serve outer suburbs.

Smaller cities have shared the construction boom. In France alone, new subway lines have been built since the 1970s in Lille, Lyon, and Marseille, and hundreds of kilometers of entirely new tracks have been laid between the country's major cities to operate a high-speed train known as the TGV (*Train à Grande Vitesse*). Growth in the suburbs has stimulated nonresidential construction, including suburban shops, industry, and offices.

Local Government Fragmentation

The fragmentation of local government in the United States makes it difficult to solve regional problems of traffic, solid-waste disposal, and building affordable housing. The number of local governments exceeds 1,400 in the New York area, 1,100 in the Chicago area, and 20,000 throughout the United States. Approximately 40 percent of these 20,000 local governments are general units, such as cities and counties, and the remainder serve special purposes, such as schools, sanitation, transportation, water, and fire districts.

Long Island, which extends for 150 kilometers (90 miles) east of New York City and is approximately 25 kilometers (15 miles) wide, contains nearly 800 local governments. The island includes 2 counties, 2 cities, 13 towns, 95 villages, 127 school districts, and more than 500 special districts, such as for garbage collection.

The multiplicity of local governments on Long Island leads to problems. When police or fire fighters are summoned to the State University of New York at Old Westbury, two or three departments sometimes respond, because the campus is in five districts. The boundary between the communities of Mineola and Garden City runs down the center of Old Country Road, a busy, four-lane route. Mineola set a 40-mile-per-hour speed limit for the eastbound lanes, while Garden City set a 30-mile-per-hour speed limit for the westbound lanes.

Metropolitan Government

The large number of local government units has led to calls for a metropolitan government that could coordinate—if not replace—the numerous local governments in an urban area.

Most U.S. metropolitan areas have a **council of government**, which is a cooperative agency consisting of

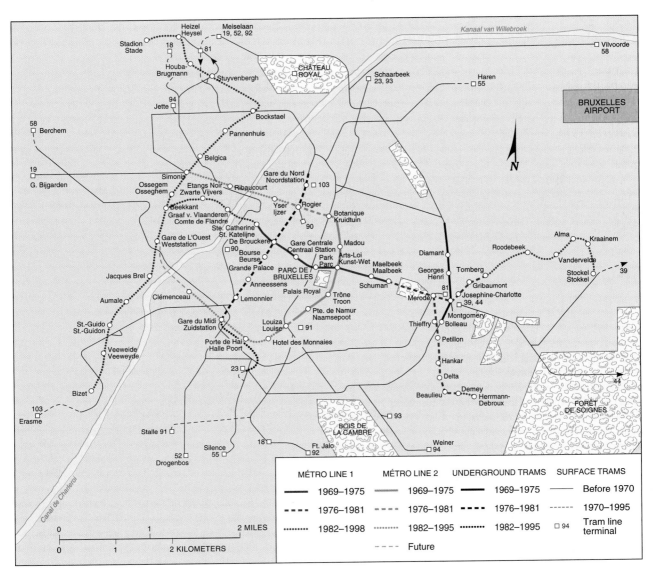

Figure 13–24 Brussels, Belgium, subway and tram lines. European cities like Brussels have invested substantially to improve public transportation in recent years. Brussels provides a good example of a public transport system that integrates heavy rail (Métro Lines 1 and 2) with light rail (trams). Trams initially used Métro tunnels, but the tunnels were large enough to convert to heavy rail lines as funds became available.

representatives of the various local governments in the region. The council of government may be empowered to do some overall planning for the area that local governments cannot logically do.

Strong metropolitan-wide governments have been established in a few places in North America. Two kinds exist, federations and consolidations.

Federations. Toronto, Ontario, has a federation system. Toronto's local government has two tiers. The region's six local governments, which range in size from 100,000 to 600,000 inhabitants, are responsible for police, fire, and tax collection services. A regional government, known as the Metropolitan Council, or Metro, sets the tax rate for the region as a whole, assesses the value of property, and borrows money for new projects. Metro shares responsibility with local governments for public

services, such as transportation, planning, parks, water, sewage, and welfare.

Consolidations. Several U.S. urban areas have consolidated metropolitan governments; Indianapolis and Miami are examples. Both have consolidated city and county governments. The boundaries of Indianapolis were changed to match those of Marion County, Indiana. Government functions that were handled separately by city and county now are combined into a joint operation in the same office building. In Florida, the city of Miami and surrounding Dade County have combined some services, but the city boundaries have not been changed to match the county's.

Metropolitan Governments in the United Kingdom. The creation of metropolitan governments has been somewhat easier in other countries, where the national government has more authority. In the United Kingdom, for example, the national government can change local government boundaries when it wishes. During the 1970s, the British government redrew the country's local government boundaries and created six new metropolitan governments, including Greater London, Greater Manchester, West Midlands around the city of Birmingham, West Yorkshire around Leeds, Merseyside around Liverpool, and Tyneside around Newcastle.

Then, in the 1980s, the British government decided to eliminate the Greater London Council, which had been created only a few years earlier to govern the London region. The national government decided that the 1,580-square-kilometer (610-square-mile) London metropolitan area would be better governed by 32 local boroughs than by one regional government.

Derek Senior, a geographer and a member of the royal commission that restructured local government, filed a minority report calling for a more radical local government restructuring in England. He identified the important cities in England and allocated the surrounding countryside to each, so that local government would truly be based on a series of urban regions. Senior's plan was rejected because the government wished to change the traditional counties as little as possible. However, in recent years, the idea has resurfaced.

■ Summary

Many people live in urban areas and never venture into inner-city neighborhoods or downtown. They live in suburbs, attend school in suburbs, work in suburbs, shop in suburbs, visit friends and family in suburbs, and attend movies and sports events in suburbs. Motor vehicles allow movement across urban areas without entering the center.

Conversely, inner-city residents may rarely venture out to suburbs. Lacking a motor vehicle, they have no access to most suburban locations. Lacking money, they do not shop in suburban malls or attend sports events at suburban arenas. The spatial segregation of inner-city residents and suburbanites lies at the heart of the stark contrasts so immediately observed in any urban area.

Here is a review of the key issues raised at the beginning of the chapter:

1. **Where have urban areas grown?** Urbanization involves increases in the percentage and in the number of people living in urban areas. More developed countries have higher percentages of urban residents, but less developed countries now have most of the largest urban areas.

2. **Where are people distributed within urban areas?** Three models explain where various groups of people live in urban areas: the concentric zone, sector, and multiple nuclei models. Combined, the three models present a useful framework for understanding the distribution of social and economic groups within urban areas. With modifications, the models also apply to cities in Europe and less developed countries.

3. **Why do inner cities have distinctive problems?** Inner-city residential areas have physical problems stemming from the high percentage of older deteriorated housing, social problems stemming from the high percentage of low-income households, and economic problems stemming from a gap between demand for services and supply of local tax revenue.

4. **Why do suburbs have distinctive problems?** The suburban lifestyle as exemplified by the detached, single-family house with surrounding yard attracts most people. Transportation improvements, most notably the railroad in the nineteenth century and the automobile in the twentieth century, have facilitated the sprawl of urban areas. Among the negative consequences of large-scale sprawl are segregation and inefficiency.

Case Study Revisited

Contrasts in the City

What is the future for cities? As this chapter has shown, contradictory trends are at work simultaneously. Why does one inner-city neighborhood become a slum and another a high-class district? Why does one city attract new shoppers and visitors while another languishes?

The Camden, New Jersey, urban area displays the strong contrasts that characterize American urban areas. The central city of Camden houses an isolated underclass, while suburban Camden County prospers. The population of the city of Camden has declined from 117,000 in 1960 to less than 90,000 today. African-Americans comprise 56 percent of the city's population, Hispanics 31 percent. The white, non-Hispanic population has declined from 90,000 in 1960 to 10,000 today.

Annual per capita income in Camden is $7,500, compared to $30,000 for the United States as a whole. More than half the population receive government assistance. Infant mortality rate for the city's African-American population is 27 per 1,000, about the level of Mexico, and four times higher than the rest of the United States.

Half of Camden's residents are under age 25, closer to the level found in LDCs than to the rest of the United States. Job prospects are not promising for these young people, because more than half have left school without obtaining a high school diploma. Camden's unemployment rate is 20 percent, four times the national average.

In the past, Camden's youths could find jobs in factories that produced Campbell's soups, Esterbrook pens, and RCA Victor records, radios, and televisions, but the city has lost 90 percent of its industrial jobs. The Esterbrook and Campbell factories in Camden are closed, although Campbell's corporate offices remain; General Electric operates the former RCA factory but with a labor force at only 15 percent of the level during the 1960s.

As Camden's population and industries decline, few shops have enough customers to remain open. The city once had 13 movie theaters, but none are left. The murder rate soared after gangs carved up the city into districts during the mid-1980s to control cocaine trafficking. Violent crimes like murder, rape, and robbery are increasing in Camden while dropping nationally. New Jersey state troopers help the city's understaffed police force deal with crime.

Meanwhile, Camden County (excluding the city) has grown from 275,000 in 1960 to about 420,000 in 1990. Cherry Hill has more than 70,000 residents today, compared to fewer than 10,000 in 1960, and will probably surpass Camden as the largest city in the county before the end of the decade. About 85 percent of Cherry Hill's high school graduates go on to college.

Cherry Hill is an example of an edge city, a large node of office and retail activities on the edge of an urban area. Despite its rapid population growth and trained labor force, an edge city like Cherry Hill has become both a residential area that commuters leave and an employment center that attracts other commuters. Cherry Hill has attracted so many new jobs that a major obstacle to further economic growth is a shortage of qualified workers. But many inner-city Camden residents lack transport to reach the jobs, or the skills to hold the jobs. Camden's mismatch among locations of people, jobs, resources, and services exemplifies the urban crisis throughout the United States, as well as in other countries. Geographers help us understand why these patterns arise, and what can be done about them.

■ Key Terms

Annexation Legally adding land area to a city in the United States.

Census tract An area delineated by the U.S. Bureau of the Census for which statistics are published; in urbanized areas, census tracts correspond roughly to neighborhoods.

Concentric zone model A model of the internal structure of cities in which social groups are spatially arranged in a series of rings.

Consolidated metropolitan statistical area (CMSA) In the United States, two or more adjacent metropolitan statistical areas with overlapping commuting patterns.

Council of government A cooperative agency consisting of representatives of local governments in a metropolitan area in the United States.

Density gradient The change in density in an urban area from the center to the periphery.

Edge city A large node of office and retail activities on the edge of an urban area.

Filtering A process of change in the use of a house, from single-family owner occupancy to abandonment.

Gentrification A process of converting an urban neighborhood from a predominantly low-income renter- to a predominantly middle-class owner-occupied area.

Greenbelt A ring of land maintained as parks, agriculture, or other types of open space to limit the sprawl of an urban area.

Metropolitan statistical area (MSA) In the United States, a central city of at least 50,000 population, the county within which the city is located, and adjacent counties meeting one of several tests indicating a functional connection to the central city.

Multiple nuclei model A model of the internal structure of cities in which social groups are arranged around a collection of nodes of activities.

Peripheral model A model of North American urban areas consisting of an inner city surrounded by large suburban residential and business areas tied together by a beltway or ring road.

Primary metropolitan statistical area (PMSA) In the United States, a metropolitan statistical area exceeding 1 million population located within a consolidated metropolitan statistical area.

Public housing Housing owned by the government; in the United States, it is rented to low-income residents, and the rents are set at 30 percent of the families' incomes.

Redlining A process by which banks draw lines on a map and refuse to lend money to purchase or improve property within the boundaries.

Rush (or peak) hour The four consecutive 15-minute periods in the morning and evening with the heaviest volumes of traffic.

Sector model A model of the internal structure of cities in which social groups are arranged around a series of sectors, or wedges, radiating out from the central business district (CBD).

Sprawl Development of new housing sites at relatively low density and at locations that are not contiguous to the existing built-up area.

Squatter settlement An area within a city in a less developed country in which people illegally establish residences on land they do not own or rent and erect home-made structures.

Underclass A group in society prevented from participating in the material benefits of a more developed society because of a variety of social and economic characteristics.

Urban renewal Program in which cities identify blighted inner-city neighborhoods, acquire the properties from private owners, relocate the residents and businesses, clear the site, build new roads and utilities, and turn the land over to private developers.

Urbanization An increase in the percentage and in the number of people living in urban settlements.

Urbanized area In the United States, a central city plus its contiguous built-up suburbs.

Zoning ordinance A law that limits the permitted uses of land and maximum density of development in a community.

■ Thinking Geographically

1. Nearly all residents of MDCs lead urban lifestyles even if they live in rural areas. In contrast, many residents in LDCs lead rural lifestyles, even though they live in large cities. They practice subsistence agriculture, raising animals or growing crops. Lacking electricity, they gather wood for fuel. Lacking running water and sewers, they dig latrines. Why do so many urban dwellers in LDCs lead rural lifestyles?

2. Draw a sketch of your community or neighborhood. In accordance with Kevin Lynch's *The Image of the City*, place five types of information on the map: districts (homogeneous areas), edges (boundaries that separate districts), paths (lines of communication), nodes (central points of interaction), and landmarks (prominent objects on the landscape). How clear an image does your community have for you?

3. Jane Jacobs wrote in *Death and Life of Great American Cities* that an attractive urban environment is one that is animated with an intermingling of a variety of people and activities, such as found in many New York City neighborhoods. What are the attractions and drawbacks to living in such environments?

4. Land-use activities in Communist cities were allocated by government rather than made by private market decisions. To what extent would the absence of a private-sector urban land market affect the form and structure of socialist cities? What impacts might Eastern European cities experience with the switch to market economies?

5. Officials of rapidly growing cities in LDCs discourage the building of houses that do not meet international standards for sanitation and construction methods. Also discouraged are privately owned transportation services, because the vehicles generally lack decent tires, brakes, and other safety features. Yet the residents prefer substandard housing to no housing, and they prefer unsafe transportation to no transportation. What would be the advantages and problems for a city if health and safety standards for housing, transportation, and other services were relaxed?

■ Further Readings

Arimah, Ben C. "The Determinants of Housing Tenure Choice in Ibadan, Nigeria." *Urban Studies* 34 (1997): 105–24.

Baldasarre, Mark. *Trouble in Paradise: The Suburban Transformation in America*. New York: Columbia University Press, 1986.

Berry, Brian J. L. *The Human Consequences of Urbanization*. New York: St. Martin's Press, 1973.

———, and John D. Kasarda. *Contemporary Urban Ecology*. New York: Macmillan, 1977.

Bertaud, Alain, and Bertrand Renaud. "Socialist Cities Without Land Markets." *Journal of Urban Economics* 41 (1997): 137–51.

Bourne, Larry S., ed. *Internal Structure of the City*, 2d ed. New York: Oxford University Press, 1982.

———. "Normative Urban Geographies: Recent Trends, Competing Visions, and New Cultures of Regulation." *Canadian Geographer* 40 (1996): 2–16.

Bratt, Rachel G. *Rebuilding a Low–Income Housing Policy*. Philadelphia: Temple University Press, 1990.

Carter, Harold. *An Introduction to Urban Historical Geography*. London: Edward Arnold, 1983.

Cervero, Robert. *America's Suburban Centers: The Land Use-Transportation Link*. Boston: Unwin and Hyman, 1989.

Chandler, Tertius, and Gerald Fox. *Three Thousand Years of Urban Growth*. New York: Academic Press, 1974.

Clawson, Marion, and Peter Hall. *Planning and Urban Growth*. Baltimore: The Johns Hopkins University Press, 1973.

Clay, Grady. *Real Places: An Unconventional Guide to America's Generic Landscape*. Chicago: University of Chicago Press, 1994.

Colwell, Peter F., and Henry J. Munneke. "The Structure of Urban Land Prices." *Journal of Urban Economics* 41 (1997): 321–36.

Cybriwsky, Roman. *Tokyo—The Changing Profile of an Urban Giant*. Boston: G.K. Hall, 1991.

Davis, Kingsley. *World Urbanization 1950–1970. Volume I*. Berkeley: University of California Institute of Environmental Studies, 1969.

Dear, Michael, and Steven Flusty. "Postmodern Urbanism." *Annals of the Association of American Geographers* 88 (1998): 50–72.

Detwyler, Thomas, and Melvin Marcus, eds. *Urbanization and Environment*. Belmont, CA: Duxbury Press, 1972.

Drakalis-Smith, David. *The Third World City*. London: Methuen, 1987.

———. "Third World Cities: Sustainable Urban Development: II—Population, Labour and Poverty." *Urban Studies* 33 (1996): 672–702.

Elliott, James R. "Cycles Within the System: Metropolitanisation and Internal Migration in the US, 1965–90." *Urban Studies* 34 (1997): 21–43.

Ford, Larry R. "Continuity and Change in the American City." *Geographical Review* 85 (1995): 552–68.

———. "A New and Improved Model of Latin American City Structure." *Geographical Review* 86 (1996): 437–40.

Foster, Richard H., and Mark K. McBeth. "Urban–Rural Influences in U.S. Environmental and Economic Development Policy." *Journal of Rural Studies* 12 (1996): 387–98.

Garreau, Joel. *Edge City: Life on the New Frontier*. New York: Doubleday, 1991.

Giles, Harry, and Bryan Brown. "'And Not a Drop to Drink': Water and Sanitation Services to the Urban Poor in the Developing World." *Geography* 82 (1997): 97–108.

Glasmeier, Amy K., and Jeff Kibler. "Power Shift: The Rising Control of Distributors and Retailers in the Supply Chain for Manufactured Goods." *Urban Geography* 17 (1996): 740–57.

Golany, Gideon, ed. *International Urban Growth Policies: New-Town Contributions*. New York: John Wiley, 1978.

Gordon, David L.A. "Managing the Changing Political Environment in Urban Waterfront Redevelopment." *Urban Studies* 34 (1997): 61–84.

Gottmann, Jean. *Megalopolis*. New York: Twentieth-Century Fund, 1961.

Green, Milford B., and Stephen P. Meyer. "An Overview of Commuting in Canada: With Special Emphasis on Rural Commuting and Employment." *Journal of Rural Studies* 13 (1997): 163–76.

Griffin, Ernest, and Larry Ford, "A Model of Latin American City Structure." *Geographical Review* 70 (1980): 387–422.

Guest, Avery M. "Population Suburbanization in American Metropolitan Areas, 1940–1970." *Geographical Analysis* 7 (1976): 267–83.

Hall, Tim, and Phil Hubbard. "The Entrepreneurial City: New Urban Politics, New Urban Geographies?" *Progress in Human Geography* 20 (1996): 153–74.

Hammel, Daniel J., and Elvin K. Wyly. "A Model for Identifying Gentrified Areas with Census Data." *Urban Geography* 17 (1996): 248–68.

Harris, Chauncy D. "Diffusion of Urban Models: A Case Study." *Urban Geography* 19 (1998): 49–67.

———, "'The Nature of Cities' and Urban Geography in the Last Half Century." *Urban Geography* 18 (1997): 15–35.

———, and Edward L. Ullman. "The Nature of Cities." *Annals of the American Academy of Political and Social Science* 143 (1945): 7–17.

Hart, John Fraser, ed. *Our Changing Cities.* Baltimore: The Johns Hopkins University Press, 1991.

Herzog, Lawrence A. *Where North Meets South: Cities, Space, and Politics on the U.S.-Mexico Border.* Austin: University of Texas, Center for Mexican-American Studies, 1990.

Hill, Edward W., and Harold L. Wolman. "Accounting for the Change in Income Disparities Between US Central Cities and Their Suburbs from 1980 to 1990." *Urban Studies* 34 (1997): 43–60.

Hodge, David C.; Richard L. Morrill, and Kiril Stanilov. "Implications of Intelligent Transportation Systems for Metropolitan Form." *Urban Studies* 17 (1996): 714–39.

Hoyt, Homer. *The Structure and Growth of Residential Neighborhoods.* Washington, DC: Federal Housing Administration, 1939.

Hsu, Mei-Ling. "China's Urban Development: A Case Study of Luoyang and Guiyang." *Urban Studies* 33 (1996): 895–910.

Jacobs, Allan B. *Looking at Cities.* Cambridge, MA: Harvard University Press, 1985.

Jacobs, Jane. *Death and Life of Great American Cities.* New York: Random House, 1961.

Jargowsky, Paul A. "Beyond the Street Corner: The Hidden Diversity of High-Poverty Neighborhoods." *Urban Geography* 17 (1996): 579–603.

Johnston, R. J. *City and Society: An Outline for Urban Geography.* London: Hutchinson Education, 1984.

Knox, Paul L. "The Restless Urban Landscape: Economic and Sociocultural Change and the Transformation of Metropolitan Washington, DC." *Annals of the Association of American Geographers* 81 (1991): 181–209.

Kristensen, Gustav. "Women's Economic Progress and the Demand for Housing: Theory, and Empirical Analyses Based on Danish Data." *Urban Studies* 34 (1997): 403–18.

Lawrence, Henry W. "The Greening of the Squares of London: Transformation of Urban Landscapes and Ideals." *Annals of the Association of American Geographers* 83 (1993): 90–118.

Lemon, James. "Liberal Dreams and Nature's Limits: Great Cities of North America Since 1600." *Annals of the Association of American Geographers* 86 (1996): 745–66.

Ley, David. *A Social Geography of the City.* New York: Harper and Row, 1983.

———. "Alternative Explanations for Inner-City Gentrification: A Canadian Assessment." *Annals of the Association of American Geographers* 76 (1986): 521–35.

Leyshon, Andrew, and Nigel Thrift. "Spatial Financial Flows and the Growth of the Modern City." *International Social Science Journal* 49 (1997): 41–54.

Lowder, Stella. *Inside Third World Cities.* London: Routledge, 1988.

Lynch, Kevin. *The Image of the City.* Cambridge, MA: M.I.T. Press, 1960.

Madden, Janice Fanning. "Changes in the Distribution of Poverty Across and Within the US Metropolitan Areas, 1979–89." *Urban Studies* 33 (1996): 1581–1600.

Mayer, Harold M., and Charles R. Hayes. *Land Uses in American Cities.* Champaign, IL: Park Press, 1983.

McMillen, Daniel P. "One Hundred Fifty Years of Land Values in Chicago: A Nonparametric Approach." *Journal of Urban Economics* 40 (1996): 100–24.

Mumford, Lewis. *The City in History.* New York: Harcourt, Brace, and World, 1961.

Nevarez, Leonard. "Just Wait Until There's a Drought: Mediating Environmental Crises for Urban Growth." *Antipode* 28 (1996): 246–72.

Newman, Peter, and Andy Thornley. *Urban Planning in Europe: International Competition, National Systems and Planning Projects.* London: Routledge, 1996.

Park, Robert E.; Ernest W. Burgess, and Roderick D. McKenzie, eds. *The City.* Chicago: University of Chicago Press, 1925.

Peach, Ceri. "Does Britain Have Ghettos?" *Transactions of the Institute of British Geographers New Series* 21 (1996): 216–35.

Puga, Diego. "Urbanization Patterns: European Versus Less Developed Countries." *Journal of Regional Science* 38 (1998): 231–52.

Satterthwaite, David. "Sustainable Cities or Cities that Contribute to Sustainable Development." *Urban Studies* 34 (1997): 1667–92.

Short, John. *The Humane City.* New York and Oxford: Basil Blackwell, 1989.

Tanaka, Atsuko; Takehito Takano, and Sachiko Takeuchi. "Health Levels Influenced by Urban Residential Conditions in a Megacity—Tokyo." *Urban Studies* 33 (1996): 879–94.

Teitz, Michael B. "American Planning in the 1990s: Evolution, Debate and Challenge." *Urban Studies* 33 (1996): 649–72.

United Nations. *World Urban Agglomerations.* New York: U.N. Population Division, 1992.

United States National Advisory Commission on Civil Disorders, Otto Kerner, chairman. *Report.* New York: Dutton, 1968.

Vance, James E., Jr. *The Continuing City: Urban Morphology in Western Civilization.* Baltimore and London: The Johns Hopkins University Press, 1990.

Wexler, Martin E. "A Comparison of Canadian and American Housing Policies." *Urban Studies* 33 (1996): 1909–22.

Wheaton, William C. "Land Use and Density in Cities with Congestion." *Journal of Urban Economics* 43 (1998): 258–72.

White, Paul. *The West European City: A Social Geography.* London and New York: Longman, 1984.

Whyte, William H. *City: Rediscovering the Center.* New York: Doubleday, 1988.

Wilson, David, ed. *Globalization and the Changing U.S. City.* Thousand Oaks, CA: Sage Publications, 1997.

Wirth, Louis. *On Cities and Social Life.* Chicago: University of Chicago Press, 1964.

Yeung, Yue-man. "Geography in the Age of Mega-cities." *International Social Science Journal* 49 (1997): 91–104.

Also consult these journals: *Environment and Planning; Journal of Housing; Journal of the American Planning Association; Land Economics; Planning; Urban Geography; Urban Land; Urban Studies.*

CHAPTER 14

Resource Issues

When you have finished drinking a soda, do you pitch it in the trash or place it in a recycling bin? In winter, if you feel cold, do you put on a sweater, or do you turn up the thermostat? Do you normally eat with disposable plates, cups, and plastic utensils, or do you use washable ceramic and stainless steel products? When you leave a room, do you turn off the lights and television?

When you buy a new motor vehicle, do you consider its fuel efficiency? Do you care that your family's sport utility vehicle gets poorer gas mileage than your grandparent's big "old-fashioned" sedan?

People have always transformed Earth's land, water, and air for their benefit. But human actions in recent years have gone far beyond the impact of the past. The magnitude of transformations is disproportionately shared by North Americans; with only one-twentieth of Earth's population, North Americans consume one-fourth of the world's energy and generate one-fourth of many pollutants. Elsewhere in the world, 2 billion people live without clean water or sewers. One billion live in cities with unsafe sulfur dioxide levels.

Future generations will pay the price if we continue to mismanage Earth's resources. Our shortsightedness could lead to shortages of energy to heat homes and operate motor vehicles. Our carelessness has already led to unsafe drinking water and toxic air in some places. Our inefficiency could lead to shortages of food.

Humans once believed Earth's resources to be infinite, or at least so vast that human actions could never harm or deplete them. But warnings from scientists, geographers, and governments are making clear that resources are indeed a problem. Earth Day 1970 alerted the world to the magnitude of damage that people have done to the environment. Three decades later, we have learned much about the processes that produce environmental problems, and the long-term consequences of environmental mismanagement.

Key Issues

1. Why are fossil fuel resources being depleted?

2. Why are resources being polluted?

3. Why are global food resources expandable?

Sorting trash and recycling in a sixth grade class. (B. Daemmrich/The Image Works)

Case Study

Pollution in Mexico City

Eight-year-old Carlos and nine-year-old Maria, residents of Mexico City, did not go to school today. Nor did many of their classmates. And many of their teachers failed to report for work. These people did not leave their homes, because they feared that breathing outside air in Mexico City would be too dangerous.

For much of the year, a stationary cloud hangs over Mexico City, producing a gray-brown fog that irritates the eyes and burns the throat. Residents report frequent conjunctivitis and other eye disorders, skin rashes, bronchitis, other respiratory diseases, and increased susceptibility to heart attacks. The health benefits of outdoor sports like soccer and running are outweighed by the health risks of breathing the air. Pregnant women are cautioned that living in Mexico City increases risk to fetal health.

This severe air pollution partly results from Mexico City's setting: it rests in a basin some 2,250 meters (about 7,400 feet) above sea level, surrounded by a semicircle of volcanic peaks as high as 5,545 meters (16,900 feet). This giant bowl is open only to the north. Prevailing winds from the north enter the basin and back polluted air against the surrounding mountains. Consequently, emissions from cars and factories are trapped close to the ground in a stationary cloud, especially in the winter, when the climate is cool and dry, and winds are calm.

Because the city is at a high altitude, the level of available oxygen is low. Consequently, fossil fuels burn less completely than at lower altitudes, and burning them produces more carbon monoxide and ozone.

Three-fourths of the emissions come from burning fuels in more than 2 million motor vehicles. Natural phenomena (such as fires) and industrial sources account for much of the remainder of the air pollution. Many larger industries are concentrated to the northern part of the valley, so their emissions are blown across the city by the prevailing winds.

Air pollution is not Mexico City's only environmental problem. Inadequately treated sewage flows into nearby rivers, and 30 percent of the city's homes are not even connected to the sewer system. Solid waste is deposited at large municipal dumps, where 17,000 people known as *pepenadores*, or garbage pickers, survive by going through rubbish and, in many cases, actually live at the dump. Dust from fecal matter in unsewered areas increases skin and eye infections.

Plants and animals live in harmony with their environment, but people often do not. Geographers study this troubled relationship between human actions and the physical environment in which we live. From our perspective, Earth offers a large menu of resources available for our use. A **resource** is a substance in the environment that is useful to people, is economically and technologically feasible to access, and is socially acceptable to use. A resource could include food, water, soil, plants, animals, and minerals.

The problem is that most resources are limited, and Earth has a tremendous number of consumers. Geographers observe three major misuses of resources:

1. We deplete scarce resources, especially petroleum, natural gas, and coal, for energy production.
2. We destroy resources through pollution of air, water, and soil.
3. We fail to use resources efficiently, especially the global food supply.

These three misuses are the basic themes of this chapter.

As with other topics, geographers look first at *where* resources are distributed. Some regions are relatively well-endowed with minerals, water, and other resources, whereas other regions have limited suppliers. The reason *why* problems arise from this uneven distribution is that resources are often located in places different from their users. Differences in demand may arise from the uneven distribution of people across Earth, or from variations in development.

Nowhere is the *globalization* trend more pronounced than in the study of resources. The global economy depends on the availability of natural resources to produce the goods and services that people demand. Global uniformity in cultural preferences means that people in different places value similar natural resources, although not everyone has the same access to them. In a global environment, misuse of a resource in one location affects the well-being of people everywhere.

To study resource problems, we also depend on our understanding of *local diversity*. As geographers, we understand that our energy problems derive from depletion of resources in particular regions and from differences in how consumers use resources in different places. We see that the pollution problem comes from the concentration of substances that harm the physical environment in particular regions. We find that the food supply problem relates in part to identifying the limits of a region's physical environment and in part to the regional distributions of food production and need.

Key Issue 1

Why Are Fossil Fuel Resources Being Depleted?

- Dependence on fossil fuels
- Alternative energy sources

We depend on abundant, low-cost energy to run our industries, transport ourselves, and keep our homes comfortable. However, we are depleting the global supply of some energy resources. More developed countries want to preserve current standards of living, and less developed countries are struggling to attain a better standard. All this demands tremendous energy resources, so as we deplete our current sources of energy we must develop alternative ones.

Historically, people relied on power supplied by themselves or by animals, known as **animate power**. Energy from burning wood or flowing water later supplemented animate power. Ever since the industrial revolution began in the late 1700s, humans have expanded their use of **inanimate power**, generated from machines. Humans have found the technology to harness the great potential energy stored in resources such as coal, oil, gas, and uranium.

Dependence on Fossil Fuels

Three of Earth's substances provide more than 80 percent of the world's energy: oil, natural gas, and coal (Figure 14–1). In MDCs, the remainder comes primarily from nuclear, solar, hydroelectric, and geothermal power. Burning wood provides much of the remaining energy in less developed societies.

Historically, the most important energy source worldwide was **biomass fuel**, such as wood, plant material, and animal waste. Biomass fuel is burned directly or converted to charcoal, alcohol, or methane gas. Biomass remains the most important source of fuel in some LDCs, but during the past 200 years MDCs have converted to other energy sources.

As a consequence of the industrial revolution, coal supplanted wood as the leading energy source in the late 1800s in North America and Western Europe. Petroleum was first pumped in 1859, but it was not an important resource until the diffusion of automobiles in the twentieth century. Natural gas was originally burned off as a waste product of oil drilling but now heats millions of homes.

Energy is used in three principal places: businesses, homes, and transportation. For U.S. businesses, the main energy resource is coal, followed by natural gas and oil. Some businesses directly burn coal in their own furnaces. Others rely on electricity, mostly generated at coal-burning power plants. At home, energy is used primarily for heating of living space and water. Natural gas is the most common source, followed by petroleum (heating oil and kerosene). Almost all transportation systems operate on petroleum products, including automobiles, trucks, buses, airplanes, and most railroads. Only subways, streetcars, and some trains run on coal-generated electricity.

Petroleum, natural gas, and coal are known as fossil fuels. A **fossil fuel** is the residue of plants and animals that became buried millions of years ago. As sediment accumulated over these remains, intense pressure and chemi-

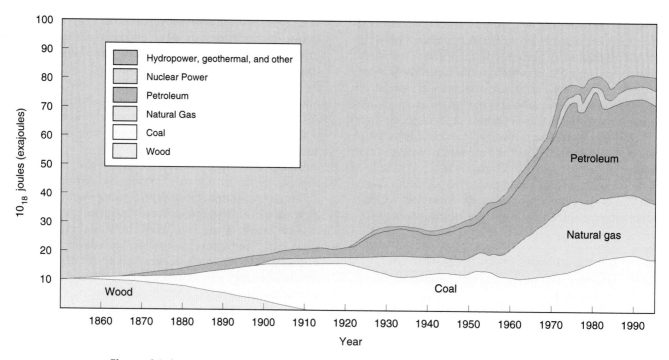

Figure 14–1 U.S. energy consumption. U.S. energy consumption increased rapidly during the 1960s, but since the early 1970s it has increased at a much slower rate. The amount of energy derived from petroleum and natural gas increased rapidly in the 1960's, while use of coal stagnated.

cal reactions slowly converted them into the fossil fuels we use today. When we burn these substances today, we are releasing energy originally stored in plants and animals millions of years ago.

Two characteristics of fossil fuels cause great concern for the future:

1. **The supply of fossil fuels is finite.** Once the present supply of fossil fuels is consumed, it is gone, and we must look to other resources for our energy. (Technically, fossil fuels are continually being formed, but the process takes millions of years, so humans must regard the current supply as essentially finite.)

2. **Fossil fuels are distributed unevenly around the globe.** Some regions enjoy a generous supply of fossil fuels, while others have little, and fossil fuels are not consumed in the same regions where they are produced.

Finiteness of Fossil Fuels

To understand Earth's resources, we distinguish between those that are renewable and those that are not:

- **Renewable energy** is replaced continually, or at least within a human lifespan: solar energy, hydroelectric, geothermal, fusion, and wind are examples. Renewable energy has an essentially unlimited supply and is not depleted when used by people.

- **Nonrenewable energy** forms so slowly that for practical purposes it cannot be renewed: the fossil fuels, as well as nuclear energy, are examples.

As nonrenewable energy sources, the three main fossil fuels, once burned, are used up for all time. The world faces an energy problem in part because we are rapidly depleting the remaining supply of the three fossil fuels, especially petroleum. Because of dwindling supplies of fossil fuels, most of the buildings in which we live, work, and study will have to be heated another way. Cars, trucks, and buses will have to operate on some other energy source. The many plastic objects that we use (because they are made from petroleum) must be made with other materials.

We can use other resources for heat, fuel, and manufacturing, but they are likely to be more expensive and less convenient to use than fossil fuels. And converting from fossil fuels will likely disrupt our daily lives and cause us hardship.

Remaining Supply of Fossil Fuels.   How much of the fossil fuel supply remains? Despite the critical importance of this question for the future, no one can answer it precisely. Because petroleum, natural gas, and coal are deposited beneath Earth's surface, considerable technology and skill are required to locate these substances and estimate their volume.

The amount of energy remaining in deposits that have been discovered is called a **proven reserve**. Proven reserves can be measured with reasonable accuracy—

about 1 trillion barrels of petroleum, about 140 trillion cubic meters of natural gas, and about 1 quadrillion metric tons of coal.

But how many deposits in the world have not yet been discovered? The energy in undiscovered deposits that are thought to exist is a **potential reserve**. When a potential reserve is actually discovered, it is reclassified as a proven reserve. The World Energy Council estimates potential oil reserves of about 500 billion barrels, with the largest fields thought to lie beneath the South China Sea and northwestern China.

To determine when remaining reserves of an energy source will be depleted, we must know the rate at which the resource is being consumed. At the current world petroleum consumption rate of about 25 billion barrels a year, Earth's proven petroleum reserves of 1 trillion barrels will last 40 years.

New petroleum deposits are being discovered each year and added to the inventory of proven reserves (thus extending the number of years of remaining supply), but petroleum is being consumed at a more rapid rate than it is being found, and world demand is increasing by more than 1 percent annually. Unless substantial new proven reserves are found—or consumption decreases sharply—the world's petroleum reserves will be depleted sometime in the twenty-first century.

Similarly, at current rates of use the world's proven reserves of natural gas will last for about 80 years. Proven reserves of natural gas are less extensive than petroleum reserves, but the remaining supply is projected to last longer because the world currently uses much more oil than gas. However, if energy users switched from petroleum to natural gas, then the proven reserves of petroleum would last longer and natural gas would be depleted more quickly.

For coal, the immediate future is less grim. At current consumption, proven coal reserves can last at least several hundred years. Today, about 55 percent of U.S. electricity comes from power plants that burn coal.

Extraction of Remaining Reserves. Although scientists differ on the volume of potential reserves, they agree that extracting proven reserves will grow harder. When it was first exploited, petroleum "gushed" from wells drilled into rock layers saturated with it. Coal was quarried in open pits.

But now, extraction is harder. Sometimes pumping is not sufficient to remove petroleum, so water or carbon dioxide may be forced into wells to push out the remaining resource. Oil companies have reduced their expenditures for new drilling by about two-thirds since the 1980s. Coal mining continues in some thick, high-quality coal seams, both in open pits and underground, but already more mining is being done in thinner, poorer-quality coal.

The problem of removing the last reserves from a proven field is comparable to wringing out a soaked

towel. It is easy to quickly remove the main volume of water, but the last few percent require more time and patience, and perhaps special technology.

The largest, most accessible deposits of petroleum, natural gas, and coal already have been exploited. Newly discovered reserves generally are smaller and more remote, such as beneath the seafloor, where extraction is costly. Exploration cost has increased because methods are more elaborate and the probability of finding new reserves is less.

Unconventional sources of petroleum and natural gas are being studied and developed, such as oil shale and tar sandstones. Oil shale is a "rock that burns" because of its tar-like content. Tar sandstones are saturated with a thick petroleum. They are called unconventional because methods currently used to extract resources won't work—instead, the rocks must be "cooked" to melt out their petroleum. These are also known as unconventional sources because we do not currently have economically feasible, environmentally sound technology to extract them.

Utah, Wyoming, and Colorado contain more than ten times the petroleum reserves of Saudi Arabia, but as oil shale. The cost of conventional oil resources must increase dramatically before these unconventional sources will become profitable. Even then, the adverse environmental impacts of using these sources is likely to be high.

Uneven Distribution of Fossil Fuels

Geographers observe two important inequalities in the global distribution of fossil fuels:

1. Some regions have abundant reserves, whereas others have little.
2. Consumption of fossil fuels is much higher in some regions than in others.

Given the centrality of fossil fuels in a society's economy and culture, unequal possession and consumption of fossil fuels have been major sources of global instability in the world.

Location of Reserves. Why do some regions have abundant reserves of one or more fossil fuels, but other regions have little? This partly reflects how fossil fuels form.

Coal forms in tropical locations, in lush, swampy areas rich in plants. Thanks to the slow movement of Earth's drifting continents, the tropical swamps of 250 million years ago have relocated to the midlatitudes. As a result, today's main reserves of coal are in midlatitude countries, rather than in the tropics. The United States and Russia each have 22 percent of proven coal reserves, European countries (especially Germany and Poland) 14 percent, China 11 percent, Australia 8 percent, and India 6 percent (Figure 14–2).

Similarly, today's sources of oil and natural gas formed millions of years ago from sediment deposited on the

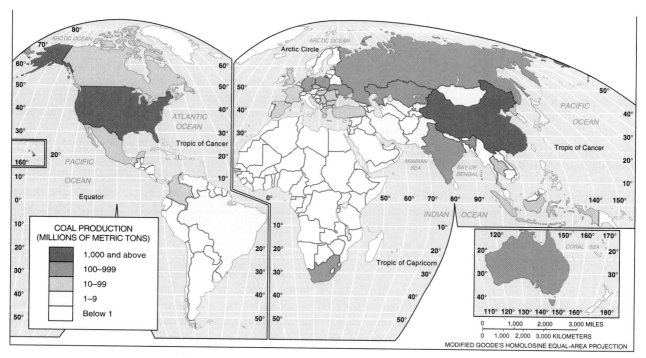

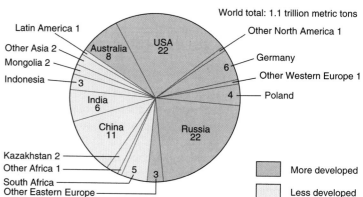

Figure 14–2 Coal production and proven reserves. China and the United States are the largest producers. The United States and Russia have the largest proven reserves of coal.

seafloor. Some oil and natural gas reserves still lie beneath such seas as the Persian Gulf and the North Sea, but other reserves are located beneath land that had been under water millions of years ago, when sea level was higher.

Nearly two-thirds of the world's oil reserves are in five Middle Eastern countries—about 26 percent in Saudi Arabia and about 10 percent each in Iran, Iraq, Kuwait, and United Arab Emirates. Mexico and Venezuela have the most extensive proven reserves in the Western Hemisphere. The United States accounts for about 10 percent of the world's annual production of petroleum but possesses less than 10 percent of the proven reserves (Figure 14–3).

Russia possesses 37 percent of the world's proven natural gas reserves. The United States currently produces 24 percent of the world's natural gas, but its proven reserves are extremely limited (Figure 14–4).

Taken as a group, more developed countries historically have possessed a disproportionately high percentage of the world's fossil fuel reserves. Europe's nineteenth-century industrial development depended on its abundant coalfields, and extensive coal and petroleum supplies helped the United States to become the leading industrial power of the twentieth century. A handful of less developed countries in Africa, Asia, and Latin America have extensive reserves of one or another of the fossil fuels, but most have little.

During the nineteenth and twentieth centuries, the MDCs produced most of the world's fossil fuels. But in the twenty-first century, this dominance is likely to end. Many of Europe's coal mines have closed in recent years, because either the coal was exhausted, or the remaining supply was too expensive to extract, and the region's petroleum and natural gas (in the North Sea) account for small percentages of worldwide reserves. The United States still has extensive coal reserves, but its petroleum and natural gas reserves are being depleted rapidly. Japan has never had significant fossil fuel reserves.

Most of the world's proven reserves (and probably potential reserves) are in a handful of Asian countries, especially China, the Middle East, and former Soviet Union republics. How these reserves are divided up between more developed and less developed countries (as well as among LDCs) is a critical issue for the world community in the twenty-first century.

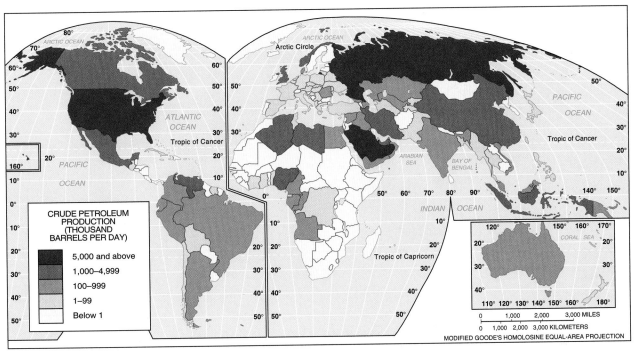

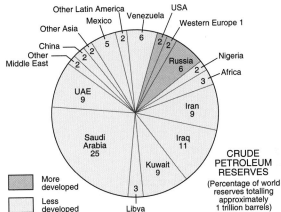

Figure 14–3 Petroleum production and proven reserves. Saudi Arabia is the largest producer and has the largest proven reserves of petroleum. The United States is a major producer but has limited reserves.

Consumption of Fossil Fuels. The global pattern of fossil fuel consumption—like production—will shift in the twenty-first century. More developed countries, with about one-fourth of the world's population, currently consume about three-fourths of the world's energy. Annual per capita consumption of energy exceeds 300 million BTUs in North America and 100 million BTUs in Western Europe, compared to less than 25 million in most less developed countries (Figure 14–5). This high energy consumption by a modest percentage of the world's population supports a lifestyle rich in food, goods, services, comfort, education, and travel.

The sharp regional difference in energy consumption has two geographic consequences for the future:

- As they promote development and cope with high population growth, LDCs will consume much more energy. As a result of increased demand in LDCs, global consumption of petroleum is expected to increase by about 50 percent during the next two decades, while both coal and natural

gas consumption are expected to double. The share of world energy consumed by LDCs will increase from about 25 percent today to 40 percent by 2010 and 60 percent by 2020.

- Because MDCs consume more energy than they produce, they must import more fossil fuels, especially petroleum, from LDCs. The United States and Western Europe import more than half their petroleum, and Japan more than 90 percent. However, because of development and population growth in LDCs, the more developed countries will face greater competition in obtaining the world's remaining supplies of fossil fuels.

Control of World Petroleum

The sharpest conflicts over energy will be centered on the world's limited proven reserves of petroleum. The United States produced more petroleum than it consumed during the first half of the twentieth century. Beginning in the 1950s, the handful of large transna-

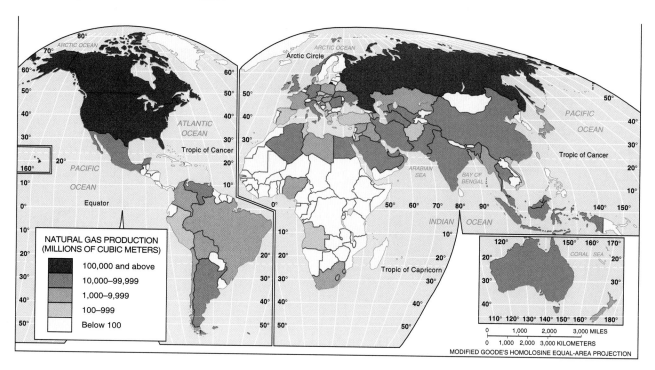

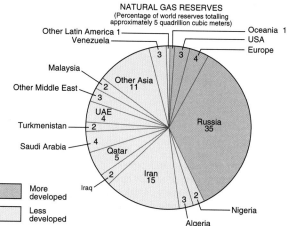

Figure 14–4 Natural gas production and proven reserves. Russia is the largest producer and has the largest proven reserves of natural gas. The United States is a major producer but has limited reserves.

tional companies then in control of international petroleum distribution determined that extracting domestic petroleum was more expensive than importing it from the Middle East. Thus, U.S. petroleum imports have increased from 14 percent of total consumption in 1954 to more than 50 percent today. European countries and Japan have always depended on foreign petroleum because of limited domestic supplies.

The MDCs import most of their petroleum from the Middle East, where most of the world's proven reserves are concentrated. Both U.S. and Western European transnational companies originally exploited Middle Eastern petroleum fields and sold the petroleum at a low price to consumers in MDCs. At first, Western companies set oil prices and paid the Middle Eastern governments only a small percentage of their oil profits. But government policies changed in the petroleum-producing countries, especially during the 1970s. Foreign-owned petroleum fields were either nationalized or more tightly controlled, and prices were set by governments rather than by petroleum companies.

OPEC Policies During the 1970s. Several LDCs possessing substantial petroleum reserves created the Organization of Petroleum Exporting Countries (OPEC) in 1960. Arab OPEC members in the Middle East are Algeria, Iraq, Kuwait, Libya, Qatar, Saudi Arabia, and United Arab Emirates. Another Middle East OPEC member, Iran, is not Arab. OPEC countries elsewhere in the world include Indonesia, Gabon, Nigeria, and Venezuela. Ecuador was a member until 1993.

OPEC's Arab members were angry at North American and Western European countries for supporting Israel during that nation's 1973 war with the Arab states of Egypt, Jordan, and Syria. So during the winter of 1973/74, they flexed their new economic muscle with a boycott—Arab OPEC states refused to sell petroleum to the nations that had supported Israel.

Soon, gasoline supplies dwindled in MDCs. Each U.S. gasoline station was rationed a small quantity of fuel, which ran out early in the day. Long lines formed at gas stations and some motorists waited all night for fuel. Gasoline was rationed by license plate number (cars with

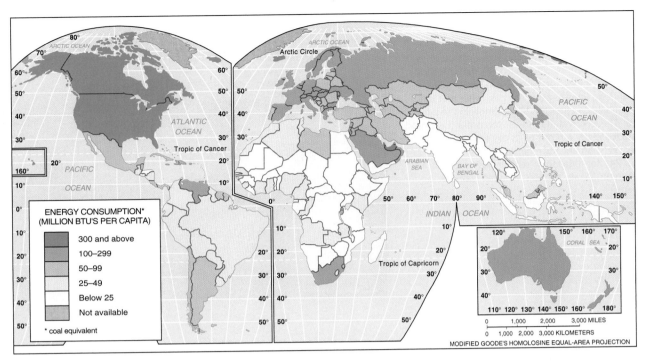

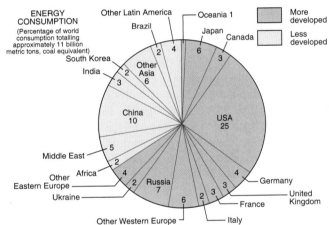

Figure 14–5 Per capita energy consumption. More developed countries consume much more energy per capita than do less developed countries. The United States, with 5 percent of the world's population, consumes about one-fourth of the world's energy.

licenses ending in an odd number could buy only on odd-numbered days). European countries took more drastic action—the Netherlands, for example, banned all but emergency motor vehicle travel on Sundays.

OPEC lifted the boycott in 1974 but raised petroleum prices from $3 per barrel to more than $35 by 1981. Prices at U.S. gas pumps soared from an average of 39¢ in 1973 to $1.38 in 1981. Thus did OPEC force a massive increase in oil cost for Western countries. To import oil, U.S. consumers spent $3 billion in 1970, but $80 billion in 1980.

The rapid escalation in petroleum prices caused severe economic problems in MDCs during the 1970s. Production of steel, motor vehicles, and other energy-dependent industries plummeted in the United States in the wake of the 1973/74 boycott and have never regained their pre-boycott levels (recall Figure 11–21, which shows declining steel production in more developed countries since the 1970s). Many manufacturers were forced out of business by soaring energy costs, and the

survivors were forced to restructure their operations to regain international competitiveness.

The LDCs were hurt even more. They depended on low-cost petroleum imports to spur industrial growth. Their fertilizer costs shot up, because many fertilizers are derived from oil. North American and Western European states cushioned themselves by creating a profitable return path for money that was going to OPEC: they encouraged OPEC countries to invest in American and European real estate, banks, and other safe and profitable investments. Comparable investment opportunities were limited in less developed countries.

Reduced Dependency on OPEC. Internal conflicts weakened OPEC's influence in the 1980s and 1990s. Iraq warred with Iran, and Libya grew more radical, supporting terrorists. By not acting together, individual OPEC members produced more petroleum than the world demanded, and MDCs stockpiled some of the surplus as protection against another boycott.

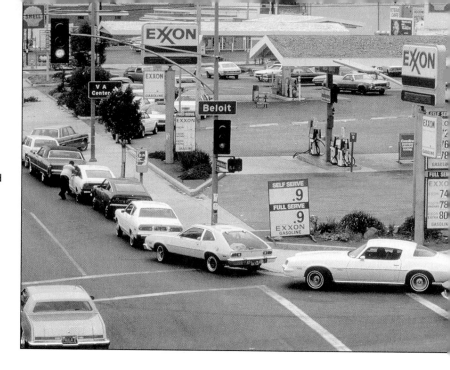

The Organization of Petroleum Exporting Countries refused to sell petroleum to North American and Western European countries for a few months during the winter of 1973–74 to protest Western countries support for Israel in the October 1973 war. Motorists in the United States, such as these in Los Angeles, waited in long lines to purchase gas. Note that the cars are in line to enter the Shell station in the background; The Exxon station in the foreground is closed because it already ran out of its day's allotment of fuel. In 1973, Americans regarded Exxon's posted price of 74¢ for a gallon of regular gas to be outrageously high. (Craig Aurness/Woodfin Camp & Associates)

The price of petroleum plummeted from over $30 to below $10 per barrel during the 1980s. During the 1991 Gulf War, the price rose briefly to about $40, then settled at about $20, less than the price before the 1973/74 boycott when accounting for inflation.

Conservation measures have also dampened demand for petroleum in developed countries. The average car sold in the United States got less than 16 miles per gallon in 1975, compared to 28 miles per gallon in 1998. However, the savings from more efficient cars have been offset by increased sales of gas-guzzling sport utility vehicles and pickup trucks that average under 20 mpg. Further, the average vehicle is being driven more in a year.

The United States has reduced petroleum imports from Arab OPEC countries by one-third since the 1970s, and only about 10 percent of total U.S. petroleum consumption now comes from these countries, mostly Saudi Arabia. Instead, the United States imports more petroleum from countries that are firm allies located in the Western Hemisphere, especially Venezuela, Canada, and Mexico. Europeans and Japanese have also decreased dependency on Arab OPEC countries. But the Middle East remains the world's principal source of petroleum (Figure 14–6), and given the global distribution of proven reserves, the Middle East is likely to account for an even higher share of trade in the future.

The world will not literally "run out" of petroleum during the twenty-first century. However, at some point extracting the remaining petroleum reserves will prove so expensive and environmentally damaging that use of alternative energy sources will accelerate, and dependency on petroleum will diminish.

The issues for the world are whether dwindling petroleum reserves are handled wisely and other energy sources are substituted peacefully. Given the massive growth in petroleum consumption expected in LDCs such as China and India, the United States and other more developed countries may have little influence over when prices rise and supplies decline.

Problems with Coal

Coal could substitute for petroleum during the next couple of centuries, especially in the United States, which possesses large proven coal reserves (refer to Figure 14–2). But problems hinder expanded use of coal: air pollution, mine safety, land subsidence and erosion, economics, and use as automotive fuel. Here is a look at each of these problems.

Air Pollution. Uncontrolled burning of coal releases several pollutants into the atmosphere: sulfur oxides, hydrocarbons, carbon dioxide, and particulates ("soot") into the atmosphere. Many communities suffered from coal-polluted air earlier in this century and encouraged their industries to switch to cleaner-burning natural gas and oil.

The U.S. Clean Air Act now requires utilities to use better-quality coal or to install "scrubbers" on smokestacks. These methods can work; Pittsburgh, Pennsylvania, once noted for terrible air pollution when coal was burned for steel mills and glass factories, today has remarkably clean air. But coal-fired power plants still pump copious carbon dioxide into the atmosphere.

Mine Safety. Historically, mining was an especially dangerous occupation. One thousand miners once died annually in the United States, especially in underground mines. Miners also are prone to "black lung" disease, for which the U.S. government pays several billion dollars per year in compensation.

Strictly enforced U.S. mine safety laws, improved mine ventilation, intensive safety programs, automation

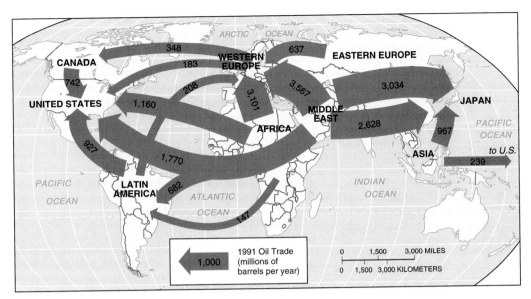

Figure 14–6 World petroleum trade. Because petroleum is not produced in the same countries where it is consumed, 21 million barrels a day are carried by tankers from one country to another. More petroleum is transported internationally than the next three commodities combined (iron ore, coal, and grain), when figured by weight.

of mining, and a smaller workforce have made the American coal industry much safer. Annual U.S. mine mortality now is below 100. But that figure could rise if mining operations expanded.

Subsidence and Erosion. Both surface mining and underground mining can cause environmental damage. Underground mining may release acidic groundwater that may pollute water by draining into streams, and subsidence or sinking of the ground can damage buildings. The removal of trees and other vegetation during surface mining can cause soil erosion. In the United States, the mining industry is highly regulated, and most companies today have a good record of "cleaning up after themselves." But less sensitive mining practices in the past have left a legacy of environmental damage.

Economics. Although heavy, bulky, and expensive to transport, coal must be shipped long distances, because most of the factories and power plants using it are far from the coalfields. Ironically, the principal methods of transporting coal—barge, rail, or truck—are all powered by petroleum. A considerable amount of energy thus is expended to mine and transport coal so that it can be used to generate energy somewhere else.

Powering Motor Vehicles. Ever since motor vehicles appeared about a century ago, petroleum has been the basis for modern transportation. To substitute for petroleum, coal must be adapted for motor vehicles. The simplest way is to generate electricity by burning coal, and use the electricity to charge batteries in electric vehicles.

The use of electric vehicles is expanding in more developed countries, primarily to reduce air pollution rather than conserve petroleum. California and several East Coast states require that, by the year 2002, 10 percent of all new vehicles must generate "zero emissions"—essentially electric vehicles, given current technology. Motor vehicle companies now offer electric vehicles for sale or lease in these areas.

In opinion polls, consumers express a willingness to pay a premium for the sake of saving energy and reducing pollution. On the other hand, the major motor vehicle producers, as well as market research firms, have evidence that few consumers are actually willing to incur the additional cost of an electric vehicle.

Even if the costs of owning and operating electric vehicles are substantially reduced, consumers will be reluctant to use them because of their limited range. Electric vehicles must be recharged frequently, especially if the air conditioning and electronic equipment are intensively used. Improvements in batteries will eventually reduce the frequency and time needed for recharging, but electric vehicles will be hard-pressed to match the cost and convenience of gasoline-powered vehicles.

Electric vehicles are likely to look attractive to motorists only if gasoline becomes hard to find and very expensive. During the 1970s, the prospect of very scarce and expensive petroleum seemed imminent, but fears proved exaggerated. Today, consumers perceive unlimited cheap petroleum, but that optimism may soon prove exaggerated as well. By the middle of the twenty-first century, large gasoline-powered vehicles will be limited to specialized tasks—or consigned to museums.

Alternative Energy Sources

Coal-generated electricity can reduce the impact of depleted petroleum reserves over the next century or two. But coal, like petroleum, is a nonrenewable fossil fuel that will eventually be depleted. It offers little hope to countries that lack substantial coal reserves. In the long run, energy problems can be solved only by converting to renewable resources. Recall that a renewable energy source has an essentially unlimited supply and is not depleted when used by people.

Two types of energy sources—nuclear and solar—figure most prominently in current energy planning. Other alternatives to fossil fuels may become more important in the future.

Nuclear Energy

The big advantage of nuclear power is the large amount of energy that is released from a small amount of material. One kilogram of enriched nuclear fuel contains more than 2 million times the energy in 1 kilogram of coal.

Nuclear power supplies about 25 percent of electricity in North America and Europe, including 86 percent in Lithuania, 76 percent in France, 55 percent in Belgium, and more than 40 percent in Bulgaria, Hungary, Slovakia, Sweden, and Switzerland (Figure 14–7). Elsewhere in the world, Japan, South Korea, and Taiwan draw more than 30 percent of their electricity from nuclear power.

Altogether, about three dozen countries make some use of nuclear power to generate electricity, three of which are responsible for about 59 percent of the total: the United States (about 30 percent), France (about 16 percent), and Japan (about 12 percent). Although the world's largest generator, the United States, depends less than some European countries on nuclear energy because of more abundant coal reserves.

Dependency on nuclear power varies widely among U.S. states. Nuclear power accounts for more than 70 percent of electricity in Maine, New Jersey, and Vermont; 60 percent in New Hampshire and South Carolina; and 50 percent in Illinois. At the other extreme, 18 states have no nuclear power (Figure 14–8). New England draws 52 percent of its electricity from nuclear power, the Southeast 27 percent, and the Midwest 23 percent, compared to about 10 percent for states west of the Mississippi River.

Nuclear power presents serious problems. These include potential accidents, radioactive waste, generation of plutonium, a limited uranium supply, geographic distribution, and cost.

Potential Accidents. A nuclear power plant produces electricity from energy released by splitting uranium atoms in a controlled environment, a process called **fission**. One product of all nuclear reactions is **radioactive waste**, certain types of which are lethal to people exposed to it. Elaborate safety precautions are taken to prevent nuclear fuel from leaking from a power plant.

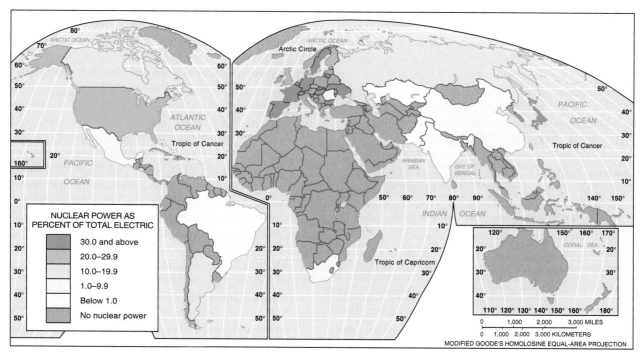

Figure 14–7 Nuclear power as percent of electricity. Nuclear power has been especially attractive to more developed countries in Europe that lack abundant reserves of either petroleum or coal.

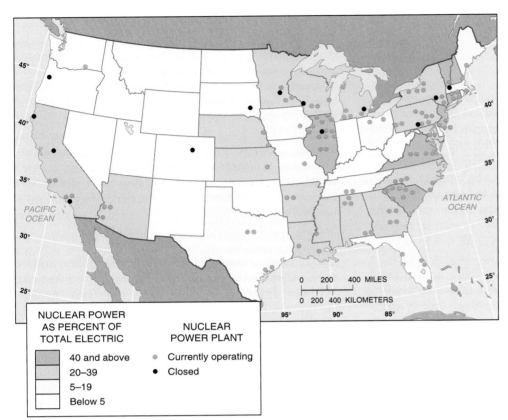

Figure 14–8 U.S. nuclear power plants and nuclear power as percent of electricity by state. Nuclear power is an important source of electricity in a number of northeastern and midwestern states. At a number of locations, more than one nuclear plant has been built.

NUCLEAR POWER AS PERCENT OF TOTAL ELECTRIC	NUCLEAR POWER PLANT
40 and above	● Currently operating
20–39	● Closed
5–19	
Below 5	

Nuclear power plants cannot explode, like a nuclear bomb, because the quantities of uranium are too small and cannot be brought together fast enough. However, it *is* possible to have a runaway reaction, which overheats the reactor, causing a meltdown, possible steam explo-

Workers carefully replace fuel rods in the reactor core at the James A. Fitzpatrick Nuclear Power Plant in Scriba, near Syracuse, New York. (Crandall/The Image Works)

sions, and scattering of radioactive material into the atmosphere. This happened in 1986 at Chernobyl, then in the Soviet Union and now in the north of Ukraine, near the Belarus border.

The Soviet Union reported at the time that the Chernobyl accident caused 31 deaths, including two at the accident site itself and 29 elsewhere to people who were exposed to severe radiation and burns. Most scientists dismiss these figures as unrealistically low, because hundreds of thousands of residents and cleanup workers were exposed to high levels of radiation. For example, officials in Belarus, where 70 percent of the fallout hit, report that cancer cases increased 45 percent in the districts closest to the plant during the first 5 years after the accident.

The impact of the Chernobyl accident extended through Europe. Most European governments temporarily banned the sale of milk and fresh vegetables, which were contaminated with radioactive fallout. Half of the eventual victims may be residents of European countries other than Ukraine and Belarus.

American nuclear plants are designed with strong, thick containment buildings surrounding the reactors. But nuclear plants built by the former Soviet Union lack containment buildings and often have defective parts. At a Soviet-built plant in East Germany, 11 of 12 cooling pumps were disabled by a fire and power failure. Had the twelfth pump failed, a meltdown, with its inevitable release of strong radioactive materials, likely would have killed the 50,000 inhabitants of the nearby city of

Greifswald. This 1975 accident went unreported for 15 years, making the case for all nuclear plants to be open for inspection.

Radioactive Waste. When nuclear fuel fissions, the waste is highly radioactive and lethal, and remains so for many years. Plutonium can be harvested from it for making nuclear weapons. Pipes, concrete, and water near the fissioning fuel also become "hot" with radioactivity.

No one has yet devised permanent storage for radioactive waste. The waste cannot be burned or chemically treated—it must be isolated for several thousand years until it loses its radioactivity. Spent fuel in the United States is stored "temporarily" in cooling tanks at nuclear power plants, but these tanks are nearly full.

The United States is Earth's third largest country in land area, yet it has failed to find a suitable underground storage site, because of worry about groundwater contamination. Proposals abound: burial at sea, in abandoned mines, and in deep layers of rock salt, or rocketing it into the Sun. But the universal response is NIMBY, which stands for "not in my back yard." People do not want a storage facility near their community.

The time required for radioactive waste to decay to a safe level is far longer than any country or civilization has existed. What government, army, or other human institution will survive for several thousand years to safeguard the stored waste?

Bomb Material. Nuclear power has been used in warfare twice, in August 1945, when the United States dropped an atomic bomb on first Hiroshima and then Nagasaki, Japan, ending World War II. Since then, the Soviet Union (now Russia), the United Kingdom, France, China, and India have tested nuclear weapons, and several other countries are actively developing them, although they have not so stated publicly. No government has dared to use them in a war, because leaders have recognized that a full-scale nuclear conflict could terminate human civilization. But the black market could provide terrorists with enough plutonium to construct nuclear weapons.

A few years ago, a Princeton University student wrote a term paper outlining how to make a nuclear weapon. Most of his information came from an encyclopedia and a few unclassified government documents. More chilling: following publicity about his paper, several organizations and foreign governments contacted him for assistance in making a bomb.

Limited Uranium Reserves. Like fossil fuels, proven uranium reserves are limited—about 60 years at current rates of use. And they are not distributed uniformly around the world—two-thirds of the world's proven uranium reserves are in Australia, the United States, South Africa, and Canada. (Russia and China probably rank among world leaders in production and proven reserves, but their levels are unknown.)

The chemical composition of natural uranium further aggravates the scarcity problem. Uranium ore naturally contains only 0.7 percent U-235, and a greater concentration is needed for power generation.

Uranium is a nonrenewable resource—the world's reserves of minable uranium ore are limited, just like coal or petroleum. Proven uranium reserves could be depleted in three more decades. A **breeder reactor** turns uranium into a renewable resource by generating plutonium, also a nuclear fuel. However, plutonium is more lethal than uranium and could cause more deaths and injuries in an accident. It is also easier to fashion into a bomb. Because of these risks, few breeder reactors have been built, and none are in the United States.

Cost. Nuclear power plants cost several billion dollars to build, primarily because of elaborate safety measures. Without double and triple backup systems, nuclear energy would be too dangerous to use. Uranium is mined in one place, refined in another, and used in still another. The complexities of safe transportation add cost. As a result, the cost of generating electricity is much higher from nuclear plants than from coal-burning plants.

The future of nuclear power has been seriously hurt by the combination of high risks and costs. Most countries in North America and Western Europe have curtailed construction of new plants. Sweden, which received nearly half of its electricity from nuclear power in the 1980s, plans to abandon its nuclear power plants completely by the year 2010. Even in France, where over three-fourths of electricity is generated from nuclear power, public opposition inhibits new development. Nuclear power will decline in other countries as older nuclear plants are closed and not replaced.

Solar Energy

Solar energy is free, does not damage the environment or cause pollution, and is quite safe, unlike nuclear energy or mining coal. The Sun's remaining life is estimated at 5 billion years, so solar energy is the ultimate renewable source. There are two general approaches to harnessing solar energy: passive and active.

Passive solar energy systems capture energy without special devices. When you sit by a window so that sunlight falls directly on you, the Sun's rays penetrate the window and are converted to heat when they strike your skin, making you a passive solar energy collector. If you wear dark clothing, you are warmed even more, because dark objects absorb more energy.

Passive solar energy systems use similar principles—south-facing windows and dark surfaces—to heat homes and buildings on sunny days. Passive solar energy can also be generated by placing on a south-facing roof an insulated glass box containing pipes filled with air or water. When heated, the air or water is piped into the home.

Active solar energy systems collect solar energy and convert it either to heat energy or to electricity (either indirectly or directly):

- In heat conversion, solar radiation is concentrated with large reflectors and lenses to heat water or rocks. These store the energy for use at night and on cloudy days. A place that receives relatively little sunlight can still use solar energy by using more reflectors and lenses and larger storage containers.

- In indirect electric conversion, solar radiation is first converted to heat, then to electricity. The Sun's rays are concentrated by reflectors onto a pipe filled with synthetic oil. The heat from the oil-filled pipe generates steam to run turbines. Several of these plants serve a quarter-million people in California's Mojave Desert.

- In direct electric conversion, solar radiation is captured with **photovoltaic cells**, which convert light energy to electrical energy. Each cell generates a small electric current, but large numbers of them wired together produce significant electricity. Solar-generated electricity is now used in spacecraft, light-powered calculators, and at remote sites where conventional power is unavailable. The cost of cells must drop and their efficiency must improve for solar power to expand.

Solar power can be generated at a central power station, as we now receive most of our power. But solar power also makes feasible individual home systems. An installation costing several thousand dollars provides a solar energy system that supplies virtually all household heat and electricity. This high initial cost is offset by very low monthly operating costs, and is economical for consumers who remain in the same house for a number of years. Individual solar energy users do not face rising electric bills from utilities that pass on their cost of purchasing fossil fuels and constructing facilities.

The United States, Israel, and Japan lead in solar use at home, mostly for water-heating. Solar energy will become more attractive as other energy sources become more expensive. A bright future for solar energy is indicated, for petroleum companies now own the major U.S. manufacturers of photovoltaic cells.

Back in the 1950s, government policy in the United States and Soviet Union was instrumental in developing peacetime use for nuclear energy. Similarly, the speed with which solar energy diffuses in the years ahead depends on the extent of government support. Tax credits for home solar systems and solar-generated electric cars would reduce their cost and encourage consumers to buy them.

Other Energy Sources

Other energy sources include hydroelectric, geothermal, biomass, and fusion. The first three are currently used but offer limited prospects for expansion. Fusion is not a practical source at this time, but may be in the future.

Hydroelectric Power. Water has been a source of mechanical power since before recorded history. It turned water wheels, and the rotational motion was used to grind grain, saw timber, pump water, and operate machines. Over the last hundred years, the energy of moving water has been used to generate electricity, called **hydroelectric power**.

Hydroelectric power is the world's second most popular source of electricity, after coal, supplying about a fourth of worldwide demand. The United States, though, obtains only about 3 percent of its energy through hydroelectric power, and little growth is anticipated, because few acceptable sites to build new dams remain.

Hydroelectric power has drawbacks. Dams may flood formerly usable land, cause erosion, and upset ecosystems. Political problems can result from building dams on rivers that flow through more than one country. Turkey's recently built dam on the Euphrates River was

Wind is a source of renewable energy. Rows of windmills with three rotating blades generate electrical power near Palm Springs, California. (Russell D. Curtis/Photo Researchers, Inc.)

strongly opposed by Syria and Iraq, through which the river also passes. The new dam diverts too much water from the river and makes its water saltier.

Geothermal Energy. Earth's interior is hot from natural nuclear reactions. Toward the surface, heat is especially pronounced in volcanic areas. The hot rocks can encounter groundwater, producing heated water or steam that can be tapped by wells. Energy from this hot water or steam is called **geothermal energy**.

Harnessing geothermal energy is most feasible at the rifts along Earth's surface where crustal plates meet. These rifts also are the sites of many earthquakes and volcanoes. Geothermal energy is being tapped in several locations, including California, Italy, New Zealand, and Japan, and other rift sites are being explored. Iceland and Indonesia make extensive use of this resource. Ironically, in Iceland, an island named for its glaciers, nearly all homes and businesses in the capital of Reykjavik are heated with geothermal steam.

Biomass. Forms of biomass, such as sugar cane, corn, and soybeans, can be processed into motor vehicle fuels. Brazil in particular makes extensive use of biomass to fuel its cars and trucks. Potential for increasing the use of biomass for fuel is limited for several reasons. Burning biomass may be inefficient, because the energy used to produce the crops may be as much as the energy supplied by the crops. When wood is burned for fuel instead of being left in the forest, the fertility of the forest may be reduced. The most important limitation on using biomass for energy is that it already serves other essential purposes: providing much of Earth's food, clothing, and shelter.

Nuclear Fusion. Some nuclear power problems could be solved with nuclear **fusion**, which is the fusing of hydrogen atoms to form helium. Fusion releases spectacular amounts of energy—a gnat-sized amount of hydrogen releases the energy of thousands of tons of coal. But fusion can occur only at very high temperatures (millions of degrees). Such high temperatures have been briefly achieved in hydrogen bomb tests, but not on a sustained basis in a power-plant reactor, given present technology.

Alternatives such as fusion do not offer immediate solutions to energy shortages in the twenty-first century, but may become more practical if the price of current energy sources substantially rises. Earth is not "running out" of energy resources, but the era of dependency on nonrenewable fossil fuels for energy will constitute a remarkably short period of human history.

Key Issue 2

Why Are Resources Being Polluted?

- Pollution sources
- Alternatives for reducing pollution

In our consideration of resources, consumption is half of the equation—waste disposal is the other half. All of the resources we use eventually are returned to the atmosphere, bodies of water, or land surface, through burning, rinsing, or discarding.

We rely on air, water, and land to remove and disperse our waste. Not all human actions harm the environment, for every resource can accept some waste. When we wash household cleaners and chemicals into a river, the river may dilute them until their concentration is insignificant. However, when more waste is added than a resource can accommodate, we have **pollution**. Pollution levels generally are greater where people are concentrated. The actions of many people in a small area are likely to exceed the capacity of the environment to absorb the waste.

When we discard something, we never really eliminate that product, but simply put it somewhere else. It may cause pollution, depending on where we placed it. Natural processes may transport pollutants from one part of the environment to another. Discharges to the air often turn up in rivers, and wastes dumped in landfills can produce gases that leak to the atmosphere.

To better understand the causes of pollution, Blair Bower of Resources for the Future, a nonprofit research organization, asks us to consider a dairy cow as an analog of a factory, with inputs, processing, and outputs. We can think of a dairy cow as a factory that uses two main inputs—feed and water—to produce a desired output or product—milk (Figure 14–9).

Like a factory, a cow also may pollute the environment. Pollution can result because a cow (like a factory) generates some unintended waste byproducts, along with the desired product. In the case of a cow, it is manure. One of two things can happen to the waste: either it is collected and reused, or it is discarded into the environment. The waste may or may not constitute pollution. It is a pollutant if the level discharged into the environment exceeds the capacity of the environment to accept it.

In the case of the dairy cow, the waste is discharged into the environment in some societies and reused in others. In South Asia, dried manure is reused for fuel, and in many societies it is used for fertilizer. Farmers in more developed countries allow the manure to lie on the ground, because collecting and treating it is more expensive than purchasing commercial fertilizer. If the amount discharged by the cow exceeds the capacity of the field and nearby river to absorb it, the manure constitutes pollution.

Not all pollution is caused by humans. Natural pollution occurs when volcanoes erupt, spewing vast quantities of ash, cinders, sulfur gases, and steam into the atmosphere. Erosion from floods can clog streams with silt. However, our focus here is on the pollution that humans cause.

In the following sections, we look at air, water, and land pollution. Each has distinctive characteristics that illustrate the close connection between human activities and environmental quality.

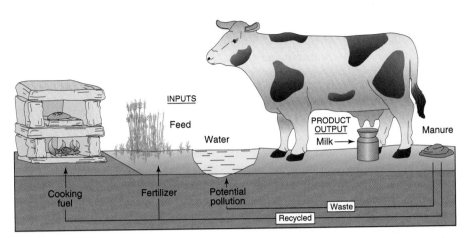

Figure 14-9 A dairy cow, a model for understanding pollution. The cow has inputs (feed and water) and a product (milk). Manure is a waste that can be recycled, in this case perhaps as fertilizer, or it may wash into the stream and become pollution.

Pollution Sources

When we discard something, we do not eliminate the material, but merely place it back into one of our three basic resources: air, water, or land. We will now look at how we pollute each.

Air Pollution

At ground level, Earth's average atmosphere comprises about 78 percent nitrogen, 21 percent oxygen, and less than 1 percent argon. The remaining 0.04 percent includes several trace gases, some of which are critical. **Air pollution** is a concentration of trace substances at a greater level than occurs in average air.

The most common air pollutants are carbon monoxide, sulfur dioxide, nitrogen oxides, hydrocarbons, and solid particulates. Concentrations of these trace gases in the air can damage property and adversely affect the health of people, other animals, and plants.

Three human activities generate most air pollution: motor vehicles, industry, and power plants. In all three cases, pollution results from the burning of fossil fuels. Burning gasoline or diesel oil in cars, trucks, buses, and motorcycles produces carbon monoxide, hydrocarbons, nitrogen oxides, and other pollutants. Factories and power plants produce sulfur dioxides and solid particulates, primarily from burning coal.

Air pollution concerns geographers at three scales—global, regional, and local:

- **Global scale.** Air pollution may contribute to global warming and damage the atmosphere's ozone layer.
- **Regional scale.** Air pollution may damage a region's vegetation and water supply through acid deposition.
- **Local scale.** Air pollution is especially severe in local areas where emission sources are concentrated, such as urban areas.

We can examine distinctive problems associated with air pollution at each scale.

Global Warming. Human actions, especially the burning of fossil fuels, may be causing Earth's temperature to rise. Earth is warmed by sunlight that passes through the atmosphere, strikes the surface, and is converted to heat. When the heat tries to pass back through the atmosphere to space, some gets through and some is trapped. This process keeps Earth's temperatures moderate and allows life to flourish on the planet.

A concentration of trace gases in the atmosphere can block or delay the return of some of the heat leaving the surface heading for space, therefore raising Earth's temperatures. When fossil fuels are burned, one of the trace gases, carbon dioxide, is discharged into the atmosphere.

Plants and oceans absorb much of the discharges, but increased fossil fuel burning during the past 200 years has caused the level of carbon dioxide in the atmosphere to rise by more than one-fourth, according to the United Nations Intergovernmental Panel on Climate Change. The level will continue to increase even if fossil fuel burning is reduced immediately, because of lingering effects of past emissions. Carbon dioxide is also increasing in the atmosphere from the burning and rotting of trees cut in the rain forests.

During the past century, the average temperature of Earth's surface has increased by 1° Celsius (2° Fahrenheit). This increase may or not have been caused by the buildup of carbon dioxide. However, unless carbon dioxide emissions are sharply curtailed in the near future, average temperatures at the surface of Earth will increase by several degrees over the next century, according to the U.N. panel. This would be comparable to the increase over the past 20,000 years, since the end of the last Ice Age.

The anticipated increase in Earth's temperature, caused by carbon dioxide trapping some of the radiation emitted by the surface, is called the **greenhouse effect**. The term is somewhat misleading, because a greenhouse does not work in the same way as trace gases in the atmosphere. In a real greenhouse, the interior gets very warm when its windows remain closed on a sunny day. The Sun's light energy passes through the glass into the greenhouse and is converted to heat, while the heat trapped inside the building is unable to escape out through the glass. Although an imprecise analogy, greenhouse effect has been a widely adopted term to evoke the anticipated warming of Earth's surface when trace gases block some of the heat trying to escape into space.

Regardless of what it is called, global warming of only a few degrees could melt the polar ice caps and raise the level of the oceans many meters. Coastal cities such as New York, Los Angeles, Rio de Janeiro, and Hong Kong would flood. Global patterns of precipitation could shift—some deserts could receive more rainfall, but currently productive agricultural regions, such as the U.S. Midwest, could become too dry for farming. Humans can adapt to a warmer planet, but the shifts in coastlines and precipitation patterns could require massive migration and be accompanied by political disputes.

Global-Scale Ozone Damage. Earth's atmosphere has zones with distinct characteristics. The stratosphere, the zone between 15 and 50 kilometers (9 to 30 miles) above Earth's surface, contains a concentration of **ozone** gas. The ozone layer absorbs dangerous ultraviolet (UV) rays from the Sun. Were it not for the ozone in the stratosphere, UV rays would damage plants, cause skin cancer, and disrupt food chains.

Earth's protective ozone layer is threatened by pollutants called **CFCs (*chlorofluorocarbons*)**. CFCs such as *Freon* were widely used as coolants in refrigerators and air conditioners. When they leak from these appliances, the CFCs are carried into the stratosphere, where they break down Earth's protective layer of ozone gas. The 1987 Montreal Protocol called for more developed countries to cease using CFCs by 2000, and for LDCs to cease by 2010.

Regional-Scale Acid Deposition. Industrialized, densely populated regions in Europe and eastern North America are especially affected by **acid deposition**. Sulfur oxides and nitrogen oxides, emitted by burning fossil fuels, enter the atmosphere, where they combine with oxygen and water. Tiny droplets of sulfuric acid and nitric acid form and return to Earth's surface as acid deposition. When dissolved in water, the acids may fall as **acid precipitation**—rain, snow, or fog. The acids can also be deposited in dust. Before they reach the surface, these acidic droplets might be carried hundreds of kilometers.

Acid precipitation has damaged lakes, killing fish and plants. Aquatic life has been completely eliminated from 4 percent of the lakes in the eastern United States and Canada; another 5 percent of the lakes in the eastern United States and 20 percent in eastern Canada have acidity levels that threaten some species.

On land, concentrations of acid in the soil can injure plants by depriving them of nutrients, and can harm soil worms and insects. Acid precipitation has contributed to the decline of the red spruce tree at higher elevations. Buildings and monuments made of marble and limestone have suffered corrosion from acid rain; engravings on old marble tombstones may be illegible as a result.

Since the 1970s, the United States has reduced sulfur dioxide emissions significantly. Many Western European countries cut theirs in half, largely by reducing coal use. Despite this progress, acid precipitation continues to damage forests and lakes. Governments are reluctant to impose the high cost of controls on their industries and consumers.

Geographers are particularly interested in the effects of acid precipitation because the worst damage is not experienced at the same location as the emission of the pollutants. Within the United States, the major generators of acid deposition are in Ohio and other industrial states along the southern Great Lakes. However, the severest effects of acid rain are felt in several areas farther east (Figure 14–10).

The problem of acid precipitation is compounded by the fact that pollutants emitted in one country cause adverse impacts in another. Acid rain falling in Ontario, Canada, for example, can be traced to emissions from coal-burning power plants in the U.S. Great Lakes. Government officials at the source of the pollution may be reluctant to impose strong controls on the offending factories because they fear damaging the local economy.

Eastern Europe has suffered especially severely from acid precipitation, a legacy of Communist policies that encouraged the construction of factories and power plants without pollution control devices. Destruction of forests is widespread because of acid rain emitted from Eastern Europe's major industrial region (southeastern Germany, southern Poland, and northern Czech Republic). Affected by acid precipitation more than any other European state, the Czech Republic has suffered severe damage in more than 80 percent of the Bohemian Forest and more than one-third of its other forests.

The destruction of trees has harmed Eastern Europe's seasonal water flow. In dense forests, snow used to melt slowly and trickle into rivers. Now, on the barren sites, it melts and drains quickly, causing flooding in the spring and water shortages in the summer.

Perhaps the most severe impact is on human life. One-third of the residents of St. Petersburg, Russia's second-largest city, suffer from upper respiratory tract ailments as a result of the intense air pollution. A 40-year-old man living in Poland's polluted southern industrial area has a life expectancy 10 years less than his father had at the same age. Poland is estimated to have between 20,000 and 50,000 additional deaths per year due to pollution.

Local-Scale Urban Air Pollution. The air above urban areas may be especially polluted because a large number of factories, motor vehicles, and other polluters emit residuals in a concentrated area. Weather conditions may make it difficult for the emissions to dissipate.

Urban air pollution has three basic components:

1. ***Carbon Monoxide.*** Proper burning in power plants and vehicles produces carbon dioxide, but improper combustion produces carbon monoxide.

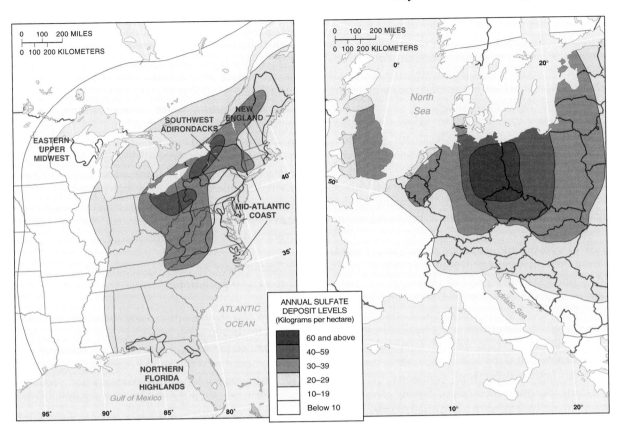

Figure 14–10 Acid precipitation in North America and Europe. Levels exceeding 20 kg/ha are considered threatening. (Left) Because of prevailing wind patterns across North America, damage is generally found to the east of the emissions. (Right) Deposition levels in eastern Germany are higher than anywhere in the United States, although elsewhere in central Europe levels are comparable to those in the eastern United States.

Breathing carbon monoxide reduces the oxygen level in blood, impairs vision and alertness, and threatens those with breathing problems.

2. *Hydrocarbons* also result from improper fuel combustion, as well as evaporation of paint solvents. Hydrocarbons and nitrogen oxides in the presence of sunlight form **photochemical smog**, which causes respiratory problems, stinging in the eyes, and an ugly haze over cities.

3. *Particulates* include dust and smoke particles. The dark plume of smoke from a factory stack and the exhaust of a diesel truck are examples of particulates being emitted.

The severity of air pollution resulting from emissions of carbon monoxide, hydrocarbons, and particulates depends on the weather. The worst urban air pollution occurs when winds are slight, skies are clear, and a temperature inversion exists. When the wind blows, it disperses pollutants, and when it is calm pollutants build. Sunlight provides the energy for the formation of smog. Air is normally cooler at higher elevations, but during temperature inversions—in which air is warmer at higher elevations—pollutants are trapped near the ground.

According to the U.S. Environmental Protection Agency, the worst U.S. city for concentrations of carbon

Acid precipitation has killed a large percentage of the trees in the Krušné Hory mountains of Czech Republic, in the far west of the country, near the border with Germany. Emissions of sulfur dioxide and nitrogen oxides from factories and power plants built without pollution control devices in the former Communist East Germany and Czechoslovakia caused this widespread death of trees.

monoxide and second worst for particulates is Denver, where residents call the smog "the brown cloud." The Rocky Mountains help trap the gases and produce a permanent temperature inversion. Ironically, the beautiful view of the mountains, which attracted so many migrants to Denver, is often obscured by smog.

The problem is not confined to MDCs. Santiago, Chile, nestled between the Pacific Ocean and the Andes Mountains, suffers severe smog problems. Motor vehicles are also responsible for much of the pollution in Santiago, especially particulates from burning diesel fuel, combined with dust kicked up from dirt streets. Mexico City's serious air pollution problem is discussed in the opening case study.

Progress in controlling urban air pollution is mixed. Air has improved in developed countries where strict clean air regulations are enforced. Changes in automobile engines, manufacturing processes, and electric generation all have helped. For example, in the quarter-century since the U.S. government has required catalytic converters on motor vehicles, carbon monoxide emissions have been reduced by more than three-fourths, nitrogen oxide and hydrocarbon emissions by more than 95 percent. But more people are driving more, offsetting gains made by emission controls. Limited emission controls in LDCs are contributing to severe urban air pollution.

Water Pollution

Water serves many human purposes. People must drink water to survive, and they cook and bathe with water. The typical U.S. urban resident consumes 680 liters (180 gallons) of water per day for drinking, cooking, and bathing. Water provides a location for boating, swimming, fishing, and other recreation activities. People consume fish and other aquatic life. These uses depend on fresh, clean, unpolluted water.

Clean water is not always available, because people also use water for purposes that pollute it. Manufacturers use water each year to process food and manufacture goods. People discharge waste down the drain and into water. Farmers let waste wash away into water. When all of these uses are included, the average American consumes nearly 10,000 liters (2,400 gallons) of water per day. By polluting water, humans harm the health of aquatic life and the health of land-based life (including people themselves).

Pollution is widespread, because it is easy to dump waste into a river and let the water carry it downstream where it becomes someone else's problem. Water can decompose some waste without adverse impact on other activities, but the volume exceeds the capacity of many rivers and lakes to accommodate it.

Pollution Sources. Three main sources generate most of the water pollution:

- ***Water-Using Industries***. Industries such as steel, chemicals, paper products, and food processing are major water polluters. Each requires a large amount of water in the manufacturing process and generates a lot of wastewater. Food processors, for example, wash pesticides and chemicals from fruit and vegetables. They also use water to remove skins, stems, and other parts. Water can also be polluted by industrial accidents, such as petroleum spills from ocean tankers and leaks from underground tanks at gasoline stations.

- ***Municipal Sewage***. In more developed countries, sewers carry wastewater from sinks, bathtubs, and toilets to a municipal treatment plant, where most—but not all—of the pollutants are removed. The treated wastewater is then typically dumped back into a river or lake. In LDCs, sewer systems are rare, and wastewater usually drains untreated into rivers and lakes.

- ***Agriculture***. Fertilizers and pesticides spread on fields to increase agricultural productivity are

Mexico City suffers from a combination of circumstances that lead to significant air pollution problems. The city lies in a mountain basin that limits dispersion of pollutants, and motor vehicle traffic is heavy. (Robert Frerck/Odyssey Productions)

carried into rivers and lakes by the irrigation system or natural runoff. Expanded use of these products may help to avoid a global food crisis, yet they destroy aquatic life by polluting rivers.

These three sources of pollution can be divided into point sources and nonpoint sources. Point-source pollution enters a stream at a specific location, whereas non-point-source pollution comes from a large diffuse area. Manufacturers and municipal sewage systems tend to pollute through point sources, such as a pipe from a wastewater treatment plant.

Farmers tend to pollute through nonpoint sources, such as by permitting fertilizer to wash from a field during a storm. Point-source pollutants are usually smaller in quantity and much easier to control. Nonpoint-sources usually pollute in greater quantities and are much harder to control.

Impact on Aquatic Life. Polluted water can harm aquatic life. Aquatic plants and animals consume oxygen, but so does the decomposing organic waste that humans dump in the water. The oxygen consumed by the decomposing organic waste constitutes the **biochemical oxygen demand (BOD)**. If too much waste is discharged into the water, the water becomes oxygen-starved, and fish die.

This condition is typical when water becomes loaded with municipal sewage or industrial waste. The sewage and industrial pollutants consume so much oxygen that the water can become unlivable for normal plants and animals, creating a "dead" stream or lake. Similarly, when runoff carries fertilizer from farm fields into streams or lakes, the fertilizer nourishes excessive aquatic plant production—a "pond scum" of algae—that consumes too much oxygen. Either type of pollution unbalances the normal oxygen level, threatening aquatic plants and animals.

Some of the residuals may become concentrated in the fish, making them unsafe for human consumption. For example, salmon from the Great Lakes became unfit to eat because of high concentrations of the pesticide DDT, which washed into streams from farm fields.

Many factories and power plants use water for cooling, and then discharge the warm water back into the river or lake. The warm water may not be polluted with chemicals, but it raises the temperature of the body of water it enters. Fish adapted to cold water, such as salmon and trout, might not be able to survive in the warmer water.

Wastewater and Disease. Since passage of the U.S. Clean Water Act and equivalent laws in other developed countries, most treatment plants meet high water-quality standards. Improved treatment procedures have resulted in cleaner rivers and lakes in more developed countries (see Globalization and Local Diversity box).

Although less developed countries generate less waste-water per person than do more developed countries, they have less capacity to treat their wastewater. In LDCs, sewage often flows untreated directly into rivers. The drinking water, usually removed from the same rivers, may be inadequately treated as well. And in squatter settlements on the edge of rapidly growing cities, running water and sewers may be totally lacking.

The combination of untreated water and poor sanitation makes drinking water deadly in LDCs. Waterborne diseases such as cholera, typhoid, and dysentery are major causes of death.

Land Pollution

When we consume a product, we also consume an unwanted byproduct—a glass, metal, paper, or plastic box, wrapper, or container in which the product is packaged. About 2 kilograms (4 pounds) of solid waste per person is generated daily in the United States, including about 60 percent from residences and 40 percent from businesses.

Paper products, such as corrugated cardboard and newspapers, account for the largest percentage of solid waste in the United States, especially among residences and retailers (Figure 14–11). Food products and rubbish clean-up from yards, such as grass clipping and leaves, are other important sources of solid waste. Manufacturers discard large quantities of metal and concrete, as well as paper.

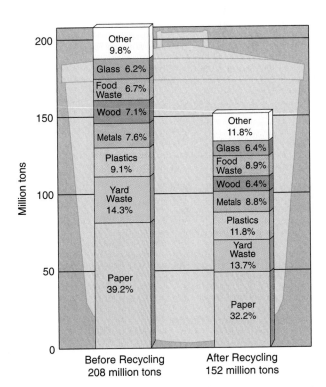

Figure 14–11 Sources of solid waste. Paper products account for the largest percentage of U.S. solid waste, followed by food products and yard rubbish. Plastics and metals are comparatively small percentages of solid waste.

Globalization and Local Diversity

Two Pollution Stories: Aral Sea and Thames River

Geographers study *where* resources such as water supplies are polluted. Water quality is improving in North America and Western Europe while deteriorating in Africa, Asia, Eastern Europe, and Latin America. Human geographers are especially concerned with *why* pollution is related to differences in development.

In less developed countries, pollution may be a small price to pay for participating in the *globalization* of the economy. Industrialization may take a higher priority than clean water. More developed countries caused most of the water pollution in the past. Now, they possess the wealth and technology to clean up polluted rivers and lakes. But even within MDCs, geographers observe considerable *local diversity* in the willingness of citizens to spend money for the clean up.

Cleanup of the Thames

Western European and North American communities have successfully cleaned up lakes and streams. One dramatic example is the River Thames, which passes through London. Prior to the industrial revolution, the Thames was a major food source for Londoners. Some apprentice workers even went on strike in the early 1800s because their masters fed them too much fish. During the industrial revolution, the Thames became the principal location for dumping waste. The fish died, and the water grew unsafe to drink. The river became so dark, murky, and smelly that novelist Charles Dickens called the Thames "London's Styx," after the underworld river that the dead had to cross in Greek mythology.

In the late 1960s, the British government began a massive cleanup to restore the Thames to health. Regulations prohibited industrial dumping, and sewage systems were modernized to improve treatment. In 1982, a salmon was caught in the Thames, just upstream from London. This event was remarkable because it was the first salmon caught in the river since 1833. Salmon are particularly sensitive to pollution, and for nearly 150 years the Thames was too polluted for salmon to survive.

To demonstrate the success of the cleanup operation, the Thames Water Authority has released 50,000 young salmon, known as parr, into the river each year since 1979. When they are approximately 2 years old, the salmon, then known as smolts, migrate from the river to the sea, where they spend 1 to 3 years. The salmon then return to the river to spawn upstream. Several dozen of the 50,000 salmon released into the river in 1979 were the first to be caught in 1982.

Actually, the first salmon may have been caught several days earlier. An Englishman produced a salmon allegedly taken from the Thames and claimed the prize offered to the person who caught the first one. However, government authorities tested his salmon and ruled that it had come from the man's freezer, not from the Thames.

Destruction of the Aral

One of the world's most extreme instances of water pollution is the Aral Sea in Kazakhstan and Uzbekistan. Once the world's fourth-largest lake, the Aral has lost nearly half of its water since 1960. Carp, sturgeon, and other fish species have disappeared, the last fish dying in 1983. Large ships lie aground in salt flats that were once the lake bed, outside of abandoned fishing villages that now lie tens of kilometers from the rapidly receding shore (Figure 1).

Some consumers demonstrate obvious unconcern for the environment by discarding waste along roadsides and sidewalks, where they cause visual pollution. But even consumers who carefully dispose of solid waste are contributing to a major pollution problem. A particularly severe threat is posed by the careless discharge of toxic waste.

About five-sixths of the solid waste generated in the United States is discharged into the environment in two ways: landfills and incineration. The remainder is recycled.

Sanitary Landfills.   The **sanitary landfill** is by far the most common strategy for disposal of solid waste in the United States: over 70 percent of the country's waste is trucked to landfills and buried under soil. This strategy is opposite our disposal of gaseous and liquid wastes: we *disperse* air and water pollutants into the atmosphere, rivers, and eventually the ocean, but we *concentrate* solid waste in thousands of landfills.

Concentration would seem to eliminate solid waste pollution, but it may only hide it—temporarily. Chemicals released by the decomposing solid waste can leak from the landfill into groundwater. This can contaminate water wells, soil, and nearby streams.

Eventually, landfills fill up. To preserve landfill space, many communities prohibit discarding bulky yard waste like grass clippings, weeds, and leaves. Many communities have closed them. Few new ones are being built, because landfills can contaminate groundwater and devalue property, and no one wants to live near one—the NIMBY principle at work again.

Some communities now pay to use landfills elsewhere. San Francisco trucks solid waste to Altamont, California, 100 kilometers (60 miles) away. Passaic County, New Jersey, hauls waste 400 kilometers (250 miles) west to Johnstown, Pennsylvania. New Jersey and New York are the two states that regularly try to dispose of their solid waste by transporting it out of state (Figure 14–12).

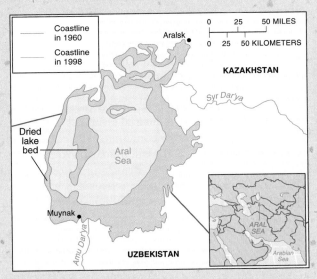

The Aral Sea. Once the world's fourth-largest freshwater lake, the Aral Sea has declined in area by nearly one-half since 1960. The principal cause was diversion of two rivers, the Amu Dar'ya and the Syr Dar'ya, to irrigate cotton fields. (Louchine Ugoniok/Sygma)

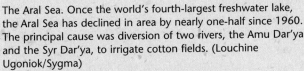

The Aral Sea died after the Soviet Union diverted its tributary rivers, the Amu Dar'ya and the Syr Dar'ya, to irrigate cotton fields. Not only did this sharply reduce the lake's source of fresh water input; pollution accelerated as chemicals sprayed on the fields flowed into the sea. Ironically, the cotton now is withering because winds pick up salt from the exposed lake bed and deposit it on the cotton fields!

Worse, most households in the surrounding Karakalpak region of Kazakhstan obtain their water directly from polluted ditches, rather than through pipes from treatment plants. Two-thirds of these households now suffer from liver disorders, typhoid, or cancer of the esophagus.

New York's problem of transporting solid waste long distances generated an absurd situation in 1987. The town of Islip, New York, sold its solid waste to a private individual who planned to transport the waste to Morehead City, North Carolina, and convert it to methane gas. The 3.1 million kilograms (3,100 tons) of refuse were trucked to Long Island City and transferred to a barge, named the *Mobro*, for the journey to North Carolina. However, officials in Morehead City obtained a court injunction prohibiting the unloading of refuse there.

The *Mobro* then continued farther south in search of a home, but communities in Alabama, Mississippi, Louisiana, Texas, and Florida all refused to accept it. After Mexico, Belize, and the Bahamas also turned it away, the barge returned to New York, after nearly 2 months and a 9,000-kilometer (5,500-mile) journey. Four months later, the refuse was burned in New York City, and the ashes were trucked back to Islip, where the trash had started its journey 6 months earlier (Figure 14–12).

Incineration. Burning the trash reduces its bulk by about three-fourths, and the remaining ash demands far less landfill space. Incineration also provides energy—the incinerator's heat can boil water to produce steam heat or to operate a turbine that generates electricity. Over 100 incinerators now burn about 10 percent of the trash in the United States. Given the anticipated shortage of space in landfills, the percentage of solid waste that is burned is likely to increase.

However, solid waste, a mixture of many materials, may burn inefficiently. Burning releases some toxics into the air, and some remain in the ash. Thus, solving one pollution problem may increase another pollution problem.

Recycling Solid Waste. As with other residuals, recycling is the only alternative to preventing solid waste from being discharged into a landfill or incinerator. Recycling of solid waste addresses problems of both

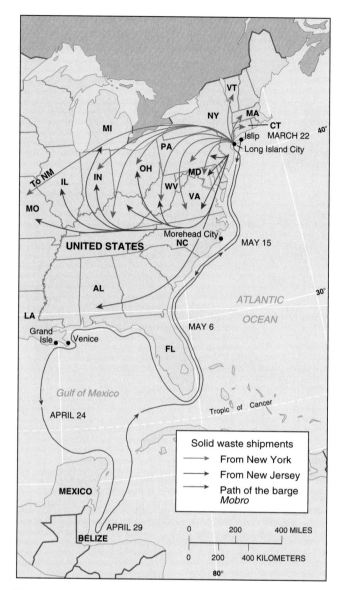

Figure 14–12 Solid-waste disposal. New York and New Jersey try to solve their solid waste disposal problems by transporting it out of state. New York's problem of solid waste disposal was graphically illustrated by the 1987 odyssey of the barge *Mobro*. The Town of Islip, New York, lacking space in its landfill for solid waste, hired a company to remove the waste by barge. The waste was to be converted to methane gas near Morehead City, North Carolina, but that state prohibited the *Mobro* from landing. The barge was towed into the Gulf of Mexico but was refused landing rights by Alabama, Florida, Louisiana, and Texas, as well as by Belize and Mexico. Finally, following a 2-month, 9,000-kilometer (5,500-mile) journey, the *Mobro* returned to the New York area. Four months later, the waste was unloaded and burned.

pollution and resource depletion. Recycling not only reduces the need for landfills and incinerators; it also reuses natural resources that already have been extracted.

Most U.S. communities have instituted some form of mandatory recycling. Newspapers, glass, plastic, and metals are separated from the solid waste destined for the landfill or incinerator. The trash collector sends newspaper to paper mills, "tin" (actually steel) cans to steel-producing minimills, aluminum cans to recycling companies, and

glass to bottlers. To encourage recycling, some communities haul recyclables for free or a small fee and charge a high fee to pick up nonrecyclables. In other places, citizens are fined for failing to comply.

Recycling has steadily increased in the United States, from 7 percent of all solid waste in 1970 to 10 percent in 1980, 17 percent in 1990, and 22 percent in 1994. The percentage of recovered materials varies widely by product; 97 percent of discarded auto batteries are recycled, compared to only 2 percent of plastics.

Toxic Pollutants. Disposing of toxic wastes is especially difficult. If poisonous industrial residuals are not carefully placed in protective containers, the chemicals may leach into the soil and contaminate groundwater or escape into the atmosphere. Breathing air or consuming water contaminated with toxic wastes can cause cancer, mutations, chronic ailments, and even immediate death.

Toxic wastes include heavy metals (including mercury, cadmium, and zinc), PCB oils from electrical equipment, cyanides, strong solvents, acids, and caustics. Burial of wastes was once believed to be sufficient to handle the disposal problem, but many of the burial sites have leaked.

One of the most notorious is Love Canal, near Niagara Falls, New York. During the 1930s, the Hooker Chemicals and Plastic Company buried toxic wastes in metal drums. A school and several hundred homes were built on the site in 1953. Erosion eventually exposed the metal drums, and in 1976 they began to give off a strong stench and slime oozed from them.

Residents at Love Canal reported a high incidence of liver ailments, nervous disorders, and other health problems. After four babies were born with birth defects on the same block, New York State officials relocated most of the families and began an expensive clean-up effort. Love Canal is not unique. Toxic wastes have been improperly disposed of at thousands of dumps.

Companies in the United States that release chemicals classified as toxic by the U.S. Environmental Protection Agency (EPA) must report the amounts released. About 2.5 billion kilograms (6 billion pounds) of toxic chemicals are discharged in the United States. About one-fourth of the discharges are by ten companies (DuPont, Monsanto, American Cyanimid, B.P. America, Renco Holdings, 3M, Vulcan Materials, General Motors, Eastman Kodak, and Phelps Dodge).

The EPA ranks U.S. counties on the amount of toxic pollutants (Figure 14–13). Counties with relatively high levels of toxic discharges are clustered in the Northeast and Midwest—essentially the U.S. manufacturing belt (refer to Figure 11–4).

As toxic-waste disposal sites become increasingly hard to find, some European and North American firms have tried to transport their waste to West Africa, often unscrupulously. Some firms have signed contracts with West African countries, while others have found isolated locations to dump waste without official consent.

A garbage scow is pulled to land in New Jersey, within sight of the New York skyline. (Ray Ellis/Photo Researchers, Inc.)

Alternatives for Reducing Pollution

Burning fossil fuels, dumping wastewater, and throwing away paper may or may not constitute pollution. Pollution occurs when the amount of the waste discharged into the environment exceeds the capacity of the air, water, or land to absorb it.

If the amount of a discharge exceeds the capacity of the environment, then only two basic alternatives can reduce pollution: the amount of waste discharged into the environment must be reduced, or the capacity of the environment to accept discharges must be expanded. The merits of the two alternatives can be examined.

Reducing Pollution by Reducing Discharges

Pollution can be prevented by reducing the amount of waste discharged into the environment. Two strategies can reduce discharges:

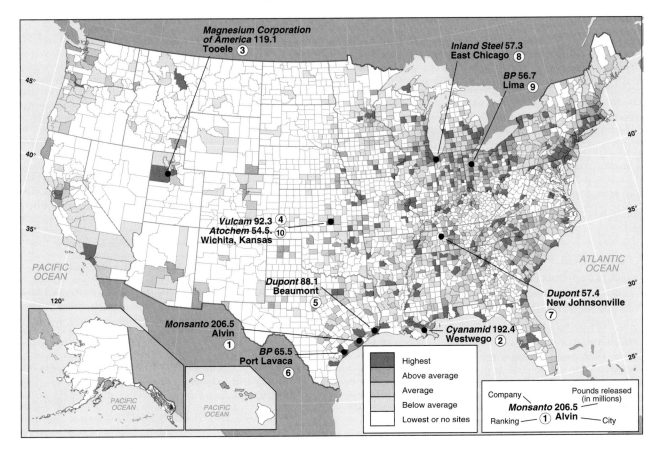

Figure 14–13 Toxic chemical concentrations per county. The highest concentrations are in the Northeast and southern Great Lakes regions, where the country's manufacturers are clustered.

Solid waste accumulates at unsanitary disposal sites in less developed countries, such as the Smokey Mountains rubbish dump, in the Philippines. Poor people carry baskets through the site picking up unclean objects thrown away by others. (Nigel Dickinson/ Tony Stone Images)

- *Reduce* the amount of waste created.
- *Recycle* the waste rather than discharge it into the environment.

Recalling the analogy of a polluter to a dairy cow (Figure 14–9), a farmer can reduce a cow's discharge of manure onto the ground in two basic ways. One is to reduce the amount of the waste (manure) that the cow generates, either by changing the inputs into the production system (the feed), or by reducing demand for the product (eat less meat or dairy products). The other is to recycle the waste, by collecting the manure and using it as fertilizer or fuel rather than let it accumulate on the ground. These alternatives may not be practical, but they illustrate the variety of strategies available to reducing the discharge of waste into the environment.

Reduce Waste Created. Pollution would be reduced if less waste were generated. Waste can be reduced in two ways:

- *Change Mix of Inputs*. What goes into a production system affects what comes out—as product and as waste. The mix of various inputs can be adjusted to produce a higher ratio of product to waste. For example, gasoline for motor vehicles once contained lead. But most of the lead was discharged through the exhaust pipe and contributed to air pollution. To reduce the generation of lead—once a significant waste—automakers modified engines so that they operate on unleaded instead of leaded gasoline.
- *Reduce Demand for Product*. If consumers purchase smaller quantities of a product, then the production system is down-sized. Factories produce less of the product—or shut altogether.

The creation of fewer products would result in the production of less waste, as well. If consumers drive less, then they will use less gasoline and therefore generate less pollution.

Recycling. Once waste is discharged into the environment, the only possible strategy for preventing it from accumulating is to recycle more of it. Recycling can take two forms:

- *Reuse Waste in Same Production System*. For example, a cow's manure can be used as fertilizer to grow more cattle feed, which generates more milk and manure. Or, rather than discharging wastewater into a river, a food processor can treat the water for reuse in production.
- *Reuse Waste in Different Production Process*. For example, the cow's manure can be used for another purpose, like fuel for heating or cooking. The vegetable pulp produced as an unwanted byproduct of food processing can be used to make pet food.

Increasing Environmental Capacity

The second way to handle pollution is to increase the capacity of the environment to accept the discharges. We can increase environmental capacity to accept waste by using two strategies:

- Make *more efficient use* of the resource currently receiving the discharge.
- *Transform* the waste and discharge it into a different part of the environment.

Again recalling the analogy of the dairy cow, a farmer can make more effective use of the environment in two basic ways. First, the farmer can disperse the manure over a wider land area, either by allowing the cow to roam over a larger area, or by spreading the manure onto large fields. Second, the manure can be converted from a solid to a gas by burning it, or it can be washed into a nearby stream.

Use Resource More Efficiently. The capacity of air, water, and land to accept waste is not fixed, but varies among places and at different times. Adding a particular amount of wastewater to a stream may or may not constitute pollution, depending on the flow of the water. A deep, fast-flowing river has a greater capacity to absorb wastewater than a shallow, slow-moving one. Wastewater can be stored when the river level is low and released when the river is high.

The same is true for air: exhaust released into a brisk wind is quickly dispersed, whereas exhaust released into stagnant air during a temperature inversion quickly accumulates to irritating levels. Industries and utilities reduce local air pollution by building taller smokestacks, which better disperse gases at greater heights. Air quality may also be improved by staggering workers' hours so pollution from cars is spread more evenly through the day.

Discharge into Different Part of Environment. The other way to increase environmental capacity is to transform the waste so that it is discharged into a resource that has the capacity to assimilate it. Matter can be transformed among gaseous, liquid, and solid states and discharged into air, water, or land.

For example, a coal-burning power plant discharges gases into the atmosphere, causing air pollution. To reduce air pollution, wet scrubbers are installed to wash particulates from the gas before it is released to the atmosphere. Wet scrubbers capture the particulates in water, which then can be discharged into a stream. If the stream is polluted by the discharge, then the wastewater can be cleaned in a settling basin where the particulates drop out. This transforms the residue into a solid waste for disposal on land.

A Coking Plant: Using All Reduction Strategies

A coking plant provides an example of applying all four pollution reduction strategies—the two ways to reduce discharges and the two ways to increase environmental capacity (Figure 14–14). The main input into a coking plant is a mixture of coal types, and the intended product is coke, which becomes an input in steel production. The coal is placed in a blast furnace and cooked at very high temperatures to form coke. Four unwanted byproducts result: gases, tars, oils, and heat.

Discharging the heat into the environment can cause air pollution. To reduce air pollution from the heat, a coking plant increases the capacity of the environment to accept discharges in two ways. First, the hot coke is taken to a quench station and doused with water to cool it. This process transforms the residual (hot gas) into a liquid (dirty water) as well as another gas (steam). In this way, the waste is transformed and discharged into different parts of the environment. Then the steam is discharged into the environment from a tall smokestack, an example

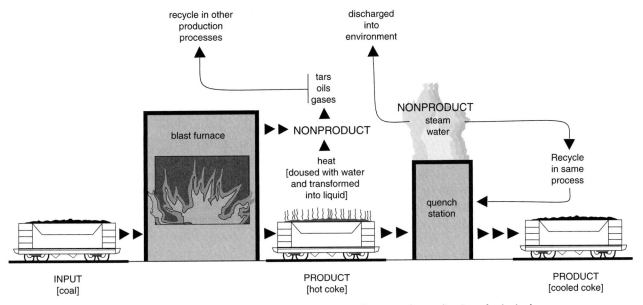

Figure 14–14 A coking plant, used for steelmaking, illustrates the application of principal alternatives for reducing pollution:

1. Reduce discharges of waste
   a. Reduce waste (by purchasing cleaner-burning coal)
   b. Recycle discharge (by reusing quenching water)
2. Increase environmental capacity
   a. Use current resources more efficiently (use taller smokestack for discharging steam)
   b. Discharge elsewhere into environment (transform heat into liquid and gaseous residuals).

of making more efficient use of whatever initially received the discharge (air).

The coking plant also minimizes pollution by reducing discharges. The dirty water produced at the quench station is reused to cool more hot coke, an example of recycling in the same production process. Meanwhile, the three unwanted byproducts from the blast furnace (other than heat)—gases, tars, and oils—are captured and sold to other companies for recycling in other processes. The other alternative for reducing discharges—changing the mix of coal used as inputs—is also employed, because the amount of gases emitted by the burning of coke varies depending on the mix of coal.

Comparing Pollution Reduction Strategies

Relying on an increase in the capacity of the environment to accept discharges is risky. Because we do not always know the environment's capacity to assimilate a particular waste, we are likely to exceed it at times. Recent history is filled with examples of wastes discharged in the environment with the belief that they would be dispersed or isolated safely: CFCs in the stratosphere, garbage off-shore, and toxic chemicals beneath Love Canal.

Dispersed wastes may remain harmful. Tall smoke-stacks built to reduce sulfur dioxide discharges around coal-burning industries and metal smelters were successful at dispersing sulfur over a larger area. But the result of the dispersal was that acid precipitation (containing sulfur) fell hundreds of kilometers away, polluting vegetation and lakes over a wide area.

Many pollutants are mobile. They often travel from air to soil, or soil to water. A pollutant like sulfur dioxide might exist at tolerable levels in the air, but it damages trees when it accumulates in the soil. In view of the many uncertainties associated with increasing environmental capacity, reducing discharges into the environment (by either changing the production process or recycling) is usually the preferred alternative.

Although the environment has the capacity to accept some discharges, consumers must learn to use this environmental capacity most efficiently. At the same time, consumers must learn to waste less, either by reducing the consumption of products that result in waste or by recycling more. With careful management, we can enjoy the benefits of both industrial development and a cleaner, safer environment.

However, environmental improvements in the more developed countries of North America and Western Europe are likely to be offset in the twenty-first century by increased pollution in less developed countries. According to economists Gene Grossman and Alan Krueger, a rising level of economic development generates increased pollution, at least until a country reaches a GDP of about $5,000 per person (Figure 14–15). In the early stages of industrialization, pollution control devices are an unpopular luxury that makes cars and other consumer goods more expensive.

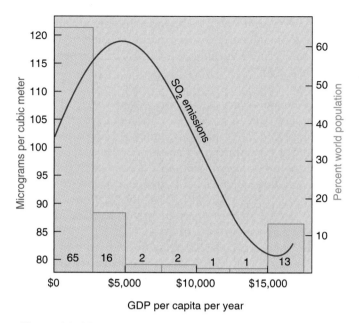

Figure 14–15 Pollution compared to a country's wealth. As a country's GDP per capita increases, discharge of sulfur dioxide increases, until GDP reaches about $5,000. Then, discharges tend to decrease, as a country begins to spend money on pollution-control devices. The green bars show the percentage of people in each GDP per capita group.

Key Issue 3

Why Are Global Food Resources Expandable?

- Alternative strategies to increase food supply
- Africa's food supply crisis

Instead of depleting and destroying resources, we can use them more efficiently. One dramatic example is our better use of Earth to provide food.

Returning to the issue of overpopulation raised in Chapter 2, we concluded that problems result when an area's population exceeds the capacity of the environment to support it at an acceptable standard of living. The ratio of population to the capacity of the environment to provide food has become more favorable in many LDCs. In Asia, especially, population growth has slowed whereas the food supply has increased. The challenge in these regions is to continue recent progress by further expanding food resources.

On the other hand, population is expanding more rapidly than food production in sub-Saharan Africa. Food supply must be expanded quickly for Africa's burgeoning population.

Alternative Strategies to Increase Food Supply

Four strategies can increase the food supply:

- Expand the land area used for agriculture.

- Increase the productivity of land now used for agriculture.
- Identify new food sources.
- Increase exports from other countries.

We will now examine each alternative.

Increase Food Supply by Expanding Agricultural Land

Historically, world food production increased primarily by expanding the amount of land devoted to agriculture. When the world's population began to increase more rapidly in the late eighteenth and early nineteenth centuries, during the industrial revolution, pioneers could migrate to uninhabited territory and cultivate the land. Sparsely inhabited land suitable for agriculture was available in western North America, central Russia, and Argentina's pampas.

Two centuries ago, people believed that good agricultural land would always be available for willing pioneers. Today, few scientists believe that further expansion of agricultural land can feed the growing world population. Beginning about 1950, the human population has increased faster than the expansion of agricultural land.

At first glance, new agricultural land appears to be available, because only 11 percent of the world's land area is currently cultivated. In fact, growth is possible in North America, where some arable land is not cultivated for economic reasons, and the tropics of Africa and South America offer some hope for new agricultural land in LDCs. However, prospects for expanding the percentage of cultivated land are poor in much of Europe, Asia, and Africa.

Land Removed from Agriculture. In some regions, farmland is abandoned for lack of water. Especially in semiarid regions, human actions are causing land to deteriorate to a desertlike condition, a process called **desertification** (more precisely, semiarid land degradation). Semiarid lands that can support only a handful of pastoral nomads are overused because of rapid population growth. Excessive crop planting, animal grazing, and tree cutting exhaust the soil's nutrients and preclude agriculture. The United Nations estimates that desertification removes 27 million hectares (104,000 square miles) of land from agricultural production each year, an area roughly equivalent to Colorado (Figure 14–16).

Excessive water threatens other agricultural areas, especially drier lands that receive water from human-built irrigation systems. If the irrigated land has inadequate drainage, the underground water level rises to the point where roots become waterlogged. The United Nations estimates that 10 percent of all irrigated land is waterlogged, mostly in Asia and South America. If the water is salty, it can damage plants. The ancient civilization of Mesopotamia may have collapsed in part because of waterlogging and excessive salinity in their agricultural lands near the Tigris and Euphrates rivers.

Urbanization can also contribute to reducing agricultural land. As urban areas grow in population and land area, farms on the periphery are replaced by homes, roads, shops, and other urban land uses. In North America, farms outside urban areas are left idle until the

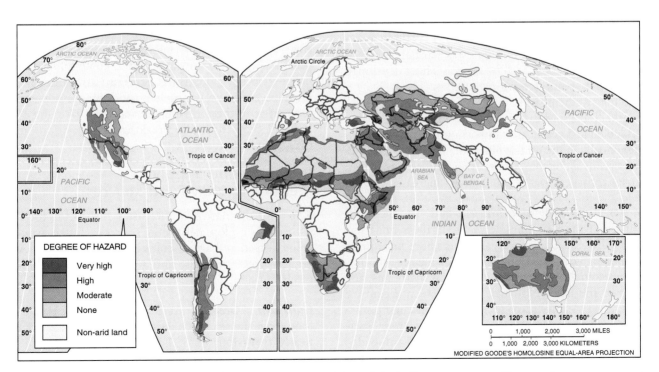

Figure 14–16 Desertification (semiarid land degradation). The most severe problems are in northern Africa, central Australia, and the southwestern parts of Africa, Asia, North America, and South America.

speculators who own them can sell them at a profit to builders and developers, who convert the land to urban uses.

Increase Food Supply Through Higher Productivity

During the 1960s, population began to grow faster than agricultural land expanded, especially in LDCs. At the time, many experts forecast massive global famine within a decade. However, these dire predictions have not come true, because new agricultural practices have permitted farmers worldwide to achieve much greater yields from the same amount of land.

The Green Revolution. The invention and rapid diffusion of more productive agricultural techniques during the 1970s and 1980s is called the **green revolution**. The green revolution involves two main practices: the introduction of new higher-yield seeds and the expanded use of fertilizers. Because of the green revolution, agricultural productivity at a global scale has increased faster than population growth.

During the 1950s, scientists began an intensive series of experiments to develop a higher-yield form of wheat. A decade later, the "miracle wheat seed" was ready. Shorter and stiffer than traditional breeds, the new wheat was less sensitive to variation in day length, responded better to fertilizers, and matured faster. The Rockefeller and Ford Foundations sponsored many of the studies, and the program's director, Dr. Norman Borlaug, won the Nobel Peace Prize in 1970.

The International Rice Research Institute, established in the Philippines by the Rockefeller and Ford Foundations, worked to create a miracle rice seed. During the 1960s, their scientists introduced a hybrid of Indonesian rice and Taiwan dwarf rice that was hardier and that increased yields. More recently, scientists have developed new high-yield maize (corn).

The new miracle seeds were diffused rapidly around the world. India's wheat production, for example, more than doubled in 5 years. After importing 10 million tons of wheat annually in the mid-1960s, India by 1971 had a surplus of several million tons. Other Asian and Latin American countries recorded similar productivity increases.

Need for Fertilizer and Machinery. To take full advantage of the new miracle seeds, farmers must use more fertilizer and machinery. Farmers have known for thousands of years that application of manure, bones, and ashes somehow increases, or at least maintains, the fertility of the land. Not until the nineteenth century did scientists identify nitrogen, phosphorus, and potassium (potash) as the critical elements in these substances that improved fertility. Today, these three elements form the basis for fertilizers—products that farmers apply on their fields to enrich the soil by restoring lost nutrients.

Nitrogen, the most important fertilizer, is a ubiquitous substance. Europeans most commonly produce a fertilizer known as urea, which contains 46 percent nitrogen. In North America, nitrogen is available as ammonia gas, which is 82 percent nitrogen but more awkward than urea to transport and store.

Both urea and ammonia gas combine nitrogen and hydrogen. The problem is that the cheapest way to produce both types of nitrogen-based fertilizers is to obtain hydrogen from natural gas or petroleum. As fossil fuel prices increase, so do the prices for nitrogen-based fertilizers, which then become too expensive for many farmers in LDCs.

In contrast with nitrogen, phosphorus and potash reserves are not distributed uniformly across Earth's surface. Two-thirds of the world's proven phosphate rock reserves are clustered in Morocco and the United States. Proven potash reserves are concentrated in Canada, Germany, Russia, and Ukraine.

Farmers need tractors, irrigation pumps, and other machinery to make most effective use of the new miracle seeds. In LDCs, farmers cannot afford such equipment, nor, in view of high energy costs, can they buy fuel to operate the equipment. To maintain the green revolution, governments in LDCs must allocate scarce funds for subsidizing the cost of seeds, fertilizers, and machinery.

The green revolution did not stop with miracle seeds. Scientists have continued to create higher-yield hybrids that are adapted to environmental conditions in specific regions. Thanks to the green revolution, Dutch scientists calculate that the maximum annual crop yield currently has reached 6,000 kilograms of grain per hectare (5,000 pounds per acre) in parts of Asia and Latin America. This, however, still is far lower than the maximum possible yields of 15,000 kilograms per hectare (13,000 pounds per acre) in Asia and 20,000 kilograms per hectare (18,000 pounds per acre) in Latin America. The green revolution was largely responsible for preventing a food crisis in these regions during the 1970s and 1980s, but will these scientific breakthroughs continue in the twenty-first century?

Increase Food Supply by Identifying New Food Sources

The third alternative for increasing the world's food supply is to develop new food sources. Three strategies being considered are to cultivate the oceans, to develop higher-protein cereals, and to improve palatability of rarely consumed foods.

Cultivating Oceans. At first glance, increased use of food from the sea is attractive. Oceans are vast, covering nearly three-fourths of Earth's surface and lying near most population concentrations. But historically, the sea has provided only a small percentage of world food supply. About two-thirds of the fish caught from the ocean is consumed directly, while the remainder is converted to fish meal and fed to poultry and hogs.

Hope grew during the 1950s and 1960s that increased fish consumption could meet the needs of a rapidly growing global population. Indeed, the world's annual fish catch increased about five times, from 22 million tons in 1954 to more than 100 million tons in 1991. However, the population of some fish species has declined because they have been harvested faster than they can reproduce. Overfishing has been particularly acute in the North Atlantic and Pacific oceans. The U.S. National Marine Fisheries Service estimates that 65 of 153 species of fish that it monitors off the Atlantic and Pacific coasts are overfished.

To protect fishing areas, many countries have claimed control of the oceans within 200 nautical miles of the coast. These countries have the right to seize foreign fishing vessels that venture into the so-called exclusive economic zone.

Peru has been especially sensitive to the overfishing problem after the country's catch of anchovies, its most important fish, declined by more than 75 percent between 1970 and 1973. To prevent further overfishing, the government nationalized its fish meal production industry, but the Peruvian experience demonstrates that the ocean is not a limitless source of fish.

Higher-Protein Cereals. A second possible new food source is higher-protein cereal grains. People in MDCs obtain protein by consuming meat, but people in LDCs generally rely on wheat, corn, and rice, which lack certain proteins. Scientists are experimenting with hybrids of the world's major cereals that have higher protein content.

People can also obtain needed nutrition by consuming foods that are fortified during processing with vitamins, minerals, and protein-carrying amino acids. This approach achieves better nutrition without changing food consumption habits. However, fortification has limited application in LDCs, where most people grow their own food rather than buy processed food.

Improve Palatability of Rarely Consumed Foods. To fulfill basic nutritional needs, people consume types of food adapted to their community's climate, soil, and other physical characteristics. People also select foods on the basis of religious values, taboos, and other social customs that are unrelated to nutritional or environmental factors. A third way to make more effective use of existing global resources is to encourage consumption of foods that are avoided for social reasons.

A prominent example of an underused food resource in North America is the soybean. Although the soybean is one of the region's leading crops, most of the output is processed into animal feed, in part because many North Americans avoid consuming tofu, sprouts, and other recognizable soybean products. However, burgers, franks, oils, and other products that do not look like soybeans are more widely accepted in North America. New food products have been created in LDCs as well. In Asia, for example, high-protein beverages made from seeds resemble popular soft drinks.

Krill (a term for a group of small crustaceans) could be an important source of food from the oceans. The krill population has increased rapidly in recent years, because overhunting has reduced the number of whales that eat krill. About 1 million tons of krill are currently harvested, most of which goes to Russia and Eastern Europe to feed chickens and livestock. The harvest could be substantially increased for human food with new processing methods, because krill deteriorates rapidly. But krill does not taste very good.

Increase Food Supply by Increasing Exports from Other Countries

The fourth alternative for increasing the world's food supply is to export more food from countries that produce surpluses. The three top export grains are wheat, maize (corn), and rice. Few countries are major exporters of food, but increased production in these countries could cover the gap elsewhere.

Before World War II, Western Europe was the only major grain-importing region. Prior to their independence, colonies of Western European countries supplied food to their parent states. Asia became a net grain importer in the 1950s, Africa and Eastern Europe in the 1960s, and Latin America in the 1970s. Population increases in these regions largely accounted for the need to import grain. By 1980, North America was the only major exporting region in the world.

In response to the increasing global demand for food imports, the United States passed Public Law 480, the Agricultural, Trade, and Assistance Act of 1954 (frequently referred to as "P.L.-480"). Title I of the act provided for the sale of grain at low interest rates, and Title II gave grants to needy groups of people.

The largest beneficiary of U.S. food aid has been India. In 1966 and 1967, when the monsoon rains failed, 60 million Indians were fed entirely by U.S. grain. At the height of the rescue, 600 ships filled with grain sailed to India, the largest maritime maneuver since the Allied invasion of Normandy on D-Day, June 6, 1944. During those years, the United States allocated 20 percent of its wheat crop to feed India's population.

The United States remains the largest grain exporter and accounts for nearly two-thirds of all corn and soybean exports, one-third of wheat, and one-fifth of rice. However, since 1980, the United States has decreased its grain exports, while other countries have increased theirs. Thailand replaced the United States as the leading rice exporter. Other Asian countries like Pakistan, Vietnam, and India account for most of the remaining rice exports. Australia and France have joined the United States and Canada as major wheat exporters (Figure 14–17).

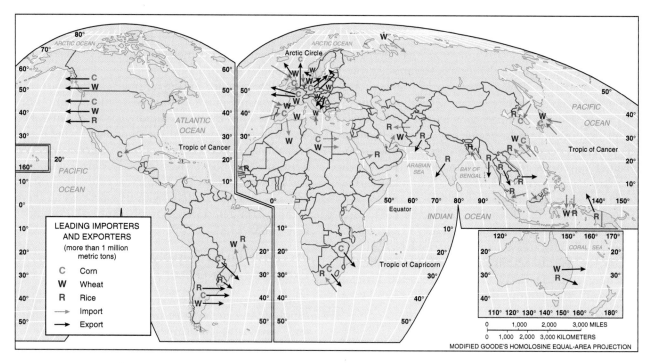

Figure 14–17 Grain imports and exports. Most countries must import more food than they export. The United States has by far the largest excess of food exports compared to imports. Argentina, Australia, Canada, and France are the other leading food exporters.

Russia is by far the leading grain importer and ranks at or near the top in wheat, corn, and rice. Russia and Japan together account for one-half of the world's corn imports, while Russia and China together account for one-fourth of the wheat imports. Asian countries account for nearly all of the rice imports.

Africa's Food Supply Crisis

Some countries that previously depended on imported grain have become self-sufficient in recent years. Higher productivity generated by the green revolution is primarily responsible for reducing dependency on imports, especially in Asia. India no longer ranks as a major wheat importer, and China no longer imports rice. As long as population growth continues to decline and agricultural productivity continues to increase, the large population concentrations of Asia can maintain the delicate balance between population and resources.

In contrast, sub-Saharan Africa is losing the race to keep food production ahead of population growth. The United Nations Food and Agricultural Organization estimates that 70 percent of Africans have too little to eat. Widespread famine exists in half of the African countries. By all estimates, the problems will grow worse.

Production of most food crops is lower today in Africa than in the 1960s. At the same time, population is increasing more rapidly than in any other world region. As a result, food production per person declined during the 1970s, 1980s and 1990s in all but a handful of the region's countries, in several cases by more than 20 per-

cent. Agriculture in sub-Saharan Africa can feed little more than half of the region's population.

The problem is particularly severe in the Horn of Africa, including Somalia, Ethiopia, and Sudan. Also facing severe food shortages are countries in the Sahel region, a 400- to 550-kilometer (250- to 350-mile) belt

In a region of Zimbabwe hit hard by drought, this woman is catching termites to consume them for protein. (Louise Gubb/The Image Works)

Blue wildebeests (gnus) and impalas cluster at a water hole in Namibia. The drylands of Africa can support limited populations, but the number of people and animals has exceeded the supply of food and water in some regions of Africa. (Leonard Lee Rue III/Photo Researchers, Inc.)

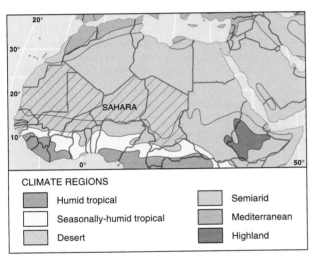

CLIMATE REGIONS

- Humid tropical
- Seasonally-humid tropical
- Desert
- Semiarid
- Mediterranean
- Highland

Figure 14–18 The Sahel. The Sahel, which lies south of the drylands of the Sahara, faces severe food supply problems, as does the Horn of Africa.

in West Africa that marks the southern border of the Sahara (Figure 14–18). The most severely affected countries in the Sahel are Gambia, Senegal, Mali, Mauritania, Burkina Faso, Niger, and Chad.

Traditionally, this region supported limited agriculture. Pastoral nomads moved their herds frequently, permitting vegetation to regenerate. Farmers grew groundnuts for export and used the receipts to import rice. With rapid population growth, herd size increased beyond the capacity of the land to support them. Animals overgrazed the limited vegetation and clustered at scarce water sources. Many died of hunger.

Farmers overplanted, exhausting soil nutrients, and reduced fallow time, during which unplanted fields can recover. Soil erosion increased after most of the remaining trees were cut for wood and charcoal, used for urban cooking and heating. Productivity declined further following several unusual drought years in the 1970s, 1980s, and 1990s.

Government policies have aggravated the food shortage crisis. To make food affordable for urban residents, governments keep agricultural prices low. Constrained by price controls, farmers are unable to sell their commodities at a profit and therefore have little incentive to increase productivity.

■ Summary

We have examined problems of depletion, degradation, and inefficient use of Earth's resources. The distribution of resources, as well as patterns of use and abuse, vary locally. But actions with regard to resources in one region can affect people everywhere.

Some scientists believe that further depletion and destruction of Earth's resources will lead to disaster in the near future. A quarter-century ago, a group of scientists known as the Club of Rome presented a particularly influential statement of this position in a report titled *The Limits to Growth*. According to these scientists, many of whom were professors at the Massachusetts Institute of Technology, the combination of population growth, resource depletion, and unrestricted use of industrial technology will disrupt the world's ecology and economy and lead to mass starvation, widespread suffering, and destruction of the physical environment.

In a recent update, the authors argued that environmental destruction is proceeding at a more rapid rate than they had originally thought. If new sets of attitudes and policies toward environmental protection are not in place within 20 years, the environment will be permanently damaged, and people's standard of living will fall.

The threat of irreparable global environmental damage is heightened by confrontation between more developed and less developed regions. The MDCs have achieved wealth in part by using large percentages of the world's resources and discharging large percentages of the world's pollutants. Now, LDCs are being asked to promote economic development with greater sensitivity to the environment than today's MDCs showed in the past. People in more developed countries are increasingly willing to allocate some of their wealth to clean up the environment. Subsistence farmers in LDCs cannot afford to invest in environmental protection.

Most geographers recognize that unrestricted industrial and demographic growth will have negative consequences, but they do not believe that the dire predictions of *The Limits to Growth* are inevitable. Human actions have depleted some resources, but substitutes may be available. Although pollution degrades the physical environment, industrial growth can be compatible with environmental protection. Demand for food is increasing, but human actions are also expanding the capacity of Earth to provide food.

Here again are the key issues in Chapter 14:

1. **Why are fossil fuel resources being depleted?** As we consume fossil fuel resources to produce energy, we are depleting Earth's supply. Fossil fuels, as well as remaining supplies, are distributed unevenly across Earth. Over the coming decades, we must turn to renewable energy sources. However, at this time, renewable energy sources are more expensive than are fossil fuels.

2. **Why are resources being polluted?** Human beings are damaging and destroying Earth's resources through pollution. Pollution is the discharge of waste at a rate that exceeds the environment's capacity to absorb it. Pollutants are discharged into the atmosphere, water, and onto land. We can reduce pollution only by decreasing the amount of waste we generate or by increasing the environment's capacity to accept it.

3. **Why are global food resources expandable?** We fail to make full use of Earth's resources to feed Earth's rapidly growing population. Four alternatives exist to increasing the food supply: expand the amount of cultivated land, increase productivity of land now used for agriculture, develop new food sources, and expand exports from productive countries. Because of increased productivity, most regions are generating enough food to feed their populations. The exception is sub-Saharan Africa, where the gap between food supply and population size is growing wider.

Case Study Revisited

Future Directions

Mexico City has taken steps to reduce air pollution. The government closed a major employer, the PEMEX oil refinery, located in the northwestern region of Mexico City, because it was responsible for 7 percent of the city's air pollution. Cars have been banned from a 50-square-block central area, and motorists are not allowed to use their cars one day each week, depending on the last digit of the license plate. Cars must now have catalytic converters and use unleaded fuel, and older buses and taxicabs have been removed from service.

However, Mexico must pay a price for implementing these pollution controls. Closing the oil refinery hurt Mexico's economy, because not only were jobs lost, but the country also had to import some fuel to replace the loss of the refinery's production. Motor vehicles are essential to economic development, because they allow people to get to work and businesses to deliver goods. And the number of vehicles in Mexico City has expanded because of the rapid population growth.

Rapid population growth means that rapidly growing cities in LDCs face pressure to expand economic opportunities and material benefits for the people, regardless of environmental impact. Stricter enforcement of pollution controls would require shutting down many businesses and eliminating jobs.

Geographers emphasize that each resource in the physical environment has a distinctive capacity for accommodating human activities. Just as a good farmer knows how many animals can be fed on a parcel of land, a scientist can pinpoint the constraints that resources place on population density or economic development in a particular region. With knowledge of these constraints, we will be able to maintain agricultural and industrial development in the future.

Future generations can maintain agricultural and industrial development without depleting and abusing resources, but they must move toward sustainable development. **Sustainable development** is the level of development that can be maintained in a country without depleting resources to the extent that future generations will be unable to achieve a comparable level of development. The United Nations defined sustainable development as "development that meets the needs of the present without compromising the ability of future generations to meet their own needs."

The concept of sustainable development is based on the current practice of sustained yield management of renewable resources, such as forests and fisheries. In some places, the amount of timber cut down in a forest or the number of fish removed from a body of water is controlled at a level that does not reduce future supplies.

In recent years, the World Bank and other international development agencies have embraced the concept of sustainable development. Planning for development involves consideration of many more environmental and social issues today than was the case in the past. However, one important recommendation of the U.N. report has not been implemented: increased international cooperation to reduce the gap between more developed and less developed countries.

Similarly, food supply can be expanded without damaging Earth's resources if farmers adopt sustainable agriculture.

Sustainable agriculture is farming methods that preserve long-term productivity of land and minimize pollution of the soil, groundwater, and streams that drain the land. In most cases, sustainable agriculture means rotating soil-restoring crops with cash crops and reducing inputs of fertilizer and pesticides.

Farmers practicing sustainable agriculture typically generate lower revenues than do conventional farmers, but they also have lower costs. On balance, sustainable agriculture may enjoy greater net benefits when costs are subtracted from revenues, especially if long-term environmental costs are included. It remains the world's best hope for increased long-term produc-tivity in LDCs, consistent with sound environmental management.

A generation ago, environmentalists coined the phrase "think global, act local" so that we would recognize that our actions in our own communities—and even in our own back-yards—could affect the entire planet. Now, geographers urge us to "think global" *and* "think local." In an age of globalization, we cannot lose sight of the importance and pleasure of the diversity of local physical conditions and human behavior. Think both global and local, and act wherever you can do some good.

■ Key Terms

Acid deposition Sulfur oxides and nitro-gen oxides, emitted by burning fossil fuels, enter the atmosphere—where they combine with oxygen and water to form sulfuric acid and nitric acid—and return to Earth's surface.

Acid precipitation Conversion of sulfur oxides and nitrogen oxides to acids that return to Earth as rain, snow, or fog.

Active solar energy systems Solar ener-gy system that collects energy through the use of mechanical devices like pho-tovoltaic cells or flat-plate collectors.

Air pollution Concentration of trace sub-stances, such as carbon monoxide, sulfur dioxide, nitrogen oxides, hydrocarbons, and solid particulates, at a greater level than occurs in average air.

Animate power Power supplied by peo-ple or animals.

Biochemical oxygen demand (BOD) Amount of oxygen required by aquatic bacteria to decompose a given load of organic waste; a measure of water pol-lution.

Biomass fuel Fuel that derives from plant material and animal waste.

Breeder reactor A nuclear power plant that creates its own fuel from pluto-nium.

Chlorofluorocarbon (CFC) A gas used as a solvent, a propellant in aerosols, a refrigerant, and in plastic foams and fire extinguishers.

Desertification Degradation of land, es-pecially in semiarid areas, primarily be-cause of human actions like excessive crop planting, animal grazing, and tree cutting.

Fission The splitting of an atomic nucleus to release energy.

Fossil fuel Energy source formed from the residue of plants and animals buried millions of years ago.

Fusion Creation of energy by joining the nuclei of two hydrogen atoms to form helium.

Geothermal energy Energy from steam or hot water produced from hot or molten underground rocks.

Green revolution Rapid diffusion of new agricultural technology, especially new high-yield seeds and fertilizers.

Greenhouse effect Anticipated increase in Earth's temperature, caused by car-bon dioxide (emitted by burning fossil fuels) trapping some of the radiation emitted by the surface.

Hydroelectric power Power generated from moving water.

Inanimate power Power supplied by ma-chines.

Nonrenewable energy A source of ener-gy that is a finite supply capable of being exhausted.

Ozone A gas that absorbs ultraviolet solar radiation, found in the strato-sphere, a zone between 15 and 50 kilo-meters (9 to 30 miles) above Earth's surface.

Passive solar energy systems Solar en-ergy system that collects energy with-out the use of mechanical devices.

Photochemical smog An atmospheric condition formed through a combina-tion of weather conditions and pollu-tion, especially from motor vehicle emissions.

Photovoltaic cell Solar energy cells, usu-ally made from silicon, that collect solar rays to generate electricity.

Pollution Addition of more waste than a resource can accommodate.

Potential reserve The amount of energy in deposits not yet identified but thought to exist.

Proven reserve The amount of a resource remaining in discovered deposits.

Radioactive waste Particles from a nu-clear reaction that emit radiation; con-tact with such particles may be harmful or lethal to people and must therefore be safely stored for thousands of years.

Renewable energy A resource that has a theoretically unlimited supply and is not depleted when used by humans.

Resource A substance in the environ-ment that is useful to people, is eco-nomically and technologically feasible to access, and is socially acceptable to use.

Sanitary landfill A place to deposit solid waste, where a layer of earth is bull-dozed over garbage each day to reduce emissions of gases and odors from the decaying trash, to minimize fires, and to discourage vermin.

Sustainable agriculture Farming meth-ods that preserve long-term productiv-ity of land and minimize pollution, typically by rotating soil-restoring crops with cash crops and reducing inputs of fertilizer and pesticides.

Sustainable development The level of development that can be maintained in a country without depleting resources to the extent that future generations will be unable to achieve a comparable level of development.

Thinking Geographically

1. What steps has your community taken to recycle solid waste and to conserve energy?

2. U.S. automakers must meet a standard for Corporate Average Fuel Efficiency (CAFE). This means that the average miles-per-gallon achieved by all models of a company's American-made cars must meet a government-mandated level. If they do not, the company must pay a stiff fine. Should the United States raise the CAFE standard to conserve fuel and reduce air pollution, even if the result is a loss of American jobs? Explain.

3. A recent study compared paper and polystyrene foam drinking cups. Conventional wisdom is that foam cups are bad for the environment, because they are made from petroleum and do not degrade in landfills. However, the manufacture of a paper cup consumes 36 times as much electricity and generates 580 times as much wastewater. Further, as they degrade in landfills, paper cups release methane gas, a contributor to the greenhouse effect. Which types of cups should companies like McDonald's be encouraged to use? Why?

4. Pollution is a byproduct of producing almost anything. How can more developed countries, which historically have been responsible for generating the most pollution, encourage less developed countries to seek to minimize the adverse effects of pollution as they improve their levels of development?

5. Malthus argued 200 years ago that overpopulation was inevitable, because population increased geometrically, while food supply increased arithmetically. Was Malthus correct? Why, or why not?

Further Readings

Barbier, Edward B., and Joanne C. Burgess. "The Economics of Tropical Forest Land Use Options." *Land Economics* 73 (1997): 174–95.

Beach, Timothy, and P. Gershmehl. "Soil Erosion, T Values, and Sustainability: A Review and Exercise." *Journal of Geography* 92 (1993): 16–22.

Bower, Blair T., and Daniel J. Basta. *Residuals-Environmental Quality Management: Applying the Concept.* Baltimore, MD: The Johns Hopkins University Center for Metropolitan Planning and Research, 1973.

Brown, Lester R. "Higher Crop Yields? Don't Bet the Farm on Them." *World Watch* 10 (1997): 8–17.

_____, et al. *State of the World.* New York and London: W. W. Norton & Co., annually since 1984.

_____, and Edward C. Wolf. *Reversing Africa's Decline.* Worldwatch Paper 65. Washington, DC: Worldwatch Institute, 1985.

_____, and Pamela Shaw. *Six Steps to a Sustainable Society.* Worldwatch Paper 48. Washington, DC: Worldwatch Institute, 1982.

Bugliarello, George; Ariel Alexandre, John Barnes, and Charles Wakstein. *The Impact of Noise Pollution.* New York: Pergamon Press, 1976.

Calzonetti, Frank J., and Barry D. Solomon. *Geographical Dimensions of Energy.* Dordrecht, Netherlands: D. Reidel Publishing Co., 1985.

Chakravarti, A. K. "Green Revolution in India." *Annals of the Association of American Geographers* 63 (1973): 319–30.

Clements, Donald W. "Recent Trends in the Geography of Coal." *Annals of the Association of American Geographers* 67 (1977): 109–25.

Cole, H.S.D.; Christopher Freeman, Marie Jahoda, and K.L.R. Pavitt. *Models of Doom: A Critique of the Limits to Growth.* New York: Universe Books, 1972.

Commoner, Barry. *Making Peace with the Planet.* New York: New Press, 1990.

_____. *The Poverty of Power.* New York: Knopf, 1976.

Costanza, Robert, and H.E. Daly. "Natural Capital and Sustainable Development." *Conservation Biology* 6 (1992): 37–46.

Cuff, David, and William J. Young. *The United States Energy Atlas*, 2d ed. New York: Free Press, 1984.

Cusack, David F., ed. *Agroclimate Information for Development: Reviving the Green Revolution.* Boulder, CO: Westview Press, 1983.

Cutter, Susan L.; Hilary Lambert Renwick, and William H. Renwick. *Exploitation, Conservation, Preservation: A Geographic Perspective on Natural Resource Use*, 2d ed. New York: John Wiley, 1991.

Dakers, Sonya. *Sustainable Agriculture: Future Dimensions.* Ottawa: Library of Parliament, Research Branch, 1992.

de Freitas, C.R. "The Greenhouse Crisis: Myths and Misconceptions." *Area* 23 (1991): 11–18.

Dodd, Jerrold L. "Desertification and Degradation in sub-Saharan Africa: The Role of Livestock." *BioScience* 44 (1994): 28–34.

Ehrlich, Anne H., and Paul R. Ehrlich. *The Earth.* New York: Franklin Watts, 1987.

Elsom, Derek. *Atmospheric Pollution: A Global Problem*, 2d ed. Cambridge, MA: Blackwell, 1992.

Falkenmark, Malin, and Carl Widstrand. "Population and Water Resources: A Delicate Balance." *Population Bulletin* 47. Washington, DC: Population Reference Bureau, November 1992.

Flavin, Christopher. *World Oil: Coping with the Dangers of Success.* Worldwatch Paper 66. Washington, DC: Worldwatch Institute, July 1985

_____. *Electricity's Future: The Shift to Efficiency and Small-Scale Power.* Worldwatch Paper 61. Washington, DC: Worldwatch Institute, November 1984.

_____. *Nuclear Power: The Market Test.* Worldwatch Paper 57. Washington, DC: Worldwatch Institute, December 1983.

Francis, Charles A., ed. *Sustainable Agriculture in Temperate Zones.* New York: John Wiley, 1990.

Greenberg, M. R.; R. Anderson, and G. W. Page. *Environmental Impact Statements.* Washington, DC: Association of American Geographers, 1978.

Grigg, David. *The World Food Problem 1950–1980.* New York: Basil Blackwell, 1985.

_____. "The World's Hunger: A Review 1930–1990." *Geography* 82 (1997): 197–206.

Grossman, Gene M., and Alan B. Krueger. *Economic Growth and the Environment.* Princeton, NJ: Princeton University, Woodrow Wilson School Discussion Paper in Economics, 1994.

Hammer, W. M., "Krill—Untapped Bounty from the Sea," *National Geographic* 165 (1984): 626–43.

James, Peter. *The Future of Coal,* 2d ed. London: Macmillan, 1984.

Kates, Robert W.; Christoph Hohenemser, and Jeanne X. Kasperson, eds. *Perilous Progress: Technology as Hazard.* Boulder, CO: Westview Press, 1984.

Knight, C. G., and P. Wilcox. *Triumph or Triage? The World Food Problem in Geographical Perspective.* Washington, DC: Association of American Geographers, 1975.

Kotlyakov, Vladimir M. "Climatic Change and the Future of the Human Environment." *International Social Science Journal* 48 (1996): 511–24

Meadows, Donnela H.; Dennis L. Meadows, Jorgen Randers, and William W. Behrens III. *The Limits to Growth,* 2d ed. New York: Universe Books, 1973.

Meadows, Donnela H.; Dennis L. Meadows, and Jorgen Randers. *Beyond the Limits.* Post Mills, VT: Chelsea Green Publishing Co., 1992.

Munton, Richard. "Engaging Sustainable Development: Some Observations on Progress in the UK." *Progress in Human Geography* 21 (1997): 147–63.

Mounfield, P. R. "Nuclear Power in Western Europe: Geographical Patterns and Political Problems." *Geography* 70 (1985): 315–27.

Murdock, Steve H.; F. Larry Leistritz, and Rita R. Hamm, eds. *Nuclear Waste: Socioeconomic Dimensions of Long-Term Storage.* Boulder, CO: Westview Press, 1983.

National Geographic. *Energy: Special Report.* Washington, DC: National Geographic Society, 1981.

National Research Council, Board on Radioactive Waste Management, Panel on Social and Economic Aspects of Radioactive Waste Management. *Social and Economic Aspects of Radioactive Waste Disposal: Considerations for Institutional Management.* Washington, DC: National Academy Press, 1984.

Openshaw, Stan. *Nuclear Power: Siting and Safety.* London: Routledge and Kegan Paul, 1986.

Pasqualetti, Martin J., and K. David Pijawka, eds. *Nuclear Power: Assessing and Managing Hazardous Technology.* Boulder, CO: Westview Press, 1984.

Pierce, John T. *The Food Resource.* New York: Longman Scientific and Technical, 1990.

Pollack, Cynthia. *Decommissioning: Nuclear Power's Missing Link.* Worldwatch Paper 69. Washington, DC: Worldwatch Institute, 1986.

Postel, Sandra. *Conserving Water: The Untapped Alternative.* Worldwatch Paper 67. Washington, DC: Worldwatch Institute, 1985.

_____. *Air Pollution, Acid Rain, and the Future of Forests.* Worldwatch Paper 58. Washington, DC: Worldwatch Institute, 1984.

Pryde, Philip R. *Environmental Management in the Soviet Union.* Cambridge: Cambridge University Press, 1991.

Rinaldi, S.; W. Sanderson, and A. Gragnani. "Pollution Control Policies and Natural Resource Dynamics: A Theoretical Analysis." *Journal of Environmental Management* 48 (1996): 357–74.

Rosegrant, Mark W., and Robert Livernash. "Growing More Food, Doing Less Damage." *Environment* 38 (1996): 6–11

Sagers, Matthew J. "Russian Crude Oil Production in 1996: Conditions and Prospects." *Post–Soviet Geography and Economics* 37 (1996): 523–87.

Sankoh, O.A. "Environmental Impact Assessment Convincible to Developing Countries." *Journal of Environmental Management* 47 (1996): 185–89.

Schmandt, Jurgen, and Hilliard Roderick, eds. *Acid Rain and Friendly Neighbors: The Policy Dispute between Canada and the U.S.* Durham, NC: Duke University Press, 1985.

Sills, David L.; C. P. Wolf, and Vivien B. Shelanski, eds. *Accident at Three Mile Island: The Human Dimensions.* Boulder, CO: Westview Press, 1982.

Smil, Vaclav. "China Shoulders the Cost of Environmental Change." *Environment* 39 (1997): 6–9.

Turner, B. L., 2d; Robert W. Kates, and William C. Clark. *The Earth as Transformed by Human Action.* Cambridge: Cambridge University Press, 1990.

U.S. Congress, Office of Technology Assessment. *Acid Rain and Transported Air Pollutants: Implications for Public Policy.* Washington, DC: U.S. Government Printing Office, 1984.

World Commission on Environment and Development. *Food 2000: Global Policies for Sustainable Agriculture.* London: Zed Books, 1987.

_____. *Our Common Future.* London: Oxford University Press, 1987.

Young, Liz. "World Hunger: A Framework for Analysis." *Geography* 81 (1996): 97–110.

_____. "Environmental Intelligence." *World Watch* 10 (1997): 7–9.

Also consult these journals: *Ecological Economics; Ecologist; Energy Journal; Energy Policy; Environment; Environmental Management; Environmental Pollution; Journal of Environmental Management; Worldwatch.*

Conclusion
CAREERS IN GEOGRAPHY

An increasing number of students recognize that geographic education is practical as well as stimulating. Employment opportunities are expanding for students trained in geography, especially in teaching, government service, and business.

Teaching. As of 1998, a doctorate in geography was offered at fifty-two U.S. and twenty-seven Canadian universities, and the master's was the highest available degree at ninety-five U.S. and five Canadian universities. Traditionally, most trained geographers became teachers in high schools, colleges, or universities.

A career as a geography teacher is promising, because schools throughout North America are expanding the amount of geography in the curriculum. Educators increasingly recognize geography's role in teaching students about global diversity.

Some university geography departments have emphasized good teaching over research; others are increasingly concerned with research. The Association of American Geographers includes several dozen specialty groups organized around research themes, including agricultural, industrial, medical, and transportation geography.

Government. Some geographers find employment with cities, states, provinces, and other units of local government. Typically, these opportunities are found in departments of planning, transportation, parks and recreation, economic development, housing, or other similarly titled government agencies. Geographers may be hired to conduct studies of local economic, social, and physical patterns; to prepare information through maps and reports; and to help to plan the community's future.

Many national government agencies also employ geographers. In the United States, the Bureau of the Census in the Department of Commerce has a geography division that studies and reports on changing national population trends. Other U.S. government agencies that employ geographers include the Department of Defense Mapping Agency, the Soil Conservation Service and the Geological Survey in the Department of Interior, and the National Aeronautics and Space Administration.

Geographers contribute their knowledge of the location of activities, the patterns underlying the distribution of various activities, and the interpretation of information from maps. For example, in the 1940s the British government hired the geographer L. Dudley Stamp to conduct a land-use study of the entire country. Stamp identified the country's best agricultural land and contributed to the development of laws that strictly protected valuable land from urban development.

In recent years, geographers have been hired by government agencies because of their ability to interpret data generated from satellite imagery. Geographers also increasingly display information and interpret information through computer-generated maps.

Another career area for geographers is government foreign service, especially if they have expertise about other parts of the world. The tradition of geographic service in foreign affairs strengthened after World War I. The American geographer Isaiah Bowman advised President Woodrow Wilson on redrawing the map of Europe after the war, so that national boundaries more closely conformed to cultural patterns. At approximately the same time, the British geographer Halford J. Mackinder advised the British government on military strategy. He argued that the world's heartland, Eastern Europe, was surrounded by a rim of maritime powers. The key to Mackinder's international military strategy was understanding the relationship of the location of countries to their opportunity to exercise either land or naval power.

Business. An increasing number of American geographers are finding jobs with private companies. The list of possibilities is long, but here are some common examples.

- Developers hire geographers to find the best locations for new shopping centers.
- Real estate firms hire geographers to assess the value of properties.
- Supermarket chains, department stores, and other retailers hire geographers to determine the potential market for new stores.
- Banks hire geographers to assess the probability that a loan applicant has planned a successful development.
- Distributors and wholesalers hire geographers to find ways to minimize transportation costs.
- Transnational corporations hire geographers to predict the behavior of consumers and officials in other countries.
- Manufacturers hire geographers to identify new sources of raw materials and markets.
- Utility companies hire geographers to determine future demand at different locations for gas, electricity, and other services.

For more information on careers in geography, contact the Association of American Geographers in Washington, D.C., or the National Council for Geographic Education at Indiana University of Pennsylvania.

Woody Allen's Final Word to the Graduates This book is supposed to end with a word about the future to the graduates of the class. Woody Allen is not a geographer, but his speech to graduates has captured effectively some of the interrelationships among various human actions and the physical environment, which form the core of human geography.

More than any other time in history, mankind faces a crossroads. One path leads to despair and utter hopelessness. The other, to total extinction. Let us pray we have the wisdom to choose correctly. . . .

Science is something we depend on all the time. If I develop a pain in the chest I must take an X ray. But what if the radiation from the X ray causes me deeper problems? Before I know it, I'm going in for surgery. Naturally, while they're giving me oxygen an intern decides to light up a cigarette. The next thing you know I'm rocketing over the World Trade Center in bed clothes. . . .

At no other time in history has man been so afraid to cut into his veal chop for fear that it will explode. Violence breeds more violence and it is predicted that...kidnapping will be the dominant mode of social interaction. Overpopulation will exacerbate problems to the breaking point. Figures tell us that there are already more people on earth than we need to move even the heaviest piano. If we do not call a halt to breeding . . . there will be no room to serve dinner unless one is willing to set the table on the heads of strangers. Then they must not move for an hour while we eat. Of course energy will be in short supply and each car owner will be allowed only enough gasoline to back up a few inches. . . .

Summing up, it is clear the future holds great opportunities. It also holds pitfalls. The trick will be to avoid the pitfalls, seize the opportunities, and get back home by six o'clock.*

Human geographers do not know how to solve all of the world's problems of population growth, cultural and political conflict, economic development, and abuse of resources. This course has tried to expose you to the need to understand differences among human actions in different regions, the interdependencies between people and the environment, and other geographic perspectives on world problems. Above all, this book's aim is to heighten your sense of global awareness, that is, an understanding that our comfort—if not survival—requires greater knowledge of Earth's human and physical processes.

---

*Reprinted with the permission of Woody Allen and Random House, Inc. Reprinted from Woody Allen, "My Speech to the Graduates," Side Effects. Copyright © Random House, Inc., and Woody Allen.

Appendix
MAP SCALE AND PROJECTIONS
Phillip C. Muercke

Unaided, our human senses provide a limited view of our surroundings. To overcome those limitations, humankind has developed powerful vehicles of thought and communication, such as language, mathematics, and graphics. Each of those tools is based on elaborate rules, each has an information bias, and each may distort its message, often in subtle ways. Consequently, to use those aids effectively, we must understand their rules, biases, and distortions. The same is true for the special form of graphics we call maps: we must master the logic behind the mapping process before we can use maps effectively.

A fundamental issue in cartography, the science and art of making maps, is the vast difference between the size and geometry of what is being mapped—the real world, we will call it—and that of the map itself. Scale and projection are the basic cartographic concepts that help us understand that difference and its effects.

Map Scale

Our senses are dwarfed by the immensity of our planet; we can sense directly only our local surroundings. Thus, we cannot possibly look at our whole state or country at one time, even though we may be able to see the entire street where we live. Cartography helps us expand what we can see at one time by letting us view the scene from some distant vantage point. The greater the imaginary distance between that position and the object of our observation, the larger the area the map can cover but the smaller the features will appear on the map. That reduction is defined by the *map scale*, the ratio of the distance on the map to the distance on the earth. Map users need to know about map scale for two reasons: (1) so that they can convert measurements on a map into meaningful real-world measures and (2) so that they can know how abstract the cartographic representation is.

Real-World Measures A map can provide a useful substitute for the real world for many analytical purposes. With the scale of a map, for instance, we can compute the actual size of its features (length, area, and volume). Such calculations are helped by three expressions of a map scale: a word statement, a graphic scale, and a representative fraction.

A *word statement* of a map scale compares X units on the map to Y units on the earth, often abbreviated "X unit to Y units." For example, the expression "1 inch to 10 miles" means that 1 inch on the map represents 10 miles on the earth (Figure A-1). Because the map is always smaller than the area that has been mapped, the ground unit is always the larger number. Both units are expressed in meaningful terms, such as inches or centimeters and miles or kilometers. Word statements are not intended for precise calculations but give the map user a rough idea of size and distance.

A *graphic scale*, such as a bar graph, is concrete and therefore overcomes the need to visualize inches and miles that is associated with a word statement of scale (see Figure A-1). A graphic scale permits direct visual comparison of feature sizes and the distances between features. No ruler is required; any measuring aid will do. It needs only to be compared with the scaled bar; if the length of 1 toothpick is equal to 2 miles on the ground and the map distance equals the length of 4 toothpicks, then the ground distance is 4 times 2, or 8 miles. Graphic scales are especially convenient in this age of copying machines, when we are more likely to be working with a copy than with the original map. If a map is reduced or enlarged as it is copied, the graphic scale will change in proportion to the change in the size of the map and thus will remain accurate.

The third form of a map scale is the *representative fraction* (RF). An RF defines the ratio between the distance on the map and the distance on the earth in fractional terms, such as 1/633,600 (also written 1/633,600 or 1 : 633,600). The numerator of the fraction always refers to the distance on the map, and the denominator always refers to the distance on the earth. No units of measurement are given, but both numbers must be expressed in the same units. Because map distances are extremely small relative to the size of the earth, it makes sense to use small units, such as inches or centimeters. Thus, the RF 1 : 633,600 might be read as "1 inch on the map to 633,600 inches on the earth."

Herein lies a problem with the RF. Meaningful map-distance units imply a denominator so large that it is impossible to visualize. Thus, in practice, reading the map scale involves an additional step of converting the denominator to a meaningful ground measure, such as miles or kilometers. The unwieldy 633,600 becomes the

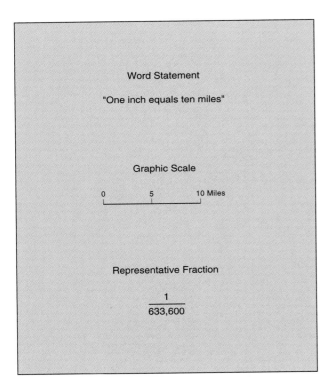

FIGURE A–1 Common expressions of map scale.

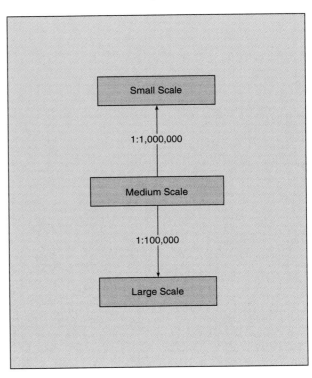

FIGURE A–2 The scale gradient can be divided into three broad categories.

more manageable 10 miles when divided by the number of inches in a mile (63,360).

On the plus side, the RF is good for calculations. In particular, the ground distance between points can be easily determined from a map with an RF. One simply multiplies the distance between the points on the map by the denominator of the RF. Thus, a distance of 5 inches on a map with an RF of 1/126,720 would signify a ground distance of 5 × 126,720, which equals 633,600. Because all units are inches and there are 63,360 inches in a mile, the ground distance is 633,600 ÷ 63,360, or 10 miles. Computation of area is equally straightforward with an RF. Computer manipulation and analysis of maps is based on the RF form of map scale.

Guides to Generalization Scales also help map users visualize the nature of the symbolic relation between the map and the real world. It is convenient here to think of maps as falling into three broad scale categories (Figure A-2). (Do not be confused by the use of the words *large* and *small* in this context; just remember that the larger the denominator, the smaller the scale ratio and the larger the area that is shown on the map.) Scale ratios greater than 1 : 100,000, such as the 1 : 24,000 scale of U.S. Geological Survey topographic quadrangles, are large-scale maps. Although those maps can cover only a local area, they can be drawn to rather rigid standards of accuracy. Thus, they are useful for a wide range of applications that require detailed and accurate maps, including zoning, navigation, and construction.

At the other extreme are maps with scale ratios of less than 1 : 1,000,000, such as maps of the world that are found in atlases. Those are small-scale maps. Because they cover large areas, the symbols on them must be highly abstract. They are therefore best suited to general reference or planning, when detail is not important. Medium- or intermediate-scale maps have scales between 1 : 100,000 and 1 : 1,000,000. They are good for regional reference and planning purposes.

Another important aspect of map scale is to give us some notion of geometric accuracy; the greater the expanse of the real world shown on a map, the less accurate the geometry of that map is. Figure A-3 shows why. If a curve is represented by straight line segments, short segments (*X*) are more similar to the curve than are long segments (*Y*). Similarly, if a plane is placed in contact with a sphere, the difference between the two surfaces is slight where they touch (*A*) but grows rapidly with increasing distance from the point of contact (*B*). In view of the large diameter and slight local curvature of the earth, distances will be well represented on large-scale maps (those with small denominators) but will be increasingly poorly represented at smaller scales. This close relationship between map scale and map geometry brings us to the topic of map projections.

Map Projections

The spherical surface of the earth is shown on flat maps by means of map projections. The process of "flattening" the earth is essentially a problem in geometry that

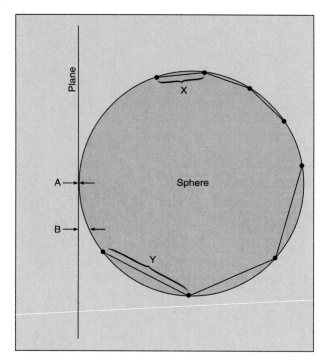

FIGURE A–3 Relationships between surfaces on the round earth and a flat map.

has captured the attention of the best mathematical minds for centuries. Yet, no one has ever found a perfect solution; there is no known way to avoid spatial distortion of one kind or another. Many map projections have been devised, but only a few have become standard. Because a single flat map cannot preserve all aspects of the earth's surface geometry, a mapmaker must be careful to match the projection with the task at hand. To map something that involves distance, for example, a projection should be used in which distance is not distorted. In addition, a map user should be able to recognize which aspects of a map's geometry are accurate and which are distortions caused by a particular projection process. Fortunately, that objective is not too difficult to achieve.

It is helpful to think of the creation of a projection as a two-step process (Figure A-4). First, the immense earth is reduced to a small globe with a scale equal to that of the desired flat map. All spatial properties on the globe are true to those on the earth. Second, the globe is flattened. Since that cannot be done without distortion, it is accomplished in such a way that the resulting map exhibits certain desirable spatial properties.

Perspective Models Early map projections were sometimes created with the aid of perspective methods, but that has changed. In the modern electronic age, projections are normally developed by strictly mathematical means and are plotted out or displayed on computer-driven graphics devices. The concept of perspective is still useful in visualizing what map projections do, however. Thus, projection methods are often illustrated by

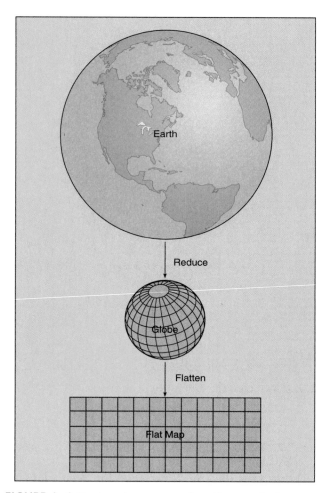

FIGURE A–4 The two-step process of creating a projection.

using strategically located light sources to cast shadows on a projection surface from a latitude/longitude net inscribed on a transparent globe.

The success of the perspective approach depends on finding a projection surface that is flat or that can be flattened without distortion. The cone, cylinder, and plane possess those attributes and serve as models for three general classes of map projections: *conic*, *cylindrical*, and *planar* (or azimuthal). Figure A-5 shows those three classes, as well as a fourth, a false cylindrical class with an oval shape. Although the oval class is not of perspective

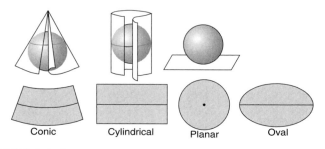

FIGURE A–5
General classes of map projections. *(Courtesy of ACSM)*

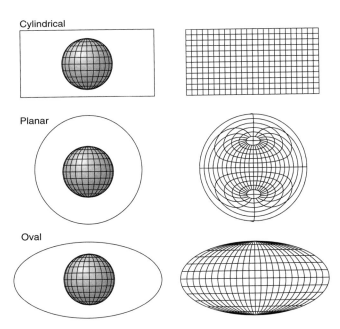

FIGURE A-6
The visual properties of cylindrical and planar projections combined in oval projections. *(Courtesy of ACSM)*

origin, it appears to combine properties of the cylindrical and planar classes (Figure A-6).

The relationship between the projection surface and the model at the point or line of contact is critical because distortion of spatial properties on the projection is symmetrical about, and increases with distance from, that point or line. That condition is illustrated for the cylindrical and planar classes of projections in Figure A-7. If the point or line of contact is changed to some other position on the globe, the distortion pattern will be recentered on the new position but will retain the same symmetrical form. Thus, centering a projection on the area of interest on the earth's surface can minimize the effects of projection distortion. And recognizing the

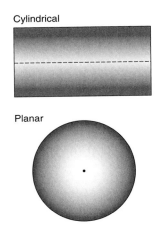

FIGURE A-7 Characteristic patterns of distortion for two projection classes. Here, darker shading implies greater distortion. *(Courtesy of ACSM)*

general projection shape, associating it with a perspective model, and recalling the characteristic distortion pattern will provide the information necessary to compensate for projection distortion.

Preserved Properties For a map projection to truthfully depict the geometry of the earth's surface, it would have to preserve the spatial attributes of *distance, direction, area, shape,* and *proximity.* That task can be readily accomplished on a globe, but it is not possible on a flat map. To preserve area, for example, a mapmaker must stretch or shear shapes; thus, area and shape cannot be preserved on the same map. To depict both direction and distance from a point, area must be distorted. Similarly, to preserve area as well as direction from a point, distance has to be distorted. Because the earth's surface is continuous in all directions from every point, discontinuities that violate proximity relationships must occur on all map projections. The trick is to place those discontinuities where they will have the least impact on the spatial relationships in which the map user is interested.

We must be careful when we use spatial terms, because the properties they refer to can be confusing. The geometry of the familiar plane is very different from that of a sphere; yet, when we refer to a flap map, we are in fact making reference to the spherical earth that was mapped. A shape-preserving projection, for example, is truthful to local shapes—such as the right-angle crossing of latitude and longitude lines—but does not preserve shapes at continental or global levels. A distance-preserving projection can preserve that property from one point on the map in all directions or from a number of points in several directions, but distance cannot be preserved in the general sense that area can be preserved. Direction can also be generally preserved from a single point or in several directions from a number of points but not from all points simultaneously. Thus, a shape-, distance-, or direction-preserving projection is truthful to those properties only in part.

Partial truths are not the only consequence of transforming a sphere into a flat surface. Some projections exploit that transformation by expressing traits that are of considerable value for specific applications. One of those is the famous shape-preserving *Mercator projection* (Figure A-8). That cylindrical projection was derived mathematically in the 1500s so that compass bearing (called rhumb lines) between any two points on the earth would plot as straight lines on the map. That trait let navigators plan, plot, and follow courses between origin and destination, but it was achieved at the expense of extreme areal distortion toward the margins of the projection (see Antarctica in Figure A-8). Although the Mercator projection is admirably suited for its intended purpose, its widespread but inappropriate use for nonnavigational purposes has drawn a great deal of criticism.

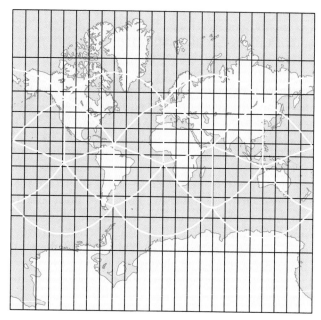

FIGURE A–8 The useful Mercator projection, showing extreme area distortion in the higher latitudes. *(Courtesy of ACSM)*

The *gnomonic projection* is also useful for navigation. It is a planar projection with the valuable characteristic of showing the shortest (or great circle) route between any two points on the earth as straight lines. Long-distance navigators first plot the great circle course between origin and destination on a gnomonic projection (Figure A-9, top). Next they transfer the straight line to a Mercator projection, where it normally appears as a curve (Figure A-9, bottom). Finally, using straight-line segments, they construct an approximation of that course on the Mercator projection. Navigating the shortest course between origin and destination then involves following the straight segments of the course and making directional corrections between segments. Like the Mercator projection, the specialized gnomonic

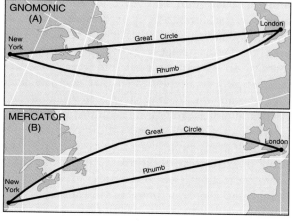

FIGURE A–9 A gnomonic projection (A) and a Mercator projection (B), both of value to long-distance navigators.

projection distorts other spatial properties so severely that it should not be used for any purpose other than navigation or communications.

Projections Used in Textbooks   Although a map projection cannot be free of distortion, it can represent one or several spatial properties of the earth's surface accurately if other properties are sacrificed. The two projections used for world maps throughout this textbook illustrate that point well. *Goode's homolosine projection*, shown in Figure A-10, belongs to the oval category and shows area accurately, although it gives the impression that the earth's surface has been torn, peeled, and flattened. The interruptions in Figure A-10 have been placed in the major oceans, giving continuity to the land masses. Ocean areas could be featured instead by placing the interruptions in the continents. Obviously, that type of interrupted projection severely distorts proximity relationships. Consequently, in different locations the properties of distance, direction, and shape are also distorted to varying degrees. The distortion pattern mimics that of cylindrical projections, with the equatorial zone the most faithfully represented (Figure A-11).

An alternative to special-property projections such as the equal-area Goode's homolosine is the compromise projection. In that case no special property is achieved at the expense of others, and distortion is rather evenly distributed among the various properties, instead of being focused on one or several properties. The *Robinson projection*, which is also used in this textbook, falls into that category (Figure A-12). Its oval projection has a global feel, somewhat like that of Goode's homolosine. But the Robinson projection shows the North Pole and the South Pole as lines that are slightly more than half the length of the equator, thus exaggerating distances and areas near the poles. Areas look larger than they really are in the high latitudes (near the poles) and smaller than they really are in the low latitudes (near the equator). In addition, not all latitude and longitude lines intersect at right angles, as they do on the earth, so we know that the Robinson projection does not preserve direction or shape either. However, it has fewer interruptions than the Goode's homolosine does, so it preserves proximity better. Overall, the Robinson projection does a good job of representing spatial relationships, especially in the low to middle latitudes and along the central meridian.

Scale and Projections in Modern Geography

Computers have drastically changed the way in which maps are made and used. In the preelectronic age, maps were so laborious, time-consuming, and expensive to make that relatively few were created. Frustrated, geographers and other scientists often found themselves trying to use maps for purposes not intended by the map designers. But today anyone with access to computer

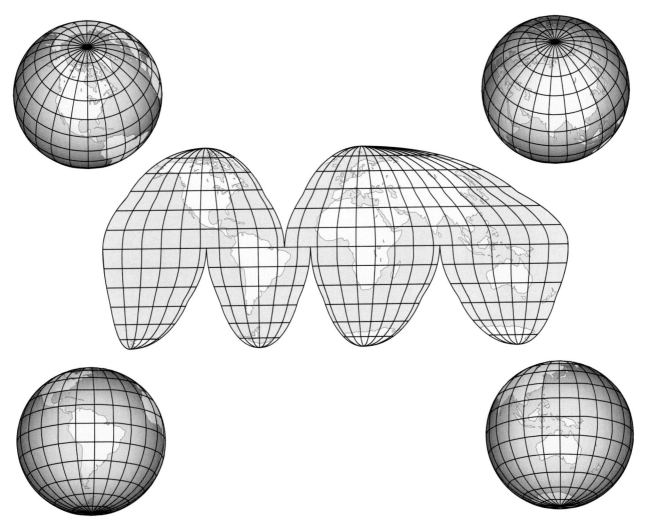

FIGURE A–10 An interrupted Goode's homolosine, an equal-area projection. *(Courtesy of ACSM)*

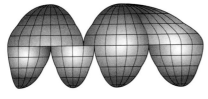

FIGURE A–11
The distortion pattern of the interrupted Goode's homolosine projection, which mimics that of cylindrical projections. *(Courtesy of ACSM)*

mapping facilities can create projections in a flash. Thus, projections will be increasingly tailored to specific needs, and more and more scientists will do their own mapping rather than have someone else guess what they want in a map.

Computer mapping creates opportunities that go far beyond the construction of projections, of course. Once maps and related geographical data are entered into computers, many types of analyses can be carried out involving map scales and projections. Distances, areas, and volumes can be computed; searches can be conducted; information from different maps can be combined; optimal routes can be selected; facilities can be allocated to the most suitable sites; and so on. The term used to describe such processes is *geographical information system*, or GIS (Figure A-13). Within a GIS, projections provide the mechanism for linking data from different sources, and scale provides the basis for size calculations of all sorts. Mastery of both projection and scale becomes the user's responsibility because the map user is also the map maker. Now more than ever, effective geography depends on knowledge of the close association between scale and projection.

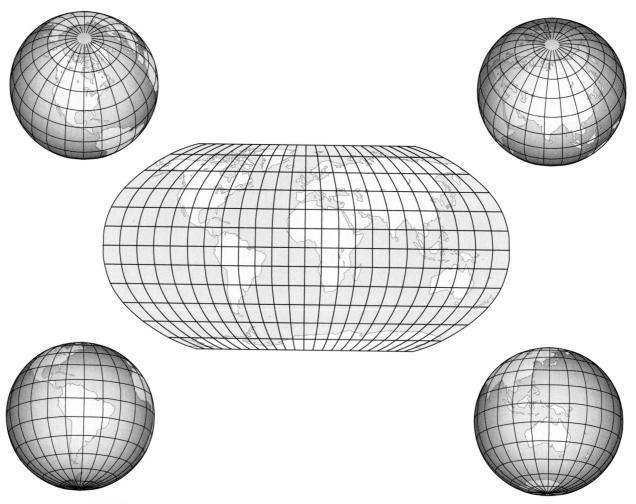

FIGURE A–12 The compromise Robinson projection, which avoids the interruptions of Goode's homolosine but preserves no special properties. *(Courtesy of ACSM)*

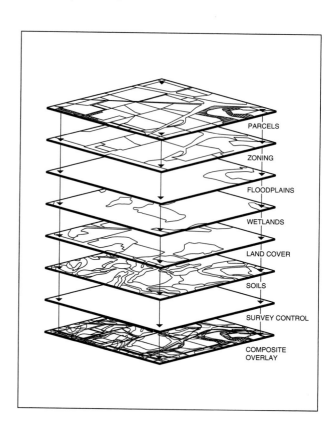

PARCELS

ZONING

FLOODPLAINS

WETLANDS

LAND COVER

SOILS

SURVEY CONTROL

COMPOSITE
OVERLAY

FIGURE A-13 Within a GIS, environmental data attached to a common terrestrial reference system, such as latitude/longitude, can be stacked in layers for spatial comparison and analysis.

Credits

FIGURE 1-3 Adapted from David Lewis, *The Voyaging Stars* (New York: W.W. Norton, 1978), drawing page 120.

FIGURE 4-1 Adapted from John F. Rooney, Jr., Wilbur Zelinsky, and Dean R. Louder, eds., *This Remarkable Continent: An Atlas of United States and Canadian Society and Culture* (College Station, TX: Texas A&M University Press), 1982, page 244, figure 11-13, George O. Carney.

FIGURE 4-3 Reproduced by permission from the *Annals of the Association of American Geographers*, Volume 68, 1978, p. 262, figure 11, W. K. Crowley.

FIGURE 4-4 Reproduced by permissions from the *Annals of the Association of American Geographers*, Volume 66, 1976, p. 490, figure 2; P. P. Karan and E. E. Mather.

FIGURE 4-6 Adapted from Jean-Paul Bourdier and Nezar Alsayyad, *Dwellings, Settlements, and Tradition* (Lanham, MD: University Press of America, 1989).

FIGURE 4-7 Adapted from Robert W. McColl, "By Their Dwellings Shall We Know Them: Home and Setting Among China's Inner Asian Ethnic Groups," *Focus*, Winter 1989, photos pp. 6 and 7, and Ronald G. Knapp, *China's Traditional Rural Architecture: A Cultural Geography of the Common House* (Honolulu: University of Hawaii Press, 1986), plate 2, p. 114.

FIGURE 4-8 and 4-9 Kniffen, Fred B. "Folk-Housing: Key to Diffusion." *Annals of the Association of American Geographers*, 55 (December 1965): p. 549–77.

FIGURE 4-10 Adapted from Virginia McAlester and Lee McAlester, *A Field Guide to American Houses* (New York: Alfred A. Knopf, 1984)

FIGURE 4-11 Adapted from John A. Jakle, Robert W. Bastian, and Douglas K. Meyer, *Common Houses in America's Small Towns* (Athens: The University of Georgia Press, 1989).

FIGURE 4-12 Adapted from John F. Rooney, Jr., and Paul L. Butt, "Beer, Bourbon, and Boone's Farm: A Geographical Examination of Alcoholic Drink in the United States," in *Journal of Popular Culture*, Vol. 11 (1968) 4:842–48. Reprinted with the permission of the editor.

FIGURE 4-13 Adapted from Barbara A. Shortridge and James R. Shortridge, "Consumption of Fresh Produce in the Metropolitan United States," *Geographical Review*, Vol. 79 (1989) 1: p.86, Table IV. Used by permission of the American Geographical Society.

FIGURE 4-14 Adapted from John F. Rooney, Jr., "American Golf Courses: A Regional Analysis of Supply," *Sport Place International*, Vol. 3 (1989) 1/2: pp. 6–8, figures 3, 4, and 6.

FIGURE 4-19 Adapted from David H. Kaplan, "Population and Politics in a Plural Society: The Changing Geography of Canada's Linguistic Groups," *Annals of the Association of American Geographers*, March 1994, Figure 3.

FIGURE 5-1 Adapted from A. Meillet and M. Cohen, *Les langues du monde*, 1952 (Paris: Centre National de la Recherche Scientifique), carte XIB.

FIGURE 5-4 Adapted from *Encyclopaedia Britannica*, 15th edition (1987), 22: 660.

FIGURE 5-5 Adapted from Antoine Meillet and Marcel Cohen, *Les langues du monde* 1952.

FIGURE 5-6 Adapted from A. K. Ramanujan and Colin Masica, "A Phonological Typology of the Indian Liguistic Area," *Current Trends in Linguistics* 5 (1969): 561, with permission of Mouton de Gruyter, a division of Walter de Gruyter & Co.

FIGURE 5-8 Adapted from *Encyclopaedia Britannica*, 15th edition, 1987, 22:766 and 769.

FIGURE 5-9 Adapted from *Encyclopaedia Britannica*, 15th edition, 1987, 29:907.

FIGURE 5-12 Reprinted from George Cardona, Henry M. Hoenigswald, and Alfred Senn, "Indo-European and Indo-Europeans," in *Proto-Indo-European Culture* by Marija Gimbutas, University of Pennsylvania Press, Philadelphia, 1970.

FIGURE 5-13 Adapted from Colin Renfrew, "The Origins of Indo-European Languages," *Scientific American*, October 1989, map, p. 112.

FIGURE 5-15 Adapted from *Children's Games in Street and Playground* by Iona and Peter Opie. Iona and Peter Opie 1969. Published by Oxford University Press 1969.

FIGURE 5-16 Adapted from Hans Kurath, *A Word Geography of the Eastern United States* (Ann Arbor: University of Michigan Press), 1949, Figure 3.

FIGURE 5-17 Adapted from *Encyclopaedia Britannica*, 15th edition (1987), 28:358.

FIGURE 5-18 Adapted from *Encyclopaedia Britannica*, 15th edition (1987) 23:346.

FIGURE 6-1 Adapted from D. Sopher, *Geography of Religions* (Englewood Cliffs, NJ.: Prentice Hall), 1967.

FIGURE 6-2 Adapted from Jan and Mel Thompson, *The R. E. Atlas: World Religions in Maps and Notes* (Sevenoaks, England: Hodder & Stoughton), 1986, p. 44; by permission of the publisher.

FIGURE 6-3 Adapted from Douglas W. Johnson, Paul R. Picard, and Bernard Quinn, *Churches and Church Membership in the United States* (Bethesda, MD; Glenmary Research Center), 1971.

FIGURE 6-5 From W. Shepherd, *Historical Atlas*, by permission of Barnes & Noble Books.

FIGURE 6-6, 6-7, 6-8, 6-9 From Ismail Ragi al Farugi and David E. Sopher, *Historical Atlas of the Religions of the World* (New York: Macmillan), 1974.

FIGURE 7-13 Adapted from George P. Murdock, *Africa: Its Peoples and Their Cultural History* (New York: McGraw Hill, 1959).

FIGURE 7-15 Adapted from *Encyclopaedia Britannica*, 15th edition (1987), 28:181.

FIGURE 9-3 Adapted from Edward B. Espenshade, Jr., editor, *Goode's World Atlas*, 18th ed. (Chicago: Rand McNally, 1990), map p. 201.

FIGURE 10-4 Reproduced by permission from the *Annals of the Association of American Geographers*, Vol. 26, 1936, p. 241, figure 1; D. Whittlesey.

FIGURE 11-2 From Peter A. Gould, "Spatial Diffusion" (Washington, D.C.: *Association of American Geographers*) Resource Publication in Geography, No. 4, 1969, page 52 (including figure 57). Reprinted by permission.

FIGURE 12-8 From Walter Christaller, *Die Zentralen Orte in Sudeutschland as found in The Central Places in Southern Germany*, a translation by Carlisle W. Baskin (Englewood Cliffs, N.J.: Prentice Hall, 1966). Used by permission.

FIGURE 12-11 Adapted from Leonardo Benevolo, *The History of the City* (Cambridge, MA: MIT Press), 1980, figures 31, 33, and 36.

FIGURE 12-15 Adapted from Paul Knox, *Urbanization: An Introduction to Urban Geography* (Englewood Cliffs, NJ: Prentice Hall, 1994), p. 61.

FIGURE 12-16 Adapted from Paul Knox, *Urbanization: An Introduction to Urban Geography* (Englewood Cliffs, NJ: Prentice Hall, 1994), p. 64.

FIGURE 12-17 Adapted from J. Clark Archer and Ellen R. White, "Service Classification of American Cities," *Urban Geography*, Vol 5 (1985): 122–151.

FIGURE 13-13 Adapted from Leonardo Benevolo, *The History of the City* (Cambridge: MIT Press, 1980), map pp. 828–29.

FIGURE 13-23 From Marion Clawson and Peter Hall, *Planning and Urban Growth* (Baltimore: The Johns Hopkins University Press), 1973, p. 131. Published for Resources for the Future, Inc. by the Johns Hopkins University Press.

FIGURE 14-9 From Blair T. Bower and Daniel J. Basta, *Residuals-Environmental Quality Management: Applying the Concept* (Baltimore: The Johns Hopkins University Center for Metropolitan Planning and Research), 1973, p.3.

FIGURE 14-10 Adapted from William K. Stevens, "Study of Acid Rain Uncovers a Threat to Far Wider Area," *The New York Times*, January 16, 1990, p. 21, map.

FIGURE 14-12 Adapted from Peter Passell, "The Garbage Problem: It May Be Politics, Not Nature," *The New York Times*, February 26, 1991, p. B5, map; and Philip S. Gutis, "New York Begins Getting Rid of the Trash No One Wanted," *The New York Times*, May 18, 1987, p. 12, map.

FIGURE 14-13 Adapted from "The Nation's Polluters—Who Emits What, and Where," *The New York Times*, October 13, 1991, p. 10, map.

FIGURE 14-15 Adapted from Edward Carr, "Power to the People," *The Economist*, June 18, 1994, chart, p. 15

Selected Map Index

Index

Index